D0114669

Progress in Inorganic Chemistry

Volume 43

Advisory Board

PROGRESS IN INORGANIC CHEMISTRY

Edited by

KENNETH D. KARLIN

DEPARTMENT OF CHEMISTRY
THE JOHNS HOPKINS UNIVERSITY
BALTIMORE, MARYLAND

VOLUME 43

AN INTERSCIENCE® PUBLICATION
JOHN WILEY & SONS, INC.
New York • Chichester • Brisbane • Toronto • Singapore

Cover Illustration of "a molecular ferric wheel" was adapted from Taft, K. L. and Lippard, S. J., *J. Am. Chem. Soc.*, **1990,** 112, 9629.

This text is printed on acid-free paper

An Interscience® Publication

Library of Congress Catalog Card Number 59-13035
ISBN 0-471-12336-6

Printed in the United States of America

10 9 8 7 6 5 4 3 2 1

Contents

Progress in Inorganic Chemistry

Volume 43

Oxovanadium and Oxomolybdenum Clusters and Solids Incorporating Oxygen-Donor Ligands

M. ISHAQUE KHAN

Department of Chemical and Biological Sciences
Illinois Institute of Technology
Chicago, IL

JON ZUBIETA

Department of Chemistry
Syracuse University
Syracuse, NY

CONTENTS

Progress in Inorganic Chemistry, Vol. 43, Edited by Kenneth D. Karlin.
ISBN 0-471-12336-6

I. INTRODUCTION

A. General Considerations

All the elements of the periodic table except the lighter noble gases form oxides, and most elements form more than one binary compound with oxygen (1). The structural chemistry of oxides is extensive, encompassing simple molecular species, complex oligomeric clusters, and solid state phases exhibiting chain, layer, and three-dimensional (3D) network structures. Thus, the metal–oxo unit {M=O} is a fundamental constituent both of soluble metal–oxo complexes of varying nuclearities and of complex metal oxide solids.

The metal oxide clusters known as the polyoxometalates or polyoxoanions constitute an important subclass of the metal oxide system. These clusters, which are characteristic of the early transition elements, exhibit an extensive chemistry including both isopolyanions $[M_xO_y]^{n-}$, which contain only a transition element together with oxygen and hydrogen, and the heteropolyanions $[X_aM_xO_y]^{n-}$, which contain one or more atoms of other elements in addition to M. While such "naked" core polyoxoanions, that is, metal oxide clusters that do not incorporate organic ligands, have provided the major focus in the development of the chemistry of metal oxide clusters, the past decade has witnessed the emergence of a considerable coordination chemistry of polyoxoanions, which is characterized by metal oxide cores incorporating a variety of organic ligands or fused to organometallic fragments.

B. Historical Perspective

While polyoxoanions have been known since the time of Berzelius (2), the first polyanions to contain covalently attached organic groups were not reported until the early twentieth century (3, 4) and their structures remained enigmatic until the mid-1970s (5, 6). The structural characterizations of $(NH_4)_6[(HCO)_2Mo_8O_{28}]\cdot 2H_2O$ (5), first described by Miolati (3) in 1908, and of $[(R_2AsO_2)Mo_4O_{13}H]^{2-}$ (6), prepared by Rosenheim and Belicki (4) in 1913 signaled the emergence of interest in the coordination chemistry of polyoxoanions. The dramatic developments in the area of polyoxoanion coordination chemistry received significant impetus from several seminal contributions:

1. The synthesis and characterization of a variety of metal–oxo-organophosphonato and metal–oxo-organoarsonato clusters by Pope and coworkers in the 1970s (7–9).
2. Systematic studies of polyoxoanions in inert, aprotic solvents by Day and Klemperer (10) in the early 1980s, which allowed the preparation of a number of novel covalent derivatives.

3. The introduction of templates to direct the self-assembly of complex metal oxide ligand shells and a new appreciation for the relationship of polyanion coordination chemistry to host–guest and host–hostage chemistry by Müller and co-workers (11–13) in the late 1980s.
4. The expansion of synthetic methodologies afforded by the exploitation of solvothermal methods introduced by Klemperer and co-workers (14) and Jacobson and co-workers (15, 16).

The dramatic expansion of the coordination chemistry of polyoxoanions in the past decade is also linked to a growing interest in these materials as models for surface bound intermediates in the selective catalytic oxidations and/or dehydrogenations of organic substrates by metal oxide solids, as versatile synthetic precursors for the preparation of complex solid state phases with designed properties, and as potential catalysts and sorbent materials. Furthermore, polyoxoanion coordination compounds are of intrinsic interest as a consequence of their diverse structures, complex magnetic properties, and rich electrochemistry.

C. Scope of This Chapter

In view of the extensive literature on polyoxoanions and the availability of a number of excellent reviews on the general subject, the scope of the discussion has been limited generally to oxovanadium and oxomolybdenum clusters incorporating simple organic ligands with oxygen-donor atoms. This choice is predicated on the dominant role of organic ligands with oxygen donors in the coordination chemistry of polyoxoanions and on the extensive chemistry of the oxovanadium and oxomolybdenum cores in comparison to other metal–oxo functionalities. To keep this chapter within manageable proportions, clusters derived from polyoxometalate cores by oxo-group substitution, such as nitrido (17), organoimido (18), nitrosyl (19), and hydrazido groups (20) have been excluded, as have the growing classes of polyoxometalates supporting organometallic subunits, whether of the transition metal or main group types (10, 21–25).

The general field of polyoxoanion chemistry is enormous, with over 1000 well-characterized complexes reported in the literature. An excellent overview of the area and a comprehensive literature review to 1983 is available in the monograph by Pope (26). The more recent literature has been discussed in updated reviews (27, 28), and the current status of the field is well represented in the book edited by Pope and Müller (29). More specialized reviews on the structural systematics of various classes of polyoxoanion clusters (30, 31) and on polyanion coordination chemistry (20, 32, 33) have also appeared in recent years.

In view of the rapid developments in the area of polyanion coordination chemistry and the absence of a comprehensive overview of the topic, this chapter is timely, exploring the literature up to the beginning of 1994. More specifically, the chapter discusses the ligand types and the rationales for their choice (see Section II), the synthetic methodologies adopted in the preparations of the clusters (see Section III), some general principles of the structural chemistry (see Section IV), and the detailed structural characteristics of the complexes (see Sections V–VII). This latter discussion is structured by ligand type, whether of the conventional organic classes (alkoxide, carboxylate, or carbon oxo anion) or of organophosphonate or organoarsonate types, and further subdivided according to cluster nuclearity (see Tables I–V). We have chosen not to discuss the enormous literature on complexes of nuclearity less than three, unless these are intimately related to the higher nuclearity species as synthetic precursors or structural prototypes. Triangular cores that possess reduced metal sites and fewer oxo groups than metal centers, such as the triangular Mo^{IV} and V^{III} carboxylate clusters, have also been excluded.

II. LIGAND TYPES

While an extensive coordination chemistry of polyoxoanions incorporating ligands with nitrogen, oxygen, and sulfur donors and supporting organometallic subunits has emerged in the past two decades, our discussion in this chapter will focus on two general classes of ligands with oxygen-donor groups: (a) the more or less conventional ‘‘organic ligands, such as alkoxides, carboxylates, acetals, and carbon oxo anions; and (b) the organophosphonate, organoarsonate, and related ligand types.

Several general features of the coordination of these ligand types with polyoxoanions merit comment. The introduction of such ligand types vastly expands the coordination chemistry of polyoxoanions both by providing charge compensation by the substitution of peripheral O^{2-} oxo groups with mononegative ligands in ‘‘classical’’ compact polyanion structures (Section III) and by allowing structure expansion and/or modification through bridging interactions. While ‘‘naked’’ polyanion structures exhibit frameworks necessarily constructed exclusively from $\{MO_5\}$ square pyramids and/or $\{MO_6\}$ octahedra, the structural diversity of the coordination clusters reflects the introduction of ligand tetrahedra into the cluster framework. In addition to providing this diversity of structural types, metal oxide coordination clusters serve as models for the interactions of substrate molecules with metal oxides in heterogeneous catalysis, as potential precursors for the synthesis of new solid materials and supramolecular assemblies, and as materials with useful catalytic and sorptive properties of their own.

TABLE I
Oxovanadium Clusters with Alkoxide, Carboxylate, Carbon Oxo Anion and Related Ligand Types

Complex	Ligand Type[a]	References
a. Trinuclear Cores		
$[(VO)_3(thf)(PhCO_2)_6]$ **(1)**	C2	111
$[(VO)_3(sal)_3(MeOH)_3]$ **(2)**	A1, sal[b]	116
$[\{RC(CH_2O)_3V_3\}P_2W_{15}O_{59}]^{6-}$ **(3)**	TA3	120
b. Tetranuclear Cores		
$[V_4O_8(C_2O_4)_4(H_2O)_2]^{4-}$ **(4)**	Terminal bidentate	121
$[V_4O_8(NO_3)(\text{thiophene-2-carboxylate})_4]^{2-}$ **(5)**	C2	122
$[V_4O_7(OH)(O_2CR)_4K]^+$ **(6)**	C2	124
$[(VO)_4(H_2O)_2(SO_4)_2\{(OCH_2)_3CR\}_2]^{2-}$ **(7)**	TA5	127
$[V_4O_6Cl_2(OCH_2CR_2CH_2OH)_4]$ **(8)**	BA1	128
$[V_2Mo_2O_8(OMe)_2\{(OCH_2)_3CR\}_2]^{2-}$ **(9)**	A1, TA3	127
$[V_4O_2(acac)_4(OMe)_6]$ **(10)**	A2, Terminal bidentate acac	129
c. Pentanuclear Core		
$[V_5O_9Cl(\text{thiophene-2-carboxylate})_4]^{2-}$ **(11)**	C2	122
d. Hexanuclear Cores		
$[V_6O_{10}(PhCO_2)_9]$ **(22)**	C2	154
$[V_6O_{12}(OMe)_7]^{1-}$ **(12a)**	A2	141
$[V_6O_{13}(OMe)_3\{(OCH_2)_3CR\}]^{2-}$ **(12)**	A2, TA3	139
$[V_6O_{13}\{(OCH_2)_3CR\}_2]^{2-}$ **(13)**	TA3	142
$[V_6O_{11}(OH)_2\{(OCH_2)_3CR\}_2]$ **(14)**	TA3	146
$[V_6O_{10}(OH)_3\{(OCH_2)_3CR\}_2]^{2-}$ **(15)**	TA3	149
$[V_6O_9(OH)_4\{(OCH_2)_3CMe\}_2]^{2-}$ **(16)**	TA3	146
$[V_6O_7(OH)_6\{(OCH_2)_3CMe\}_2]^{2-}$ **(17)**	TA3	146
$[V_6O_8\{(OCH_2)_3CEt\}_2\{(OCH_2)_2CEt(CH_2OH)\}_4]^{2-}$ **(13a)**	BA2, TA5	139
$Ba[V_6O_7(OH)_3\{(OCH_2)_3CMe\}_3]$ **(18)**	TA3	152
$(Me_3NH)[V_6O_7(OH)_3\{(OCH_2)_3CMe\}_3]$ **(19)**	TA3	152, 153
$Na_2[V_6O_7\{(OCH_2)_3CEt\}_4]$ **(20)**	TA3	152, 153
$Na[(VO)_6F(OH)_3\{(OCH_2)_3CMe\}_3]$ **(21)**	TA3	152
$[V_6O_{10}(PhCO_2)_9]$ **(22)**	C2	154
$[(VO)_6(CO_3)_4(OH)_9]^{5-}$ **(23)**	CA1, CA2	155, 156
e. Octanuclear and Larger Nuclearity Cores		
$[(VO)_8(OMe)_{16}(C_2O_4)]^{2-}$ **(24)**	A2, O4	160
$[V_9O_{16}(bdta)_4]^{7-}$ **(28)**	bdta[c]	162
$[V^{IV}_{10}O_{13}\{(OCH_2)_3CMe\}_5]^{1-}$ **(32)**	TA3	163
$[V^{IV}_{10}O_{16}\{(OCH_2)_3CR\}_4]^{4-}$ **(29)**	TA3	163, 164
$[V^{IV}_8V^{V}_2O_{16}\{(OCH_2)_3CR\}_4]^{2-}$ **(30)**	TA3	165
$[V^{IV}_{10}O_{14}(OH)_2\{(OCH_2)_3CR\}_4]^{2-}$ **(31)**	TA3	165
$[V_{15}O_{36}(CO_3)]^{7-}$ **(25)**	CA2	161
$[H_6V_{10}O_{22}(RCO_2)_6]^{2-}$ **(26)**	C2	11
$[H_2V_{22}O_{54}(MeCO_2)]^{7-}$ **(27)**	Encapsulated RCO_2^-	11
$[V_{16}O_{20}\{(OCH_2)_3CR\}_8(H_2O)_4]$ **(33)**	TA3	166

TABLE I (*Continued*)

Complex	Ligand Type[a]	References
f. Polymeric Materials		
$(Ph_4P)[VOCl(C_2O_4)]$ (**36**)	O2	172
$K_2[VO(HCO_2)_4]$ (**35**)	C1 and anti-anti form of C2	171
$VO(OMe)_3$ (**34**)	A1, A2	167

[a]Ligand types refer to the common binding modes illustrated in Figs. 1, unless otherwise noted.
[b]Salicylhydroximate = sal, $-OC_6H_4C(O)NHO^-$.
[c]Butanediaminetetracetic acid = bdta, $\{(O_2CCH_2)_2N(CH_2)_4N(CH_2CO_2)_2\}^{4-}$.

TABLE II
Oxovanadium Organophosphonate, Organoarsonate, and Related Clusters

Complex	Ligand Type	References
a. Tetranuclear Cores		
$[V_4O_6(PhPO_3)_4F]^{1-/2-}$ (**39, 40**)	P4	178
$[(VO)_4\{PhP(O)_2OP(O)_2Ph\}_4Cl]$ (**41**)	See text	179
b. Pentanuclear Core		
$[V_5O_7(OMe)_2(PhPO_3)_5]^{1-}$ (**42**)	A1, P4	89
c. Hexanuclear Cores		
$[V_6O_{10}(PhAsO_3)_4(PhAsO_3H)_2]^{2-}$ (**45**)	P3, P4	180
$[V_6O_{10}(PhPO_3)_4(PhPO_3H)_2]^{2-}$ (**44**)	P3, P4	180
$[(VO)_6(t\text{-}BuPO_3)_8Cl]$ (**43**)	P4	179
d. Heptanuclear Core		
$[V_7O_{12}(PhPO_3)_6Cl]^{2-}$ (**46**)	P4	182
e. Decanuclear Core		
$[V_{10}O_{24}(H_2NC_6H_4AsO_3)_3]^{4-}$ (**47**)	P4	180
f. Dodecanuclear Cores		
$[H_{12}(VO_2)_{12}(PhPO_3)_8(H_2O)_4]^{4-}$ (**48**)	P4	15
$[2(MeOH)V_{12}O_{14}(OH)_4(PhAsO_3)_{10}]^{4-}$ (**51**)	P4	131
$[(VO)_{12}(OH)_2(PhAsO_3)_{10}(PhAsO_3H)_4]^{4-}$ (**50**)	P3, P4	131
$[(V_4O_8)_2\{(VO)_4(H_2O)_{12}\}(PhPO_3)_4Cl_2]^{2-}$ (**49**)	P4	184
g. Tetradecanuclear Cores		
$[2(NH_4Cl)V_{14}O_{22}(OH)_4(H_2O)_2(PhPO_3)_8]^{6-}$ (**52**)	P4	12
$[2(MeCN)_2V_{14}O_{22}(OH)_4(PhPO_3)_8]^{6-}$ (**53**)	P4	131
h. Hexadecanuclear Cores		
$[H_6(VO_2)_{16}(MePO_3)_8]^{8-}$ (**54**)	P4	15
$[H_8(VO_2)_{16}(PhAsO_3)_8]^{6-}$ (**55**)	P4	185
i. Octadecanuclear Core		
$[V_{18}O_{25}(H_2O)_2(PhPO_3)_{20}Cl_4]^{4-}$ (**56**)	P4	179

TABLE III
Oxovanadium Organophosphonate Solid State Phases

Complex	Ligand Type	References
a. One-Dimensional Phases		
$[H_2N(CH_2CH_2)_2NH_2][VO\{CH_2(PO_3)_2\}]$ (**57**)	P1	191
b. Two-Dimensional Phases		
$[VO(PhPO_3)(H_2O)]$ (**58**)	P4	186
$[Et_2NH_2][Me_2NH_2][(VO)_4(OH)_2(PhPO_3)_4]$ (**60**)	P4	188
$[Et_4N]_2[(VO)_6(OH)_2(H_2O)_2(EtPO_3)_6]$ (**61**)	P4	189
$[EtNH_3][(VO_3)(H_2O)(PhPO_3)_4]$ (**59**)	P4	187
$[(VO)_2\{CH_2(PO_3)_2\}(H_2O)_4]$ (**63**)	P4	190
$[H_3NCH_2CH_2NH_3][(VO)(O_3PCH_2CH_2PO_3)]$ (**64**)	P1	191
$[H_2N(CH_2CH_2)_2NH_2][(VO)_2(O_3PCH_2CH_2CH_2PO_3)_2]$ (**66**)	P4	*a*
$[V_2O_4(PhAsO_3H)(H_2O)]$ (**62**)	P3	181
c. Pillared Two-Dimensional Phases		
$(H_3NCH_2CH_2NH_3)[(VO)_4(OH)_2\{O_3PCH_2CH_2CH_2PO_3\}_2]$ (**65**)	P4	191

[a]V. Soghomonian, R.C. Haushalter, and J. Zubieta, unpublished results.

A. Alkoxides, Carboxylates, Carbon Oxo Anions, and Related Ligand Types

The major products of heterogeneous catalysis are formed by O_2 oxidations that employ transition metal oxides to effect dehydrogenation and/or oxygen atom transfer to the substrate (34). Given the versatility of metal oxides in catalyzing organic transformations, the chemical processes that occur on these surfaces are of considerable interest. Since oxide surfaces adopt complex structures that are difficult to characterize on a molecular level (35), particularly with respect to surface bound intermediates, the relatively simple soluble metal oxides have received increasing attention as models for the more complex systems. As Day and Klemperer (10) pointed out, the relationship between polyoxoanions and solid oxides extends beyond simple size considerations to fundamental aspects of structure and bonding, such that the chemical information elicited from the molecular systems is not only of intrinsic interest but may provide insights into the complex solid systems.

The recently described polyoxoalkoxymetalates $[M_aO_b(OR)_c]^{x-}$ represent a particularly prominent subclass of polyanion coordination compounds that are of considerable interest by virtue of their relationship to the oligomeric metal alkoxides, a major class of inorganic materials (36–41). Metal alkoxides display a range of reactivity patterns (42–48), including catalytic activity (48) and in their high oxidation states are versatile molecular precursors of oxides of high purity and a variety of "high tech" materials, including catalyst supports, biomaterials, and ceramics (49–58). The relationship of polyoxoalkoxymetalate to both the metal oxide and metal alkoxide systems identifies these clusters as

TABLE IV
Oxomolybdenum Clusters with Alkoxide, Carboxylate, Carbon Oxo Anion, and Related Ligand Types

Complex	Ligand Type	References
a. Trinuclear Cores		
$[Mo_3O_8(OMe)(C_4O_4)_2]^{3-}$ (**67**)	A3, SA1	192
$[Mo_3O_8(pinacolate)_2]^{2-}$ (**68**)	BA1	193
$[Mo_3O_7\{Me(OCH_2)_3CMe\}_2]^{2-}$ (**69**)	TA2, TA3	83
$[Mo_3O_6(OMe)\{(OCH_2)_3CMe\}_2]^{1-}$ (**70**)	A1	83
$[Mo_3O_9(dmso)_4]$ (**71**)	See text	196
b. Tetranuclear Cores		
$[Mo_4O_{10}(OMe)_6]^{2-}$ (**72**)	A1, A2, A3	84, 199
$[Mo_4O_{10}(OMe)_4Cl_2]^{2-}$ (**73**)	A2, A3	84, 199
$[Mo_4O_{10}(C_6H_2O_4)_2]^{2-}$ (**74**)	See text	217, 218
$[Mo_4O_8(OEt)_2\{(OCH_2)_3CR\}_2]$ (**75**)	A1, TA5	200
$[Mo_4O_{10}(OMe)_2(OC_6H_4O)_2]^{2-}$ (**76**)	A2, bidentate catecholate	199
$[Mo_4O_8(OMe)_2(HOMe)_2Cl_4]^{2-}$ (**77**)	A2	199
$[Mo_4O_8(OEt)_2(HOEt)_2Cl_4]^{2-}$ (**78**)	A2	200
$[Mo_4O_6(OPr)_4(HOPr)_2Cl_4]$ (**79**)	A1, A2	203
$[Mo_4O_8(OPr)_4(py)_4]$ (**80**)[a]	A2	204
$[Mo_4O_{11}(malate)_2]^{2-}$ (**81**)	C1, C3	210, 211
$[Mo_4O_{11}(citrate)_2\}^{4-}$ (**82**)	C1, C3	212, 213
$[Mo_4O_6Cl_2(O_2CR)_6][R = O_2CMe$ (**83a**) or $O_2C_6H_4Me$ (**83b**)	C3	214, 215
$[CH_2Mo_4O_{15}H]^{3-}$ (**84**)	CA3	207
$[(HCCH)Mo_4O_{15}X]^{3-}$ [X= F (**85a**) or HCO_2 (**85b**)]	CA4	208
$[(C_9H_4O)Mo_4O_{15}(OMe)]^{3-}$ (**86**)	A2, CA4	209
$[(C_{14}H_{10})Mo_4O_{15}(PRCO_2)]^{3-}$ (**87**)	C3, CA4	209
$[(C_{14}H_8)Mo_4O_{15}(OH)]^{3-}$ (**88**)	CA4	209
$[Mo_4O_8(OMe)_2(C_4O_4)_2(C_4O_4H)_2]^{4-}$ (**89**)	A2, SA1, and monodentate $HC_4O_4^-$	206
$[Mo_4O_6(OH)_4(CO_3)(CO)_2(PMe_3)_6]$ (**90**)	CA3	216
c. Pentanuclear Core		
$[Mo_5O_{16}(OMe)]^{3-}$ (**91**)	A2	192
d. Hexanuclear Core		
$[Mo_6O_{10}(OPr)_{12}]$ (**92**)	A1, A2	219
e. Octanuclear Cores		
$[Mo_8O_{26}(HCO)_2]^{6-}$ (**93**)	Monodentate	5
$[Mo_8O_{24}(OH)_4(sal)_2]^{2-}$ (**94**)	Monodentate through oxygen donor	220
$[Mo_8O_{24}(OH)_2(met)_2]^{4-}$ (**95**)	Monodentate through carboxylate oxygen	202
$[Mo_8O_{24}(OMe)_4]^{4-}$ (**96**)	A2	195
$[Mo_8O_{20}(OMe)_4\{(OCH_2)_3CR\}_2]^{2-}$ (**97**)	A1, A2, TA5	221
$[H_2Mo_8O_{24}(OMe)_2]^{4-}$ (**98**)	A2	205
$[Mg_2Mo_8O_{22}(OMe)_6(HOMe)_4]^{2-}$ (**99**)	A2	222
$[Mo_8O_{16}(OMe)_8(C_2O_4)]^{2-}$ (**100**)	A2, O4	223
$[Mo_8O_{16}(OMe)_8(PR_3)_4]$ (**101**)	A2	224
f. Dodecanuclear Cores		
$[Mo_{12}O_{36}(C_4O_4H)_4]^{4-}$ (**102**)	SA3	225
g. Superclusters		
$[Na(H_2O)_3H_{15}Mo_{42}O_{109}\{(OCH_2)_3CCH_2OH\}_7]^{7-}$ (**103**)	TA3	227a
$[Mo_{43}H_{13}O_{112}\{(OCH_2)_3CMe\}_7]^{10-}$ (**104**)	TA3	227b

[a]Pyridine = py.

TABLE V
Oxomolybdenum Clusters with Organophosphonate, Organoarsonate, and Related Ligand Types

Complex	Ligand Type	References
a. Trinuclear Core		
$[Mo_3O_9(O_3PCMe(O)PO_3)]^{5-}$ **(105)**	XXIV	228
b. Tetranuclear Cores		
$[Mo_4O_8\{MeC_6H_4AsO_2(Cl)\}_4Cl]$ **(106)**	See text (P1)	230
$[Mo_4O_8\{PhAsO_2(Cl)\}_4Cl][MoOCl_4]$ **(106a)**		230
$[Mo_4O_{13}H(Me_2AsO_2)]^{2-}$ **(107)**	P3	6
$[Mo_4O_{10}(PhPO_3)_4]^{4-}$ **(108)**	P2	230
α − $[Mo_4O_{10}(PhAsO_3)_4]^{4-}$ **(109)**	P1, P5	230
β − $[Mo_4O_{10}(C_7H_7AsO_3)_4]^{4-}$ **(110)**	P1, P5	230
b. Pentanuclear Cores		
$[Mo_5O_{15}(RPO_3)_2]^{n-}$ **(111)**	P7	7,231
R = —Me, —Ph; $n = 4$		
R = $-CH_2N(CH_2)_4O$; $n = 2$		232
$[Mo_5O_{15}(C_3H_7AsO_3)_2]^{4-}$ **(111a)**	P7	100
c. Hexanuclear Cores		
$[Mo_6O_{18}(RAsO_3)_2]^{4-}$ **(112)**	P6	9
$[Mo_6O_{18}(RPO_3)_2]^{4-}$ **(118)**	P7	
$[Mo_6O_{18}(H_2O)(PRAsO_3)_2]^{4-}$ **(113)**	P5, P6	8
$[Mo_6O_{18}(H_2O)_6(MeAsO_3)]^{2-}$ **(114)**	P6	238
$[Mo_6O_{17}\{HOCH(PO_3)_2\}_2]^{6-}$ **(115)**	P2	234
$[Mo_6O_{17}\{OCH(PO_3)_2\}_2]^{8-}$ **(116)**	P2	228
$\{M[Mo_6O_{15}(PhPO_3H)_3(PhPO_3)]_2\}^{9-}$ **(117)** M = Na^+ or K^+	P1, P6	236–237
d. Dodecanuclear Core		
$[Mo_{12}O_{34}(H_3NC_6H_4AsO_3)_4]$ **(119)**	P6	6(d)

models for the initial stages in the sol–gel process (59). As such, the development of the chemistry of the polyoxoalkoxymetalates may provide insights into controlling the aggregation process and providing useful synthetic precursors amenable to some degree of molecular engineering.

B. Organophosphonates, Organoarsonates, and Related Ligand Types

The organophosphonates and organoarsonates provide effective bridging groups linking metal–oxo fragments $\{MO\}_n$ in unusual oligomeric species and even solid phases of various dimensionalities. Indeed, the metal–organophosphonate system specifically spans a rich variety of coordination chemistry including mononuclear coordination compounds (60, 61), linear one-dimensional (ID) complexes (62), and layered solid compounds (63–65). The layered metal organophosphonates exhibit unusual sorptive and catalytic properties and serve as catalyst supports and ion exchangers (66–70). Most specifically, the oxovanadium organophosphonate solids, exemplified by $[VO(PhPO_3)]\cdot H_2O$, possess structurally well-defined internal void spaces and coordination sites that intercalate alcohols by coordination of the substrate molecule to the vanadium centers of the inorganic V/P/O layer. The shape selectivity of these phases in absorbing alcohols is related to the steric constraints imposed by the organic residues of the phosphonates surrounding the metal sites and to the interplay of hydrophobic and hydrophilic domains in the structure (71). Since the functional groups of these solids may be introduced through the organic moiety of the

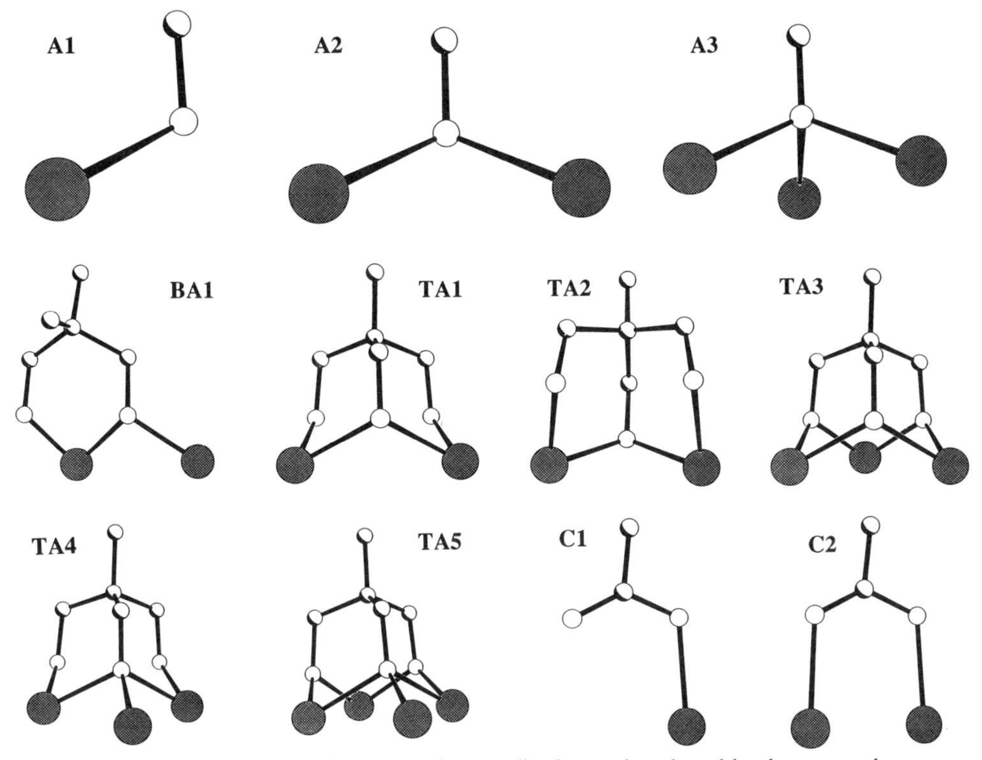

Figure 1. Schematic views of representative coordination modes adopted by the oxygen-donor ligands discussed in this chapter. The atom-labeling scheme, which is consistently maintained throughout this chapter, depicts vanadium or molybdenum atoms as cross-hatched circles, phosphorus or arsenic as circles lined from bottom left to top right, oxygen atoms as open circles, and carbon atoms as highly highlighted circles. A1: terminal alkoxide; A2: μ^2-bridging alkoxide; A3, μ^3-bridging alkoxide; BA1, mixed bridging and terminal mode adopted by bisalkoxides; TA1–TA5, coordination modes adopted by trisalkoxide ligand types; C1–C3, coordination modes adopted by carboxylate ligands; SA1–SA3, common coordination modes for squaric acid and hydrogen squarate; O1–O4, oxalate coordination modes; CA1 and CA2, common coordination modes for carbonate; CA3 and CA4, acetal or ketal coordination types; P1–P7, organoarsonate and organophosphonate coordination modes.

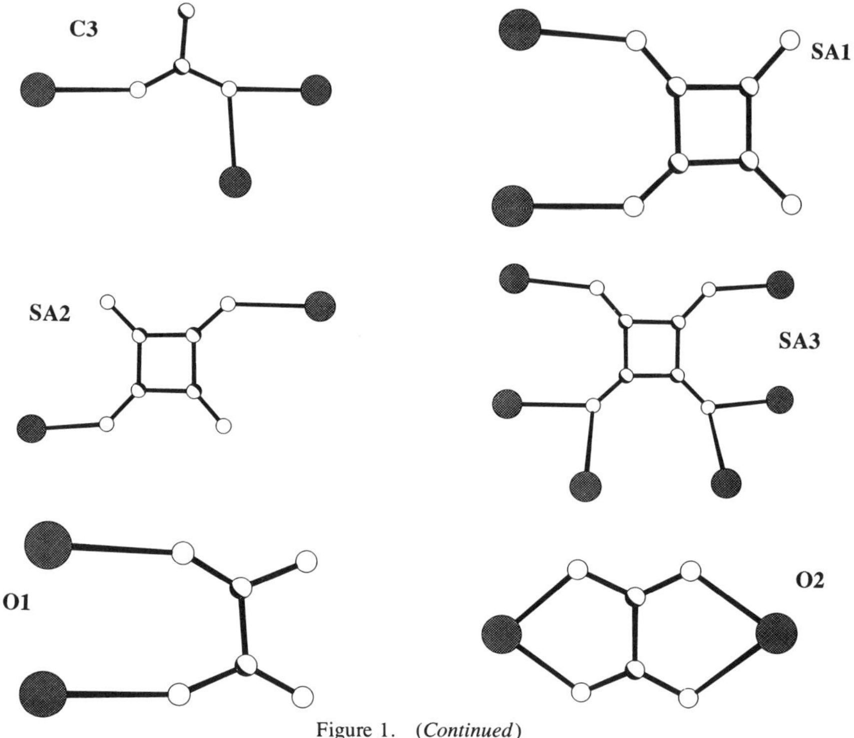

Figure 1. *(Continued)*

phosphonate or incorporated by postsynthesis reactions, a wealth of chemistry may be exploited in the preparation of new materials. Furthermore, since layered phosphonates undergo both Brønsted and Lewis intercalative chemistry (72), the optical and electronic properties of the host may be modified. While the chemistry of the molecular systems of the M/O/REO_3^{2-} system (E = P or As) remains relatively undeveloped, it is tempting to speculate on the potential role of such molecular species in the design of solid phases with tailored properties (73, 74) for applications as catalyst supports, sorbent materials, and electronic or optical devices.

C. Classification of Ligand Types and Coordination Modes

The representative members of the two general classes of ligands discussed in this chapter and the major coordination types adopted are summarized in Fig. 1. Although other coordination modes are possible, the polyanion coordination compounds are generally limited to those of the figure. Several unique cases are discussed in subsequent sections. Although examples of ligands with mixed-

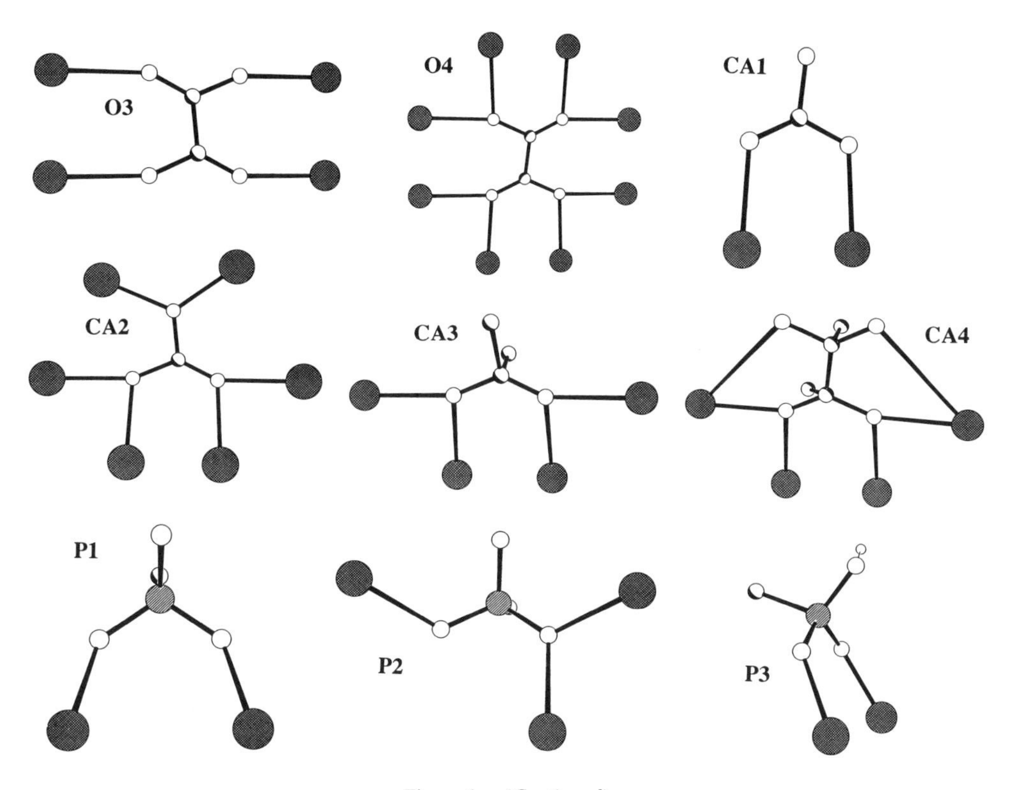

Figure 1. *(Continued)*

donor groups (O and N, and O and S) are known (75, 76): these groups have been excluded from this discussion.

III. SYNTHETIC METHODOLOGIES

A. General Features

There are no general synthetic routes for preparing cluster compounds and solid materials of predictable features. Rational synthetic methods available for molecular and soluble complexes are more developed compared to somewhat primitive approaches that have thus far been adopted for preparing solids. Since organic ligands do not survive under the typical solid state reaction conditions, which often involve highly refractory amorphous materials that diffuse together very slowly, even at high temperatures, the scope of these methods is generally

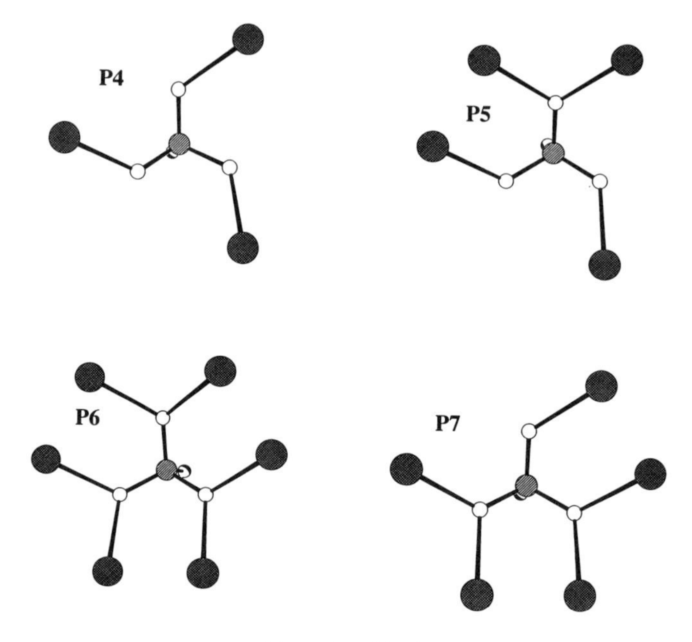

Figure 1. *(Continued)*

limited to the preparation of metal oxides and refractory materials. Furthermore, under solid state reaction conditions, only the most stable reaction product(s) are isolated, and, therefore the possibility of achieving analogous syntheses from simple molecular precursors of materials with desired features appears impractical. In other terms, the solid state analogue of organic functional group chemistry is still in an embryonic state.

Reaction fluxes or molten salts, which behave as high-temperature solvents to provide reaction media at elevated temperatures, promote the reactions of starting materials and crystallization of final product by enhancing diffusion rates and particle mobility of the reactants. Such conditions have been widely employed for preparing metal oxides and metal–chalcogenide based solid materials (77). However, this technique poses serious problems for the preparation of molecular species containing conventional organic ligands and even for solid phase materials incorporating organic substituents. Not only are most organic precursors decomposed at the high temperatures required, but the activation energy for diffusion is a substantial fraction of the individual bond strengths in the products and exceeds the activation energy for the conversion of metastable products into the thermodynamically favored product. Thus, there is no access to the kinetic products that are stable at lower temperatures.

Another common preparative method, the sol–gel technique, consists of making a homogenous solution of precursors in a suitable solvent, followed by controlled hydrolysis to form a gel. The technique has found wide applications in the preparation of oxide glasses and ceramics by the controlled hydrolyses of solvents (usually alcohols), to produce the gel, which after drying and pyrolysis produces the final products as glassy solids or ceramics (40). However, the products of sol–gel synthesis are characteristically amorphous polymeric materials, a distinct disadvantage in the design of complex molecular clusters and related solid phases with ordered arrays.

Conventional synthetic methods, on the other hand, involve hydrolysis and fragment condensation reactions or ''self-assembly'' of complex products from simpler molecular precursors in solution. Under these conditions molecular species present at the initial stages of hydrolysis–condensation processes may be isolated, and the structures and properties of these species may be related to those of solid materials formed under more forcing conditions. Furthermore, these simpler molecular products may in turn serve as the precursors for some of the low-temperature, nonconventional reactions, to be discussed in the following sections. Also, by bringing together these soluble building blocks, supramolecular materials and designer solids can be synthesized (13). The molecular building block approach is quite attractive since it provides not only a conceptional framework for the design of topologically controlled materials but also several synthetic advantages:

1. There is an extensive chemistry of coordination complexes and clusters that provide precursors for the synthesis of more complex assemblies.
2. Functional group chemistry and molecular symmetry of the precursors may be exploited in the synthesis of the product cluster.
3. The use of such soluble starting materials allows low-reaction temperatures, favoring the kinetic trapping of interesting metastable structures.
4. Reactions proceed by transformation of some of the functional groups of the organic and inorganic starting materials while retaining the covalent bonding relationships between most of the atoms, a feature that tends to direct the assembly of the more complex frameworks.

Furthermore, the structure-directing character of ionic bonds, through which these individual units may also be linked, provides an alternative approach to conventional solution synthesis of supramolecular materials, as evidenced by the success of ''Lego chemistry'' (73). However, the high degree of success that conventional synthetic methods enjoy in preparing simple molecular species is not uniformly paralleled in the preparation of supramolecular materials and solid phases. The current interest in giant clusters, and supramolecular, lamellar, and zeolitic materials has resulted in significant activity directed towards development of more effective synthetic procedures. Consequently, relatively low-temperature (100–400 °C) nonconventional synthetic techniques suitable for a variety of both molecular and nonmolecular materials have evolved over the past decade.

Nonconventional hydrothermal syntheses are conveniently employed for molecular complexes of varying degrees of complexity, 1D and lamellar solids, as well as 3D open-framework solid materials. Hydrothermal reactions, typically carried out at a temperature range 140–260 °C under autogenous pressure, exploit the self-assembly of the product from soluble precursors (78). The reduced viscosity of the solvent under these conditions results in enhanced rates of solvent extraction of solids and crystal growth from solution. Since differential solubility problems are minimized, a variety of simple precursors may be introduced, as well as a number of organic and/or inorganic structure-directing (templating) agents from which those of appropriate shape(s) and size(s) may be selected for efficient crystal packing during the crystallization process. Use of nonaqueous solvents such as alcohols, acetonitrile, acids, and mixtures of organic solvents in solvothermal synthesis (79), has received considerable attention in applications requiring the exclusion of water to prevent hydrolysis of desired products or for which effective solubility of reactants in nonaqueous solvents is practical or in which the solvent is itself a reactant (78). Solvothermal reactions are typically carried out at lower temperature (100–160 °C) than hydrothermal counterparts in sealed, thick-walled quartz or glass tubes. The technique is particularly suitable for air and moisture sensitive reactions.

While there are examples of the synthesis of large molecular clusters by stepwise fragment condensation processes, generally through hydrolysis–condensation reactions of simple precursors, the most common synthetic routes rely upon "self-assembly," the spontaneous formation of higher ordered structures (80). Self-assembly may involve simply the spontaneous formation of product upon combining the component parts under the proper conditions or may be regulated in a variety of ways such as precursor modification, postsynthesis modification of functional groups introduced into the product, direct self-assembly involving a buttressing agent, which itself does not appear in the final product, the introduction of structure directing templates, which often appear as guests in host–guest assemblies of molecular clusters or as charge compensating and space-filling entities in solid phase materials.

B. Representative Examples of Syntheses of Clusters and Solid Phase Materials

These general synthetic principles may be illustrated by several selected examples of the preparations of clusters and solid-phase materials of the general classes discussed in this chapter. Although the principles underlying polyanion accretion are not understood in detail, fragment condensation processes have been invoked (81, 82). The aggregation process is demonstrated most convincingly in the synthesis of higher oligomers from $[Mo_2O_4\{(OCH_2)_3CR\}_2]^{2-}$ (83–88). The structural interrelationships of the bi-, tri-, and tetranuclear species are shown in Fig. 2, which illustrates schematically the course of the aggregation process viewed as successive condensations of $[MoO_2(OR)]^+$ units.

$$[Mo_2O_7]^{2-} + 2H_3L \longrightarrow [Mo_2O_4L_2]^{2-} + 3H_2O$$

$$2[Mo_2O_4L_2]^{2-} + [Mo_2O_7]^{2-} + H_3L + MeOH \rightarrow [Mo_3O_6(OMe)L_2]^- + [Mo_3O_7L_2]^{2-} + 2H_2O + L^{3-}$$

$$2[Mo_3O_6(OMe)L_2]^- + [Mo_2O_7]^{2-} + 6MeOH \longrightarrow [Mo_4O_8(OMe)_2L_2] + [Mo_4O_{10}(OMe)_6]^{2-} + H_2O + 2H_2L^-$$

$$2[Mo_4O_8(OMe)_2L_2] + 4H_2O \longrightarrow [H_2Mo_8O_{20}(OMe)_4L_2] + 2H_3L$$

Similar fragment condensation processes are apparent in the vanadium chemistry. The reaction of tris(hydroxymethyl)alkane ligands with $[VO_2Cl_2]^{2-}$ under oxidizing conditions yields the binuclear $[(VO)_2Cl_2\{(OCH_2)_2CR(CH_2OH)\}_2]$, as shown in Fig. 3. The structure is reminiscent of that of $[Mo_2O_4\{(OCH_2)_3CR\}_2]^{2-}$ discussed above in exhibiting two sets of three alkoxy oxygen donors, each so disposed as to provide facial tridentate coordination. Further aggregation to produce higher oligomers should then be possible by condensation of appropriate units onto these faces. This expectation was realized in the preparation of a variety of tetranuclear species.

$$[(VO_2)Cl_2\{(OCH_2)_2CR(CH_2OH)\}_2] + Mo_2O_7^{2-} + 2MeOH \longrightarrow [V_2Mo_2O_8(OMe)_2\{(OCH_2)_3CR\}_2]^{2-} + 2HCl + H_2O$$

$$2[(VO)_2Cl_2\{(OCH_2)_2CR(CH_2OH)\}_2] + 2HSO_4^- + 2H_2O + 2MeOH \longrightarrow [(VO)_4(H_2O)_2(SO_4)_2\{(OCH_2)_3CR\}_2]^{2-} + 4HCl + 2H_2CO + 2\{(HOCH_2)_3CR\}$$

$$2[(VO)_2Cl_2\{(OCH_2)_2CR(CH_2OH)\}_2] + 3HOR \longrightarrow [(VO)_4(OR)_3\{(OCH_2)_3CR\}_3] + 4HCl + (HOCH_2)_3CR$$

The tetranuclear species in turn undergo hydrolysis–condensation reactions to form higher oligomers.

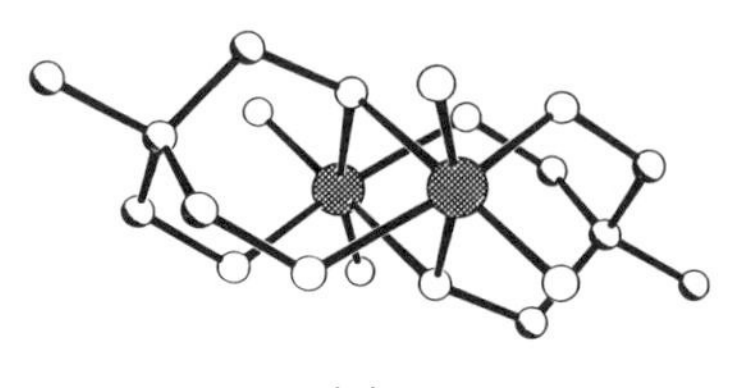

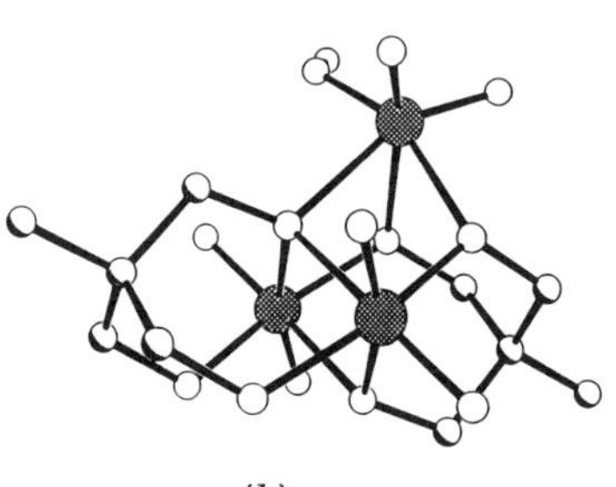

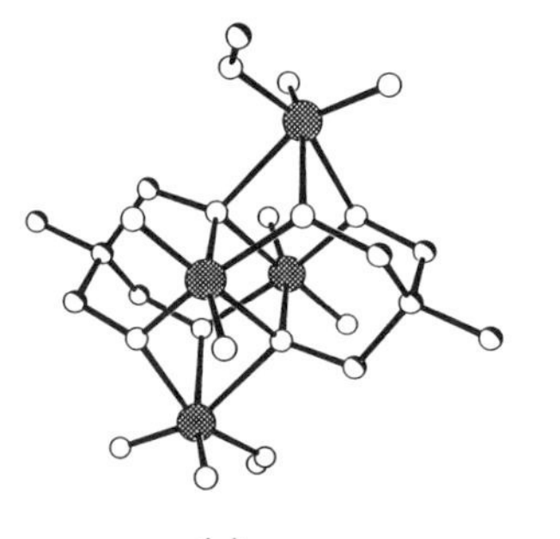

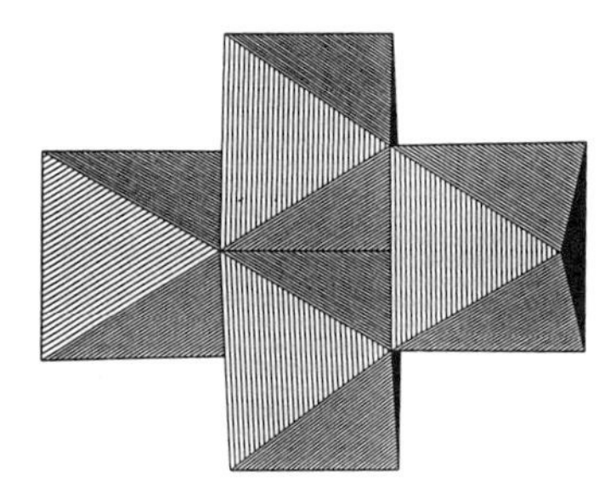

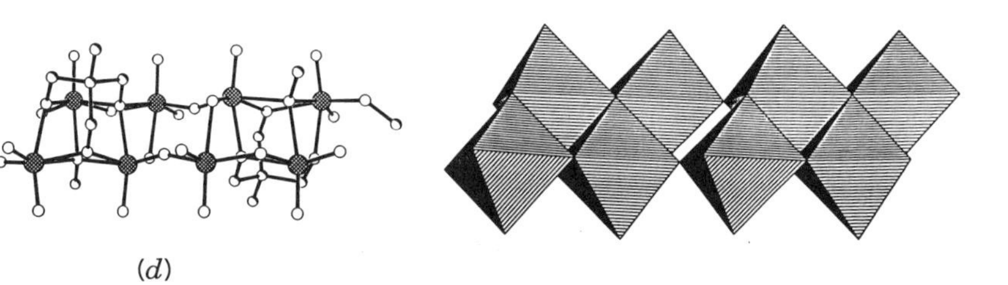

(*d*)

Figure 2. Ball-and-stick and polyhedral representations of the structures of (*a*) $[Mo_2O_4\{RC(CH_2O)_3\}_2]^{2-}$, (*b*) $[Mo_3O_7\{RC(CH_2O)_3\}_2]^{2-}$, (*c*) $[Mo_4O_8(OR)_2\{RC(CH_2O)_3\}_2]$, and (*d*) $[H_2Mo_8O_{20}(OR)_4\{RC(CH_2O)_3\}]_2$.

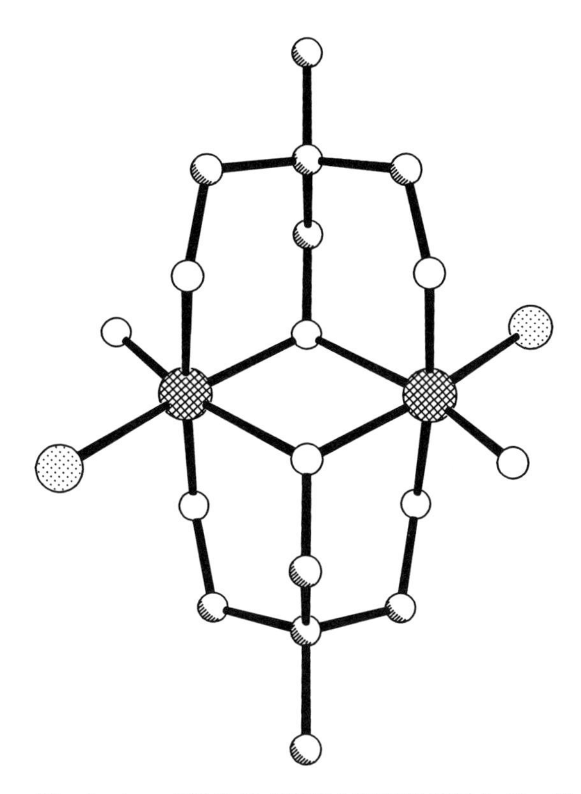

Figure 3. A view of the structure of $[V_2O_2Cl_2\{(OCH_2)_2CR(CH_2OH)\}_2]$. The chloride donors are represented as stippled spheres.

$$2[(VO)_4(OR)_3\{(OCH_2)_3CR\}_2] + 2[V_5O_{14}]^{3-} + 3H_2O \longrightarrow$$
$$3[V_2O_{13}\{(OCH_2)_3CR\}_2]^{2-} + 6HOR$$

$$3[(VO)_4(OR)_3\{(OCH_2)_3CR\}_2] + 4H_2O + 4MeOH + 4R_3N \longrightarrow$$
$$2[V_6O_8\{(OCH_2)_3CR\}_2\{(OCH_2)_2CR(CH_2OH)\}_4]^{2-} + 9MeOH$$
$$+ 4H_2CO + (HOCH_2)_3CR + 4R_3NH^+$$

The isolation of reduced and mixed-valence species, $[(VO)_4(H_2O)_2(SO_4)_2\text{-}\{(OCH_2)_3CR\}_2]^{2-}$ and $[V_6O_8\{(OCH_2)_3CR\}_2\{(OCH_2)_2CR(CH_2OH)\}_4]^{2-}$, respectively, reflects the facile reduction of V^V species in alcoholic solutions and demonstrates the additional structural complexity, which is introduced by the rich redox chemistry of the metal oxides.

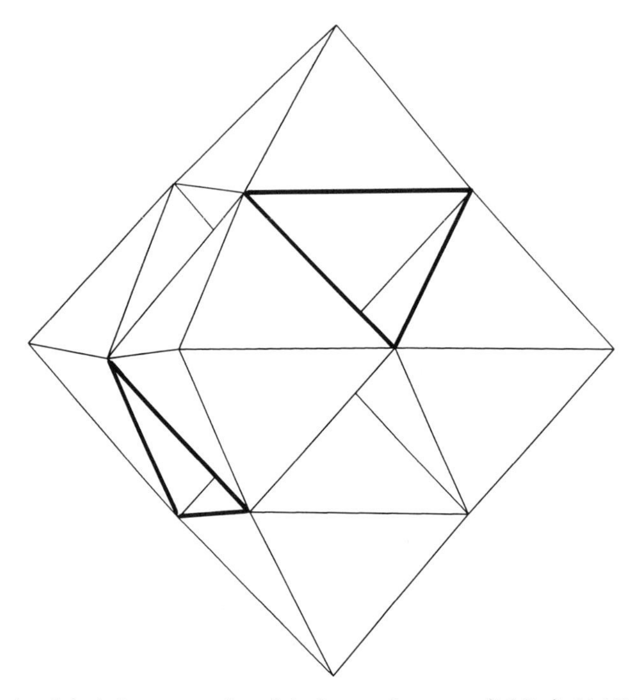

Figure 4. A polyhedral representation of the hexanuclear core, $\{M_6O_{19}\}$, highlighting the triangular faces commonly occupied by the oxygen donors of trisalkoxide ligand types.

Two features of the hexametalate core are noteworthy. As shown in Fig. 4, the tris(hydroxymethyl)alkane ligands occupy the triangular faces of the tetrahedral cavities of the hexametalate framework. Although there are eight triangular faces of this type associated with the $\{V_6O_{19}\}$ core, only four may be occupied at any one time by the trisalkoxy ligands. Under the conditions of conventional synthesis, the only derivatives isolated were those of the classes with one or two trisalkoxy ligands, $[V_6O_{13}(MeO)_3\{(OCH_2)_3CR\}]^{2-}$ and $[V_6O_{13}\{(OCH_2)_3CR\}_2]^{2-}$ respectively. Similarly, the $\{V_6O_{19}\}$ core should serve as a framework for further condensation to yield higher oligomers. Again, under the conditions of conventional synthesis, no clusters of higher nuclearity could be isolated. However, by adopting the conditions of hydrothermal synthesis both the progressively substituted hexametalate cores and higher oligomers were readily isolated. Thus, in the presence of Ba^{2+} cations, vanadate, and $(HOCH_2)_3CMe$ react at 150°C in H_2O to give $Ba[V_6O_7(OH)_3\{(OCH_2)_3CMe\}_3]$, while the reaction under similar conditions with Na^+ yields the tetra-substituted product $Na_2[V_6O_7\{(OCH_2)_3CMe\}_4]$. By raising the temperature and increasing reactions times, decanuclear cores of the classes $[V_{10}O_{16}\{(OCH_2)_3CR\}_4]^{n-}$ and $[V_{10}O_{13}\{(OCH_2)_3CR\}_5]^{1-}$ are isolated.

The interplay of fragment condensation, self-assembly, templating effects, and solvothermal synthesis is demonstrated in the preparation of $[V_{18}O_{25}(H_2O)_2(PhPO_3)_{20}Cl_4]^{4-}$. The reaction of $[V_5O_{14}]^{3-}$ with $PhPO_3H_2$ in methanol spontaneously yields the pentanuclear cluster $[V_5O_7(OMe)_2(PhPO_3)_5]^{1-}$ [Fig. 5(*b*)] (89). Controlled hydrolysis with HCl/H_2O in acetonitrile at 100°C of the pentanuclear species yields the octadecanuclear $[V_{18}O_{25}(H_2O)_2(PhPO_3)_{20}Cl_4]^{4-}$, a "cluster of clusters" formed in the condensation of five pentanuclear building blocks and encapsulating four chloride anions in a host–guest type assembly [Fig. 5(*a*)].

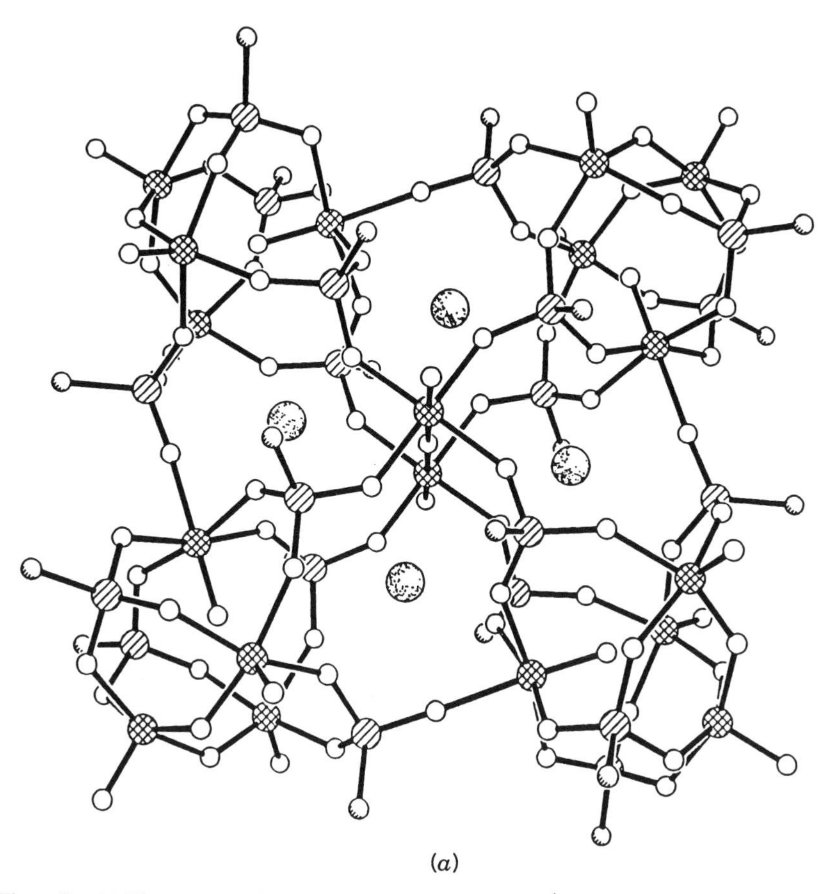

Figure 5. (*a*) The structure of the $[V_{18}O_{25}(H_2O)_2(PhPO_3)_{20}Cl_4]^{4-}$ host–guest assembly. (*b*) Schematic representation of the structure of the $[V_5O_7(OMe)_2(PhPO_3)_5]^{1-}$ core, which provides the structural motif for the octadecanuclear cluster by loss of the alkoxy group and the pendant $\{VO(OR)\}^{2+}$ unit shown as dashed spheres.

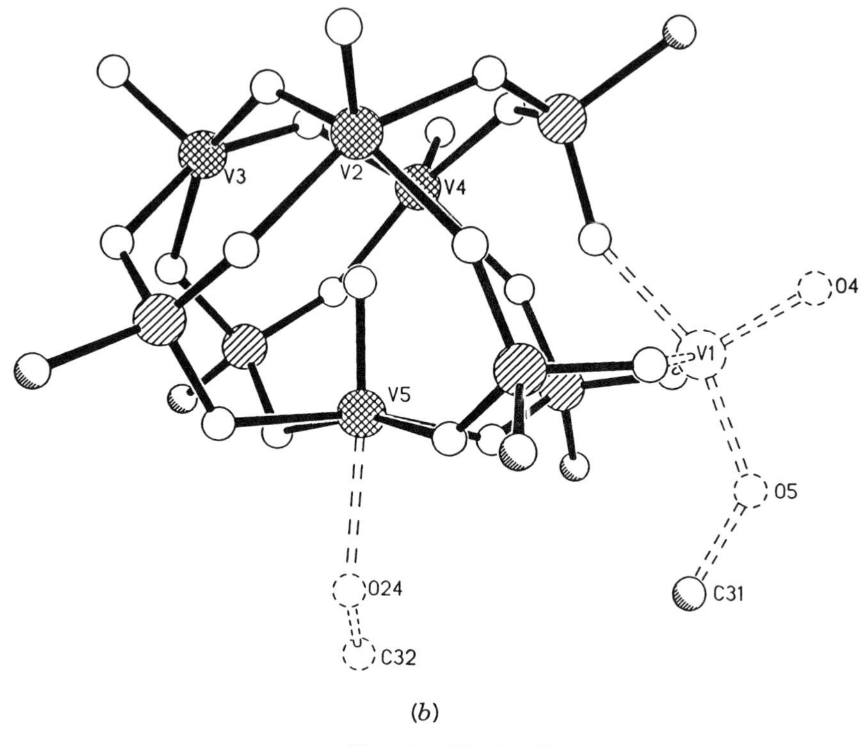

(*b*)

Figure 5. (*Continued*)

The principles of self-assembly under hydrothermal conditions to produce nanometer sized clusters are also applicable to the isolation of the ''supercluster'' $[Na(H_2O)_3H_{15}Mo_{42}O_{109}\{(OCH_2)_3CR\}_7]^{7-}$. While the synthesis may be effected directly by hydrothermal reaction of the precursors Na_2MoO_4, Mo metal, and ligand, stepwise synthesis has also been effected. The latter route relies on the self-assembly of the structural core $[H_nMo_{16}O_{52}]^{(20-n)-}$ (90) shown in Fig. 6. This reactive, highly negatively charged and nucleophilic intermediate can then be linked to appropriate subunits to produce the supercluster. In this fashion, the hydrothermal reaction of $[H_{14}Mo_{16}O_{52}]^{6-}$ with Na_2MoO_4 and $(HOCH_2)_3CR$ yields the unusual $[Na(H_2O)_3H_{15}Mo_{42}O_{109}\{(OCH_2)_3CR\}_7]^{7-}$. Although this fragment condensation process, shown schematically below, provides a satisfactory conceptual framework for the preparation of tailor-made clusters by control of inorganic hydrolysis–condensation and aggregation reactions, structural preassembly of the constituent subunits relies on self-assembly.

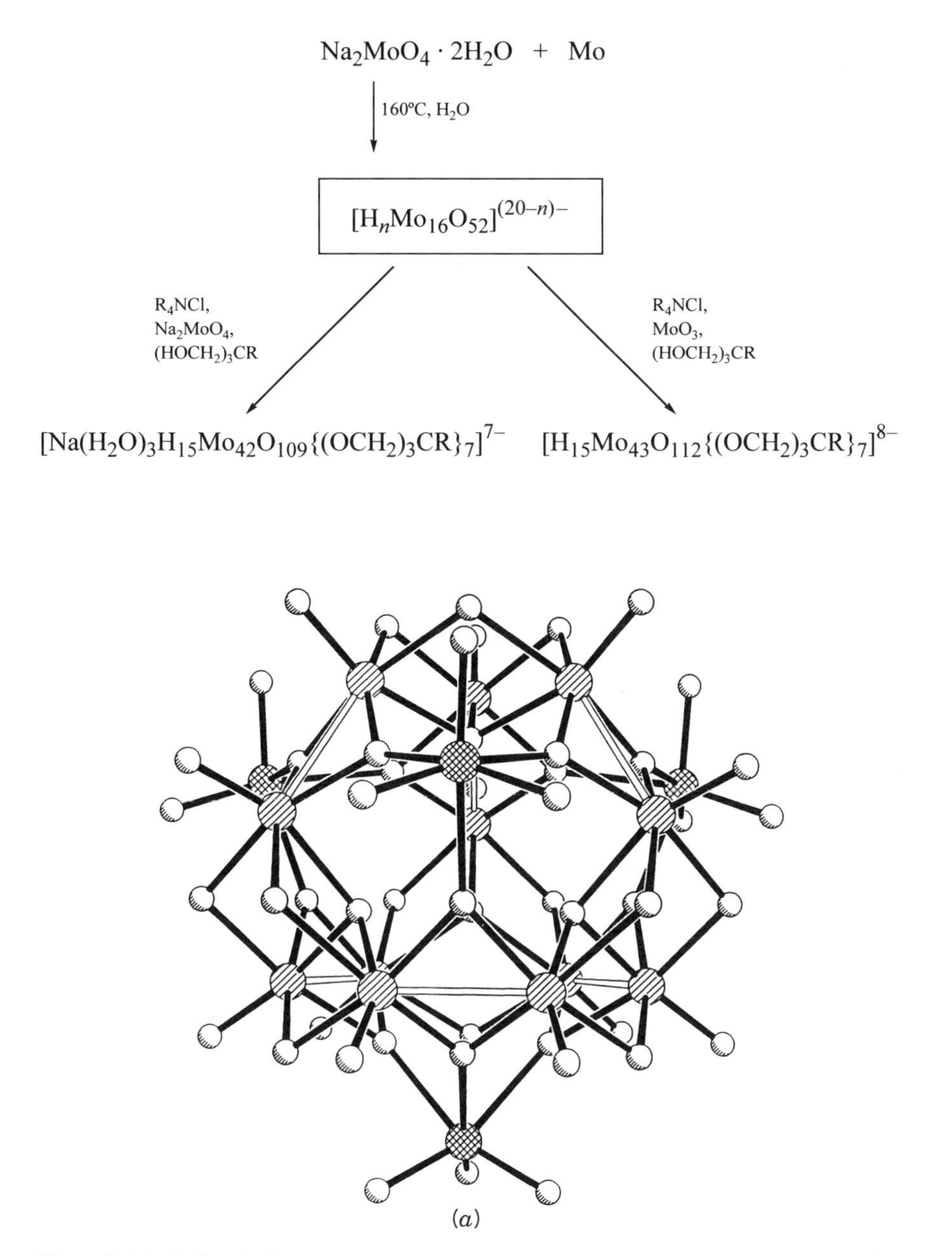

Figure 6. (*a*) Ball-and-stick and polyhedral representations of the structure of $[H_{14}Mo_{16}O_{52}]^{6-}$, which consists of an ε-Keggin core of 12 edge-sharing octahedra capped by four {MoO} groups. The twelve Mo^V sites of the Keggin core form six pairs of Mo^V-Mo^V dimers with short Mo—Mo distances. The four $\{Mo^{VI}O_3\}$ sites are shown as cross-hatched spheres and octahedra. (*b*) A polyhedral representation of the anion cluster.

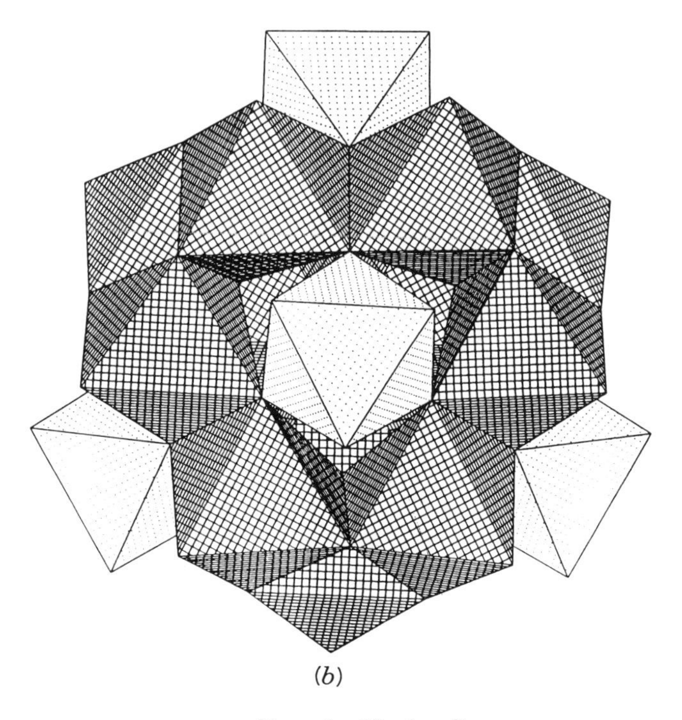

(*b*)

Figure 6. (*Continued*)

That the structures of metal oxide containing materials may be influenced by the choice of templating agent is a noteworthy observation and one amply demonstrated in the solid state phases of the $V/O/RPO_3^{2-}$ system. In the absence of organic cations, the reaction of V_2O_5 and $PhPO_3H_2$ in H_2O yields the prototypical layered material $[VO(PhPO_3)(H_2O)]$. In contrast, layered materials of significantly different structures within the inorganic V/P/O layer and even with respect to the registry of inorganic and organic layers are isolated upon addition of appropriate templates, as illustrated in the structures of $[EtNH_3][(VO)_3(H_2O)(PhPO_3)_4]$ and $[Et_2NH_2][Me_2NH_2][(VO)_4(OH)_2(PhPO_3)_4]$, as described in Section V1.A.

In the presence of the same templating reagent, modifications of the organic groups of the organophosphonates may also result in dramatic structural changes. For example, ethylenediammonium phases of the V/O/diphosphonate systems with 1D, 2D, and 3D structures may be prepared under similar conditions simply by varying the tether length n of the diphosphonate $\{O_3P(CH_2)_nPO_3\}^{4-}$, as illustrated by the phases $[H_3NCH_2CH_2NH_3][VO\{(O_3P)_2CH_2\}]$, $[H_3NCH_2CH_2NH_3][(VO)(O_3PCH_2CH_2PO_3\}]$, and $[H_3NCH_2CH_2NH_3]$-

$[(VO)_4(OH)_2\{O_3PCH_2CH_2CH_2PO_3\}]$, respectively. While the templating mechanism is not well understood, judicious choices of templating agents may be exploited to fashion new and often unexpected structures completely.

IV. GENERAL STRUCTURAL CHARACTERISTICS OF POLYOXOANION COORDINATION COMPLEXES

A. Cores Containing Molybdenum(VI) Sites

As pointed out by Pope (26), the formation of polyoxoanions depends both on the appropriate relationship of Coulombic factors of ionic radius and charge and on the accessibility of empty *d* orbitals for metal–oxygen bonding. Polyanion structures can often be represented as clusters of edge-sharing MO_6 octahedra in which the metal ions are displaced as a consequence of M—O π-bonding toward the vertices at the surface of the structure (27). Two other general principles are associated with classical polyoxoanion structures of molybdenum(VI) and tungsten(VI):

1. Polyanion structures that contain three or more terminal oxo groups are not observed, a restriction that has been explained in terms of the strong trans influence of terminal M—O bonds facilitating dissociation of the cluster (91).
2. The shapes of polyanions can be predicted by linking octahedra in an edge-sharing fashion such that each octahedron shares two edges on the same face.

The first restriction, "Lipscomb's principle," is not absolute as evidenced by the existence of several mononuclear *fac*-trioxomolybdenum complexes (92–96) and even "anti-Lipscomb" polyoxomolybdate clusters (97–100). In the case of polyoxoanion coordination complexes, ligand influences often modify such general principles (97, 98).

The second principle, the "rule of compactness" (101), allows the construction of polyanion structural cores from the fundamental structural motif, the pair of edge-sharing octahedra, by aggregation through edge-sharing of additional octahedra (102), as shown in Fig. 7. Of these potential structural prototypes, only the hexanuclear and the octanuclear structures have been structurally characterized for the underivatized polyoxomolybdates. In nonaqueous solvents, $[Mo_6O_{19}]^{2-}$ (103) and β-$[Mo_8O_{26}]^{4-}$ (104) may be isolated, while the heptamolybdate $[Mo_7O_{24}]^{6-}$ (105) has been confirmed as the predominant species in aqueous solution at pH 3–5.5. The decanuclear core $[V_{10}O_{28}]^{6-}$ is common to both the aqueous and nonaqueous solution chemistry of vanadium(V)

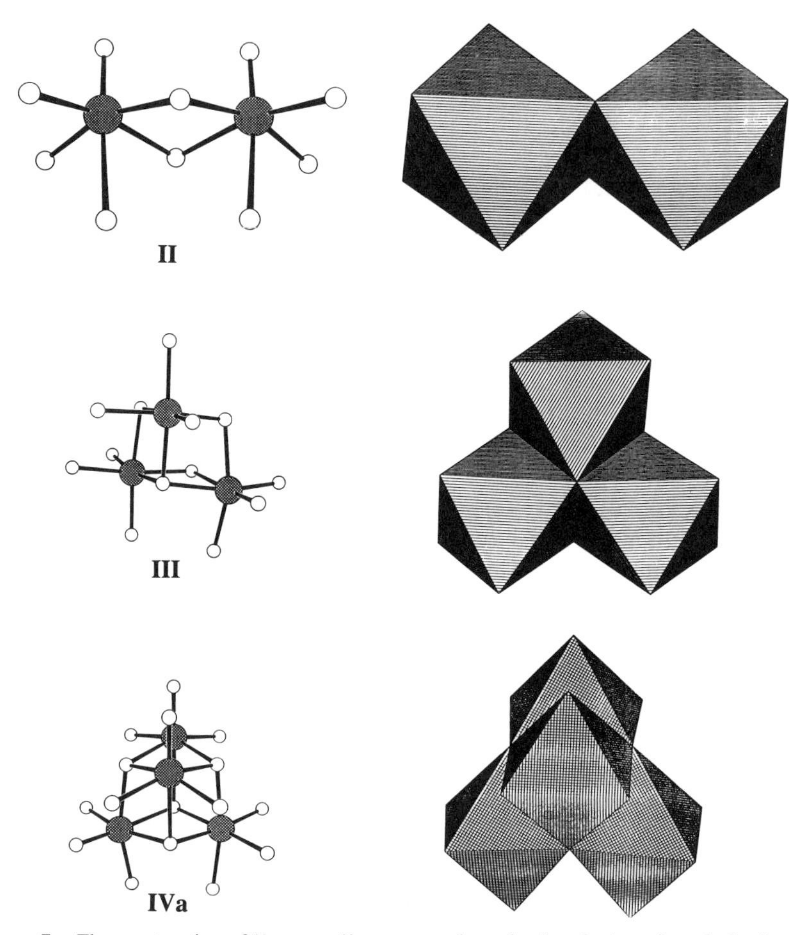

Figure 7. The construction of ''compact'' structures through edge sharing of octahedra for common aggregates formed for Mo^{VI} and some V^{V}/V^{VI} cluster types: II, binuclear core; III, trinuclear core; IVa and IVb, two possible packing geometries for the tetranuclear core; V, pentanuclear core; VI, hexanuclear core; VII, heptanuclear core; VIII, octanuclear core; X, decanuclear core.

and numerous structural determinations have been carried out (106). The tetranuclear core is known only in the solid phase $Li_{14}(WO_4)_3W_4O_{16} \cdot 4H_2O$ (107). However, the introduction of common organic ligands expands the structural chemistry of the polyoxomolybdates to include not only such prototypes but a variety of novel geometries constructed from corner sharing and face sharing of octahedra, in addition to the more common edge sharing, and from tetrahedral and square pyramidal units, as well as the more common octahedral motifs.

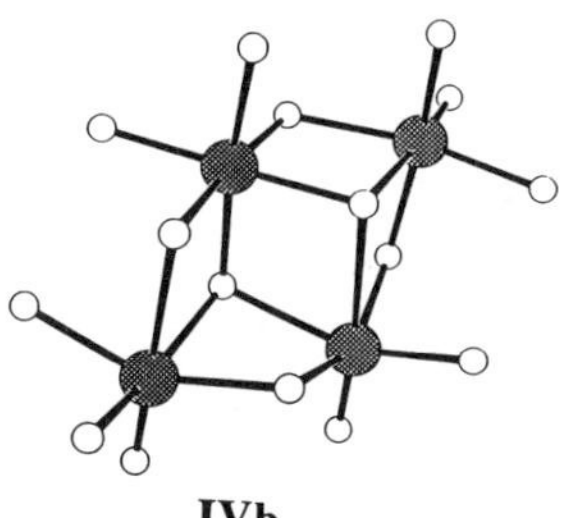

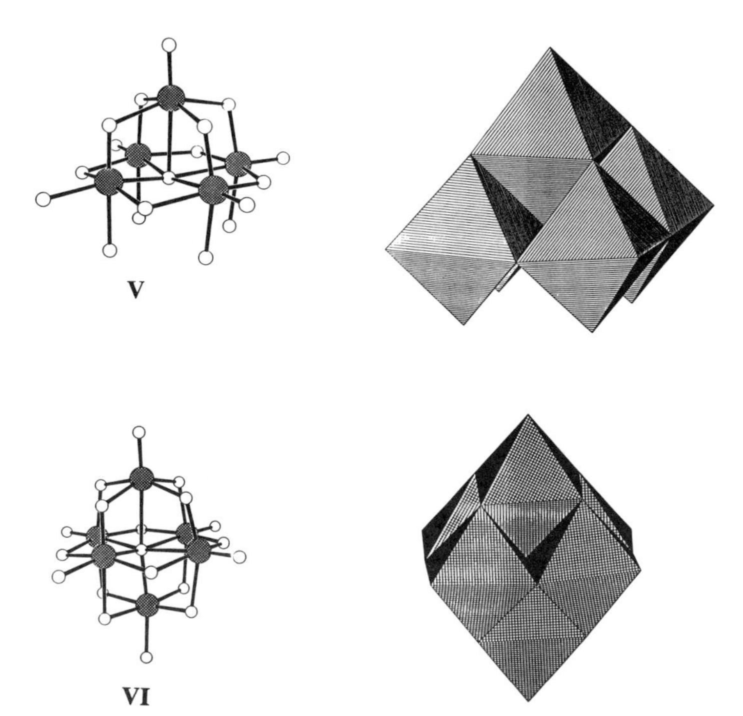

Figure 7. *(Continued)*

B. Binuclear Molybdenum(V)–Molybdenum(V) Cores

While the classical isopolymolybdates conforming to the structural models of Fig. 7 contain Mo^{VI} centers, in recent years a new class of high nuclearity oxomolybdenum complexes has emerged, one characterized by Mo^{V} sites with localized Mo^{V}–Mo^{V} single bonds (31). In these instances, the fundamental structural motif may be described as the $\{Mo_2O_8\}$ unit of Fig. 8, consisting of

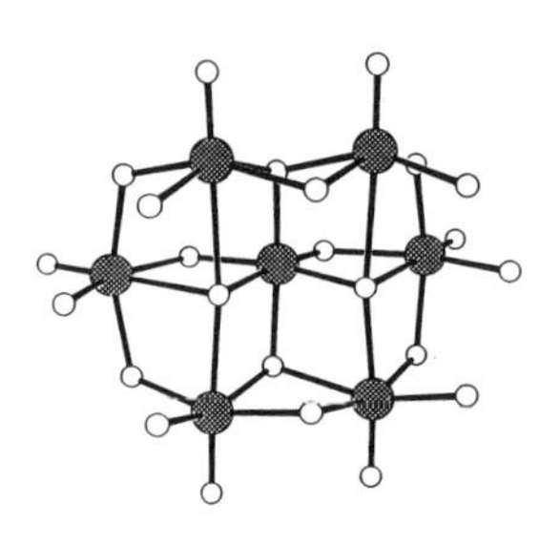

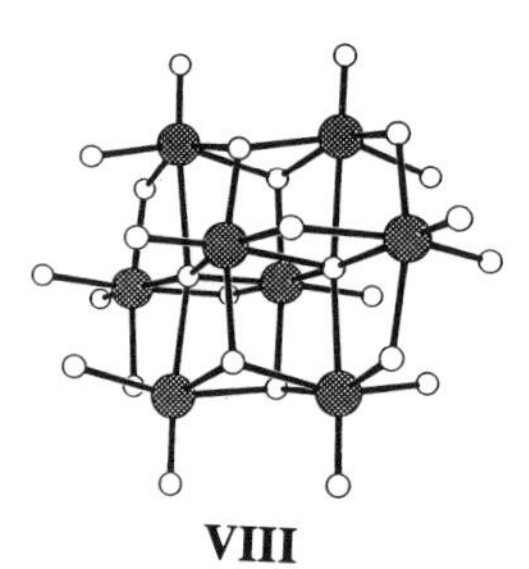

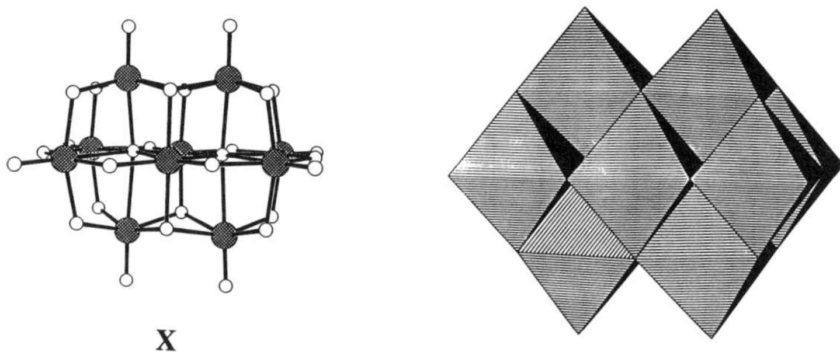

Figure 7. (*Continued*)

edge-sharing square pyramids with a short Mo—Mo distance in the range 2.5–2.7 Å. In fact, structures of this class generally exhibit additional weak axial interactions trans to the multiply bonded terminal oxo groups to produce a highly distorted octahedral environment about the Mo^V centers. Whichever view of the bonding is adopted, the structures of this subgroup may be rationalized by linking $\{Mo_2O_8\}$ units to form assemblies of corner- and/or edge-sharing square pyramids or octahedra.

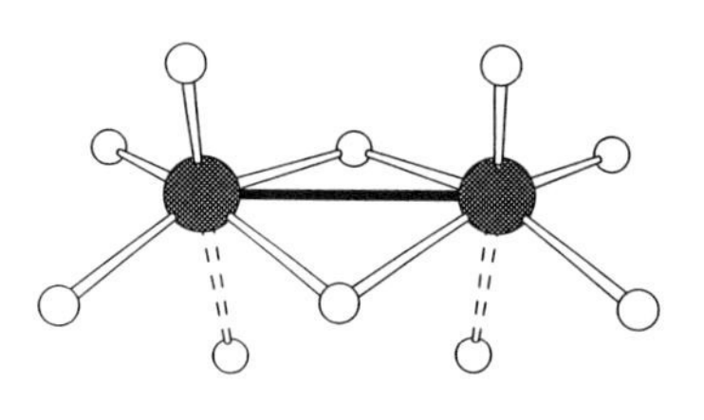

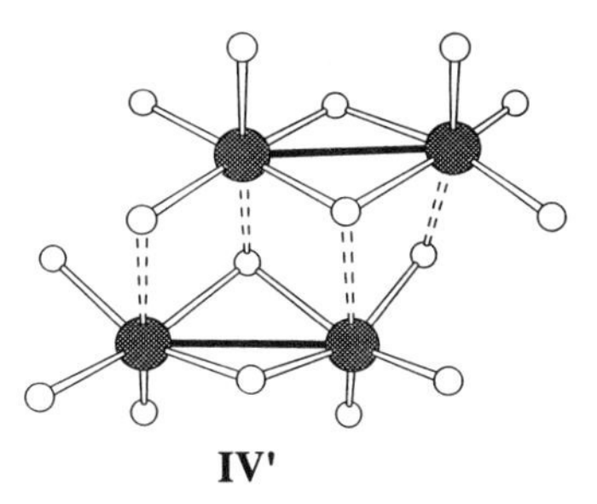

IV'

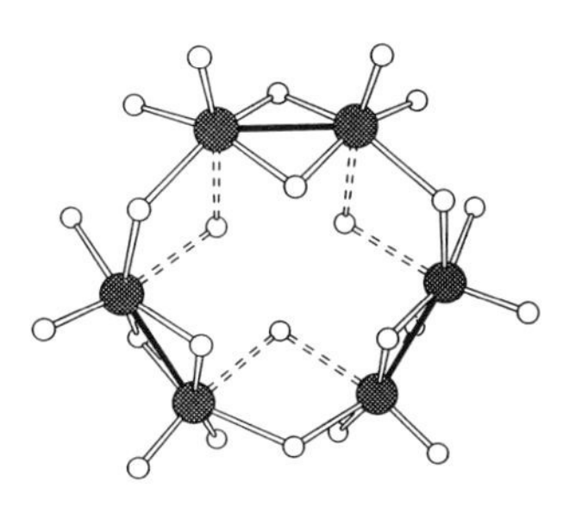

VI'

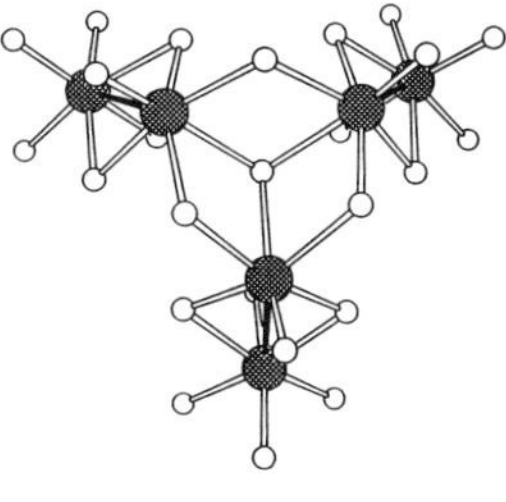

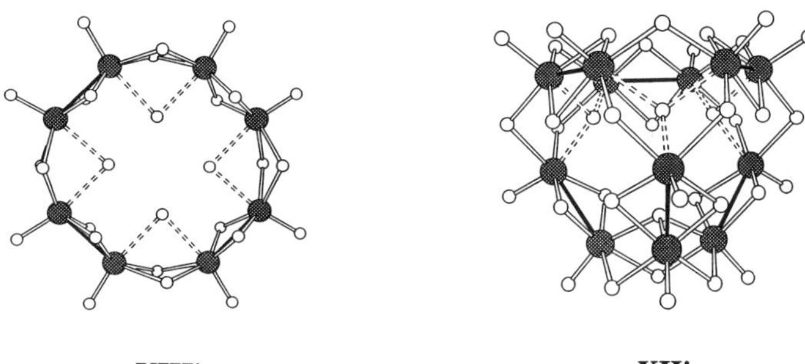

Figure 8. Cluster cores based on the $\{Mo^{V}_{2}O_{10}\}$ motif, which is characterized by short $Mo^{V}—Mo^{V}$ distances and a relatively weak Mo—O distance to the sixth ligand site trans to the multiply bonded terminal oxo group. Binuclear core II′; tetranuclear core, IV′; hexanuclear cores VI′ and VI″; octanuclear core VIII′; and dodecanuclear core XII′.

C. Vanadium(V) and Vanadium(IV) Cores

Although only four high nuclearity, "naked" core vanadium(V) isopolyanions are known, considerable structural variability is observed. While $[V_{10}O_{28}]^{6-}$ (106) and $[V_{13}O_{34}]^{3-}$ (108) conform to the general principles of classical polyanion structural chemistry, $[V_{12}O_{32}]^{4-}$ (109) exhibits a bowl-like structure constructed from edge-sharing square pyramids and $[V_{15}O_{42}]^{9-}$ (110) consists of a central Keggin core with two square pyramidal capping units, resulting in a structure with tetrahedral, square pyramidal, and octahedral vanadium centers.

Introduction of reduced V^{IV} sites into polyoxovanadium frameworks further expands the structural chemistry. Nearly all ratios of V^{IV}/V^{V} have been observed (27). However, unlike the Mo^{V}-containing clusters, metal–metal interactions are generally not observed. Furthermore, the variability in the coordination polyhedra adopted by vanadium (tetrahedral, square pyramidal, trigonal bipyramidal, octahedral) allows the construction of structural types not observed for the molybdenum oxide clusters. A recurrent theme of the chemistry of reduced and mixed-valence, naked core polyoxovanadates is the formation of "hollow spheres" capable of encapsulating a variety of templates (11).

The structural versatility of the polyoxovanadium clusters is enhanced by the introduction of oxygen-donor ligands, which may serve to substitute for peripheral oxo groups of the parent polyvanadate core or which may bridge $\{V_xO_y\}$ units in the construction of novel cluster frameworks. The influence of templates (anionic, cationic, and neutral) is also pronounced in the isolation of polyoxovanadium coordination compounds, resulting in a new family of polyanion aggregates with host–guest characteristics (13).

V. OXOVANADIUM CLUSTERS

A. Oxovanadium Clusters with Alkoxide, Carboxylate, Carbon Oxo, and Related Ligand Types

1. *Trinuclear Clusters*

The reaction of $VCl_3 \cdot 3thf$ [where thf = tetrahydrofuran (ligand)] with sodium benzoate in dichloromethane yields the trinuclear V^{IV} species $[(VO)_3(thf)(PhCO_2)_6]$ (**1**) (111). As shown in Fig. 9, the structure consists of a trinuclear, oxo-centered carboxylate-type cluster (112–115). The structure does not conform to the "rule of compactness," which would result in a structure of type III, but rather exhibits three V^{IV} octahedra sharing a central vertex and bridged through symmetrically coordinated carboxylates of type C2. A curious feature of the structure is the geometry of the triply bridging oxo group. In contrast to other structures of the $\{M_3O\}$ core, this oxo group forms a short, multiply bonded $\{V{=}O\}$ interaction at 1.62 Å to one vanadium site and long $V \cdot \cdot \cdot O$ distances of about 2.40 Å to the remaining two vanadium centers. This structural feature defines the cluster as an oxovanadium(IV) trinuclear species $\{(VO)_3\}$ rather than a normal oxo-centered trinuclear complex $\{V_3O\}$, a

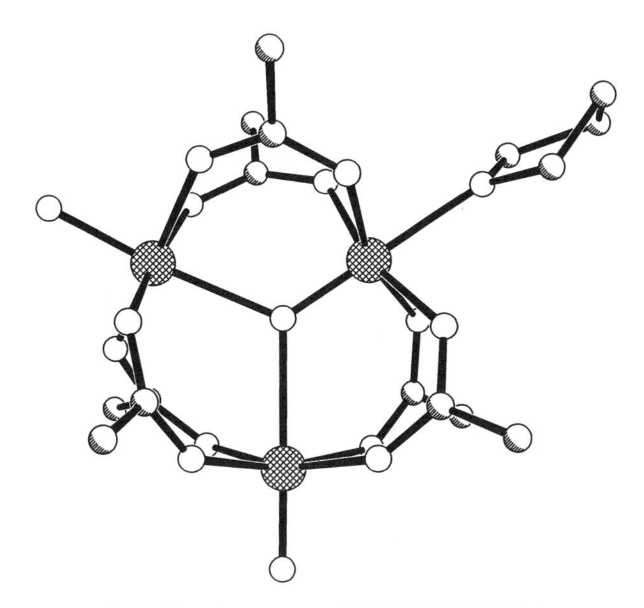

Figure 9. The structure of $[(VO)_3(thf)(PhCO_2)_6]$.

characteristic presumably associated with the high-oxidation state of the metals. Other examples of oxo-centered trinuclear complexes exhibit mean oxidation states of III or lower. The higher oxidation state of vanadium centers of $[(VO)_3(thf)(PhCO_2)_6]$ (**1**) provides strong covalent bonding that reduces the effective positive charge in the metal centers, thus favoring the formation of oxovanadium(IV) units. It is noteworthy that $[(VO_3)(thf)(PhCO_2)_6]$ provides the lowest molecularity example of a terminal oxovanadium moiety directed toward the interior of the cluster cavity, rather than toward the exterior as almost universally observed for $\{M{=}O\}$ units. Thus, this simple trinuclear core provides a structural prototype for the pentanuclear $[V_5O_7(OMe)_2(PhPO_3)_5]^{1-}$ (**42**) and the dodecanuclear $[(VO)_{12}(OH)_2(PhAsO_3)_{10}(PhAsO_3H)_4]^{4-}$ (**50**), discussed in Section V.B.

In contrast to the oxo-bridged core of $[(VO)_3(thf)(PhCO_2)_6]$, the structure of $[(VO)_3(sal)_3(MeOH)_3]$ (**2**) (116) exhibits "isolated" vanadium octahedra, that is, there is no edge or corner sharing between vanadium centers through bridging oxo groups. The $\{VO_5N\}$ vanadium octahedra are lined through the salicylhydroximate groups that are present in the triply deprotonated form, identifying the cluster as a V^V species. The deprotonated hydroximate nitrogen and phenolate oxygen bind to one vanadium site while the two remaining oxygen atoms bind to an adjacent vanadium. Repetition of this motif produces a triangular unit with the $\{{-}V{-}N{-}O{-}\}_3$ core.

The compact arrangement of three corner-sharing octahedra is represented by the trinuclear $[(VO)_3\{OCH_2)_3CR\}]^{6+}$ unit of $[RC(OCH_2)_3V_3P_2W_{15}O_{59}]^{6+}$, which may be regarded as a trinuclear $[(VO_3)\{(OCH_2)_3CR\}L]^{6-}$ complex with the "ligand" L identified as the lacunary Dawson cluster $[P_2W_{15}O_{56}]^{12-}$ (117, 118). The tris(hydroxymethyl)ethane ligand adopts the common coordination mode TA3 with the alkoxy oxygen donors bridging three metals in a triangulo arrangement. Such ligands are generally observed to cap the triangular faces of the tetrahedral cavities formed in the compact structures of oxide clusters exhibiting corner-sharing octahedral motifs. The synthesis of $[(VO)_3\{(OCH_2)_3CR\}L]^{6-}$ exploits this tendency and proceeds directly from the reaction of the tris-alkoxy ligand and the mixed-addenda Dawson cluster (119).

$$RC(CH_2OH)_3 + H_4[V_3P_2W_{15}O_{62}]^{5-} \longrightarrow$$

$$[RC(OCH_2)_3V_3P_2W_{15}O_{59}]^{5-} + 3H_2O$$

Curiously, the cluster exhibits substantial kinetic stability with respect to hydrolysis, a feature of some interest in the design of polyoxometalates with catalytic applications (120).

2. *Tetranuclear Clusters*

The tetranuclear clusters of the oxovanadium unit with organic oxygen-donor ligands exhibits two fundamental structural cores, the compact edge-sharing arrangement denoted as type V1a of Fig. 7 and the more open framework based on the corner-sharing $\{V_4O_8\}$ unit shown in Fig. 10.

The V^V species $K_4[V_4O_8(H_2O)_2(C_2O_4)_4]$ (121), shown in Fig. 10, illustrates the latter coordination type. The structural core consists of a cyclic $\{V_4O_8\}^{8+}$ unit with the central V_4O_4 array arranged in a crown conformation with the vanadium atoms and the bridging oxo groups in distinct planes. The distorted octahedral coordination geometry about each vanadium center is completed by the chelating oxalate ligand in type O2 coordination mode, a terminal oxo group, and an aqua ligand bridging alternating pairs of vanadium sites. The presence of these bridging aqua groups results in an overall structure that may be described as two pairs of edge-sharing octahedra linked via two corner-sharing interactions. The resulting open framework structure possesses a distinct central cavity, a characteristic that is exploited in accommodating templating cations or anions in the structures of $[V_4O_7(OH)(O_2CR)_4K]^+$ and $[V_4O_8(O_2CR)_4(NO_3)]^{2-}$. The $\{V_4O_8\}^{n+}$ grouping also provides a recurring structural motif in larger oligomers, a feature that is apparent in the structure of $[(V_4O_8)_2\{(VO)_4(H_2O)_{12}\}(PhPO_3)_8Cl_2]^{2-}$ described in Section V.B. It is also noteworthy that capping the $\{V_4O_8]^{n+}$ unit with a $\{VO\}^{3+}$ or an $\{AsR\}^{4+}$ moiety generates the $\{V_5O_9\}^{n+}$ and $\{V_4O_8(AsR)\}^{n+}$ assemblies, respectively, which provide com-

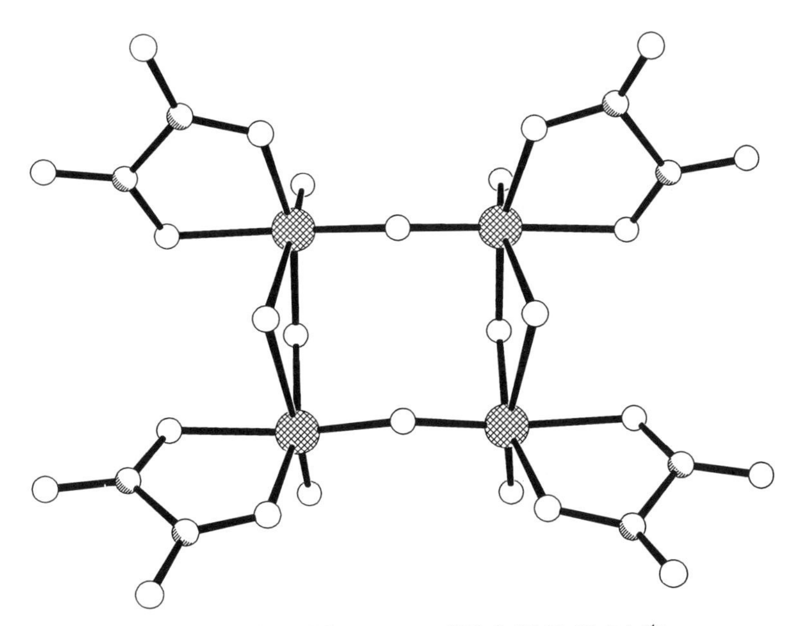

Figure 10. A view of the structure of $[V_4O_8(H_2O)_2(C_2O_4)_4]^{4-}$.

mon structural motifs in higher oligomers of the vanadium oxide system, as described in Section V.B.

The bowl-like cavity associated with the $\{V_4O_8L_4\}$ framework is exploited in the isolation of $[V_4O_8(O_2CR)_4(NO_3)]^{2-}$ (122), a $V^{IV}/3V^{V}$ species whose structure is shown in Fig. 11. The cluster is representative of the growing class of mixed-valence vanadium oxide clusters (123) whose chemistry is now beginning to emerge (27). While $[V_4O_8(O_2CR)_4(NO_3)]^{2-}$ remains the central $\{V_4O_8\}$ core previously described for $[V_4O_8(C_2O_4)_4(H_2O)_2]^{4-}$, the replacement of the oxalate ligands by carboxylate groups and capture of the nitrate group in the molecular cavity results in significant structural modification. Whereas the oxalate ligands of $[V_4O_8(C_2O_4)_4(H_2O)_2]^{4-}$ adopt the chelating mode O2, the carboxylate groups of $[V_4O_8(O_2CR)_4(NO_3)]^{2-}$ symmetrically bridge adjacent vanadium centers. Furthermore, the oxygen donor of the nitrate guest molecule bridges the four vanadium sites, resulting in a ring of four vanadium octahedra sharing a common central vertex. Alternatively, the structure may be viewed as a ring of corner-sharing square pyramids providing a cavity for a weakly associated guest molecule.

The structural flexibility of the $\{V_4O_8\}^{n-}$ core is demonstrated by the remarkable $V^{IV}/3V^{V}$ cluster $[V_4O_7(OH)(O_2CR)_4K]^+$ (124). While the gross features of the $V/O/RCO_2^-$ cores of $[V_4O_8(O_2CR)_4(NO_3)]^{2-}$ and $[V_4O_7(OH)(O_2CR)_4K]^+$ are similar, the latter structure exhibits a protonated bridging

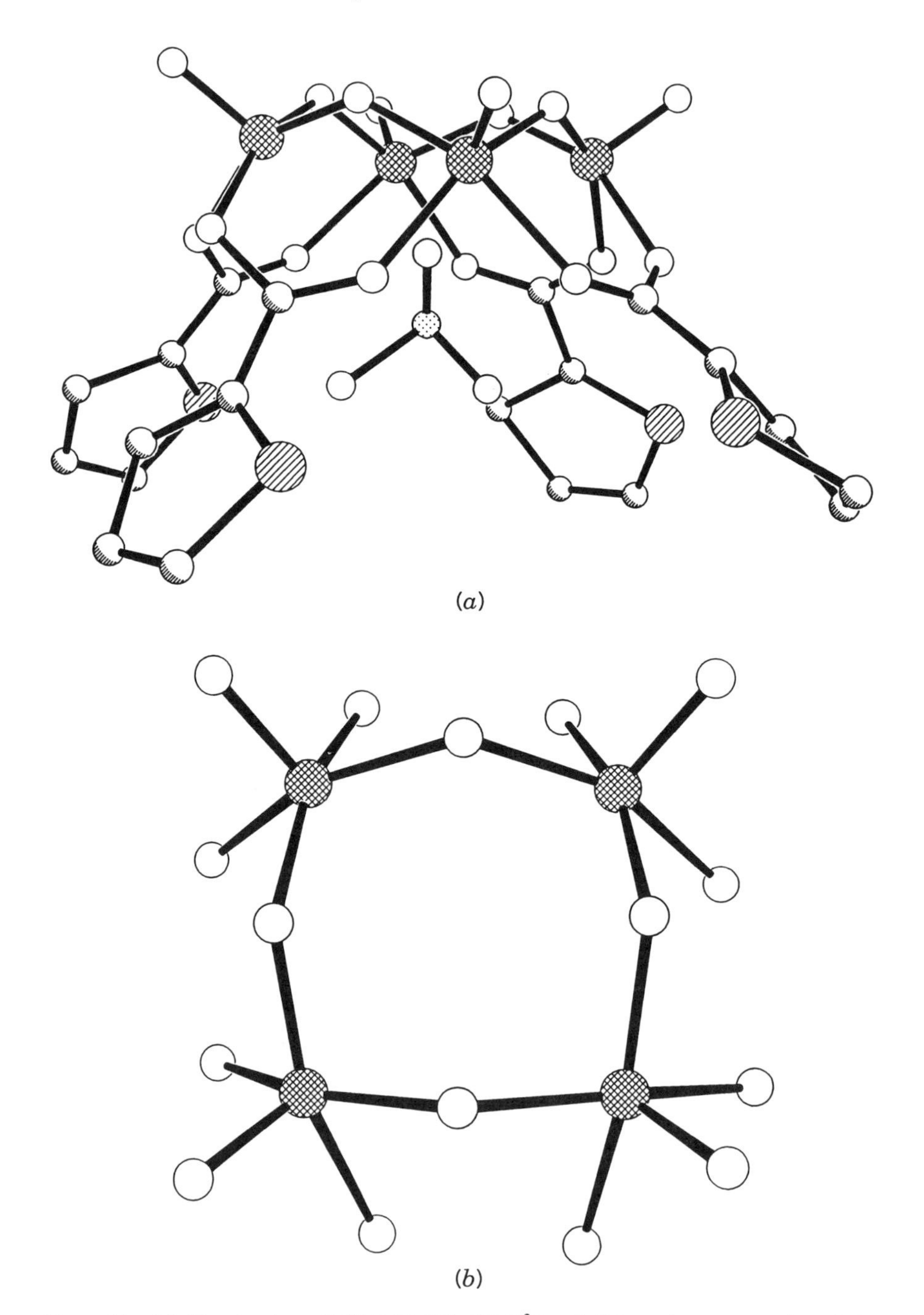

Figure 11. (*a*) The structure of $[V_4O_8(O_2CR)_4(NO_3)]^{2-}$. (*b*) Schematic representation of the $\{V_4O_4\}$ ring produced by corner sharing of four vanadium square pyramids.

oxo group and a K^+ cation replacing the NO_3^- group in the molecular cavity. As illustrated in Tables VI and VII, the identity of bridging —OH groups is clearly revealed by the metrical parameters when compared to those for bridging oxo groups. Valence sum calculations (125) on the oxygen atoms are consistent with these arguments and also serve to identify protonated oxo groups in metal oxide coordination complexes. While "naked" core metal oxides $\{M_xO_y\}^{n-}$ are invariably anions, $[V_4O_7(OH)(O_2CR)_4K]^+$ demonstrates an intriguing characteristic of the coordination chemistry of polyoxoanion-based clusters—the introduction of organic ligands into the core structure allows the isolation of neutral and even cationic clusters, $[M_xO_yL_z]^0$ and $[M_xO_yL_z]^{n+}$ types, respectively.

While the tetranuclear clusters, **4**, **5**, and **6** of Table I exhibit the $\{V_4O_4\}^{n-}$ core reminiscent of the "naked" core cluster $[V_4O_{12}]^{4-}$ (126), the latter cyclic tetravanadate is distinctly different in detail, as the structure consists of corner-sharing tetrahedra rather than octahedra or square pyramids. This observation is consistent with the variability of coordination geometries available to vanadium sites that may exhibit tetrahedral geometry for V^V, square pyramidal, or trigonal bipyramidal for V^V and V^{IV}, and octahedral for V^V and V^{IV} (27). When combined with the facile V^V/V^{IV} redox couple, an enormous molecular and solid phase structural chemistry may be actualized.

The remaining clusters of the tetravanadium core adopt the compact structure based on four edge-sharing octahedra, classified as type IVa in Fig. 7. As discussed in Section III.B, the prototype structure $[(VO)_4(H_2O)_2(SO_4)_2\{(OCH_2)_3CR\}_2]^{2-}$ (**7**) (127) is synthesized from the binuclear $[(VO)_2Cl_2\{(OCH_2)_2C(R)CH_2OH\}_2]$ in the presence of vanadate and $[Bu_4N]HSO_4$. As shown in Fig. 12, the structure consists of a compact arrangement of edge-sharing $\{MO_6\}$ octahedra with the characteristic tetrametalate geometry. The remnant of the binuclear precursor is retained in the $[(VO)_2\{(OCH_2)_3CR\}_2]^{2-}$ binuclear core that provides two faces, each defined by a triangular arrangement of three alkoxy oxygen donors, which fuse with the two $[VO(H_2O)(SO_4)]$ moieties to complete the aggregate. The trisalkoxy ligands assume the type TA5 coordination mode with two alkoxy oxygen donors of each ligand in the μ^2-bridging mode and the third adopting the μ^3-mode. The sulfate ligands bridge vanadium sites of the binuclear core to each of the two $\{VO(H_2O)\}$ fragments condensed onto the faces of the core. The vanadium sites are in the IV oxidation state, as revealed by the metrical parameters and valence sums collected in Table VI.

The propanediolate complex $[V_4O_6Cl_2(OCH_2CR_2CH_2OH)_4]$ (**8**) (128) prepared from the reaction of $VOCl_3$ and the appropriate diol, exhibits the same core structure as **7**. However, the triply bridging alkoxy oxygen donors of **7** are replaced by triply bridging oxo groups in **8**, the terminal aqua ligands by chlorides, and the bridging sulfato oxygen atoms by alkoxide oxygen donors. In contrast to the "fully reduced" state of **7** with all vanadium sites in the IV oxidation state, cluster **8** is "fully oxidized" with four V^V sites. It is a char-

TABLE VI
Comparison of Selected Structural Parameters for Representative Vanadium Oxide Ligand Clusters

Compound	V^{IV}/V^{V}	V—O							
		Ligand Terminal Oxygen (L)	μ_2-oxo	μ_3-oxo	μ_2-OH	μ_3-OH	μ_2-OH_2	μ_2-OR	μ_3-OR
$[(VO)_2Cl_2\{(OCH_2)_2CRR^1\}_2]$	0/2	1.783(OR) 2.057(HOR)						2.045	
$Na[(VO)_2(OH_2)_3(OH)_2(C_4O_4)_2]$	1/1	1.97(C_4O_4) 2.04(H_2O)			1.98		2.40		
$Bu_4N[V_2O_3(OH_2)_3(C_4O_4)_2]$	1/1	1.989(C_4C_4) 2.033(H_2O)	1.822				2.460		
$[V_4O_8(O_2CR)_4(NO_3)]^{2-}$	1/3	1.96–207(RCO_2)	1.78–1.85						
$[V_4O_7(OH)(O_2CR)_4K]^{+}$	1/3	1.95–2.00(RCO_2)	1.77–1.87		1.95				
$[V_4O_8(H_2O)_2(C_2O_4)_4]^{4-}$	0/4	2.02/(C_2O_4)	1.83				2.44		
$[(VO)_4(H_2O)_2(SO_4)_2\{(OCH_2)_3CR\}_2]^{2-}$	4/0	2.001(SO_4) 2.013(H_2O)						1.991	2.186
$[V_4O_6Cl_2(OCH_2CR_2CH_2OH)_4]$	0/4	1.75(OR) 1.06(HOR)		2.01				1.83–2.31	
$[V_4O_8(OR)_4(bpy)_2]$	0/4	1.82(OR)	1.84						2.204
$[V_5O_9(O_2CR)_4Cl]^{2-}$	4/1	2.02(RCO_2)		1.94					
$[V_6O_{13}\{(O_2CH_2)_3CR\}_2]^{2-}$	0/6		1.823					2.022	
$[V_6O_{11}(OH)_2\{(OCH_2)_3CR\}_2]$	0/6		1.823		1.879			2.011	
$[V_{10}O_{14}(OH)_2\{(OCH_2)_3CR\}_4]^{2-}$	10/0		1.848	1.967	1.970			1.997	2.086
$[H_6V_{10}O_{22}(O_2CR)_6]^{2-}$	8/2	2.01(RCO_2)		2.015	1.97	2.15			

TABLE VII

Comparison of Structural Parameters for Clusters with the $\{V_6O_{19}\}$ and $\{V_6FO_{18}\}$ Cores

Complex	Vanadium Oxidation States	V—O: Bridging Oxo	Bridging Hydroxy	Bridging Alkoxy	Central Oxo (Fluoro)
$[V_6O_{12}(OMe)_7]^{1-}$	$6V^V$	1.826		1.974	2.212
$[V_6O_{13}(OMe)_3\{(OCH_2)_3CMe\}]^{2-}$	$6V^V$	1.826		2.011	2.241
$[V_6O_{13}\{(OCH_2)_3CMe\}_2]^{2-}$	$6V^V$	1.823		2.022	2.242
$[V_6O_{11}(OH)_2\{(OCH_2)_3CMe\}_2]$	$6V^V$	1.823	1.879	1.994, 2.019	2.219, 2.253
$[V_6O_{10}(OH)_3\{(OCH_2)_3CNO_2\}_2]^{2-}$	$3V^{IV}/3V^V$	1.86	1.94	2.017	2.28
$[V_6O_9(OH)_4\{(OCH_2)_3CMe\}_2]^{2-}$	$4V^{IV}/2V^V$	1.869	1.946	2.004, 2.038	2.287
$[V_6O_7(OH)_6\{(OCH_2)_3CMe\}_2]^{2-}$	$6V^{IV}$		1.994	2.004	2.038
$[V_6O_7(OH)_3\{(OCH_2)_3CMe\}_3]^{2-}$	$6V^{IV}$		1.999	1.973, 2.074	2.272, 2.328
$[V_6O_7\{(OCH_2)_3CCH_2Me\}_4]^{2-}$	$6V^{IV}$			2.007	2.315
$[V_6O_7(OH)_3\{(OCH_2)_3CMe\}_3]^{1-}$	$IV^V/5V^{IV}$		1.974	1.976, 2.006	2.36, 2.382
$[(VO)_6F(OH)_3\{(OCH_2)_3CMe\}_3]^{1-}$	$6V^{IV}$		2.037	1.955, 2.045	2.245, 2.522 (F)

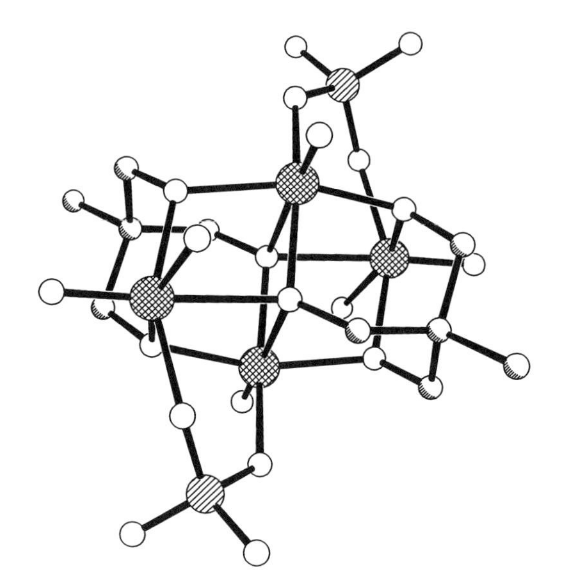

Figure 12. A view of the structure of $[(VO)_4(H_2O)_2(SO_4)_2\{(OCH_2)_3CR\}_2]^{2-}$ (**7**).

acteristic feature of the clusters of the metal oxide ligand system that the same structural core may accommodate different metal oxidation states, a variety of peripheral ligand types and even mixed-metal cores. It should be noted that the structural analyses of Tables VI and VII establish two of the terminally coordinated alkoxy oxygen donors as the protonation sites, rather than the triply bridging oxo groups as initially formulated. Some care must be taken in assigning the protonation positions of such structures, but a careful analysis of the metal–oxygen bond distances will generally allow a unambiguous assignment.

The complex $[V_2Mo_2O_8(OMe)_2\{(OCH_2)_3CR\}_2]^{2-}$ (**9**) (127) provides an example of this latter type of mixed-metal core. The existence of an analogous series of binuclear and tetranuclear metal oxo–alkoxide clusters in the molybdenum system, as discussed in Section III.B, suggested that the mixed-metal complexes should also be accessible. This expectation was realized in the isolation of **9** from the aggregation reaction between $[(VO)_2Cl_2\{(OCH_2)_3CR\}_2]$ and $[Mo_2O_7]^{2-}$ in alcohol. The structure consists of the tetranuclear core IVb constructed from the condensation of two $[MoO_2(OMe)]^+$ units into two triangular alkoxide oxygen faces of the $[V_2O_4\{(OCH_2)_3CR\}_2]^{4-}$ core. The structure of **9** is isotypic with that of $[Mo_4O_8(OEt)_2\{(OCH_2)_3CR\}_2]$ (**75**) with the two central Mo^{VI} centers replaced by V^V. The cluster represents another example of a full-oxidized species with Mo^{VI} and V^V sites.

The ability of a parent core structure to accommodate a range of oxidation states is dramatically illustrated by the structure of $[V_4O_2(acac)_4(OMe)_6]$ (where acac = acetylacetonate) (**10**) (129), a V^{III}/V^{IV} cluster. The complex is prepared in the reduction by thiolate of $[VO(acac)_2]$ in $MeOH/CH_2Cl_2$. The binuclear core of **10** consists of the $[V_2(acac)_2(OMe)_6]^{2-}$ unit, in place of the more common $\{M_2O_4(OR)_4\}^n$ or $\{M_2O_2(OR)_6\}^n$ units, to which two $\{VO(acac)\}^{1+}$ groups are condensed. The vanadium centers are thus bridged exclusively by μ^{2-} and μ^{3-} methoxy groups. As consequences of introducing the V^{III} sites, the central vanadium centers do not possess terminal oxo coordination, a characteristic feature of V^V and V^{IV} but not of V^{III}, and the triply bridging methoxy groups are only weakly associated with the V^{IV} sites, resulting in a more open structural framework based on the type IVb core.

While the compact tetranuclear core IVb is common to a number of metal oxide coordination compounds, the core is known only in the solid phase $Li_{14}(WO_4)_3(W_4O_{16})\cdot 4H_2O$ (107) in the metal oxide chemistry. However, other tetranuclear cores, such as type IVa, have been reported in oxometalate phosphate phases (130). Furthermore, as mentioned previously, the introduction of organic ligands expands the structural chemistry of the metal oxide cores with the ligands providing structural support for cores otherwise not readily accessible.

3. *Pentanuclear Clusters*

Pentanuclear species are relatively rare in the M/O/ligand system for M = V or Mo, an observation that reflects the tendency of polymetalates to assume approximately spherical structures. Thus, the addition of an additional {M = O} vertex via condensation onto a compact pentanuclear core V will generate the favored hexametalate structure VI.

There exists a single example of a pentanuclear coordination complex of the vanadium oxide system, $[V_5O_9Cl(\text{thiophene-2-carboxylate})_4]^{2-}$ (**11**) (122), shown in Fig. 13. The cluster forms by self-assembly in the reaction of $[VOCl_4]^{2-}$ with thiophene-2-carboxylate in acetonitrile. The structure of **11** is clearly related to that of **5** by capping of the $\{V_4O_8\}$ cyclic unit of the latter with a $\{VO\}^{3+}$ group and replacement of the NO_3^- guest by a chloride group. The $\{V_5O_9\}^{3+}$ assembly exhibits a domed structure, providing a well-defined interior cavity for the accommodation of a guest molecule; in this instance Cl^-. Complex **11** is a mixed-valence species $V^V/4V^{IV}$ with a trapped valence of +5 at the apical vanadium site of the $\{V_5O_9\}^{3+}$ core. The topological relationship of **5** and **11** extends to their chemical behavior. Thus, **5** may be prepared from **11** by treatment with $AgNO_3$, which displaces the weakly associated Cl^- of **11**, but curiously also results in loss of the apical {VO} unit.

The isolation of **11** suggests that the observed polyanion core may be a consequence of the template effect (13), the organization of the host framework

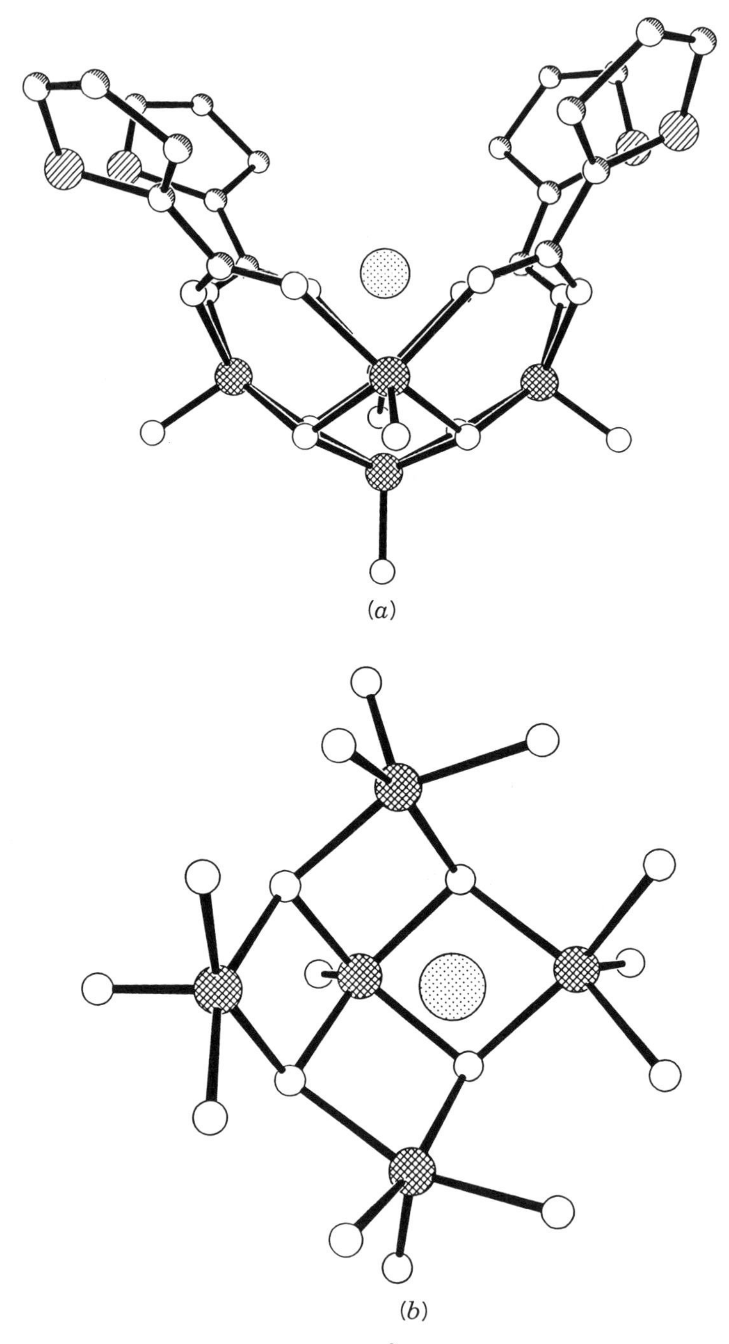

Figure 13. (*a*) The structure of $[V_5O_9(O_2CR)_4Cl]^{2-}$. (*b*) The $\{V_5O_9L_8\}$ core and the location of the Cl^- anion in the bowl formed by condensing the five vanadium square pyramids in an edge-sharing arrangement.

around the guest in the condensation reaction. While **11** represents a unique example of the pentanuclear core, the $\{V_5O_9\}^{n-}$ moiety is a common structural motif, recently identified in the higher oligomers of the class $[V_{14}O_{22}(OH)_4(H_2O)_2(PhPO_3)_8(Y)_n]^{6-}$, where Y is a guest cation, anion, or neutral molecule (12, 131). The incorporation of the guest moiety into these structures again reflects the templating influence of these reagents. It is also noteworthy that the $\{V_5O_9\}$ unit is found in complex 3D frameworks of the V/O/PO_4^{3-} system, such as $(A)_x(B)_y[(V_5O_9)(PO_4)_4]$, where A and B are NH_4^+, alkali metal cations, and/or alkylammonium cations (132), and that the domed structure of the $\{V_5O_9\}^{n+}$ unit once again acts as a host for the cationic templates that are presumably responsible for the organization of the V/O/PO_4^{3-} shells of these materials. It is tempting to speculate on the topological and chemical relationships of polynuclear complexes to inorganic materials. While molecular units are of limited 3D extension, their molecular structure and solid state packing should reflect some features of similarly composed solids (133). While the notion that molecular clusters are structural images of extended structure materials is no doubt superficial, the modeling of topological relationships and local geometric features can result in useful insights, as demonstrated by the metal alkoxides that serve as models for metal oxide structures and reactivity (134, 135) and by the vanadium oxide clusters and their relationships to the solid phases (136). We shall return to this theme in the discussion of the clusters of the V/O/RPO_3^{2-} systems and their relationships to the solid phase materials.

4. *Hexanuclear Clusters*

Clusters with hexanuclear cores are the most commonly observed for the vanadium oxide coordination compounds, and of the possible polyhedral arrangements for six metal centers the hexametalate core VI is by far the most representative. The $[M_6O_{19}]^{n-}$ core is known for $[Nb_6O_{19}]^{8-}$, $[Ta_6O_{19}]^{8-}$, $[Mo_6O_{19}]^{2-}$, and $[W_6O_{19}]^{2-}$ (26, 28). The naked $[V_6O_{19}]^{8-}$ cluster has not been isolated, possibly as a consequence of the high charge/volume ratio. However, the $[V_6O_{19}]^{8-}$ core may be stabilized by appropriate electropositive groups, as in the organometallic vanadium oxide–rhodium cyclopentadienyl cluster $[(C_5Me_5)Rh]_4V_6O_{19}$ (137, 138). Likewise, the hexavanadate core may be stabilized by reducing the cluster charge through the expedient of substituting alkoxy oxygen donors for bridging oxo groups of the parent core. The tris(hydroxymethyl)alkane class of ligands appears to be particularly effective in this regard as a consequence of their ability to stabilize trinuclear alkoxy-bridged fragments of the type represented by structure TA3. Conceptually, the hexanuclear core is derived from tetranuclear species of the type $[(VO)_4(H_2O)_2(SO_4)_2\{(OCH_2)_3CR\}_2]^{2-}$ (**7**) by displacement of the terminal aqua ligands and the peripheral SO_4^{2-} groups and condensation of two additional {VO} groups onto opposite faces of the tetranuclear core.

In practice, the prototypical hexavanadate complex $[V_6O_{13}(OMe)_3\{(OCH_2)_3CR\}]^{2-}$ (**12**) (139) is isolated from the reaction of $[V_5O_{14}]^{3-}$ (140) with $(HOCH_2)_3CR$ in MeOH. The structure of **12**, shown in Fig. 14, consists of six vanadium $\{VO_6\}$ octahedra in a compact edge-sharing arrangement to produce an arrangement of vanadium centers disposed at the vertices of a "superoctahedron." In contrast to the $\{M_{16}O_{19}\}^{n-}$ naked cores, six doubly bridging oxo groups have been replaced by the bridging alkoxy donors of the $\{(OCH_2)_2CR\}^{3-}$ ligand in the type TA3 coordination mode and by three methoxy ligands. Alternatively, the structure may be described as a hexametalate core $\{V_6O_{19}\}^{8-}$ supporting a $\{(OCH_2)_3CR\}^{3+}$ and three CH_3^+ subunits. The core is maintained in the analogous methoxy derivative $[V_6O_{12}(OMe)_7]^{1-}$ (**12a**) (141).

Minor modifications of reaction stoichiometries allow isolation of the analogous cluster with two tris(alkoxymethyl)alkane ligands, $[V_6O_{13}\{(OCH_2)_3CR\}_2]^{2-}$ (**13**) (142). The structure shown in Fig. 15 is derived from that of **12** by replacing the three methoxy ligands by the second tris(hydroxymethyl)alkane group. The trisalkoxides bond to opposite triangular faces of the $\{V_6O_{19}\}$ cage.

The presence of the trisalkoxy ligand serves to deform the regular hexametalate core. Thus, whereas the highly symmetrical hexametalate structures,

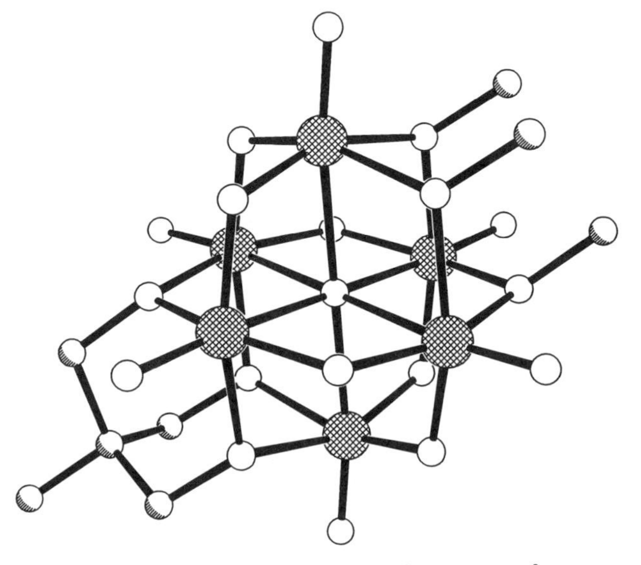

Figure 14. The structure of $[V_6O_{13}(OMe)_3\{(OCH_2)_3CR\}]^{2-}$ (**12**).

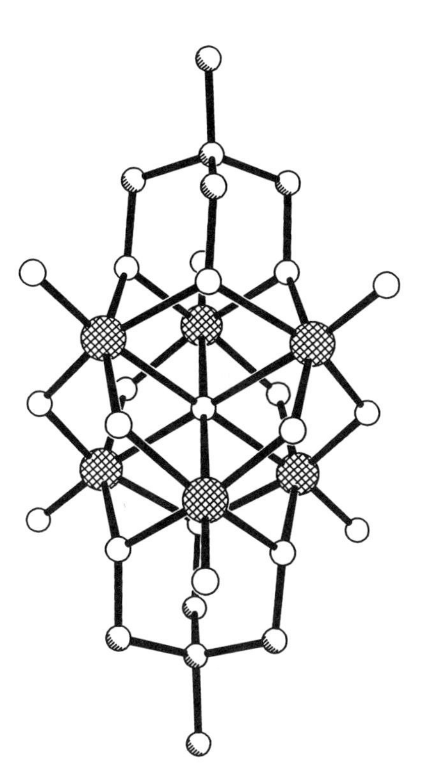

Figure 15. A view of the structure of $[V_6O_{13}\{(OCH_2)_3CR\}_2]^{2-}$ (**13**).

such as $[Mo_6O_{19}]^{2-}$ (143), $[W_6O_{19}]^{2-}$ (144), and $[Ta_6O_{19}]^{8-}$, each exhibit 12 doubly bridging oxo groups with essentially equivalent M—O distances, the hexavanadate structures $[V_6O_{13}\{(OCH_2)_3CR\}_2]^{2-}$ possess six doubly bridging oxo groups and six doubly bridging alkoxy groups with average V—O distances of 1.823(4) and 2.022(4) Å, respectively, as shown in Table VI.

Since the doubly bridging oxo groups of **13** are basic and readily protonated (145), the neutral hexametalate $[V_6O_{11}(OH)_2\{(OCH_2)_3CR\}_2]$ (**14**) (146) may be isolated from solutions of **13** upon addition of HBF_4. The structural consequence of protonation are expressed in an expansion of V—O(H) distances relative to the unprotonated species and a small expansion of the cluster volume, a feature that will subsequently be discussed in more detail.

In addition to the basicity of the bridging oxo groups allowing facilie protonation, **13** is also readily reduced, both electrochemically and chemically. The hexametalate cluster **13** exhibits metal sites that correspond grossly to mononuclear VOL_5 species with the vanadate centers in approximately tetragonal sites with a single terminal oxo group. As such, these species are classified as type I polyoxoanions (147) and are expected to exhibit well-behaved reduction processes, in common with other hexametalate species (148). By exploiting the protonation of bridging oxo groups and the facile reduction of the hexametalate core, successive coupled reduction–protonation of **13** has been employed to synthesize both mixed-valence V^{IV}/V^{V} clusters and fully reduced V^{IV} species (146, 149).

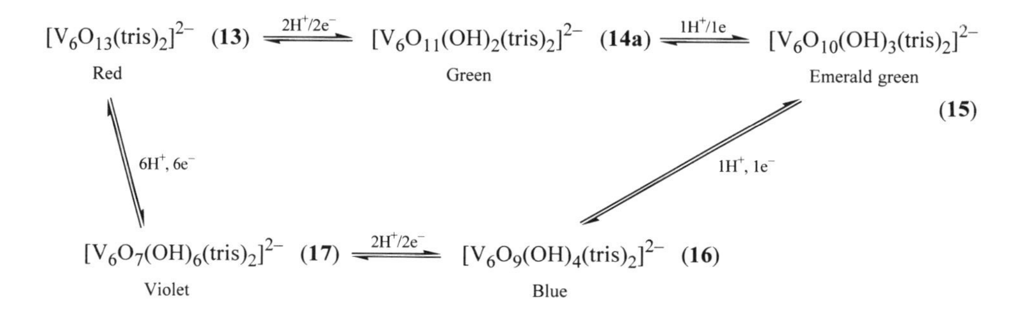

where tris = tris(hydroxymethyl)ethane, $\{(OCH_2)_3CMe\}^{3-}$.

Structural studies revealed the doubly bridging oxo groups of the hexametalate core as the protonation sites in all cases. The structural consequences of reduction and protonation are apparent in the parameters listed in Tables VII and VIII.

The most extreme structural changes are observed in the fully reduced or $6e^-$, six hydroxy group cluster $[V_6O_7(OH)_6\{(OCH_2)_3CMe\}_2]^{2-}$ (**17**), whose structure is shown in Fig. 16. Complex **17** retains the hexavanadate core, but with all six doubly bridging oxo groups protonated and all six vanadium centers in the V^{IV} oxidation state. These changes are most obviously reflected in the

TABLE VIII
Distances between Planes Defined in Fig. 17 for Structures with the $\{V_6O_{19}\}$ and $\{V_6FO_{18}\}$ Cores

Complex	Vanadium Oxidation States	a	b	c
$[V_6O_{19}\{Rh(C_5Me_5)\}_4]$	$6V^V$	0.89	2.59	4.40
$[V_6O_{13}(OMe)_3\{(OCH_2)_3CMe\}]^{2-}$	$6V^V$	0.99	2.35	4.35
$[V_6O_{13}\{(OCH_2)_3CMe\}_2]^{2-}$	$6V^V$	1.01	2.36	4.36
$[V_6O_{11}(OH)_2\{(OCH_2)_3CMe\}_2]$	$6V^V$	1.00	2.40	4.39
$[V_6O_{10}(OH)_3\{(OCH_2)_3CNO_2\}_2]^{2-}$	$3V^V/3V^{IV}$	0.98	2.54	4.49
$[V_6O_9(OH)_4\{(OCH_2)_3CMe\}_2]^{2-}$	$2V^V/4V^{IV}$	0.97	2.56	4.50
$[V_6O_7(OH)_6\{(OCH_2)_3CMe\}_2]^{2-}$	$6V^{IV}$	0.95	2.68	4.59
$[V_6O_7(OH)_3\{(OCH_2)_3CMe\}_3]^{2-}$	$6V^{IV}$	0.94	2.68	4.59
$[V_6O_7\{(OCH_2)_3CCH_2Me\}_4]^{2-}$	$6V^{IV}$	0.94	2.67	4.54
$[V_6O_7(OH)_3\{(OCH_2)_3CMe\}_3]^{1-}$	$1V^V/5V^{IV}$	0.93	2.69	4.54
$[(VO)_6F(OH)_3\{(OCH_2)_3CMe\}_3]^{1-}$	$6V^{IV}$	0.94	2.67	4.54

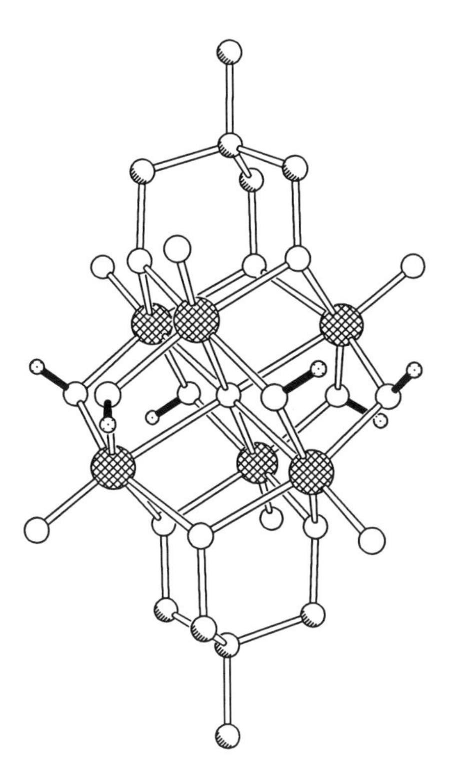

Figure 16. The structure of the V^{IV} species $[V_6O_7(OH)_6\{(OCH_2)_3CMe\}_2]^{2-}$ (**17**), highlighting the protonation sites on doubly bridging oxo groups forming the girdle of the cluster.

bond lengths of the vanadium-bridging hydroxy group interactions. As illustrated by Table VII, the average V—O(H) distance for **17** is 1.994(4) Å, an expansion of +0.171 Å compared to the average vanadium-bridging oxo group distance of 1.823(4) Å for **13–17**. Since the structural parameters for protonation of doubly bridged oxo groups in the absence of reduction are known from the structure of **13**, the structural consequences of reduction relative to protonation may be assessed. While protonation alone results in an increase of the protonated V—O (bridge) distances by an average of +0.056 Å, protonation coupled to reduction results in a further bond lengthening of +0.115 Å, to give an average increase of +0.171 Å in the structure of **17**. It would appear that reduction alone would be manifest by a bond lengthening of the V—O (bridge) distance of twice the order of magnitude of that produced by protonation alone.

The bond distances observed for the V—O bonds of **17** are also consistent with valence sum calculations shown in Table VII. Bond length–bond strength correlations of the form $s = (R/R_1)^{-N}$, where s is Pauling's bond strength in valence units, R is the metal-oxygen bond length, and R_1 and N are empirical parameters, have been derived by several authors (125) and are used in deriving the valence sums that are consistent with the oxidation state assignments of

Table VII. While structures **11–14** yield valence sums of approximately 5.0, Structure **17** clearly conforms to the valence for a cluster with exclusively V^{IV} d^1 centers. In contrast, the structure of $[V_6O_9(OH)_4\{(OCH_2)_3CMe\}_2]^{2-}$ (**16**) is clearly a mixed-valence species. The calculations indicate an average vanadium valence of 4.36 for **16**, which is consistent with the $V_2^VV_4^{IV}$ formalism. Similarly, complex **15** yields an average vanadium valence of 4.52, which conforms to the $V_3^VV_3^{IV}$ formalism.

The structural effects of reduction and/or protonation can also be demonstrated by comparing the spacing between approximately planar layers of negatively charged and close-packed oxygen atoms separated by layers of cationic vanadium centers (106) in the oxidized clusters **12** and **13**, the protonated species **14**, the mixed-valence polyanions **15** and **16**, and the fully reduced six proton cluster **17**, as shown in Fig. 17 and Table VIII. In all cases, the plane containing the central six-coordinate oxygen atom and the six doubly bridging oxo and/or hydroxy groups defines the reference plane from which the spacings to the other parallel layers (to within 1.2°) have been calculated. There are two types of planes above and below the reference plane: two sets of three terminal oxo groups, and three bridging alkoxy oxygen donors and two sets of three vanadium atoms. The outermost layers of oxygen atoms are displaced 2.186 Å

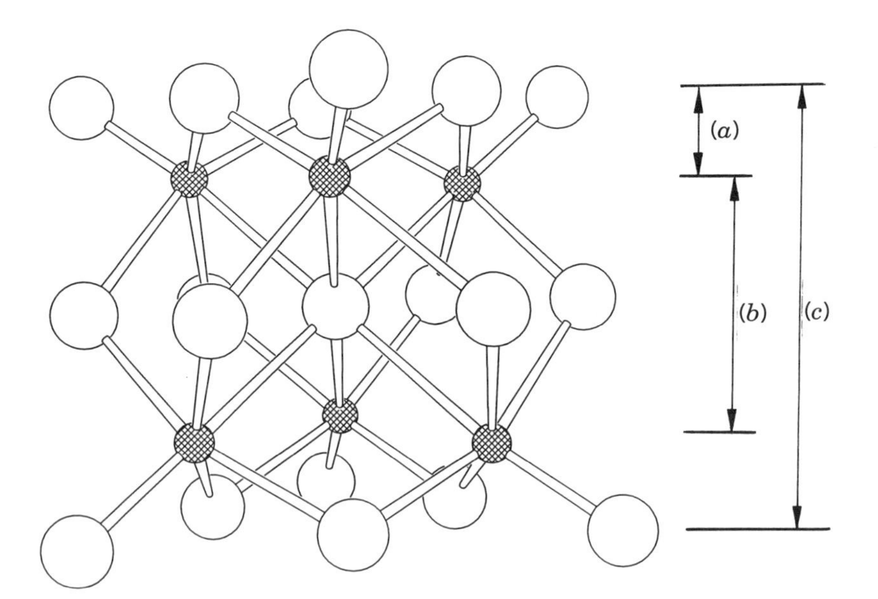

Figure 17. Schematic representation of the hexametalate core viewed as parallel oxide layers with vanadium layers in the interstitial positions. The distances between such layers for various compounds are tabulated in Table VIII.

from the reference plane, while the vanadium layers are situated 1.176 Å from this plane. As anticipated, the vanadium layer is considerably displaced toward the outermost layer of oxygen atoms, away from the reference plane: 1.009 Å to the outermost oxygen layer versus 1.176 Å to the reference plane. This difference reflects the usual displacement of the metal centers in polyanions from the center of the MO_6 octahedron in the direction of the terminal oxo group. Protonation of two bridging oxo groups in **14** results in an overall expansion of the cluster volume. Closer inspection reveals that the predominant effects reside in the location of the vanadium planes relative to the oxygen layers. Formation of two O—H bonds removes charge from the central layer of oxygen atoms and weakens the bonds between these oxygen atoms and the vanadium layers. Consequently, the vanadium layers move away from the reference plane by 0.023 Å and move 0.011 Å closer to the outermost layer of negatively charged oxygen atoms. Such effects would be expected to be more pronounced upon coupled protonation and reduction of the core, as a result of the larger covalent radius of V^{IV} relative to V^{V}. This expectation is confirmed in the Structures **15–17**, which exhibit increased displacements of the vanadium planes from the reference plane. The structural changes concomitant to protonation and reduction of the hexavanadate core are also evident in an overall expansion of the core volume as reflected in an increase in the distances between the outermost planes of oxygen atoms in the various structures. The effects of coupled protonation–reduction are most significant, as illustrated by a distance of 4.59 Å for the fully reduced cluster **17**.

The consequences of successive reduction of vanadium sites in these hexametalate cores is also reflected in the magnetic properties. The magnetic behavior of the $2e^-$ reduced species $[V_6O_{11}(OH)_2\{(OCH_2)_3Cr\}_2]^{2-}$ and **15–17** is qualitatively similar to that reported for other vanadium–oxygen clusters with varying numbers of V^{IV} d^1 centers (Table IX). In species with a small number of spins, V^{IV} d^1 sites, the spins may be trapped and remain as far apart as possible, resulting in nearly spin-only values for μ_{eff} per V^{IV}. Spin–spin coupling increases with an increasing number of V^{IV} centers, which necessarily results in close contacts. As shown in Table IX, the variations in the magnetic properties of vanadium oxide clusters span weak exchange interactions, through very strong exchange interactions, to relatively strongly covalently bonded pairs of V^{IV}–V^{IV} sites.

When viewing the hexametalate core as an assembly of edge-sharing octahedra, the trisalkoxy ligands of structures **12–17** are seen to occupy triangular faces of tetrahedral cavities produced by the close-packing arrangement of metal octahedra. Although there are eight of these faces, a maximum of four may be occupied at any one time by trisalkoxy ligand types, as shown schematically in Fig. 4. Yet, by conventional synthetic techniques, a maximum of two trisalkoxides could be linked to the hexametalate core. The existence of polyoxoalkoxide clusters with the hexametalate core, such as $[Fe_6O(OR)_{18}]^{2-}$ (150)

TABLE IX

Magnetic Properties of Selected Vanadium Oxide Clusters and Solids[a]

Compound	V^{IV}/V^{V}	μ_{eff}/μ_B per V^{IV}	$V^{IV}-V^{IV}$ Distances (Å)	No. of V^{IV} Trapped Pairs
$[V_{14}AsO_{40}]^{7-}$	2/12	1.77	8.72	0
$[H_3KV_{12}As_3O_{39}(AsO_4)]^{6-}$	4/8	1.75	5.68	0
$[V_{12}As_8O_{40}(HCO_2)]^{3-}$	6/6	1.75	5.25	0
$[V_{12}As_8O_{40}(HCO_2)]^{5-}$	8/4	1.39	*b*	*b*
$[V_{12}As_8O_{40}(H_2O)]^{4-}$	8/4	1.38	*b*	*b*
$[V_{18}O_{42}(SO_4)]^{8-}$	12/6	1.14	*b*	*b*
$[V_{15}As_6O_{42}(H_2O)]^{6-}$	15/0	1.08	2.96	0
$[H_4V_{18}O_{42}(I)]^{9-}$	18/0	1.06	2.91	0
$[V_{34}O_{82}]^{10-}$	16/18	1.2	2.65	2
$[V_{19}O_{41}(OH)_9]^{8-}$	12/7	0.9	2.95	6
VO_2, monoclinic	1/0	0.4	2.62	All
$[V_6O_{11}(OH)_2\{(OCH_2)_3CR\}_2]^{2-}$	2/4	1.84	*b*	0
$[V_6O_{10}(OH)_3\{(OCH_2)_3CR\}_2]^{2-}$	3/3	1.75	3.2–4.6	0
$[V_6O_9(OH)_4\{(OCH_2)_3CR\}_2]^{2-}$	4/2	1.74	3.2–4.5	0
$[V_6O_7(OH)_6\{(OCH_2)_3CR\}_2]^{2-}$	6/0	1.25	3.25–4.6	0
$[V_{10}O_{14}(OH)_2\{(OCH_2)_3CR\}_4]^{2-}$	10/0	1.41	3.3–3.6	0
$[V_{10}O_{16}\{(OCH_2)_3CR\}_4]^{2-}$	8/2	1.50	3.2–3.6	0
$[V_{16}O_{20}\{(OCH_2)_3CR\}_8(H_2O)_4]$	16/0	1.50	2.68 3.0–3.3	2

[a]Discussions of the magnetic properties of V/O and V/O/As clusters may be found in A. Müller, J. Döring, and H. Boggë, *J. Chem. Soc. Chem. Commun.*, 274 (1991); A. Müller, R. Rohlfing, J. Döring, and M. Penk, *Angew. Chem. Int. Ed. Engl.*, *30*, 588 (1991).

[b]When the number of V^{IV} centers is greater than one half of the total number of V sites, both trapped and delocalized V^{IV} centers are observed.

and of a variety of species containing the central $\{M_6(\mu\text{-}O_6)\}$ unit (151) suggested that further substitution of alkoxide ligands in the vanadium oxide system should be possible. This expectation was realized by exploiting the potential of hydrothermal synthesis to induce crystal growth of reduced phases.

Minor modifications in the conditions of the hydrothermal syntheses can result in important variations in cluster composition, which may be reflected exclusively in cluster oxidation states or degree of substitution about a given core, or more dramatically in the isolation of higher nuclearity cores, as summarized in Table X. The former aspects are demonstrated in the hexametalate series in the structures of **18–21** (152, 153). Thus, introduction of various cationic templates can influence the cluster oxidation states, as in **18** and **19**, which are formally $6V^{IV}$ and $V^{V}/5V^{IV}$ clusters, respectively, or the number of trisalkoxide ligands associated with the $\{V_6O_{19}\}$ core, as in **20**. The schematic representation of Fig. 18 illustrates the substitution pattern about the $\{V_6O_{19}\}$ for the structures with two, three, and four trisalkoxide ligands. While the overall geometries of the $\{V_6O_{19}\}$ cores are grossly similar, the metrical parameters listed in Tables VII and VIII illustrate the structural consequences of substitution of doubly bridging oxo groups by alkoxide donors and of changes in the oxidation states of the vanadium centers.

The most unusual member of this class of hexametalate clusters is $Na[(VO)_6F(OH)_3\{(OCH_2)_3CCH_3\}_3]$ (**21**), the first example of this core type possessing a central anion other than oxide. The trapping of the fluoride anion reflects the organization of the metal oxide framework about a templating agent, a significant feature of the hydrothermal process and of the preparation of several classes of metal oxide coordination clusters. The most significant structural distortion consequent to the replacement of the central oxide by fluoride is the displacement of the fluoride toward one triangular metal face, resulting in two sets of V—F distances of 2.245 and 2.522 Å. This behavior of the $\{V_6(\mu_3\text{-}F)O_{18}\}$ core contrasts dramatically with that of the $\{V_6(\mu_6\text{-}O)O_{18}\}$ cores when the V—O (central) bond distances are essentially identical. Of the remaining examples of hexanuclear species of the {V/O/carbon–oxygen ligand} class of clusters, two members, **22** and **23**, possess cyclic structures while the third is related to the compact tetranuclear core.

The benzoate cluster, $[V_6O_{10}(PhCO_2)_9]$ (**22**) (154), formed in the reaction of benzoate with $VOCl_3$ under anaerobic conditions, possesses the cyclic structure shown in Fig. 19. The compound is a mixed-valence $V^{IV}/5V^{V}$ species with the vanadium sites arranged in a flat twist-boat conformation. The cluster consists of a dinuclear $\{(VO)_2O(RCO_2)_2\}$ unit linked to a tetranuclear $\{(VO)_4O_3(RCO_2)_3\}$ unit connected by four bridging benzoate ligands. The vanadium centers of the binuclear unit are best described as corner-sharing octahedra, while the vanadium sites of the linear tetranuclear moiety provide a corner-sharing arrangement of octahedra when the weak interaction of the vanadium

TABLE X

Summary of Experimental Conditions for the Hydrothermal Syntheses of V/O/alkoxide and Mo/O/alkoxide Phases

Compound	Metal Oxidation States	Reactant Composition	Temperature (°C)	Duration (h)
$Ba[V_6O_7(OH)_3\{(OCH_2)_3CMe\}_3]$ (**18**)	$6 \times V^{IV}$	V_2O_3, KVO_3, $(HOCH_2)_3CR$, $BaCl_2$, H_2O 3:6:10:20:300	150	50
$(Me_3NH)[V_6O_7(OH)_3\{(OCH_2)_3CMe\}_3]$ (**19**)	$5 \times V^{IV}$, $1 \times V^{V}$	V_2O_3, V_2O_5, $(HOCH_2)_3CR$, Me_3NHCl, H_2O 1.25:1.25:2.5:5:300	210	17
$Na_2[V_6O_7\{(OCH_2)_3CEt\}_4]$ (**20**)	$6 \times V^{IV}$	V_2O_3, $NaVO_3$, $(HOCH_2)_3CR$, $NaCl$, H_2O 3:6:10:5:300	150	21
$Na[(VO)_6F(OH)_3\{(OCH_2)_3CMe\}_3]$ (**21**)	$5 \times V^{IV}$	V_2O_3, $NaVO_3$, $(HOCH_2)_3CR$, $NaBF_4$, $NaCl$, H_2O 2.5:5:5:10:10:5:300	150	24
$(NH_4)_4[V_{10}O_{16}\{(OCH_2)_3CR\}_4]$	$10 \times V^{IV}$	V_2O_3, NH_4VO_3, $(HOCH_2)_3CR$, NH_4Cl, H_2O 3:6:10:5:300	150	20
$Na_2[V_{10}O_{16}\{(OCH_2)_3CR\}_4]$	$8 \times V^{IV}$, $2 \times V^{V}$	V_2O_3, $NaVO_3$, $(HOCH_2)_3CR$, $NaCl$, $CuBr$, H_2O 3:6:5:10:10:300	150	50
$(Me_3NH)_2[V_{10}O_{14}(OH)_2\{(OCH_2)_3CR\}_4]$	$10 \times V^{IV}$	V_2O_3, V_2O_5, $NaVO_3$, $(HOCH_2)_3CR$, Et_2NHCl, H_2O 5:5:5:10:20:300	150	20
$(Et_4N)[V_{10}O_{13}\{(OCH_2)_3CR\}_5]$	$10 \times V^{IV}$	V_2O_3, V_2O_5, $(HOCH_2)_3CR$, Et_4NBr, H_2O 2.5:2.5:5:4:600	200	22
$[V_{16}O_{20}(H_2O)_4\{(OCH_2)_3CR\}_8]$	$16 \times V^{IV}$	V_2O_3, V_2O_5, $(HOCH_2)_3CR$, Me_2NH_2Cl, H_2O 2:2:10:10:300	170	48
$(Me_3NH)_2(Et_4N)Na_4[Na(H_2O)_3H_{15}Mo_{42}$-$O_{109}\{(OCH_2)_3CCH_2OH\}_7]$	$36 \times Mo^{V}$, $6 \times Mo^{VI}$	$Na_2MoO_4 \cdot 2H_2O$, MoO_3, Mo, $(HOCH_2)_3CR$, Me_3NHCl, Et_4NCl, H_2O 6:6:4:10:10:10:300	160	72
$(Me_3NH)_2Na_6[Na(H_2O)_3H_{15}Mo_{42}O_{109}$-$\{(OCH_2)_3CCH_2OH\}_7]$	$36 \times Mo^{V}$, $6 \times Mo^{VI}$	$Na_2MoO_4 \cdot 2H_2O$, MoO_3, $(HOCH_2)_3CR$, Me_3-$NHCl$, H_2O 3:3:5:5:150	160	72
$Na_9[(MoO_3)H_{14}Mo_{42}O_{109}\{(OCH_2)_3$-$CCH_2OH\}_7]$	$36 \times Mo^{V}$, $7 \times Mo^{VI}$	$Na_2MoO_4 \cdot 2H_2O$, MoO_3, $(HOCH_2)_3CR$, $NaCl$, H_2O 3:3:5:10:170	160	72

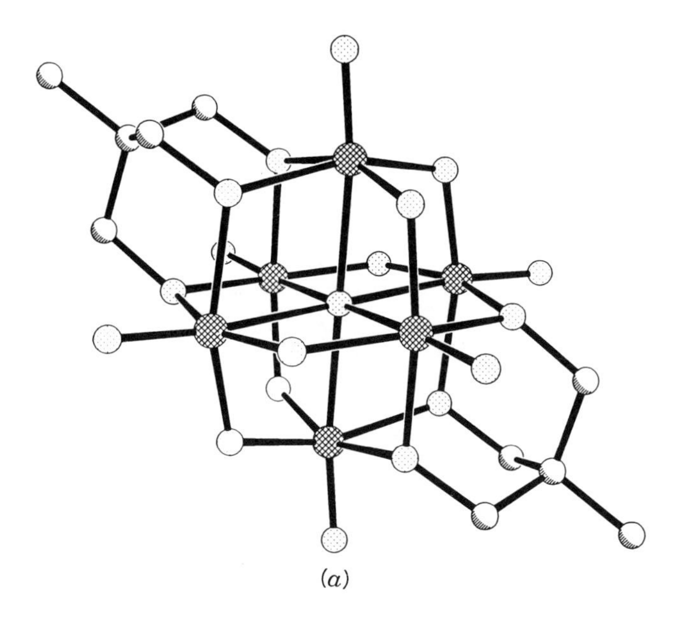

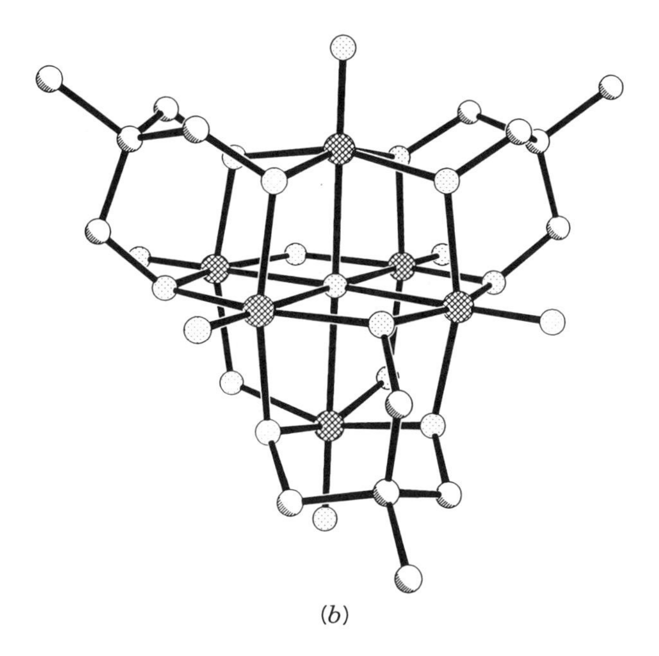

Figure 18. Schematic representations of the $\{M_6O_{19}\}$ core with (*a*) two, (*b*) three, and (*c*) four trisalkoxy ligands.

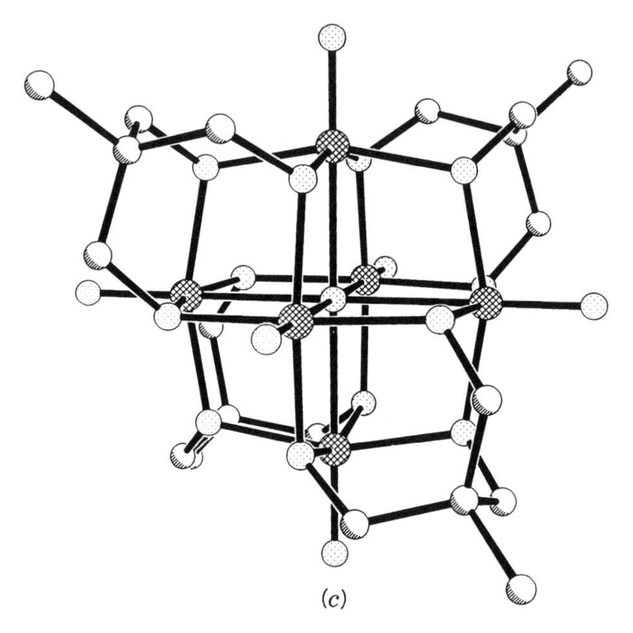

Figure 18. *(Continued)*

centers to the oxo groups of the binuclear site is taken into account. Cluster **22** provides an additional example of ''terminal'' $\{V{=}O\}$ moieties directed toward the interior of the cluster cavity, in effect providing a template for cluster organization. A similar effect has been noted in the structure of $[(VO)_3(thf)(PhCO_2)_6]$ (**1**).

A much more regular geometry is exhibited by the V^{IV} cluster $[(VO)_6(CO_3)_4(OH)_9]^{5-}$ (**23**) (155, 156), formed in the reaction of $VOCl_3$ with NH_4HCO_3 under CO_2. As shown in Fig. 20, the anion consists of a crown-shaped hexanuclear aggregate with bridging hydroxo and carbonato groups. The central carbonato ligand functions in the μ_6 mode, while the exterior carbonato units adopt the more common μ_2 mode. The templating role of the carbonato group is evident, and the structure of **23** should be compared to that of **25**, discussed in Section V.B.e.

The trisalkoxy ligand type is again prominent in the structure of $[V_6O_8\{(OCH_2)_3CEt\}_2\{(OCH_2)_2CEt(CH_2OH)\}_4]^{2-}$ (**13a**) (139). As shown in Fig. 21, the structure of the anion of **24** presents a novel hexavanadium framework based on a tetranuclear $\{V_4O_{16}\}$ core of edge-sharing octahedra linked via edge sharing to two peripheral vanadium square pyramids. The geometry of the tetranuclear framework of **13a** is reminiscent of that observed for the tetranuclear species **7–9**, which also possess the common core type IVa. The

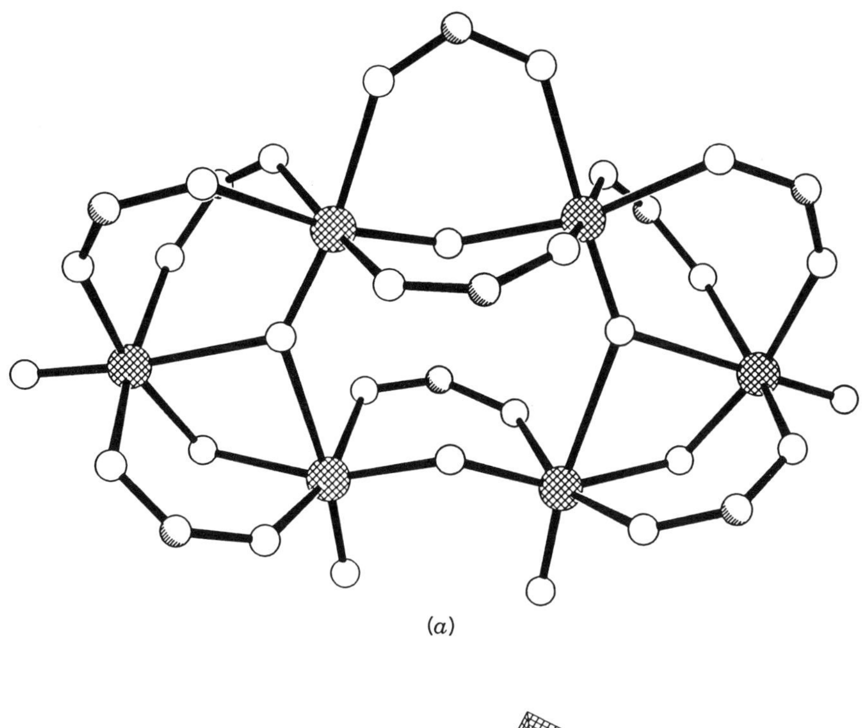

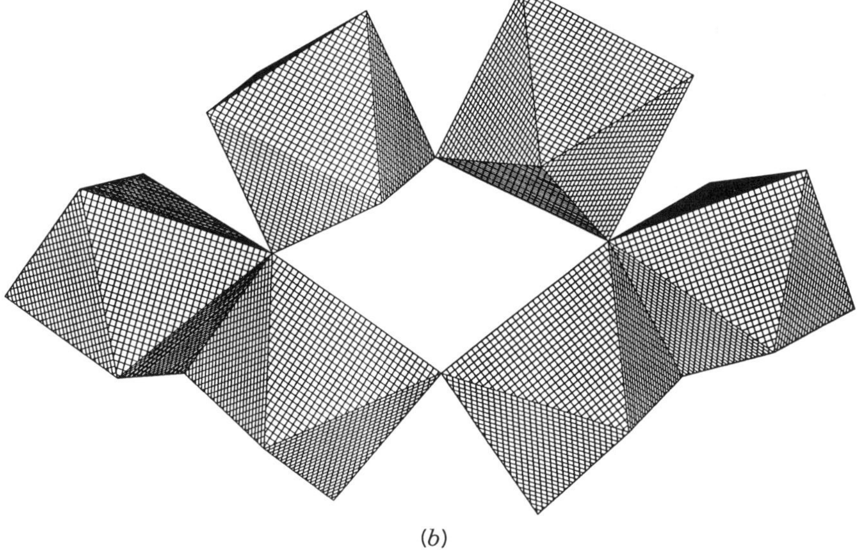

Figure 19. (*a*) The structure of $[V_6O_{10}(PhCO_2)_9]$. (*b*) A polyhedral representation showing the edge- and corner-sharing octahedra.

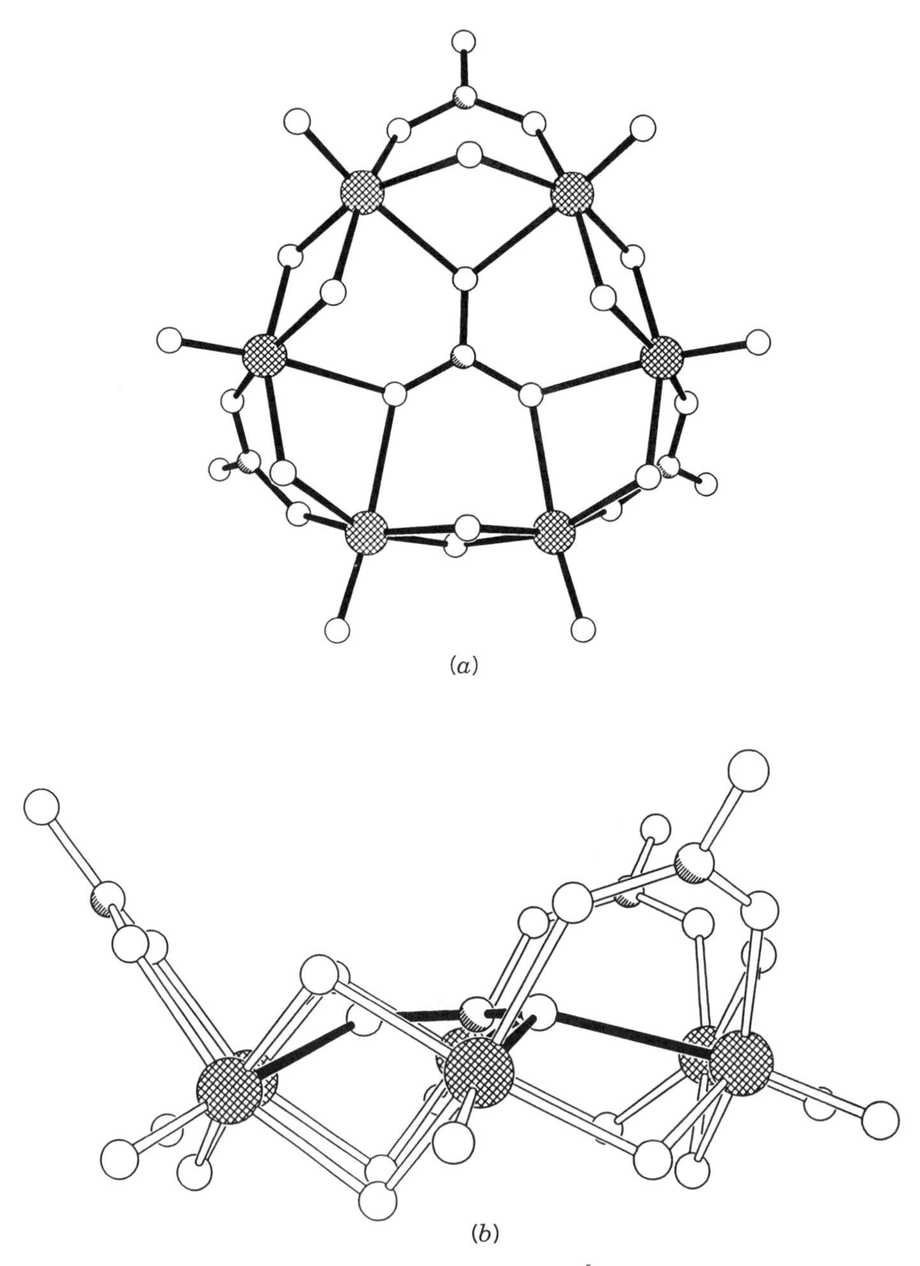

Figure 20. (*a*) A view of the structure of $[(VO)_6(CO_3)_4(OH)_9]^{5-}$ (**23**), normal to the V_6 plane. (*b*) A view with the V_6 plane normal to the page, illustrating the location of the CO_3^{2-} group above the V_6 plane. This contrasts with the structure of $[(VO)_8(OR)_{16}(C_2O_4)]^{2-}$ where the organic structure $C_2O_4^{2-}$ is coplanar with the V_8 ring.

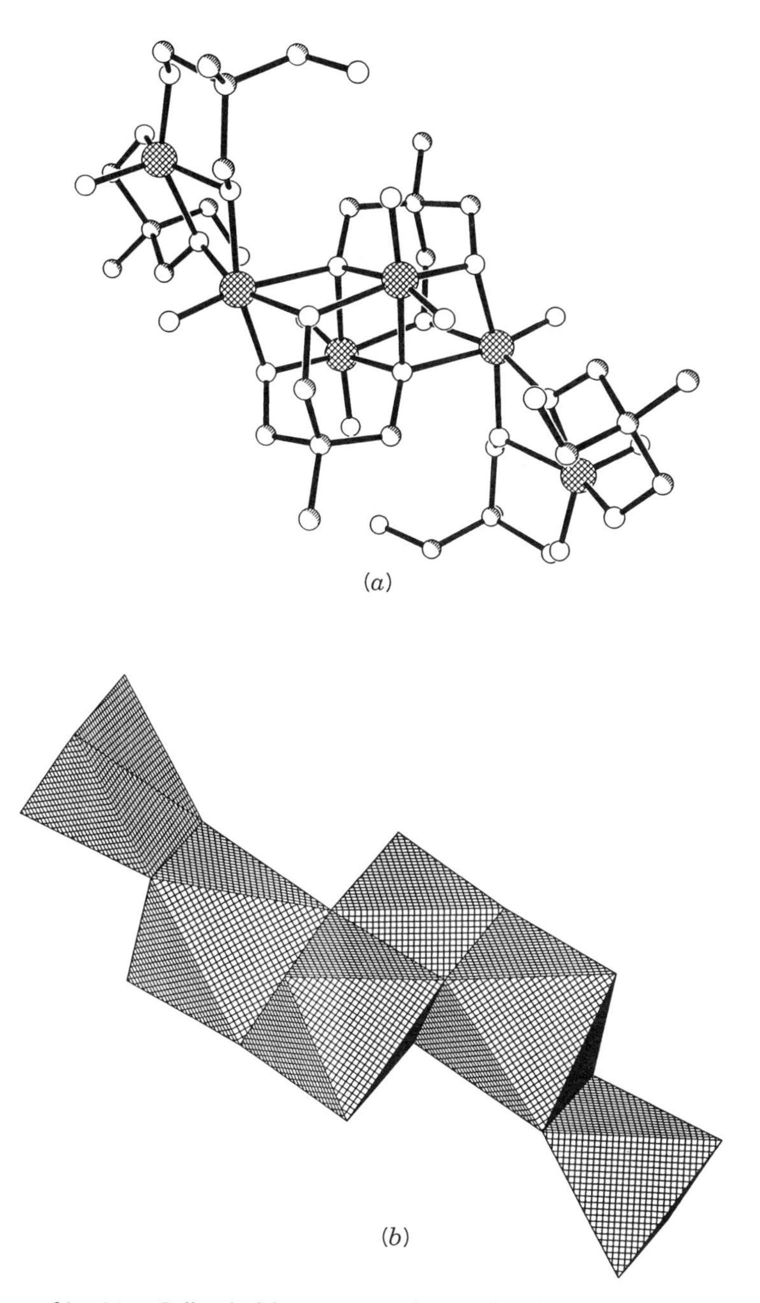

Figure 21. (*a*) Ball-and-stick representation of the structure of $[V_6O_8\text{-}\{(OCH_2)_3CEt\}_2\{(OCH_2)_2CEt(CH_2OH)\}_4]^{2-}$ (**13a**). (*b*) A polyhedral representation showing the central Type IVa core and the terminal edge-sharing vanadium square pyramids.

structure of the anion cluster of **24** is realized by fusing two $\{VO_5\}$ square pyramids to the two exterior octahedra of the central core. The resultant hexavanadium cluster exhibits three distinct vanadium geometries: one distinguished by the cis dioxo $[VO_2]^+$ unit, an unusual instance of this structural motif in polyvanadate clusters; the second, which participates in bonding to four polyalkoxide ligands, two which bridge sites with the tetranuclear core and two which bridge core vanadium sites to the exterior vanadium centers; and the square pyramidal site, which bonds to two bidentate (HtrisEt) ligands. Valence sum calculations confirm that the first site is in the +5 oxidation state while the latter two exhibit an intermediate oxidation state +4.5, suggesting that the electrons associated with the reduced metal sites are delocalized. Mixed-valence V^V/V^{IV} species are not uncommon and examples include binuclear species of the $\{V_2O_3\}$ core (157), as well as high nuclearity clusters, such as $[V_{10}O_{26}]^{4-}$ (158) and $[V_{15}O_{36}]^{5-}$ (159).

5. *Octanuclear and Higher Nuclearity Cores*

Cores of nuclearity greater than six are relatively rare for the {V/O/carbon–oxygen ligand} class of materials and when observed they are generally limited to the compact decavanadate $\{V_{10}O_{28}\}$ core. However, a number of unusual examples of nuclearity other than 10 have been reported recently.

The reduction of vanadate by rhodizonic acid, $H_2C_6O_6$, in alcohols yields the octanuclear V^{IV} species $[(VO)_8(OMe)_{16}(C_2O_4)]^{2-}$ (**24**) (160). The structure shown in Fig. 22 exhibits an octagonal array of $\{VO\}^{2+}$ centers, doubly bridged by methoxy groups to produce a cyclic tiara framework $[(VO)_8(OMe)_{16}]^0$. The cavity produced by this unit is occupied by an oxalate group $[C_2O_4]^{2-}$ with each oxygen donor bridging two V centers to give an arrangement of alternating face-sharing and edge-sharing $\{VO_6\}$ octahedra. The cyclic structure of **24** is reminiscent of the structure of $[(VO)_6(CO_3)_4(OH)_9]^{5-}$ (**23**) and once again demonstrates the templating influence of organic anions in directing the shell organization. Apparently, the ring expansion from six vanadium sites in **23** to eight in **24** is dictated by the coordination requirements of the template.

While the control of the linkage of fragments of inorganic complexes to form oligomers or infinite 3D structures is of interest in the design of materials by molecular engineering, the underlying chemical principles remain obscure. However, the structures of **23** and **24** suggest that $\{VO_n\}$ polyhedra may be linked in a more or less controlled fashion by judicious choice of anionic reagents as templates. This observation has been exploited in the construction of a variety of novel cluster shells using organic anionic reagents that either occupy the cavity of the shell or occupy positions as the outer surface of the shell, as illustrated by clusters **25–27** (11, 161).

The structure of the mixed-valence $8V^{IV}/7V^V$ species $[V_{15}O_{36}(CO_3)]^{7-}$ (**25**)

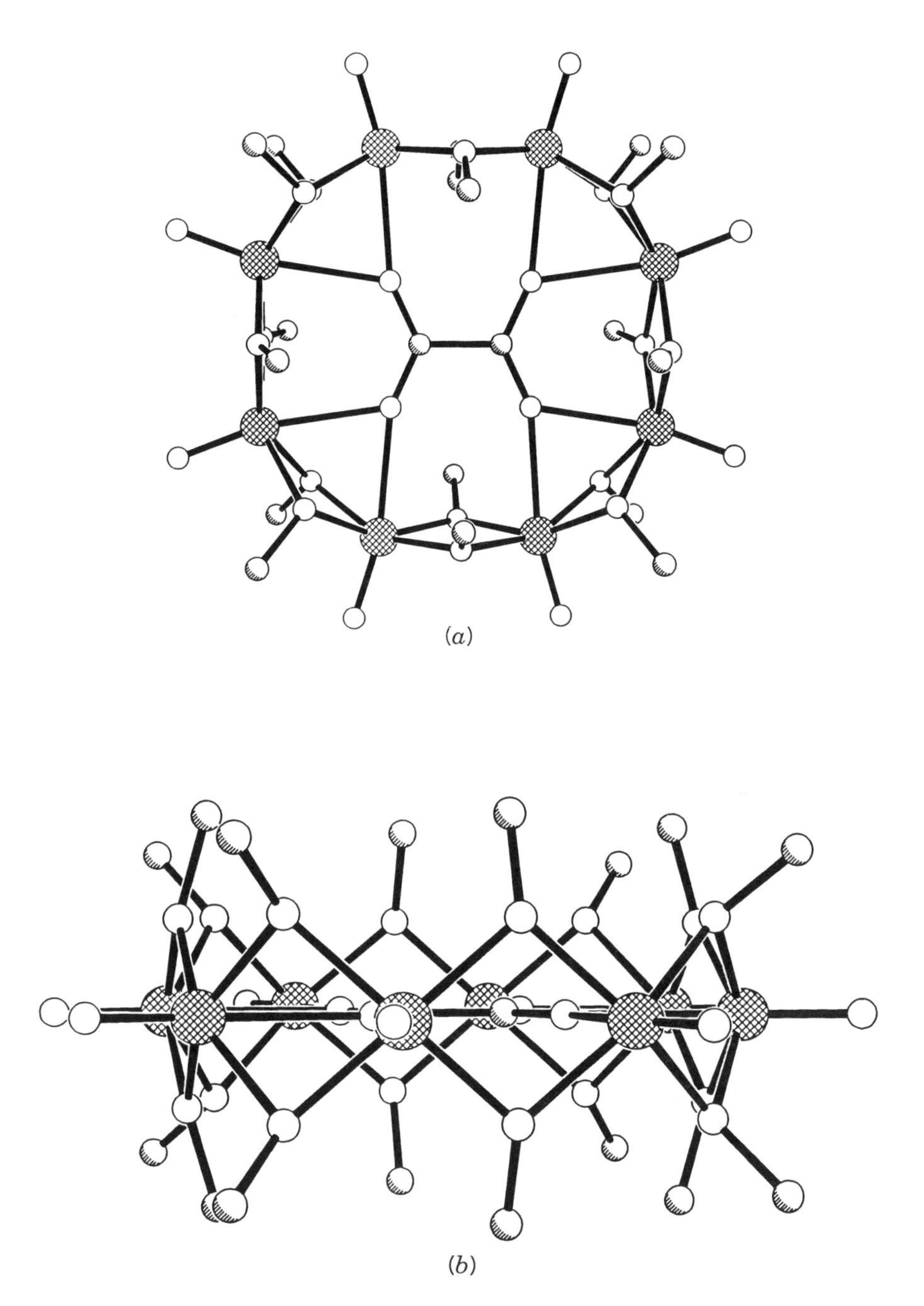

Figure 22. (*a*) A view of the structure of $[(VO)_8(OMe)_{16}(C_2O_4)]^{2-}$ (**24**) normal to the V_8 plane. (*b*) The "tiara" structure of the core and the coplanarity of the V_8 plane and the $[C_2O_4]^{2-}$ substrate.

(161) consists of a spherical shell of corner-sharing $\{VO_5\}$ square pyramids encapsulating the CO_3^{2-} anion. The $[V_{15}O_{36}]^{n-}$ shell may accommodate a variety of anions, including Cl^- and Br^-. When the attractive type of interaction between the central nucleophilic anion, as in the cases of Cl^- and Br^-, and the electrophilic vanadium centers is weak, square pyramids are linked to form the carcerand. If the interaction is strong, as for CO_3^{2-}, octahedral sites also occur in the structures, as shown in Fig. 23.

Introduction of a sterically bulkier template, such as a carboxylate, requires either expansion of the cavity or positioning of the ligand on the exterior surface of the cluster. Both structural types are observed for the $V/O/RCO_2^-$ system. The structure of $[H_2V_{22}O_{54}(MeCO_2)]^{7-}$ (**27**) (11), a $10V^{IV}/12V^{V}$ mixed-valence cluster shown in Fig. 24, consists of a shell of $\{VO_5\}$ square pyramids encapsulating the $MeCO_2^-$ group. In the case of $[H_6V_{10}O_{22}(RCO_2)_6]^{2-}$ (**26**) (11), a mixed-valence $8V^{IV}/2V^{V}$ cluster shown in Fig. 25, the ligands occupy the exterior of the cluster so as to link $\{V^{IV}O_6\}$ octahedra. The structure may be best described as two sets of four edge-sharing octahedra linked through two $\{V^{V}O_4\}$ tetrahedra. In this case the spins are clearly localized on unique vanadium centers.

The reaction of vanadate with the multidentate butanediaminetetraacetic acid (bdta) in water yields the curious mixed-valence $5V^{V}/4V^{IV}$ cluster $[V_9O_{16}(bdta)_4]^{7-}$ (**28**) (162), shown in Fig. 26. Unlike the previous examples of higher nuclearity members of the {V/O/carbon–oxygen ligand} class that

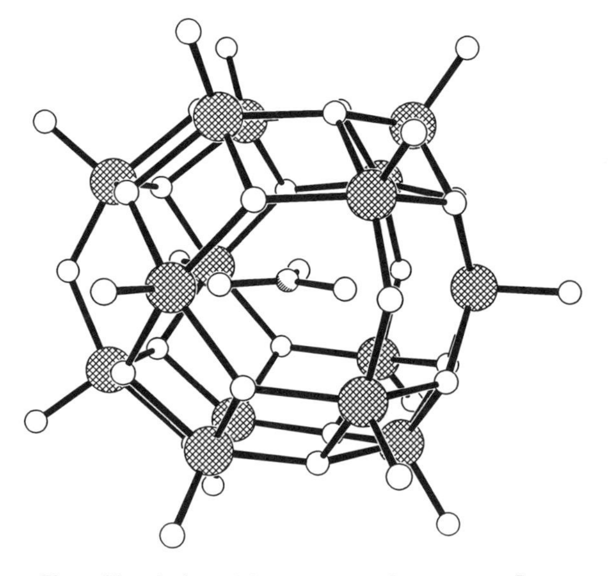

Figure 23. A view of the structure of $\{V_{15}O_{36}(CO_3)\}^{7-}$ (**25**).

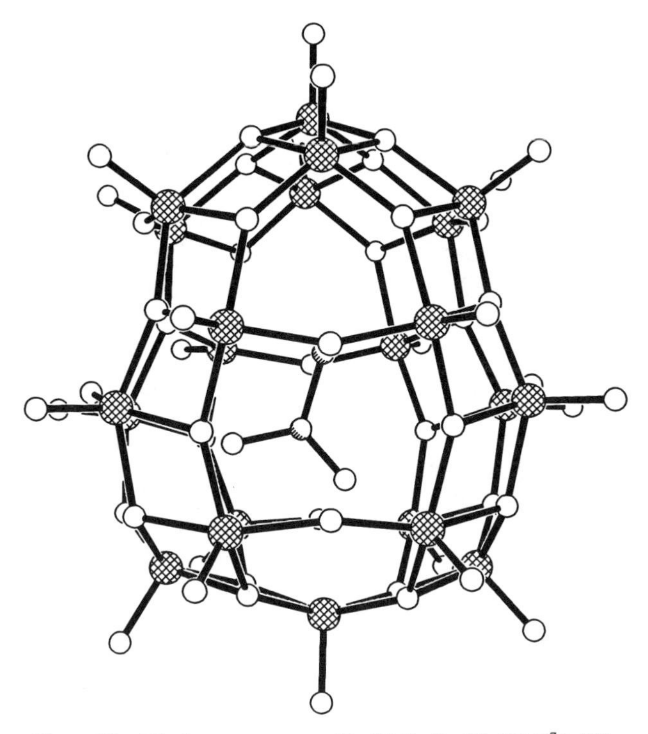

Figure 24. The host–guest assembly $[H_2V_{22}O_{54}(MeCO_2)]^{7-}$ (**27**).

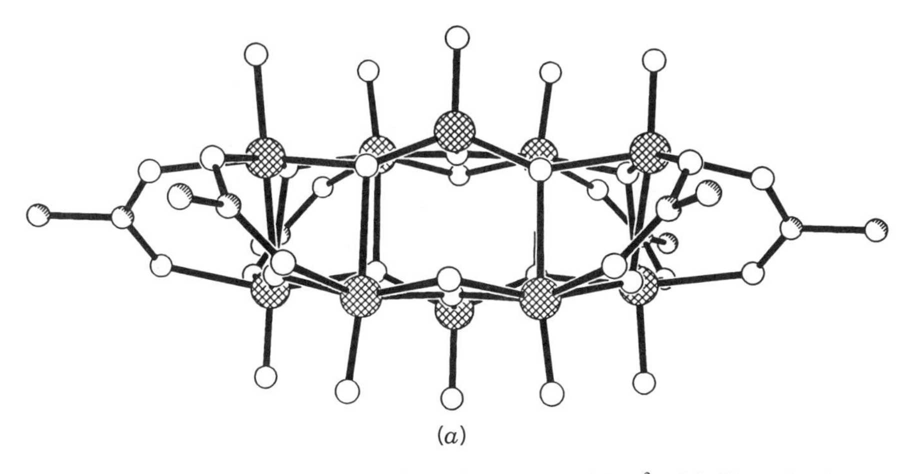

Figure 25. (*a* and *b*) Two ball-and-stick views of $[H_6V_{10}O_{22}(RCO_2)_6]^{2-}$ (**26**), illustrating the two V_5 units linked through bridging RCO_2^- groups and the tetrahedral $\{VO_4\}$ groups. (*c*) A polyhedral view showing the pattern of edge- and corner-sharing octahedra and tetrahedra.

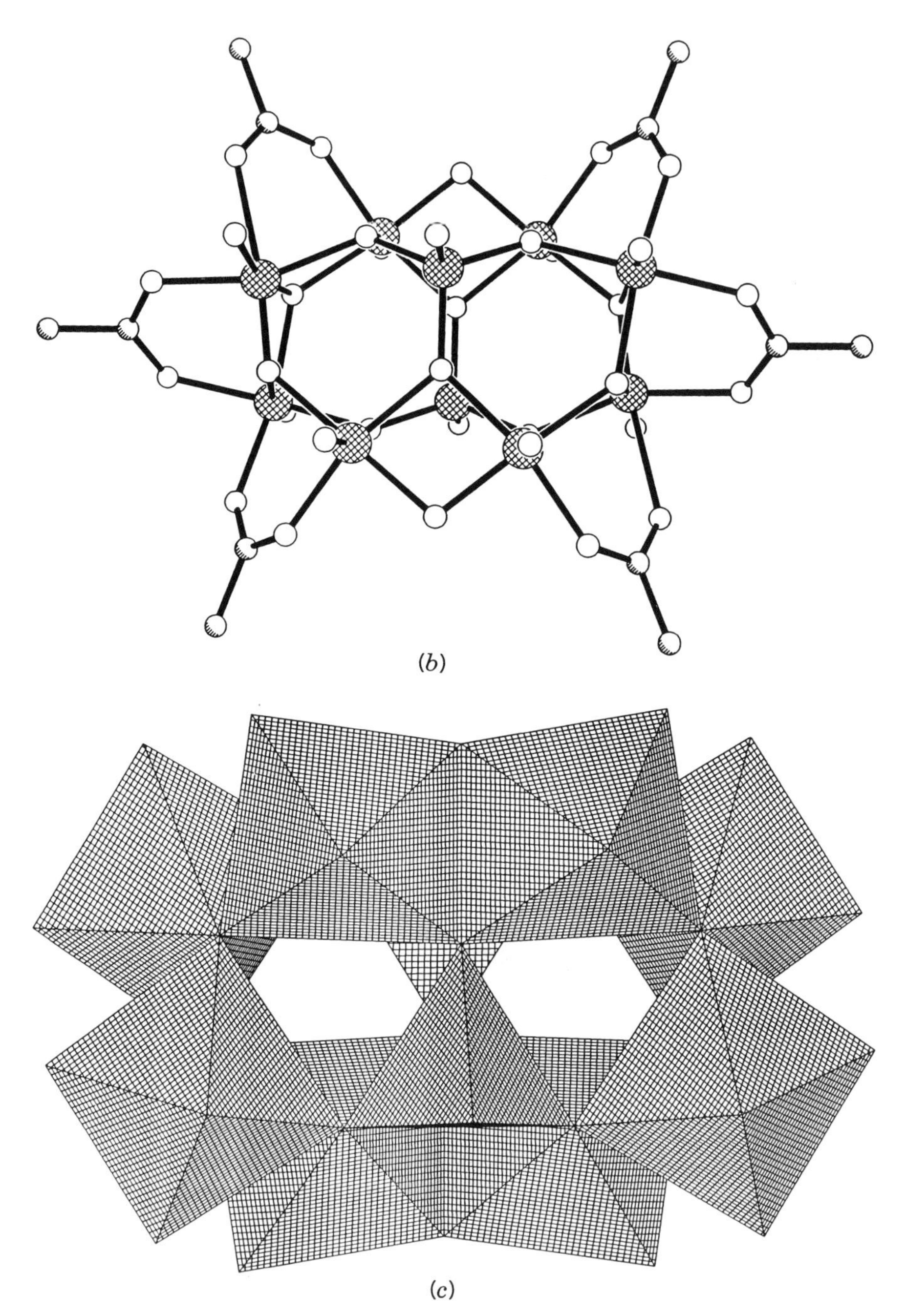

(b)

(c)

Figure 25. (*Continued*)

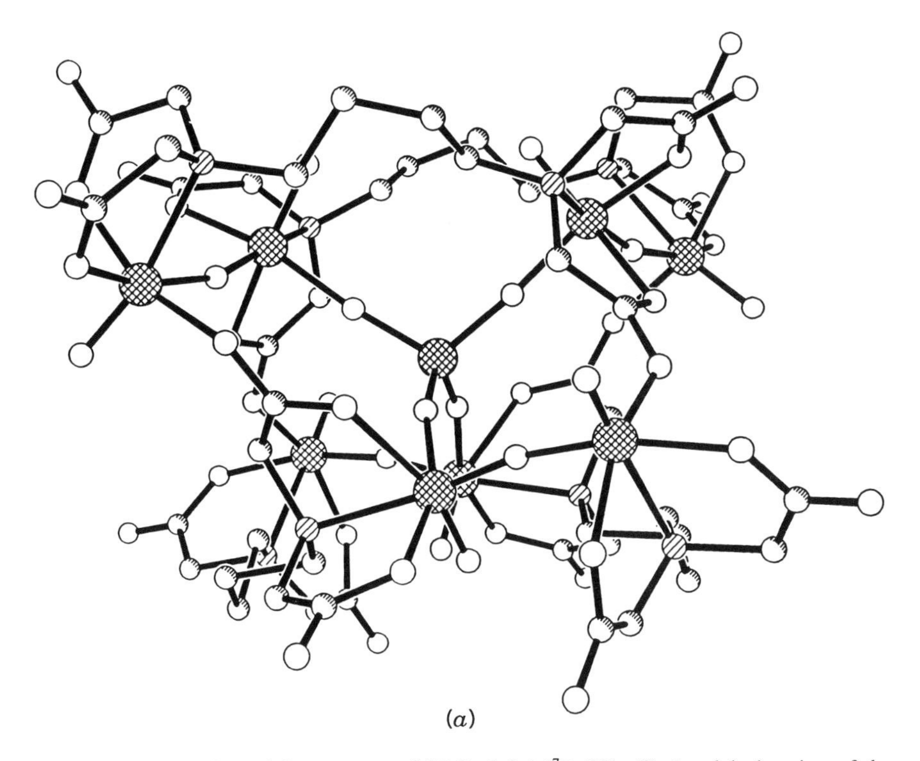

Figure 26. (*a*) A view of the structure of $[V_9O_{16}(bdta)_4]^{7-}$ (**28**). (*b*) A polyhedra view of the vanadium sites, illustrating the unique central tetrahedron linking the four binuclear corner-sharing octahedral units.

form ring, spherical, or bilayer structures, **28** assumes an open framework, characterized by a central $\{V^VO_4\}$ tetrahedron linking four binuclear units, constructed of corner linked square pyramids with the common $\{V_2O_3\}^{3+}$ mixed-valence core (156).

The remaining members of the high nuclearity group of clusters belong to the V/O/trisalkoxide family and illustrate the exploitation of hydrothermal techniques to induce crystallization of larger aggregation. The hexavanadate core $[V_6O_{19}]$ may be considered a fragment of larger clusters. Thus, condensation of the $\{V_6O_{19}\}$ unit with four additional vanadium octahedra provides the decavanadate core shown in Fig. 27. While the $[V_{10}O_{28}]^{6-}$ is well established, no coordination compounds of this core have been isolated by conventional methods. However, as shown in Table X, appropriate conditions for the isolation of ligand substituted $\{V_{10}O_{28}\}$ core materials have been explored in the hydrothermal domain. In this fashion, the series of decavanadates $[V^{IV}_8V^V_2O_{16}(tris)_4]^{4-}$ (**29**), $[V^{IV}_{10}O_{14}(OH)_2(tris)_4]^{2-}$ (**31**), $[V^{IV}_8V^V_2O_{16}(tris)_4]^{2-}$

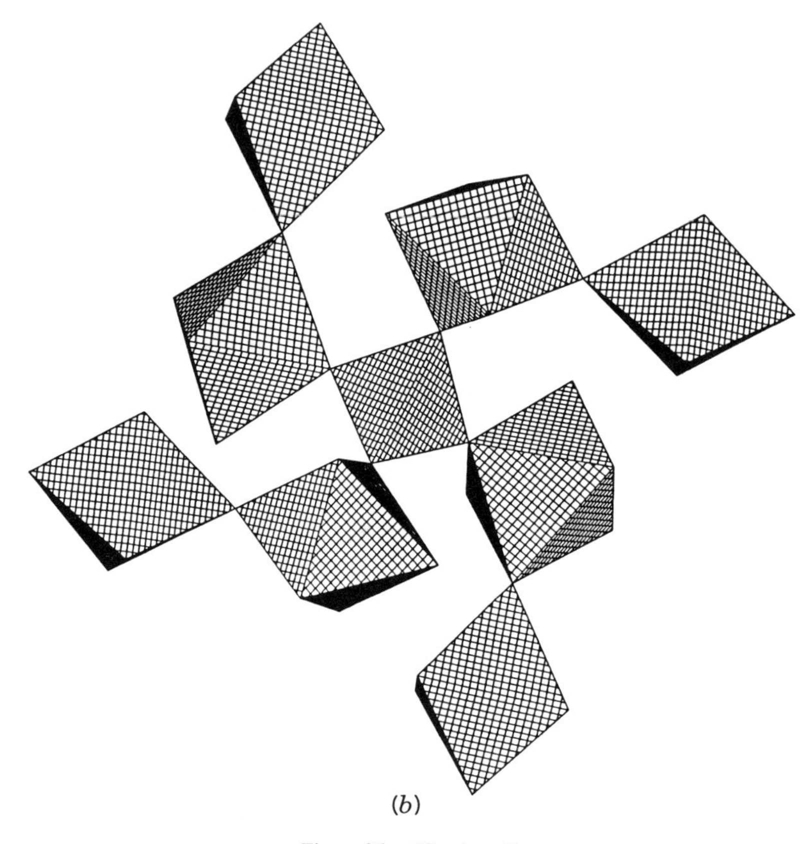

(*b*)

Figure 26. (*Continued*)

(**30**), and $[V^{IV}_{10}O_{13}(tris)_5]^{1-}$ (**32**) were isolated (163–165). Once again as in the hexavanadate species, the pattern of substitution adopted by the trisalkoxy ligands is dictated by the bonding requirements of the ligands that will bridge between three metals in a triangulo arrangement when possible. Thus, the ligands cap the triangular faces of the tetrahedral cavities of the decametalate framework as shown in Fig. 27. Although there are 12 of these sites associated with the cluster, only 6 may be occupied by bridging tridentate ligands, as indicated by the highlighting. In the structures of **29–31**, four sites are occupied while in **32** a fifth site is also occupied.

In common with observations on both naked core $\{V_xO_y\}^{n-}$ clusters and those incorporating a variety of templates $\{V_xO_yL\}^{n-}$ (12), structures with different electronic populations and degrees of protonation are readily isolated. As in the case of the hexavanadates, such changes in cluster oxidation states are

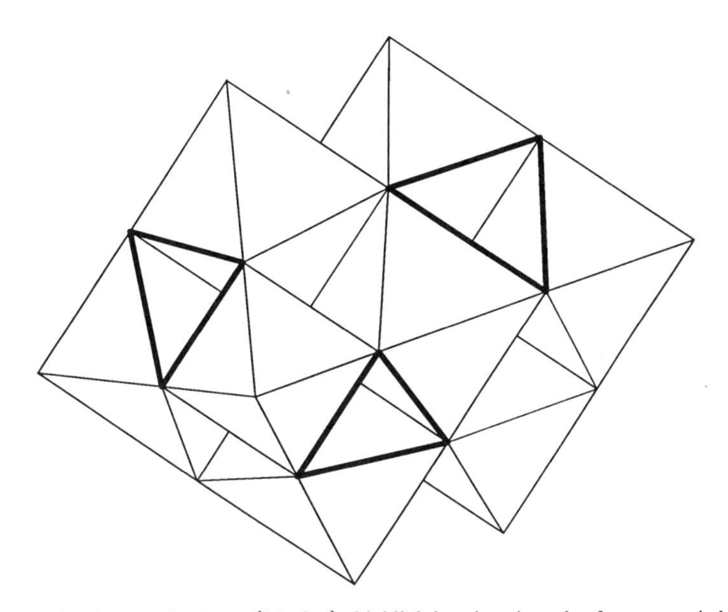

Figure 27. The decametalate core $\{M_{10}O_{28}\}$, highlighting the triangular faces occupied by trisalkoxy ligand types.

reflected by only minor structural perturbations. In fact, the structures of **29–31** are virtually identical but for the V(4)–O(10) bond lengths. The doubly bridging O(10) is the protonation site, while in the mixed species V(4) is the localized V^{V} center. The structures **29–32** also demonstrate the general principle that when the ligands occupy positions on the exterior surface of the cluster, octahedral geometry appears to be favored and consequently compact structural cores based on corner-sharing $\{VO_6\}$ octahedra are adopted.

The influence of minor variations in hydrothermal reaction conditions is demonstrated in the isolation of the hexadecanuclear V^{IV} cluster $[V_{16}O_{20}\{(OCH_2)_3CR\}_8(H_2O)_4]$ (**33**) (166), an unusual example of neutral cluster of the V/O/alkoxide family. As shown in Fig. 28, the structure of **33** consists of two $(VO)_8\{(OCH_2)_3CCH_2OH\}_4(H_2O)_2]^{4+}$ units connected through four μ^2-oxo groups. The topology of each $[(VO)_8\{(OCH_2)_3CCH_2OH\}_4(H_2O)_2]^{4+}$ motif is clearly related to the decametalate core $\{V_{10}O_{28}\}$ by removal of two of the $\{(\mu\text{-O})V{=}O\}$ groups occupying the polar capping positions of the latter. Face-to-face condensation of two of the $[(VO)_8\{(OCH_2)_3CCH_2OH\}_4(H_2O)_2]^{4+}$ units through four μ^2-oxo-groups produces the neutral core of **33**. Twenty doubly bridging and four triply bridging oxo groups of the hypothetical $\{V_{16}O_{48}\}$ core have been replaced by alkoxy donors from the eight trisalkoxy ligands. In ad-

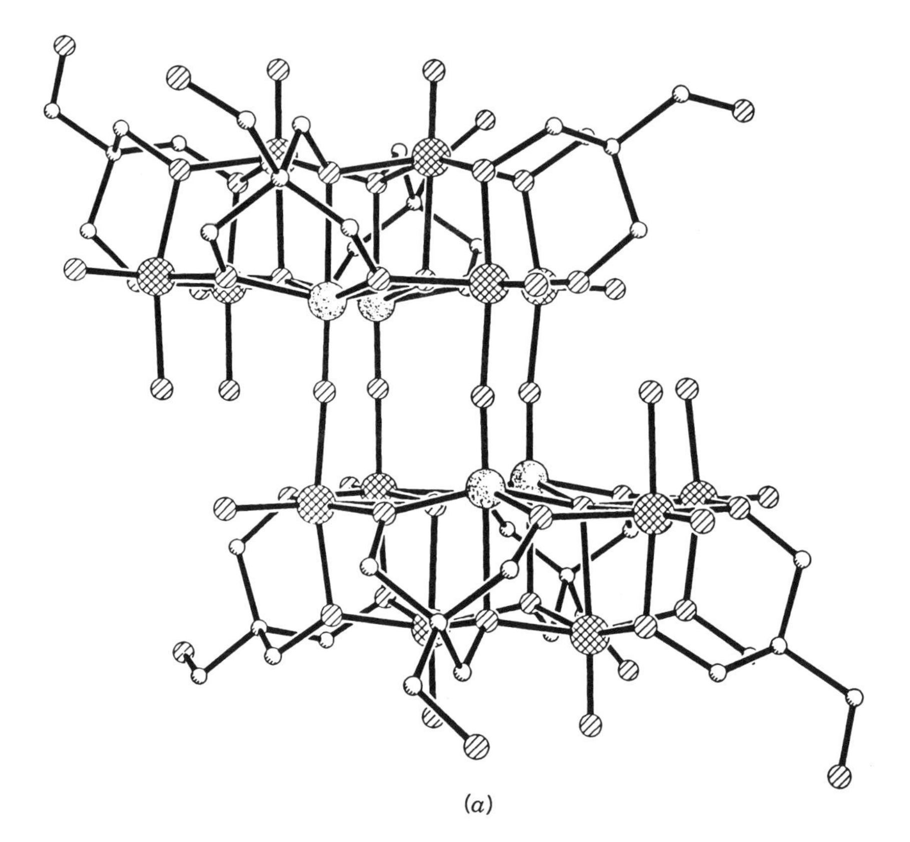

Figure 28. (*a*) A view of the structure of $[V_{16}O_{20}\{(OCH_2)_3CR\}_8(H_2O)_4]$ (**33**). (*b*) A polyhedral representation.

dition, there are four aqua ligands sited on the two exterior vanadium centers of the central hexavanadium ring of each $[(VO)_8\{(OCH_2)_3CCH_2OH\}_4(H_2O)_2]^{4+}$ layer and directed toward the interlayer region. One consequence of the aquo coordination is that the two $[(VO)_8\{(OCH_2)_3CCH_2OH\}_4(H_2O)_2]^{4+}$ layers do not align with the hexavanadium ring of one unit directly above the hexavanadium face of the second but with the layers sheared via a parallel displacement to produce four {V—O=V} interactions with long–short V—O distances.

Within each $[(VO)_8\{(OCH_2)_3CCH_2OH\}_4(H_2O)_2]^{4+}$ subunit, the $\{VO_6\}$ octahedra aggregate by edge-sharing into a compact arrangement, while the two octavanadium units are interconnected by corner sharing of four pairs of vanadium octahedra. This arrangement produces an unusual central $\{V_4O_6\}$ parallelepiped with two pairs of short V—V distances [2.671(5) and 2.684(5) Å

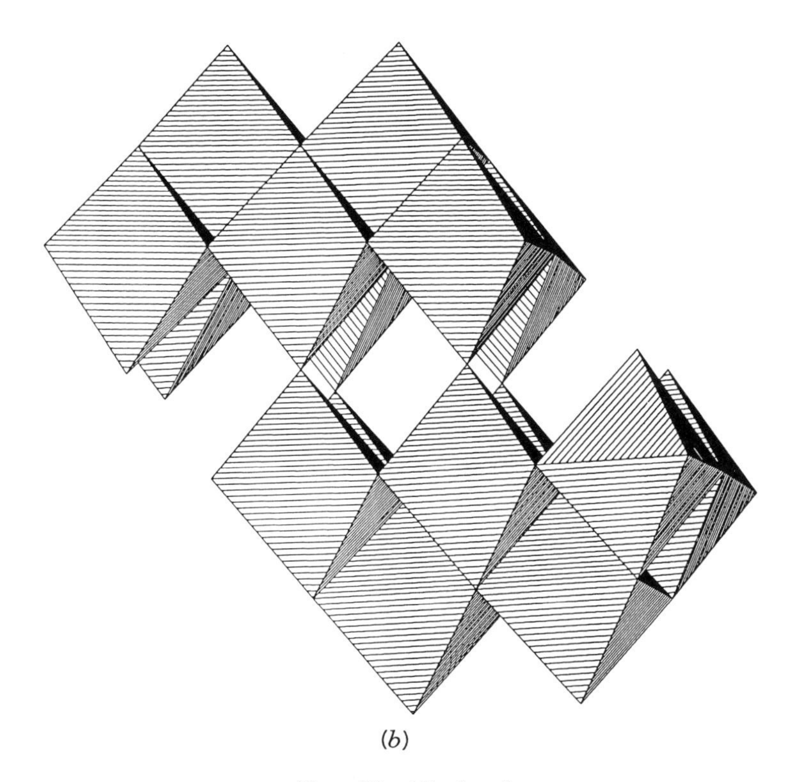

(b)

Figure 28. (*Continued*)

compared to a range of 0.303–3.30 Å for the distances between other V—V pairs]. The vanadium centers involved in these short interactions are also displaced from the planes through the hexavanadium rings into the region between two octanuclear structural units, so as to produce short V—O distances to the four oxo groups bridging these units.

We have noted previously in the structures of the $[V_6O_{19-3n}\{(OCH_2)_3CR\}_n]^{x-}$ and $[V_{10}O_{28-3n}\{(OCH_2)_3CR\}_n]^{x-}$ classes of polyoxoalkoxyvanadium clusters that the preferred substitution pattern of the trisalkoxide ligands adopts a bridging mode between three metals in a triangulo arrangement, so as to cap the triangular faces of the tetrahedral cavities of the metal oxide framework. Such an arrangement in the case of **33** would limit trisalkoxide occupancy to six sites. However, by adopting a ligation mode that alternate occupancy of triangular cavities, which bridge three V sites, and of triangular faces of the central vanadium sites of each hexavanadium ring, thus bridging such vanadium sites to four adjacent vanadium centers, eight alkoxide ligands are accommodated about the framework. This observation raises the possibility of isomeric structures for a given polyoxoalkoxyvanadium cluster, depending on the disposition of trisalkoxide ligands over triangular faces defined by three neighboring V sites or by a single V center of the cluster.

6. *Polymeric Materials*

The V/O/ligand system also forms polymeric materials, exploiting the potential of either multidentate oxygen-donor ligands or of the vanadium–oxo unit to act as bridging groups. However, as these units are somewhat limited in their spatial extension, the polymeric materials are limited to 1D phases.

The reaction of $VOCl_3$ with methanol yields the polymeric V^V alkyl ester $[VO(OMe)_3]_\infty$ (**34**) (167), shown in Fig. 29. The structure consists of edge-sharing binuclear units, linked into infinite chains. The vanadium coordination geometry is effectively distorted octahedral, in common with the tetranuclear and hexanuclear cores of the V/O/alkoxide system. In contrast, the dimeric vanadate esters $[VO(OR)_3]_2$, R = CH_2CH_2Cl (168) and *cyclo*- C_5H_9 (169) exhibit distorted trigonal bipyramidal geometries, while $[VOCl(OCH_2CH_2O)]_2$ (170) exhibits tetrahedral geometry. The V/O/alkoxide system clearly possesses extensive structural diversity, which has yet to be fully developed.

The V^{IV} material $K_2[VO(HCO_2)_4]$ (**35**) (171) also exhibits distorted octahedral geometry about the vanadium centers. The structure of **35** consists of infinite zigzag chains of V^{IV} centers linked in an axial–equatorial manner by formate bridges. It is noteworthy that the magnetic moment of 1.79 BM indicates little or no magnetic interaction between the V atoms.

The structure of the V^{IV} species $(Ph_4P)[VOCl(C_2O_4)]$ (**36**) (172), formed in the reaction of $(Ph_4P)_2[VO_2Cl_2]$ with $Na_2C_2O_4$, is unique for this class of polymeric materials in exhibiting the V^{IV} square pyramid as the structural motif. Vanadium polyhedral are linked through bridging oxalate ligands into infinite zigzag chains, separated by the bulky interstrand cations, as shown in Fig. 30.

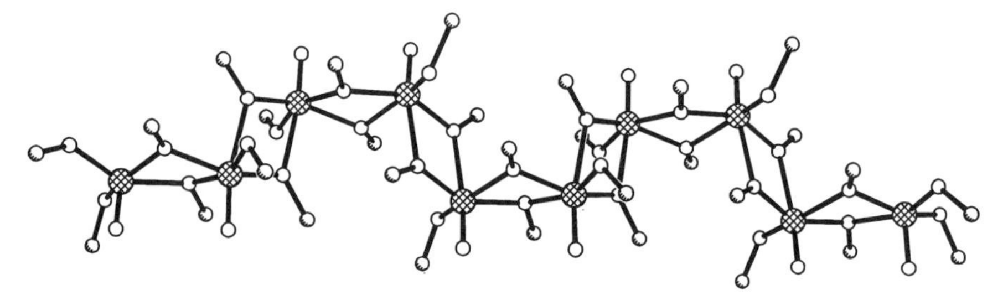

Figure 29. The infinite chain structure of $[VO(OMe)_3]_\infty$ (**34**).

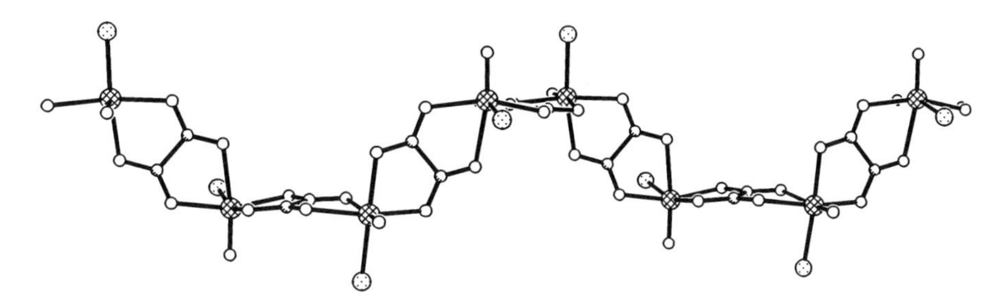

Figure 30. The 1D chain structure of $[VOCl(C_2O_4)]_n^{n-}$.

B. Oxovanadium Clusters with Organophosphonate, Organoarsonate, and Related Ligand Types

1. Structural Prototypes: The Binuclear Core and the Layered Vanadyl Organophosphonate, [VO(RPO$_3$)(H$_2$O)]

The contemporary interest in the $V/O/RPO_3^{2-}$ system reflects not only the extensive and fundamental molecular chemistry spanning binuclear species to "superclusters" of nanometer dimensions but the applications of solid phases, particularly layered materials, as sorbents, catalyst supports, and optical devices. Catalytic oxidation of alkanes in layered solids of the type $[(VO_2)P_2O_7]$ (173) provided the impetus for the development of the metal–organic version of this catalyst, $[VO(RPO_3)(H_2O)]R'OH$ (135). The layered solid loses R'OH upon heating and the resultant material recognizes primary alcohols in preferences to secondary and tertiary alcohols. The structure of $[VO(PhPO_3)(H_2O)]$ (64), shown in Fig. 31, exhibits several features that are persistent structural motifs reflected in the structures of the molecular clusters. Prominent among these are the presence of $\{V{-}O{-}V\}$ linkages and of the $\{(VO)_2(\mu_2\text{-}RPO_3)_2\}$ units, containing the cyclic $\{V{-}O{-}P{-}O\}_2$ moiety. Also worthy of note is the extended structure of the solid, which consists of alternating V/O/P inorganic layers and organic bilayers.

The simplest constituent building block for the solid phase $[VO(RPO_3)(H_2O)]$ is provided by the binuclear complexes of the class $[(VO)_2Cl_2(H_2O)_2(PhPO_3H)_2]$ (**37**) (174). As shown in Fig. 32(*a*), the overall structure of (**37**) may be described in terms of two vanadium(IV) square pyramids and two phosphonate tetrahedra in a corner-sharing arrangement. Each vanadium site is coordinated to a terminal oxo group, two phosphonate oxygen donors, and exocyclic chloride and H_2O ligands. The crystallographically imposed symmetry dictates that the vanadium oxo groups adopt an anti orientation. Each organophosphonate ligand bridges vanadium centers through two oxygen donors, while the third ox-

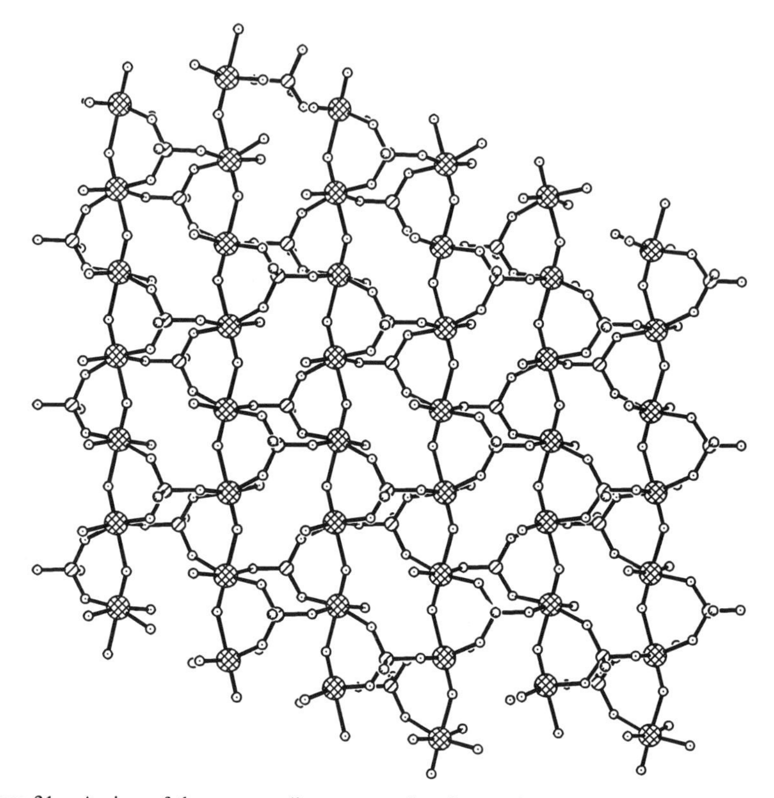

Figure 31. A view of the oxovanadium–organophosphonate layer of $[VO(PhPO_3)(H_2O)]$ (phosphonate groups not shown) (**58**).

ygen is protonated and pendant. The resultant eight-membered ring $[V-O-P-O-]_2$ is a structural motif common to vanadium organophosphonate solids (64) and also to layered vanadium oxide phosphate solids (175–177).

The structures of **37** may be viewed as a fragment of the $VO(O_3PR) \cdot H_2O$ layered materials, which may be constructed by linking binuclear units, in such a way that $\{V-O\cdot\cdot\cdot V\}$ interactions with alternating short and long distances are formed along the chains and the pendant {P—OH} groups coordinate to adjacent units. The syn orientation of the vanadyl groups is also observed upon modification of the R group, as in $[(VO)_2Cl_2(H_2O)_2(MePO_3H)_2]$ (**38**), shown in Fig. 32(*b*). The syn configuration is required for the structural motif for clusters of the $V/O/RPO_3^{2-}$ system, which require {V=O} groups directed to the exterior of the cluster shell.

The topological relationship of the $[(VO)_2Cl_2(H_2O)_2(RPO_3H)_2]$ species to the infinite $[VO(RPO_3H)(H_2O)]$ sheets of the solid, shown in Fig. 33, suggests that the binuclear units of **37** and **38** may serve as building blocks for the prep-

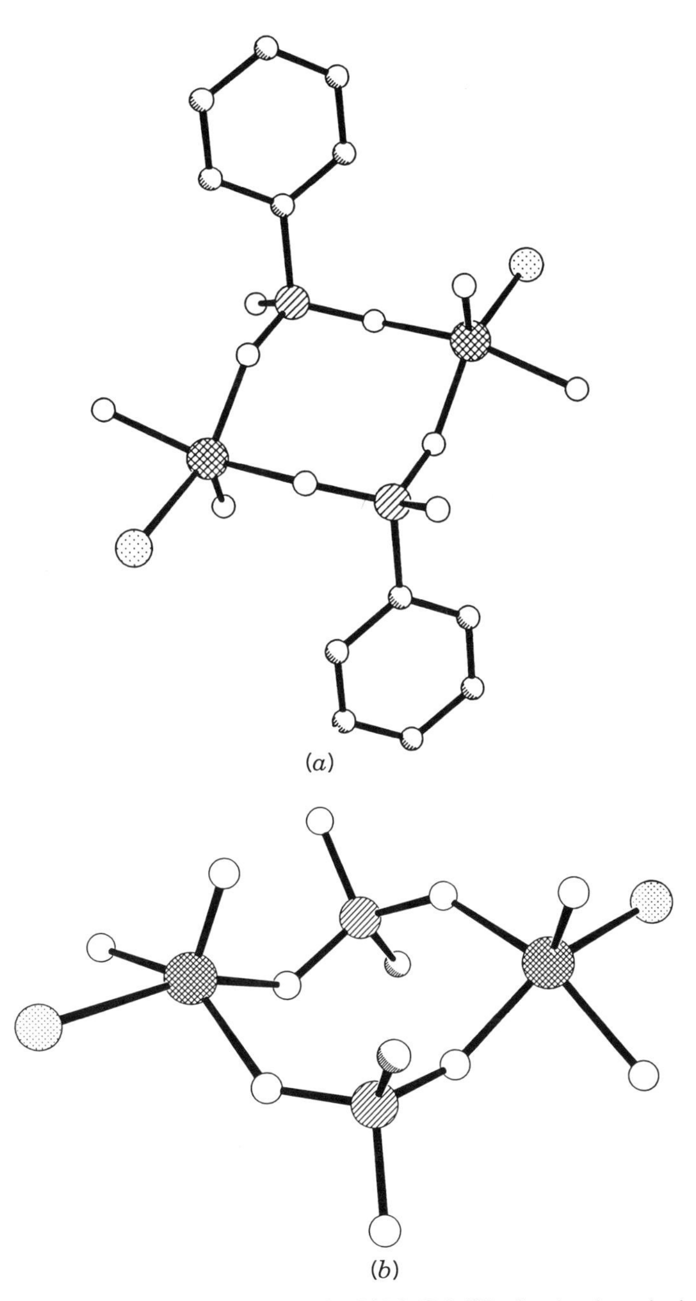

Figure 32. (*a*) The structure of $[(VO)_2Cl_2(H_2O)_2(PhPO_3H)_2]$ (**37**), showing the anti orientation of the V=O groups relative to the $\{V_2P_2O_4\}$ ring. (*b*) The structure of $[(VO)_2Cl_2(H_2O)_2(MePO_3H)_2]$ (**38**), illustrating the syn orientation of the V=O groups.

aration of molecular clusters and solids by condensation–hydrolysis processes involving the ligation of the pendant oxygen atoms of the organophosphonate groups and concomitant loss of HCl and displacement of the aqua ligand. As anticipated, these binuclear species serve as precursors for condensation into larger oligomers and inorganic acids (Scheme 1).

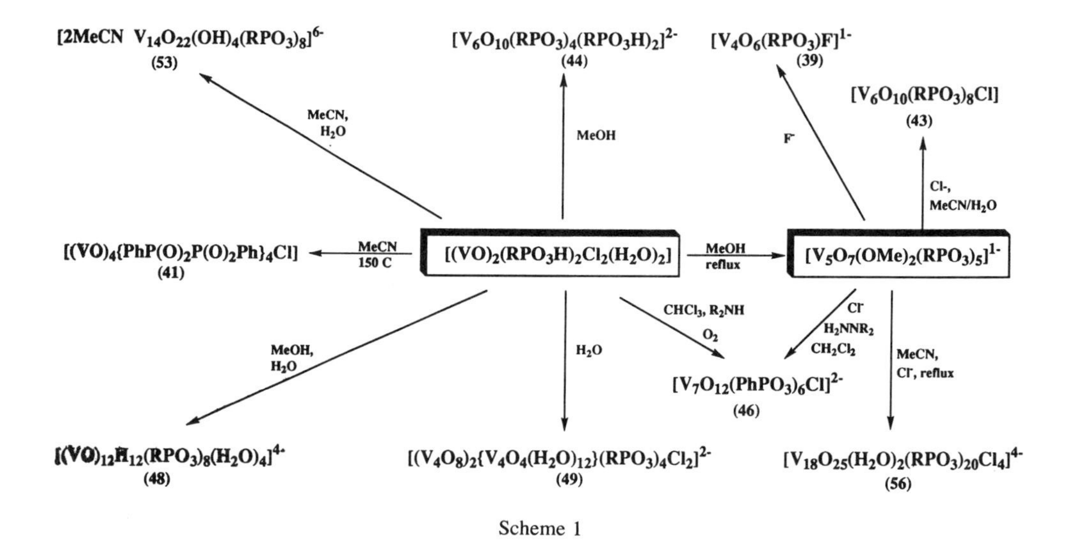

Scheme 1

2. *Tetranuclear Cores*

The tetranuclear species $[V_4O_6(PhPO_3)_4F]^{1-}$ (**39**) (178) is isolated from the reaction of $[V_5O_{14}]^{3-}$ with $PhPO_3H_2$ in ethanol in the presence of HBF_4 or from the reaction of $[VO_2F_2]^{2-}$ with $PhPO_3H_2$ in organic solvents. As shown in Fig. 34, the structure of **39** consists of two $\{V_2O_3\}^{4-}$ units bridged by four $[PhPO_3]^{2-}$ groups, each linked through all three oxygen donors, and encapsulating the F^- anion. Certain features of the structure are noteworthy and provide recurrent themes of the structural chemistry of the $V/O/RPO_3^{2-}$ system. The structure may be described in terms of two $\{(VO)_2(\mu_2\text{-}RPO_3)_2\}$ groups linked through the third $\{P{-}O\}$ arm of the organophosphonate ligands and by the introduction of two doubly bridging oxo groups. The organization of the cluster shell about an anionic template is a characteristic feature of the chemistry. Finally, the presence of $\{V{-}O{-}V\}$ interactions also relates the structure of **39** to that of the prototypical layered material. As illustrated in Fig. 33, cutting of several $V{-}O(P)$ bonds, removing the F^-, and flattening the structure of **39** into a sheet produces a fragment of the dehydrated $[VO(RPO)_3]$ layer.

Cluster **39** undergoes successive, reversible $1e^-$ reductions to yield mixed-

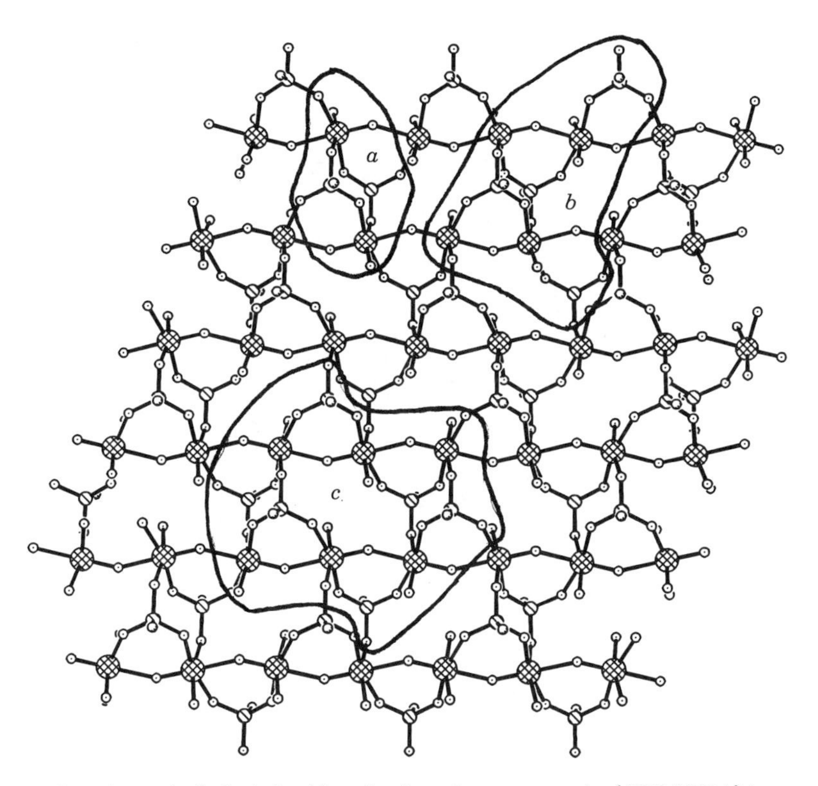

Figure 33. The topological relationships of various cluster cores to the $\{(VO)(RPO_3)\}$ layer structure. (*a*) A binuclear $[(VO)_2(RPO_3)_2]$ unit; (*b*) A tetranuclear $[(VO)_4(RPO_3)_4]$ unit; (*c*) The putative hexanuclear unit $[(VO)_6(RPO_3)_6]$.

valence V^V/V^{IV} species. Chemical reduction of a yellow solution of **39** with hydrazine yields bright green crystals the $3V^V/V^{IV}$ species $[V_4O_6(PhPO_3)_4F]^{2-}$ (**40**). The electron is delocalized throughout the cluster, although the valence sums associated with the vanadium sites clearly establish the total metal valency.

The consequence of introducing a bulkier anionic template are demonstrated by the structure of $[(VO)_4\{Ph(O_2)POP(O_2)Ph\}_4Cl]$ (**41**) (179), shown in Fig. 35. The metrical parameters indicate that the F^- anion of **39** is tightly held and that the shell dimensions are incapable of accommodating the larger Cl^- template. When the V^{IV} species $[VO_2Cl_2]^{2-}$ is used as a starting material, cluster expansion is achieved not by introducing additional vanadyl units into the shell but rather by utilizing the $[PhP(O)_2OP(O)_2Ph]^{2-}$ groups that are formed in a metal mediated, thermally induced condensation process. The structure of **41** contrasts with that of **39** in possessing isolated V^{IV} square pyramids with no

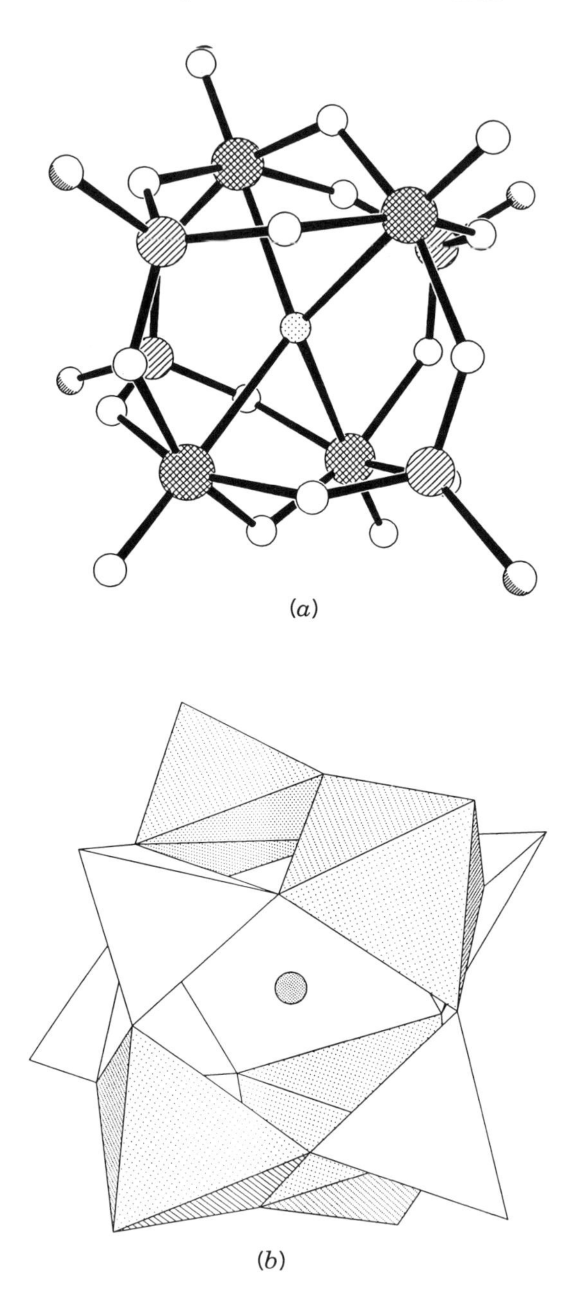

Figure 34. (*a*) Ball-and-stick representation of the structure of $[V_4O_6(PhPO_3)_4F]^{1-}$ (**39**). (*b*) Polyhedral representation illustrating the corner sharing of vanadium square pyramids and phosphorus tetrahedra.

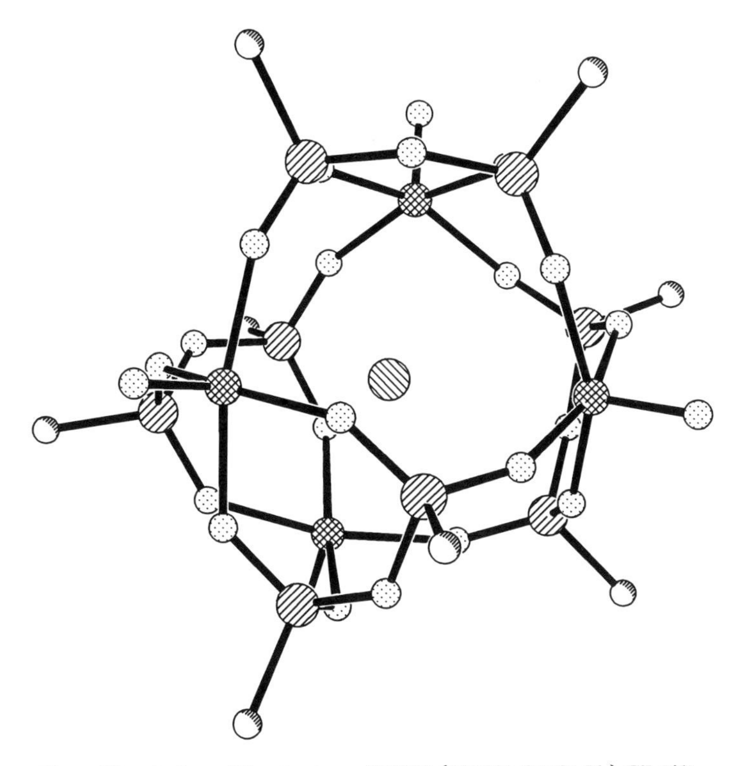

Figure 35. A view of the structure of $[(VO)_4\{PhP(O)_2OP(O)_2Ph\}_4Cl]$ (**41**).

{V—O—V} interactions. Each vanadium center is bonded to the oxygen donors of three neighboring pyrophosphonate ligands; two ligands are each bonded through one O atom and the third through two O atoms to form a {V—O—P—O—P—O—} ring. The V/P/O shell of **41** acts as a cage for a Cl^- ion. The V· · ·Cl distances in **41** are 3.07 Å compared to an average of 2.35 Å for the V—F distances in **39**. The shell associated with **41** acts as a hollow sphere with cryptand properties encapsulating a Cl^- anion in a close packing arrangement, while in **39** the F^- group forms a weak covalent interaction with the vanadium centers in a fashion reminiscent of the V/O/alkoxide species $Na[(VO)_6F(OH)_3\{(OCH_2)_3CMe\}_3]$ (**21**).

3. Pentanuclear Core

The reaction of $[V_5O_{14}]^{3-}$ with $PhPO_3H_2$ in methanol yields the pentanuclear V^V cluster $[V_5O_7(OMe)_2(PhPO_3)_5]^{1-}$ (**42**) (89), shown in Fig. 36. In contrast to spherical structure exhibited by the tetranuclear cluster **39**, the structure of

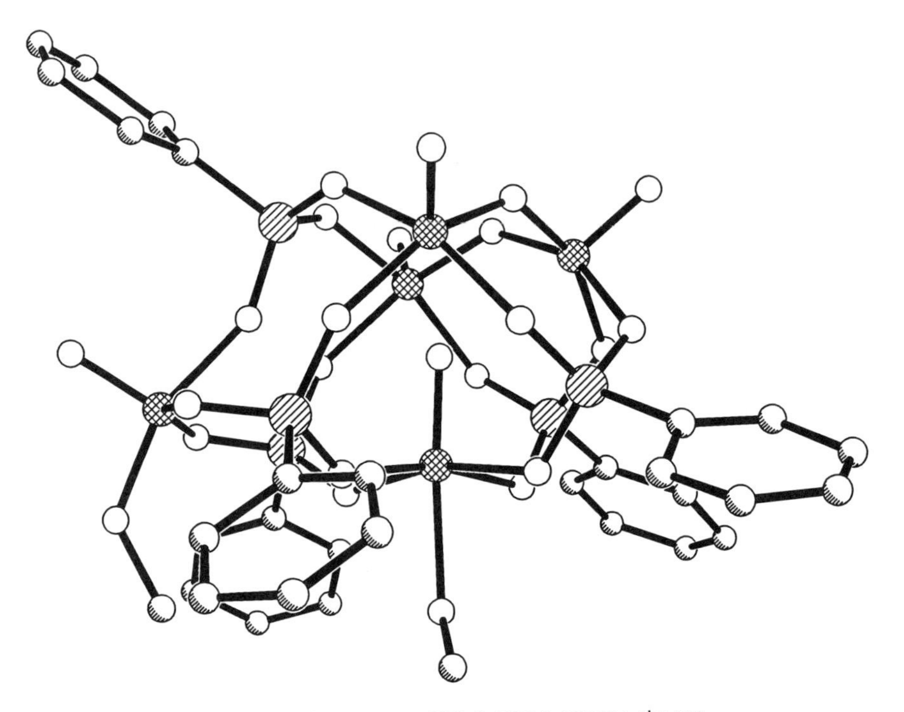

Figure 36. The structure of $[V_5O_7(OMe)_2(PhPO_3)_5]^{1-}$ (**42**).

42 is distinctly irregular. The core of the structure consists of a trinuclear $[(VO)_3(\mu_2\text{-}O)_2]$ unit connected through four of the five phenylphosphonate ligands to the *endo*-[(VO)(OMe)] moiety and through three phenylphosphonate ligands to the *exo*-[(VO)(OMe)] moiety. While four vanadium centers exhibit distorted square pyramidal geometries, the fifth displays distorted octahedral geometry with four oxygen donors of the phosphonato ligands occupying the equatorial positions and a terminal oxo group and a methoxy group in the axial positions.

The structure of **42** exhibits the $[(VO)_2(\mu_2\text{-}RPO_3)_2\}$ motif common to both the $[VO(RPO_3)(H_2O)]$ solid phase and to other molecular species. In common with the molecular clusters of the $V/O/RPO_3^{2-}$ system, **42** may be unfolded and flattened to produce a fragment of the $[VO(RPO_3)(H_2O)]$ structure. A pseudo-mirror plane passes through the structure resulting in four structurally distinct vanadium centers. Two square pyramidal vanadium sites form the termini of the $[(VO)_3(\mu\text{-}O)_2]$ fragment and are each coordinated to a terminal and a bridging oxo group and to three $(\mu\text{-}O_3PPh)$ ligands. The central vanadium atom of this trinuclear unit is bound to a terminal and two bridging oxo groups and two $(\mu_3\text{-}O_3PPh)$ units. The unique octahedrally coordinated vanadium atom caps the bowlike fragment found by the fusing of the square pyramidal geometries of the vanadium atoms of the central $[(VO)_3(\mu\text{-}O)_2]$ moiety and four (O_3PPh) tetrahedra. The $\{VO(OMe)O_3\}$ center has "condensed" onto the surface of the cluster so as to satisfy the coordination requirements of three of the phosphonato ligands.

The most unusual feature of the structure of **42** is the folding of the shell so as to direct an oxo group into the molecular cavity. In a sense, the oxide group may be regarded as the template for the organization of the V/O/P framework of **42**. This unusual orientation of the oxo group is similar to that previously noted for $[(VO)_3(thf)(PhCO_2)_6]$ (**1**) and $[V_6O_{10}(PhCO_2)_9]$ (**22**). Another curious feature of the structure is the presence of the exterior vanadium center, which is weakly attached to the cluster surface, a point that will be invoked in the discussion of the "cluster of clusters" $[V_{18}O_{25}(H_2O)_2(RPO_3)_{20}Cl_4]^{4-}$.

4. *Hexanuclear Cores*

The hexanuclear cores adopted by clusters of the $V/O/RPO_3^{2-}$ system reflect the presence or absence of guest anions and the vanadium oxidation states. Thus, a spherical and symmetrical shell is observed for the $5V^V/V^{IV}$ cluster $[(VO)_6(t\text{-}BuPO_3)_8Cl]$ (**43**) (179), while the $4V^V/2V^{IV}$ cluster $[V_6O_{10}(PhPO_3)_4(PhPO_3H)_2]^{2-}$ (**44**) (180) adopts an unusual cyclic structure.

The solvothermal reaction of $t\text{-}BuPO_3H_2$ with $[Ph_4P][VO_2Cl_2]$ (14) in acetonitrile yields lustrous dark green crystals of the mixed-valence cluster $[(V^VO)_5(V^{IV}O)(t\text{-}BuPO_3)_8Cl]$ (**43**). As shown in Fig. 37, the structure consists of a spherical V/P/O shell of corner-sharing vanadium-centered square pyramids and organophosphonate tetrahedra encapsulating a chloride anion, which seems to serve as a template for cluster formation. Although the structure of **43** is related to those of $[V_7O_{12}(PhPO_3)_6Cl]^{2-}$ (**46**) and $[V_{15}O_{36}Cl]^{6-}$, the topological details are distinct. The structure of $[V_{15}O_{36}Cl]^{6-}$ is constructed exclusively from $\{VO_5\}$ square pyramids in a highly symmetrical arrangement of approximate D_{3h} symmetry; the faces of the sphere consist of two $\{V_3O_3\}$, three $\{V_4O_4\}$, and eighteen $\{V_2O_2\}$ rings. The structure of $[V_7O_{12}(PhPO_3)_6]^-$ is much less symmetrical, exhibiting C_2 symmetry and faces constructed from $\{V_2PO_3\}$, $\{V_2P_2O_4\}$, $\{V_3PO_4\}$, and $\{V_3P_2O_5\}$ rings. The structure of **43** exhibits idealized D_{4h} symmetry, and the shell is constructed exclusively from twelve $\{V_2P_2O_4\}$ faces. In contrast to the structure of $[V_7O_{12}(PhPO_3)_6]^{1-}$, **43** does not possess $\{V{-}O{-}V\}$ bridges, and the vanadium centers are isolated from each other. The structural relationship of **43** to the dinuclear complexes of the type $[(VO)_2Cl_4(t\text{-}BuPO_3H)_2]^{2-}$ is also apparent. Formally, the shell of **43** is constructed from the condensation of three such units with concomitant loss of HCl and incorporation of two additional $t\text{-}BuPO_3^{2-}$ groups. The structure of **43** is

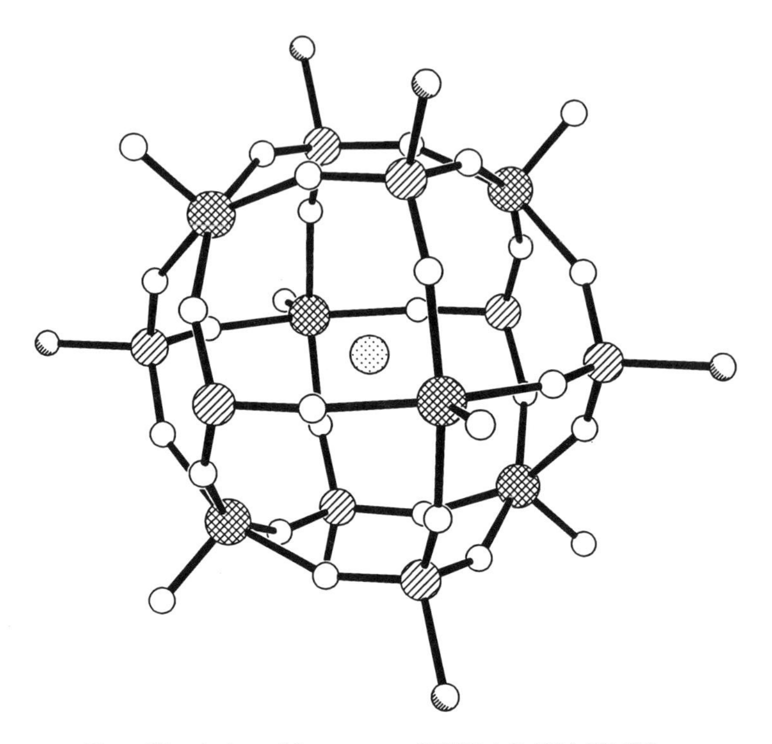

Figure 37. A view of the structure of $[(VO)_6(t\text{-}BuPO_3)_8Cl]$ (**43**).

also related to that of the layered phase $[VO(PhPO_3)H_2O]$, since cleavage of three V—O(phosphonate) bonds in **43** and flattening of the cage produces a motif seen in the layered phase.

Although the $4V^{V}/2V^{IV}$ species $[V_6O_{10}(PhPO_3)_4(PhPO_3H)_2]^{2-}$ (**44**) does not yield crystals of sufficient quality for X-ray analysis, the structure of the analogous organoarsonate species $[V_6O_{10}(PhAsO_3)_4(PhAsO_3H)_2]^{2-}$ (**45**), isolated in the reaction of $[V_{10}O_{28}]^{6-}$ with $PhAsO_3H_2$ in acetonitrile, was confirmed crystallographically. The structure of **45**, shown in Fig. 38, was revealed to exhibit a twisted 24-membered ring $\{V_6As_6O_{12}\}$, which is highly distorted by the presence of additional intraring V—O—V and As—O—V bridges; the As-phenyl substituents, the terminal oxo groups of the vanadium atoms, and the hydroxyl groups of two $\{RAsO_3H\}$ groups project from this central ring. Alternatively, the structure may be described in terms of a layer structure of three $\{V_2As_2O_4\}$ rings connected in a stacking fashion by μ^2-oxo groups.

The vanadium centers exhibit square pyramidal geometry with the standard apical disposition of the terminal oxo groups. Each of the organoarsonate groups is bound to two oxygen atoms and contributes to the formation of the $\{V_2As_2O_4\}$

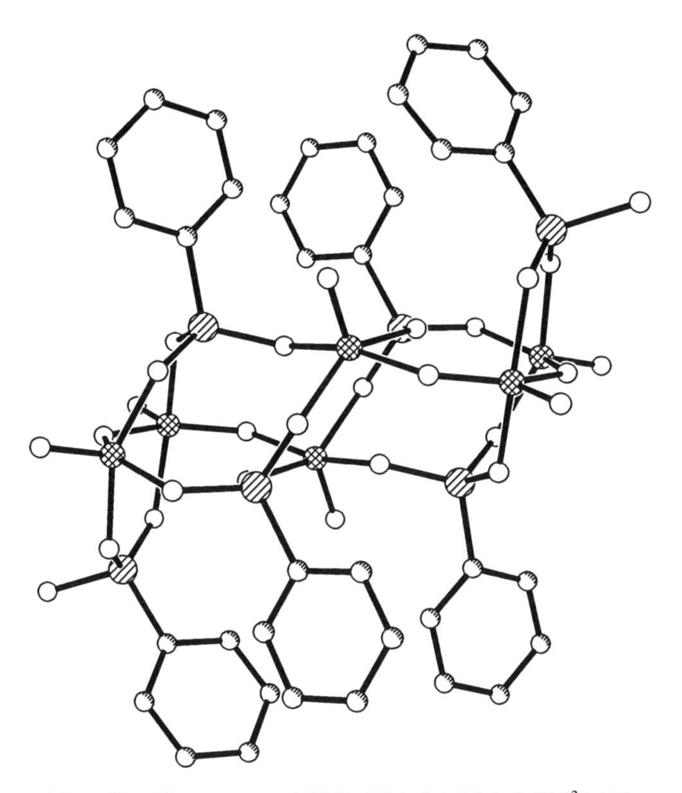

Figure 38. The structure of $[V_6O_{10}(PhAsO_3)_4(PhAsO_3H)_2]^{2-}$ (**45**).

rings: Four of the groups employ a third oxygen atom to coordinate to an adjacent ring. The remaining two organoaronate groups participate in the outermost $\{V_2As_2O_4\}$ rings, while the third oxygen atom of each is a terminal ligand. The latter oxygen atoms are protonated. The overall anionic charge of 2− requires the presence of four V^V centers. On the basis of valence sum calculations, the reduced sites are identified as the central vanadium centers, that is, those that do not participate in {V—O—V} bonding. Unfolding and flattening of the structure of **45** reveals a topological relationship to both $[VO(RPO_3)(H_2O)]$ and to the unique V/O/As layered phase, $[V_2O_4(RAsO_3H)(H_2O)]$ (181), shown in Fig. 39. The $[V_2O_4(RAsO_3H)(H_2O)]$ structure, unlike that of $[VO(RPO_3)(H_2O)]$, exhibits a protonated and pendant —OH group on each $\{HO_3AsPh\}$ moiety and undulating V/O/As layers with phenyl groups projecting toward both the upper and lower adjacent layers. Once again, by cutting V—O(As) bonds of **45**, the cyclic $\{V_6As_2O_{12}\}$ repeating motif of the layer structure is obtained.

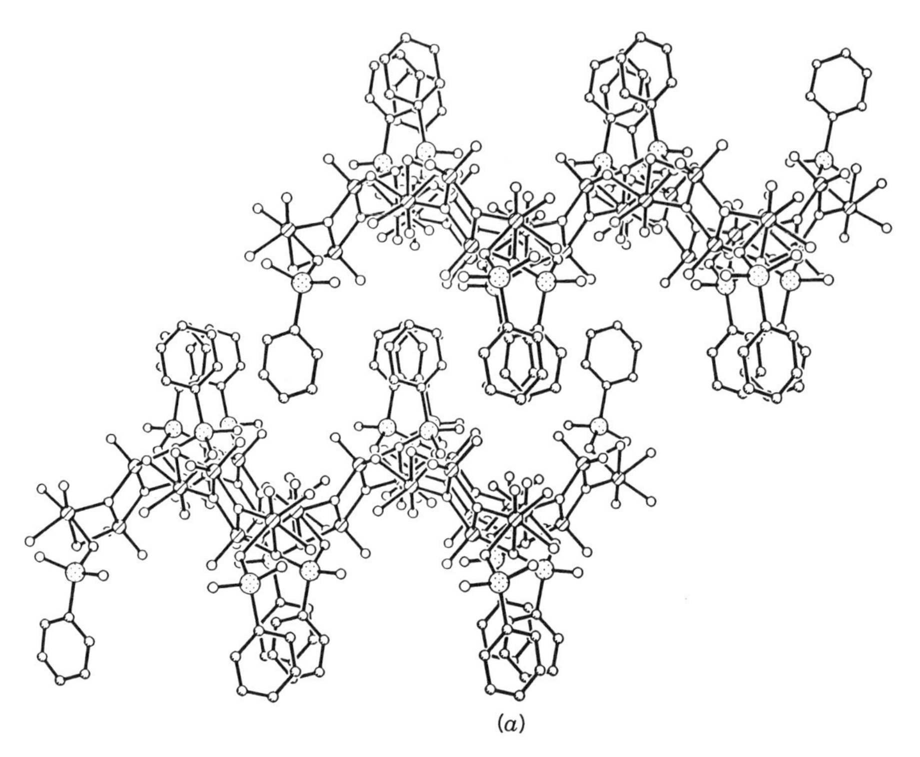

Figure 39. (*a*) A view of the undulating layer structure of $[V_2O_4(PhAsO_3H)(H_2O)]$ (**62**), illustrating the alternation of organic and inorganic layers. (*b*) A view of the inorganic layer highlighting the unfolded hexavanadate core of **45**.

5. *The Heptanuclear Core*

The V^V cluster $[V_7O_{12}(PhPO_3)_6Cl]^{2-}$ (**46**) (182) represents a unique example of the heptanuclear core in the $V/O/RPO_3^{2-}$ system. Red crystals of **46** are prepared in the reaction of $[VO_2Cl_2]^{2-}$ with $PhPO_3H_2$ in acetonitrile. An X-ray structural analysis of **46** revealed the structure shown in Fig. 40.

The V/P/O framework consists of a spherical shell of corner-sharing vanadium/oxygen square pyramids and organophosphonate tetrahedra. This $[V_7O_{12}(O_3PPh)_6]^-$ cage acts as a host for the chloride ion, which serves as a template for the organization of the cluster during the condensation process. In common with the shell of $[(VO)_6(t\text{-}BuPO_3)_8Cl]$ (**43**), the structural features of **46** may be compared and contrasted to those of $[V_{15}O_{36}Cl]^{6-}$.

The "facets" of the $[V_{15}O_{36}Cl]^{6-}$ spheroid consist of two six-membered

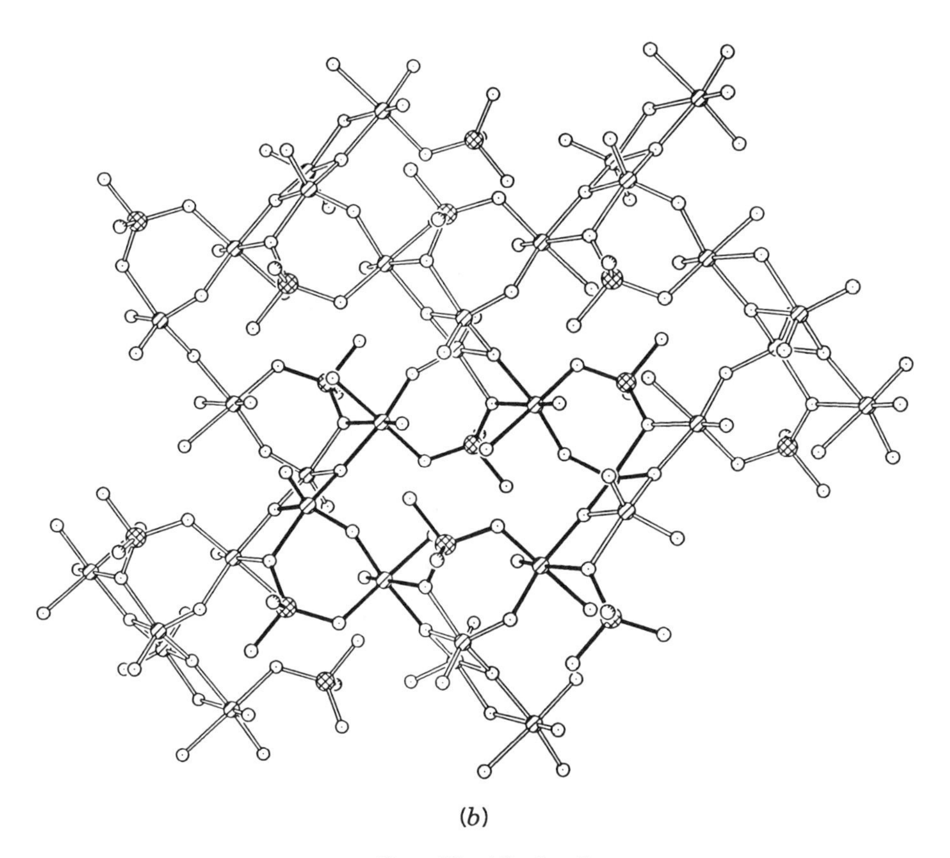

(*b*)

Figure 39. (*Continued*)

$\{V_3O_3\}$ rings, three $\{V_4O_4\}$ rings, and eighteen four-membered $\{V_2O_2\}$ rings. In contrast, cluster **46**, which is constructed from vanadium/oxygen square pyramids and phosphorus/oxygen/carbon tetrahedra and displays irregular geometry with approximate C_2 symmetry, exhibits 6-membered $\{V_2PO_3\}$ rings, 8-membered $\{V_2P_2O_4\}$, and $\{V_3PO_4\}$ rings, and 10-membered $\{V_3P_2O_5\}$ rings defining four, four, two, and two faces of the cage, respectively. The cage is constructed from the fusing of a trinuclear $\{V_3O_5(PhPO_3)_6\}$ unit to a tetranuclear $\{V_4O_7\}$ moiety through bridging organophosphonate groups. While neither unit exists independently in the V/O/organophosphonate system, the trinuclear unit is reminiscent of the $\{V_3O_5(PhPO_3)_5\}$ unit in $[V_5O_7(OMe)_2(PhPO_3)_5]^{1-}$ (**42**). The presence of the persistent $\{(VO)_2(\mu_2\text{-}O_3PPh)_2\}$ structural motif is apparent, as is the topological relationship to the $[VO(RPO_3)(H_2O)]$ layer structure, shown in Fig. 32.

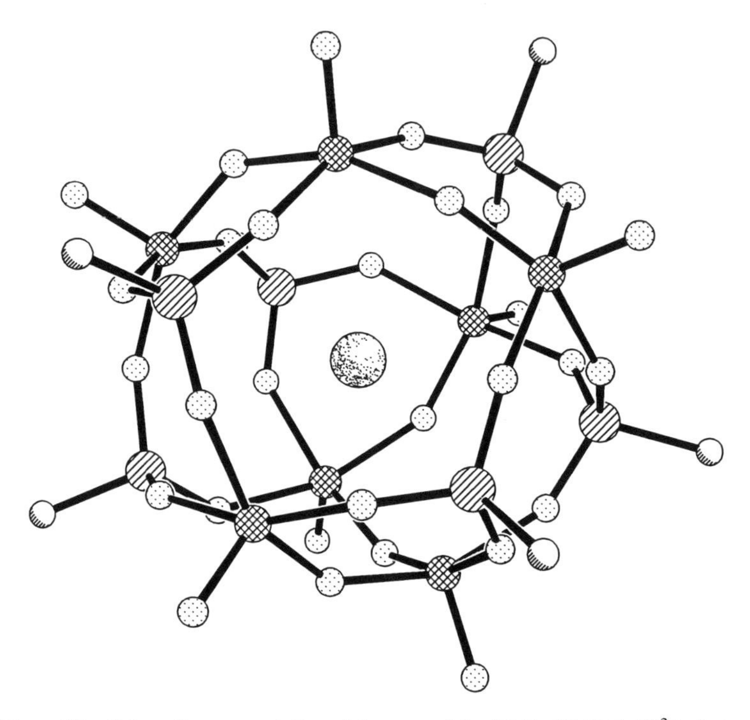

Figure 40. Schematic representation of the core of the $[V_7O_{12}(PhPO_3)_6Cl]^{2-}$ cluster.

6. *Decanuclear Cores*

The decanuclear core is represented by a single example in the V/O/$RAsO_3^{2-}$ system, $[V_{10}O_{24}(H_2NC_6H_4AsO_3)_3]^{4-}$ (**47**) (180), a V^V cluster formed in the reaction of $[V_{10}O_{28}]^{6-}$ with arsanilic acid in methanol. As shown in Fig. 41, the overall structure of the anion may be described as a $[V_9O_2(H_2NC_6H_4AsO_3)]^{3-}$ toroid encapsulating a VO_3^- moiety. The toroid consists of three trinuclear $\{V_3O_{13}\}$ units of edge-sharing octahedra with the conventional 60° angle formed by the V centers, which are linked through bridging oxo groups and the arsonato ligands. The central vanadium atom is in a highly distorted octahedral environment, in which the metal atom is displaced toward one triangular face of the octahedron, forming three short V—O distances of 1.70(2) Å and three long distances of 2.14(2) Å. This unusual oxometalate arrangement in **47** gives rise to four distinct types of oxo groups: terminal O atoms associated with the V centers of the nine-membered ring, μ^2-oxo groups that bridge V atoms of the $\{V_3O_{13}\}$ units, μ^3 types that link the $\{V_3O_{13}\}$ units to the central vanadium atom, and μ^4-oxo groups that provide both the common vertex for the V octahedra of the trinuclear units and the link to the encapsulated

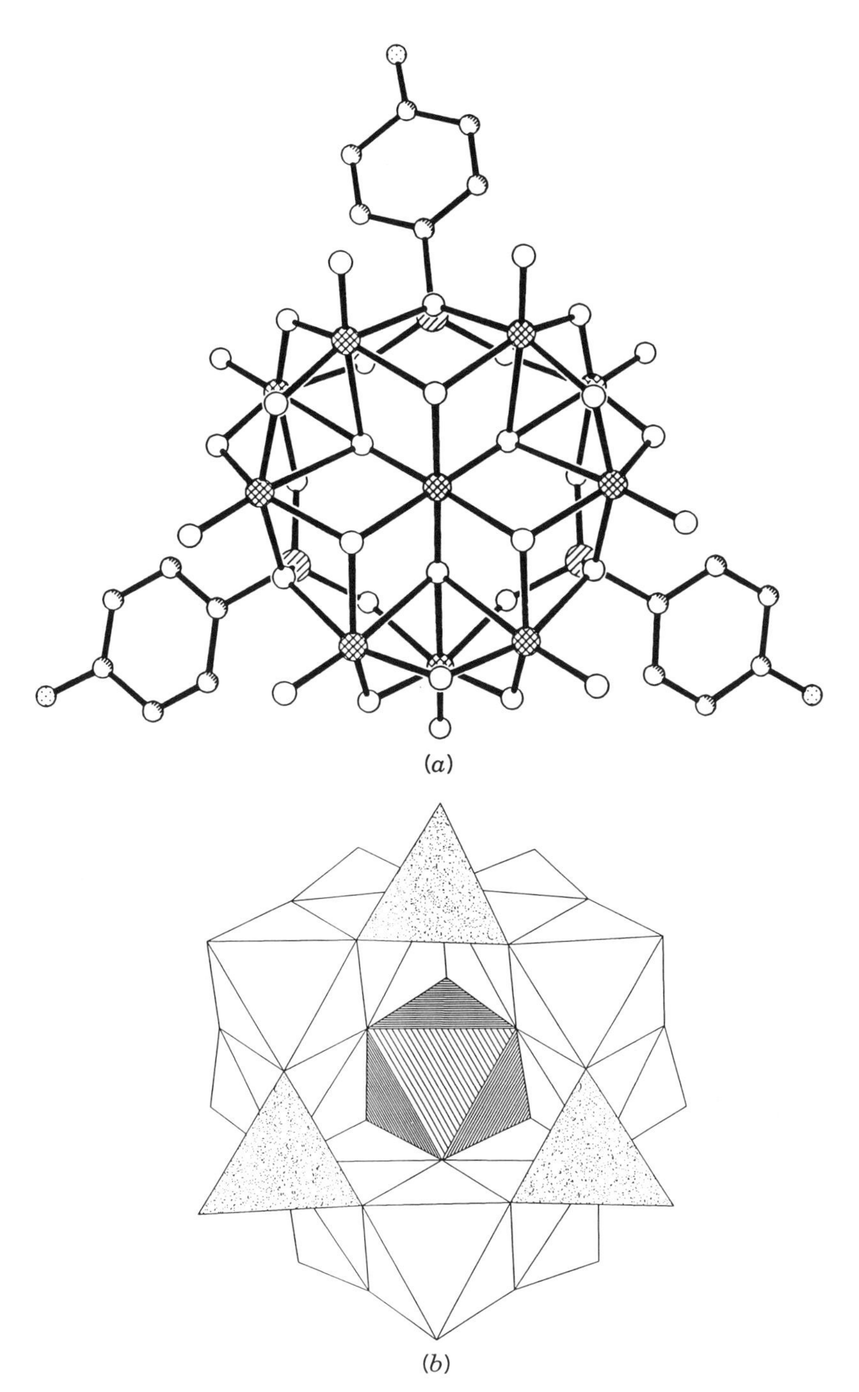

Figure 41. (*a*) Ball and stick. (*b*) Polyhedral representations of the structure of $[V_{10}O_{24}(H_2NC_6H_4AsO_3)_3]^{4-}$ (**47**).

V atom. Each organoarsonate group shares two of its oxygen atoms with vanadium centers of adjacent trinuclear units, while the third is bound to two vanadium centers in neighboring trinuclear units. The anion **47** exhibits structural features common to polyanions exhibiting a central MO_6 octahedron (26) and to ϵ-Keggin types. The polyhedral representation clearly shows the relationship of **47** to the Anderson $\{M_7O_{24}\}$ clusters (26). The structure of the anion in **47** may be best described as a double layer of polyhedra, one layer composed of the $[V_7O_{24}]^{13-}$ Anderson core and the second consisting of a $\{V_3(RAsO_3)_3\}^{9+}$ ring that rests on one face of the Anderson core. The structure of **47** appears to provide a unique instance in the $V/O/REO_3^{2-}$ (E = P or As) chemistry where the cluster is topologically unrelated to the layered phases but rather exhibits structural characteristics more akin to classical polyoxoanion species.

7. *Dodecanuclear Cores*

The higher oligomers of the $V/O/REO_3^{2-}$ systems exhibit the common feature of encapsulated guest molecules; however, these clusters also exhibit a remarkable versatility in providing shells capable of accommodating anions, cations, or even neutral molecules. Furthermore, the topological relationships of these clusters to the layered solids of the family are less obvious and, in certain cases, absent.

By exploiting the conditions of hydrothermal synthesis, several inclusion compounds of the V/O/organophosphonate system have been characterized. The structure of one of these, the spherical anion $[H_{12}(VO_2)_{12}(PhPO_3)_8(H_2O)_4]^{4-}$ (**48**) (15), is shown in Fig. 42. The anion consists of six condensed vanadium–oxygen dimers covalently connected by eight $\{O_3PMe\}$ tetrahedra through corner-sharing oxygen atoms. The vanadium centers exhibit the usual square pyramidal geometry with one terminal oxygen donor, two phosphonate oxygen donors, and two doubly bridging hydroxy oxygen atoms, V—(OH)—V. The vanadium(IV) square pyramids share edges to form bishydroxy bridged binuclear units $\{(VO)_2(OH)_2\}$. The hydroxy protons form hydrogen bonds to stabilize the cluster. The anion sphere encloses water molecules at four partially disordered positions.

It is noteworthy that the $\{(VO)_2(\mu_2\text{-}O_3PR)_2\}$ structural motif does **not** appear in the structure **48**, nor is the structure related to the $[VO(RPO_3)(H_2O)]$ layer network. The binuclear unit $\{V_2(OH)_2(XPO_3)_4\}$ based on edge-sharing V^{IV} square pyramids is, however, a constituent of the VOPO layer of $(H_3NCH_2CH_2N(CH_2CH_2)_2NCH_2CH_2NH_3)[(VO)_5(OH)_2(PO_4)_3]\cdot 2H_2O$ (183). The structural variability of the layered and 3D phases of the $V/O/PO_4^{3-}$ family is remarkable (183), suggesting that corresponding molecular clusters may be accessible under appropriate reaction conditions and employing judicious choices of templates.

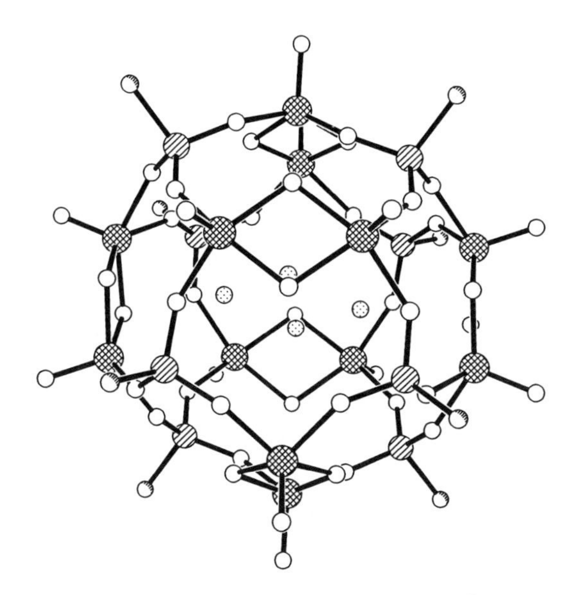

Figure 42. A view of the structure of $[H_{12}(VO_2)_{12}(PhPO_3)_8(H_2O)_4]^{4-}$ (**48**), illustrating the positions of four water guest molecules.

It is a curious feature of the various dodecanuclear cores that they bear no structural kinship to each other. The mixed-valence species $[(V_4O_8)_2\{(VO)_4(H_2O)_{12}\}(PhPO_3)_4Cl_2]^{2-}$ (**49**) (184) illustrates this point. Cluster **49** is formed in the hydrolysis/condensation reaction of $[V_5O_7(OMe)_2(PhPO_3)_5]^{1-}$ (**42**) with HCl in methanol. The structure of the anion **49**, shown in Fig. 43 consists of two $\{V_4O_8\}^{4+}$ caps each bridged through four $(PhPO_3)^{2-}$ tetrahedra to a central girdle of four $\{VO(H_2O)_3\}^{2+}$ groups. Each tetranuclear cap exhibits a ring of V^V square pyramids, corner sharing through bridging oxo groups and linked in a pairwise fashion by corner sharing to $(RPO_3)^{2-}$ tetrahedra adopting the symmetrically bridging bidentate mode. The remaining oxygen donors of the eight organophosphonate groups serve to bridge the $\{V_4O_8\}$ caps to the four central V^{IV} octahedra. These sites are most unusual in that the coordination about each of the V^{IV} centers consists of a terminal oxo group, two organophosphonate oxygen atoms, and three aquo ligands, one of which projects into the interior of the molecular cavity so as to partition the interior volume into two compartments, each of which is occupied by a chloride anion. Charge requirement suggest that **49** is a mixed-valence species $V^V_8V^{IV}_4$ and valence sum calculations (125) clearly identify the capping vanadium centers as V^V while the four vanadium atoms of the central girdle are in the +4 oxidation state. The room temperature magnetic moment of 2.60 μ_B is close to the spin-only value

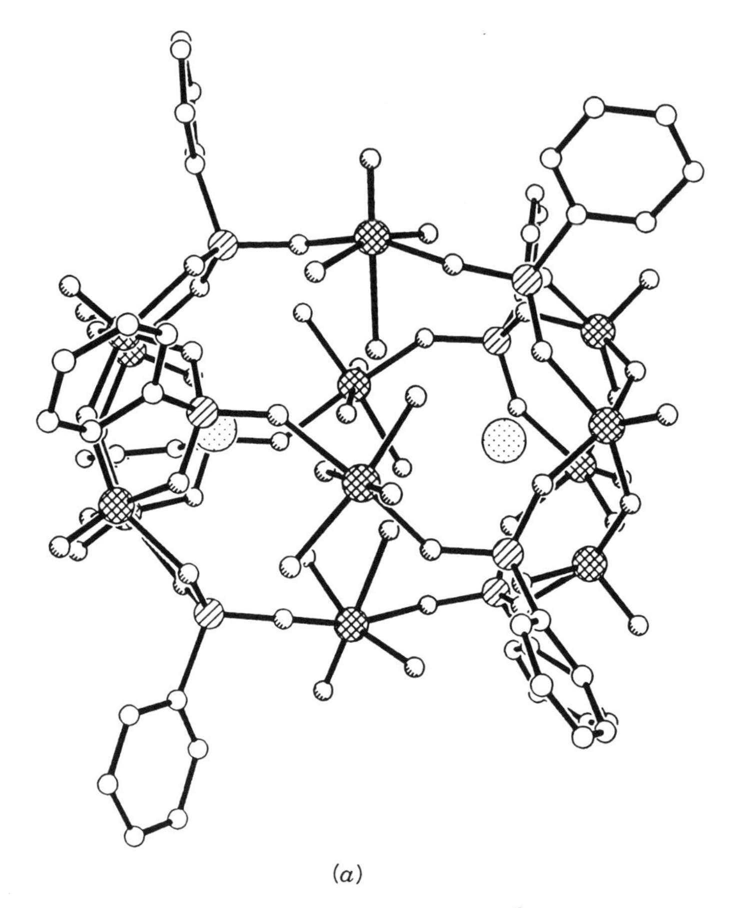

(a)

Figure 43. (*a*) The cluster $[(V_4O_8)_2\{(VO)_4(H_2O)_{12}\}(PhPO_3)_4Cl_2]^{2-}$ (**49**). (*b*) The $\{V_4O_8(RPO_3)_4\}$ capping unit. Note the similarity to the $\{V_4O_8\}$ core of $[V_4O_8(O_2CR)_4(NO_3)]^{2-}$, Fig. 11.

for four isolated V^{IV} sites. It is also noteworthy that the $\{V_4O_8\}^{4+}$ units of **49** are topologically related to the $\{V_5O_9\}^{3+}$ groups observed in $[2(NH_4Cl)V_{14}O_{22}(OH)_4(H_2O)_2(PhPO_3)_8]^{6-}$ (**52**) (see below) and $[H_4V_{18}O_{42}(X)]^{9-}$ by removal of the central $\{VO\}^{3+}$ and flattening of the $\{V_4O_4\}$ ring, an observation that reveals the structural relationship of **49** and $[V_{14}O_{22}(OH)_4(H_2O)_2(PhPO_3)_8]^{6-}$. The latter is formally constructed from **49** by reduction of the V^V centers, capping of each $\{V_4O_8\}$ unit by a $\{VO\}^{3+}$ group, and pairwise condensation of the four $\{VO(H_2O)_3\}^{2+}$ units to give two bridging $\{V_2O_2(OH)_2(H_2O)\}^{2+}$ groups.

The dodecanuclear clusters of the $V/O/RAsO_3^{2-}$ family are represented by $[(VO)_{12}(OH)_2)(PhAsO_3)_{10}(PhAsO_3H)_4]^{4-}$ (**50**) and $[2(MeOH)V_{12}O_{14}(OH)_4$-

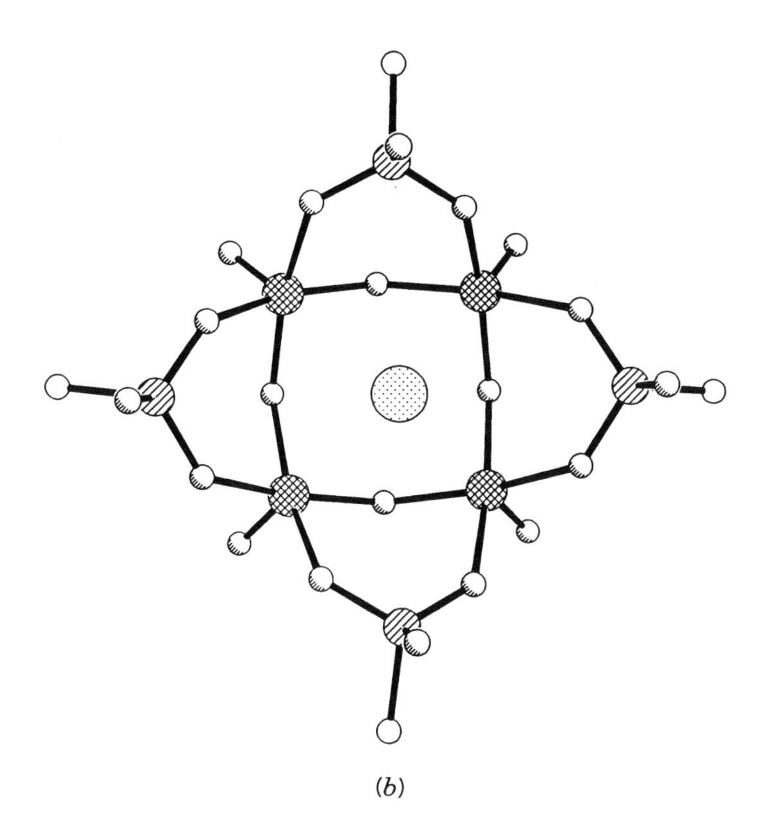

(*b*)

Figure 43. (*Continued*)

$(PhAsO_3)_{10}]^{4-}$ (**51**) (131). The solvothermal reaction of $[H_3V_{10}O_{28}]^{3-}$, Ph-AsO_3H_2, and $MeOH/H_2O$ yields green crystals of the V^{IV} species **50**, whose structure is shown in Fig. 44. This structure is constructed from vanadium-centered square pyramids and octahedra, and organoarsonate tetrahedra and square pyramids. The dominant structural motifs are two $\{V^{IV}O_6(PhAsO_3)_3\}^{2+}$ units bridged by two $\{V^{IV}(OH)(H_2O)(PhAsO_3)_2(PhAsO_3H_2\}^{3-}$ groups. The shell encloses two water molecules, which are confined to the rather restricted free volume of the interior and strongly hydrogen bonded to the hydroxo groups of the central bridging units. These hydroxo groups, as well as the oxo groups of one vanadium center in each $\{V_5O_6(PhAsO_3)_3]^{2+}$ unit, are directed toward the interior of the cavity and thus reduce the volume accessible to guest molecules. The core of the $\{V_5O_6(PhAsO_3)_3\}^{2+}$ unit consists of four vanadium-centered square pyramids that share edges with a central $\{PhAsO_4\}$ square pyramid. The central $\{V_4O_8AsPh\}^{4+}$ unit is analogous to the $\{V_5O_9\}^{3+}$ unit of

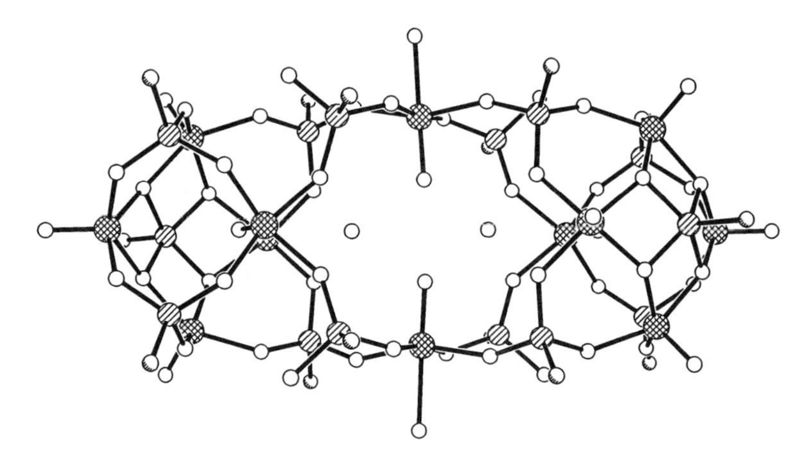

Figure 44. The dodecanuclear cluster, $[(VO)_{12}(OH)_2(PhAsO_3)_{10}(PhAsO_3H)_4]^{4-}$ (**50**).

$[V_5O_9Cl(RCO_2)_4]^{2-}$ (**11**) and of the higher oligomers of the class characterized by $[V_{14}O_{22}(OH)_4(H_2O)_2(RPO_3)_8]^{6-}$ (**52** and **53**). The $\{AsR\}^{4+}$ unit is thus topologically identical to the $\{VO\}^{3+}$ group and may substitute for the latter in structural motifs. In fact, valence bond arguments indicate that the lengths of $As^V—O$ and $V^V—O$ single bonds should be nearly identical, which confirms the geometric equivalence of these groups. Since the $P^V—O$ bond is 0.15 Å shorter than the $As^V—O$, organophosphonate analogues of **50** containing the hypothetical $\{V_4O_5(PhPO_3)\}^{4+}$ unit will be inaccessible. Two $\{PhAsO_3\}^{2-}$ tetrahedra each bridge three vanadium centers of this core and provide linkage to the fifth vanadium center. Reminiscent of the structures of $[(VO)_3(thf)(PhCO_2)_6]$ (**1**), $[V_6O_{10}(PhCO_2)_9]$ (**22**), and $[V_5O_7(OMe)_2(PhPO_3)_5]^{1-}$ (**42**), the oxo group of this latter vanadium site penetrates the interior of the shallow bowl formed by the $\{V_4O_5(PhAsO_3)\}$ fragment and interacts weakly with the four vanadium atoms of this fragment. This appears to be a common structural motif in the chemistry of the V/O/organophosphonate and V/O/organoarsonate systems. The central $\{V(OH)(H_2O)\}$ fragments are each linked to these $\{V_5O_6(PhAsO_3)_3\}^{2+}$ units by two $(PhAsO_3)^{2-}$ and two $(PhAsO_3H)^-$ groups. The protonated oxygen atoms are identified by the relatively long As—O distances and their pendant position.

Several of the structural motifs adopted by **50** persist in the structure of $[2(MeOH)V_{12}O_{14}(OH)_4(PhAsO_3)_{10}]^{4-}$ (**51**) (131). The structure is constructed from edge- and corner-sharing vanadium-centered square pyramids, organoarsonate tetrahedra, and arsonate square pyramids, and may be described as two $\{V^{IV}O_5(PhAsO_3)\}$ units bridged by two dinuclear $[V^{IV}O_2(OH)_2(PhAsO_3)_4]^{6-}$ moieties. The cavity formed encloses two methanol molecules.

While the structure of **51** is also related to that of $[V_{14}O_{22}(OH)_4(H_2O)_2(PhPO_3)_8]^{6-}$ (**52** and **53**) (12, 131), there are a number of significant differences. In the organophosphonate cluster the vanadium centers in the bridging dinuclear $[(VO)_2(OH)_2(H_2O)(PhPO_3)_4]^{6-}$ units have octahedral coordination because of a bridging aqua ligand; since **51** lacks this bridging water molecule, the vanadium centers of the corresponding dinuclear units have square pyramidal coordination. Consequently, the interior of the cluster shell may be less polar in **51** than in the organophosphonate analogue, a situation that favors enclosure of neutral organic molecules; in contrast inorganic cation/anion pairs are encapsulated in the hydrophilic cavity of $[V_{14}O_{22}(OH)_4(H_2O)_2(PhPO_3)_8]^{6-}$. It is also noteworthy that the pentanuclear $\{V_5O_9\}^{3+}$ caps of **52** and **53** are replaced by the topologically equivalent $\{V_4O_8AsR\}^{4+}$ units in **51**.

8. *Tetradecanuclear Cores*

The tetradecanuclear core is represented by the $[V_{14}O_{22}(OH)_4(H_2O)_2(PhPO_3)_8]^{6-}$ shell, first identified as a host network providing a cavity for the encapsulation of both NH_4^+ cations and Cl^- anions (12). The cluster **52** was prepared by the hydrazine reduction in water of NH_4VO_3 followed by addition of $PhPO_3H_2$ and the appropriate alkylammonium cation. The structure of **52**, shown in Fig. 45, is closely related to that of **51**. While the topologies of the

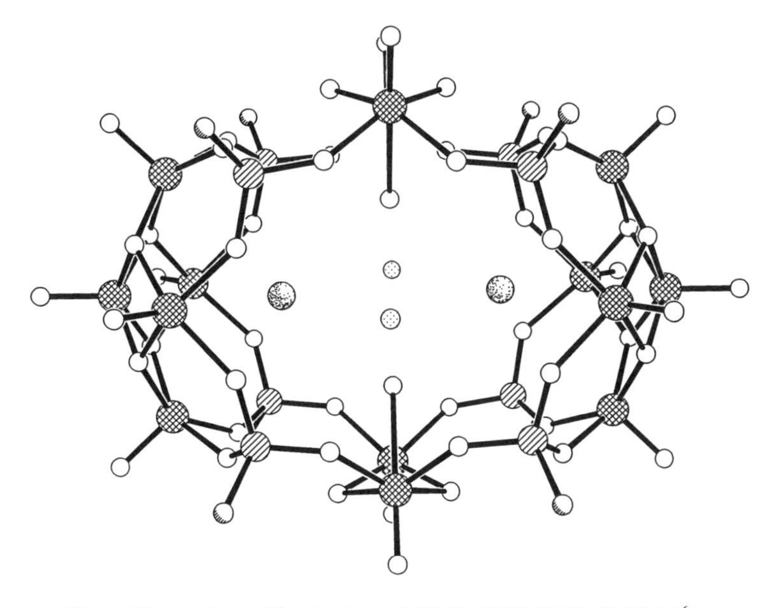

Figure 45. A view of the structure of $[V_{14}O_{22}(OH)_4(H_2O)_2(PhPO_3)_8]^{6-}$.

shells of **51** and **52** are similar, it is noteworthy that **52** possesses $\{V_5O_9\}^{3+}$ caps in place of the $\{V_4O_8(AsR)\}^{4+}$ units in **51** and that the central binuclear units of **52**, $\{(VO)_2(\mu\text{-}OH)_2(\mu\text{-}OH_2)\}^{2+}$, are constructed from face-sharing octahedra rather than edge-sharing $\{V_2O_2(\mu\text{-}OH)_2\}^{2+}$ square pyramids, as observed for **51**. The interior of the cavity produced by this arrangement of vanadium and phosphorus polyhedra is occupied by two NH_4^+ cations and two Cl^- anions. The organization of the shell appears to be directed by the oppositely charged ions inducing generation of nucleophilic and electrophilic fragments that are linked in the host assembly.

However, the structure of **51** suggested that the analogous $[V_{14}O_{22}(OH)_4(PhPO_3)_8]^{6-}$ shell should be formed about neutral organic templates in the absence of inorganic cations and in nonaqueous reaction media. This expectation was realized in the preparation of $[2(MeCN)_2V_{14}O_{22}(OH)_4\text{-}(PhPO_3)_8]^{6-}$ (**53**) (131) in the solvothermal reaction of $[n\text{-}Bu_4N]_3[H_3V_{10}O_{28}]$, $PhPO_3H_2$, and $PhCH_2NEt_3Cl$ in MeCN/MeOH at 120°C. The structure of the anion shell of **53** is identical to that of **52** except that the bridging aqua ligands in the central binuclear moieties of **52** are absent in **53**. Furthermore, in **53**, the shell has organized about two MeCN molecules, which are aligned within the cavity so as to nestle the nitrogen in the polar cavity of the $\{V_5O_9\}$ basket and direct the methyl group toward the relatively open and nonpolar central region of the framework. The acetonitrile molecules interact only weakly with the vanadium centers of the $\{V_5O_9\}$ unit (average V· · ·N distance 3.21 Å), which is reminiscent of the $[MeCNV_{12}O_{32}]^{4-}$ inclusion complex (109).

The synthesis of **53** establishes that the same shell may organize about different templates. While the template effect is well known for cluster shells of this type, the mechanism is not well understood. Both charge-compensation and space-filling effects play a role (22), and the relative importance of these influences may reflect the synthetic conditions. It is noteworthy that the same shell may enclose species as diverse as neutral organic molecules and inorganic cations and anions.

9. *The Hexadecanuclear Core*

A most unusual anionic ''tire'' is isolated from the hydrothermal reaction of vanadium oxide with methylphosphonic acid. As shown in Fig. 46, the structure of $[H_6(VO_2)_{16}(MePO_3)_8]^{8-}$ (**54**) (16) consists of a barrel enclosing a single Me_4N^+ cation. The anion is best described as four condensed $(VO_2)_4$ tetramers linked by four $\{O_3PMe\}$ tetrahedra through corner-sharing oxygen atoms. While the vanadium oxide motif $\{H_2V_4O_8\}$ is similar in composition to the $\{V_4O_8\}$ unit of $[V_4O_8(NO_3)(\text{thiopene-2-carboxylate})_4]^{2-}$ (**5**), the latter exhibits a cyclic $\{V_4O_4\}$ framework of corner-sharing $\{VO_5\}$ square pyramids; in contrast, the

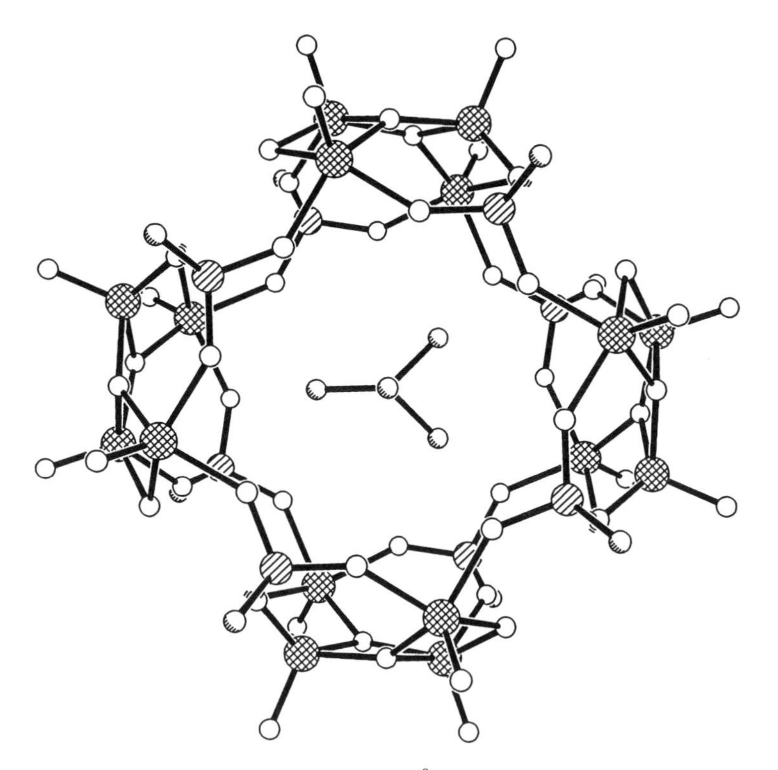

Figure 46. The structure of $[H_6(VO_2)_{16}(MePO_3)_8]^{8-}$ (**54**), showing the encapsulated $(Me)_4N^+$ cation.

$\{H_2V_4O_8\}$ core of **54** is constructed by edge-sharing square pyramids in a unique tetranuclear arrangement.

The analogous organoarsonate shell is represented by $[H_8(VO_2)_{16}$-$(PhAsO_3)_8]^{6-}$ (**55**) (185). A feature of the chemistry of the higher oligomers is the existence of cores with variable cluster oxidation states, consequences either of different degrees of protonation or of variable oxidation states of the vanadium centers. While **54** and **55** are both formally $14V^{IV}/2V^V$ mixed-valence species, **55** exhibits protonation at two additional bridging oxo groups of the $\{V_4O_8\}$ units.

10. The Octadecanuclear Core

When the pentanuclear cluster $[V_5O_7(OMe)_2(PhPO_3)_5]^{1-}$ (**42**) is heated in acetrotrile at reflux, the "cluster of clusters" $[V_{18}O_{25}(H_2O)_2(PhPO_3)_{20}Cl_4]^{4-}$ (**56**) (179) is isolated. The structure of this V^V cluster, shown in Fig. 5, may be described as a cyclic arrangement of four $\{V_4O_6(PhPO_3)_5\}^{2-}$ clusters linked through four P—O—V interactions and fused through eight V—O(phosphonate) bonds to a central $\{V_2O(H_2O)_2\}^{8+}$ moiety. The four chloride ions occupy the resulting four cavities.

The structural relationship of the anion of **56** to $\{V_5O_7(PhPO_3)_5]^{1-}$, shown in the lower part of Fig. 5, (*b*) is apparent. Removal of the pendant {VO(OMe)} group of **42** frees three phosphonate oxygen atoms to condense to additional vanadium atoms. One phosphonate oxygen atom displaces a methoxy group from the site trans to the interior-directed V=O unit of a $\{VO(OMe)(PhPO_3)_4\}$ center of a second $\{V_4O_6(OMe)(PhPO_3)_5\}$ unit; the remaining two P—O groups are utilized in bonding to the central $\{V_2O(H_2O)_2\}^{8+}$ moiety. Two of these $[\{V_4O_6(PhPO_3)_5\}_2]^{4-}$ units are then organized about the central $\{V_2O(H_2O)_2\}^{8+}$ fragment to form the cluster shell. It is noteworthy that the unusual {V=O} units, which are directed toward the interior of the subunit cavities and that have been noted in the structures of **1**, **22**, **42**, and **52** are retained in the structure of the supercluster. The anion in **56** may be considered a ''cluster of clusters'' and demonstrates that small oligomers may be linked by appropriate ligand types to form clusters of nanometer dimensions by more or less rational synthetic routes.

VI. OXOVANADIUM ORGANOPHOSPHONATE SOLID PHASES

A. Oxovanadium Monophosphonate Phases: $V/O/RPO_3^{2-}$

The structural prototype for the solid phases of the $V/O/RPO_3^{2-}$ system is $[VO(PhPO_3)(H_2O)]$ (**58**) (186), whose structure is shown in Fig. 31. The structure consists of layers of corner-sharing $\{VO_6\}$ octahedra linked through $\{CPO_3\}$ tetrahedra. Phenyl groups extend from both faces of the inorganic V/O/P layer, to produce a repeating motif of alternating inorganic layers and organic bilayers with a layer repeat distance of 14.14 Å between V/O/P layers. Within the inorganic layer, the vanadium octahedra share an axial oxygen to form infinite ··[···V=O···V=O···]·· chains with alternating short-long V—O distances. The chains are separated by phosphate tetrahedra linking two adjacent vanadium sites in one chain and one vanadium center in the neighboring chain. The interchain interactions produce the common $\{(VO)_2(\mu_2\text{-}O_3PC)_2\}$ structural motif. The network of vanadium octahedra and phosphorus tetrahedra also results in a 12 membered interchain heterocycle $\{P—O—V—O—P—O\}_2$ and an intrastrand motif $[V_2O_3(\mu_2\text{-}O_3PC)]$ characterized by the presence of the $\{V_2O_3\}^{2+}$ unit and a symmetrically bridging $\{\mu_2\text{-}O_3PC\}$ group.

The chemistry of the $V/O/RPO_3^{2-}$ phases may be expanded dramatically by introducing organic cations, which may not only occupy interlamellar voids, thus increasing the separation between planes, but also influence the connectivities between vanadium polyhedra and organophosphonate tetrahedra. Hydrothermal synthesis provides an expedient method for introducing a variety of potential templates to direct the organization of new phases and for adequate crystal growth. The optimal synthetic conditions summarized in Table XI reflect the vastness of the hydrothermal parameter space and certainly provide no sense of rational design of solid phases by well-defined reaction routes. On the other hand, by exploiting hydrothermal techniques and judicious choice of organic cations, it is evident that major structural modifications may be accomplished and that a significant structural chemistry of these phases may be developed. Furthermore, since the functional group of the organophosphonate is also subject to variation, presynthesis and/or postsynthesis modification may afford routes to more rational synthetic design. This feature of the chemistry will be revisited in the discussion of the diphosphonate, $V/O/R(PO_3)_2^{4-}$ phases in Section VI.

The phase most closely related to the prototype structure $[VO(PhPO_3)(H_2O)]$ (**58**) is the organically modified $(EtNH_3)[(VO)_3(H_2O)(PhPO_3)_4]$ (**59**) (187). As shown in Fig. 47, the structure of **59** may be described as layers of corner-sharing $\{VO_6\}$ octahedra and $\{PO_3C\}$ tetrahedra, with the phenyl groups extending from both sides of the V/O/P layer. The pattern of alternating organic and inorganic layers is reminiscent of the structure of $[VO(PhPO_3)(H_2O)]$. However, the structural similarities do not extend to the detailed structure of the oxide layers, which distort in **59** to accommodate the presence of the organic cationic templates. Whereas the structure of $[VO(PhPO_3)(H_2O)]$ exhibits an oxide layer with a vanadium–oxo to phenylphosphonate composition of 1 : 1 and exhibiting infinite $\{-V{=}O-V-\}$ chains with alternating short and long V—O bonds, **59** features a $\{VO\}/(RPO_3)^{2-}$ composition of 3:4 with discrete trinuclear $\{V_3O_3(H_2O)\}$ units bridged through $\{PhPO_3\}^{2-}$ groups (Fig. 48). The $\{V_3O_3(H_2O)\}$ trinuclear $\{V_3O_3(H_2O)\}$ units bridge through $\{PhPO_3\}^{2-}$ groups (Fig. 48). The $\{V_3O_3(H_2)\}$ trinuclear units exhibit zigzag $\{-V{=}O-V-\}$ chains with short–long alternation of V—O bonds, such that each $\{(VO)_3(H_2O)\}$ chain terminates in an oxo-group atom at one end and an aqua ligand at the other. The kinks in the $\{-V{=}O-V-\}$ chains are a consequence of edge sharing between the three vanadium octahedra of the trinuclear unit and four bridging $(RPO_3)^{2-}$ groups. Thus, the central vanadium octahedron of the unit coordinates to four bridging organophosphonates, while each terminal vanadium octahedron shares two of these bridging $(RPO)_3{}^{2-}$ tetrahedra with the central vanadium and forms two additional edge-sharing interactions to phosphate tetrahedra that bridge to an adjacent trinuclear unit.

The trinuclear structural motifs fuse in such a fashion as to generate cavities in the V/P/O layer, defined by rings constructed from the edge sharing of six vanadium octahedra and four $(RPO_3)^{2-}$ tetrahedra. These cavities are polar in character with aquo groups and vanadyl oxygen atoms projecting into the void

TABLE XI
Summary of Experimental Conditions for the Hydrothermal Syntheses of $V/O/REO_3^{2-}$ (R = P or As) Phases

Compound	Reactant Composition	Temperature (°C)	Duration (h)
$[VO(PhPO_3)(H_2O)\}$	V_2O_3, $PhPO_3H_2$, H_2O 1:3:500	200	96
$[(VO)_2\{(CH_2(PO_3)_2\}(H_2O)_4]$	V_2O_3, $CH_2(PO_3H_2)_2$, H_2O 1:1:250	200	24
$(EtNH_3)_2[(VO)_3(PhPO_3)_4(H_2O)]$	NH_4VO_3, $pHPO_3H_2$, NH_4Cl, $EtNH_3Cl$, H_2O 0.3:6:3:6:200	200	48
$(Et_2NH_2)(Me_2NH_2)[(VO)_4(OH)_2(PhPO_3)_4]$	$RbVO_3$, $PhPO_3H_2$, Et_2NH_2, Me_2NH_2, H_2O 4:6:4:4:300	160	96
$(Et_4N)_2[(VO)_6(OH)_2(EtPO_3)_6(H_2O)_2]$	$NaVO_3$, $EtPO_3H_2$, $(Et_4NCl$, H_2O 3:4:6:300	140	120
$[H_2N(CH_2CH_2)_2NH_2][VO\{CH_2(PO_3)_2\}]$	VCl_4, $CH_2(PO_3H_2)_2$, $HN(CH_2CH_2)_2NH$, H_2O 1:1:1.93:945	200	50
$[H_3NCH_2CH_2NH_3][VO(O_3PCH_2CH_2CH_3PO_3)]$	VCl_4, $H_2PO_3(CH_2)_2PO_3H_2$, $H_2NCH_2CH_2NH_2$, H_2O 1:3.11:5.1:1890	200	96
$[H_3NCH_2CH_2NH_3][(VO)_4(OH)_2(H_2O)_2$- $(O_3PCH_2CH_2CH_2PO_3)_2]$	VCl_4, $H_2PO_3(CH_2)_3PO_3H_2$, $HN(CH_2CH_2)_2NH_2H_2O$ 1:1.47:4.8:9.45	200	80
$[V_2O_4(PhAsO_3)(H_2O)]$	V_2O_3, $PhAsO_3H_2$, H_2O 1:3:250	200	120

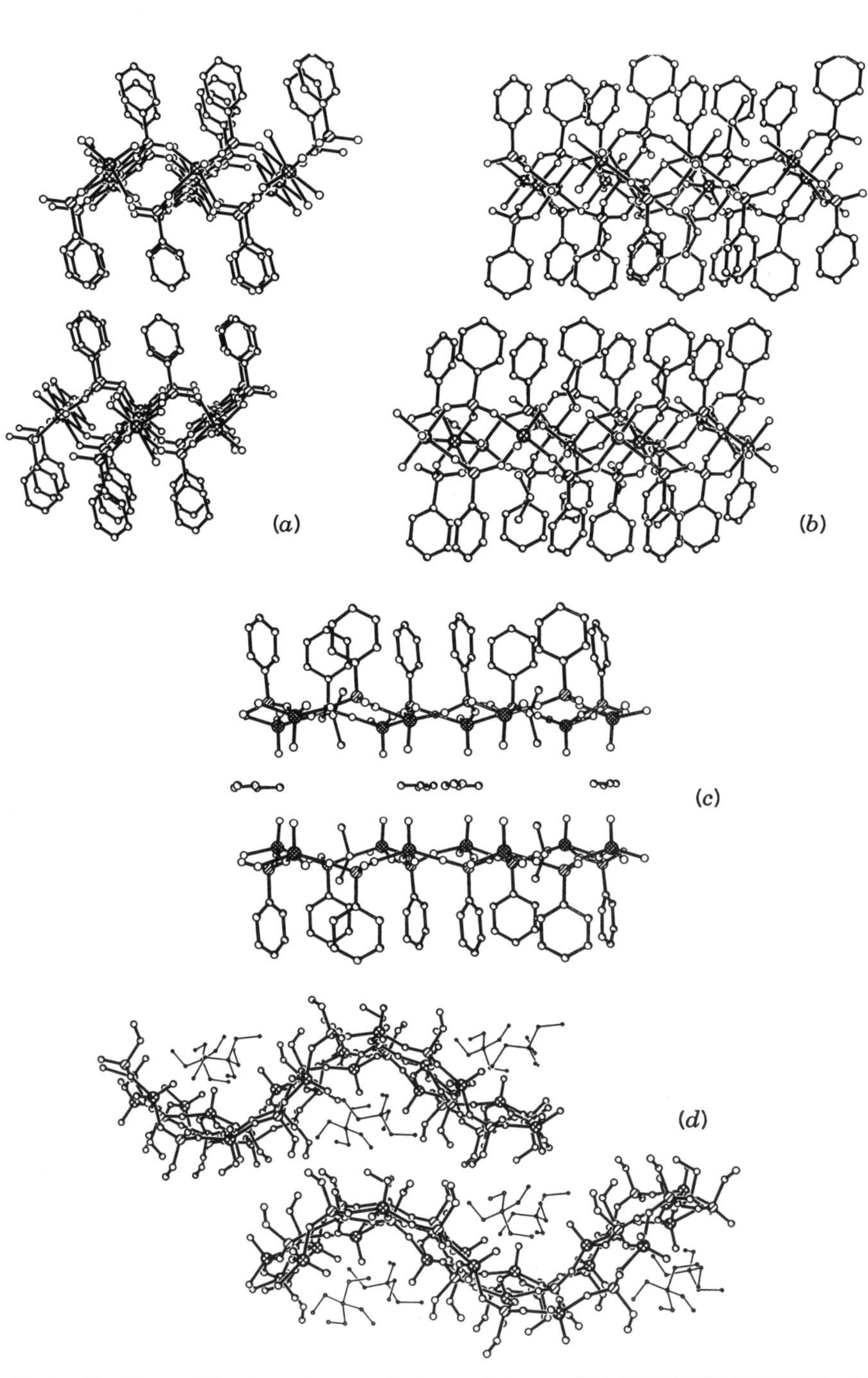

Figure 47. Views of the alternating organic–inorganic layers of (*a*) $[VO(PhPO_3)(H_2O)]$ (**58**), (*b*) $(EtNH_3)[(VO)_3(H_2O)(PhPO_3)_4]$ (**59**), (*c*) $(Et_2NH_2)(Me_2NH_2)[(VO)_4(OH)_2(PhPO_3)_4]$ (**60**), and (*d*) $[Et_4N]_2[(VO)_6(OH)_2(H_2O)_2(EtPO_3)_6]$.

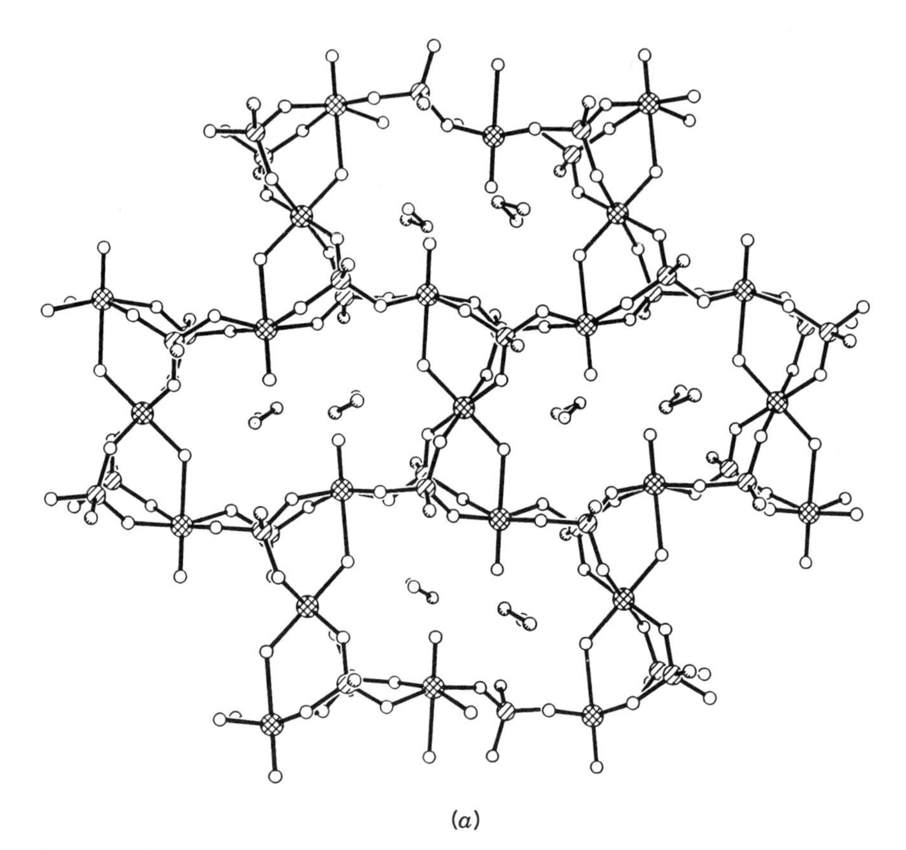

Figure 48. The oxovanadium–phosphonate planes of (*a*) $(EtNH_3)[(VO)_3(H_2O)(PhPO_3)_4]$; (*b*) $(Et_2NH_2)(Me_2NH_2)[V_4O_4(OH)_2(PhPO_3)_4]$; (*c*) $[(Et)_4N][(VO)_3(OH)(EtPO_3)_3{\cdot}H_2O$.

space. The V/P/O layers are stacked to produce tunnels parallel to the cell *a* axis, which are occupied by the $(EtNH_3)^+$ cations. The walls of these tunnels are defined by the organophosphonate phenyl groups that project above and below the V/P/O planes. The organic cations are oriented in the cavities with the ammonium group, $-NH_3^+$, directed toward the hydrophilic holes in the V/P/O layer, while the —Et group trails into the hydrophobic region generated by the phenyl substituents. The incorporation of the organic template into the V/O/$PhPO_3^{2-}$ system results in dramatic structural rearrangement within the V/P/O layer of 1 as compared to $[VO(PhPO_3)(H_2O)]$, on contrast to a relatively minor distortion of the gross alternating pattern of inorganic and organic layers. The structure may be described as amphiphilic and rationalized on the basis of partitioning into hydrophilic and hydrophobic domains. While the templating mechanism remains enigmatic, it is clear that charge-compensating and space-

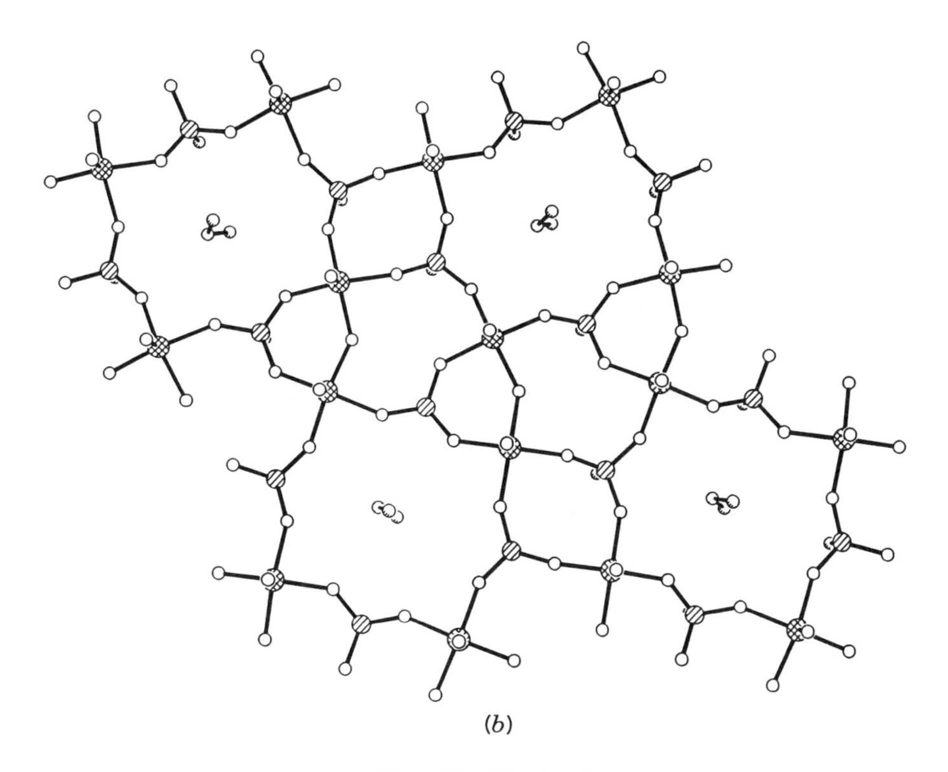

Figure 48. *(Continued)*

filling effects contribute to the process and that different template molecules will provide different distortions of the V/P/O layer, as in (Et_2NH_2)-$(Me_2NH_2)[(VO)_4(OH)_2(PhPO_3)_4]$ (**60**) (188) and $[Et_4N]_2[(VO)_6(OH)_2(H_2O)_2$-$(EtPO_3)_6]$ (**61**) (189).

Additional template modification forces more severe structural modification, as observed for $[Et_2NH_2][Me_2NH_2][(VO)_4(OH)_2(PhPO_3)_4]$ (**60**) (188). The introduction of the intercalating organic cations has disrupted the registry of layers such that the structure of **60** exhibits a trilayer repeat represented by a layer of $(Et)_2NH_2^+$ cations sandwiched between inorganic V/P/O layers that are in turn bounded by organic bilayers consisting of phenyl group from adjacent $V/O/PhPO_3$ slabs. The structure of $[VO(PhPO_3)(H_2O)]$ is constructed from layers of corner-sharing $\{VO_6\}$ octahedra and $\{CPO_3\}$ tetrahedra, with *phenyl groups extending from both sides of the metal oxide layers*. In contrast, the structure of **60** consists of corner-sharing $\{VO_5\}$ square pyramids and $\{CPO_3\}$ tetrahedra, with *phenyl groups directed exclusively to one face of the layer* while the $\{V{=}O\}$ groups are directed to the other. Since the $\{V{=}O\}$ groups

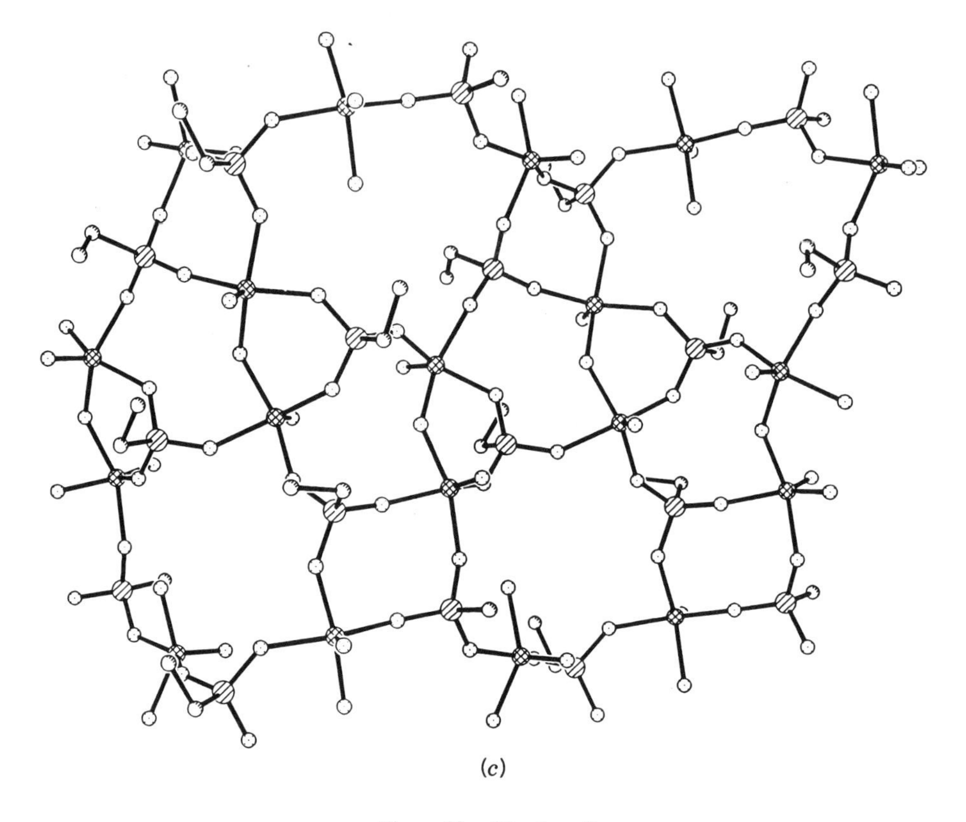

Figure 48. (*Continued*)

of the adjacent layer are directed toward the vanadyl face of the neighboring layer, a highly hydrophilic region is produced that accommodates the layer of $(Et)_2NH_2^+$ cationic templates. Whereas the structure of $[VO(PhPO_3)(H_2O)]$ (**58**) is defined by an interlayer repeat of 14.14 Å, the registry of layers in **60** is such as to produce two repeat distances: 8.89 Å between V/P/O layers sandwiching the $(Et)_2NH_2^+$ intercalators and 12.29 Å between V/P/O layers sandwiching phenyl group bilayers. The partitioning of the structure into polar and nonpolar domains is evident, and **60** appears to be amphiphilic with hydrophobic interactions determining the layer packing.

The structural differences between **58** and **60** are further accentuated in the oxide layer, shown in Fig. 48(*b*). The vanadium octahedra of $[VO(PhPO_3)(H_2O)]$ share axial oxygen atoms, forming infinite $\{-V{=}O-V{=}O-\}$ chains with alternating short–long interactions. In **60**, however, discrete $\{V_2O_2(OH)\}^{3+}$ binuclear units are bridged through phosphonate tetrahedra to form the layer motif. The vanadium centers of each binuclear unit are bridged by a phosphonate group that directs the third oxygen donor to an adjacent binuclear unit. The coordination at each vanadium site of the binuclear unit is completed by bonding to two oxygen donors from two phosphonate groups each of which in turn bridges to an adjacent binuclear unit. Thus, while the largest cavity in the V/P/O layer of **58** is constructed from the corner sharing of four vanadium octahedra and two phosphonate tetrahedra, producing a 12-membered ring, the cavity size in **60** is expanded by corner sharing of four vanadium square pyramids and four phosphonate tetrahedra, resulting in a 16-membered ring with a V—V diagonal distance of about 10.5 Å. The distortion of the V/P/O layer in **60** from that in $[VO(PhPO_3)(H_2O)]$ reflects the necessity of accommodating the $(Me)_2NH_2^+$ cation, which projects through the V/P/O inorganic layer. The structure of **60** demonstrates the dramatic topological flexibility of vanadium polyhedra and phosphonate tetrahedra in incorporating a variety of substrates. The structure of **60** is thus able to accommodate intercalation of organic cations both between layers and within layers, by suitable modification of the $[VO(PhPO_3)(H_2O)]$ parent structure.

The steric constraints of bulky organic cations will also cause structural reorganization relative to the prototype structure. The structure of $(Et_4N)_2[(VO)_6(OH)_2(H_2O)_2(EtPO_3)_6]$ (**61**) (189) exhibits layers of corner-sharing vanadium(IV) pyramids and $\{PO_3C\}$ tetrahedra, with the phosphate —Et groups and the vanadyl $\{V{=}O\}$ moieties projecting from both surfaces of the V/P/O layer, as shown in Fig. 47*d*. In contrast to the prototypical structure of $[VO(PhPO_3)(H_2O)]$, which exhibits planar V/P/O layers and alternating inorganic layers and phenyl bilayers, the V/P/O layers of **61** undulate in such a fashion as to produce interlamellar cavities occupied by the organic cationic templates. The organic cations produce channels defined by the crests and troughs of adjacent V/P/O layers, which exhibit an interlayer repeat distance of 10.73 Å. These channels are bounded on two flanks by the —Et substituents of the organophosphonate groups, on a third side by the concave surface of the V/P/O layer, which projects both —Et groups and $\{V{-}O\}$ units into the interlamellar region, and on the remaining side by the convex surface of the adjacent layer, which projects vanadyl groups exclusively. Thus, through the interplay of hydrophobic–hydrophilic interactions, nonpolar interlamellar regions bounded by organic substituents and relatively nonpolar vanadyl groups have been created.

As shown in Fig. 48(*c*), there are two distinct vanadium environment associated with the V/P/O planes of **61**. Two vanadium centers form a hydroxy-bridged binuclear unit with the terminal oxo groups adopting the anti-configuration relative to the $\{V_2O_7\}$ plane. On the other hand, the third site consists of isolated $\{VO_5\}$ square pyramids with the coordination geometry at the vanadium defined by a terminal oxo group, three oxygen donors from each of three phosphonate units, and an aqua ligand. The geometries adopted by the vanadium centers of **60** or **61** may be contrasted to those observed for $[VO(PhPO_3)(H_2O)]$ (**58**), which contains infinite 1D {—V=O—V=O—} chains and for **58**, which is characterized by isolated $\{VO_6\}$ octahedra and $\{RPO_3\}$ tetrahedra in a corner-sharing array.

Ethylphosphonate groups in three distinct environments serve to bridge the vanadium centers and to link the structure into its unique 2D array. Each organophosphonate group associated with two of the phosphonate sites adopts the symmetrical bridge mode between V atoms of the $\{V_2(\mu^2\text{-}OH)(\mu^2\text{-}O_2P(Et)O)\}$ binuclear units, as well as serving as a monodentate ligand to an adjacent binuclear unit. The second type of organophosphonate group bridges vanadium centers from each of two adjacent binuclear units to a mononuclear vanadium site, while the third type links a vanadium site of a binuclear unit to two adjacent mononuclear vanadium centers. The complexity of the connectivity pattern is reflected in the presence of 6-membered $\{V_2PO_3\}$ rings, 8-membered $\{V_2P_2O_4\}$ rings, 10-membered $\{V_3P_2O_5\}$ rings, and 16-membered $\{V_4P_4O_8\}$ rings formed by the corner sharing of the $\{VO_5\}$ square pyramids and the $\{O_3PC\}$ tetrahedra.

The single example reported to date of a phase of the $V/O/RAsO_3^{2-}$ system is represented by $[V_2O_4(PhAsO_3H)(H_2O)]$ (**62**) (181), whose structure is shown in Fig. 39. The inorganic layers are formed from pairs of edge-sharing $\{VO_5\}$ square pyramids linked through corner-sharing to $\{VO_6\}$ octahedra and $\{CAsO_3\}$ tetrahedra. The inorganic layers undulate and project phenyl groups above and below the layer to give an alternating pattern of inorganic and organic layers with a layer repeat distance of 11.51 Å. A curious feature of the structure is the presence of both doubly binding —OH groups and triply bridging oxo groups producing a network of ∤V—O(H)—V—O—V—O(H)∤ linked vanadium sites extending through the inorganic layer. A structurally related $V/O/RPO_3^{2-}$ phase has yet to be described.

B. Oxovanadium Diphosphonate Phases: $V/O/R(PO_3)_2^{4-}$

The structural expansion induced by the straightforward modification of replacing the monophosphonate ligand RPO_3^{2-} by the diphosphonate group $R(PO_3)_2^{4-}$ is quite dramatic. The parent structure is again represented by the $V/O/R(PO_3)_2^{4-}$ phase unconstrained by introduction of organic cations: $[(VO)_2\{CH_2(PO_3)_2\}(H_2O)_4]$ (**63**) (190). As shown in Fig. 49, the structure possesses undulating layers of corner-sharing $\{VO_6\}$ octahedra and diphosphonate polyhedra stacked with a repeat distance of 15.08 Å. There are four coordinated water molecules, three go to one vanadium center and one goes to the second unique vanadium site; these all project into the interlamellar void. The diphosphonate ligand employs all six oxygen donors in coordination to vanadium centers, acting as a bidantate ligand toward two vanadium sites and a

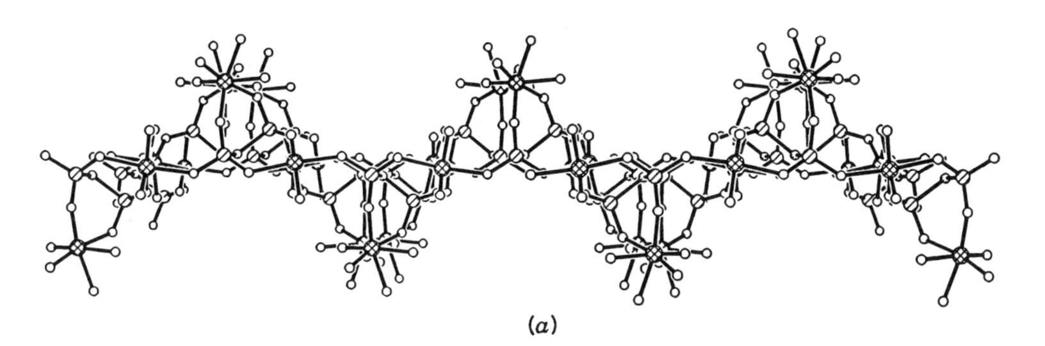

(a)

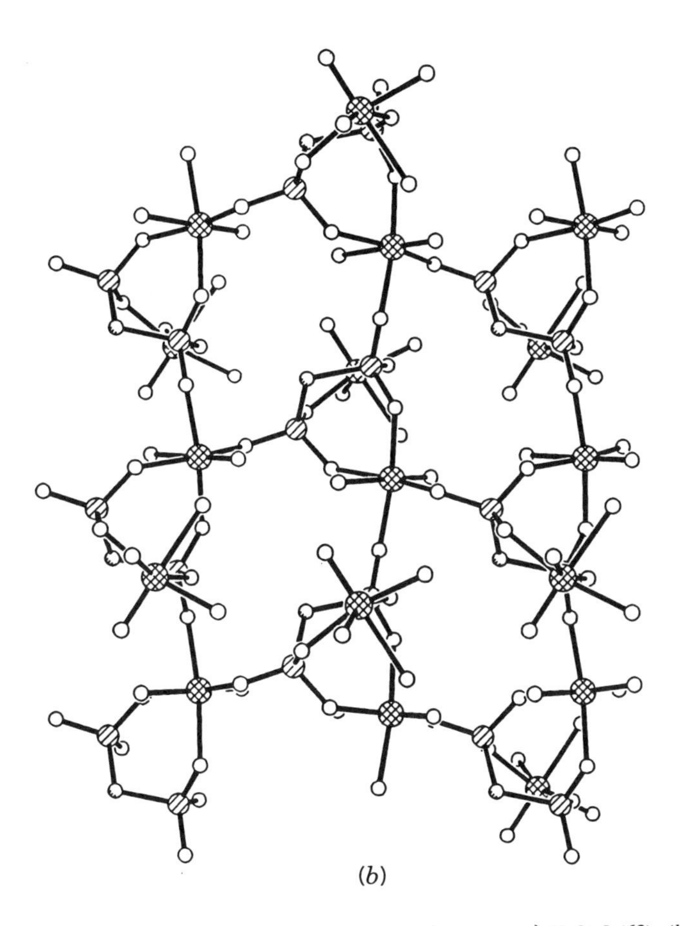

(b)

Figure 49. (a) A view of one undulating layer of $[(VO)_2\{CH_2(PO_3)_2\}(H_2O)_4]$ (**63**). (b) A view of the layer showing the linking of four vanadium sites by each diphosphonate ligand.

monodentate ligand for two additional vanadium centers. One consequence of this coordination mode and the presence of extensive aqua ligation is the absence of direct interactions between $\{VO_6\}$ octahedra by $\{V-O-V\}$ linkages.

By introducing organic cations as potential templates, $V/O/R(PO_3)_2^{4-}$ phases with 1-, 2-, and 3D structures have been isolated, the dimensionality paralleling the number of methylene group spacers in the diphosphonate group.

As shown in Fig. 50, the structure of $[H_2N(CH_2)_4NH_2][VO\{CH_2(PO_3)_2\}]$ (**57**) (191) consists of infinite puckered chains of $[(VO)(O_3PCH_2PO_3)]^{1-}$ units with piperazinium cations occupying the interstrand regions. The V^{IV} sites exhibit square pyramidal coordination geometry with the apical position occupied by a terminal oxo group and the four equatorial positions defined by phosphon-

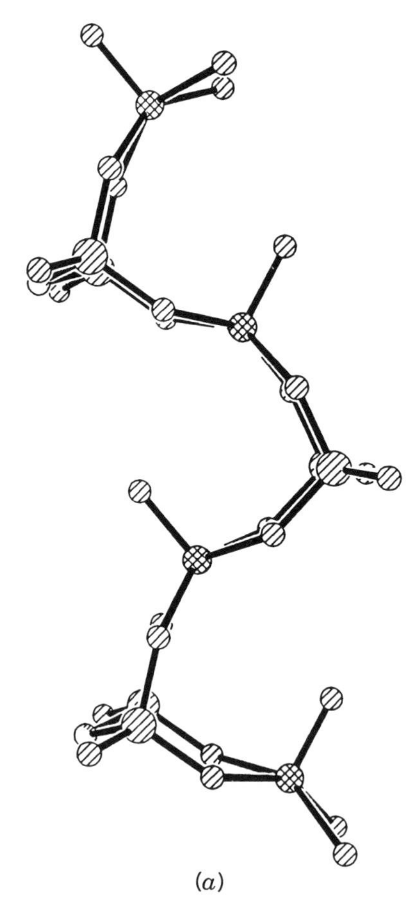

(*a*)

Figure 50. Two views of the $[VO\{O_3PCH_2PO_3\}_n^{n-}$ chains of **57**.

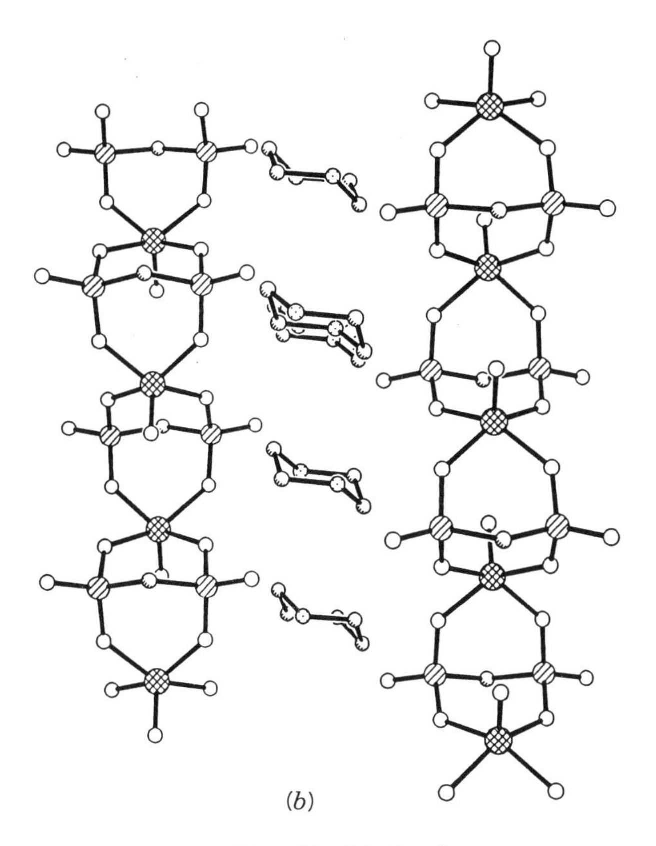

Figure 50. *(Continued)*

ate oxygen atoms, two from each of two diphosphonate ligands. Each diphosphonate bridges two vanadium centers, acting as a symmetrically bridging bidentate ligand to each metal site. Each phosphonate unit of the ligand retains a pendant {P=O} group, confirmed by the short P—O distance of 1.518(6) Å. While pendant {P—OH} groups are more common in the organophosphonate coordination chemistry, the uncoordinated {P—O} group has been demonstrated in the structure of $(TBA)_2[V_2O_4(RPO_3)_2]$ [where TBA = tetra-*n*-butylammonium]. This diphosphonate coordination mode contrasts with that of the prototype layered solid $[(VO)_2\{CH_2(PO_3)_2\}(H_2O)_4]$ (**63**), wherein all phosphonate oxygen donors are coordinated to metal centers. On the other hand, both **63** and **57** exhibit the symmetrically bridging bidentate ligation of two {VO} sites. This ligand coordination mode is responsible for the puckering of the chain in **57** and in undulation of the layers in **62**.

The structure of $[H_3NCH_2CH_2NH_3][(VO)(O_3PCH_2CH_2PO_3)]$ (**64**) (191) may be described as layers of vanadium(IV) square pyramids corner sharing with diphosphonate tetrahedra to produce $[(VO)(O_3PCH_2CH_2PO_3)]_n^{n-}$ layers with ethylenediammonium cations occupying the interlamellar regions (Fig. 51). In addition to the apical oxo group, each vanadium(IV) site is bonded in the equatorial positions to four diphosphonate oxygen donors, one from each of four diphosphonate groups. Hence, each diphosphonate ligand bridges four vanadium centers in a monodentate fashion through two oxygen donors of each $\{PO_3\}$ unit. In a fashion akin to **57**, each phosphate group exhibits a pendant $\{P{=}O\}$ group, which projects into the interlamellar space, as do the vanadyl

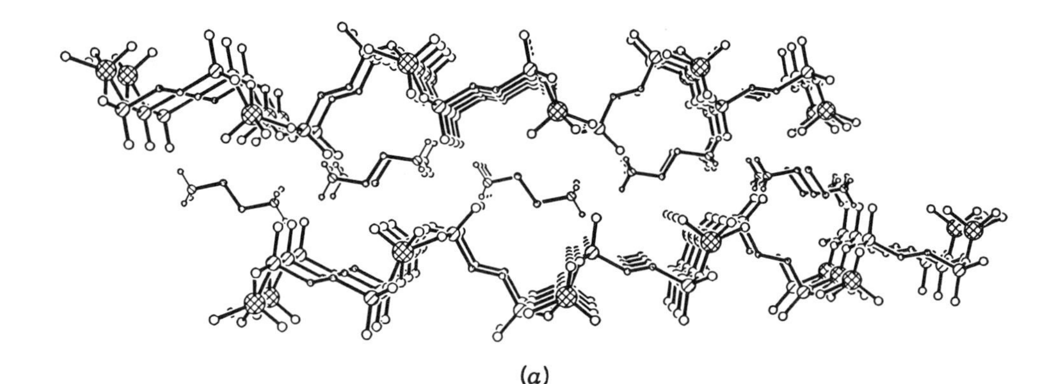

(*a*)

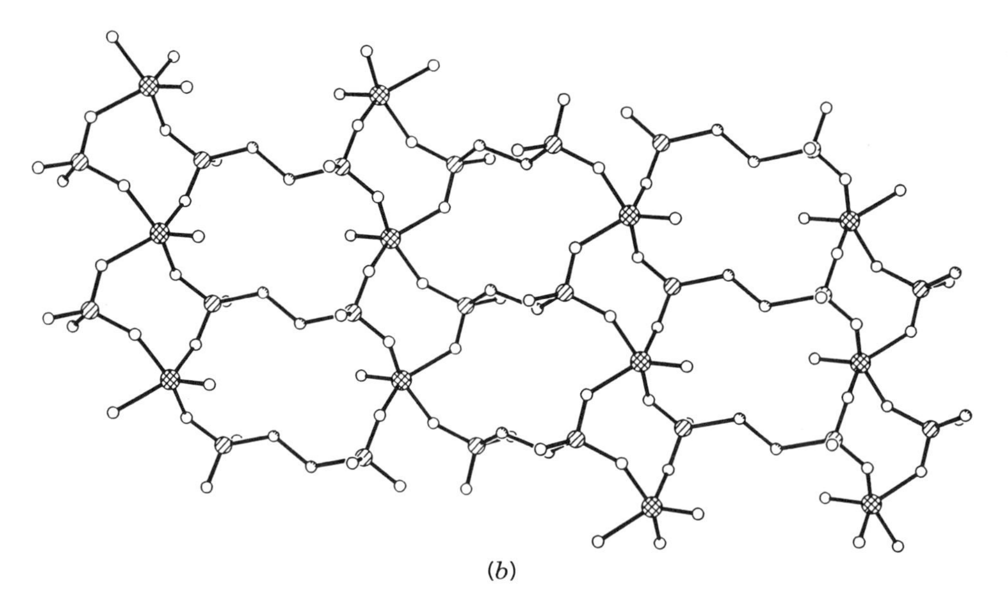

(*b*)

Figure 51. (*a*) The alternating organic–inorganic layers of $[H_3NCH_2CH_2NH_3]$-$[(VO)(O_3PCH_2CH_2PO_3)]$ (**64**). (*b*) A view of the oxovanadium–phosphonate layer of **64**.

oxo groups. In comparison to the prototypical layered structures **58** and **63** the V/O/RPO_3 layer of **64** exhibits a more open net as a consequence of the presence of 14-membered {V—O—P—C—C—P—O}$_2$ rings (Fig. 51). The layer also exhibits the 8-membered {V—O—P—O—V—O—P—O—} ring, characteristic of both molecular and solid phase structures of the V/O/RPO_3^{2-} system.

The dramatic structural variations attendant to expansion of the phosphonate tether group are evident in the "pillared" structure adopted by $[H_3NCH_2CH_2NH_3][(VO)_4(OH)_2(O_3PCH_2CH_2CH_2PO_3)_2]$ (**65**) (191). As shown in Fig. 52, the structure of **65** may be described in terms of inorganic V/P/O layers joined covalently by the trimethylene bridges of the diphosphonate groups. Since the covalently linked framework includes the ligand carbon framework, the solid may alternatively be described as a 3D V/P/O/C framework with ethylenediammonium cations occupying the channels parallel to the crystallographic *a* axis. There are two distinct vanadium(IV) environments: the first is square pyramidal with bonds to an apical oxo group, to three phosphon-

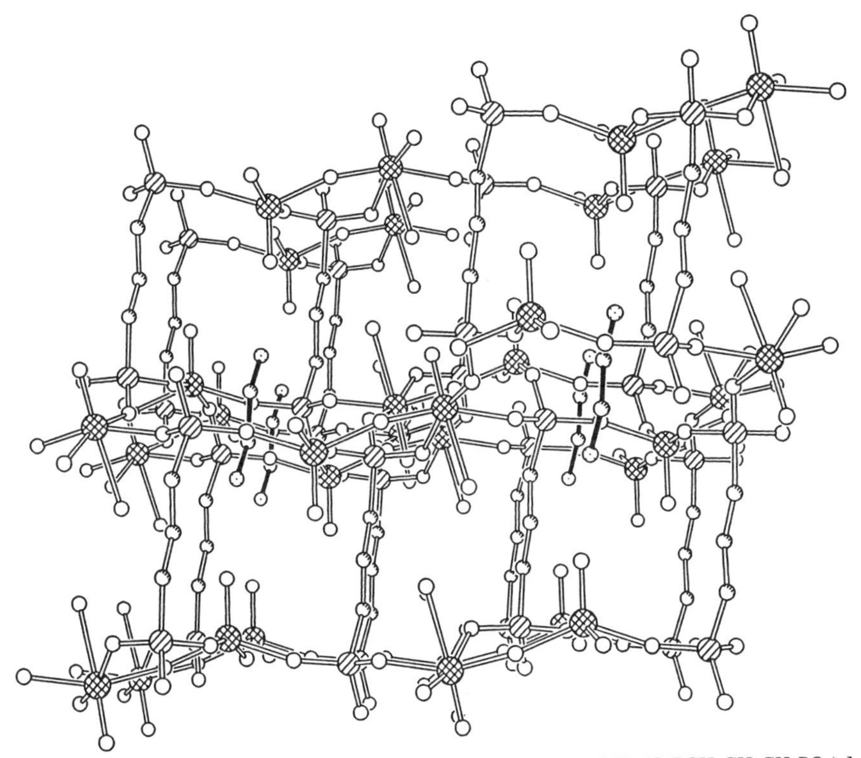

Figure 52. A view of the structure of $[H_3NCH_2CH_2NH_3][(VO)_4(OH)_2(O_3PCH_2CH_2CH_2PO_3)_2]$ (**65**), highlighting the positions of the organic cations.

ate oxygen donors, and to bridging hydroxy group, while the second site is distorted octahedral with bonding to an apical oxo group, three phosphonate oxygen atoms, the bridging hydroxy group, and a terminal aqua ligand. The binuclear $\{V_2^{IV}\ O_2(OH)\}$ unit provides an unusual structural motif within the V/P/O layers. The diphosphonate ligands span adjacent layers such that one $\{PO_3\}$ group bridges the V sites of a $\{V_2O_2(OH)\}$ unit in a symmetrical monodentate mode and coordinates through the remaining oxygen donor to the V(2) site of an adjacent binuclear unit, while the second $\{PO_3\}$ group bonds to two V(1) sites and one V(2) site of three proximal $\{V_2O_2(OH)\}$ units from a neighboring layer. In contrast to the structures of **57** and **64**, all the oxygen donors of the diphosphonate ligand coordinate to metal centers.

An unusual feature of the V/O/RPO_3^{2-} system is the range of connectivity patterns that are beginning to emerge for vanadium polyhedra and phosphonate tetrahedra within the V/P/O layers. As shown in Figs. 48, Compound **58** exhibits vanadium octahedra corner sharing through short–long V—O interactions into infinite chains, while $[(VO)_2\{CH_2(PO_3)_2\}(H_2O)_4]$ (**63**) contains isolated $[VO_6]$ octahedra, that is, no V—O—V bonding. While **64** resembles the latter in the absence of V—O—V bonds, the vanadium sites are square pyramidal and as noted previously the details of the polyhedral connectivity are quite distinct. The V/P/O structure of **65** is unique in presenting $\{V_2O_2(OH)\}$ binuclear units with both octahedral and square pyramidal vanadium sites. While the structure of $[Et_4N]_2[(VO)_6(OH)_2(H_2O)_2(EtPO_3)_6]$ (**61**) also displays the $[V_2O_2(OH)]$ motif, the detailed connectivity within the plane is quite distinct from that of **65**.

While the tether length n of the diphosphonate ligands $(H_2PO_3)(CH_2)_n$-(H_2PO_3) enhances the ligand flexibility and may thus influence the dimensionality of the resultant phase, the isolation of both 1D and 2D phases with methylenediphosphonate, $[H_2N(C_2H_4)_2NH_2][VO\{CH_2(PO_3)_2\}]$ (**57**) and $[(VO)_2\{CH_2(PO_3)_2\}(H_2O)_4]$ (**63**), suggests that reaction conditions, introduction of templates, and other factors are also critical determinants of structure. In this vein, minor modification of reaction conditions for the vanadium oxide phases of the propylenediphosphonate yields the "stepped" 2D phase $[H_2N(CH_2CH_2)_2NH_2][(VO)_2(O_3PCH_2CH_2CH_2PO_3)_2]$ (**66**) in addition to the 3D or "pillared" 2D phase **65**, described above. The structure of **66** is shown in Fig. 53 and seen to consist of stepped layers constructed of vanadium(IV) square pyramids and diphosphonate polyhedra. Adjacent vanadium sites are bridged by phosphonate oxygen atoms from two different diphosphonate ligands to give the recurrent motif $\{(VO)_2(\mu_2\text{-}O_3PC\text{—})\}$. The $\{V_2P_2O_4\}$ eight-membered rings formed by this connectivity pattern fuse along common edges to produce ribbons of linked vanadium square pyramids and phosphonate tetrahedra. Each propylenediphosphonate ligand bridges adjacent strips. However, while one $\{\text{—}PO_3\}$ group employs all three oxygen donors in bonding to the vanadium centers of one strip, the other $\{\text{—}PO_3\}$ terminus of the diphosphonate group

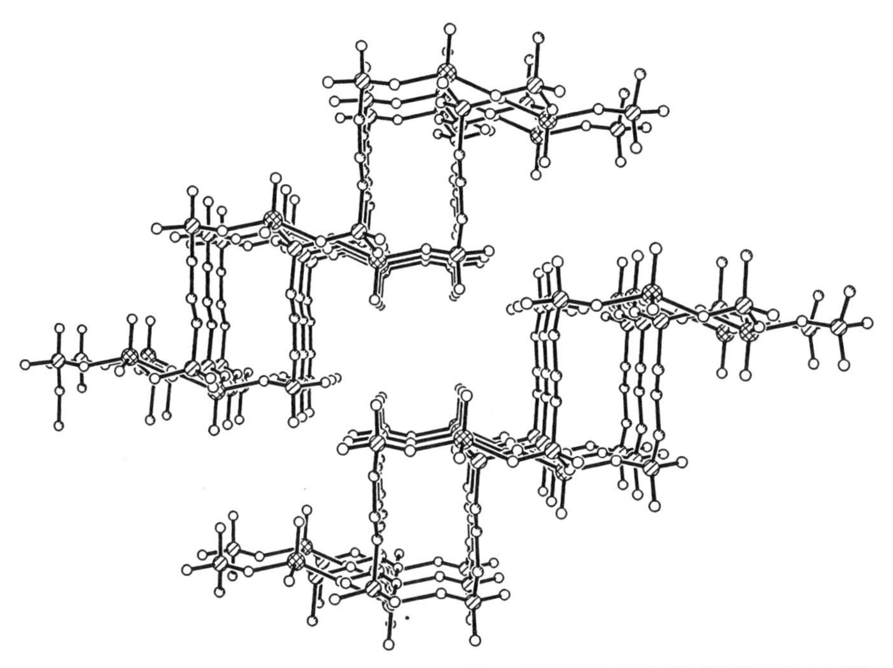

Figure 53. A view of the structure of $[H_2N(CH_2CH_2)_2NH_2][(VO)_2(O_3PCH_2CH_2CH_2PO_3)_2]$ (**66**).

bonds through a single oxygen donor, leaving two pendant {P=O} units on the same phosphorus. This coordination mode has the effect of terminating the connectivity and producing ribbons two vanadium and four phosphate polyhedra in width. The propylene bridges are disposed in pairs above and below the plane of each strip, so as to connect neighboring ribbons in a "stepped" orientation. The vanadyl oxygen atoms and the terminal {P=O} groups project into the interlamellar region such that adjacent "stepped" layers generate well-defined tunnels occupied by the organic cations.

VII. OXOMOLYBDENUM CLUSTERS

A. Oxomolybdenum Clusters with Alkoxide, Carboxylate, Oxocarbon, and Related Ligand Types

1. Trinuclear Cores

Figure 54 represents four different ways of constructing a trinuclear core, based on three edge-sharing octahedra, which is observed as a constituent unit in several polyoxometalates. Synthesis of the isolated unit has been achieved

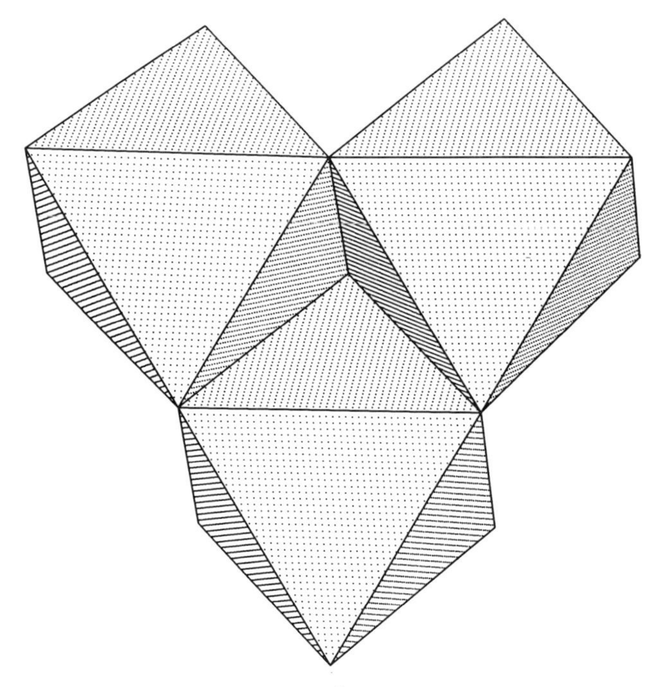

(a)

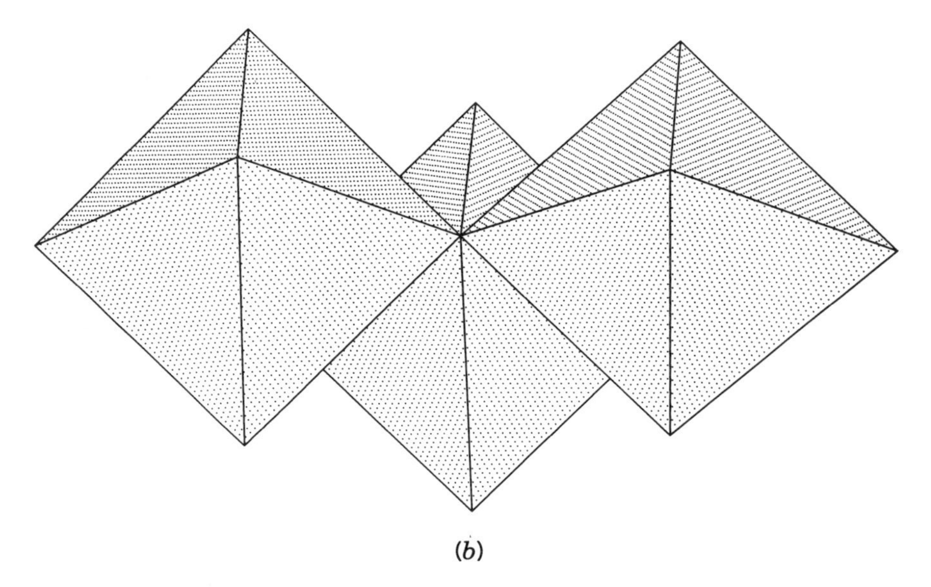

(b)

Figure 54. Four different ways of constructing a trinuclear core, based on three edge-sharing octahedra.

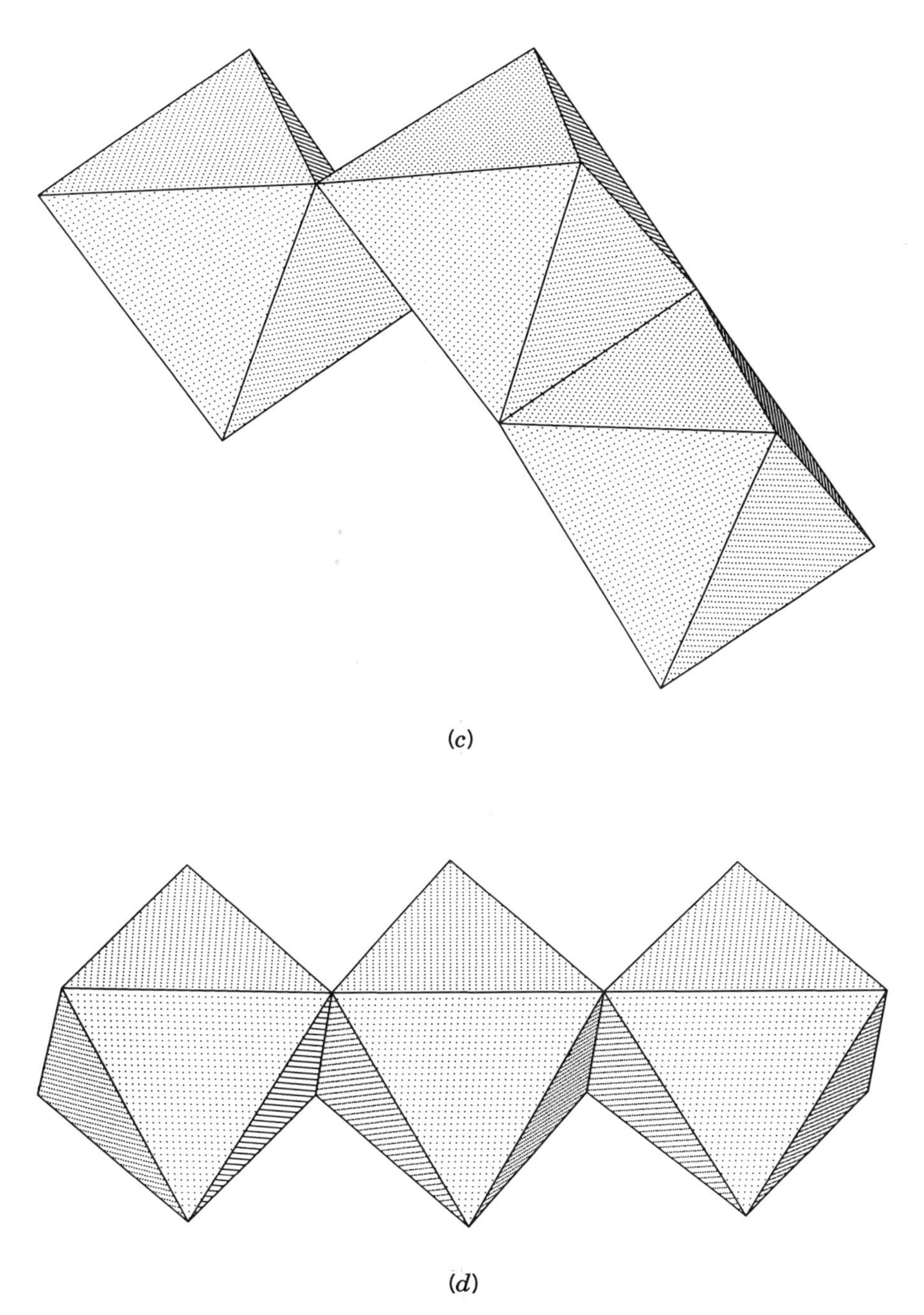

Figure 54. (*Continued*)

by the reactions of $[Mo_5O_{16}(OMe)]^{3-}$ (**91**) with either squarate or pinacolate ligands resulting in the formation of $[Mo_3O_8(OMe)(C_4O_4)_2]^{3-}$ (**67**) (192) (Fig. 55) and $[Mo_3O_8$ (pinacolate)$_2]^{2-}$ (**68**) (193), respectively. While the squarate ligand effects the direct displacement of $[MoO_4]^{2-}$ groups with the retention of the central triangular core of structural type shown in Fig. 54(*a*), the pinacolate ligand causes structural rearrangement during the reaction leading to the product with the polyhedral configuration of Fig. 54(*c*). The pronounced effect of ligand geometry on structure is marked by the coordination chemistry of polyoxomolybdate with trisalkoxy type ligands, $[RC(CH_2O)_3]^{3-}$, whose chelating sites favor some degree of bridging across metal centers. The cluster $[Mo_3O_7\{(OCH_2)_3CMe\}_2]^{2-}$ (**69**) (83), shown in Fig. 56, displays triangular core arrangement of Fig. 54(*a*) with two of its molybdenum centers each bonded to two terminal oxo groups, one terminal alkoxo, and two bridging alkoxo groups. The remaining molybdenum center, which bonds to three bridging alkoxo groups, also exhibits three terminal oxo groups, thus displaying the less commonly observed anti-Lipscomb MoO_3 unit, to achieve octahedral geometry. The rarely observed structural moiety $\{MoO_3\}$ dissociates readily in solution, as observed in the case of polynuclear clusters $[Mo_{10}O_{34}]^{8-}$ (194) and $[Mo_8O_{24}(OMe)_4]^{4-}$ (195) where it functions as part of a corner sharing $[MoO_4]^{2-}$ tetrahedron; the unit is also sufficiently basic/nucleophilic to provide a site for further condensation reactions, a feature exemplified in the synthesis

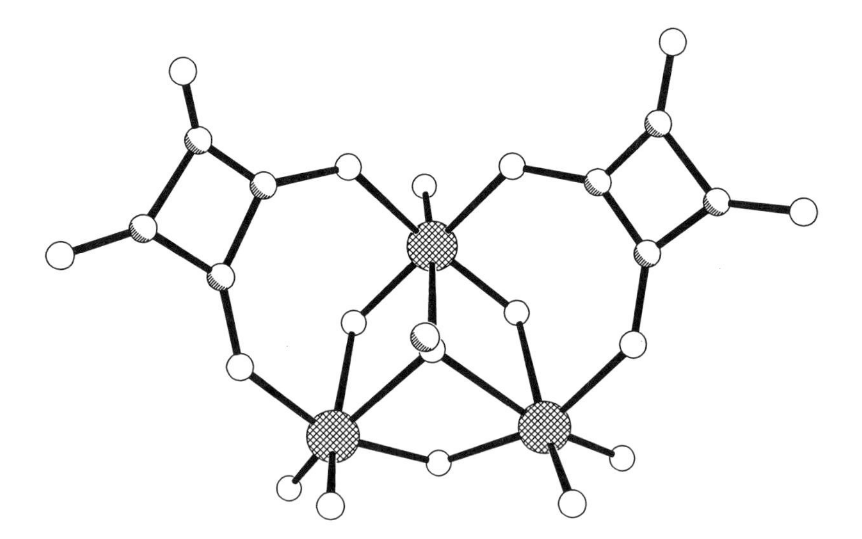

Figure 55. The structure of $[Mo_3O_8(OMe)(C_4O_4)_2]^{3-}$ (**67**).

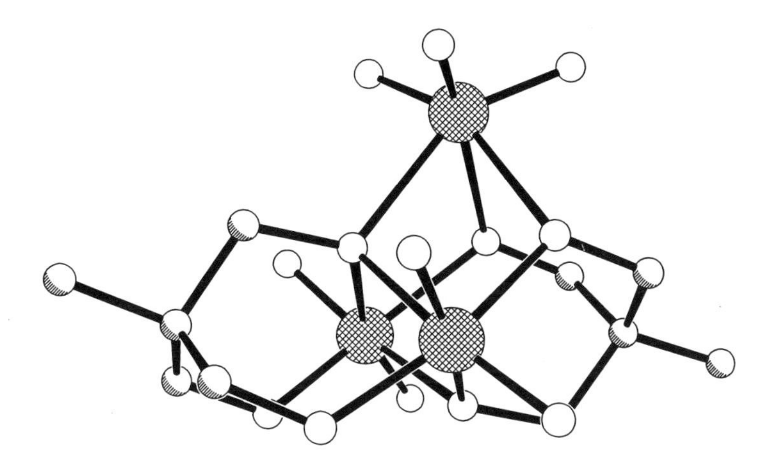

Figure 56. A view of the structure of $[Mo_3O_7\{(OCH_2)_3CMe\}_2]^{2-}$ (**69**).

of $[Mo_3O_6(OMe)\{(OCH_2)_3CMe\}_2]^{1-}$ (**70**) from the direct reaction of $[Mo_3O_7\{(OCH_2)_3CMe\}_2]^{2-}$ with MeOH.

An interesting adduct containing a trinuclear repeat unit is found in the species $\{Mo_3O_9(dmso)_4]$ (**71**) (196) (Fig. 57) [where dmso = dimethylsulfoxide], which consists of infinite chains of both octahedral and tetrahedral molybdenum(VI) sites repeating in a (oct—tet—oct—oct—tet—oct)$_n$ motif. The unique 1D nature of the structure is closely related to the structures of inorganic materials $Na_2Mo_2O_7$ (197) and $K_2Mo_2O_7$ (198), which contain infinite chains of $\{MoO_4\}$ tetrahedra and $\{MoO_6\}$ octahedra in a somewhat different pattern.

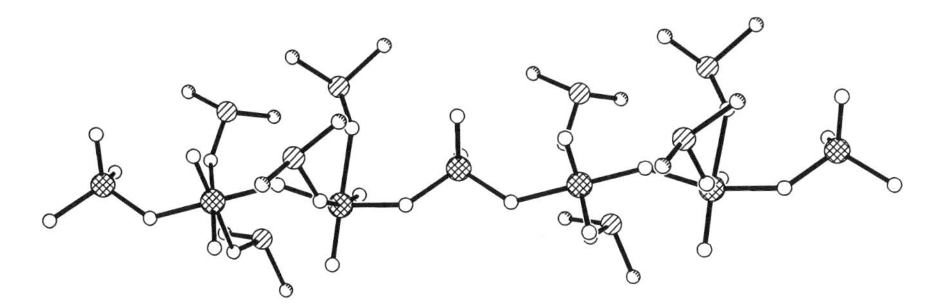

Figure 57. A view of the structure of $[Mo_3O_9(dmso)_4]$ (**71**).

2. *Tetranuclear Clusters*

Two possible octahedral edge-sharing arrangements for the tetranuclear core are illustrated in Fig. 7, namely, IVa and b. The tetranuclear core is the most commonly observed motif in the coordination chemistry of polyoxomolybdates, and the most frequently encountered arrangement is exemplified by the cluster $[Mo_4O_{10}(OMe)_6]^{2-}$ (**72**) (84, 199), which possesses four edge-sharing octahedra in the compact cluster of type IVa. As shown in Fig. 58, the structure displays two unique Mo centers with incorporation of alkoxy groups into the cluster as commonly observed in alcoholic reaction media. Rather than forming a highly charged naked cluster, $[Mo_4O_{16}]^{8-}$, alkoxy ligands substitute varying numbers of oxo groups to reduce the charge and stabilize the cluster in alcoholic solvents. This forms the basis of synthesis of a variety of clusters: $[Mo_4O_{10}(OMe)_4Cl_2]^{2-}$ (**73**) (84, 199), $[Mo_4O_{10}(OMe)_2(OC_6H_4O)_2]^{2-}$ (**76**) (199), $[Mo_4O_8(OEt)_2\{(OCH_2)_3CR\}_2]$ (**75**) (200), and $[Mo_4O_{10}(OMe)_2$-$(C_6H_4CONO)_2]^{2-}$ (201). The tetranuclear core sustains electron-transfer processes to produce analogous Mo^V species. Thus, the cluster $[Mo_4O_{10}$-$(OMe)_4Cl_2]^{2-}$ may be reduced into $[Mo_4O_8(OMe)_2(HOMe)_2Cl_4]^{2-}$ (**77**), shown in Fig. 59, containing two short Mo· · ·Mo distances 2.595(1) Å, consistent with a significant degree of metal–metal bonding. Other members of this fully reduced class of compounds include $[Mo_4O_8(OEt)_2(HOEt)_2Cl_4]^{2-}$ (**78**) (202) and $[Mo_4O_6(OPr)_4(HOPr)_2Cl_4]$ (**79**) (203), which were originally described as mixed-valence Mo^V/Mo^{VI} species and later reformulated in their correct oxidation states.

These reduced clusters (**78–80**) (202–204) undergo ligand substitution reactions such that chloride and alcohol ligands are readily displaced by other

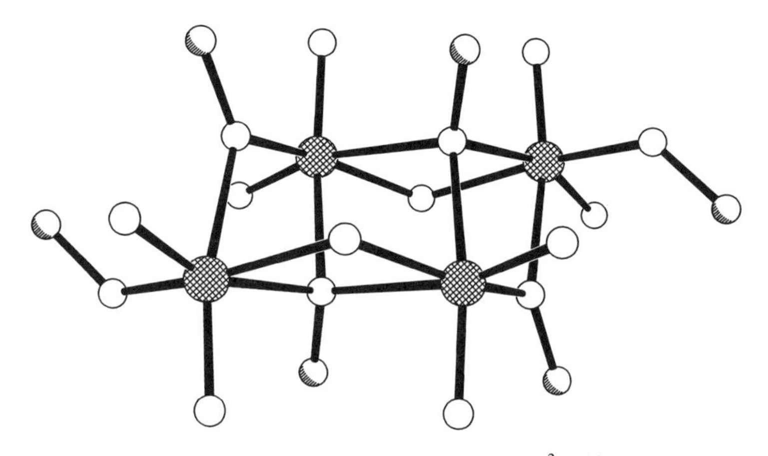

Figure 58. The structure of $[Mo_4O_{10}(OMe)_6]^{2-}$ (**72**).

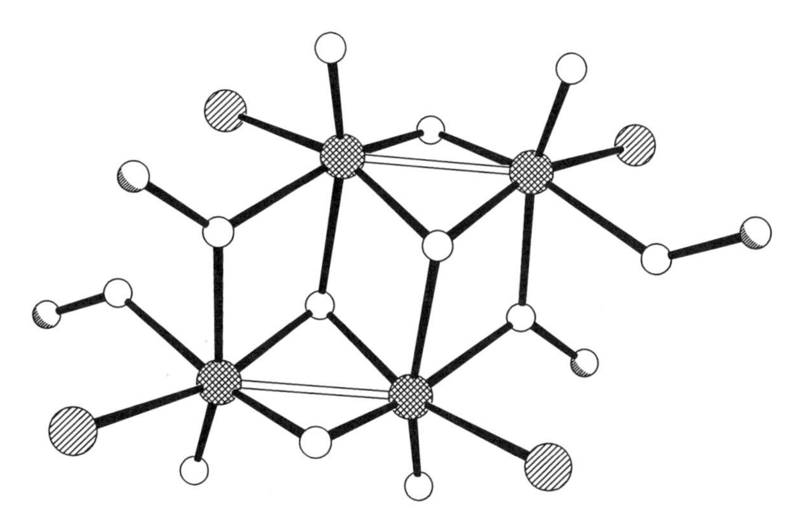

Figure 59. A view of the structure of $[Mo_4O_8(OMe)_2(HOMe)_2Cl_4]^{2-}$ (**77**), which exhibits the IV′ core of two pairs of Mo^V-Mo^V units.

appropriate ligands to generate higher nuclearity clusters retaining the robust $\{Mo_4O_8(OMe)_2\}^{2+}$ core, as demonstrated by the synthesis and structure of the mixed-valence cluster $[H_2Mo_8O_{24}(OMe)_2]^{4-}$ (**98**) (205) and the squarate cluster $[Mo_4O_8(OMe)_2(C_4O_4)_2(C_4O_4H)_2]^{4-}$ (**89**) (206), shown in Fig. 60.

Intramolecular carbonyl insertion is an attractive synthetic route (207) for several complexes including $[CH_2Mo_4O_{15}H]^{3-}$ (**84**), which has a ring of oxo-bridged *cis*-dioxomolybdate units, $\{MoO_2(\mu\text{-}O)\}_4$, capped by acetal and hydroxide groups from above and below, respectively. An identical core structure is present in the species $[(R_2As)Mo_4O_{15}H]^{2-}$ (6). Other related structures with different capping groups are the formylated and fluoro derivatives $[(HCCH)Mo_4O_{15}(HCO_2)]^{3-}$ (**85b**) and $[(HCCH)Mo_4O_{15}F]^{3-}$ (**85a**) (208) illustrated schematically in Fig. 61. The most pronounced structural effects of introducing the diacetal moiety on the $\{Mo_4O_4\}^{2+}$ ring are the incorporation of one of the bridging oxo groups of $\{Mo_4O_4\}$ unit into the diacetal linkage with concomitant increase in the Mo· · ·Mo distances and replacement of the quadruply bridging OH^- group by the doubly bridging F^- or HCO_2^- groups. As a result, the structures display nonequivalent molybdenum centers. Similarly, incorporation of organic substrates containing the α-diketone subunit into tetramolybdate framework yields diketal derivatives of the type $[RMo_4O_{15}X]^{3-}$. Examples of such derivatives are $[(C_9H_4O)Mo_4O_{15}(OMe)]^{3-}$ (**86**), $[(C_{14}H_{10})$-$Mo_4O_{15}(PhCO_2)]^{3-}$ (**87**), and $[(C_{14}H_8)Mo_4O_{15}(OH)]^{3-}$ (**88**), which have been prepared by the incorporation of ninhydrin, benzil, and phenanthraquinone, respectively (209) (Fig. 61). While the methoxy derivative $[(C_9H_4O)$-

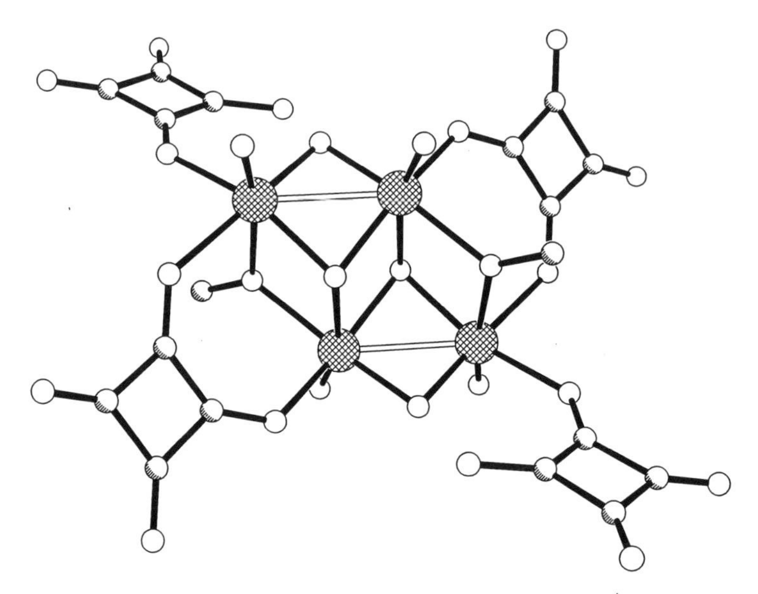

Figure 60. The structure of $[Mo_4O_8(OMe)_2(C_4O_4)_2(C_4O_4H)_2]^{4-}$ (**89**).

$[(C_9H_4O)Mo_4O_{15}(OMe)]^{3-}$ is structurally analogous to $[(HCCH)Mo_4O_{15}F]^{3-}$, Structures **87** and **88** demonstrate the flexibility of this core. The phenanthraquinone derivative $[(C_{14}H_8)Mo_4O_{15}(OH)]^{3-}$ exhibits a triply bridging OH^- group, leading to a pentacoordinate Mo, two octahedral Mo centers bonded to two bridging oxo groups, and a third octahedral center at the open edge of the $\{Mo_4(\mu\text{-}O)_3\}$ core. In $[(C_{14}H_{10})Mo_4O_{15}(PhCO_2)]^{3-}$, the benzoate group bridges two Mo centers through one carboxylate oxygen and interacts with a third Mo through the second carboxylate oxygen to produce an unsymmetrical bonding mode.

The structures of the tetranuclear compounds $[Mo_4O_{11}(malate)_2]^{2-}$ (**81**) (210, 211) and $[Mo_4O_{11}(citrate)_2]^{4-}$ (**82**) (212, 213), consist of pairs of edge-sharing octahedra connected through corner-sharing interactions. The citrate group uses both oxygen atoms of one carboxylate group, a single oxygen of a second carboxylate group and the central hydroxy group to produce quadradentate behavior, leaving the third carboxylate group as a pendant arm. A similar bonding pattern is present in the malate derivative.

Carboxylate derivatives of the reduced polyoxomolybdates of the type $[Mo_4O_6Cl_2(O_2CR)_6]$ [R = O_2CMe (**83a**) or $O_2C_6H_4Me$ (**83b**)] (214, 215) containing pairs of Mo^V centers with short $Mo^V{-}Mo^V$ interactions have been prepared. While two $\{MoO_6\}$ octahedra are fused through edge sharing, the remaining two $\{MoO_5Cl\}$ moieties each share a vertex with the $[Mo_2O_2]$ rhombus

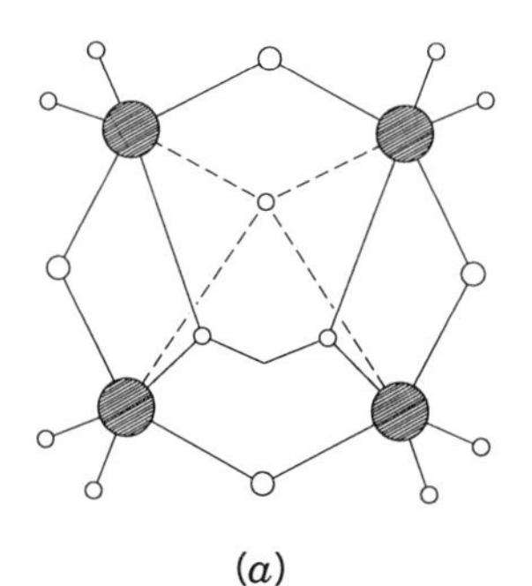

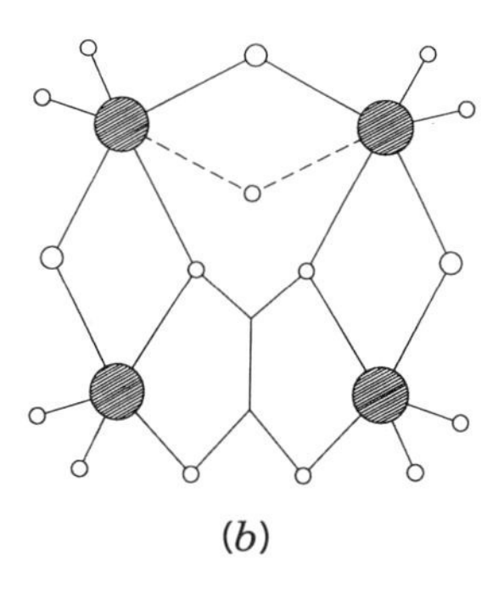

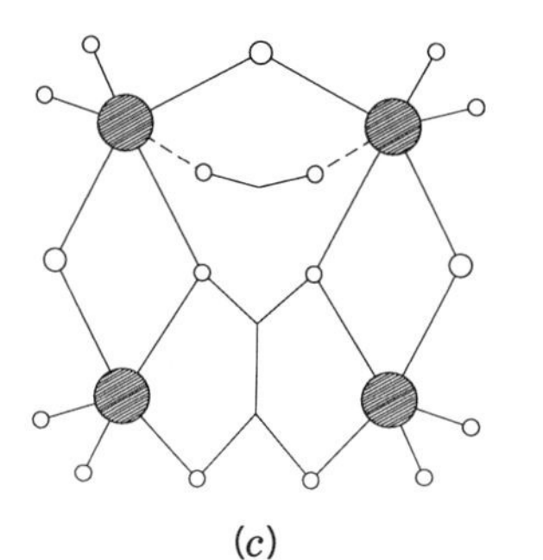

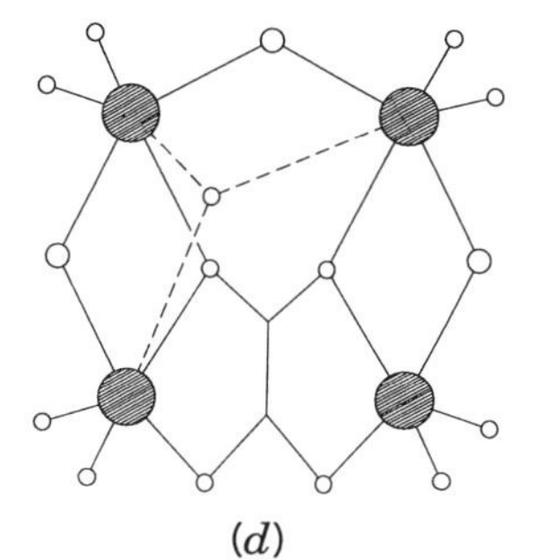

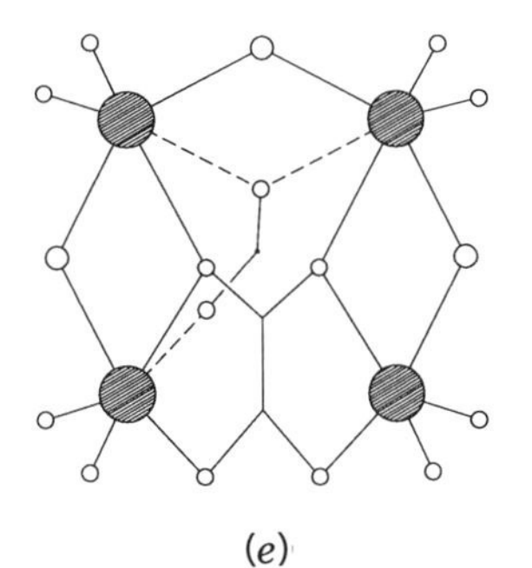

Figure 61. Schematic representation of the $[RMo_4O_{15}X]^{3-}$ core as adapted for the structures of (*a*) $[(CH_2)Mo_4O_{15}]^{3-}$, (*b*) $[(HCCH)Mo_4O_{15}F]^{3-}$, (*c*) $[(HCCH)Mo_4O_{15}(HCO_2)]^{3-}$, (*d*) $[(C_{14}H_8)Mo_4O_{15}(OH)]^{3-}$, (*e*) $[(C_{14}H_{10})Mo_4O_{15}(C_7H_5O_2)]^{3-}$.

of this binuclear fragment. The six carboxylate ligands adopt the bridging bidentate mode to complete the coordination about the Mo centers.

A most unusual structure is provided by the cluster $[Mo_4O_6(OH)_4(CO_3)(CO)_2(PMe_3)_6]$ (**90**) (216), which is schematically illustrated in Fig. 62. Each Mo^V center of the $(\mu\text{-}O)_2$ bridged core, $\{Mo_2O_4\}$, is bridged through two

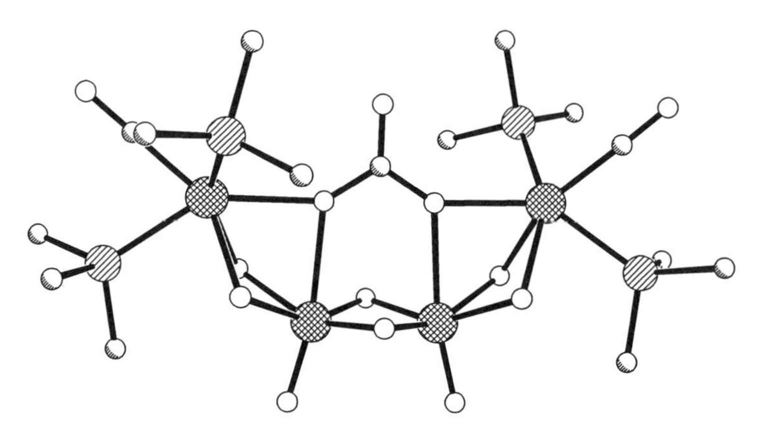

Figure 62. A view of the structure of $[Mo_4O_6(OH)_4(CO_3)(CO)_2(PMe_3)_6]$ (**90**).

hydroxy groups, to a seven-coordinate Mo^{II} site. The CO_3^{2-} group coordinates through two oxygen donors in μ-4-coordination mode to bridge all of the four molybdenum centers.

The compound $[Mo_4O_{10}(C_6H_2O_4)_2]^{2-}$ (**74**) (217, 218) exhibits a structure based on the well-known $[Mo_2O_5]^{2+}$ core, bridged through the semiquinone ligand group.

3. Pentanuclear Clusters

The unique example of the pentanuclear core, $[Mo_5O_{16}(OMe)]^{3-}$ (**91**) (192), prepared from a methanolic solution of $[Mo_2O_7]^{2-}$ in the presence of Lewis bases (192), reflects the strong tendency of polyoxoanions to adapt spherical structures. Thus, condensation of an additional {Mo=O} unit on the pentanuclear core to generate favored hexanuclear core $[Mo_6O_{19}]^{2-}$ is readily achieved. The noncompact structure of **91**, similar to that of the previously described compound $[Mo_3O_8(OMe)(C_4O_4)_2]^{3-}$, consists of a trinuclear $\{Mo_3O_8(OMe)\}^+$ unit bridged by two bidentate $\{MoO_4\}^{2-}$ groups. While two of the Mo sites adopt a tetrahedral coordination environment, the remaining sites exhibit distorted octahedral geometries.

4. Hexanuclear Clusters

Although the commonly observed hexametalate core $[Mo_6O_{19}]^{n-}$ and its derivatives $[Mo_6O_{19-n}L_n]^{m-}$ favor spherical and pseudospherical structures, the mixed-valence complex, $[Mo_6O_{10}(OPr)_{12}]$ (**92**) (219) displays an extended structure consisting of a zigzag chain of edge-sharing square pyramids, with

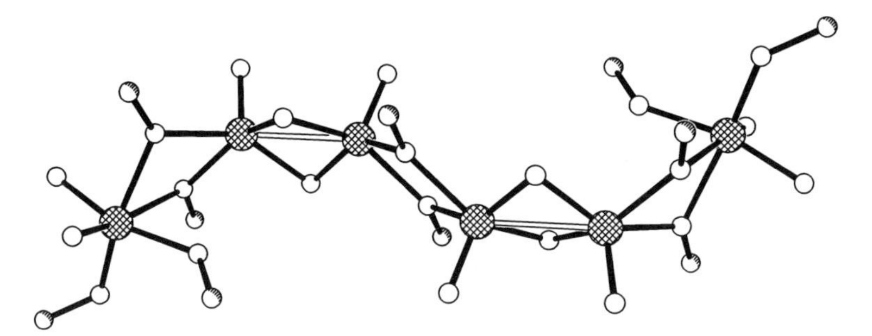

Figure 63. A view of the chain of edge-sharing octahedra and square pyramids adopted by $[Mo_6O_{10}(OPr)_{12}]$ (**92**). The Mo^V sites are trapped as pairs of $Mo^V{-}Mo^V$ units.

alternating (μ-OR) and (μ-O) bridges, terminated at either end by edge-sharing $[MoO(OR)_5]$ octahedra, shown in Fig. 63.

5. *Octanuclear Clusters*

Formation of the frequently encountered octanuclear core, such as that found in $[Mo_8O_{24}(OMe)_4]^{4-}$ (**96**) (195), may be described in terms of the condensation of two tetranuclear species $[Mo_4O_{10}(OMe)_6]^{2-}$ along two edges of each unit and concomitant loss of eight methoxy groups. While four sites are occupied by $(OMe)^-$ groups in $[Mo_8O_{24}(OMe)_4]^{4-}$, the formylated derivative, $[Mo_8O_{26}(HCO_2)_2]^{6-}$ (**93**) (5) has a different ligand occupation pattern with two distinct positions occupied by the HCO_2^- groups, shown in Fig. 64. On the other hand, $[Mo_8O_{24}(sal)_2]^{2-}$ (**94**) (sal = salicylidenepropyliminato, $PrN{=}CHC_6H_4O^-$) and $[Mo_8O_{24}(OH)_2(met)_2]^{4-}$ (**95**) [met = methionato, $MeSCH_2CH_2CH(NH_2)CO_2^-$] (220) provide variants of this structural type.

Another structural variation is exhibited by the polyoxoalkomolybdate cluster $[Mo_8O_{20}(OMe)_4\{(OCH_2)_3CR\}_2]^{2-}$ (**97**) (221) which consists of two tetranuclear unit $[Mo_4O_{10}(OMe)_2\{(OCH_2)_3CMe\}]^{-1}$ related by a center of symmetry. While each tetranuclear unit is composed of edge-sharing $\{MoO_6\}$ octahedra, the entire structure is generated by the sharing of two vertices between the tetranuclear moieties, as shown in Fig. 2.

Another example of octanuclear core is observed in the mixed-valence heterometallic oxoalkoxide cluster $[Mg_2Mo_8O_{22}(OMe)_6(HOMe)_4]^{2-}$ (**99**) (222). This structure consists of two tetranuclear cores fused along two edges. The central core consists of two pairs of two edge-sharing Mo^V octahedra each with short Mo· · ·Mo distances, joined by two corner-sharing interactions. The cavities formed by the stacking of Mo octahedra are occupied by two $\{Mg(OMe)_2\}$ moieties, as illustrated schematically in Fig. 65.

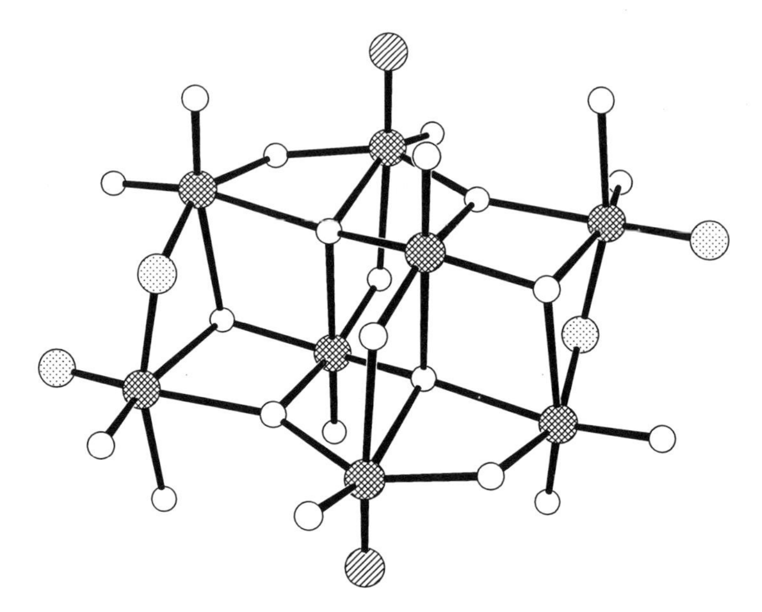

Figure 64. A schematic representation of the octametalate core $\{M_8O_{28}\}$ and the positions most commonly occupied by oxygen donors of organic ligands. In $[Mo_8O_{26}(HCO_2)_2]^{6-}$ (**93**) only the A sites are occupied, while the B sites are occupied in $[Mo_8O_{24}(OMe)_4]^{4-}$ (**96**).

Two members of this octanuclear series of complexes, $[Mo_8O_{16}(OMe)_8(C_2O_4)]^{2-}$ (**100**) (223) and $[Mo_8O_{16}(OMe)_8(PR_3)_4]$ (**101**) (224), exhibit the cyclic $\{Mo_8O_{16}(OMe)_8\}^0$ framework. The former is prepared by the reduction in alcoholic solution of polyoxomolybdate by rhodizonic acid, which is itself oxidized to produce oxalate anions. The structure, shown in Fig. 66, displays an octagonal array of Mo centers, alternatively bridged by two methoxy and two oxo groups, in a planar ring with alternatively short (2.578 Å) and long (3.282 Å) Mo—Mo distances associated with Mo_2O_2 and $Mo_2(OMe)_2$ units, respectively. Each oxygen of the oxalate moiety, which occupies the central cavity, bridges two Mo centers to produce an arrangement of alternating edge-sharing and face-sharing $\{MoO_6\}$ octahedra. The notable differences in the structure of $[Mo_8O_{16}(OMe)_8(PR_3)_4]$ are the vacant cavity and a puckered $[Mo_8O_{16}(OMe)_8]$ ring.

6. *Dodecanuclear Clusters*

The cluster $[Mo_{12}O_{36}(C_4O_4H)_4]^{4-}$ (**102**) (225) displays an open cyclic structure, shown in Fig. 67, rather than the compact close-packed oxide assembly generally favored by high nuclearity molybdenum oxides. Each of the

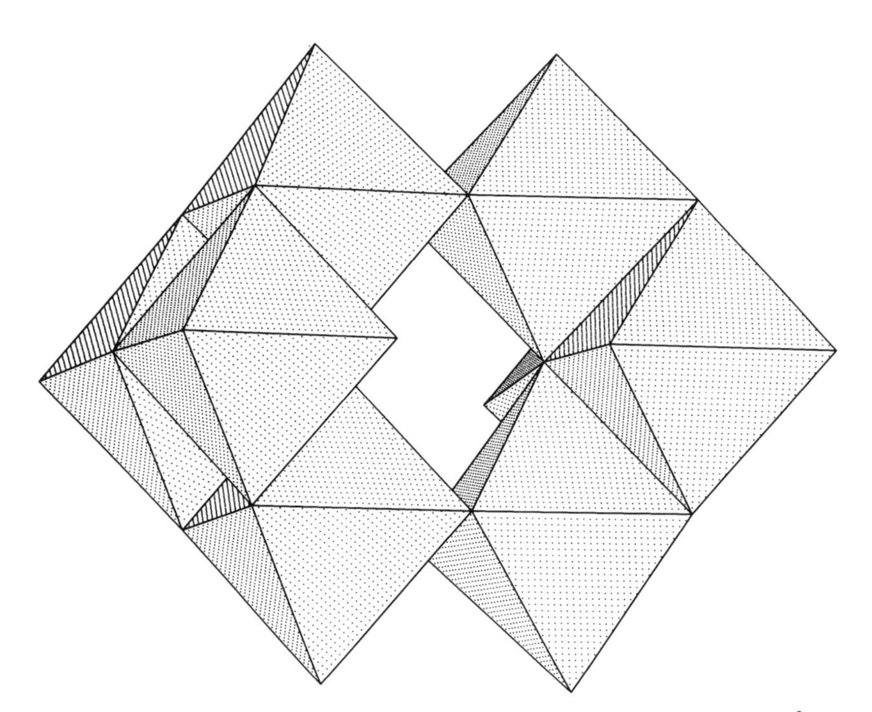

Figure 65. The arrangement of $\{MoO_6\}$ octahedra in $[Mg_2Mo_8O_{22}(OMe)_6(HOMe)_4]^{2-}$ (**99**), showing the cavities occupied by the $\{Mg(OMe)_2\}$ moieties. The magnesium $\{MgO_6\}$ octahedra complete the decametalate core $\{M_{10}O_{28}\}$.

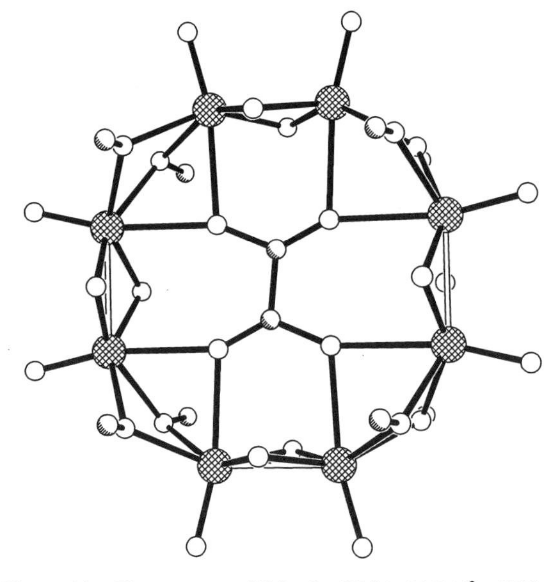

Figure 66. The structure of $[Mo_8O_{16}(OMe)_8(C_2O_4)]^{2-}$ (**100**).

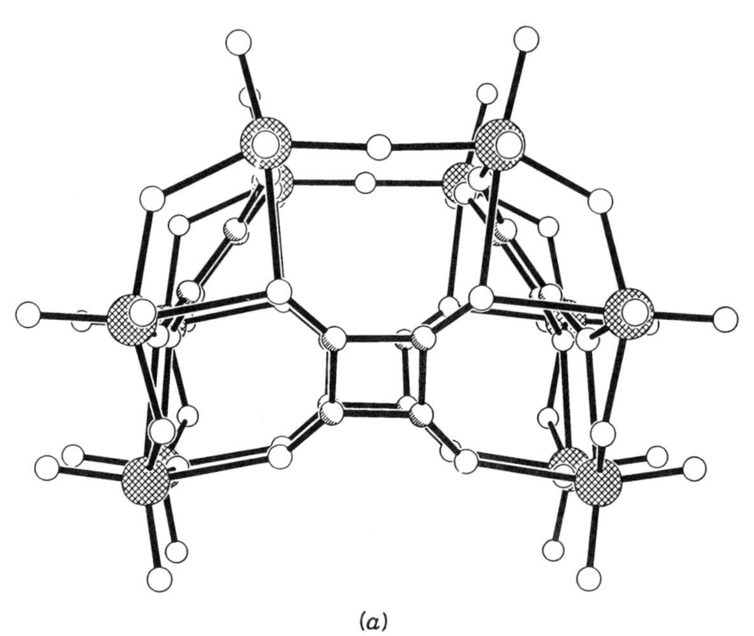

(a)

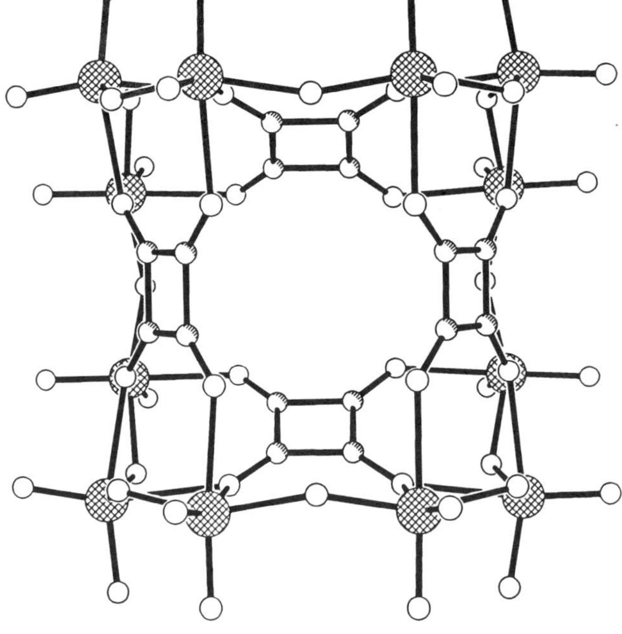

(b)

Figure 67. (*a* and *b*) Two-ball-and-stick representations of the structure of $[Mo_{12}O_{36}(C_4O_4H)_4]^{4-}$ (**102**) and the corresponding polyhedral views (*c* and *d*).

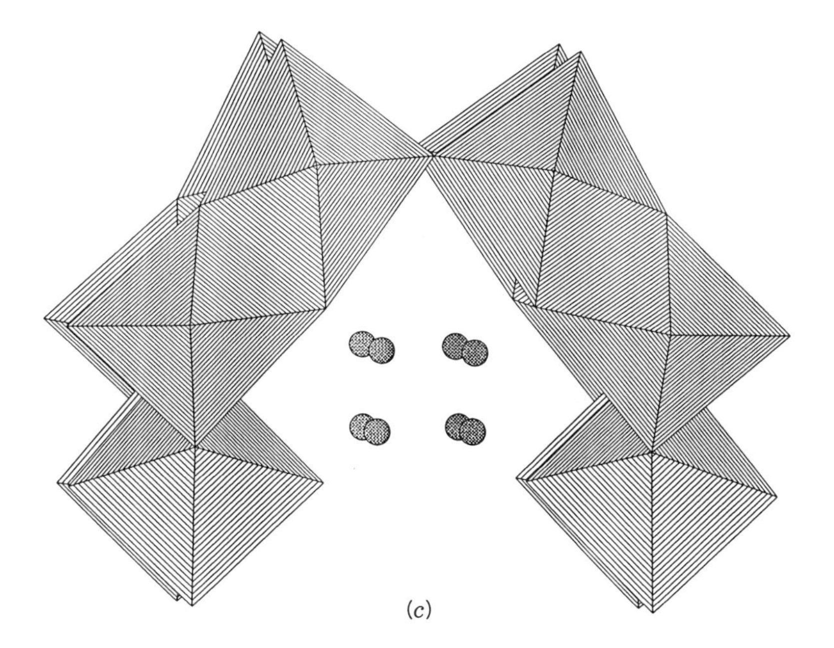

(*c*)

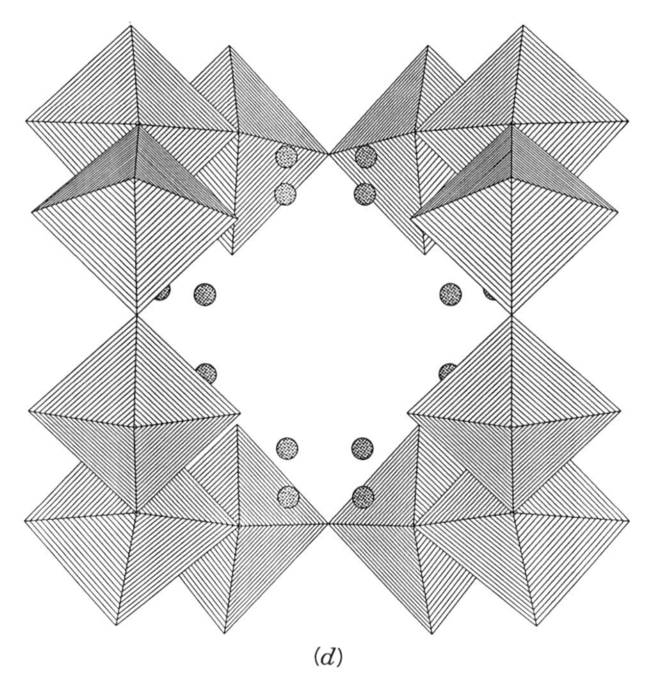

(*d*)

Figure 67. (*Continued*)

$[HC_4O_4]^{1-}$ ligands bridges six Mo^{VI} sites, such that two of the oxygen donors adopt bridging modes and two assume terminal monodentate coordination. The resultant $[MoO_6]$ octahedra form an edge- and corner-sharing framework that encapsulates the organic moieties. The structure may also be described as four trinuclear units in a corner-sharing arrangement of the type illustrated in Fig. 54(*c*). A similar core has been observed for the W^{VI}-diphosphonate cluster $[(O_3PCH_2PO_3)_4W_{12}O_{36}]^{16-}$ (226).

7. *Superclusters*

As described in Section III.B, hydrothermal reactions of simple molybdate precursors with tris(hydroxymethyl)alkanes, at 160°C, resulted in the syntheses of superamolecular clusters of which $[Na(H_2O)_3H_{15}Mo_{42}O_{109}\{(OCH_2)_3CCH_2OH\}_7]^{7-}$ (**103**) (227a) is a prototype. The structure of the Goliath cluster, illustrated in Fig. 68, consists of a framework of edge- and corner-

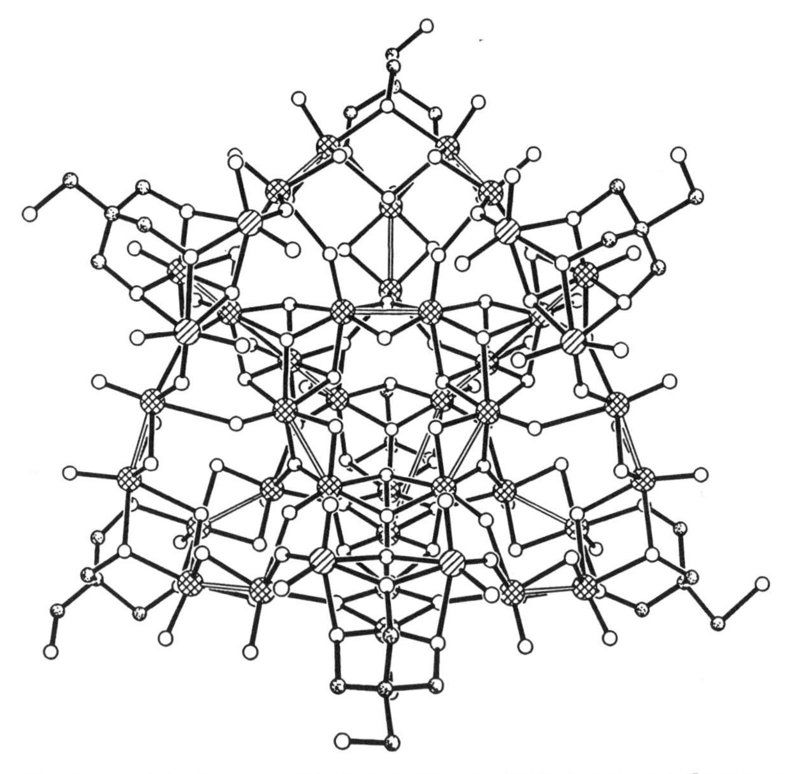

Figure 68. A view of the structure of $[Na(H_2O)_3H_{15}Mo_{42}O_{109}\{(OCH_2)_3CCH_2OH\}_7]^{7-}$ (**103**); Mo^V sites are cross-hatched; Mo^{VI} sites lined lower left to upper right.

sharing $\{MoO_6\}$ octahedra with the pendant groups of the ligands projecting outward from the central core. Three structural motifs of the cluster are shown in Fig. 69. The first unit (Fig. 69*b*) of which four are present in the cluster, consists of the tridentate ligand, $\{(OCH_2)_3CCH_2OH\}^{3-}$, bridging a triangular arrangement of Mo^V centers, which in turn are each associated through edge sharing of oxo groups to an adjacent Mo^V site; the Mo—Mo distances within these binuclear units are in the range 2.55–2.65 Å, consistent with typical Mo^V—Mo^V bonding interactions. The second structural motif (Fig. 69*a*), of which there are three in the cluster, consists of a tridentate ligand bridging a triangular core of one Mo^V and two Mo^{VI} centers, the Mo^V center in turn associated through edge sharing by two oxo-groups with an exocyclic Mo^V unit and exhibiting the usual Mo^V—Mo^V single-bond distance. The last unit (Fig. 69*c*) is a cyclic ring $\{Mo_6O_{24}\}$ of edge-sharing Mo^V octahedra with alternating short–long Mo—Mo distances. The four hexanuclear and three tetranuclear structural motifs of the cluster are disposed about this central ring through edge

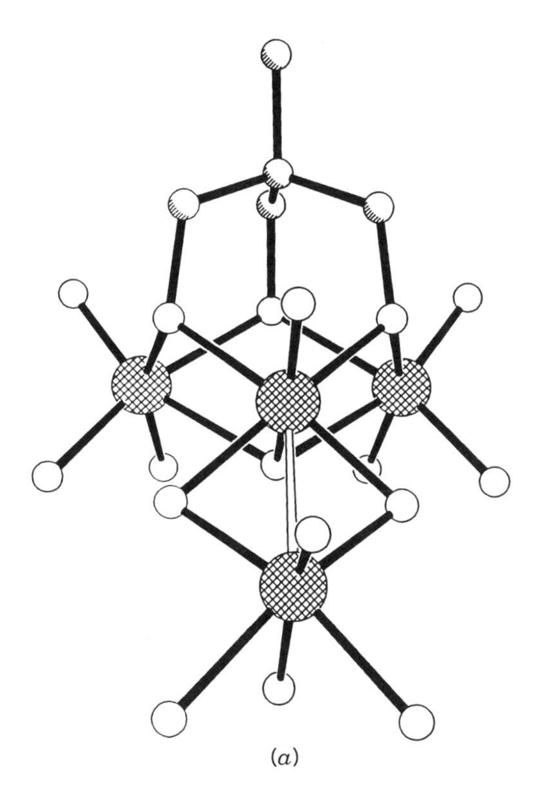

Figure 69. (*a–c*) The three structural motifs that combine to produce the Goliath cluster. (*d*) A polyhedral view of the structure, highlighting the locations of the structural building blocks.

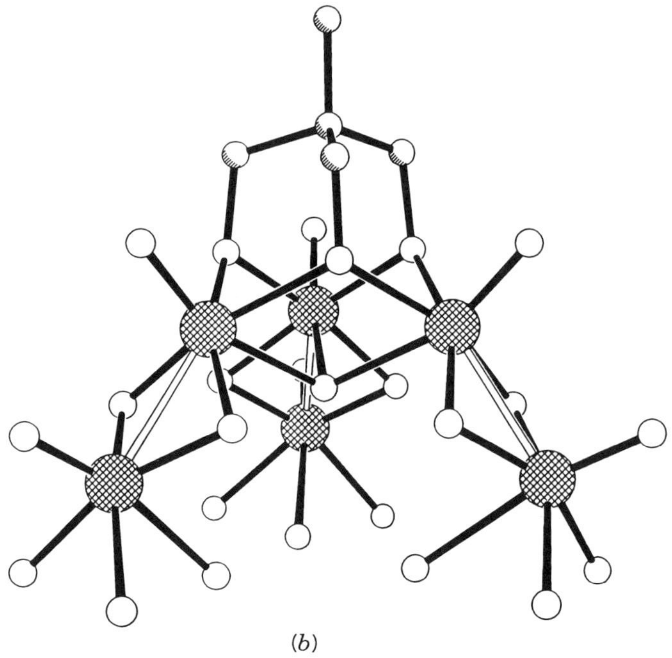

(*b*)

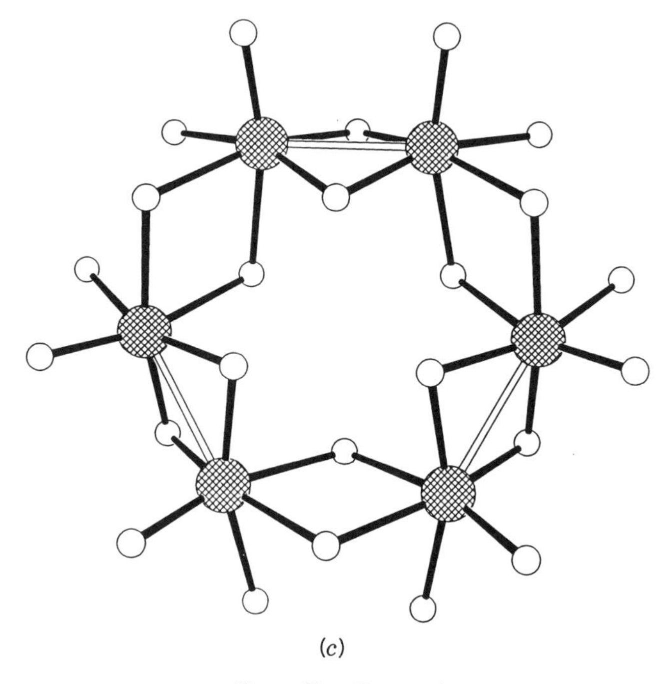

(*c*)

Figure 69. (*Continued*)

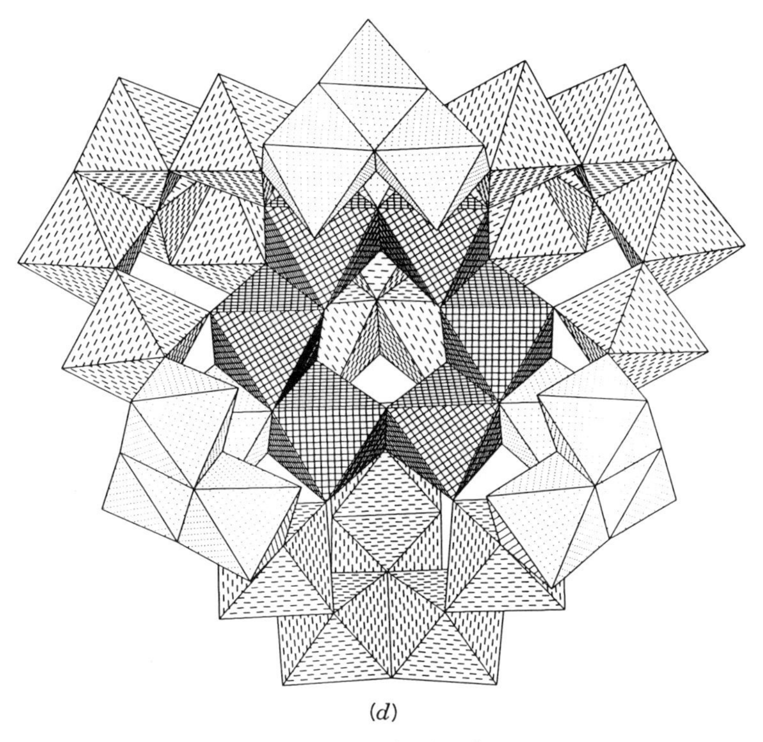

(*d*)

Figure 69. (*Continued*)

and corner sharing as shown in Fig. 68. The tetranuclear units project outward from the cluster to produce a cavity, the sides of which are defined by the octahedra of the central ring and 12 oxygen atoms associated with cis-dioxo groups of the six Mo^{VI} sites. The cavity is occupied by the $\{Na(H_2O)\}^+$ moiety. Slight modifications in the hydrothermal reaction conditions resulted in the isolation of a second supercluster type, $[Mo_{43}H_{13}O_{112}\{(OCH_2)_3CMe\}_7]^{10-}$ (**104**) (227b). The basic core of the structure is similar to that of the first supercluster, with thirty six Mo^V sites, and six Mo^{VI} sites. The notable difference here is the occupation of the central cavity by a neutral $\{MoO_3\}$ group. Thus, the structure of the cluster in this case may be considered to have assembled around the "template" cluster core $\{Mo_{16}O_{52}\}$, as schematically shown in Figs. 6 and 70, respectively.

B. Oxomolybdenum Clusters with Organophosphonate, Organoarsonate, and Related Ligand Types

In contrast to the corresponding vanadium systems, which exhibit chemistry for fully oxidized V^V, fully reduced V^{IV}, and mixed-valence V^V/V^{IV} species,

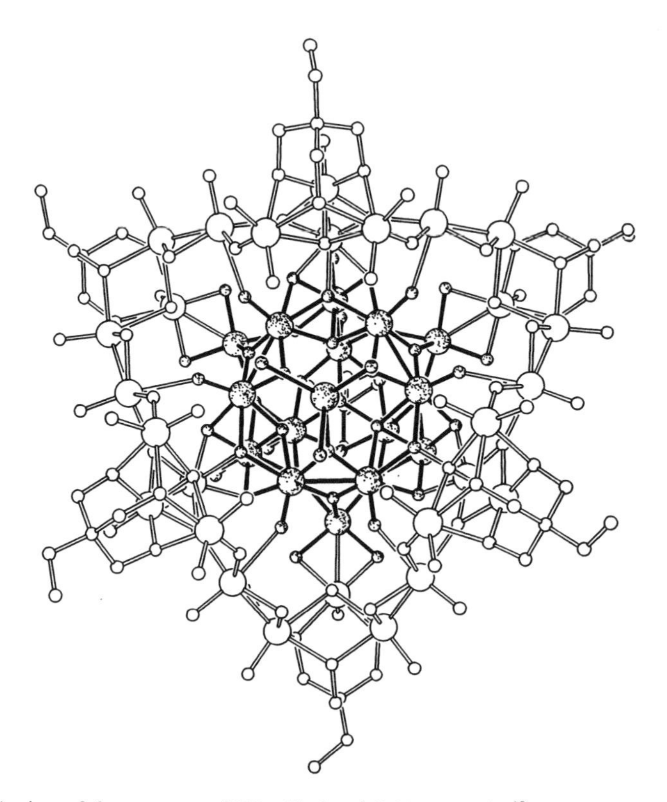

Figure 70. A view of the structure of $[Mo_{43}H_{13}O_{112}\{(OCH_2)_3CMe\}_7]^{10-}$ (**104**) highlighting the $\{Mo_{16}O_{52}\}$ core.

the $Mo/O/RPO_3$ and $Mo/O/RAsO_3$ systems are mainly limited to fully oxidized Mo^{VI} state. This observation may in part reflect the characteristic of the reduced Mo^V species to form binuclear units with short Mo· · ·Mo distances, a feature rarely displayed by motifs incorporating V^{IV} sites.

1. Trinuclear Clusters

The fully oxidized compound $[Mo_3O_9\{O_3PCMe(O)PO_3\}]^{5-}$ (**105**) (228) has been isolated by the interaction of an aqueous molybdate solution with the phosphonate ligand at pH 6. The structure of the anion shown in Fig. 71 consists of triangular arrangement of three $\{MoO_6\}$ octahedra, two of which are fused through a common face with the remaining unit linked through a common vertex. The pentadentate chelating ligand is tridentate with respect to one Mo center through two of its oxygen donors on one $-PO_3^{2-}$ unit and one oxygen atom

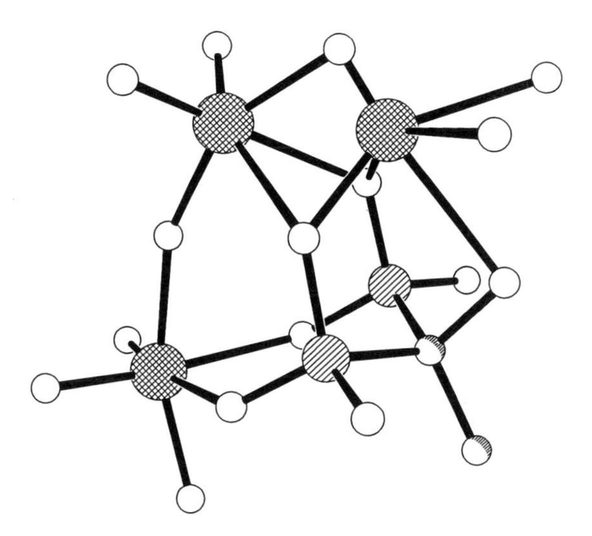

Figure 71. A view of the structure of $[Mo_3O_9\{O_3PCMe(O)PO_3\}]^{5-}$ (**105**).

of the α-hydroxy group, and bidentate with respect to the other two Mo centers. The coordination sphere of two Mo centers contain two terminal oxo groups while the remaining site has three terminal oxo groups.

2. *Tetranuclear Clusters*

While the coordination chemistry of molybdenum organoarsonate and related ligand systems remains underdeveloped, the synthesis of the classical cluster $[Mo_4O_{13}H(Me_2AsO_2)]^{2-}$ (**107**), originally by Gibbs (229) and later by Rosenheim and Bilecki (4), marked the beginning of a new era in the coordination chemistry of polyoxoanions. The structure of the anion shown in Fig. 72 (6) consists of a flat Mo_4O_{15} group resulting from the fusion of four octahedral $\{MoO_6\}$ moieties through two face-sharing and two edge-sharing interactions. Each donor oxygen of the ligand bridges both Mo centers of each of the two Mo_2O_5 moieties, which in turn are also linked to each other through two μ^2-O and one μ^4-O groups.

The influence of organic substituents on the reaction product(s) is reflected in the structures of a series of tetramolybdate clusters, prepared by the reactions of $[MoO_2(acac)_2]$ with appropriate organophosphonate–organoarsonate ligands, in refluxing acetonitrile in the presence of excess triethylamine (230). The structure of tetranuclear species $[Mo_4O_{10}(PhPO_3)_4]^{4-}$ (**108**), shown in Fig. 73, consists of two identical $\{Mo_2O_5\}$ units symmetrically bridged by two organophosphonate ligands. This arrangement generates a ring that is capped above

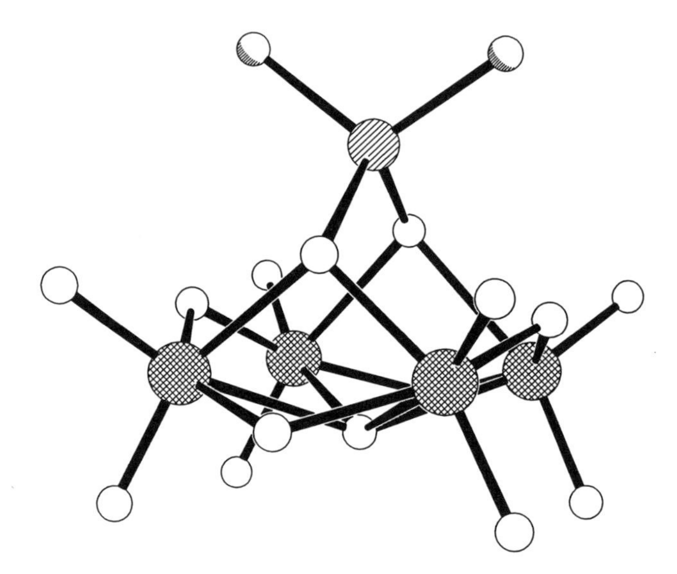

Figure 72. A view of the structure of $[Mo_4O_{13}H(Me_2AsO_2)]^{2-}$ (**107**). Note the resemblance to the common core of Fig. 61.

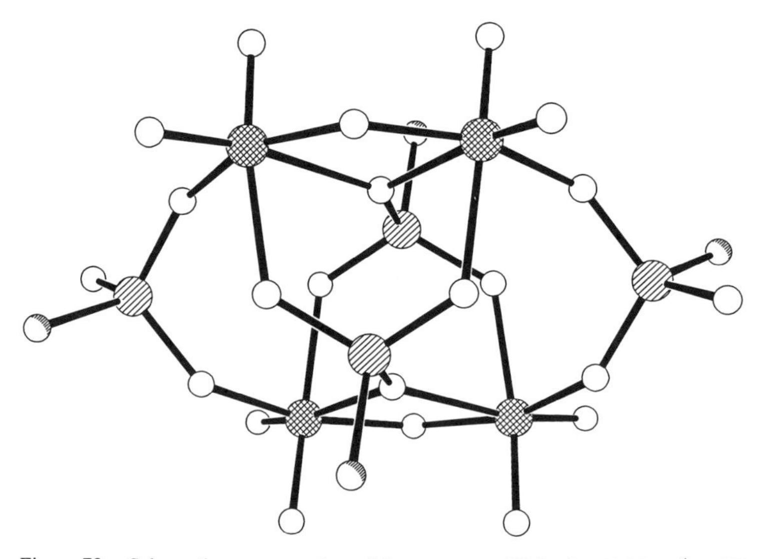

Figure 73. Schematic representation of the structure of $[Mo_4O_{10}(PhPO_3)_4]^{4-}$ (**108**).

and below by the remaining two RPO_3^{2-} ligands, which function in a tridentate mode in such a manner that two of the donor oxygen atoms are each coordinated to one Mo center and the third one acts in the μ^2-O bridging mode. The structure is based on two pairs of face-sharing $\{MoO_6\}$ octahedra, each of which in turn shares three vertices with three RPO_3 tetrahedra. The corresponding arsonate analogue α-$[Mo_4O_{10}(PhAsO_3)_4]^{4-}$ (**109**) has an identical core. Introduction of the *p*-toluenearsonate ligand causes structural changes that are illustrated by the species β-$[Mo_4O_{10}(C_7H_7AsO_3)_4]^{4-}$ (**110**), shown in Fig. 74. The structure consists of two binuclear units $\{Mo_2O_3\}$ and $\{Mo_2O_5\}$, linked through two μ^2-bridging oxo groups to give a $\{Mo_4O_4\}$ cyclic core. Two organoarsonate groups serve to bridge the $\{Mo_2O_3\}$ and $\{Mo_2O_5\}$ units in a symmetrical bidentate mode, while a third symmetrically bridges the Mo sites of the $\{Mo_2O_3\}$ unit. The fourth ligand acts as a tridentate moiety symmetrically bridging the Mo centers of the $\{Mo_2O_5\}$ unit and using the third oxygen donor in a bridging mode between the Mo sites of the $\{Mo_2O_3\}$ grouping. The structure is unique in presenting two distinct binuclear Mo cores and three distinct ligand-binding modes.

In contrast to the products of reactions of Mo^{VI} precursors with REO_3H_2 (E

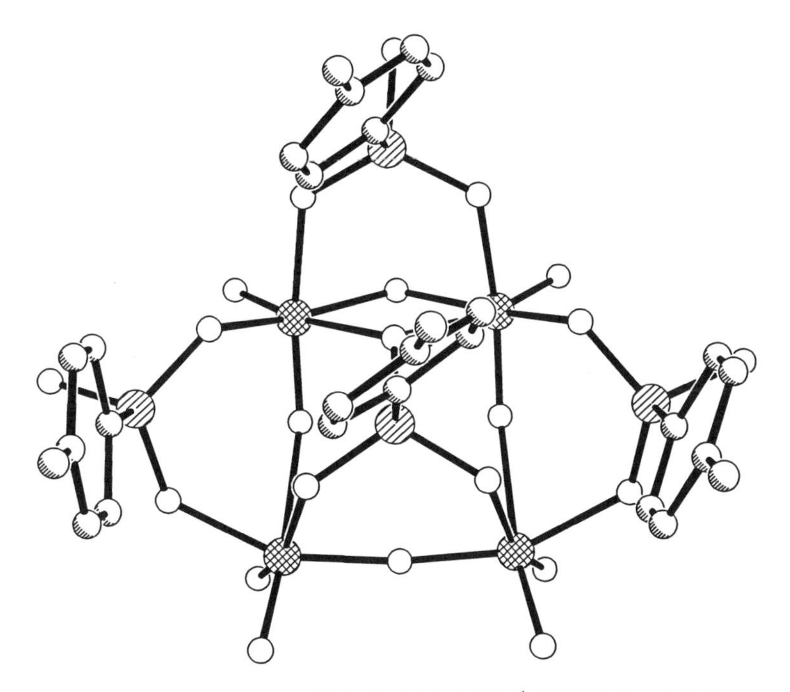

Figure 74. The structure of $[Mo_4O_{10}(C_7H_7AsO_3)_4]^{4-}$ (**110**) ball-and-stick.

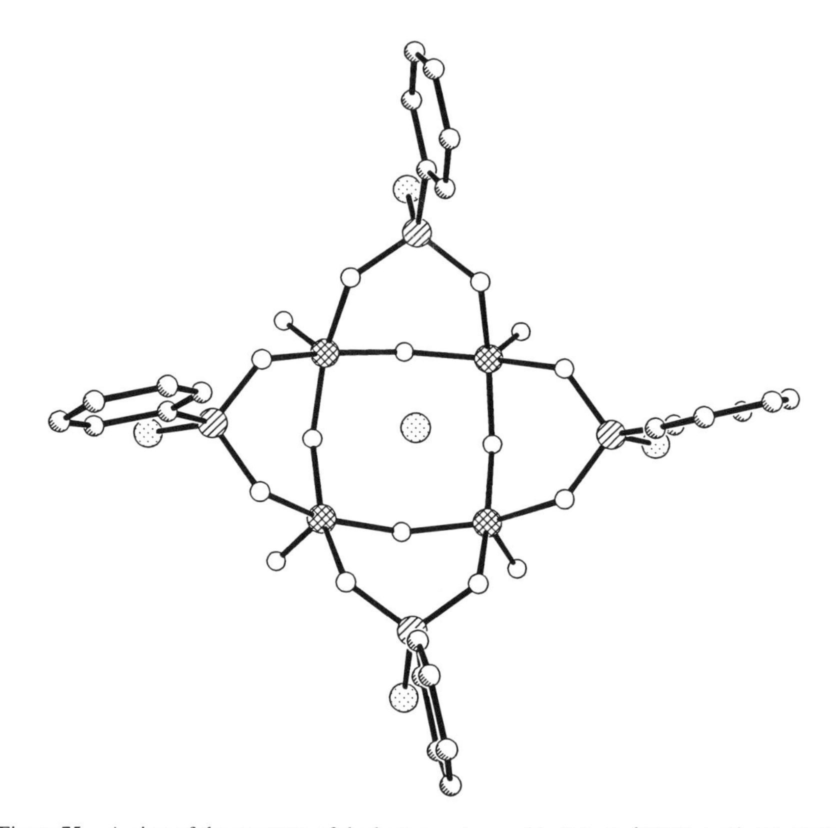

Figure 75. A view of the structure of the host–guest assembly $[Mo_4O_8\{(Cl)O_2AsR\}_4Cl]$ (**106**).

= P or As) ligand types, the reactions of $MoCl_5$ with p-$MeC_6H_4AsO_3H_2$ resulted in two unusual cluster types. The structure of the cluster $[Mo_4O_8\{MeC_6H_4AsO_2(Cl)\}_4Cl]$ (**106**), shown in Fig. 75, consists of a ring of four corner-sharing $\{MoO_5\}$ square pyramids forming a $\{Mo_4O_4\}$ unit. Each $RAsO_2(Cl)$ ligand bridges two Mo centers as a symmetrical bidentate donor. All four ligands project in the same direction to produce a bowl-like structure, which is occupied by a central Cl^- anion. The ligand $(p$-$MeC_6H_4AsO_2Cl)^-$, is apparently synthesized in a metal-mediated chlorination during the course of the reaction. The analogous cluster $[Mo_4O_8\{PhAsO_2(Cl)\}_4Cl]$ $[MoOCl_4]$ (**106a**) has an essentially identical structure, but possesses, in addition, a capping $[MoOCl_4]$ group.

3. *Pentanuclear Clusters*

The various pentamolybdate clusters incorporating different organophosphonate ligands that have been synthesized can be represented by the general

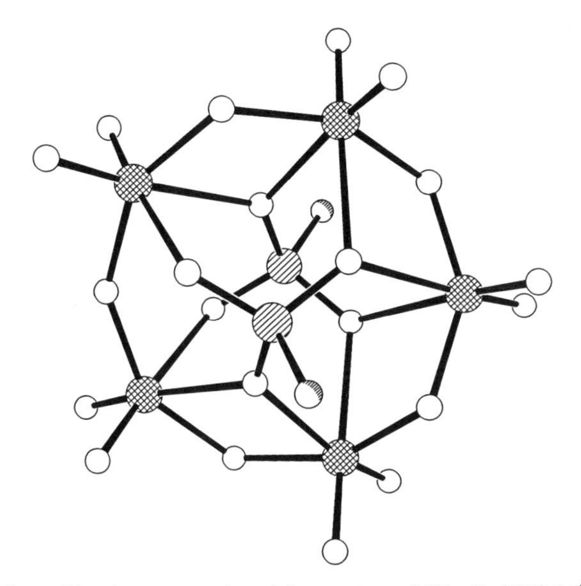

Figure 76. A representation of the structure of $[Mo_5O_{15}(RPO_3)_2]^{4-}$.

formulation $[Mo_5O_{15}(RPO_3)_2]^{n-}$ (**111**) (231) (R = $-$Me, $-$Ph: $n = 4$; R = $-C_2H_4NH_3$, $-CH_2N(CH_2)_4O$: $n = 2$) (7, 231, 232). The structure of the prototypical cluster for this class of compound, $[Mo_5O_{15}(PhPO_3)_2]^{4-}$, is illustrated in Fig. 76. The anion, which is chiral, is composed of five $\{MoO_6\}$ octahedra joined together by edge and corner sharing. This arrangement generates a cyclic ring, each face of which is capped by a tetrahedral RPO_3 group, two of whose donor oxygen atoms are shared by two adjacent Mo centers, and one oxygen shared by one Mo center of the ring. The organic groups of the ligands project out of the surface of spherical structure. This common structural core in heteropolyanion chemistry is shared by a variety of species of the $[Mo_5O_{21}X]$ type.

On the basis of the larger covalent radius of arsenic (9), it had been argued that the arsonate analogue of $[Mo_5O_{15}(RPO_3)_2]^{4-}$ would not exist. Contrary to the argument, the cluster $[Mo_5O_{15}(C_3H_2AsO_3)_2]^{4-}$ (**111a**) has been isolated and its structure has been shown to be analogous to the phosphonate derivative (100). While the O· · ·O distances around the $\{Mo_5O_{21}\}$ unit were considered to be too small to accommodate the arsonate group (233), the structure of the arsonate analogue demonstrates that the expansion of the $\{Mo_5O_{21}\}$ ring accompanied by the distortion of the $RAsO_3$ tetrahedron satisfies the steric requirements for the formation of the organoheteropolyanion.

4. *Hexanuclear Clusters*

Variations in the synthetic conditions such as the pH of the reacting mixture, the nature of the substituent, and the linking of the phosphonate groups lead

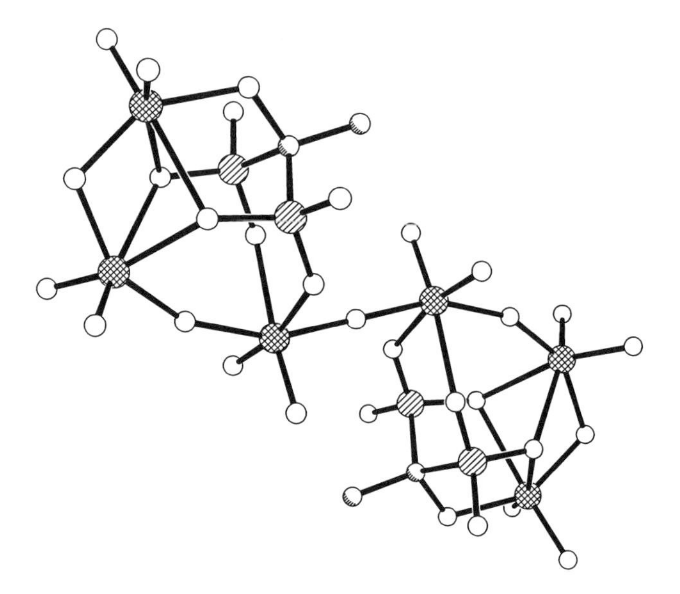

Figure 77. The structure of $[Mo_6O_{17}\{HOCH(PO_3)_2\}_2]^{6-}$ (**115**).

to further aggregation to produce hexamolydophosphonates $[Mo_6O_{17}\{HOCH(PO_3)_2\}_2]^{6-}$ (**115**) (234) and $[Mo_6O_{17}\{OCH(PO_3)_2\}_2]^{8-}$ (**116**) (228). As schematically represented in Fig. 77, the hexanuclear species consists of two trinuclear $\{Mo_3O_8\{(H)OCH(PO_3)_2\}\}$ fragments, similar to that present in $[Mo_3O_9\{(O)CMe(PO_3)_2\}]$ (228) described previously, bridged through an $\{Mo{-}O{-}Mo\}$ interaction.

The pronounced influence of synthetic conditions on the nature of the final products is amply demonstrated by the use of hydrothermal synthetic methods to prepare a series of phenylphosphonatohexamolybdate clusters $[M\{Mo_6O_{15}(PhPO_3H)_3(PhPO_3)]_2\}^{9-}$ (**117**) (M = Na or K) (235–237) containing reduced binuclear Mo^V sites hitherto unknown in $Mo/O/O_3PR$ systems. The structure of the anion $[Na\{Mo_6O_{12}(OH)_3(PhPO_3)_4\}_2]^{9-}$, shown in Fig. 78, consists of two hexanuclear oxomolybdenum(V)–organophosphonate clusters linked by a sodium cation. The hexametallic moiety $\{Mo_6O_{24}\}$ forms a cyclic core, constructed from edge sharing of $\{MoO_6\}$ octahedra. The Mo^V centers of the core are associated in strongly interacting pairs so as to produce an alternating pattern of short and long $Mo\cdots Mo$ contacts, characterized by distances of 2.58 and 3.58 Å, respectively. This cyclic core is reminiscent of the Anderson structure with the central atom removed (28) and appears to be a recurrent theme of the structural chemistry of reduced oxomolybdenum clusters, having been previously observed in the oxomolybdenum alkoxide cluster $(Me_3NH)_2$-

$(Et_4N)Na_4[Na(H_2O)_3H_{15}Mo_{42}O_{109}(OCH_2)_3CCH_2OH\}_7]\cdot 15H_2O$ (227). As shown in Fig. 78 there are two types of organophosphonate groups associated with this hexamolybdenum core in the $[Mo_6O_{12}(OH)_3(PhPO_3)_4)]^{5-}$ unit. The central RPO_3^{2-} group provides three oxygen bridges so as to span each pair of Mo centers with long Mo· · ·Mo distances. The three remaining organophosphonate groups assume μ_2-$O_2P(O)R$ geometry so as to bridge pairs of nonbonded Mo centers; the uncoordinated oxygen of these units is not protonated but present as a pendant {P=O} group. Careful consideration of bond lengths and valence sums identifies the doubly bridging oxo groups as the protonation sites. As shown in Fig. 78(*b*) the organophosphonate groups assume a syn conformation relative to the hexamolybdenum plane, an orientation that results in a double layer of polyhedra: one layer of phosphonate tetrahedra and a second of molybdenum octahedra.

One Na^+ cation is sandwiched between a pair of $[Mo_6O_{12}(OH)_3(PhPO_3)_4]^{5-}$ moieties, which present the faces defined by the hexamolybdenum rings in a

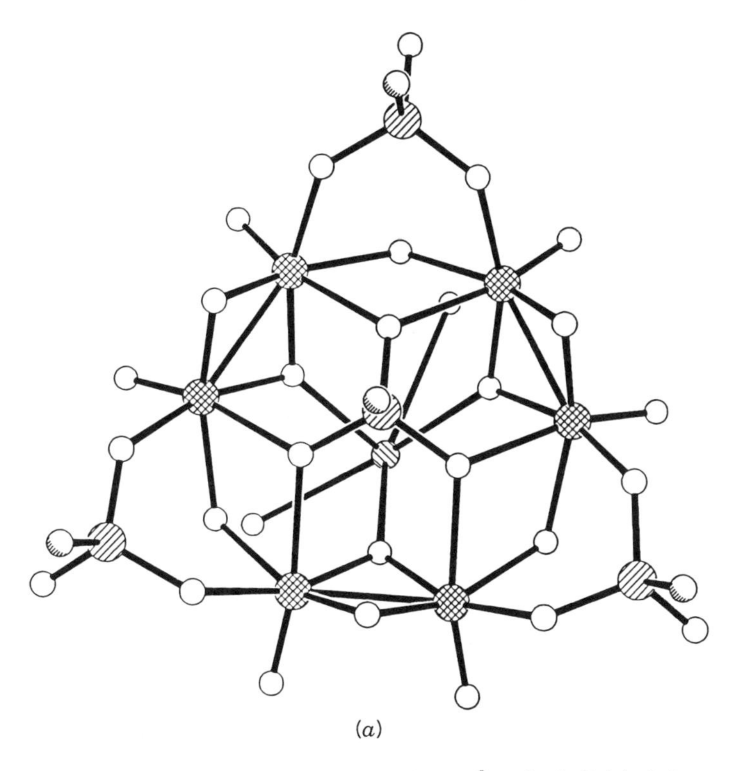

Figure 78. (*a*) The structure of the $[Mo_6O_{12}(OH)_3(PhPO_3)_4]^{5-}$ unit. (*b*) Polyhedral representation of the "sandwich" cluster $\{Na[Mo_6O_{12}(OH)_3(PhPO_3)_4]_2\}^{9-}$.

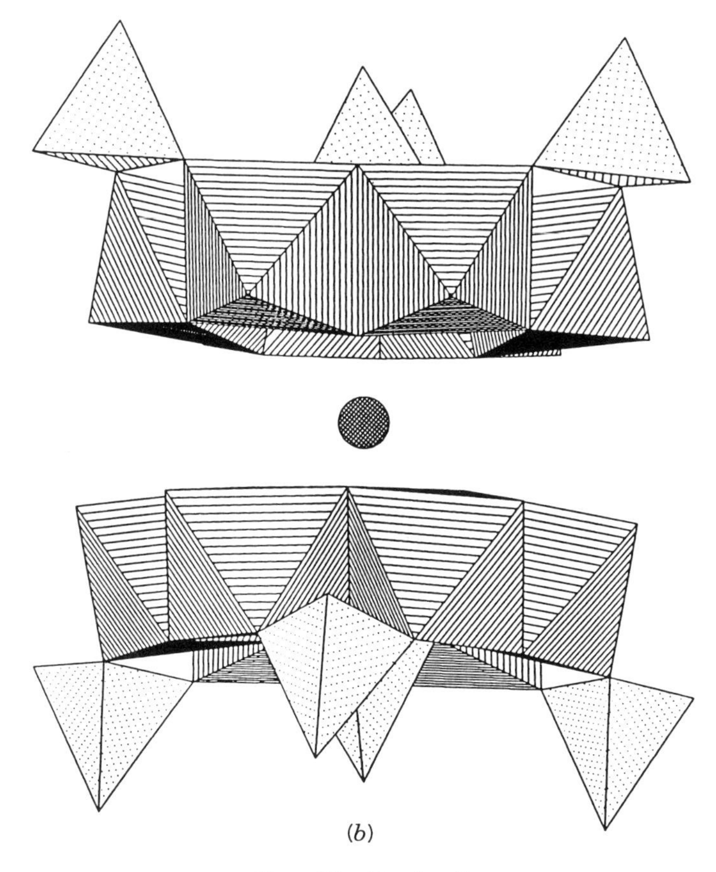

(*b*)

Figure 78. (*Continued*)

staggered orientation so as to provide an octahedral cavity for the Na^+ cation. The Na—O distances associated with this site are relatively short, suggesting that the formation of the sandwiched cation core is crucial to isolation of the polyanion. As illustrated in Fig. 78(*b*), a second Na^+ cation serves to bridge adjacent pairs of $\{Na[Mo_6O_{12}(OH)_3(PhPO_3)_4]_2\}^{9-}$ units to generate double ribbons with the phenyl groups disposed about the exterior.

The potassium analogue (237) exhibits interesting modifications in the crystal packing arrangements attributable to the large ionic radius of K^+ as compared to that of Na^+. While the structure of the hexanuclear units remains unchanged, the K^+ cation is no longer sandwiched between two perfectly overlapping hexanuclear units. Instead one of the hexanuclear units slides off

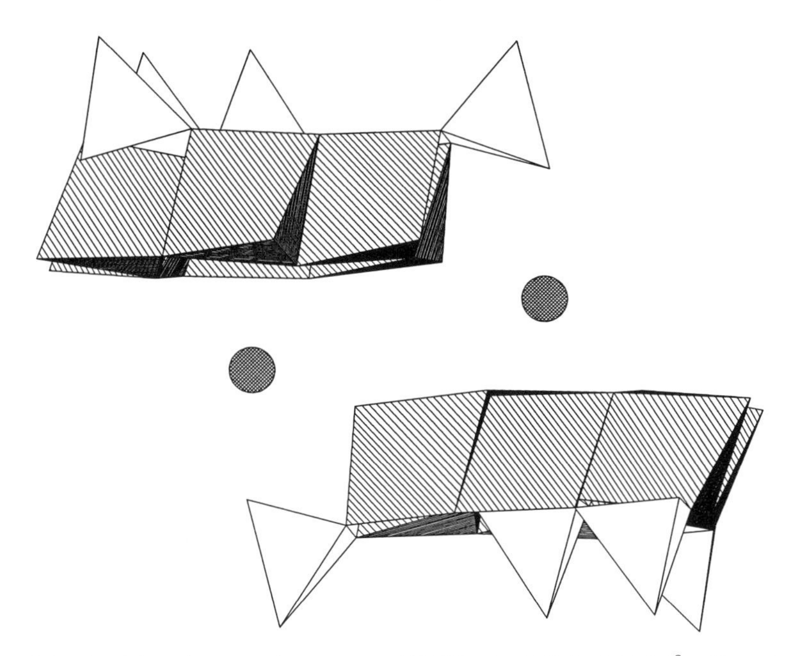

Figure 79. A polyhedral view of the structure of $\{K_2[Mo_6O_{12}(OH)_3(PhPO_3)_4]_2\}^{8-}$, illustrating the "slippage" of the $[Mo_6O_{12}(OH)_3(PhPO_3)_4]^{5-}$ planes to accommodate one K^+ cation per unit.

slightly to accommodate the larger size of K^+, which in this case is also bonded to two water molecules, as shown in Fig. 79. Long-range interactions with other six oxygen atoms arise from the proximity of three oxygen donors on the faces of each of the two hexanuclear units.

The reaction conditions employed to prepare organophosphonate pentamolydate clusters can be exploited to synthesize hexametalate species of types $[Mo_6O_{18}(RAsO_3)_2]^{4-}$ (**112**) (9) (R = —Me, —Ph, *p*-$C_6H_4NH_2$). The prototypical structure of $[Mo_6O_{18}(RAsO_3)_2]^{2-}$, shown in Fig. 80 consists of a ring of six edge-sharing $\{MoO_6\}$ octahedra, capped above and below by $\{O_3AsR\}$ tetrahedra. A hydrated form of this cluster $[Mo_6O_{18}(H_2O)(PhAsO_3)_2]^{4-}$ (**113**) (R = —Ph), in which a bridging aquo ligand is inserted between two of the molybdenum atoms (Fig. 81), has been isolated and characterized (8). Consequently, the structure consists of four edge-sharing $\{MoO_6\}$ octahedra linked by two corner-sharing interactions to the aqua bridged binuclear unit of face-sharing $\{MoO_6\}$ octahedra. The unusual cluster $[Mo_6O_{18}(H_2O)_6(MeAsO_3)]^{2-}$ (**114**) has been prepared at pH 2 (238). As shown in Fig. 82 the structure consists of three pairs of edge-sharing molybdenum octahedra, fused by corner-sharing interactions into a $\{Mo_6O_6\}$ ring. The donor oxygen atoms of the

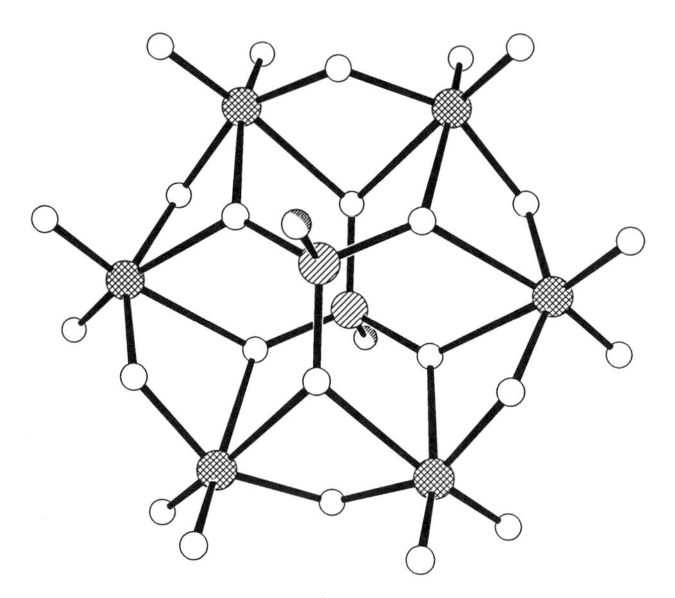

Figure 80. The prototypical $[Mo_6O_{18}(REO_3)_2]^{4-}$ (E = P or As) structure, consisting of a ring of edge-sharing $\{MoO_6\}$ octahedra.

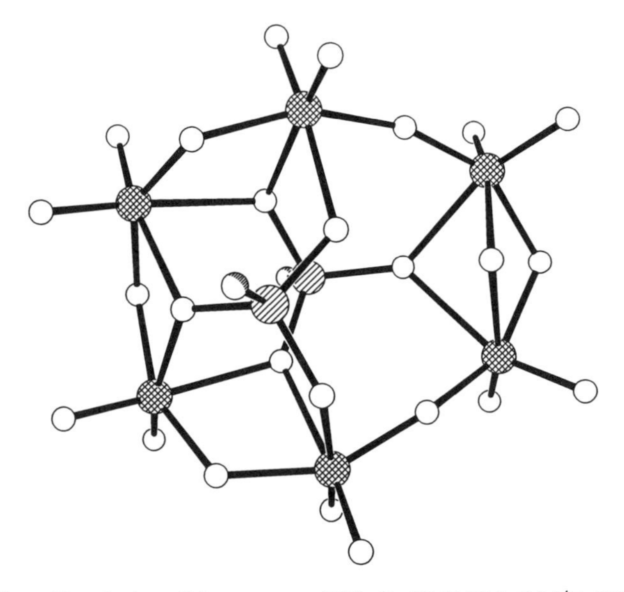

Figure 81. A view of the structure of $[Mo_6O_{18}(H_2O)(PhAsO_3)_2]^{4-}$ (**113**).

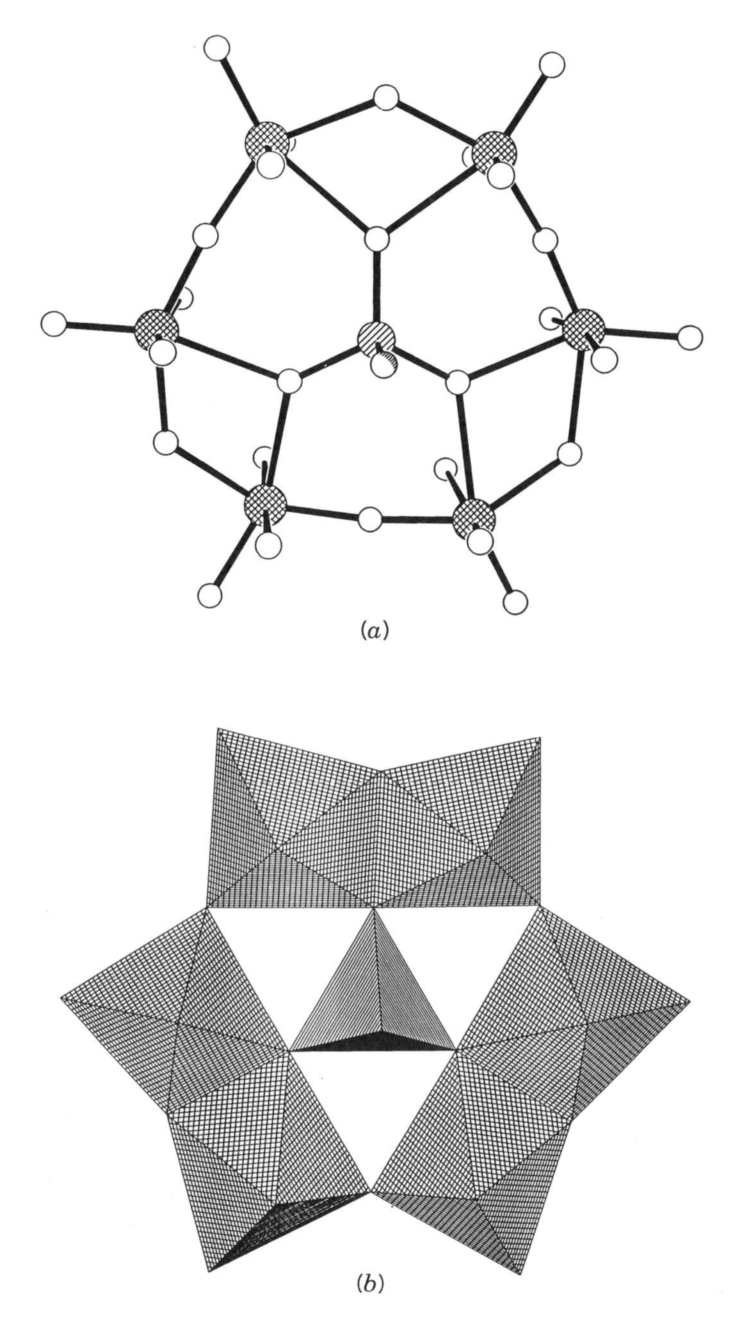

Figure 82. (*a*) Ball-and-stick and (*b*) polyhedral representations of the structure of $[Mo_6O_{18}(H_2O)_6(MeAsO_3)]^{2-}$ (**114**).

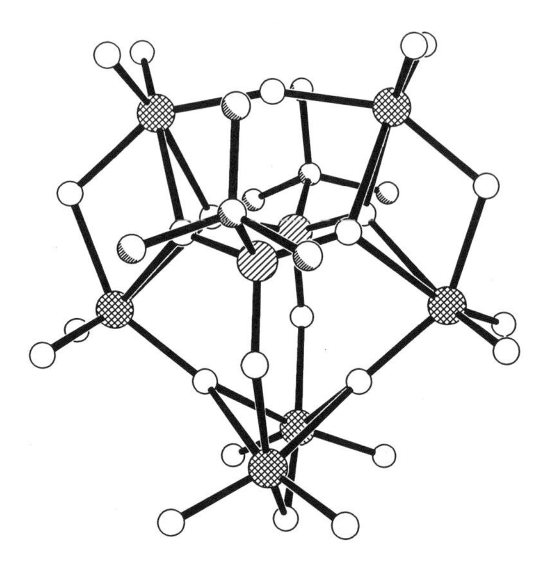

Figure 83. A view of the structure of $[Mo_6O_{18}(t\text{-}BuPO_3)_2]^{4-}$.

$MeAsO_3^{2-}$ group assume triply bridging modes and provide one bridging oxygen to each pair of edge-sharing Mo octahedra. A curious feature of the chemistry of the $Mo/O/REO_3$ system is the dependence of the structure on the substituent. Thus, when $t\text{-}BuPO_3^{2-}$ is employed as the ligand, the cluster $[Mo_6O_{18}(t\text{-}BuPO_3)_2]^{2-}$ exhibits the structure shown in Fig. 83, which is quite distinct from that of the $[Mo_6O_{18}(RAsO_3)_2]^{2-}$ prototype of Fig. 80. Rather than a ring of edge-sharing octahedra, $[Mo_6O_{18}(t\text{-}BuPO_3)_2]^{2-}$ consists of three sets of face-sharing octahedra linked by corner-sharing interactions.

Further lowering the pH of the reacting solution (pH < 1) yields the dodecamolybdate cluster $[Mo_{12}O_{34}(H_3NC_6H_4AsO_3)_4]$ (**119**), isolated as a zwitterionic species (6d). Its structure, illustrated in Fig. 84, is made up of four groups of three edge-sharing $\{MoO_6\}$ octahedra, bridged by four $\{RAsO_3\}$ units. The core of the cluster $[Mo_{12}As_4O_{48}]$ adopts a structure similar to that observed in $[Mo_{12}O_{34}(AsO_4H)_4]^{4-}$ (239) and may be described as an inverted Keggin structure.

While the chemistry of $Mo/O/REO_3$ system (E = P or As) is still not well developed, the fragmentary reports on the synthesis and characterization of novel structural cores indicate a very rich coordination chemistry. Furthermore, the polymolybdate cores incorporating organodiphosphonates and diarsonates are virtually unknown. The emergent chemistry of the vanadium system strongly suggests the possibility of similar diversity of structural types for the diphosphonate and diarsonates of the polymolybdate system.

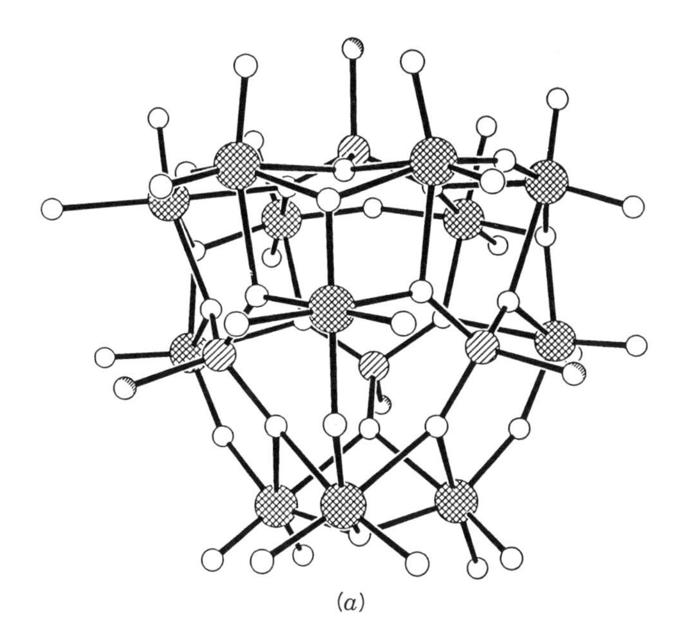

(*a*)

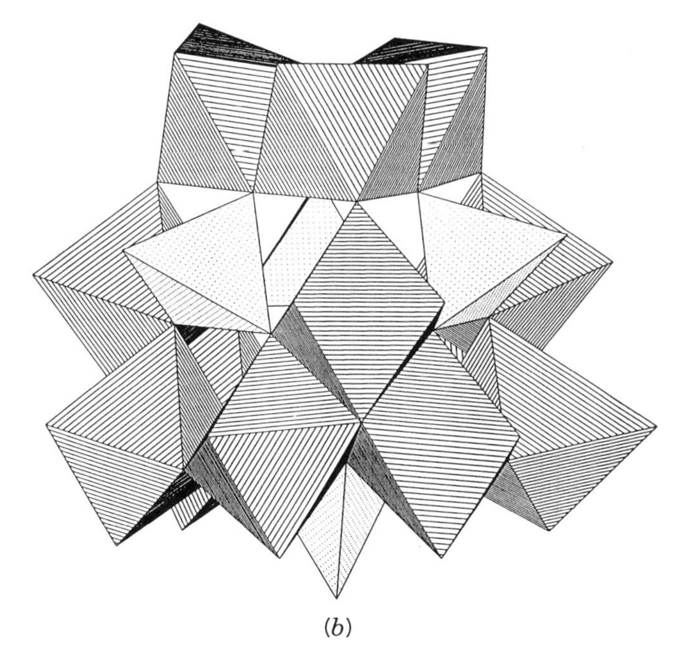

(*b*)

Figure 84. (*a*) A view of the core of the $[Mo_{12}O_{34}(RAsO_3)_4]$ cluster. (*b*) A polyhedral representation.

VIII. CONCLUSIONS

The necessarily narrow focus of this chapter serves to emphasize the scope of oxometalate chemistry. The oxovanadium and oxomolybdenum compounds with oxygen-donor ligands that comprise the subject of this chapter represent only a minor fraction of the coordination compounds of polyoxometalates. The "classical" isopolyanions and heteropolyanions provide thousands of additional examples of clusters containing the characteristic multiple metal–oxygen terminal bonds. Furthermore, a rapidly expanding chemistry of oxometalate clusters supporting organometallic subunits has begun to emerge in the past two decades.

In addition to a range of compositional and structural types, the general class of polyoxometalate materials possesses properties and applications of relevance to catalysis, medicine, and materials sciences. The remarkable extensiveness of the field can only be appreciated by examining the numerous recent reviews that address various aspects of structure, reactivity, and applications (28–33).

Although systematic studies of the subfield of polyoxometalate coordination chemistry are of recent origin, the growth of the area in the past decade has been dramatic. However, there are yet great opportunities for the realization of interesting chemistry. While a multitude of new structural types have been characterized, the process of formation of polyoxometalates remains enigmatic and is often referred to as "self-assembly." Even for relatively simple systems as aqueous vanadate hydrolysis, spectator counterions may participate in and control equilibria. Similarly, control of aggregation has not been achieved to allow the construction of clusters of a given nuclearity or molecular architecture. Perhaps the time is ripe for the design of ligands that constrain the aggregation process or extend the nuclearity. Ligands developed from arboral technology or from the bifunctional ligand design, for example, may provide a starting point for the molecular engineering of novel polyoxometalate clusters and even oxometal containing solid phases.

Several emergent themes require further elaboration. Oxometalate cluster chemistry has revealed a new direction in the development of structural types from the "classical" cages and clusters to novel baskets, belts, bowls, barrels, and soccer balls. These unusual structural types often reflect the organizational influence of a template, providing either shell-like clusters with the directing agent situated within the cavity or aggregates with the template acting as a ligand incorporated on the outer surface of the shell. Templates as varied as halides and pseudohalides, oxoanions, alkylammonium cations, carboxylates, alkali metal cations, water, and neutral organic molecules have been encapsulated within oxometalate cluster shells. While the mechanism of the template effect is not well understood, the organization of such shells depends on the size and shape of the enclosed species and on charge balance considerations.

Systematic studies of the consequences of introducing various templating reagents have only recently been undertaken.

While the mixed-valence chemistry of polyoxometalate coordination clusters has developed sporadically, recent results suggest a rich chemistry. The wide range of oxidation states available to vanadium and molybdenum and their variable coordination numbers and geometries provide an unusual potential for the existence of novel mixed-valence clusters. Furthermore, the recent description of a family of clusters containing oxomolybdenum $Mo^{V}—Mo^{V}$ binuclear units presents a synthetic challenge in the development of clusters and even solid state materials based on the $[Mo_2O_8L_2]^{n-}$ structural motif.

The development of the chemistry of the $M/O/EO_4^{n-}$ (E = P or As) and $M/O/REO_3^{n-}$ systems has revealed that heterometal centers may be incorporated to provide both clusters, $[M_xM'_yO_z(EO_4)_n]$, and solid phases of unusual compositions and properties. While such results are encouraging, the systematic chemistry remains unexplored.

The exploitation of hydrothermal techniques in the preparation of oxometalate ligand compounds has allowed a dramatic expansion of both the cluster and solid state chemistry of these materials. Unusual high nuclearity superclusters and "clusters of clusters" have been characterized and naturally pose the question of the relationship of such materials to extended solid phases. While molecular clusters are of limited 3D extension, their molecular structures and solid state packing should reflect some features of similarly composed solids. There is the necessary caveat that the consideration of clusters as structural models of extended structure phases may be superficial and restricted to the consideration of local geometry and nearest-neighbor interactions. However, such a view does provide a useful model for the structures of metal oxides and their reactivities with small substrate molecules (38, 240).

The question of the potential role of molecular clusters as a synthetic precursor in the design of new materials, exploiting hydrothermal or solvothermal techniques or hydrolysis–condensation processes naturally also arises. A major challenge facing the synthetic chemist is the preparation of solid materials designed to display a specific set of physical properties. While it is tempting to propose that preformed clusters may be incorporated into or fused via hydrolysis–condensation reactions into extended phases of predictable structures, the evidence is sketchy for such structural preassembly in solvothermal reactions. On the other hand, it is evident that the use of solvothermal techniques, template organization of cluster shells or solid frameworks, and introduction of appropriate ligand types has produced a variety of novel solid phase materials. It is intriguing that many of these extended phases contain small oxometalate aggregates $[M_xO_y]^{n-}$ (with y in the range 2–5) reminiscent of oxometalate cluster structures. Curiously, the solids of the M/O/oxoligand classes are characterized by either infinite {M—O—M} chains or $[M_xO_y]$ clusters of nuclearity 5 or less.

Since nuclearities of 18 and greater are common to oxometalate structures, can larger clusters be incorporated into the extended frameworks? Can oxometalate clusters be linked by appropriate ligands into solid networks? Since the functional groups and side chains of the ligand connectors can be modified, the properties of such materials could be fine-tuned in a more or less rational fashion to provide materials with donor–acceptor interactions, electronic conductivity, unusual optical properties, and extended multicentered magnetic interactions (241).

ACKNOWLEDGMENTS

Much of the work described in this chapter was funded in our laboratories by NSF Grants CHE9119910 and CHE9318824 to JZ. We thank Dr. Qin Chen, Victoria Soghomonian, José Salta, and Yuan Da Chang, without whose skills in synthesis and structural characterization this chapter would have proved narrow indeed, and Dr. Robert C. Haushalter of NEC Research Institute, our hydrothermal guru, coinvestigator in the development of the solid-phase chemistry, and constant source of new ideas and directions. We also thank Dr. E. M. McCarron III for providing the coordinates of his early and prototypical structural types.

REFERENCES

1. N. N. Greenwood and A. Earnshaw, *Chemistry of the Elements*, Pergamon, New York, 1984.
2. J. Berzelius, *Pogg. Ann.*, *6*, 369, 380 (1826).
3. A. J. Miolati, *J. Prakt. Chem.*, *77*, 439 (1908).
4. A. Rosenheim and R. Bilecki, *Berichte*, *4*, 543 (1913).
5. R. D. Adams, W. G. Klemperer, and R.-S. Liu, *J. Am. Chem. Soc. Chem. Commun.* 256 (1979).
6. (a) K. M. Barkigia, L. M. Rajkovic-Blazer, M. T. Pope, and C. O. Quicksall, *J. Am. Chem. Soc.*, *97*, 4146 (1975); (b) K. M. Barkigia, L. M. Rajkovic-Blazer, M. T. Pope, E. Prince, and C. O. Quicksall, *Inorg. Chem.*, *19*, 2531 (1980); (c) K. Y. Matsumoto, *Bull. Chem. Soc. Jpn.*, *52*, 3284 (1979); (d) K. M. Barkigia, L. M. Rajkovic-Blazer, M. T. Pope, and C. O. Quicksall, *Inorg. Chem.*, *20*, 3318 (1981).
7. W. Kwak, M. T. Pope, and T. F. Scully, *J. Am. Chem. Soc.*, *97*, 5735 (1975).
8. W. Kwak, L. M. Rajkovic, M. T. Pope, C. O. Quicksall, K. Y. Matsumoto, and Y. Sasaki, *J. Am. Chem. Soc.*, *99*, 6463 (1977).
9. W. Kwak, L. M. Rajkovic, J. K. Stalick, M. T. Pope, and C. O. Quicksall, *Inorg. Chem.*, *15*, 2778 (1976).

10. V. W. Day and W. G. Klemperer, *Science*, *228*, 533 (1985), and references cited therein.
11. A. Müller, R. Rohlfing, E. Krickemeyer, and H. Boggë, *Angew. Chem. Int. Ed. Engl.*, *32*, 909 (1993), and references cited therein.
12. A. Müller, K. Hovemeier, and R. Rohlfing, *Angew. Chem. Int. Ed. Engl.*, *31*, 1192 (1992).
13. H. Reuter, *Angew. Chem. Int. Ed. Engl.*, *31*, 1185 (1992).
14. H. K. Chae, W. G. Klemperer, D. E. Páez Loyo, V. W. Day, and T. A. Eberspache, *Inorg. Chem.*, *31*, 3182 (1992).
15. G. H. Huan, A. J. Jacobson, and V. W. Day, *Angew. Chem.*, *103*, 426 (1991); *Angew. Chem., Int. Ed. Engl.*, *30*, 422 (1991).
16. G. Huan, V. W. Day, A. J. Jacobson, and D. P. Goshorn, *J. Am. Chem. Soc.*, *113*, 3188 (1991).
17. M. J. Abrams, C. E. Costello, S. N. Shaikh, and J. Zubieta, *Inorg. Chim. Acta*, *180*, 9 (1991).
18. Y. Du, A. L. Rheingold, and E. A. Maatta, *J. Am. Chem. Soc.*, *114*, 345 (1992).
19. A. Proust, P. Gouzerh, and F. Robert, *Inorg. Chem.*, *32*, 5291 (1993), and references cited therein.
20. Q. Chen and J. Zubieta, *Coord. Chem. Rev.*, *114*, 107 (1992).
21. T. M. Che, V. W. Day, L. C. Francesconi, W. G. Klemperer, D. J. Main, A. Yagosaki, and O. M. Yaghi, *Inorg. Chem.*, *31*, 2920 (1992) and references cited therein.
22. F. Bottomley and P. D. Boyle, *J. Organometal. Chem.*, *13*, 370 (1994), and references cited therein.
23. Y. Do, X.-Z. You, C. Zhang, Y. Ozawa, and K. Isobe, *J. Am. Chem. Soc.*, *113*, 5892 (1991).
24. A. R. Siedle and R. A. Newmark, *J. Am. Chem. Soc.*, *111*, 2058 (1989), and references cited therein.
25. Y. Lin, K. Nomiya, and R. G. Finke, *Inorg. Chem.*, *32*, 6040 (1993), and references cited therein.
26. M. T. Pope, *Heteropoly and Isopoly Oxometalates*, Springer, New York, 1983.
27. M. T. Pope and A. Müller, *Angew. Chem. Int. Ed. Engl.*, *30*, 34 (1991).
28. M. T. Pope, *Progress in Inorganic Chemistry*, Wiley-Interscience, New York, 1991, Vol. 39, p. 181; M. T. Pope, in *Comprehensive Coordination Chemistry*, G. Wilkinson, R. D. Gillard, and J. A. McCleverty, Eds., Pergamon, Oxford, UK, 1987, Chapter 38, p. 1023.
29. M. T. Pope and A. Müller, Eds., *Polyoxometalates: From Platonic Solids to Anti-Retroviral Activity*, Kluwer Academic Press, Dordrecht, The Netherlands, 1994.
30. W. G. Klemperer, T. A. Marguart, and O. M. Yaghi, *Angew. Chem. Int. Ed. Engl.*, *31*, 49 (1992).
31. H. K. Chae, W. G. Klemperer, and T. A. Marquart, *Coord. Chem. Rev.*, *128*, 209 (1993).

32. J. Zubieta, in *Polyoxometalates: From Platonic Solids to Anti-Retroviral Activity*, M. T. Pope and A. Müller, Eds., Kluwer Academic Press, Dordrecht, The Netherlands, 1994.
33. J. Zubieta, *Comments Inorg. Chem.*, *15*, 193 (1994).
34. I. M. Campbell, *Catalysis at Surfaces*, Chapman & Hall: New York, 1988.
35. F. Delanny, Ed., *Characterization of Heterogeneous Catalysts*, Marcel-Dekker, New York, 1984.
36. D. C. Bradley, R. C. Mehrotra, and D. P. Gaur, *Metal Alkoxides*, Academic, New York, 1978.
37. R. C. Mehrotra and D. P. Gaur, *Metal Alkoxides*, in *Advances in Inorganic Radiochemistry*, H. G. Emeleus and A. G. Sharpe, Eds., Academic, New York, 1983, Vol. 26, p. 269.
38. M. H. Chisholm, in *Chemistry Toward the 21st Century*, ACS, Washington, DC, 1983, p. 243.
39. M. H. Chisholm, in *Comprehensive Coordination Chemistry*, G. Wilkinson and J. McCleverty, Eds., Pergamon, Oxford, UK, 1987.
40. D. C. Bradley, *Chem. Rev.*, *89*, 1317 (1989).
41. K. G. Caulton and J. G. Hubert-Pfalzgraf, *Chem. Rev.*, *90*, 969 (1990).
42. P. P. Power, *Comments Inorg. Chem.*, *8*, 117 (1989).
43. I. P. Rothwell, *Acc. Chem. Res.*, *21*, 153 (1988).
44. R. E. LaPointe, P. T. Wolczanski, and J. F. Mitchell, *J. Am. Chem. Soc.*, *108*, 6382 (1986).
45. L. G. McCullough, P. R. Schrock, J. C. Dewan, and J. C. Murdzek, *J. Am. Chem. Soc.*, *107*, 5907 (1985).
46. W. E. Buhro and M. H. Chisholm, *Adv. Organomet. Chem.*, *27*, 311 (1987).
47. H. E. Bryndze and W. Tam, *Chem. Rev.*, *26*, 269 (1983).
48. See, for example: C. E. Rehbert, *Org. Syn.*, *26*, 18 (1946).
49. R. C. Mehrotra, *J. Non-Cryst. Solids*, *100*, 1 (1988).
50. C. J. Brinker, *J. Non-Cryst. Solids*, *100*, 31 (1988).
51. M. Guglidmin and G. Cartman, *J. Non-Cryst. Solids*, *100*, 16 (1988).
52. H. Schmidt, *J. Non-Cryst. Solids*, *100*, 51 (1988).
53. C. Sanchez, J. Livage, M. Henry, and F. Babonneau, *J. Non-Cryst. Solids*, *100*, 65 (1988).
54. C. N. R. Rao and J. Gopalakrishnan, *Acc. Chem. Res.*, *20*, 228 (1987).
55. F. Aldinger and H.-J. Kalz, *Angew. Chem., Int. Ed. Engl.*, *26*, 37 (1987).
56. D. R. Ulrich, in *Transformation of Organometallics into Common and Exotic Materials: Design and Activation*, R. Laine, Ed., Martinus Nijhoff: Dordrecht, The Netherlands, 1988, p. 207.
57. H. Dislich, *Angew. Chem., Int. Ed. Engl.*, *10*, 363 (1971).
58. C. G. Hubert-Pfalzgraf, *New J. Chem.*, 11, 663 (1987).
59. L. L. Hench and J. K. West, *Chem. Rev.*, *90*, 33 (1990).

60. E. T. Clark, P. R. Rudolf, A. E. Martell, and A. Clearfield, *Inorg. Chim. Acta*, *164*, 59 (1989).
61. P. R. Rudolf, E. T. Clark, A. E. Martell, and A. Clearfield, *J. Coord. Chem.*, *14*, 139 (1985).
62. B. Bujoli, P. Palvadeau, and J. Rouxel, *Chem. Mater.*, *2*, 582 (1990).
63. G. Cao, H. Lee, V. Lynch, and T. E. Mallouk, *Inorg. Chem.*, *27*, 2781 (1988).
64. G. H. Huan, A. J. Jacobson, J. W. Johnson, and E. W. Corcoran, Jr., *Chem. Mater.*, *2*, 92 (1990).
65. Y. Zhang and A. Clearfield, *Inorg. Chem.*, *31*, 2821 (1992).
66. M. B. Dines, R. E. Cooksey, and P. C. Griffith, *Inorg. Chem.*, *22*, 567 (1983).
67. G. Cao and T. E. Mallouk, *Inorg. Chem.*, *30*, 1434 (1991).
68. G. L. Rosenthal and J. Caruso, *Inorg. Chem.*, *31*, 3104 (1992).
69. D. A. Burwell, K. G. Valentine, J. H. Timmermans, and M. E. Thompson, *J. Am. Chem. Soc.*, *114*, 4144 (1992).
70. A. Clearfield, *Chem. Rev.*, *88*, 125 (1988).
71. R. C. Haushalter and L. A. Mundi, *Chem. Mater.*, *4*, 31 (1992).
72. K. J. Frink, R.-C. Wang, J. L. Colos, and A. Clearfield, *Inorg. Chem.*, *20*, 1438 (1991).
73. T. E. Mallouk and H. Lee, *J. Chem. Educ.*, *67*, 829 (1990).
74. A. Stein, S. W. Keller, and T. E. Mallouk, *Science*, *259*, 1558 (1993).
75. For examples of clusters incorporating ligands with both N and O donor groups see S.-G. Roh, A. Proust, P. Gouzerh, and F. Robert, *J. Chem. Soc. Chem. Commun.*, 837 (1993) and references cited therein.
76. For examples of cluster incorporating ligands with both S and O donor groups see S. Liu, X. Sun, and J. Zubieta, *J. Am. Chem. Soc.*, *110*, 3324 (1988).
77. P. Hagenmuller, *Preparative Methods in Solid State Chemistry*; Academic, New York, 1972; P. M. Keane, Y.-J. Lu, and J. A. Ibers, *Acc. Chem. Res.*, *24*, 223 (1991).
78. A. Rabenau, *Angew. Chem. Int. Ed. Engl.*, *24*, 1026 (1985).
79. W. S. Sheldrick and H.-G. Braunbeck, *Z. Naturforsch. B: Anorg. Chem.*, *47b*, 151 (1992) and references cited therein.
80. J. S. Lindsey, *New. J. Chem.*, *15*, 153 (1991).
81. D. L. Kepert, *Inorg. Chem.*, *8*, 1556 (1969).
82. K. H. Tytko and O. Glemser, *Adv. Inorg. Chem. Radiochem.*, *19*, 239 (1976).
83. L. Ma, S. Liu, and J. Zubieta, *Inorg. Chem.*, *28*, 175 (1989).
84. S. Liu, S. N. Shaikh, and J. Zubieta, *Inorg. Chem.*, *26*, 4305 (1987).
85. V. McKee and C. J. Wilkins, *J. Chem. Soc. Dalton Trans.*, 523 (1987).
86. R. N. Hilder and C. J. Wilkins, *J. Chem. Soc. Dalton Trans.*, 495 (1984).
87. C. B. Knobler, B. R. Penfold, W. T. Robinson, C. J. Wilkins, and S. M. Yong, *J. Chem. Soc. Dalton Trans.*, 284 (1980).

88. E. Gumaer, K. Lettko, L. Ma., D. Macherone, and J. Zubieta, *Inorg. Chem. Acta*, *179*, 47 (1991).
89. Q. Chen and J. Zubieta, *Angew. Chem. Int. Ed. Engl.*, *32*, 261 (1993).
90. M. I. Khan, A. Müller, S. Dillinger, H. Bögge, Q. Chen, and J. Zubieta, *Angew. Chem., Int. Ed. Engl.*, *32*, 1780 (1993).
91. W. N. Lipscomb, *Inorg. Chem.*, *4*, 132 (1965).
92. R. J. Butcher and B. R. Penfold, *J. Cryst. Mol. Struct.*, *6*, 13 (1976).
93. F. A. Cotton and R. C. Elder, *Inorg. Chem.*, *3*, 397 (1964).
94. J. Park, M. D. Glick, and J. L. Hoard, *J. Am. Chem. Soc.*, *91*, 301 (1969).
95. W. Hermann and K. Weighardt, *Polyhedron*, *5*, 513 (1986).
96. P. Schreiber, K. Weighardt, B. Nuber, and J. Weiss, *Polyhedron*, *8*, 1675 (1989).
97. L. Ma, S. Liu, and J. Zubieta, *Inorg. Chem.*, *28*, 175 (1975).
98. S. Liu, L. Ma, D. McGorwitz, and J. Zubieta, *Polyhedron*, *9*, 1541 (1990).
99. A. Müller, E. Krickemeyer, M. Penk, V. Wittenben, and J. Döring, *Angew. Chem. Int. Ed. Engl.*, *29*, 88 (1990).
100. W. Ming, Z. Peiju, L. Benyao, and G. Yidong, *Acta Crystallogr. Ser. C*, *44*, 1503 (1988).
101. A. Goiffen and B. Spinner, *Rev. Chim. Minér*, *12*, 316 (1975).
102. D. L. Kepert, *Inorg. Chem.*, *8*, 1556 (1969).
103. O. Nagano and Y. Sasaki, *Acta Crystallogr. Ser. B*, 35, 2387 (1979).
104. I. Lindquist, *Ark. Kemi*, *2*, 349 (1950).
105. H. T. Evans, *J. Am. Chem. Soc.*, *90*, 3275 (1968).
106. V. W. Day, W. G. Klemperer, and D. J. Maltbie, *J. Am. Chem. Soc.*, *109*, 2991 (1987), and references cited therein.
107. A. Hüllen, *Angew. Chem.*, *76*, 588 (1964).
108. D. Hou, K. S. Hagen, and C. L. Hill, *J. Am. Chem. Soc.*, *114*, 5864 (1992).
109. V. W. Day, W. G. Klemperer, and O. M. Yaghi, *J. Am. Chem. Soc.*, *111*, 5959 (1989).
110. D. Hou, K. S. Hagen, and C. L. Hill, *J. Chem. Soc. Chem. Commun.*, 427 (1993).
111. F. A. Cotton, G. E. Lewis, and G. N. Mott, *Inorg. Chem.*, *21*, 3127 (1982).
112. F. A. Cotton, G. E. Lewis, and G. N. Mott, *Inorg. Chem.*, *21*, 3316 (1982).
113. F. A. Cotton and W. Wang, *Inorg. Chem.*, *21*, 2675 (1982).
114. F. A. Cotton, M. W. Extine, L. R. Falvellow, D. B. Lewis, G. E. Lewis, C. A. Murillo, W. Schwotzer, M. Tomas, and J. M. Troup, *Inorg. Chem.*, *25*, 3525 (1986).
115. J. Catterick and P. Thorntos, *Adv. Inorg. Chem. Radiochem.*, *20*, 291 (1977).
116. V. L. Pecoraro, *Inorg. Chim. Acta*, *155*, 171 (1989).
117. R. Massart, R. Contant, J. M. Fruchart, J. P. Ciabrini, and M. Fournier, *Inorg. Chem.*, *16*, 2916 (1977).
118. R. Contant and J. P. Ciabrini, *J. Inorg. Nucl. Chem.*, *43*, 1525 (1981).

119. S. P. Harmallar, M. T. Pope, and M. A. Leparulo, *J. Am. Chem. Soc.*, *105*, 4286 (1983).
120. Y. Hou and C. L. Hill, *J. Am. Chem. Soc.*, *115*, 11823 (1993).
121. H. Reiskamp, P. Gietz, and R. Mattes, *Chem. Ber.*, *109*, 2090 (1976).
122. D. D. Heinrich, K. Folting, W. E. Streib, J. C. Huffmann, and G. Christou, *J. Chem. Soc. Chem. Commun.*, 1411 (1989).
123. C. G. Young, *Coord. Chem. Rev.*, *96*, 89 (1989).
124. W. Priebsch, D. Rehder, and M. von Oeynhausen, *Chem. Ber.*, *124*, 761 (1991).
125. I. D. Brown and K. K. Wu, *Acta Crystallogr. Ser. B*, *32*, 1957 (1976); I. D. Brown and R. D. Shannon, *Acta Crystallogr. Ser. A*, *29*, 266 (1973).
126. J. Fuchs, S. Mahjour, and J. Pickardt, *Angew. Chem. Int. Ed. Engl.*, *30*, 148 (1991).
127. Y.-D. Chang, Q. Chen, M. I. Khan, J. Salta, and J. Zubieta, *J. Chem. Soc. Chem. Commun.*, 1872 (1993).
128. D. C. Crans, R. W. Marshman, M. S. Gottlieb, O. P. Anderson, and M. M. Miller, *Inorg. Chem.*, *31*, 4939 (1992).
129. M. Mikuriya, T. Kotera, F. Adachi, and S. Bandoro, *Chem. Lett.*, 945 (1993).
130. R. C. Haushalter and L. A. Mundi, *J. Am. Chem. Soc.*, *113*, 6340 (1991).
131. M. I. Khan and J. Zubieta, *Angew. Chem. Int. Ed. Engl.*, *33*, 760 (1994).
132. V. Soghomonian, R. C. Haushalter, and J. Zubieta, unpublished results.
133. W. Bidell, V. Shklover, and H. Berke, *Inorg. Chem.*, *31*, 556 (1992).
134. D. C. Bradley, *Nature (London)*, *182*, 1211 (1958).
135. J. W. Johnson, A. J. Jacobson, W. M. Butler, S. E. Rosenthal, J. F. Brody, and J. T. Lewandowski, *J. Am. Chem. Soc.* *111*, 381 (1989).
136. W. G. Klemperer, T. A. Marquand, and O. M. Yaghi, *Angew. Chem. Int. Ed. Engl.*, *31*, 49 (1992).
137. H. K. Chae, W. G. Klemperer, and V. W. Day, *Inorg. Chem.*, *28*, 1424 (1989).
138. Y. Hazashi, Y. Ozawa, and K. Isobe, *Chem. Lett.*, 425 (1989).
139. Q. Chen and J. Zubieta, *J. Chem. Soc. Chem. Commun.*, 1180 (1993).
140. V. W. Day, W. G. Klemperer, and O. M. Yaghi, *J. Am. Chem. Soc.*, *111*, 4518 (1989).
141. D. Hou, G.-S. Kim, K. S. Hagen, and C. L. Hill, *Inorg. Chim. Acta*, *211*, 127 (1993).
142. Q. Chen and J. Zubieta, *Inorg. Chem.*, *29*, 1456 (1990).
143. H. R. Allcock, E. C. Bissell, and E. T. Shawl, *Inorg. Chem.* *12*, 2963 (1973).
144. J. Fuchs, W. Freiwald, and H. Hartl, *Acta Crystallogr. Ser. B*, *34*, 1764 (1978).
145. J.-Y. Kempf, M.-M. Rohmer, J.-M. Poblet, C. Bo, and M. Bénard, *J. Am. Chem. Soc.*, *114*, 1136 (1992).
146. Q. Chen, D. P. Goshorn, C. P. Scholes, X. Tan, and J. Zubieta, *J. Am. Chem. Soc.*, *114*, 4667 (1992).
147. M. T. Pope, *Inorg. Chem.*, *11*, 1973 (1972).

148. H. So and C. W. Lee, *Bull. Korean Chem. Soc.*, *11*, 115 (1990), and references cited therein.
149. Q. Chen and J. Zubieta, *Inorg. Chim. Acta*, *198–200*, 95 (1992).
150. K. Hegstschweiler, H. W. Schnalle, H. M. Streit, V. Gramlich, H. U. Hunt, and I. Erni, *Inorg. Chem.* *31*, 1299 (1992).
151. R. Schmid, A. Mosset, and J. Galy, *Inorg. Chim. Acta*, *179*, 167 (1991) and references cited therein.
152. M. I. Khan, Q. Chen, H. Höpe, S. Parkin, C. J. O'Connor, and J. Zubieta, *Inorg. Chem.*, *32*, 2929 (1993).
153. M. I. Khan, Q. Chen, and J. Zubieta, *Inorg. Chem.*, *31*, 1556 (1992).
154. D. Rehder, W. Priebsch, and M. von Oeynhausen, *Angew. Chem. Int. Ed. Engl.*, *28*, 1221 (1989).
155. T. C. Mak, P. Li, C. Zheng, and K. Huang, *J. Chem. Soc. Chem. Commun.*, 1597 (1986).
156. L. V. Boas and P. C. Pessoa, in *Comprehensive Coordination Chemistry*, G. Wilkinson, Ed., Oxford, UK, 1987, Vol. 3, p. 453.
157. M. I. Khan, Y.-D. Chang, Q. Chen, J. Salta, Y. S. Lee, C. J. O'Connor, and J. Zubieta, *Inorg. Chem.*, *33*, 6340 (1994), and references cited therein.
158. A. Bino, S. Cohen, and C. Hirtner-Wirguin, *Inorg. Chem.*, *21*, 429 (1982).
159. A. Müller, E. Krickemeyer, M. Penk, H.-J. Walberg, and H. Bögge, *Angew. Chem. Int. Ed. Engl.*, *26*, 1045 (1987).
160. Q. Chen, S. Liu, and J. Zubieta, *Inorg. Chem.*, *28*, 4434 (1989).
161. A. Müller, M. Penk, R. Rohlfing, E. Krickemeyer, and J. Döring, *Angew. Chem. Int. Ed. Engl.*, *29*, 926 (1990).
162. J.-P. Launay, Y. Jeannin, and M. Daoudi, *Inorg. Chem.*, *24*, 1052 (1985).
163. M. I. Khan, Q. Chen, D. P. Goshorn, H. Höpe, S. Parkin, and J. Zubieta, *J. Am. Chem. Soc.*, *114*, 3341 (1992).
164. M. I. Khan, Q. Chen, and J. Zubieta, *J. Chem. Soc. Chem. Commun.*, 305 (1992).
165. M. I. Khan, Q. Chen, D. P. Goshorn, and J. Zubieta, *Inorg. Chem.*, *32*, 672 (1993).
166. M. I. Khan, Y.-S. Lee, C. J. O'Connor, and J. Zubieta, *J. Am. Chem. Soc.*, *116*, 5001 (1994).
167. C. N. Caughlin, H. M. Smith, and K. Watenpaugh, *Inorg. Chem.*, *5*, 2131 (1966).
168. W. Priebsch and D. Rehder, *Inorg. Chem.*, *29*, 3013 (1990).
169. F. Hillerns, F. Olbrich, U. Behrens, and D. Rehder, *Angew. Chem. Int. Ed. Engl.*, *31*, 447 (1992).
170. D. C. Crans, R. A. Feltz, O. P. Anderson, and M. M. Miller, *Inorg. Chem.*, *32*, 247 (1993).
171. T. R. Gibson, I. M. Thom-Postlethwaite, and M. Webster, *J. Chem. Soc. Dalton Trans.*, 895 (1986).
172. J. Salta and J. Zubieta, unpublished results.

173. G. Centi, F. Trifuro, J. R. Ebner, and V. M. Franchetti, *Chem. Rev.*, *88*, 55 (1988).
174. Q. Chen, J. Salta, and J. Zubieta, *Inorg. Chem.*, *32*, 4485 (1993).
175. A. LeBail, G. Ferez, P. Amoros, and D. Beltran, *Eur. J. Solid State Chem.*, *26*, 419 (1989).
176. D. Beltran-Porter, P. Amoros, R. Ibaney, E. Martinez, A. Beltran-Porter, A. LeBail, G. Ferey, and G. Villeneuve, *Solid State Iornicz*, *32/33*, 57 (1989) and references cited therein.
177. G. Villeneuve, K. S. Suh, P. Amoros, N. Casan-Pastor, and D. Beltran-Porter, *Chem. Mater.*, *4*, 108 (1992).
178. Q. Chen and J. Zubieta, *J. Chem. Soc. Chem. Commun.*, 2663 (1994).
179. J. Salta, Q. Chen, Y.-D. Chang, and J. Zubieta, *Angew. Chem. Int. Ed. Engl.*, *33*, 757 (1994).
180. M. I. Khan, Y.-D. Chang, Q. Chen, H. Höpe, S. Parkin, D. P. Goshorn, and J. Zubieta, *Angew. Chem. Int. Ed. Engl.*, *31*, 1197 (1992).
181. G. Huan, J. W. Johnson, A. J. Jacobson, and J. S. Merola, *Chem. Mater.*, *2*, 719 (1990).
182. Y.-D. Chang, J. Salta, and J. Zubieta, *Angew. Chem. Int. Ed. Engl.*, *33*, 325 (1994).
183. V. Soghomonian, R. C. Haushalter, and J. Zubieta, *Inorg. Chem.*, in press.
184. Q. Chen and J. Zubieta, *J. Chem. Soc. Chem. Commun.* 1635 (1994).
185. M. I. Khan and J. Zubieta, unpublished results.
186. G. Huan, A. J. Jacobson, J. W. Johnson, and E. W. Corcoran, Jr., *Chem. Mater.*, *2*, 91 (1990).
187. M. I. Khan, Y.-S. Lee, C. J. O'Connor, R. S. Haushalter, and J. Zubieta, *Inorg. Chem.*, *33*, 3855 (1994).
188. M. I. Khan, Y.-S. Lee, C. J. O'Connor, R. C. Haushalter, and J. Zubieta, *J. Am. Chem. Soc.*, *116*, 4525 (1994).
189. M. I. Khan, Y.-S. Lee, C. J. O'Connor, R. C. Haushalter, and J. Zubieta, *Chem. Mater.*, *6*, 721 (1994).
190. G. Huan, J. W. Johnson, A. J. Jacobson, and J. S. Merola, *J. Solid State Chem.*, *89*, 220 (1990).
191. V. Soghomonian, Q. Chen, R. C. Haushalter, and J. Zubieta, *Angew. Chem. Int. Ed. Engl.*, *34*, 223 (1995).
192. Q. Chen, L. Ma, S. Liu, and J. Zubieta, *J. Am. Chem. Soc.*, *111*, 5944 (1989).
193. M. Filowitz, W. G. Klemperer, and W. Shum, *J. Am. Chem. Soc.*, *100*, 2580 (1978).
194. J. L. Garin and J. A. Costamanga, *Acta Crystallogr. Sect. C*, *44*, 799 (1988).
195. E. M. McCarron, III and R. L. Harlow, *J. Am. Chem. Soc.*, *105*, 6179 (1983).
196. E. M. McCarron, III and R. L. Harlow, *J. Chem. Soc. Chem. Commun.*, 90 (1983).
197. M. Seleborg, *Acta Chem. Scand.*, *20*, 2195 (1966).

198. S. A. Magarill and R. F. Klesetsova, *Sov. Phys. Crystallogr.*, *16*, 645 (1972).
199. H. Kang, S. Liu, S. N. Shaikh, T. Nicholson, and J. Zubieta, *Inorg. Chem.*, *28*, 920 (1989).
200. A. J. Wilson, W. T. Robinson, and C. J. Wilkins, *Acta Crystallogr. Sect. C*, *39*, 54 (1983).
201. P. Gouzerh and Y. Jeannin, personal communication.
202. M. F. Belicchi, G. G. Fava, and C. Pelizzi, *J. Chem. Soc. Dalton Trans.*, 65 (1983).
203. J. A. Beaver and M. G. B. Drew, *J. Chem. Soc. Dalton Trans.*, 1376 (1973).
204. M. H. Chisholm, J. C. Huffman, C. C. Kirkpatrick, J. Leonelli, and K. Folting, *J. Am. Chem. Soc.*, *103*, 6093 (1981).
205. S. Liu and J. Zubieta, *Polyhedron*, *8*, 537 (1989).
206. Q. Chen, S. Liu, and J. Zubieta, *Inorg. Chim. Acta*, *164*, 115 (1989).
207. V. W. Day, M. F. Fredrick, W. G. Klemperer, and R.-S. Liu, *J. Am. Chem. Soc.*, *101*, 491 (1979).
208. V. W. Day, M. R. Thompson, C. S. Day, W. G. Klemperer, and R.-S. Liu, *J. Am. Chem. Soc.*, *102*, 5973 (1980).
209. Q. Chen, S. Liu, H. Zhu, and J. Zubieta, *Polyhedron*, *8*, 2915 (1989).
210. M. A. Porai-Koshits, L. A. Aslanov, G. V. Ivanova, and T. N. Polynova, *J. Struct. Chem. (Engl. Trans.)*, *9*, 401 (1968).
211. C. B. Knobler, A. J. Wilson, R. N. Hilder, I. W. Jensen, B. R. Penfold, W. T. Robinson, and C. J. Wilkins, *J. Chem. Soc. Dalton Trans.*, 1299 (1983).
212. L. R. Nassimbeni, M. L. Niven, J. J. Cruywager, and J. B. B. Heyns, *J. Crystallogr. Spect. Res.*, *17*, 373 (1987).
213. N. W. Alcock, M. Dudik, R. Grybos, E. Hodorowicz, A. Kanas, and A. Samotus, *J. Chem. Soc. Dalton Trans.*, 707 (1990).
214. B. Kamenar, B. Korpar-Colig, and M. Penavić, *J. Chem. Soc. Dalton Trans.*, 311 (1981).
215. S. Chunting, G. Chunxiao, J. Yan, L. Tiejin, Y. Ling, and F. Yuguo, *Eur. J. Solid State Inorg. Chem.*, *26*, 231 (1989).
216. E. Carmona, F. Gonzalez, M. L. Poveda, J. L. Atwood, and R. D. Rogers, *J. Am. Chem. Soc.*, *105*, 3365 (1983).
217. S. Lou, S. N. Shaikh, and J. Zubieta, *J. Chem. Soc. Chem. Commun.*, 1017 (1988).
218. S. Liu, S. N. Shaikh, and J. Zubieta, *Inorg. Chem.*, *28*, 723 (1989).
219. M. H. Chisholm, K. Folting, J. C. Huffman, and C. C. Kirkpatrick, *Inorg. Chem.*, *23*, 1021 (1984).
220. B. Kamenar, B. Korpar-Colig, M. Penavić, and M. Cindrić, *J. Chem. Soc. Dalton Trans.*, 1125 (1990).
221. L. Ma, S. Liu, and J. Zubieta, *J. Chem. Soc. Chem. Commun.*, 440 (1989).
222. M. Yu Antipin, L. P. Didenko, L. M. Kachapina, A. E. Shilov, A. K. Shilova, and Y. T. Struchkou, *J. Chem. Soc. Chem. Commun.*, 1467 (1989).

223. Q. Chen, S. Liu, and J. Zubieta, *Angew. Chem. Int. Ed. Eng.*, *27*, 1724 (1988).
224. L. J. Darensbourg, R. L. Gray, and T. Delord, *Inorg. Chim. Acta*, *98*, 239 (1985).
225. Q. Chen, S. Liu, and J. Zubieta, *Angew. Chem. Int. Ed. Engl.*, *29*, 70 (1990).
226. U. Kortz, B. Jameson, and M. T. Pope, *J. Am. Chem. Soc.*, *116*, 2659 (1994).
227. (a) M. I. Khan, J. Zubieta, *J. Am. Chem. Soc.*, *114*, 10058 (1992); (b) M. I. Khan and J. Zubieta, unpublished results.
228. V. S. Sergienko, E. O. Tolkacheva, A. B. Ilyukhin, Z. A. Starikova, and I. A. Krol, *J. Chem. Soc. Chem. Commun.*, 144 (1992).
229. W. Gibbs, *Am. Chem. J.*, *5*, 363 (1883).
230. Y.-D. Chang and J. Zubieta, unpublished results.
231. J. K. Stalick and C. O. Quicksall, *Inorg. Chem.*, *15*, 1577 (1976).
232. M. P. Lowe, J. C. Lockhart, W. Clegg, and K. A. Fraser, *Angew. Chem. Int. Ed. Engl.*, *33*, 451 (1994).
233. M. Filowitz and W. G. Klemperer, *J. Chem. Soc. Chem. Commun.* 233, 1976.
234. V. S. Sergienko, E. O. Tolkacheva, A. B. Ilyukhin, Z. A. Starikova, and I. A. Krol, *Mendelew Commun.*, 144 (1992).
235. G. Cao, R. C. Haushalter, and K. G. Strohmaier, *Inorg. Chem. 32*, 127 (1993).
236. M. I. Khan, Q. Chen, and J. Zubieta, *Inorg. Chim. Acta*, *206*, 131 (1993).
237. M. I. Khan, Q. Chen, and J. Zubieta, *Inorg. Chim. Acta*, *231*, 13 (1995).
238. K. Y. Matsumoto, *Bull. Chem. Soc. Jpn.*, *52*, 3284 (1979).
239. T. Nishikawa and Y. Sasaki, *Chem. Lett.*, 1185 (1975).
240. W. Bidell, V. Shklover, and H. Berke, *Inorg. Chem.*, *31*, 5561 (1992).
241. D. Gatteschi, L. Pardi, A. L. Barra, A. Müller, and J. Döring, *Nature* (*London*), *354*, 463 (1991).

The Application of Polychalcogenide Salts to the Exploratory Synthesis of Solid State Multinary Chalcogenides at Intermediate Temperatures

MERCOURI G. KANATZIDIS and ANTHONY C. SUTORIK

Department of Chemistry and Center for Fundamental Materials Research
Michigan State University
East Lansing, Michigan

Nature is a rag-merchant, who works up every shred and ort and end into new creations; like a good chemist whom I found, the other day, in his laboratory, converting his old shirts into pure white sugar.

Emerson, *Conduct of Life: Considerations by the Way*

CONTENTS

Progress in Inorganic Chemistry, Vol. 43, Edited by Kenneth D. Karlin.
ISBN 0-471-12336-6

I. INTRODUCTION

The critical role of solid state chemistry and physics in modern technology is in little doubt. Solid state compounds have been the foundation of the entire electronics industry for many years, and many emerging technologies, such as nonlinear optics, superconductivity, high-energy density storage batteries, and photovoltaic energy conversion, to name just a few, will hinge on developments in the economical processing of existing solid state materials and the discovery of new materials with new or enhanced properties. In this context, the importance of exploratory solid state synthesis cannot be over emphasized.

Perhaps no other area of chemical synthesis deserves the title ''exploratory'' more than solid state synthesis. The majority of synthetic chemists have a measure of predictability in that the molecular units they work with remain relatively intact throughout their reactions, and so their goal is mainly to link one molecule to the next or to perform specific changes on molecular functional groups. The solid state synthetic chemist has almost no predictability in his reactions save for the simplest cases of elemental substitution, and even then his predictions can still be frustrated.

This lack of predictability arises in part from the high reaction temperatures (> 600°C) used in typical solid state syntheses. Because the starting materials used in such reactions are usually solids themselves, very high temperatures have been necessary to cause sufficient diffusion for a reaction to take place. These high temperatures give rise to two important synthetic limitations. First, the reactions almost always proceed to the most thermodynamically stable products; the high energies involved often leave little room for kinetic control. These thermodynamically stable products are typically the simplest of binary or ternary compounds, and because of their high lattice stability, they become synthetic road blocks that often take a considerable investment of effort to circumvent, if they can be circumvented at all. Second, the high reaction temperatures also dictate that only the simplest chemical building blocks can be used; that is, elements on the atomic level. Attempts at synthesis using molecules of known structure are doomed because the high temperatures used sunder all bonds and reduce the system to atoms rushing to a thermodynamic minimum. Hence, multinary compounds are more difficult to form, with the preference lying with the more stable binary and ternary compounds. Being almost totally at the mercy of thermodynamics, the solid state chemist has traditionally relied on experience and intuition, rather than a set of predictable rules.

Are there any ways to increase the odds for new compound formation and against the thermodynamic traps? Consider Fig. 1. Plot A shows a conceptual trace of a typical high-temperature solid state reaction. Note that the solid reactants encounter a large activation barrier before proceeding to the highly stable end products. The large E_{act} in this case is mainly due to the diffusion requirement of the solid reactants; hence solid state reactions are diffusion limited. If one were to develop methodology that would allow for greater diffusion of reactants, the activation energy barrier would be considerably lowered. Very high temperatures would no longer be needed, and the reaction could proceed at lower temperatures to some other outcome. Such a scenario is shown in plot B. Note that the relative energies of the products in A and B are totally arbitrary; B could be more stable than A just as easily as it could be metastable with respect to A. Certain systems may go to A, their original thermodynamic end

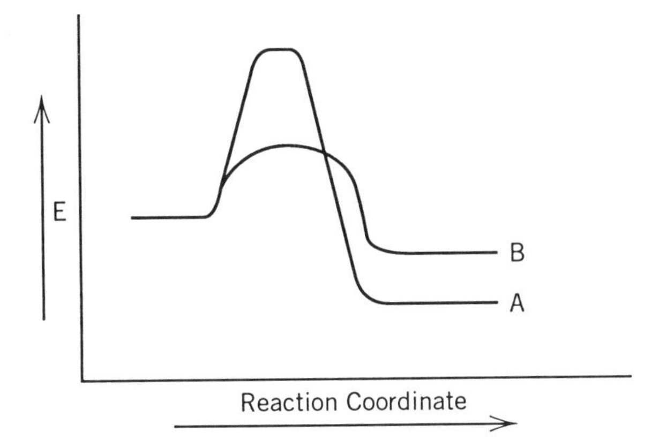

Figure 1. Hypothetical reaction coordinate plot with two traces: (A) a typical solid state reaction with low diffusion of reactants contributing to a high activation energy barrier and (B) a reaction in which enhanced diffusion of reactants lowers the activation energy barrier.

products, even with reduced activation energy. There is no way to know beforehand. *Although thermodynamic influences are not entirely avoided, it is clear that by enhancing the diffusion of the reactants, solid state reactions can be performed at lower temperatures, thereby increasing the likelihood of forming new compounds.* Lower temperatures would also make possible the use of molecular assemblies as building blocks for incorporation into solid state structures.

The amount of diffusion conducive to solid state reactions is not arbitrary. ''Infinite diffusion'' (from a solid state perspective) could be achieved by simply reacting soluble starting materials in some solvent, but when it comes to extended solids, such solution techniques are prone to produce products in the form of intractable powders. For adequate structural and physical characterization of new phases, single crystals, for use in X-ray diffraction studies, are the requirement. Methodologies for enhanced diffusion at lower temperatures ($< 600°C$) must have some mechanism for crystal growth built in or else generate products in a form amenable to other analytical techniques.

Over the past few years, work has been underway on many fronts to move away from the high temperatures of classical solid state synthesis, towards techniques that take advantage of lower reaction temperatures. One technique well known to chemists is chemical vapor deposition (CVD). Here the synthesis of solid state compounds, usually as thin films or other technologically useful forms, proceeds by the intimate gas-phase mixing of volatile precursors leading, upon pyrolysis, to deposition of solid state intermetallic materials on various substrates (1). This technique is intended for the synthesis of known solid state compounds rather than the discovery of new ones. Deposition of thin layers (15–50 Å) of starting materials, one on top of the other, has been a method used to achieve intimate mixing of reactants by reducing diffusion distances (2a and b). Although new phases have been successfully synthesized by these techniques, the deposition methods used are amenable only to elemental reactants, while synthesis using molecular species would be limited. Also noteworthy is the low-temperature solid state metathesis technique that achieves rapid synthesis of pure binary and ternary chalcogenides (3). The hydro(solvo)thermal technique has proven advantageous in that polyatomic building blocks have been used in the synthesis of solid state frameworks. This method uses solvents heated in closed containers above their boiling point (but below their critical point) as reaction media (4). The solubility of the reactants are increased by virtue of the unusually high temperature and pressure within the container. Diffusion and crystal growth are further enhanced by the inclusion of mineralizers, species that can aid in the solvation and reprecipitation of solid state reactants and products, analogous to the function of the transport agent in chemical vapor transport crystal growth. Solvothermal synthesis has been the method of choice for the synthesis of zeolites (5) and in recent years has broadened in scope to include the formation of new metal phosphates (6), metal selenates and selenites (7), alkoxopolyoxovanadium clusters (8), and metal polychalcogenides (9), to name a few. The question of synthetic outcome from reactions performed in supercritical solvents has begun to be explored only recently (10) although the solvation of normally insoluble materials and subsequent superior diffusion in such a media makes this an especially exciting approach. Performing reactions using molten salts as solvents is also a highly attractive option. Such media have been employed for well over 100 years for high-temperature single-crystal growth (11). Although many salts are high-melting species, eutectic combinations of binary salts and salts of polyatomic species often have melting points well below the temperatures of classical solid state synthesis, making possible their use in the exploration of new chemistry at intermediate temperatures. In some cases, such salts act not only as solvents, but also as reactants, providing species that can be incorporated into the final product.

One class of low-melting salts that are especially intriguing candidates for solid state synthesis is the alkali metal polychalcogenides, A_2Q_x (A = alkali metal; Q = chalcogenide). This occurs because there already exist many metal chalcogenide compounds that have technical importance. Polysulfide ligands have been implied in the hydrodesulfurization of crude oil, where they are thought to be present at the surface of metal sulfide catalysts (12). Polychalcogenide glasses are important as materials for nonlinear optics, infrared (IR) wave guides, photoconductivity, optical switching, and optical information storage, but the structure–property relationships have not been well determined due to the amorphous nature of the compounds (13). Well-characterized crystalline polychalcogenides could serve as models that would aid in illuminating such structure–property relationships. Solid state compounds with Q—Q bonds undergo reversible redox reactions with Li as shown in Eq. 1

$$2Li + (Q{-}Q)^{2-} \longrightarrow 2Li^+ + 2\,Q^{2-} \quad (1)$$

and so have been tested as rechargeable cathodes for high-energy density batteries (14). Also, reactions between metals and alkali metal polychalcogenides provide insights into the corrosion problems associated with Na/S high-energy density batteries (15). Several metal chalcogenides are photoconductive (16a and b) and $CuInSe_2$ in particular has a high radiation tolerance making it a candidate for extended extraterrestrial photovoltaic applications (16c). Group 15(IIIA) chalcogenides of the formula M_2Q_3 (M = As, Sb, or Bi; Q = S, Se, or Te) have unique applications as thermoelectric cooling materials (17). From these few examples, there is ample motivation for exploring metal reactivity in molten polychalcogenide salts both as a route to new materials with technologically interesting properties and as a way to answer long-standing corrosion problems in the molten salt area.

Over the last several years, the use of molten alkali metal–polychalcogenide salts as a low-temperature (200–600°C) solid state synthesis media has been under investigation in our laboratory. The remainder of this chapter will be devoted to this technique and the myriad new compounds that ourselves and others have discovered through its use.

II. THE NATURE OF THE POLYCHALCOGENIDE FLUX

A. Physicochemical Characteristics

Molten salts of all sorts have been used for over 100 years as high-temperature recrystallization media for a variety of binary and some ternary compounds (18). In the case of alkali metal polychalcogenides, Na_2S_x melts have been used to recrystallize binary and ternary metal sulfides at temperatures greater than 700°C by Scheel (18) and others (19). Examples of such recrystallized materials include ZnS, CdS, MnS, PbS, $NaCrS_2$, $KCrS_2$, $NaInS_2$, $KFeS_2$, FeS_2, NiS_2, CoS_2, MoS_2, NbS_2, LaS_{2-x}, Cu_3VS_4, and HgS. The use of other polychalcogenide congeners was speculated on by the investigators but only in the context of crystal growth, rather than new compound synthesis. They did recognize, however, that lower reaction temperatures could facilitate the formation of such compounds. Such an application was finally first demonstrated in 1987 by Sunshine et al. (20).

Molten A_2Q_x salts are especially well suited as solvents for intermediate temperature reactions. Figure 2 shows a phase diagram for K_2S/S, which has been adapted from studies done in connection with sodium/β-alumina/sulfur rechargeable batteries (21). It can be seen that the melting points of K_2S_x species range between the extremes of K_2S at about 800°C to K_2S_4 at 145°C, with the majority of compositions melting at less than 300°C. Although few detailed phase diagram studies of the other congeners have been performed, the melting points of many polychalcogenide salts of specific compositions have been determined and are shown in Table I (22). Minimum melting points are higher for the heavier chalcogenides, being about 250°C for the polyselenides and 350°C for the polytellurides. Although low melting, A_2Q_x fluxes remain nonvolatile over a wide temperature range, and so once above the melting point, reaction temperatures can be varied considerably without concern for solvent loss. Polychalcogenide fluxes have the added benefit of being soluble in many common polar solvents. This allows for easy isolation of products after a reaction since excess flux can simply be dissolved away. Naturally, this also requires that any products formed be insoluble. With several solvents to choose from, the most commonly used being water, methanol, or *N,N*,dimethylformamide (DMF), it is usually possible to find at least one that leaves the products intact.

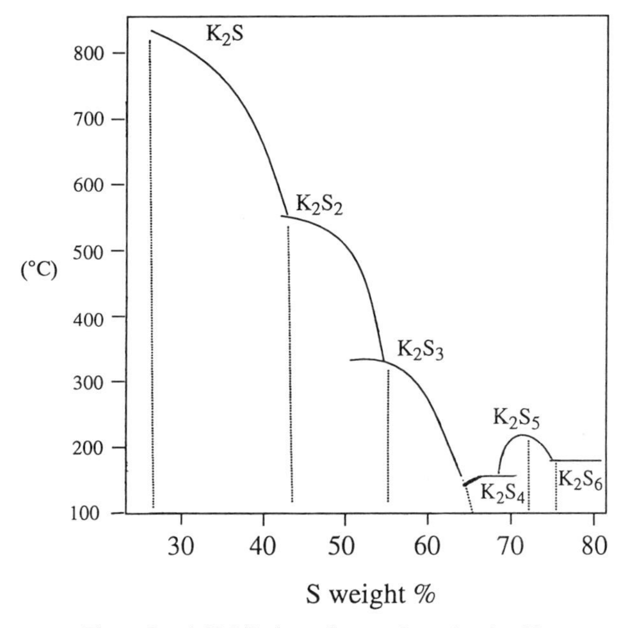

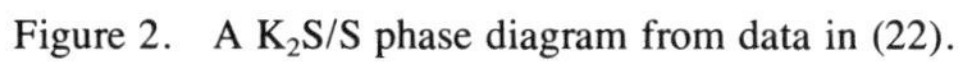
Figure 2. A K_2S/S phase diagram from data in (22).

TABLE I
Melting Points (°C) for Known Alkali Metal–Polychalcogenide Salts

Li_2S (900–975)	Li_2S_2 (369.5)				
Na_2S (1180)	Na_2S_2 (490)	Na_2S_3 (228.8)	Na_2Se_4 (275)	Na_2S_5 (251.8)	
Na_2Se (>875)	Na_2Se_2 (495)	Na_2Se_3 (313)	Na_2Se_4 (290)		Na_2Se_6 (258)
Na_2Te (953)	Na_2Te_2 (348)				Na_2Te_6 (436)
K_2S (840)	K_2S_2 (470)	K_2S_3 (252)	K_2S_4 (145)	K_2S_5 (206)	K_2S_6 (189)
	K_2Se_2 (460)	K_2Se_3 (380)	K_2Se_4 (205)	K_2Se_5 (190)	
Rb_2S (530)	Rb_2S_2 (420)	Rb_2S_3 (213)	Rb_2S_4 (160)	Rb_2S_5 (225)	Rb_2S_6 (201)
	Cs_2S_2 (460)	Cs_2S_3 (217)	Cs_2S_4 (160)	Cs_2S_5 (210)	Cs_2S_6 (186)

The exact compositional nature of a molten A_2Q_x flux remains unclear. Polychalcogenides form chain fragments of various lengths, and it is known that in solution these $(Q_x)^{2-}$ chains undergo complex self-redox equilibria of the type shown in Eq. 2 (23)

$$\left[\begin{matrix}Q\\|\\Q\end{matrix}\right]^{2-} + \left[\begin{matrix}Q\diagup Q\\|\\Q\diagdown Q\end{matrix}\right]^{2-} \rightleftharpoons \left[\begin{matrix}Q\\|\\Q\diagdown Q\end{matrix}\right]^{2} + \left[\begin{matrix}Q\diagup Q\\|\\Q\end{matrix}\right]^{2-} \qquad (2)$$

a situation that is presumably mimicked to some extent in the molten salt state. As a general phenomenon, the larger the $(Q_x)^{2-}$ fragment, the more prone it is to fall apart at higher temperatures, and so at low temperatures (250–400°C) the self-redox equilibria of the flux favors long polychalcogenide chains. Also, the inclusion of elemental Q in a reaction mixture dilutes the basicity of the polychalcogenide flux somewhat, helping to promote longer $(Q_x)^{2-}$ chains. Although providing no guarantees, working under conditions that promote the presence of long $(Q_x)^{2-}$ chains has led to the discovery of many new solid state compounds that contain such ligands in extended structures, a feature that until a few years ago had been a rarity for $(Q_x)^{2-}$ species of length greater than 2. Of course, new phases are not required to possess long $(Q_x)^{2-}$ chains; many novel phases containing either partially or exclusively monochalcogenides are also encountered in molten salt synthesis at intermediate temperatures.

B. Reactivity

Using molten A_2Q_x fluxes as reactive solvents is especially intriguing when one considers the bonding diversity of $(Q_x)^{2-}$ ligands and the wide variety of new phases that can be potentially accessed. Simple monochalcogenides already possess a high degree of flexibility in their coordination to metal cations. They can be terminal or bridging and have common coordination numbers between one and eight. Polychalcogenides retain all of the bonding versatility of monochalcogenides and then some. Table II shows some common binding modes found for $(Q_x)^{2-}$ chains along with a few examples of each (24). Most metal coordination to $(Q_x)^{2-}$ ligands occurs at the terminal atoms because of all the members of the chain, they are the most basic, being formally $1-$ in the same way as the S atom on an organic thiolate. There are several cases where an internal Q atom will also coordinate to a metal center, although this is comparatively rare. Polychalcogenides also have extensive chelation chemistry, and examples have been seen in the solid state from $(Q_2)^{2-}$ all the way to $(Q_6)^{2-}$. Solid state compounds incorporating $(Q_x)^{2-}$ are especially intriguing in that

TABLE II
Examples of Metal (M)/Polychalcogenide (Q_x^{2-}) Coordination Complexes

Type	Example	References
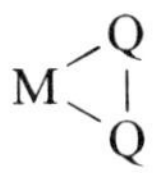 /Q M │ \Q	$(PPh_4)_2[W_2Se_4(Se_2)(Se_4)]$	24a
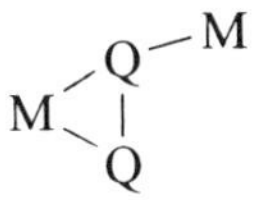 /Q—M M │ \Q	$(\eta^5\text{-}C_5H_5)Fe(CO)_2(Se_2)Cr(CO)_2(\eta_5\text{-}C_5H_5)$	24b
 /Q\ M │ M \Q/	$(\eta^5\text{-}C_5H_5)Cr_2(CO)_4Se_2$	24c
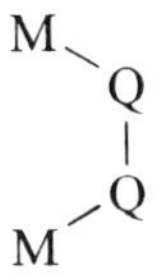 M\ Q │ Q M/	$(\eta^5\text{-}C_5H_4Me)_2V_2Se(Se_2)_2$	24d
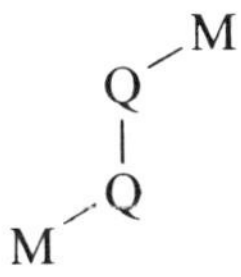 /M Q │ Q M/	$[(\eta^5\text{-}C_5H_4Me)_2Ti]_2(Se_2)_2$	24e
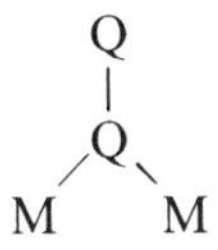 Q │ Q / \ M M	$[K\text{-}2,2,2\text{-crypt}]_2[Mo_4(Te_2)_5(Te_3)_2(en)_4]$	24f
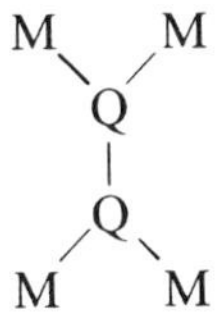 M M \ / Q │ Q / \ M M	$[NBu_4]_4[Hg_4Te_{12}]$	24g
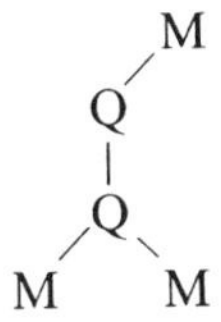 /M Q │ Q / \ M M	$[\{(\eta_5\text{-}C_5H_5)(CO)_2Fe\}_3(Se_2)][BF_4]$	24h

TABLE II. (*Continued*)

Type	Example	References
M(–Q–M)(–Q–M), Q–Q bonded	$[(CO)_5W(\mu\text{-}Te_2)][W(CO)_5]_2$	24i
M–Q_x–M	$x = 3$; $(PPh_4)_2[Au_2(Se_3)(Se_2)]$ $x = 4$; $(PPh_4)_2[Au_2(Se_2)(Se_4)]$ $x = 5$; $(PPh_4)_2[In_2(Se_4)_4(Se_5)]$	24i 24j 24k
M(Q–Q_{x-2}–Q)	$x = 3$; $(PPh_4)_2[W_2Se_4(Se_3)_2]$ $x = 4$; $[(PPh_3)_2N]_2[Au_2Se_2(Se_4)_2]$ $x = 5$; $(PPh_4)_2[Fe_2Se_2(Se_5)_2]$	24a 24l 24m
M(Q–Q_{x-2}–Q)M	$x = 4$; $(PPh_4)_2[Hg_2(Se_4)_3]$	24n
M–Q(–Q_{x-2}–)Q–M, both Q bonded to central M	$x = 4$; $(Pr_4N)_2[Ag_4(Se_4)_3]$	24o
M–Q(–Q_{x-2}–)Q–Q–M, both ring Q bonded to central M	$x = 5$; $(Me_4N)[Ag(Se_5)]$	24o

when they function as bridging ligands between two metal centers, large cages or pores can potentially be formed in the solid state. The flexibility of $(Q_x)^{2-}$ chains leads to a variety of conformational possibilities as well. The adaptability of $(Q_x)^{2-}$ ligands means that they can easily accommodate other parameters of solid state compound formation, such as the coordination number and geometry of the metal ion and the size of charge balancing countercations. Anionic M/Q frameworks form at all levels of dimensionality: one-dimensional (1D) chains, two-dimensional (2D) layers, and three-dimensional (3D) lattices. Even discrete anionic clusters have been synthesized. Compounds resulting from molten $(Q_x)^{2-}$ reactions are often metastable, either in forming only in a limited temperature window or in being dependent on the molten A_2Q_x flux for the appropriate reaction conditions. Many times, however, materials first discovered from a molten flux reaction turn out to be highly stable and can be subsequently prepared from high temperature direct combination reactions of the elements or binary chalcogenides.

Reactions between metals and molten A_2Q_x are performed *in situ*. The powdered reagents (polychalcogenide and metal or metal chalcogenide) are mixed under an inert atmosphere and loaded into reaction vessels of either pyrex or quartz. The tubes are evacuated to a pressure appropriate for the temperature used ($< 3 \times 10^{-3}$ mbar for $<600°C$; $<1 \times 10^{-4}$ mbar for $>600°C$). Once evacuated, the tubes are flame sealed and subjected to the desired heating program in a computer controlled furnace.

The polychalcogenide salt loaded into the reactions can be made in one of two ways, and the method used may have a profound effect on the reaction. First, by reacting stoichiometric amounts of elemental chalcogenide with alkali metal in liquid NH_3 (25), a direct redox reaction takes place and A_2Q_x is formed. Then *in situ* reactions with the metal may be performed. When using this method, each different length of A_2Q_x must be made separately. The second approach is to prepare one length of $(Q_x)^{2-}$ from the above NH_3 reaction and then form different polychalcogenides *in situ* by combining various ratios of A_2Q_x and Q. Upon heating, these combinations fuse together, forming the polychalcogenide flux via the self-redox equilibria previously mentioned. *Caution*: Rapid heating of these mixtures may result in an explosion. The metal, which has been *in situ* with the A_2Q_x/Q mixture all along, then proceeds to react with the flux. *In situ* polychalcogenide formation is convenient for accessing fluxes over a wide compositional range without making large quantities of various starting materials; however, this method might not be appropriate for certain very reactive metals. Such metals may react with the original starting materials before the polychalcogenide flux has a chance to form. If this premature reaction leads to phases with high thermodynamic stability, the reaction is dead in the water. Under such circumstances, new phases may be more readily accessible if the metal is reacted with already formed polychalcogenide.

Polychalcogenide fluxes are highly reactive towards metals because they are very strong oxidants. This phenomenon arises from the fact that, although the terminal atoms of the $(Q_x)^{2-}$ chains are formally $1-$, the internal atoms are all formally zero valent. Metals are drawn into solution by being oxidized by the internal atoms of the $(Q_x)^{2-}$ chains, which, upon being reduced, split into smaller fragments. The solvated metal cations are then coordinated by the basic atoms of the $(Q_x)^{2-}$ ligands, forming some manner of soluble intermediate, which is followed by nucleation into solid crystals. Presumably, the nucleated species are in equilibrium with the soluble intermediates, especially if the flux is present in excess, and hence a solvation–reprecipitation effect (often referred to as the mineralizer effect) occurs. This aids in the growth of high-quality single crystals because the flux can redissolve small or poorly formed crystallites and then reprecipitate the species onto larger, well-formed crystals. Because these molten salts are so strongly oxidizing, the phases isolated invariably possess metal cations in the highest oxidation states achievable for a given chalcogenide electronegativity. That does put some of the most oxidized metal cations beyond the reach of this method, but that is a function of the natural limits of the elements involved.

The composition of a polychalcogenide flux is one of the most important variables that can be manipulated to affect the outcome of a reaction. Such manipulations lie in altering the basicity of the fluxes leading to different reactivities with the metals. The guiding principles in this are (1) the alkali metal involved (basicity of the A_2Q_x fluxes increases with increasing cation size), and (2) the ratio of alkali metal to chalcogenide (i.e., the $(Q_x)^{2-}$ chain length). The second point is, in fact, complex and requires further elaboration. Since charge neutrality in the A_2Q_x flux must be preserved, as A is increased and Q is kept constant, the system is driven to shorter and shorter $(Q_x)^{2-}$ chains. This increases the amount of terminal basic atoms relative to the internal oxidizing atoms, increasing the overall basicity of the flux. As the internal atoms disappear, the oxidizing power of the flux reduces somewhat as well. Hence, not only are the size of the ligands changed but also a metal that is reactive in a flux of long polychalcogenide chains may be unreactive towards one with shorter chains (although such cases may be driven to reaction with increases in temperature). Varying the composition of the polychalcogenide flux is thus akin to varying the pH in aqueous solutions, and in effect amounts to having a large variety of solvents in which to perform reactions, each one having different reactivity. Changes in flux basicity initiate complex changes in reactivity and solubility that are not fully understood and that vary from metal to metal. At this stage of development in this method, each metal explored must be treated as a new and unique system, and similarities to the reactivity of other metals can only be drawn afterwards.

III. HIGH-TEMPERATURE REACTIONS (>600°C)

Although the bulk of this chapter concerns the use of molten A_2Q_x fluxes as solvents at intermediate temperatures (200–600°C), a substantial body of work exists in which reactions between metals and A_2Q_x are performed in the more traditional temperature regime of solid state synthesis. Many of the flux properties already discussed are retained at high temperatures, chief among them being nonvolatility and corrosive solvation of metallic starting materials. Although less stable at high temperatures, some polychalcogenide containing compounds have, in fact, been stabilized. As further demonstration of the general utility of these solvents, materials from high-temperature flux reactions will be surveyed in this section.

The compound $K_4Ta_2S_{11}$ (26) is an air and moisture sensitive material and was isolated from a very basic mixture of K_2S/Ta/S in a 3:2:5 molar ratio (K/S = 3:4) heated at 800°C for 48 h. in a quartz tube. The $[Ta_2S_{11}]^{4-}$ fragment is a discrete anionic cluster with no intercluster bonding. A view of the cluster is shown in Fig. 3. Each Ta^{5+} is coordinated in an irregular seven-coordinate polyhedra composed of one terminal monosulfide, one η^2-$(S_2)^{2-}$, one μ^2-S^{2-} bridging the two Ta and one more $(S_2)^{2-}$ which is η^2 to one Ta while one S atom of the unit also serves as a second bridging atom to the opposite Ta. This cluster is closely related to the neutral species $Mo_2S_9O_2$ (27) although the bridging interaction in this compound between metal and η^2-$(S_2)^{2-}$ is only weakly present. The $[W_2S_{11}]^{2-}$ complex is also similar (28).

A second compound from the reaction in a polysulfide flux at high temperature is $Na_4Ga_2S_5$ (29). Formed from a stoichiometric combination of Na_2S/Ga/S heated at 727°C, the structure features $[GaS_4]$ tetrahedra linked by a combination of edge and corner sharing into 2D layers. Although the synthesis is, stoichiometrically, a direct combination, some amount of polysulfide flux no doubt forms as an intermediate in this reaction.

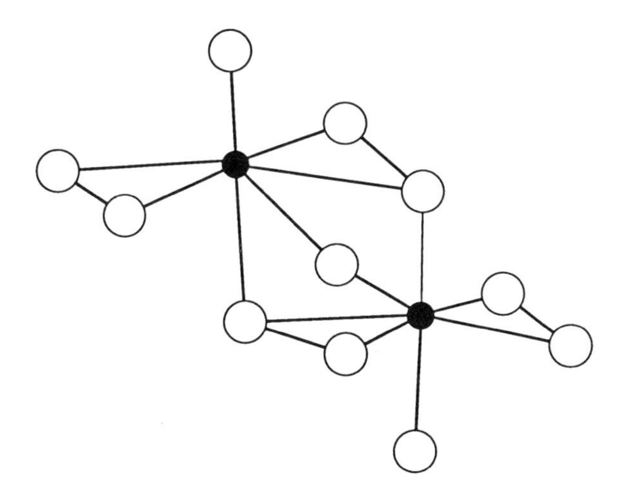

Figure 3. The structure of the $[Ta_2S_{11}]^{4-}$ anion. (Filled circles = Ta; open circles = S) [adapted from (26)].

Two ternary tellurides have been synthesized from high-temperature molten salt fluxes. The compound $KErTe_2$ was synthesized from a 2 : 1 : 6 mixture of K_2Te_3/Er/Te heated at 600°C for 4 days, followed by 4 more days at 900°C (30). The compound crystallizes in the α-$NaFeO_2$ structure type common to $ALnQ_2$ compounds (Ln = lanthanide) and is simply a NaCl derivative where the K^+ and Er^{3+} occupy the Na^+ sites and are segmented into alternating layers throughout the lattice. A new structure type was found in the compound $K_4M_3Te_{17}$ (M = Zr or Hf) (31). The phase was formed from the highly basic mixture of K_2Te/M/Te in a molar ratio of about 2 : 1 : 1.4 heated at 650°C for 6 days plus 900°C for another 4 days. The black needle shaped crystals were manually extracted from the melt surface. The structure of $K_4M_3Te_{17}$ possesses 1 D anionic chains running in the [101] crystallographic direction (Fig. 4). Coordinated about the metal is a distorted bicapped trigonal prism that shares trigonal faces with other prisms to form the chain. The capping Te atoms are one end of a $(Te_2)^{2-}$ ligand, which is η^2 on a single M atom. The compound also features a ditelluride with one nonbonding atom [Te(15)] and a $(Te_3)^{2-}$ unit in which the central atom [Te(14)] does not bond to any M cations. The Te atoms are known to form partial bonding interactions at distances as great as 4.10 Å (32), but a cutoff for an actual Te—Te bond was made by the authors at 2.94 Å to appropriately charge balance the M^{4+} centers. Hence, the formulation $K_4Hf_3(Te_3)(Te_2)_7$ was proposed. The $HfTe_5$ (33) compound was also noted to be a polytelluride with 1D chains of face-sharing bicapped trigonal prisms, albeit with different polytelluride ligands.

Examples of quaternary sulfides containing two different main group metals were reported, having the formula $KM^aM^bS_4$, where M^a = Ga for M^b = Sn, M^a = In for M^b = Ge, and M^a = Ga for M^b = Ge (34). Each compound was formed from a mixture of $K_2S_5/M^a/M^b/S$ in a 1 : 2 : 2 : 3 ratio, which was heated

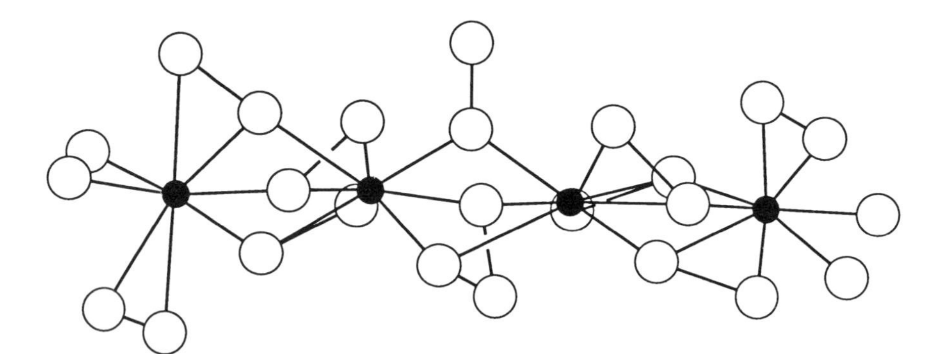

Figure 4. A fragment of an anionic chain from $K_4Hf_3Te_{17}$. (Filled circles = Hf; open circles = Te) [adapted from (31)].

at 700°C for 24 h plus 900°C for 100 h. Although possessing different structure types, all of them share several features. In each phase the two different metals are disordered over all metal sites in the structure, owing mainly to the similar coordination chemistries and sizes of the metals used. Also, all are layered compounds and are built from chains of $[MS_4]$ corner-sharing tetrahedra cross-linked by discrete $[M_2S_6]$ units, composed of two edge-sharing tetrahedra. In $KGaSnS_4$ the $[M_2S_6]$ units link the chains such that there is a dimer at the same point on opposite sides of a chain. Meanwhile, $KInGeS_4$ possesses $[M_2S_6]$ units that stagger themselves along the sides of the chain. The interchain links in $KGaGeS_4$ are the same as in $KInGeS_4$, but the former experiences conformational twisting along the chains of tetrahedra, leading to a monoclinic rather than triclinic unit cell. Figure 5 shows polyhedral representations of the compounds, all of which crystallize as colorless plates and are electric insulators.

Infinite chains of tetrahedra form the basis for another family of compounds synthesized from high-temperature flux reactions. These materials conform to the formula scheme of $K_xCu_y(MQ_4)_z$, where the sum of x and y must be equal to $3z$. Each compound possesses metal-centered tetrahedra that link into infinite chains via edge sharing. The Cu and second metal were refined as residing in alternate tetrahedral sites, although some amount of disorder could not be ruled out. When $x = 2$, $y = 1$, and $z = 1$ in the above formula, the stoichiometry K_2CuMQ_4 results (35). This phase was found for M = Nb and Q = Se and is simply the 1 D chain of edge-sharing tetrahedra surrounded by noninteracting K^+ ions (Fig. 6). The phase crystallizes as red needles from the reaction of K_2Se_5/Nb/Cu/Se in a 1:2:2:3 molar ratio heated at 800°C for 4 days. Upon doubling the amount of Cu in the reaction mixture and increasing the temperature to 870°C for 4 days, KCu_2NbSe_4 results (36). In this phase the $CuNbSe_4]_n^{2n-}$ chains are now linked into layers by $[CuSe_4]$ tetrahedra that edge-share between $[NbSe_4]$ tetrahedra on neighboring chains. The layers are perpendicular to the *a* axis and form corrugated sheets as shown in Fig. 7. The sulfide analogue was also isolated from similar reaction conditions, as was the Ta containing selenide (37). The third member of this family is intermediate between the previous two compounds. It is $K_3Cu_3M_2Q_8$ (M = Nb or Ta; Q = S or Se) (38) and features two $[CuMQ_4]_n^{2n-}$ chains linked into one large chain by $[CuQ_4]$ tetrahedra (Fig. 8). All of the analogous forms of $K_3Cu_3M_2Q_8$ were synthesized from somewhat more basic conditions than were used for $K_2CuNbSe_4$ and KCu_2NbSe_4. A 3:4:6:1 molar ratio mixture of K_2Q_5/M/Cu/Q was heated initially at 500°C for 24 h followed by 4 days at 850°C. Of note in this chemistry is the unusual tetrahedral geometry about the Nb and Ta atoms. A trigonal prismatic geometry is common for both metals in MS_2 (39) and most of their other binary chalcogenides (40), although the discrete tetrahedral anion, $(MQ_4)^{3-}$, is known for the sulfides and selenides of both metals (41).

Another example of a 1D bimetallic chain is seen in $K_3CuNb_2Se_{12}$, which

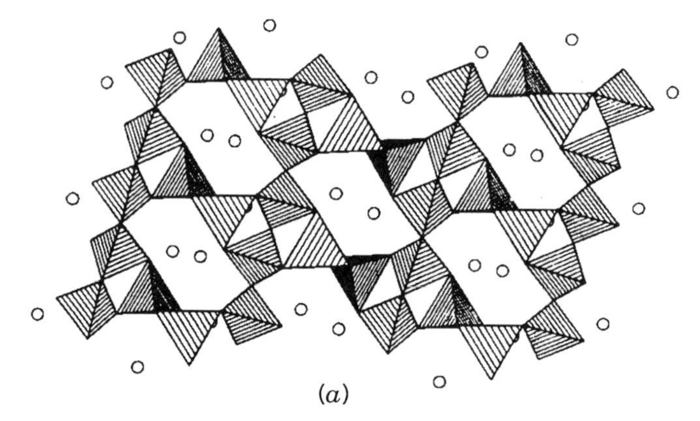

(a)

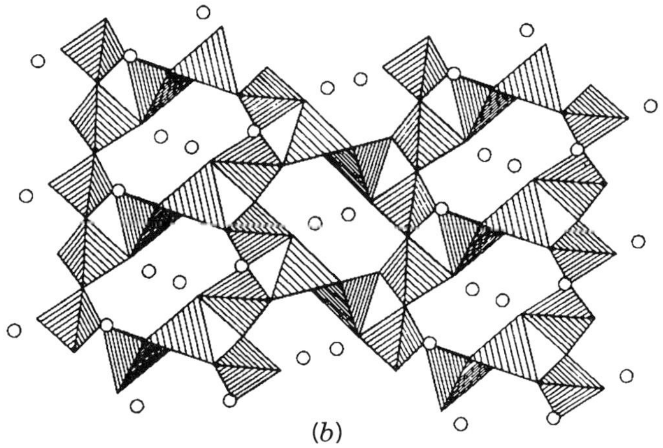

(b)

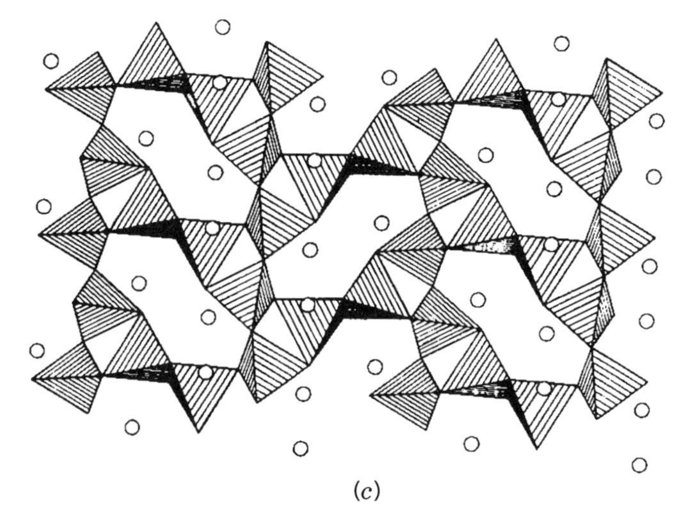

(c)

Figure 5. Polyhedral representation of (*a*) $KGaSnS_4$, (*b*) $KInGeS_4$, and (*c*) $KGaGeS_4$ [adapted from (34)].

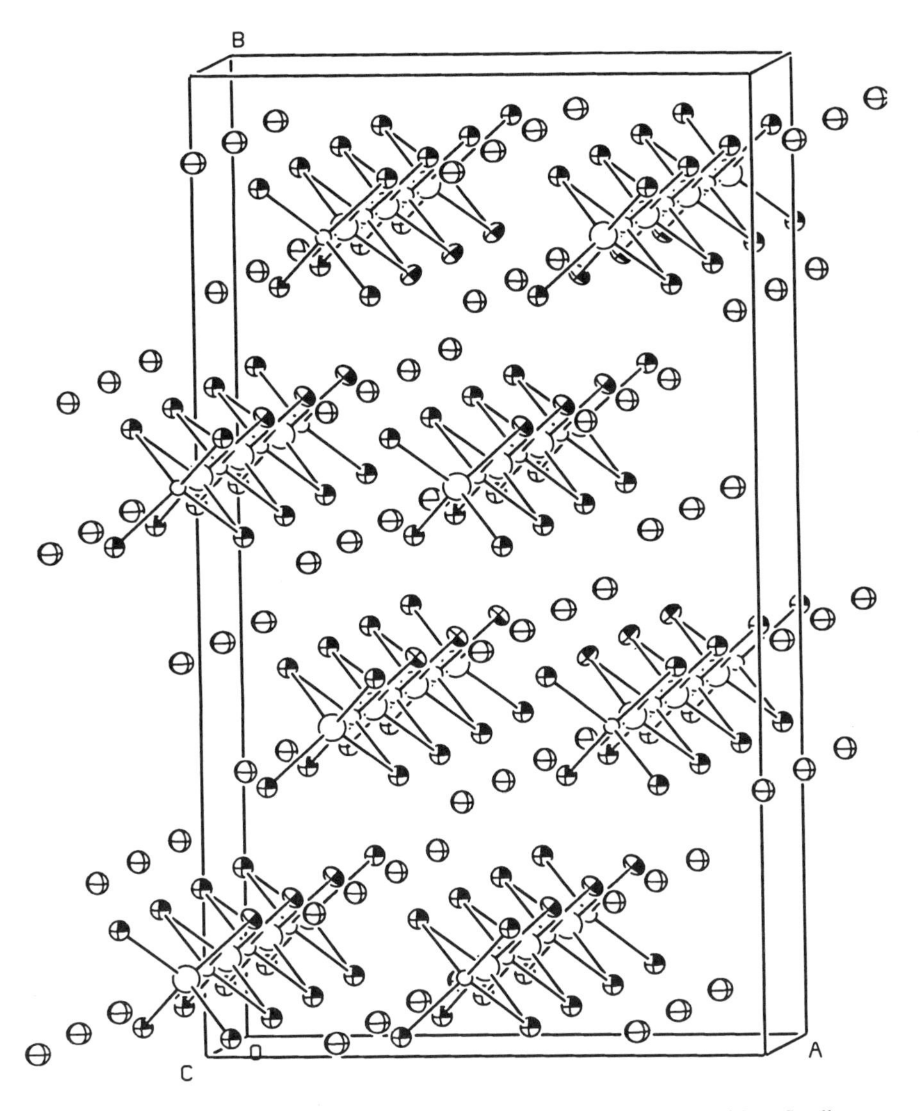

Figure 6. A view of the chains of edge-sharing $[MSe_4]$ tetrahedra of $K_2CuNbSe_4$ (Small open circles = Nb; large open circles = Cu; circles with nonshaded octants = K; circles with shaded octants = Se). [adapted from (35)].

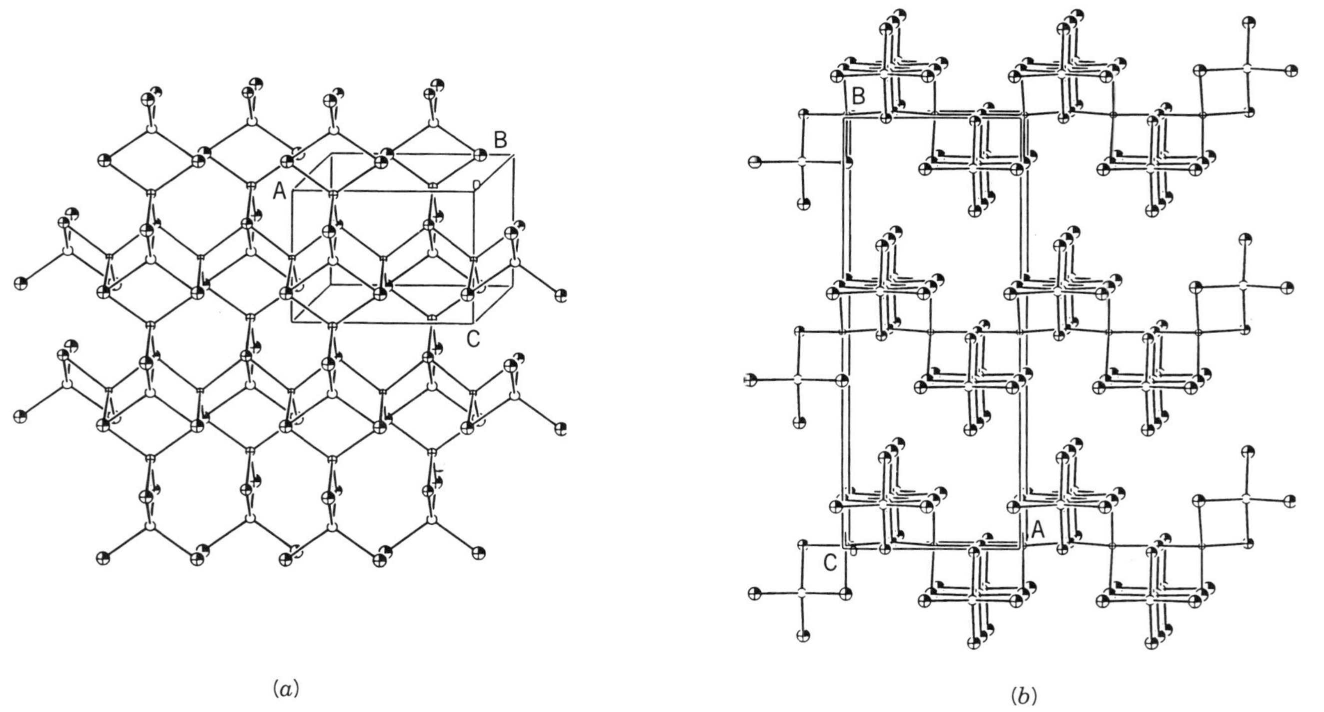

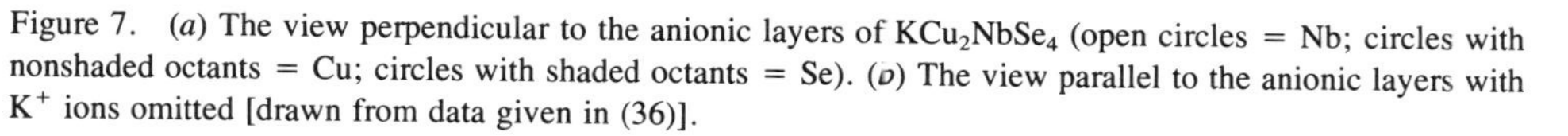

Figure 7. (*a*) The view perpendicular to the anionic layers of KCu_2NbSe_4 (open circles = Nb; circles with nonshaded octants = Cu; circles with shaded octants = Se). (*b*) The view parallel to the anionic layers with K^+ ions omitted [drawn from data given in (36)].

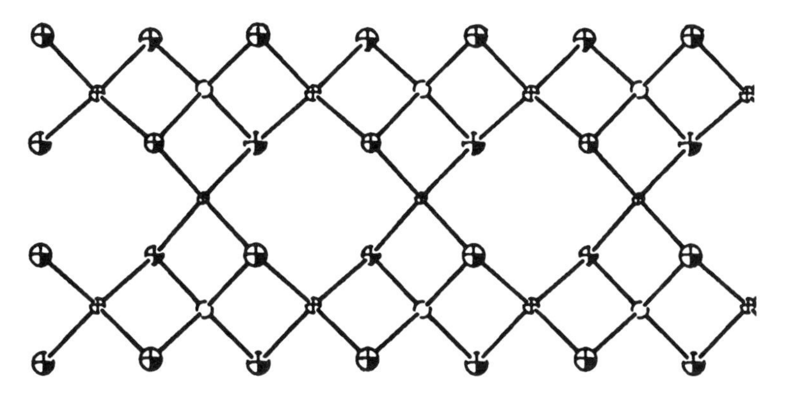

Figure 8. The anionic chains of $K_3Cu_3Nb_2Se_8$ (open circles = Nb; circles with nonshaded octants = Cu; circles with shaded octants = Se) [adapted from (38)].

was synthesized from a mixture of K_2Se_5/Nb/Cu/Se in a 1.5:2:1:4.5 molar ratio heated at 870°C for 4 days (35). In this reaction the flux is made more chalcogenide rich, hence less basic, than was used in the synthesis of the $K_xCu_y(MQ_4)_z$ compounds. As a result, it features several polyselenide ligands. In Fig. 9 the anionic chain, which has tetrahedral Cu atoms and two distinct Nb atoms in irregular seven-coordinate polyhedra, possesses two monoselenides, three diselenides, and one tetraselenide. Such a formalism would necessitate Cu^{1+} and two Nb^{4+} oxidation states. However, one of the bonds within the $(Se_4)^{2-}$ unit is appreciably longer than the others in the compound; the Se(1)—Se(6) bond is 2.726(3) Å while the other Se—Se bonds range from 2.376(4)–2.542(3) Å. Formalizing the $(Se_4)^{2-}$ unit as Se^{2-} and $(Se_3)^{2-}$ would require two Nb^{5+} oxidation states (Cu^{2+} does not form in chalcogenide envi-

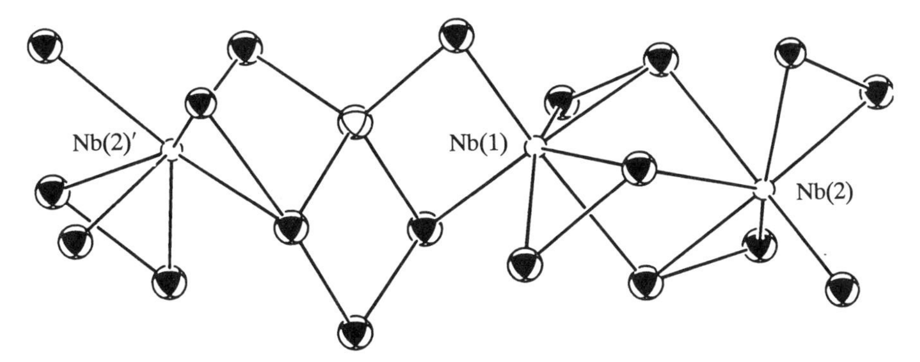

Figure 9. A fragment of the anionic chains of $K_3CuNb_2Se_{12}$ (Open circles = Nb; circles with nonshaded octants = Cu; circles with shaded octants = Se) [Adapted from (35)].

ronments). No experimental results were reported that resolved this ambiguity. The short Se(1)—Se(6) contact may imply a partial bond formation related to partial oxidation of the Se^{2-} and $(Se_3)^{2-}$ to an $(Se_4)^{2-}$ species. If so, this phenomenon may be accompanied by mixed valence on the Nb atoms, which may in turn impart unusual properties to the compound.

A family of compounds with the formula $ACuMQ_3$ (Q = S, Se, or Te) has also been synthesized (42, 43), and Table III lists the synthetic conditions for each phase. All are isostructural and feature anionic layers containing Cu, M, and Q, with A^+ in trigonal prismatic coordination in the interlayer gallery. The layers are composed of $[MQ_6]$ octahedra that share equatorial edges down the *a* axis. These chains then corner share with each other at the axial positions, forming a corrugated layer. Within the folds of these layers, the Cu atoms reside in tetrahedral sites (Fig. 10). Without the Cu^{1+}, the layers belong to the anti-Pd_3Te_2 structure type (44) in which the Te atoms occupy both Zr and K sites. An interesting change in properties was noted in these compounds as the chalcogenide was changed. The sulfur analogue is an insulator; the Se compound possesses a metal-to-semiconductor transition at 50 K; and with Te, the compound is metallic at all temperatures with a room temperature conductivity of 433 Ω^{-1} cm^{-1}. As the orbital size increases from S to Te, the resulting improved overlap increases the width of the orbital bands in the solid (45). Hence the S analogue, with the smallest orbitals and narrowest bands, is an insulator; however, Te, with the largest orbitals, must possess cross-band overlap, giving it metallic conductivity. The intermediate spacing between the bands of the Se analogue leads to some sort of unusual semiconducting behavior. Recently, a second family of phases with the same general formula of $ACuMQ_3$ have been isolated as well (43). Found for A = Na, M = Ti, and Q = S or A = Na, M = Zr, Q = Se or Te, this structure features double chains of $[MQ_6]$ edge-sharing octahedra and double chains of edge-sharing $[CuQ_4]$ tetrahedra fused into anionic layers.

Quaternaries from high temperatures are not limited to those containing Cu. The compound $CsCdAuS_2$ has been synthesized from the reaction of Cs_2S/Cd/

TABLE III
Synthetic Conditions for the Formation of $ACuMQ_3$ Phases

Compound	A_2Q/Cu/M/S Ratio	Heating Program	Cooling Rate
$KCuZrS_3$	1.51/1/1/6.53	500°C (1 day) + 800°C (4 days)	4°C h^{-1}
$KCuHfS_3$	1.5/1/2/6.6	500°C (1 day) + 800°C (4 days)	4°C h^{-1}
$KCuZrSe_3$	1/1/1/3	500°C (1 day) + 800°C (4 days)	4°C h^{-1}
$KCuZrTe_3$	1/2/2/5	650°C (6 days) + 900°C (4 days)	3°C h^{-1}
$NaCuZrS_3$	1.98/1/1/8.93	500°C (1 day) + 700°C (4 days)	4°C h^{-1}

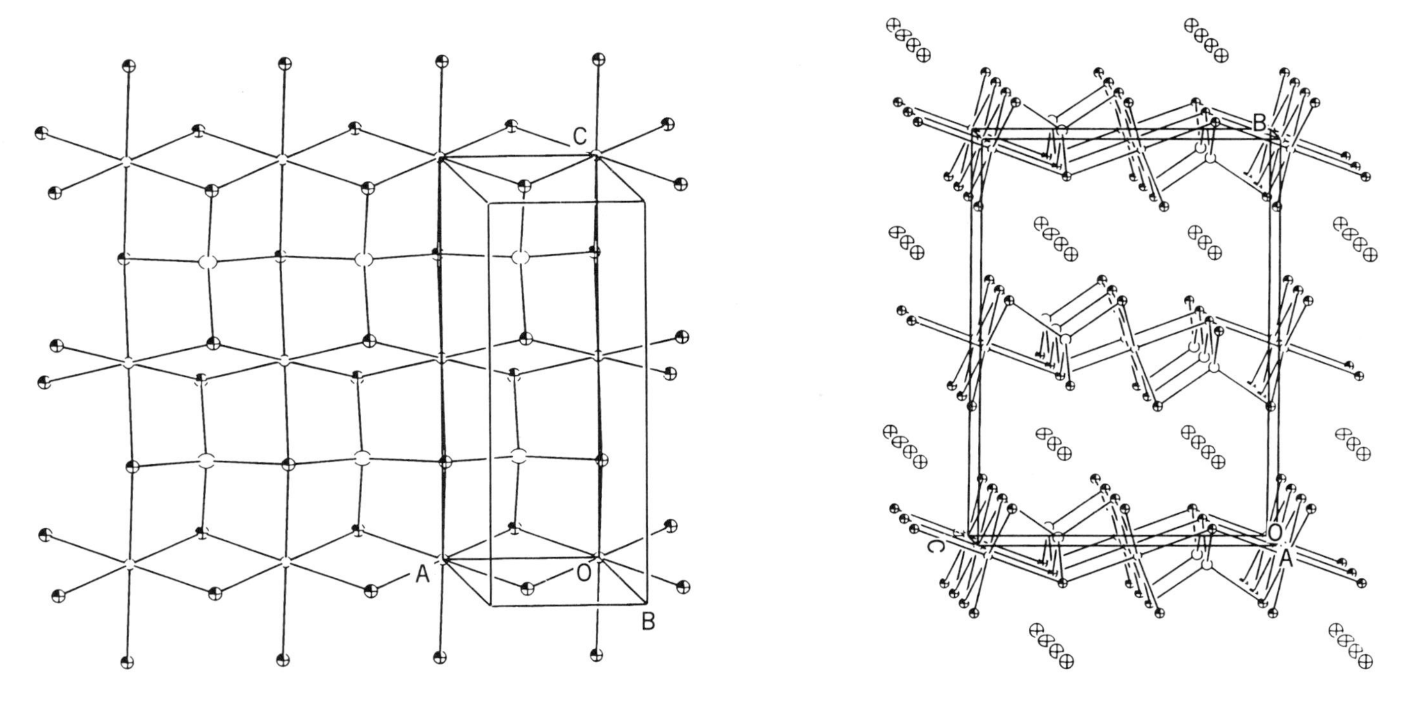

Figure 10. Two views of the $ACuMQ_3$ structure type: perpendicular (*a*) and parallel (*b*) to the anionic layers (small open circles = M; large open circles = Cu; circles with nonshaded octants = A; circles with shaded octants = Q) (42, 97).

Au/S in a 8 : 1 : 1 : 16 molar ratio heated at 650°C for 4 days (46). Its layered structure is built from chains of edge-sharing [CdS_4] tetrahedra that are linked into layers via linear [AuS_2] units (Fig. 11). A band gap of 3.0 eV has been measured for the material.

Another series of high-temperature phases is found in the $KLnMQ_4$ family (Ln = La, Nd, Ga, or Y; M = Si or Ge; and Q = S or Se) (47). All have been prepared from stoichiometric mixtures of $K_2Q_5/M/Ln_2Q_3$, which were heated at three different temperatures, successively (500°C for 1 day, followed by 700°C for 1 day, and 1000°C for 150 h). This structure type features Ln centered monocapped trigonal prisms that share the edges of their rectangular faces to form a 1D chain. These chains then are linked into layers by (a) corner sharing between capped trigonal prisms of opposite chains and (b) [MQ_4] tetrahedra that share one edge with one chain and two edges with the chain opposite.

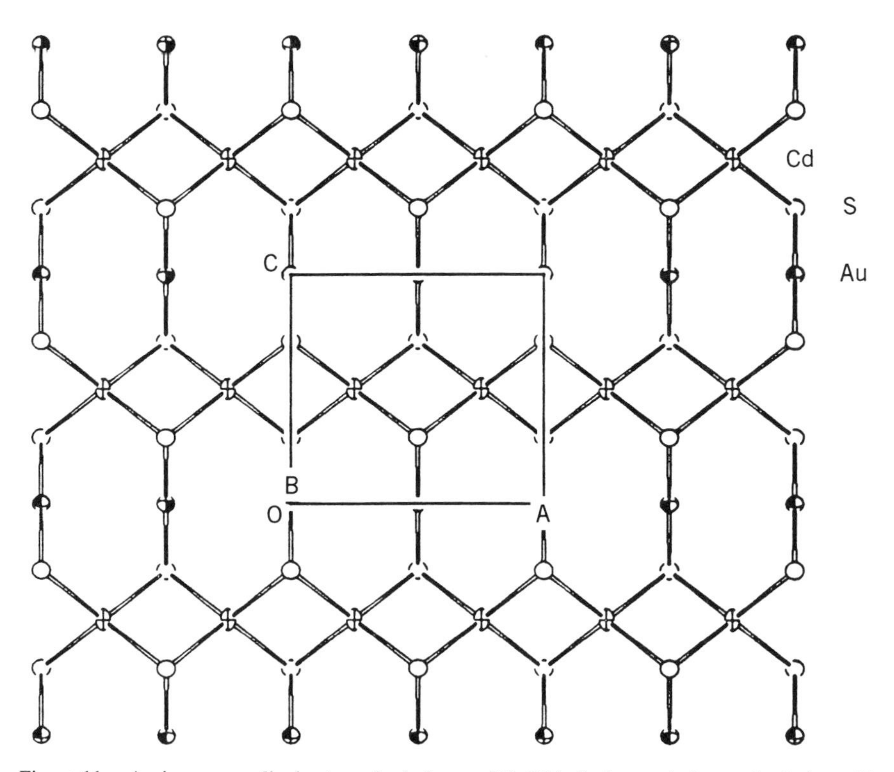

Figure 11. A view perpendicular to a single layer of $CsCdAuS_2$ (open circles = S; circles with nonshaded octants = Cd; circles with shaded octants = Au (46)).

IV. INTERMEDIATE TEMPERATURE REACTIONS (200–600°C) IN TERNARY SYSTEMS (A/M/Q) WITH A_2Q_x FLUXES

The use of reactive polychalcogenide fluxes at high temperatures can no doubt lead to the discovery of many new compounds, and as such, it is a valuable modification on traditional solid state synthetic techniques. However, the true novelty of using these salts for synthesis lies in the fact that they remain viable solvents at temperatures well below the typical solid state regime. This allows for synthetic investigations to be performed under the largely unexplored conditions present at intermediate temperatures, hence accessing all the advantages and possibilities discussed in the Introduction. The remainder of this chapter will focus on the results obtained when polychalcogenide salts are the special glasses used to observe metal–chalcogenide chemistry at intermediate reaction temperatures.

A. When M = Ti

Examples of new ternary compounds containing early transition metal elements are still few. This may lie in the fact that the binaries containing these metals possess a great deal of lattice stability and so once formed are difficult for the flux to redigest for further reaction. Despite this, some of the first compounds synthesized using A_2Q_x fluxes at intermediate temperatures contained early transition metals.

One of those compounds was $K_4Ti_3S_{14}$; the very first in fact (21). It was isolated from a reaction of $K_2S/Ti/S$ in a 2:1:6 molar ratio at 375°C for 50 h. It crystallized as black needles with hexagonal ends and has the structure shown in Fig. 12. There are two crystallographically distinct Ti atoms, the seven-

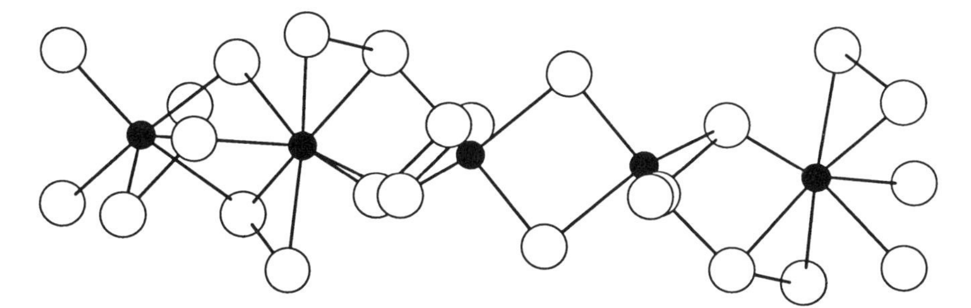

Figure 12. A fragment of the anionic chains of $K_4Ti_3S_{14}$ (solid circles = Ti; open circles = S) [adapted from (21)].

coordinate Ti(1) of which there are two per formula, and the eight-coordinate Ti(2). Each Ti(1) is linked to a Ti(2) through three bridging S atoms, each of which is part of a disulfide ligand, while connections are made to a second Ti(1) atom through two monosulfides. Thus, the compound can be represented as $K_4[Ti_3(S_2)_6S_2]$. The Ti(1) atom exists in a distorted pentagonal bipyramid with $2\eta^2$-$(S_2)^{2-}$ units in the pentagon. The Ti(2) atom is coordinated in an irregular polyhedron containing two η^2-$(S_2)^{2-}$, two bridging monosulfides, and $2\eta^1$-$(S_2)^{2-}$, which are also involved in chelating the Ti(1) atom. Both of these environments are unusual for titanium sulfides that usually exhibit octahedral coordination, although TiS_3 possesses Ti atoms in a bicapped trigonal prismatic geometry (48).

A second titanium compound, $Na_2Ti_2Se_8$ (49), was isolated from a reaction of Na_2Se/Ti/Se in a 2 : 1 : 10 molar ratio heated at 345°C for 100 h. The compound was isolated with water yielding purple–brown crystals, in 18% yield, which exhibited slight moisture sensitivity. This material is also composed of anionic polymers, a view of which is shown in Fig. 13. There are two crystallographically distinct Ti atoms; both are seven coordinate. The Ti(1) atom exhibits distorted pentagonal bipyramidal geometry in which two η^2-$(Se_2)^{2-}$ and the Se(6) atom comprise the pentagon and Se(2) and Se(5) reside in the axial positions. The Ti(2) atom possesses an even more distorted coordination environment, but a pentagonal bipyramidal geometry can be approximated here as well. As such, Se(3) and Se(5) lie in the axial positions while Se(2) and $2\eta^2$-$(Se_2)^{2-}$ occupy the equatorial positions. The chain is formed upon face sharing of the polyhedra, where the face is defined either by an axial Se^{2-} and one equatorial $(Se_2)^{2-}$ or one axial Se^{2-} and two Se atoms from two separate equatorial $(Se_2)^{2-}$ units on two separate Ti atoms.

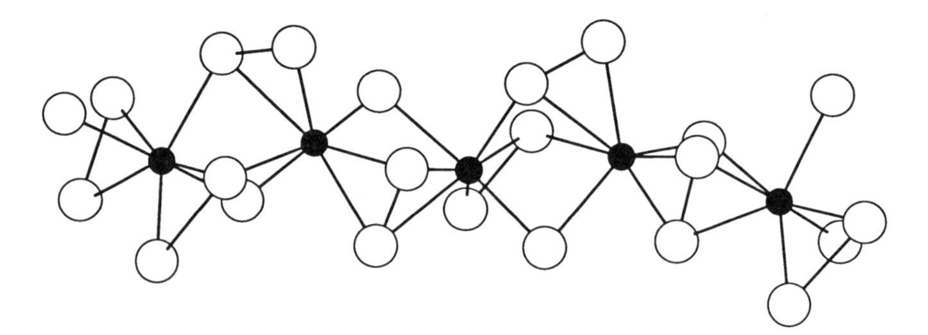

Figure 13. The anionic chain of $Na_2Ti_2Se_8$ (solid circles = Ti; open circles = Se) [adapted from (49)].

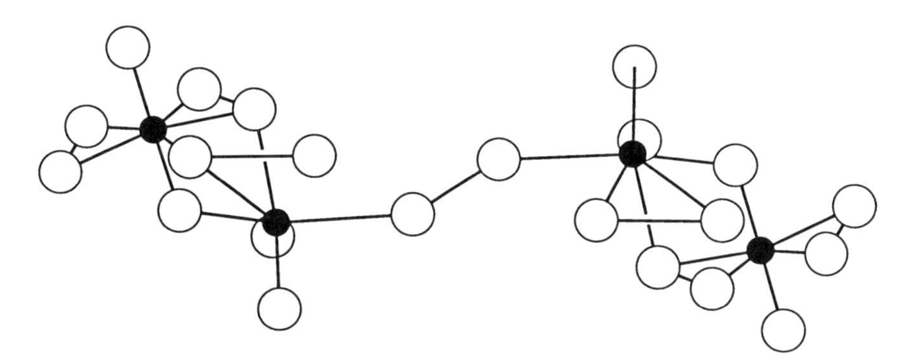

Figure 14. The structure of the anionic cluster $[Nb_4Se_{22}]^{6-}$ (solid circles = Nb; open circles = Se) [adapted from (26)].

B. When M = Nb

One phase containing Nb has been synthesized from A_2Q_x fluxes in the intermediate temperature regime: $K_3Nb_2Se_{11}$ (26). Formed from a 3 : 1 : 10 molar ratio of K_2Se/Nb/Se heated at 375°C for 100 h, the compound is composed of discrete anionic clusters. The structure is similar to that seen in the previously discussed phase $K_4Ta_2S_{11}$ (see Section III), and a view of the anion is shown in Fig. 14. The major difference between the two structures is that the monosulfide bridge in $K_4Ta_2S_{11}$ is replaced by a diselenide bridge in $K_3Nb_2Se_{11}$. Otherwise the complex bonding pattern is retained despite the change in chalcogenide and the decreased reaction temperature.

C. When M = Cu

Copper has shown an extensive and elegant reactivity towards molten polychalcogenide salts, both by itself and in combination with other metals. The naturally high chalcogenophilicity of copper is evident in mineralogy in that chalcogenides, especially Se and Te, are often found as copper containing ores. Indeed, the word *chalcogenide* has its roots in two Greek words: *chalkos* meaning copper and *genes* meaning forming or born. Under the conditions of intermediate temperatures, many new copper containing compounds have been synthesized, most of which contain polychalcogenide ligands.

Two phases have been found to exist for the stoichiometry $KCuS_4$ (50). Synthesized from a 3 : 1 molar ratio of K_2S_4/Cu heated at 215°C for 4 days, α-$KCuS_4$ is a 1D material that features Cu coordinated exclusively by $(S_4)^{2-}$ ligands. The structure of the anionic chains is shown in Fig. 15(*a*). Copper is

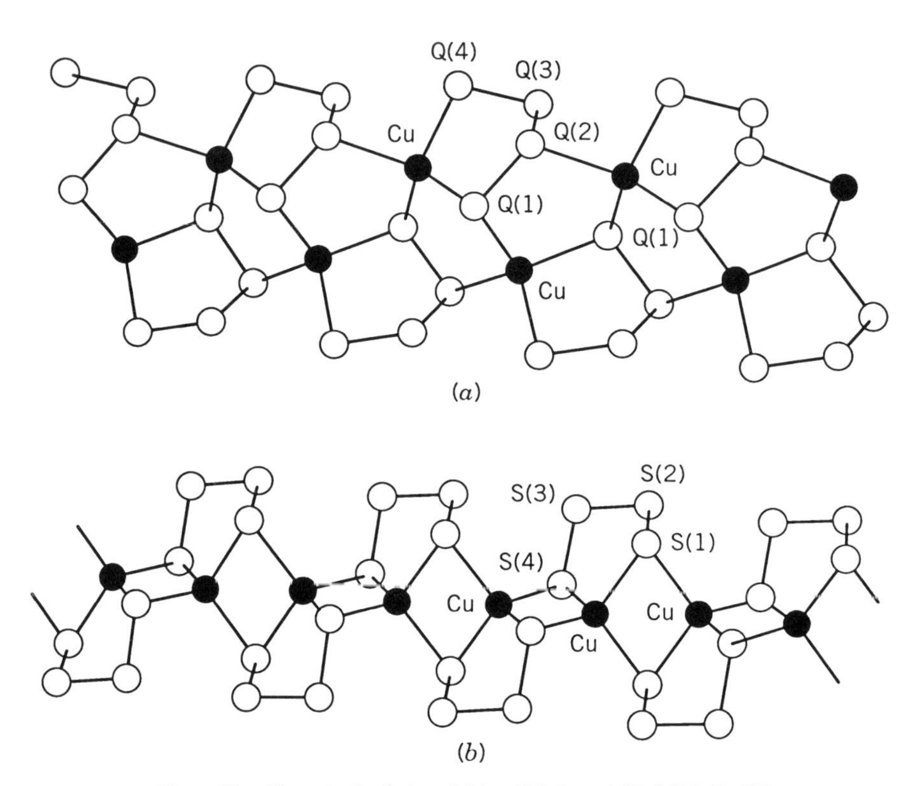

Figure 15. The anionic chains of (*a*) α-$KCuS_4$ and (*b*) β-$KCuS_4$ (50).

in tetrahedral coordination with one $(S_4)^{2-}$ ligand bonding in a chelating mode and the remaining coordination sites filled by one internal S atom and one terminal S atom, each from two separate $(S_4)^{2-}$ neighbors. The chains run along the *a* axis and are noncentrosymmetric, wrapping around a crystallographic twofold axis. These $[CuS_4]_n^{n-}$ chains have been previously seen in $[NH_4]$ $[CuS_4]$ (51), a compound synthesized at room temperature, and the selenium analogues containing K^+ and Cs^+ countercations have also been synthesized (52). The apparent stability of α-$KCuS_4$ is limited, however. If the reaction temperature is increased to 250°C, a second phase is isolated: β-$KCuS_4$. The chain nature of the anion is retained, but where α-$KCuS_4$ has a bond to an internal S atom of a $(S_4)^{2-}$ chain, β-$KCuS_4$ has bonds only to terminal S atoms. This leads to a centrosymmetric chain structure whose core of $[CuS_4]$ tetrahedra now form simple edge-sharing connections along the *a* axis [Fig. 15(*b*)]. If the temperature of the synthesis is raised still higher neither phase is stabilized; at 400°C, KCu_4S_3 is formed while at 450°C CuS results. Although synthetically related, there is no evidence for a direct phase transition between the α and β forms.

Heating of α-$KCuS_4$ to 250°C in vacuum decomposes the compound to CuS and K_2S_x. Both α- and β-$KCuS_4$ also require slow cooling; quenching the reaction of either compound from isotherm temperatures does not lead to their formation. The synthetic requirements of these compounds provide an illustrative example of just how sensitive a metastable compound can be to the conditions of molten salt synthesis.

Although other congeners of β-$KCuS_4$ have yet to be found, a related phase has been isolated in which the $(S_4)^{2-}$ ligands have been replaced with $(S_6)^{2-}$ species (53). Synthesized from a 0.8 : 1 : 8 molar ratio of Cs_2S/Cu/S heated at 270°C for 96 h, $CsCuS_6$ is a 1D material in which, as in β-$KCuS_4$, the Cu atoms are chelated by $(S_6)^{2-}$ ligands and bridged into chains by terminal S atoms (Fig. 16). Again the central core of the chain is edge-sharing $[CuS_4]$ tetrahedra. The material forms as red plates, and a band gap of 2.2 eV was measured. Although air and water stable, the compound has thermal stability only until 300°C.

In exploring the reactivity of Cu in Na_2Se_x fluxes, the new phase $Na_3Cu_4Se_4$ was isolated (52). A 58% yield of the black, needlelike compound was achieved from a reaction of Na_2Se/Cu/Se in a 4 : 1.5 : 8 molar ratio at 350°C for 4 days. Less basic fluxes were found to yield the selenide analogue of α-$KCuS_4$ as salts of K and Cs. The $Na_3Cu_4Se_4$ compound is isostructural with the sulfide of the same stoichiometry (54) and features $[Cu_4Se_4]$ rings, in a pseudochair conformation, fused together into 1D chains (Fig. 17). The Cu and Se atoms possess trigonal planar and pyramidal coordination respectively. An oxidation state formalism of $Na_3(Cu^{1+})_4(Se^{2-})_3(Se^{1-})$ has been proposed, in keeping with the fact that Cu^{2+} has proven too oxidizing to exist in chalcogenide environments (55). Hence the material is expected to have *p*-type metallic behavior due to the holes on the Se 4*p* band.

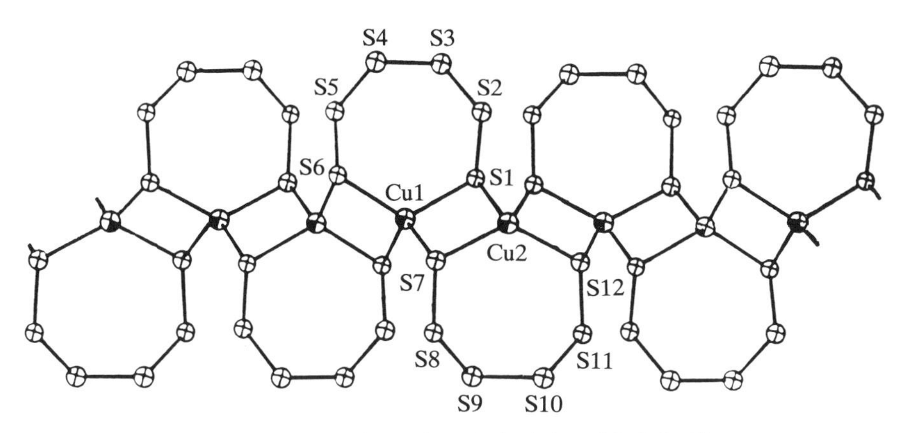

Figure 16. A portion of the anionic chains of $CsCuS_6$ (53).

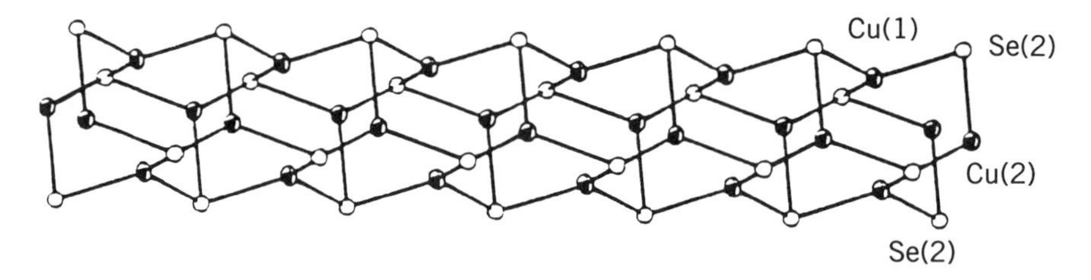

Figure 17. A view of the anionic chain of $Na_3Cu_4Se_4$ (52).

Initial experiments with Na_2Se_x fluxes afforded not only $Na_3Cu_4Se_4$ but also a second phase, crystallizing as thin black plates. Single-crystal studies revealed this phase to be $Na_{1.9}Cu_2Se_2 \cdot Cu_2O$ (56). The oxygen source was determined to be contaminated Na_2Se, and subsequent optimization of this synthesis was achieved through the reaction of freshly made Na_2Se with Cu_2O and Se in a 1.5 : 1 : 4 molar ratio at 330°C for 7 days (84% yield based on Cu). The structure is shown in Fig. 18 and consists of two alternating anionic layers of $[CuSe]_n^{n-}$ and Cu_2O with Na^+ in between. The $[CuSe]_n^{n-}$ layer possesses the anti-PbO structure type with Cu in tetrahedral geometry and Se, square pyramidal. The Cu_2O layer represents an ordering never before seen in Cu/O chemistry. The Cu atoms are linearly coordinated to two O atoms while the O atoms have square planar geometry; this results in a layer that is anti to the CuO_2 layers found in many high Tc superconductors. The Cu—O distance of 1.957(1) Å is in the typical range for such bonds (1.85–1.98 Å). Although initially formulated with two full Na atoms per formula, discovery of *p*-type metallic behavior in what should have been a semiconductor prompted a reexamination of the single-crystal data. The nonstoichiometry in the Na cations was subsequently found from refinement of the occupancies. Although other defects could not be ruled out, a deficiency of Na^+ would result in holes in the material's valence band and subsequent *p*-type conductivity. A room temperature resistivity of 1.1×10^{-4} Ω cm was measured. Not only novel in its own right, $Na_{1.9}Cu_2Se_2 \cdot Cu_2O$ indicates that molten polychalcogenide salts may also provide access to new oxychalcogenide materials as well.

Polytelluride fluxes also lead to ample new chemistry with Cu. The compound NaCuTe was synthesized from a 5 : 1 : 8 molar ratio of Na_2Te/Cu/Te heated at 400°C for 5 days (52). Single-crystal studies on the black, platelike crystals revealed the structure shown in Fig. 19. The anionic layers are in an ideal anti-PbO structure type: $[CuTe_4]$ tetrahedra edge sharing in two dimensions. Sodium cations lie between the layers. The NaCuTe compound can be thought of as a reduced derivative of CuTe, which has a distorted anti-PbO structure wherein the tetrahedra are squeezed by the presence of Te—Te bonds. Upon reduction of the CuTe layers by a full electron, the Te—Te bonds are

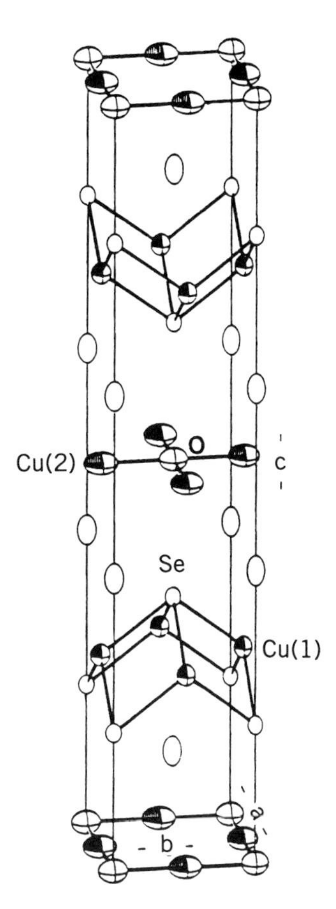

Figure 18. ORTEP representation of the unit cell of $Na_{1.9}Cu_2Se_2 \cdot Cu_2O$ (large open ellipses = Na; small open ellipses = Se; ellipses with nonshaded octants = O; ellipses with shaded octants Cu) (56).

broken, leaving the $[CuTe]_n^{n-}$ layers free of distortions. Curiously, NaCuTe possesses a very dissimilar structure from that of KCuTe. The K^+ analogue has $[CuTe]_n^{n-}$ layers that form BN-type sheets of six membered $[Cu_3Te_3]$ rings (57). With the addition to this family of the Na containing phase, a clear picture forms of an anionic structure changing in response to changing cation size. Since the Na analogue has anionic layers with more Cu—Te connectivity than the K analogue, it is reasonable that if a Li phase were possible, it would possess an even more interconnected anionic lattice, possibly resulting in some 3D structure. Similar arguments would predict a Cs^+ phase having reduced Cu—Te connectivity, perhaps resulting in a structure based on 1D anionic chains.

Upon moving to K_2Te_x fluxes, a new phase is found which, like NaCuTe, can be thought of as related to CuTe. This time the compound is $K_2Cu_5Te_5$ (58), which resulted from a reaction at 350°C for 3 days of a 3 : 1 : 8 molar ratio of $K_2Te/Cu/Te$. Where NaCuTe represented a CuTe layer that had been re-

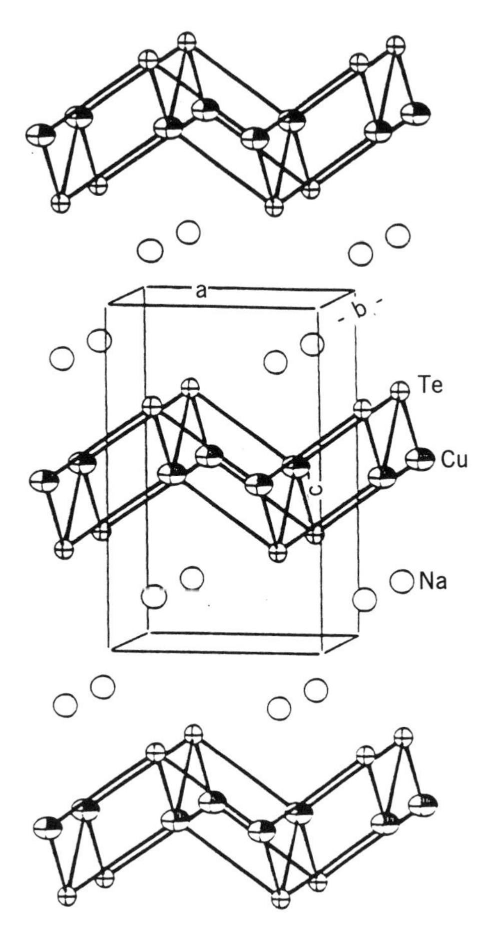

Figure 19. The view parallel to the anionic layers of NaCuTe (52).

duced by one full electron, $K_2Cu_5Te_5$ is an intermediate case: a CuTe layer that has been reduced by only 0.4 electrons. The anionic layers of $K_2Cu_5Te_5$ form an anti-PbO structure type distorted by the presence of Te—Te bonds; however, being more reduced than the layers of CuTe, there are fewer Te—Te bonds to distort the layers. Only one Te—Te contact of appropriate size for a bonding interaction is present per formula in $K_2Cu_5Te_5$, and so the layers form as shown in Fig. 20. Even taking into account a $(Te_2)^{2-}$ unit does not fully balance the formal charges of the compound. If all Cu atoms are considered 1+ and the Te_2 fragment kept at 2−, then the remaining monotellurides would require an average charge of −1.67. This model is supported, as *p*-type metallic behavior is exhibited by the compound with a room temperature conductivity of 1.5×10^4 S cm^{-1}. Also characteristic of metals, $K_2Cu_5Te_5$ exhibits Pauli-type temperature independent paramagnetism.

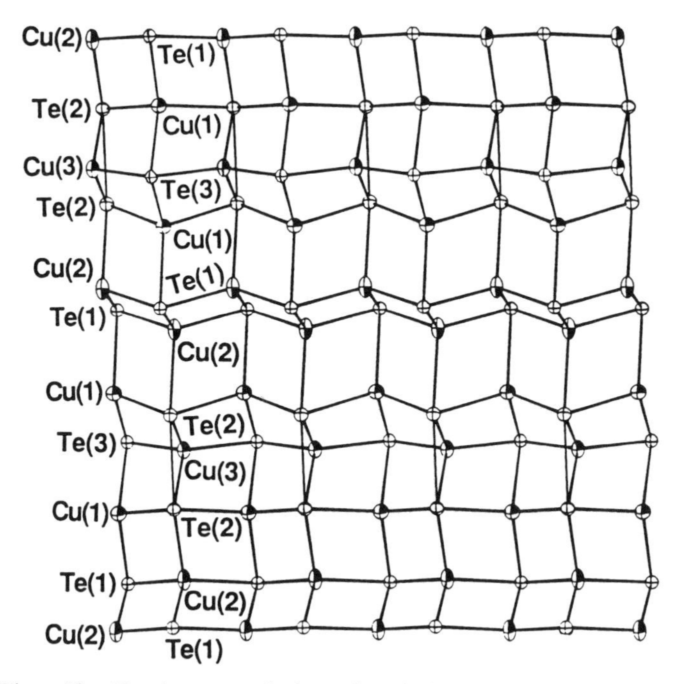

Figure 20. The view perpendicular to the anionic layers of $K_2Cu_5Te_5$ (58).

By altering the reaction conditions of $K_2Cu_5Te_5$ with a simple change in the molar ratio of K_2Te/Cu/Te to 3:2:12, a new compound with a totally new structure type emerged: $K_4Cu_8Te_{11}$ (59). The basic unit of this 3D structure is a novel pentagonal dodecahedral cluster $[Cu_8(Te_2)_6]$ [Fig. 21(*a*)]. The cluster is not void, but encapsulates a lone K^+ ion, suggesting both a template effect in the growth of the cluster and a necessity that the large negative charge of the cluster be somehow partially stabilized in order for formation to take place. Each face of the cluster is formed from a $[Cu_2Te_3]$ pentagon in which a $(Te_2)^{2-}$ lies on one of the sides. The faces are arranged such that the $(Te_2)^{2-}$ units form mutually perpendicular sets on the boundaries of the dodecahedron. The extended structure is formed from this basic cluster. First, the $[Cu_8(Te_2)_6]$ clusters form fused dimers by sharing one of the $(Te_2)^{2-}$ edges and one monotelluride [Fig. 21(*b*)]. The dimers link into chains by sharing the Te(1)—Te(2) edge, and layers are formed by the bonds between Cu(2) and Te(6) atoms in opposite chains [Fig. 21(*c*)]. Finally, the layers are connected via Te(8)—Te(8)′ bonds, neither atom of which participates in the bonding of the dodecahedron (Fig. 22). The remaining K^+ ions reside in channels parallel to the *b* axis lined by 24-member rings. The Cu atoms all have distorted tetrahedral geometry while the Te atoms are in either four-coordinated square pyramids (Te at the apex),

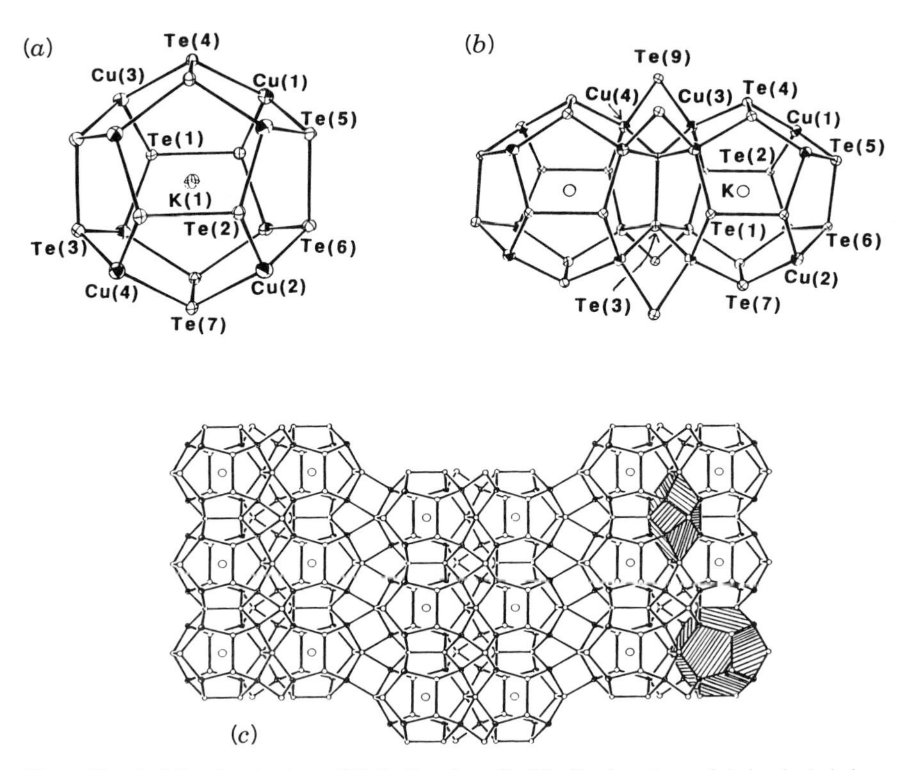

Figure 21. Building the structure of $K_4Cu_8Te_{11}$ from the $[Cu_8Te_{12}]$ pentagonal dodecahedral cluster (*a*) which forms a dimer to a second cluster via bridging monotellurides and one ditelluride (*b*). The dimers then further join into 1D stacks that further connect into layers (*c*) before being linked into the extended $K_4Cu_8Te_{11}$ structure (see Fig. 22) (59).

five-coordinate square pyramids (Cu at the apex), or trigonal pyramids. The material is valence precise and so is semiconducting, although it possesses an unusually high room temperature conductivity for a semiconductor (~ 160 S cm^{-1}).

Since a potassium cation appears to be playing a role in the stabilization of the $[Cu_8(Te_2)_6]$ cluster, the logical next question would be what might occur upon substitution for a larger cation. In moving to the next largest alkali metal cation, Rb^+, the cluster reoccurs, this time manifesting itself in a different structure type. The $Rb_3Cu_8Te_{10}$ compound was isolated (60) from a reaction of 2 : 1 : 8 molar ratio of Rb_2Te/Cu/Te heated at 350°C for 4 days. As in $K_4Cu_8Te_{11}$, a lone alkali ion is again encapsulated in the $[Cu_8(Te_2)_6]$ cluster, but now the larger size of the Rb^+ forces a 2D structure rather than the previous 3D one. Within the anionic layers of $Rb_3Cu_8Te_{10}$, each cluster is bridged to four others by sharing four of its $(Te_2)^{2-}$ edges, leaving the remaining pair of $(Te_2)^{2-}$ edges

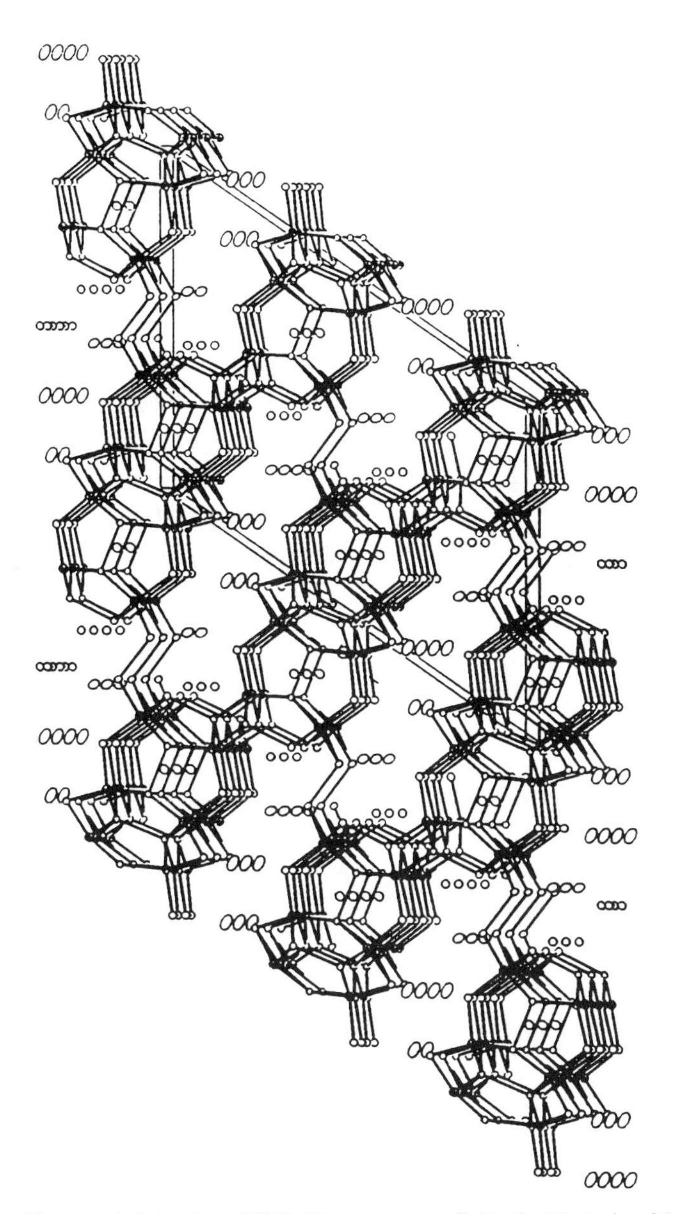

Figure 22. The extended structure of $K_4Cu_8Te_{11}$ as seen parallel to the 1D stacks of dodecahedral clusters (see Fig. 21) (59).

directed towards the interlayer gallery (Fig. 23). The tetrahedral geometry of the Cu atoms is completed by μ^4-Te^{2-} ions positioned above and below the layers. The cluster is also formed when Cs_2Te_x is employed as the flux. The compound $Cs_3Cu_8Te_{10}$ (52) possesses layers identical to those in $Rb_3Cu_8Te_{10}$, but where the later is in a monoclinic space group, the layers of the former are shifted into a more symmetrical orthorhombic modification (Fig. 24). Surpris-

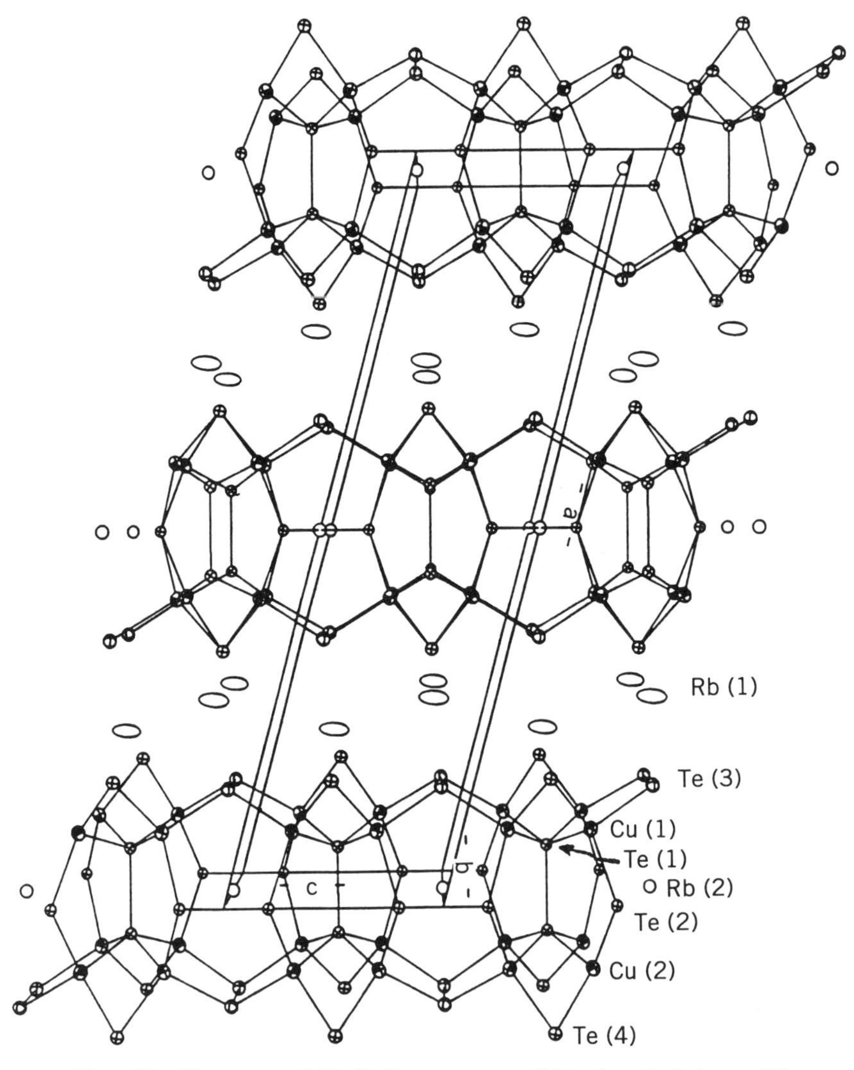

Figure 23. The compound $Rb_3Cu_8Te_{10}$ as seen parallel to the anionic layers (60).

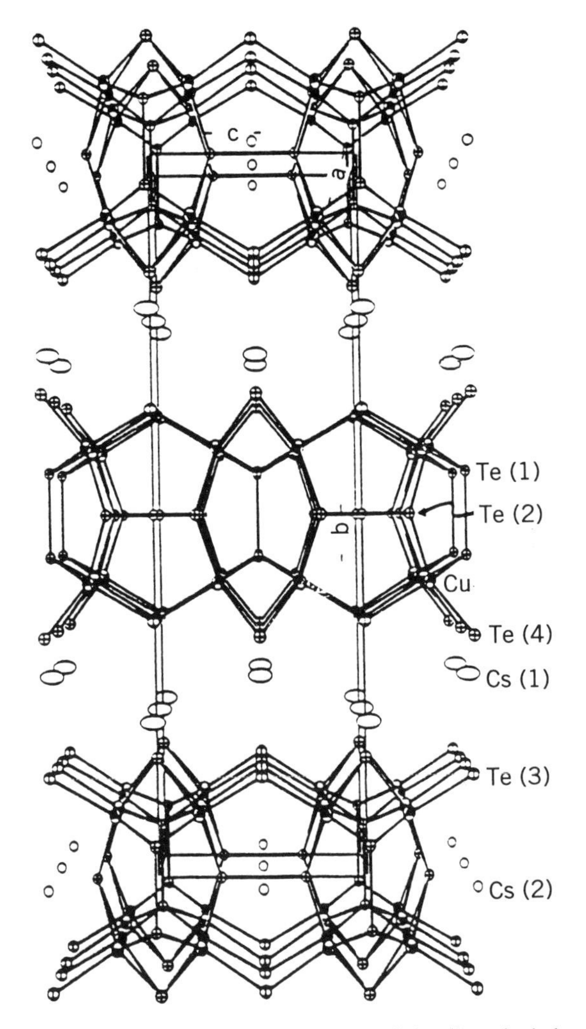

Figure 24. The compound $Cs_3Cu_8Te_{10}$ as seen parallel to the anionic layers (52).

ingly, there is little volume change in the cluster as the size of the encapsulated cation is increased, implying a substantial amount of rigidity and bond stability. The larger cations simply fit more snugly in the cavity. In addition to structural differences, the $A_3Cu_8Te_{10}$ phases also differ from $K_4Cu_8Te_{11}$ in that they are not valence precise. With the Cu atoms formally $1+$ and the ditellurides left as $2-$, a negative charge of $3-$ is left to be distributed over the two remaining monotellurides. Although a simple model, this formalism does predict *p*-type metallic behavior resulting from a partially oxidized Te 5*p* band. Such behavior is indeed observed for $Cs_3Cu_8Te_{10}$ in thermopower studies, and conductivity measurements revealed a room temperature resistivity of 1.1×10^{-3} Ω cm.

Given the fact that the framework built from the $[Cu_8(Te_2)_6]$ clusters varies with alkali cation size, the possibility of accessing new structure types by using mixed alkali metal fluxes was investigated. From a reaction of $K_2Te/Cs_2Te/Cu/Te$ in a 2:1:2:12 molar ratio heated as 420°C for 4 days, $K_{0.9}Cs_{2.1}Cu_8Te_{10}$ was successfully isolated and characterized (60). The compound is isostructural with the orthorhombic $Cs_3Cu_8Te_{10}$, except that the different cations exhibit positional preferences within the structure. The interlayer galleries are filled exclusively with Cs^+, accounting for two of the alkali atoms in the formula. The last position, that within the cage cluster, experiences an occupancy disorder in which 90% of the time K^+ is present, 10% of the time, Cs^+. That K^+ is not at full occupancy is curious, but since it is the major resident, the cluster does seem to be exhibiting some extent of size selectivity. This finding suggests that a cation such as Ba^{2+}, which has an ionic radius approximately equal to that of K^+, could also be encapsulated within the cluster. Investigations into such chemistry will be discussed in a later section, Section VI.E.

D. When M = Au

The chemistry of Au in molten polychalcogenide salts illustrates that substantially different products are possible simply by working with a different chalcogenide. This arises from differences in their electronegativities, chain stabilities, redox potentials, and the stability of known binaries. In polysulfide fluxes only one new ternary compound, $KAuS_5$, has been uncovered, and reactions of Au with polytellurides lead invariably to the highly stable binary phase, $AuTe_2$, even at very low temperatures. The bulk of new compounds containing Au have resulted from reactions in polyselenides melts.

From a reaction of $Na_2Se/Au/Se$ in a 0.9:1.02:4 molar ratio heated at 290°C for 99 h, black needles of $NaAuSe_2$ were isolated in 78% yield (61). The structure is shown in Fig. 25 and features square planar $[AuSe_4]$ units and all monoselenides. Two $[AuSe_4]$ square plane edges are shared to form a $[Au_2Se_6]$ dimer, which subsequently shares corners with other dimers in building a 2D corrugated layer. This structure represents a never before seen structure type for an MX_2 combination.

Polyselenides were successfully incorporated into a second sodium salt of gold: Na_3AuSe_8 (61). The key synthetic modification, in forming this compound, lies in changing the ratio of $Na_2Se/Au/Se$ to 1:1:8. The increased selenium content promotes a flux that contains longer polyselenide units, hence increasing the probability of their incorporation into the final product. The reaction temperature was only modestly increased to 310°C, resulting in purple needles of Na_3AuSe_8 in approximately 67% yield. Its structure proved to be 1D (Fig. 26). Gold atoms are again square planar but now are linked into chains

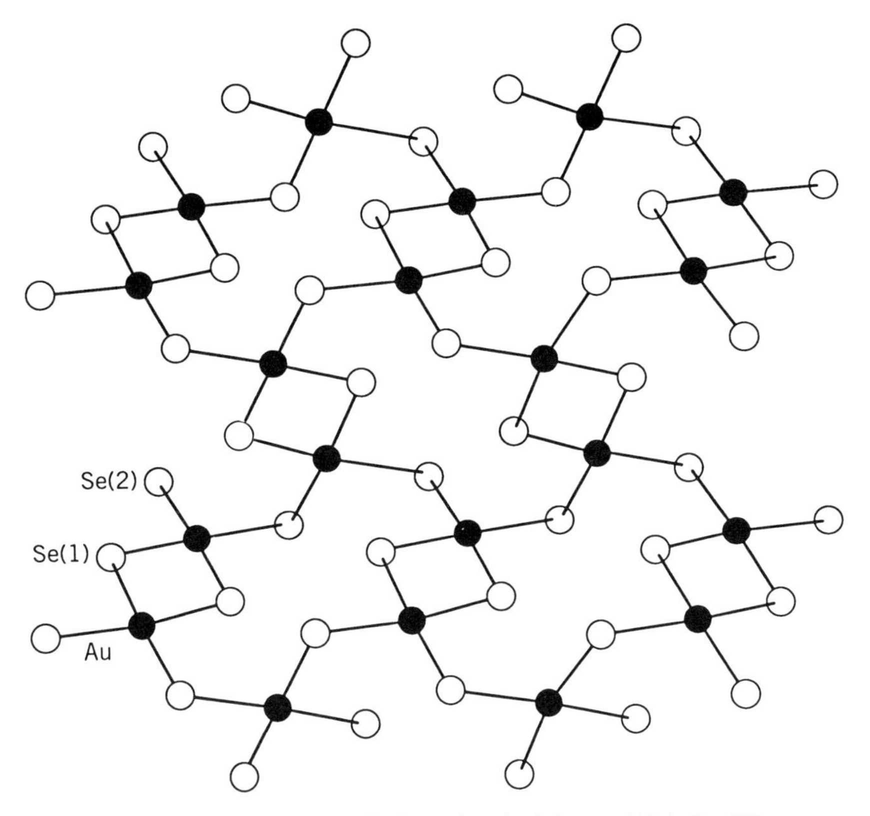

Figure 25. The view perpendicular to the anionic layers of $NaAuSe_2$ (61).

via $(Se_2)^{2-}$ units bridging the Au^{3+} centers. The diselenides lie trans to each other across the square plane, and in the remaining two sites are terminal $(Se_3)^{2-}$ ligands. These ligands have the curious feature of only bonding to a metal center at one end of the polychalcogenide chain while the other end is totally noninteracting. This sharply contrasts with the usual bridging or chelating modes seen

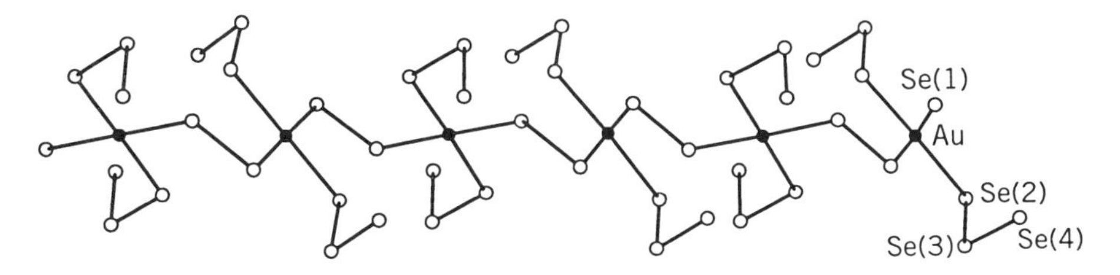

Figure 26. The anionic chain of Na_3AuSe_8 (61).

in other $(Q_x)^{2-}$ bonding. The Na^+ cations apparently play a role in stabilizing the dangling end of the $(Se_3)^{2-}$ as a portion of the compound's countercations are positioned in a pocket formed by the $(Se_3)^{2-}$ of neighboring chains.

An equally unusual dangling $(Se_5)^{2-}$ occurs when the alkali cation is changed to potassium. The K_3AuSe_{13} compound was synthesized from a 1.8 : 1 : 8 molar ratio of $K_2Se/Au/Se$ heated at only 250°C for 99 h (57% yield) (62). The material forms 1D chains analogous to those in Na_3AuSe_8, albeit containing polyselenides of different lengths. In K_3AuSe_{13} the $[AuSe_4]$ square planes are bridged by $(Se_3)^{2-}$ ligands, while the length of the dangling ligands has increased to $(Se_5)^{2-}$ (Fig. 27). The role of the K^+ cations in stabilizing the $(Se_5)^{2-}$ ligand is somewhat more evident in this compound as the chain is seen to curve about, forming a pocket containing a K^+. The occurrence of a pentaselenide in the solid state is a rarity even in the more conventional bridging [as in $[V_2Se_{13}]^{2-}$ (63) and $[In_2Se_{21}]^{4-}$ (24k)] or chelating [as in $[Fe_2Se_{12}]^{2-}$ (24m) or Cp_2TiSe_5 (64)] binding modes.

In one of the most striking examples of the dependence of product stabilization on flux composition, when the molar ratio is changed from the 1.8 : 1 : 8 used in the K_3AuSe_{13} synthesis to 2 : 1 : 8 under the same heating conditions, an entirely different phase is isolated: $KAuSe_5$ (62). The structure is shown in Fig. 28. Pentaselenide ligands are again present but now are in a more conventional bridging mode. A more startling difference from K_3AuSe_{13} is that these ligands now bridge linearly coordinated Au^+ centers. That a change in the Au oxidation state of the final product would occur on such a modest change in flux composition is remarkable, but it may be more of an indication of a very sensitive

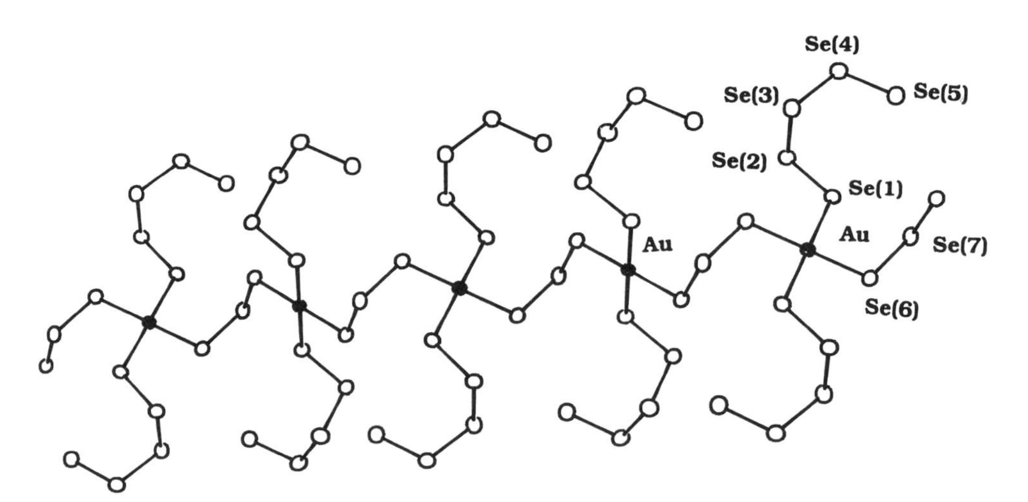

Figure 27. The anionic chain of K_3AuSe_{13} (62).

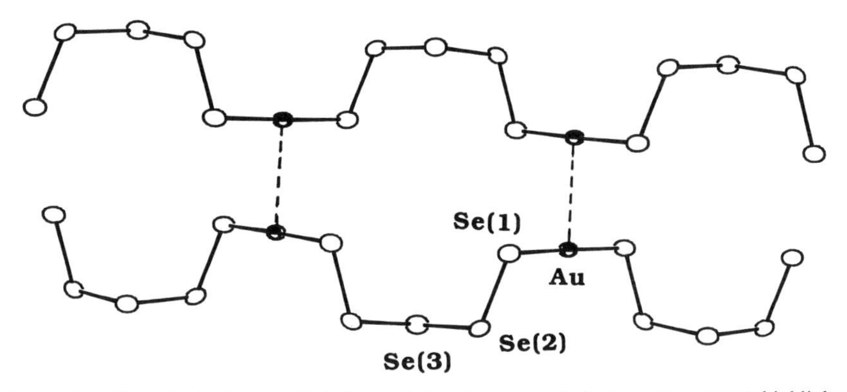

Figure 28. The anionic chains of $KAuSe_5$ with the short, interchain Au—Au contacts highlighted by dashed lines (62).

redox equilibrium between Au^+ and Au^{3+} in the polyselenide flux rather than of any real change in its reactivity. Interestingly, the sulfide analogue of $KAuSe_5$ has also been synthesized (K_2S/Au/S = 1.8 : 1 : 8 heated at 250°C for 99 h) (52), but under the synthetic conditions of K_3AuSe_{13}, which itself has no sulfide analogue yet discovered.

Upon increasing the Au concentration in the flux, a new phase containing only monoselenides was isolated. Apparently, the need to oxidize more Au metal drives the flux to shorter $(Se_x)^{2-}$ fragments. The compound $KAuSe_2$ was synthesized from similar synthetic conditions as were used for $NaAuSe_2$ (K_2Se/Au/Se = 1.8 : 2.04 : 8 heated for 99 h at 290°C) (52), but the larger size of the potassium cations drives a change in the structure away from that of the Na salt. Shown in Fig. 29, the centrosymmetric anionic chains of edge-sharing $[AuSe_4]$ square planes are of the $PdCl_2$ structure type (65) and isostructural with those seen in $(PtS_2)_n^{2n-}$ (66).

To date only one new structure has been seen in the reaction of Au with Cs_2Se_x flux. The compound $CsAuSe_3$ was synthesized from a 4 : 0.96 : 8 molar ratio of Cs_2Se/Au/Se heated at 350°C for 54 h (52). The structure is merely build from linear Au^+ bridged by $(Se_3)^{2-}$ ligands (Fig. 30). The chains are noncentrosymmetric, running parallel to the *a* axis in a helical fashion.

E. When M = Cd

The binary chalcogenides of cadmium (CdQ) have the zincblende structure type and are exceedingly stable, thermodynamically. These compounds form readily under the relatively mild synthetic conditions of intermediate temperature A_2Q_x fluxes even when the flux composition is one that has been seen to promote polychalcogenide inclusion in other systems. Despite this thermody-

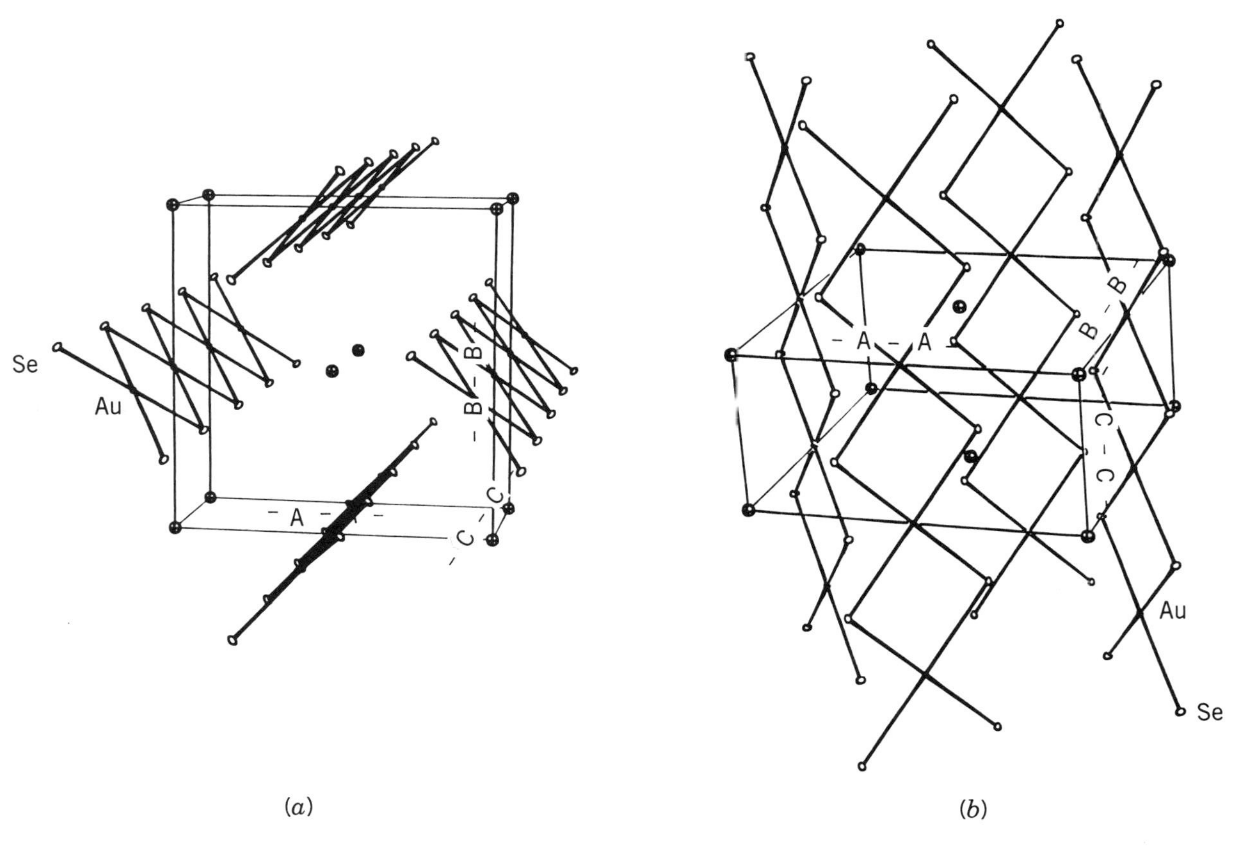

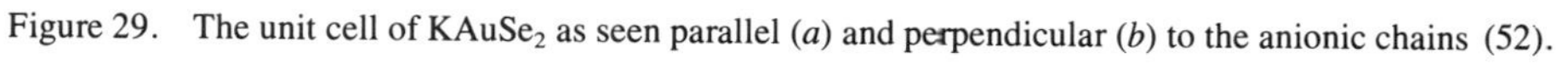

Figure 29. The unit cell of $KAuSe_2$ as seen parallel (*a*) and perpendicular (*b*) to the anionic chains (52).

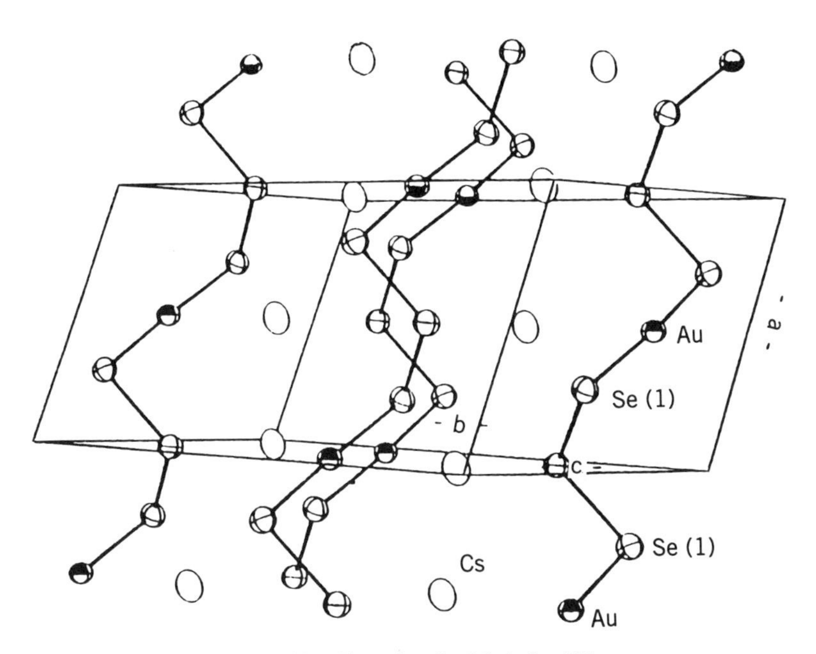

Figure 30. The unit cell of $CsAuSe_3$ (52).

namic road block, several new ternary phases containing Cd have been synthesized under conditions in which high concentrations of alkali metal cations are present in the flux.

Several compounds have been isolated that possess the formula $A_2Cd_3Q_4$ (A = K, Rb; Q = S, Se, Te) (67), and the conditions of their synthesis are given in Table IV. All of them employ fluxes that are predominantly $(Q_2)^{2-}$ in composition and as was shown in Table I, dichalcogenide fluxes require much higher melting points. Hence, $K_2Cd_3S_4$ must be synthesized at 600°C, and the selenide analogues require temperatures just beyond the region defined as "intermediate" in this chapter. The structure determined from single-crystal studies on $K_2Cd_3S_4$ is shown in Fig. 31. It possesses 2D anionic layers composed

TABLE IV
Synthetic Conditions for the Formation of $A_2Cd_3Q_4$ Phases

Compound	A_2Q/Cd/Q Ratio	Heating Program	Cooling Rate
$K_2Cd_3S_4$	7.5/1/10	600°C (2 days)	5°C h^{-1}
$K_2Cd_3Se_4$	3/1/4	650°C (2 days)	2°C h^{-1}
$Rb_2Cd_3Se_4$	6/1/6	650°C (2 days)	5°C h^{-1}
$K_2Cd_3Te_4$	1/3/3	800°C (2 days)	10°C h^{-1}

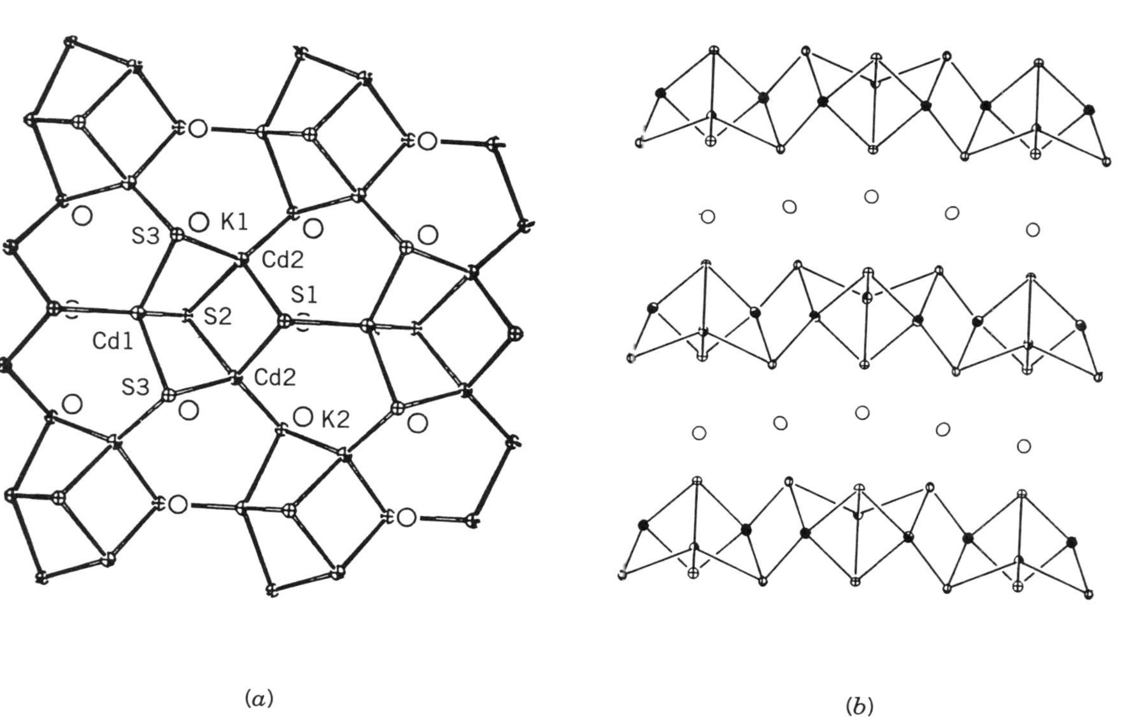

Figure 31. Two views of the structure of $K_2Cd_3S_4$: (*a*) perpendicular and (*b*) parallel to the anionic layers (67).

of interconnecting $[Cd_3S_4]$ truncated cubes. Each Cd is in characteristic tetrahedral geometry, and the K^+ resides between the layers. The breaking of a CdQ structure into the layers of $A_2Cd_3Q_4$ has an effect similar to the quantum size effect seen in nanometer size CdS particles. Specifically, with less Cd—Q bonding in $A_2Cd_3Q_4$, the Cd—Q bands become narrower, and the band gap increases relative to the binaries. For $K_2Cd_3S_4$, the band gap was estimated at 2.75 eV (CdS band gap = 2.44 eV) and for $Rb_2Cd_3Se_4$ and $K_2Cd_3Se_4$ the band gaps were 2.39 and 2.36 eV, respectively (CdSe band gap = 1.77 eV). The $K_2Cd_3Te_4$ has a band gap of 2.26 eV (CdTe band gap = 1.5 eV).

An even more alkali metal rich phase was found in $K_2Cd_2S_3$ (68). Synthesized from a 6 : 1 : 6 molar ratio of K_2S/Cd/S heated at 600°C for 6 days, the compound has a 3D structure with K^+ residing in tunnels running through the lattice. The basic repeating unit is built from two tetrahedra that edge share to form a $[Cd_2S_6]$ dimer. These dimers then corner share with other dimers, forming the walls of the structure's tunnels (Fig. 32). As with $A_2Cd_3Q_4$, the decreased Cd—S overlap in $K_2Cd_2S_3$ increases the band gap relative to the binary. The compound $K_2Cd_2S_3$ has a band gap of 2.89 eV.

A third new structure type has been found from reaction in Na_2Se_4 flux. The compound $Na_4Cd_3Se_5$ was synthesized using a 1 : 1 ratio of CdSe and Na_2Se_4 heated at 350°C for 2 days (69). Its structure is shown in Fig. 33. The Cd atoms are again in tetrahedral coordination, and the basic unit of the 2D structure is a trimer of corner sharing $[CdSe_4]$. The trimers then stack on top of each other via corner sharing along the *b* axis, and further corner sharing between the stacks leads to the formation of the anionic layers. It was noted by the authors that under similar heating conditions, stoichiometric mixtures of the elements led only to the formation of binary compounds. Although the mechanism of formation is not clear, some unfavorable reaction is apparently avoided by directly reacting the binary with polyselenide rather than forming the polyselenide *in situ*. Again, this serves to demonstrate the profound, and often surprising, effect flux composition has on the mechanism of these reactions.

F. When M = Hg

To date only two new structure types have been isolated in the A/Hg/Q system: $A_2Hg_6Q_7$ (A = K, Q = S or A = Cs, Q = Se) and $A_2Hg_3S_4$ (A = K, Q = S, Se or A = Cs, Q = Se). Both possess only monochalcogenide ligands even though they were synthesized under conditions that have generated many of the $(Q_x)^{2-}$ containing compounds discussed previously. This finding illustrates that the individual chemistry of the metal investigated often takes precedence over the reactive conditions present in the flux.

The compound $K_2Hg_6S_7$ crystallized as black needles from a reaction of K_2S_3

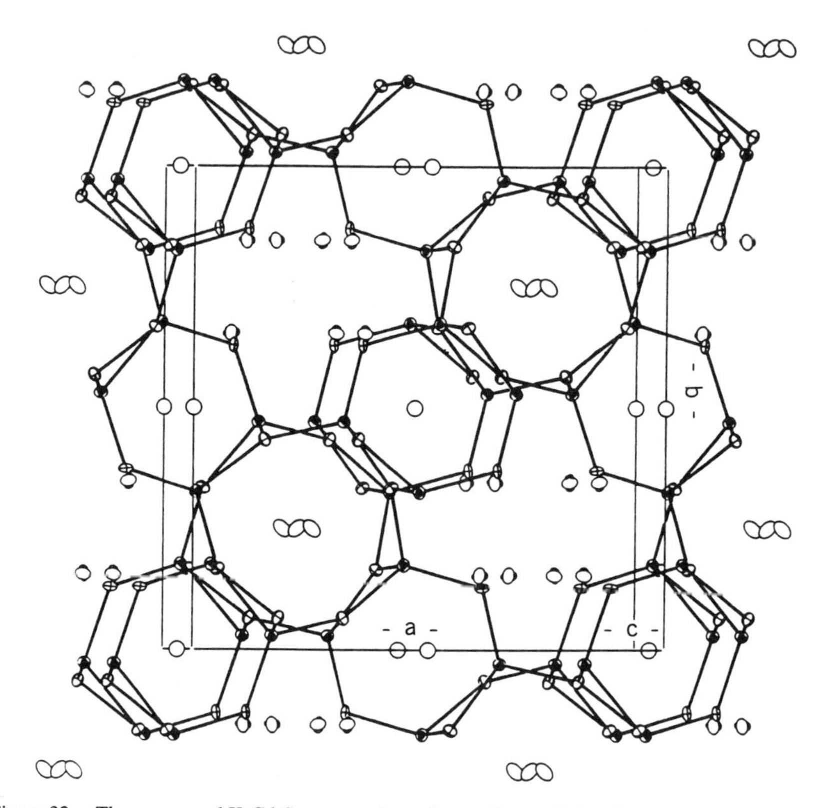

Figure 32. The compound $K_2Cd_2S_3$ as seen down the *c* axis, parallel to the tunnels running through the 3D anionic network (open ellipses = K; ellipses with octant shading = Cd; ellipses without octant shading = S) (68).

and HgS in a 4 : 1 ratio heated at 210°C for 3 days (70). The material is insoluable in water and common organic solvents. Its structure is 3D (Fig. 34) and features two different types of 1D tunnels running through the lattice. The first tunnel possesses an octagonal cross-section formed from distorted [HgS_4] tetrahedra bridged by trigonal pyramidal S^{2-} and has an approximate diameter of 4.77 Å. The second type of tunnel is larger and T-shaped, formed from Hg^{2+} in both tetrahedral and linear coordination and μ_2- and μ_3-S^{2-}. This compound is particularly noteworthy because the K^+ counterions reside only in the second, larger tunnels, leaving the smaller tunnels vacant. The stereochemically active lone pairs on the μ_3-S^{2-} atoms are directed into the interior of the larger tunnels and presumably exerts enhanced Coulombic attraction for the K^+. Hence, the anionic framework of $K_2Hg_6S_7$ may exhibit intriguing ion-exchange properties. The related selenide phase was isolated as the Cs^+ salt but could only be pre-

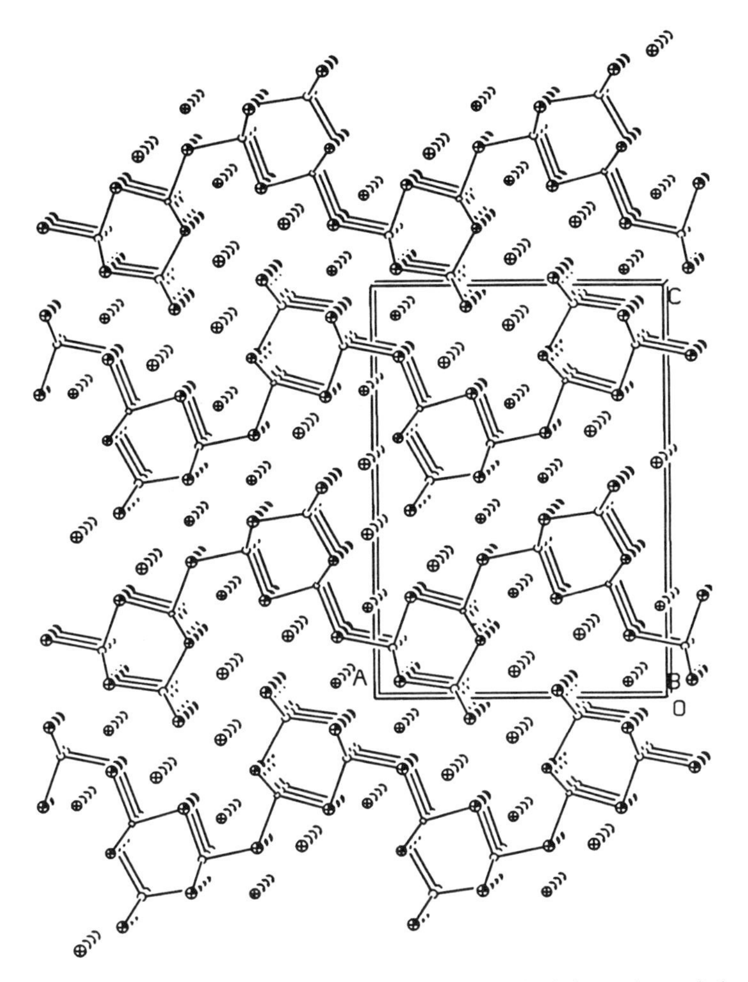

Figure 33. The structure of $Na_4Cd_3Se_5$ as seen parallel to the anionic layers (open circles = Cd; circles with nonshaded octants = Na; circles with shaded octants = S) [drawn from data in (69)].

pared by a stoichiometric combination of $Cs_2Se/HgSe$ (1:6) heated at 375°C for 3 days (52). It is crystallographically different from the sulfide in that a fourfold screw axis runs down the smaller vacant tunnels rather than the fourfold inversion axis present in the sulfide.

When a 3:1 molar ratio of K_2S_4 and HgS is heated at 220°C for 4 days, yellow hexagons of $K_2Hg_3S_4$ result (70). This material is light and moisture sensitive, degrading to HgS upon exposure. As with $K_2Hg_6S_7$, its structure is built from S^{2-} and both tetrahedral and linear Hg^{2+}. This phase, however, is 1D, and a view of the anionic chains is shown in Fig. 35. The chains, which

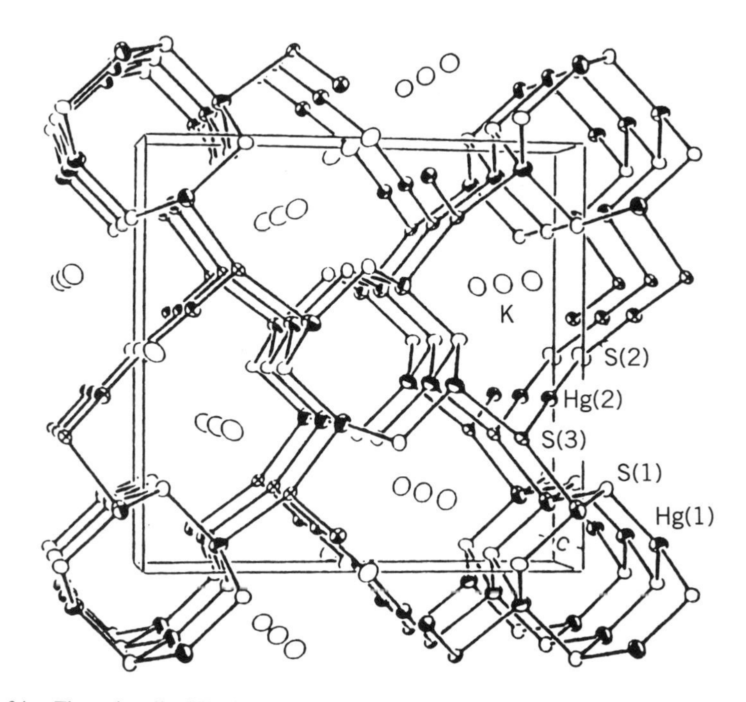

Figure 34. The unit cell of $K_2Hg_6S_7$ as seen down the *c* axis, parallel to the 1D channels of the lattice (70).

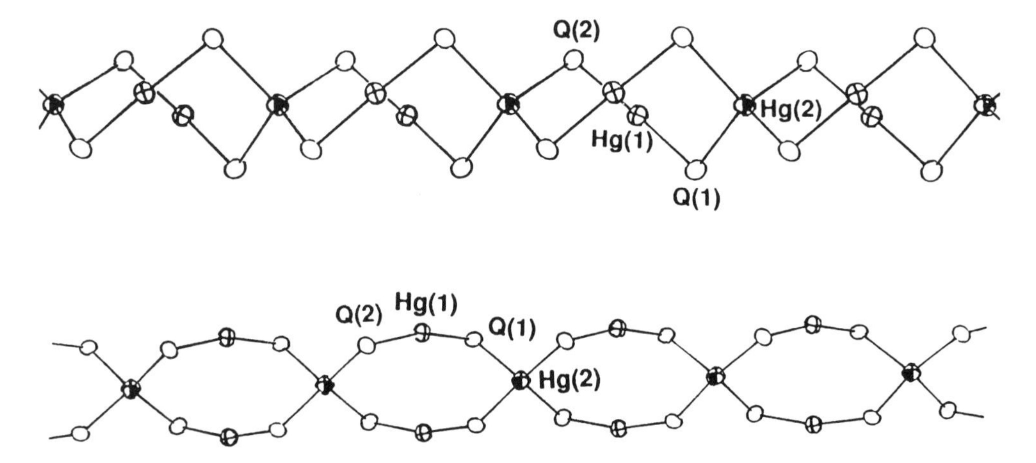

Figure 35. Two views of the anionic chains of $K_2Hg_3Q_4$ (Q = S or Se), the second being a 90° rotation of the first (70).

run parallel to the *b* axis, are built from [HgS_4] tetrahedra linked by linearly coordinated Hg^{2+}, forming a spiral of [Hg_4S_4] fused rings. The selenium analogue was subsequently synthesized from a similar flux composition heated at 250°C (70), and $Cs_2Hg_3Se_4$ has also been prepared, but under conditions of *in situ* Cs_2Se_x formation (Cs_2Se/HgSe/Se in a 2 : 1 : 8 ratio heated at 250°C for 99 h) (52).

The chemistry of Hg in A_2Q_x fluxes is similar to that of Cd in that the resulting ternary compounds do not possess polychalcogenides but can be thought of as the binary chalcogenide broken down by incorporating varying amounts of A_2Q into the lattice. In the case of the Hg systems, this may, in fact, be close to a mechanistic approximation of the flux reaction since HgS is used as the Hg starting source. As such, $A_2Hg_3Q_4$ can be formulated as A_2Q + 3 HgQ, and $A_2Hg_6Q_7$, as A_2Q + 6 HgQ. Indeed, the later was used as the synthetic ratio for forming $Cs_2Hg_6Se_7$. This concept has often been used in various ternary solid state systems, forming whole families of compounds, such as $(ZnS)_n(In_2S_3)_m$ (71). A recent example in Hg chemistry is the formation of $Na_2Hg_3S_4$ from a stoichiometric combination of Na_2S and HgS reacted at 597°C (72). Using this idea as a guide may yet lead to new compound formation in both the Hg and Cd systems.

G. When M = Sn

The chemistry of Sn in A_2S_x fluxes has proven to be very diverse. This diversity is in spite of the high stability of the $(Sn_2S_6)^{4-}$ anion, which is simply a dimer of two edge-sharing tetrahedra (73). This species readily forms from fluxes in which the A_2S/S ratio is 3 : 8 or greater, but under less basic conditions [i.e., fluxes containing longer $(S_x)^{2-}$], the anion has been either incorporated into some manner of extended structure or avoided all together.

One new phase that incorporates the [Sn_2S_6] unit is $A_2Sn_2S_8$ (A = K or Rb) (74). The potassium salt was synthesized from a 2 : 1 : 8 molar ratio of K_2S/Sn/S heated at either 250 or 275°C for 4–6 days. Upon shifting to Rb, two crystallographic modifications are seen: an α form, synthesized from a 2 : 1 : 12 ratio of Rb_2S/Sn/S heated for 4 days at 330°C, and a β form isolated from a 1 : 1 : 8 ratio heated for 4 days at 450°C. All three compounds possess the same anionic layers, but while $K_2Sn_2S_8$ and α-$Rb_2Sn_2S_8$ are isostructural with a monoclinic space group, the layers in β-$Rb_2Sn_2S_8$ are merely shifted into a more symmetrical orthorhombic space group. The structure of the layers is built from chains of [SnS_4] tetrahedra and [SnS_6] octahedra linked into layers via bridging $(S_4)^{2-}$ ligands (Fig. 36). The polyhedra within the chains are segmented into two alternating dimers: the previously mentioned edge-sharing tetrahedra of [Sn_2S_6] and edge-sharing octahedra of [Sn_2S_{10}]. The two dimers link via corner sharing between the [Sn_2S_6] and the axial corners of the [Sn_2S_{10}] octahedra. The chains

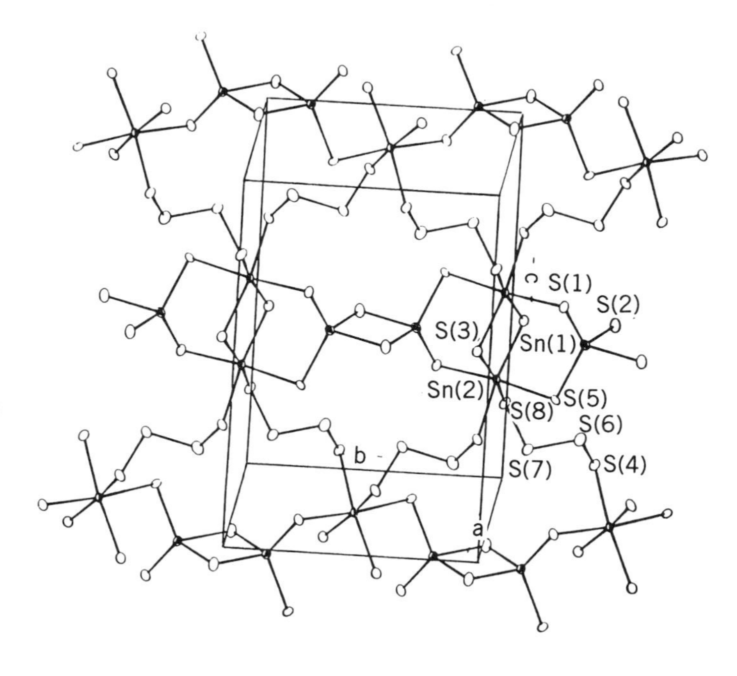

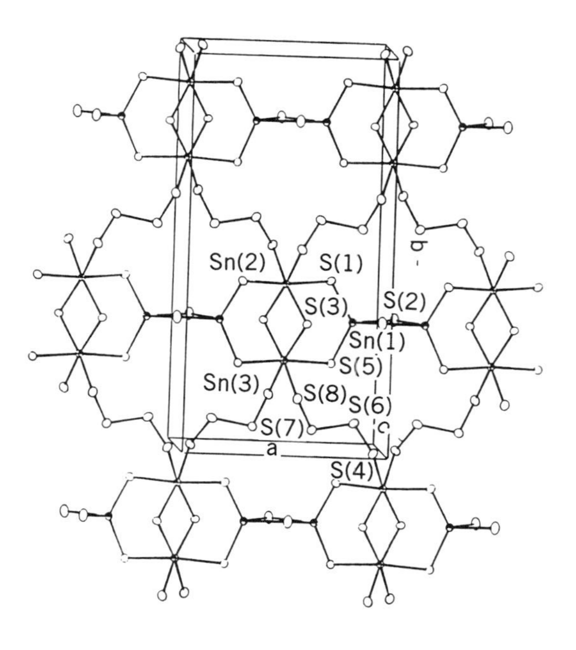

Figure 36. (*a*) The anionic layers of the monoclinic α-$Rb_2Sn_2S_8$. (*b*) The anionic layers of the orthorhombic β-$Rb_2Sn_2S_8$ (74).

are bridged into layers by $(S_4)^{2-}$ ligands that join two $[Sn_2S_{10}]$ dimers from opposite chains at the equatorial S atoms not already involved in metal bridging. The geometry about the Sn octahedra is fairly regular; however, the $[SnS_4]$ tetrahedra experience a high degree of distortion. This arises from a long contact between the tetrahedral Sn atoms and the bridging equatorial S atoms in the $[Sn_2S_{10}]$ dimer [2.934(5), 2.865(2), and 2.875(3) Å in the K, α-Rb, and β-Rb salts, respectively]. If this long contact is considered in the coordination, then a distorted trigonal bipyramidal geometry about the formerly tetrahedral Sn could be approximated, with the long contact being one of the axial bonds. This interaction apparently elongates the bond trans from the long contact, making it 2.534(6) Å while the normal Sn—S bond for a tetrahedral coordination is 2.45 Å [as in $Na_4Sn_2S_6$ (73)].

The ratio of K_2S/Sn/S employed in synthesizing $K_2Sn_2S_8$ (2 : 1 : 8) leads to an entirely different phase simply upon increasing the reaction temperature to 320°C for 4–6 days. Under such conditions, $K_2Sn_2S_5$ is formed in 81% yield (74). This compound is stable in a limited temperature window, and its yield decreases drastically at 400°C to 10%, with the major phase now being $K_4Sn_2S_6$. This temperature dependence is, however, more an example of the complex equilibria present in the flux rather than any phase metastability because $K_2Sn_2S_5$ has already been prepared at 797°C from a stoichiometric combination of the elements (75). Indeed, the isostructural $Tl_2Sn_2S_5$ (76) $K_2Sn_2Se_5$ (75), and $Rb_2Sn_2Se_5$ (77) are also known, suggesting the structure type has a fair amount of thermodynamic stability. The structure, shown in Fig. 37, is a 3D assembly of $[SnS_5]$ distorted trigonal bipyramids. The units form chains by sharing edges bounded by axial and equatorial atoms and stack such that two chains run along the [110] direction for every one that runs along the [−110], forming a criss-cross pattern. The chains are linked via sharing of the equatorial S atoms not involved in intrachain bridging. The framework forms helical tunnels parallel to the *b* axis as the interchain bonds spiral from chain to chain.

The new compound $Cs_2Sn_2S_6$ (74) can be thought of as the $K_2Sn_2S_5$ anionic framework forced open by the different atom packing requirements of a larger cation. The chains of edge-sharing trigonal bipyramids seen in $K_2Sn_2S_5$ are also present in $Cs_2Sn_2S_6$ but are now connected into layers via the formation of S—S bonds between the nonbridging equatorial S atoms on neighboring chains. Within the layers, all chains are parallel rather than criss-crossed as in $K_2Sn_2S_5$, and the disulfide linkages lead to the formation of large cavities within the layers themselves (Fig. 38). The compound $Cs_2Sn_2S_6$ forms as thin yellow plates in 80% yield from the reaction of Cs_2S/Sn/S in a 1 : 1 : 8 molar ratio heated at 400°C for 4 days.

A second compound in the Cs_2S/Sn/S system is encountered when a molar ratio of 2 : 1 : 6, 8, or 10 is heated at 250–275°C for 4–6 days. This synthesis yields a mixture of $Cs_2Sn_2S_6$ and red crystals of Cs_2SnS_{14} (74). The latter is a

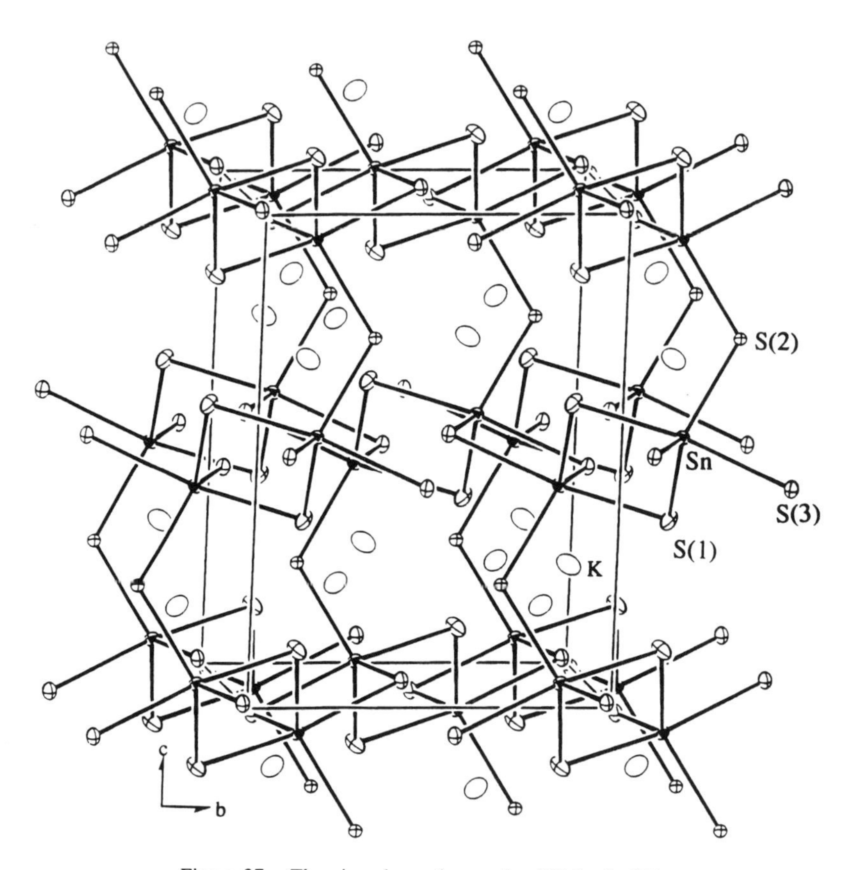

Figure 37. The view down the *a* axis of $K_2Sn_2S_5$ (74).

discrete anion surrounded by Cs^+ (Fig. 39). The Sn^{4+} is octahedrally coordinated by three chelating polysulfide chains, two $(S_4)^{2-}$ and one $(S_6)^{2-}$. The material is stable in dry air, but is slowly solubilized in water. The presence of the $(S_6)^{2-}$ ligand makes this a new species, but an all $(S_4)^{2-}$ analogue is known as the $(Me_4N)^+$ salt (77) as well as the polyselenide anion $(PPh_4)_2Sn(Se_4)_3$ (78, 79). The compound Cs_2SnS_{14} is also another example of structure formation based on cation size dependence; the exact same synthetic conditions with a potassium flux yielded $K_2Sn_2S_8$. This does not rule out the formation of K_2SnS_{14} entirely; the anion could still be present, but being a K^+ salt it may be soluble in the solvents used to remove the excess flux.

All of the A/Sn/S compounds have been found to be semiconducting with very low room temperature conductivities ($< 10^{-7}$ S cm^{-1}). Band gaps have been measured as 2.16 eV for the $A_2Sn_2S_8$ compounds, 2.36 eV for $K_2Sn_2S_5$, and 2.44 eV for Cs_2SnS_{14}.

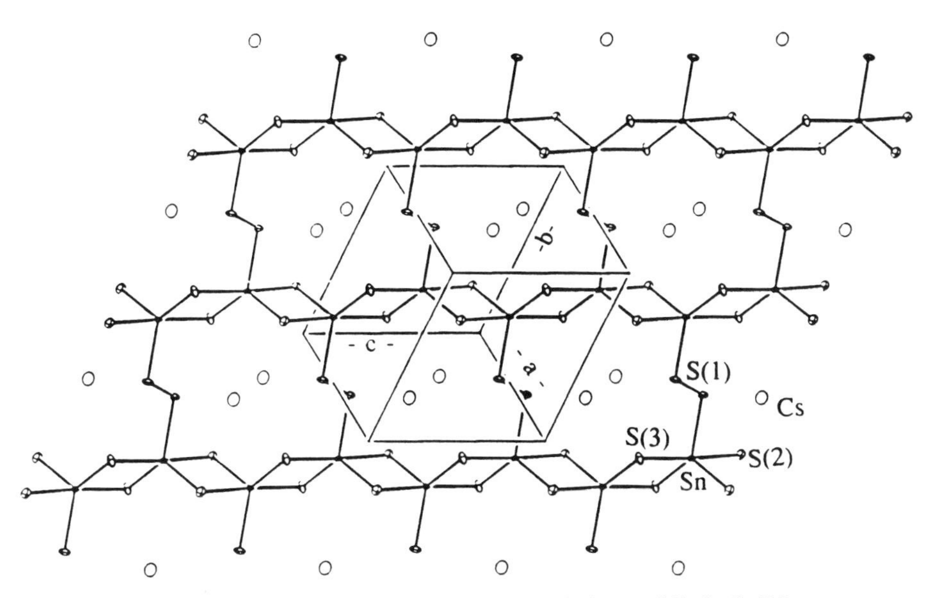

Figure 38. The view perpendicular to the anionic layers of $Cs_2Sn_2S_6$ (74).

H. When M = Sb

Investigations into the reactivity of Sb in polychalcogenide salts have lead to two new ternary compounds. Both were synthesized from Cs_2S_x fluxes, and both exhibit extended structures formed under the influence of a stereochemically active lone pair on the Sb^{3+}.

The compound $CsSbS_6$ was isolated (80) from a reaction of Cs_2S/Sb/S in a 1:1:9 molar ratio at 260°C for 5 days. Its structure is 1D in which $[Sb_2S_2]$ rhombic units are linked into a chain by bridging $(S_5)^{2-}$ ligands (Fig. 40). Each Sb atom is bonded to four S atoms with distances that are typical for Sb—S bonds [2.390(2)–2.737(2) Å] and has a stereochemically active lone pair. The geometry about the Sb atoms is trigonal bipyramidal with the lone pair in an equatorial position exerting the expected distortions from ideality on the bond angles. The material, which forms as dark orange crystals, is stable in air and water, soluble in ethylenediamine, and possesses a band gap of 2.25 eV.

The second phase is found upon changing the molar ratio of the reactants to 1.1:1:8 and increasing the reaction temperature to 280°C for 5 days. These changes result in an orange powder, presumably $CsSbS_6$, which is removed by washing the sample with ethylenediamine, and insoluble red-orange crystals, which proved to be the new phase $Cs_2Sb_4S_8$ (80). This second structure is 2D. As in $CsSbS_6$, rhombi of $[Sb_2S_2]$ are present in this new structure, but are now linked by trigonal pyramidal $[SbS_3]$ units that bridge the rhombi into chains

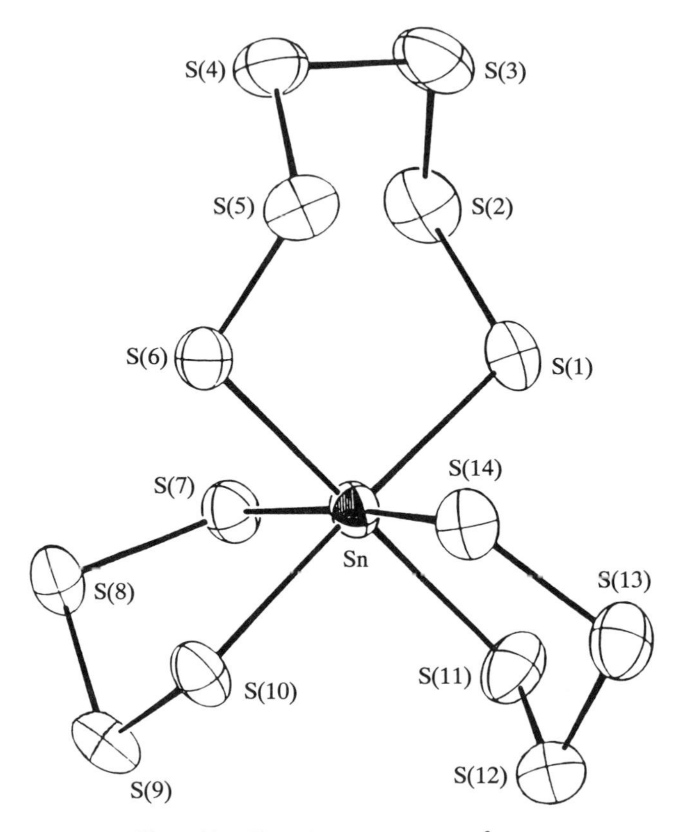

Figure 39. The anionic cluster $[SnS_{14}]^{2-}$ (74).

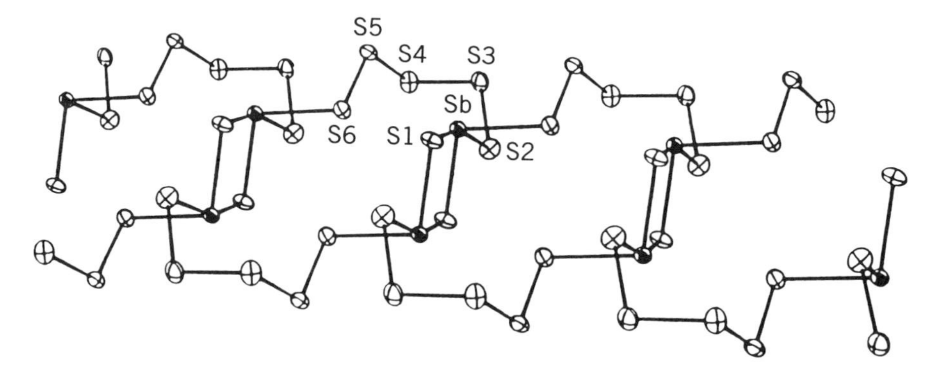

Figure 40. The anionic chains of $CsSbS_6$ (80).

along the [110] direction. These chains are then linked into layers via $(S_2)^{2-}$ ligands that bridge [SbS_3] from neighboring chains. The Sb atoms of both units have a mixture of short and long contacts to neighboring S atoms, which are shown in Fig. 41 as solid and dashed lines, respectively. The Sb(1) atom has four normal bonds to S ranging from 2.418(2)–2.783(2) Å and one long contact at 3.277(2) Å, which is well below the sum of the van der Waals radii. This leads to a square pyramidal geometry of S atoms about Sb(1) with the lone pair presumably occupying the position trans to the axial S. Meanwhile, Sb(2) possesses three normal bonds to S ranging from 2.440(2)–2.592(2) Å, one contact at a distance of 3.022(2) Å, and a very long contact to a S atom in a neighboring layer at 3.456(2) Å. The overall geometry about Sb(2) is hence a distorted

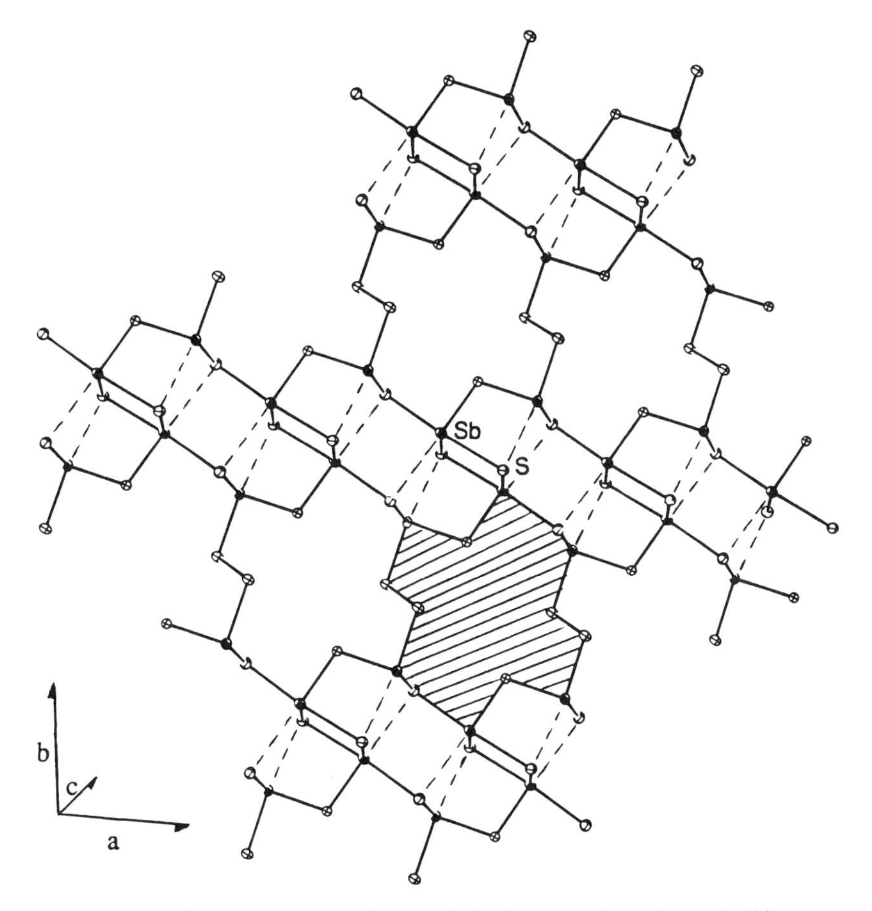

Figure 41. A single anionic layer of $Cs_2Sb_4S_8$ as seen down the *c* axis (80).

trigonal bipyramid. Within the anionic layers of $Cs_2Sb_4S_8$ are 14-member rings that stack in registry from layer to layer, forming tunnels parallel to the *c* axis.

I. When M = Bi

Although similar to Sb^{3+} in that a lone pair of electrons is sometimes stereochemically active, Bi^{3+} also exhibits coordination environments in which its lone pair does not seem to be playing a role. This has led to a very different set of new phases from the reactions of Bi in A_2Q_x fluxes.

In the compound KBi_3S_5 the stereochemical effect of the Bi^{3+} lone pair is minimal (81). Isolated from a flux containing $K_2S/Bi/S$ in a 2.6 : 1 : 12 molar ratio heated at 300°C for 5 days, KBi_3S_5 forms a 3D anionic framework built entirely from edge-sharing $[BiS_6]$ octahedra (Fig. 42). The geometry about the

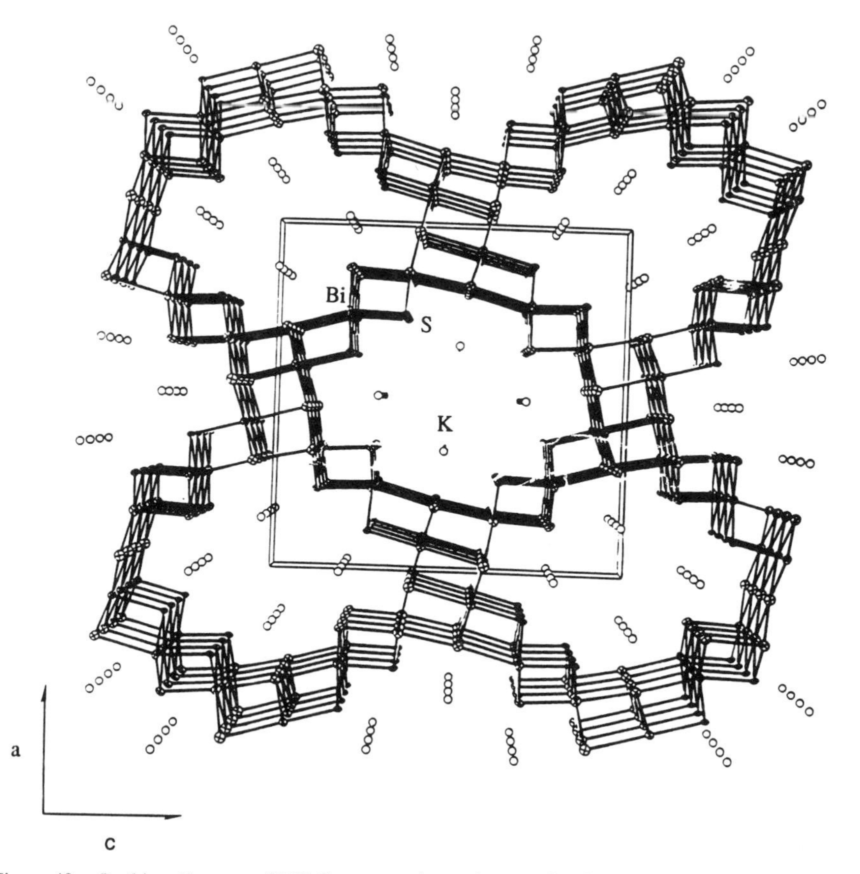

Figure 42. Packing diagram of KBi_3S_5 as seen down the tunnels of the 3D anionic network (parallel to the *b* axis) (81).

Bi atoms is somewhat distorted due to alternating long and short bonds in the axial positions, but the Bi^{3+} lone pair remains effectively delocalized. The main feature of the material is the presence of large tunnels running parallel to the *b* axis. These tunnels are lined with a 20-member ring of alternating Bi and S atoms, and the distances across the tunnel (measured from S to S) range from 9.45 Å at the shortest point to 14.02 Å at the longest. Within these tunnels reside the K^+; however, only one full cation is actually present, being disordered over two crystallographic sites. The compounds $RbBi_3S_5$ and $CsBi_3S_5$, although of different structure types, also have alkali cations residing in the tunnels of a 3D anionic framework. Their cations, however, show no disorder, and the tunnels are smaller than those in KBi_3S_5. Hence, the disorder of the K^+ must be helping to stabilize the larger tunnels of KBi_3S_5. The large size of the tunnels and the partial occupancies of the K^+ sites should allow for appreciable ion mobility through the lattice, and in fact, the material has shown a limited capacity to ion exchange for species such as H_3O^+ and Rb^+. The compound is thermally unstable, decomposing to unknown phases at about 520°C. The compound KBi_3S_5 was also found to be a semiconductor with an approximate band gap of 1.26 eV.

In moving to a Cs_2S_x flux, the compound β-$CsBiS_2$ was isolated from a 4.5 : 1 : 8 molar ratio of Cs_2S/Bi/S heated at 290°C for 4 days (82). This red crystalline material is moisture sensitive, forming a purple coating on the surface of the crystals after prolonged exposure. The structure is shown in Fig. 43 and is of the $CsSbS_2$ structure type (83). The Bi^{3+} is coordinated in a trigonal pyramid of S atoms with the lone pair presumably occupying the site of a fourth tetrahedral corner. The $[BiS_3]$ units simply corner share in 1D forming a $[BiS_2]_n^{n-}$ polymer. The Cs^+ cations are directed towards the terminal S atoms of the chain while the Bi^{3+} lone pairs are oriented towards neighboring chains. This material exhibits semiconducting behavior with a band gap of 1.43 eV.

Typically, the $ABiS_2$ compounds form with anionic layers of the $CdCl_2$ structure type with alkali between the layers (84), and a variation of this was seen in γ-$CsBiS_2$ (82). Structural data could not be successfully refined for this phase, likely due to a high probability of crystal twinning, but the cell obtained possesses a *c* axis that is a multiple of the $RbBiS_2$ cell, suggesting the layers

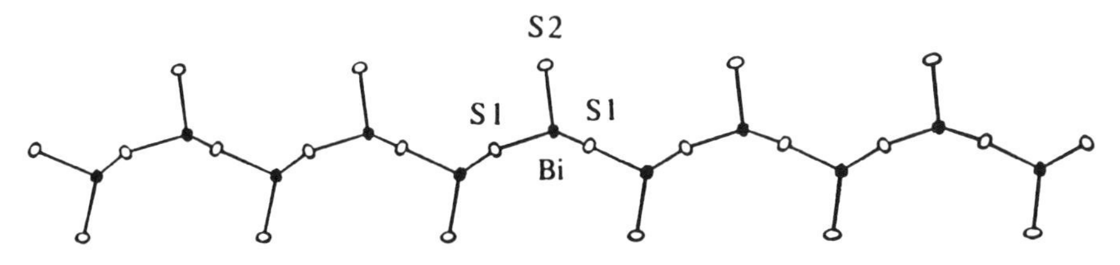

Figure 43. The anionic chains of β-$CsBiS_2$ (82).

may simply be twisted out of registry from that structure type. The compound's band gap was measured as 1.1 eV.

So far only one new ternary compound containing Bi has been discovered from reactions in polyselenide flux. The $K_2Bi_8Se_{13}$ compound was synthesized from a 2 : 1 : 8 molar ratio of $K_2Se/Bi/Se$ heated at 330°C for 10 days (82). This compound possesses a complex 3D anionic network built from $[BiSe_6]$ octahedra and $[BiSe_5]$ square pyramids and can actually be conceptualized as a hybrid of the known structures of Bi_2Te_3 (85), Sb_2S_3 (86), and CdI_2 (87). The structures of Bi_2Te_3 and CdI_2 are NaCl related: CdI_2 is a monolayer structure of face-sharing octahedra, while Bi_2Te_3 is a bilayer. The compound $K_2Bi_8Se_{13}$ possesses layers in which alternating fragments of these monolayer and bilayer structure types are fused together. These fragments are then linked into a 3D lattice by fragments resembling a third structure type, Sb_2S_3. The connecting fragments are composed of $[BiSe_5]$ square pyramids which, by sharing axial-to-equatorial edges, form 1D double chains. A view of the full structure is shown in Fig. 44 with the K^+ residing in the lattice tunnels. The mixing of these structure types has also been seen in the structures of $Cs_3Bi_7Se_{12}$ (88) and $Sr_4Bi_6S_{13}$ (89). The octahedral $[BiSe_6]$ units of $K_2Bi_8Se_{13}$ are nearly ideal, having as their main distortion a long axial bond trans to a short one, as was also present in KBi_3S_5. Meanwhile, the $[BiSe_5]$ units are more regular in their bond lengths [av equatorial bond = 2.95(3) Å; axial bond = 2.704(8) Å]. Although the Bi^{3+} lone pair is presumably stereochemically active at the base of the square pyramid, two long Bi—Se contacts are still made in that direction [average distance = 3.552(1) Å]. The compound $K_2Bi_8Se_{13}$ was formed in 88% yield as black needles, and a band gap for the material was measured at 0.76 eV. The room temperature conductivity was 10^{-2} S cm^{-1}.

J. When M = *f* Block Metal

Molten polychalcogenide salts provide a unique methodology for the study of lanthanide and actinide–chalcogenide reactivity. This result is due to the limits that the high oxophilicity of these elements imposes on typical low-temperature synthetic methods. The cations of the lanthanides and actinides are such hard acids that even oxygen atoms present on solvent molecules are preferred over the soft donating ability of the larger chalcogenides. This problem is largely eliminated by using A_2Q_x fluxes as nearly oxygen-free reaction media, but small amounts of oxygen adsorbed on the surface of the reactants may still influence the outcome. For example, $[UO_2(S_2)_3]^{4-}$ (Fig. 45), which crystallized as a sodium salt with free $(S_3)^{2-}$ fragments in the lattice, was prepared from a reaction using elemental U and an aged batch of Na_2S to make the Na_2S_x flux. Fresh Na_2S does not yield the compound, and although the Na^+ salt has

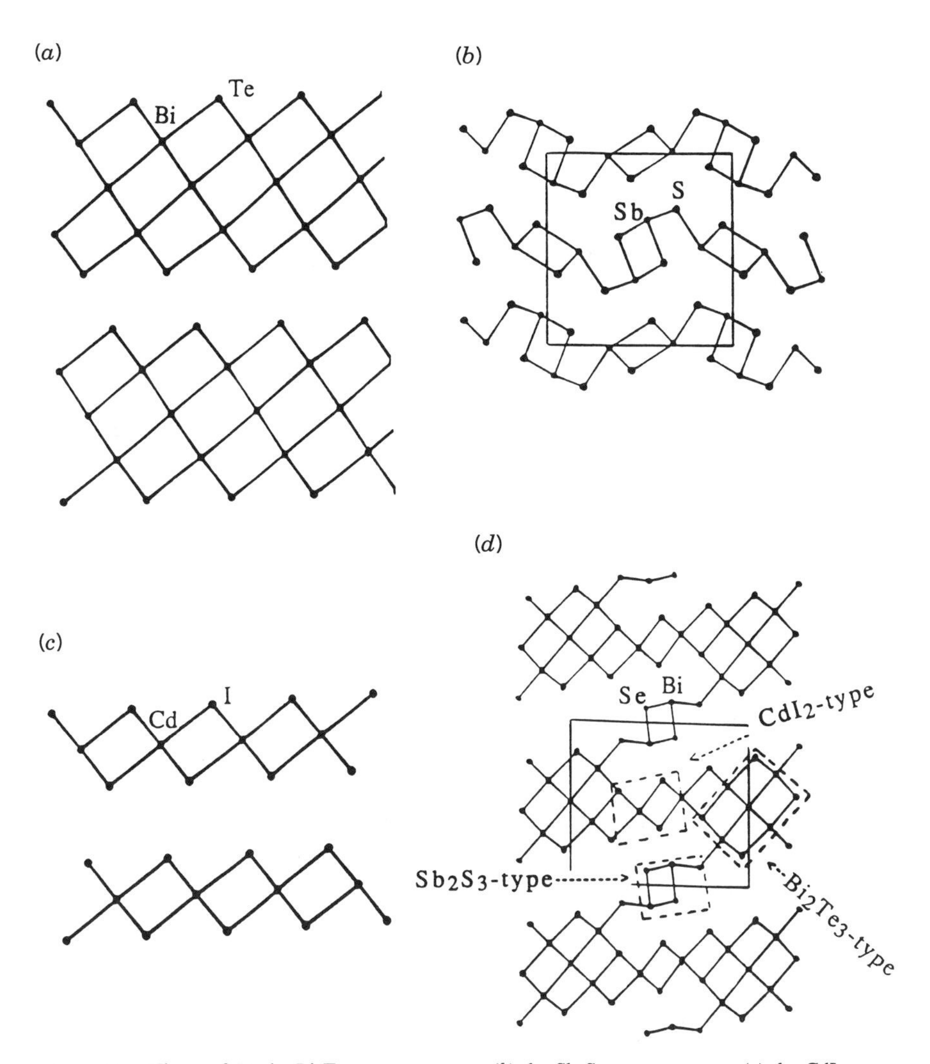

Figure 44. Views of (*a*) the Bi_2Te_3 structure type, (*b*) the Sb_2S_3 structure type, (*c*) the CdI_2 structure type, and the anionic layers of $K_2Bi_8Se_{13}$ (*d*) with each of the preceding structural features highlighted (82).

yet to be reproduced rationally, the Cs^+ analogue has recently been isolated (90). Hence, fresh reactants and oxygen-free preparation environments are especially crucial for this chemistry.

A fully reproducible, nonoxygen contaminated phase was discovered in $KCeSe_4$ (91). The material was formed from the reaction of $K_2Se/Ce/Se$ in a

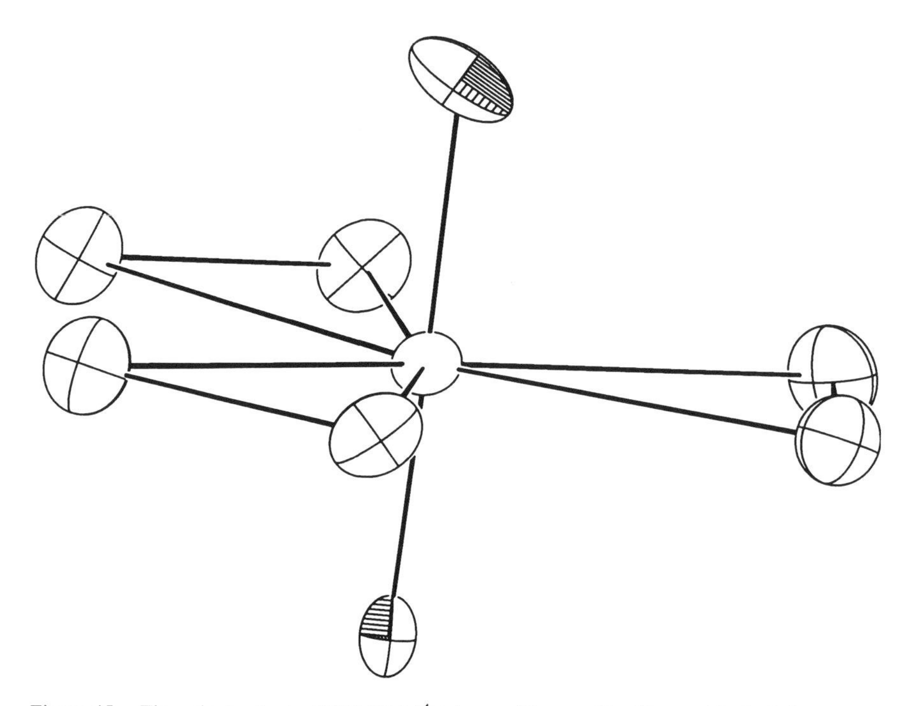

Figure 45. The anionic cluster $[(UO_2)(S_2)_3]^{4-}$ (open ellipse = U; ellipse with shaded octants = O; ellipse with nonshaded octants = S) (90).

2 : 1 : 8 molar ratio heated at 300°C for 6 days. It possesses a morphology of dark blue to black square chunks, and the yields are typically 61%. The compound is layered, and a view of the anonic component is shown in Fig. 46. The Ce^{3+} is in a square antiprismatic geometry, coordinated to a Se atom from eight different $(Se_2)^{2-}$ ligands. Each Se atom bridges two Ce atoms, and so the entire $(Se_2)^{2-}$ unit is η^4. The atoms are highly segregated, with the $(Se_2)^{2-}$ ligands all coplanar and sandwiching an equally coplanar layer of Ce^{3+}. The K^+ lie in the interlayer gallery. Magnetic studies confirmed the presence of paramagnetic Ce^{3+} with no transitions to magnetically ordered states seen down to 2 K. The band gap was measured as 1.59 eV. Recently, a Tb analogue was also isolated, indicating that the phase may be commonly formed among all the lanthanides.

The K_4USe_8 phase was synthesized by reacting U metal in a polyselenide flux of composition similar to that used for $KCeSe_4$ ($K_2Se/U/Se$ = 2 : 1 : 8.1) at 300°C for 10 days (92). Rather than an extended structure, the compound is composed of discrete anions of $[U(Se_2)_4]^{4-}$ charge balanced by K^+ ions. The U^{4+} is surrounded by four chelating $(Se_2)^{2-}$ units forming a triangulated dodecahedral geometry (Fig. 47). This cluster is distorted from an ideal dodeca-

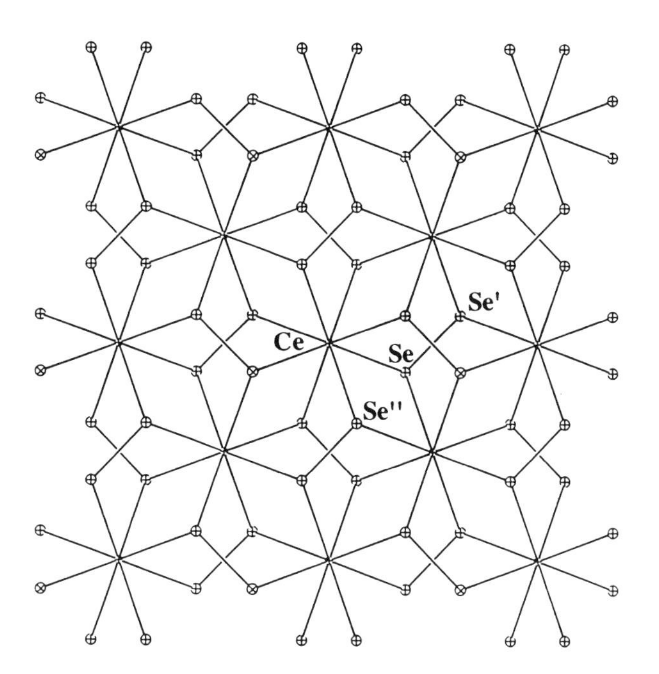

Figure 46. The view perpendicular to a single anionic layer of the $KCeSe_4$ (91).

hedron in two respects: (1) the U^{4+} is nearly coplanar with the four equatorial Se atoms where an ideal dodecahedron would have the Se atoms alternating above and below the equatorial plane and (2) the Se—Se bonds enforce a smaller bond angle than is present in the corresponding ideal solid. The material is soluble in water, DMF, and ethylenediamine(en) but insoluble in acetonitrile unless complexants, such as 222-cryptand (crypt) and 15-crown-5 (15C5) or, 18-crown-6 (18C6) crown ethers, are used to draw the K^+ into solution. If isolated from air, the $[U(Se_2)_4]^{4-}$ solutions in the organic solvents give characteristic UV–vis spectra and are stable for 3 days, after which they decompose and show spectral features for only polyselenides. The solubility of the anion may in fact lead to its use as a starting material for further solution chemistry.

The reaction of either Ce or La in Na_2S_x flux ($Na_2S/Ln/S$ = 1:1:4 heated at 370°C for 2 days) gives high yield of the phase, $NaLnS_3$ (Ln = La or Ce) (93). The Ln/S anionic framework is analogous to that of the known binary structure type, $ZrSe_3$ (94). Two monosulfides and two disulfides coordinate to the Ln atoms in a trigonal prism, and the prisms then stack in 1D by sharing trigonal faces. The stacks further link into layers as the monosulfides cap the trigonal prisms of neighboring stacks (Fig. 48). While a Zr^{4+} has no problem charge balancing such a network of S^{2-} and $(S_2)^{2-}$, the trivalent lanthanides

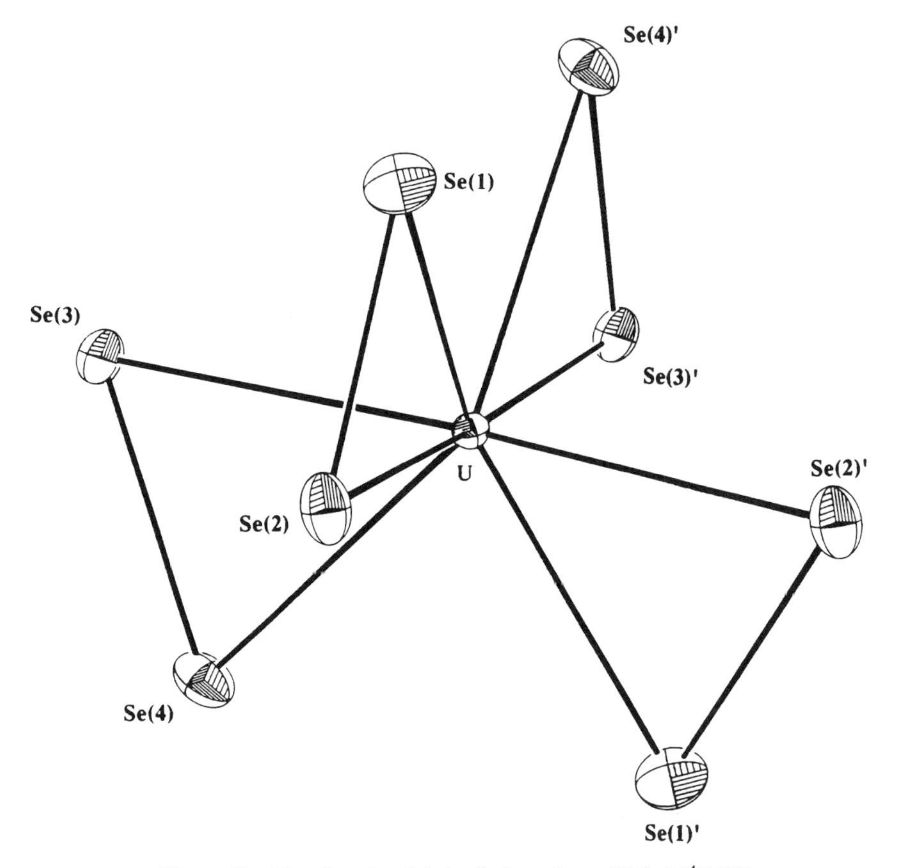

Figure 47. The triangulated dodecahedron cluster $[U(Se_2)_4]^{4-}$ (92).

need a little help from Na cations, which form a staggered bilayer in the interlayer gallery. The La analogue is a pale yellow color with a band gap of 2.6 eV, but substitution with Ce leads to a red phase with a 2.1-eV band gap.

V. $AuCuSe_4$: MAKING TERNARIES IN A_2Q_x THAT DO NOT INCORPORATE A^+

Although the alkali metal cations of the A_2Q_x flux are typically incorporated into the new compounds, there is no inherent reason why that should be so in every reaction. Despite the low-reaction temperatures used, thermodynamics is still a major force in these reactions, meaning that if a certain phase has sufficient stability under given conditions it will certainly form. As a prelude to our

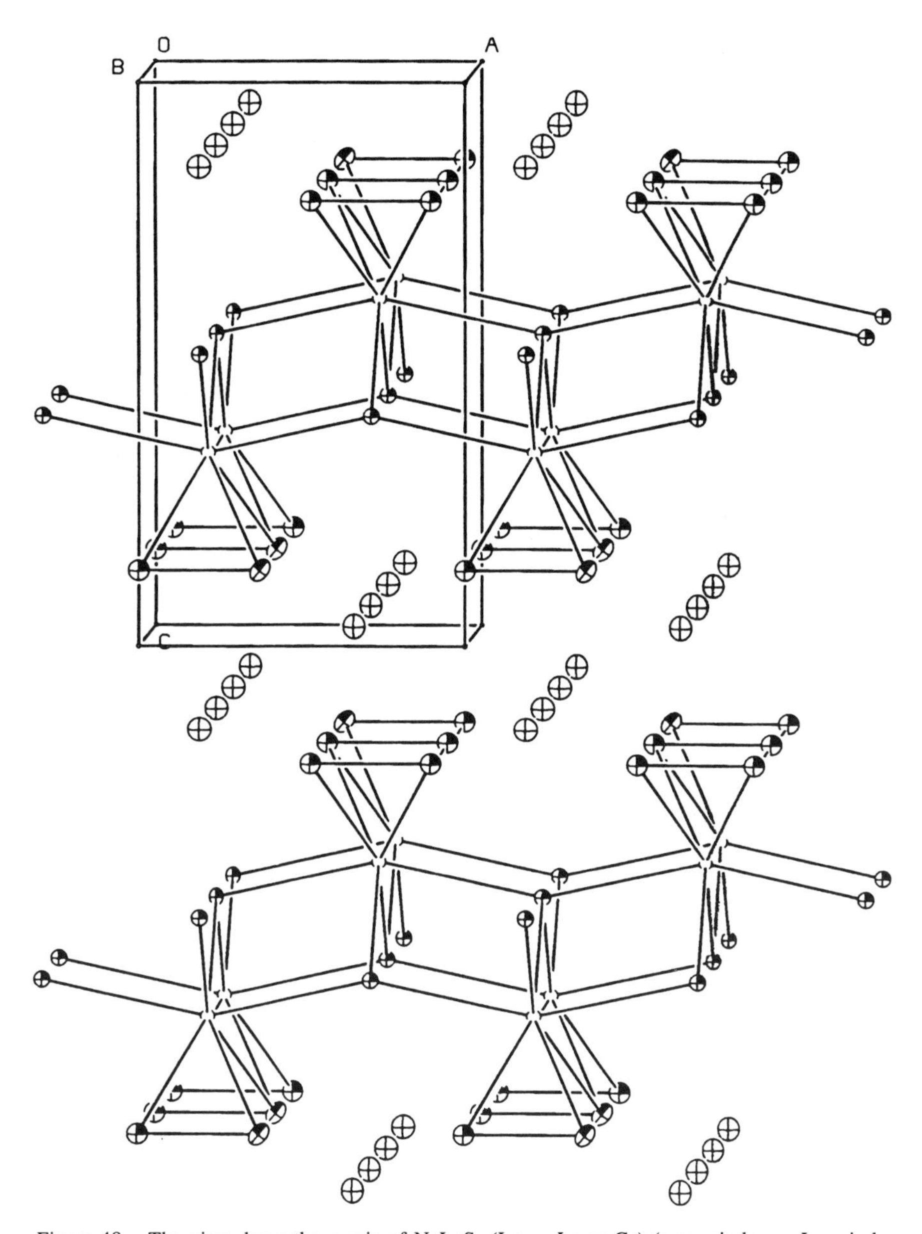

Figure 48. The view down the c-axis of $NaLnS_3$ (Ln = La or Ce) (open circles = Ln; circles with octant shading = S; circles without octant shading = Na) (93).

review of quaternary compounds, we now discuss $AuCuSe_4$, a new phase that resulted from a "failed" mixed-metal reaction designed to yield a quaternary compound.

The compound $AuCuSe_4$ was synthesized from a reaction of K_2Se/Au/Cu/Se in a 1.2:1:1:16 molar ratio at 310°C for 4 days (52). The material was isolated along with $KAuSe_5$, which was about 20% of the total product. Single-crystal studies on $AuCuSe_4$ revealed the 3D structure shown in Fig. 49. The Au^{3+} are in square planar $[AuSe_4]$ units that form chains via trans corner sharing along the *b* axis. The bridging Se atoms of the chains are monoselenides while the other positions of the square plane are occupied by the terminal ends of $(Se_3)^{2-}$ ligands. The $[Au(Se_3)Se]_n^{n-}$ chains are linked into layers by Cu^{1+}. Each Cu atom has two bonds to terminal Se atoms [Se(1) of the $(Se_3)^{2-}$] on the same chain and an internal Se atom [Se(2) on the $(Se_3)^{2-}$] of the neighboring chain. The final coordination site on the tetrahedral Cu atoms is filled by a monoselenide in the neighboring layer, leading to 3D connectivity.

The isolation of $AuCuSe_4$ suggests that other new mixed-metal chalcogenides may also be synthesized from molten A_2Q_x salts, and we may be able to draw some lessons from its synthesis. The flux used to form $AuCuSe_4$ had a low K_2Se/Se ratio (1.2:16) and so contained primarily a mixture of long polyselenides and elemental Se. Hence, the concentration of zero-valent (oxi-

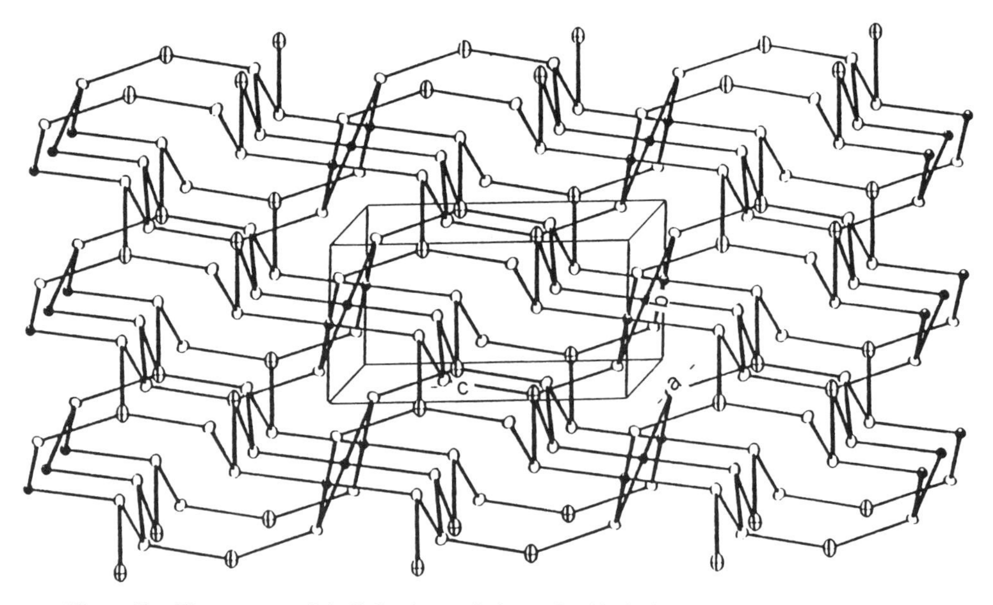

Figure 49. The structure of $AuCuSe_4$ (open circles = Se; black circles = Au; crossed circles = Cu) (52).

dizing) Se was very high while that of K^+ was very low. Apparently, under such conditions the mixture provides a good enough flux for the reaction of Au and Cu metal without having enough alkali cations to push the outcome towards a quaternary product. The reaction of different metals in fluxes of low alkali content thus would be a logical next step in the extension of this chemistry. Copper would be uniquely adaptable to this synthetic approach as the element's chalcophilicity and high mobility within the flux were presumably great aids to the growth of $AuCuSe_4$ single crystals. Hence, initial success may come from various M/Cu/Q systems. Investigations into the use of Cu_xQ_y mixtures as high-temperature fluxes have been proceeding (95), and the inclusion of small amounts of A_2Q may lead to similar investigation in the intermediate temperature regime.

VI. INTERMEDIATE TEMPERATURE REACTIONS (200–600°C) IN QUATERNARY SYSTEMS

The same advantages that led to the formation of new ternary compounds in A_2Q_x fluxes have allowed this method to be extended to several quaternary systems. One can categorize these systems into three general variations: (1) A/M/M′/Q systems in which two different elements incorporate into the anionic chalcogenide framework, (2) A/M/Q/Q′ systems involving two different chalcogenides in the same framework, and (3) A/B/M/Q systems where mixtures of counterions with different sizes and charges influence the final anionic structure. While low temperatures can still promote metastable new phases, the question of how thermally stable the known phases are is again key. In quaternary systems this is compounded by the possibilities that differences in the reactivities, solubilities, and mobilities of the various components may preclude their incorporation into quaternary compounds and lead to ternary or binary phases. For example, a material that is quickly solubilized by the flux may form a final product well before the other material even begins to react. Each system presents its own set of challenges and surprises, and as the survey of compounds in this section progresses, we shall attempt to draw some common conclusions among the disparate systems.

A. The A/Cu/M/Q System, Where M = *f* Block Element

Investigations into quaternary systems with lanthanides and actinides are of particular interest because of the intriguing structural and physical properties that have been known to result from the interplay of covalent transition metal bonding and the more ionic lanthanide and actinide bonding. As in the case of $AuCuSe_4$ formation (Section V), the unique chalcophilicity and mobility of Cu is apparently playing an important role in the formation of these compounds. Reactions with other metals have lead largely to phase separation into known materials. We suspect the limitation here is largely diffusion related; in quaternary reactions, the starting materials, once solubilized, must migrate through the flux to find each other prior to nucleation. If this migration is limited, then solubilized metals will find their own kind close to their own point of solubilization well before encountering other metals. Whatever soluble Cu species are present in the flux appear to have adequate mobility to reach the other reacting metal. Based on this argument, one would think Ag, also known to have a high ionic mobility, would make another likely candidate for easy incorporation into quaternaries. This is not the case with the *f*-block elements in intermediate temperature flux reactions, however, and phase separation still dominates. Clearly, other factors are at work in the Ag containing systems (i.e., differing rates of solubilization between the two metal species), or perhaps Ag^+ is too mobile, making any potential phase extremely kinetically sensitive. More examples of quaternary phases and the conditions under which they are formed would be needed to form a clearer mechanistic picture.

Three different quaternary structure types have been observed so far from intermediate temperature flux reactions, and we begin with $KCuCe_2S_6$ (96). This material was synthesized from a mixture of $K_2S/Cu/Ce/S$ in a 4:1:2:16 molar ratio heated between 270–310°C for 6 days. At temperatures above 310°C the product is no longer a pure phase, being contaminated by CeS_2 powder. The structure that forms is layered with K^+ residing between anionic sheets of $[CuCe_2S_6]_n^{n-}$. The Ce/S portion of the framework is analogous to both the $ZrSe_3$ structure type (94) and the previously described $[CeS_3]_n^{n-}$ framework of $NaCeS_3$ (Section IV.J) The Ce atoms are in bicapped trigonal prismatic coordination with the short basal sides of the triangular faces being $(S_2)^{2-}$ units and the rest of the sites occupied by monosulfides. The trigonal prisms face share in the [101] direction, forming prismatic columns, and further link into layers by sharing monosulfides such that the apex atoms of one chain are the capping atoms of the next. The grooves between the chains possess tetrahedral sites that accommodate the structure's Cu^{1+} ions. One Cu atom is disordered over two crystallographic sites that alternate down the groove, indicating that Cu^{1+} conductivity may be possible through Cu hopping from a filled to an empty site. A view of the structure parallel to the groove direction is shown in Fig. 50. The compound, which forms as red needles, is valence precise (Cu^{1+} and Ce^{3+} as confirmed by magnetic studies) and possess a band gap of approximately 2.0 eV. This structure type has also been found for $KCuLa_2S_6$ (isostructural via X-ray powder diffraction) and $CsCuCe_2S_6$ and $KCuCe_2Se_6$ (both of which crystallize in an orthorhombic space group), indicating an appreciable stability for this structure type (97).

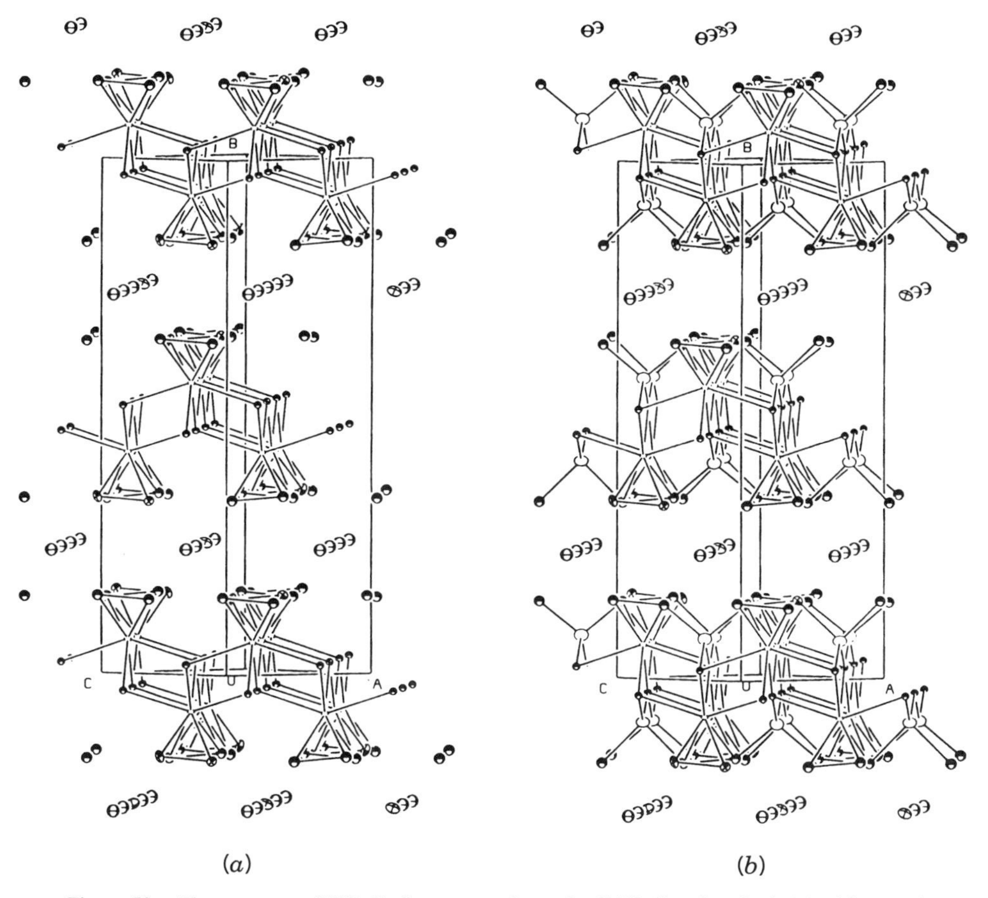

Figure 50. The structure of $KCuCe_2S_6$, as seen down the [101] direction, both (*a*) without and (*b*) with the compound's Cu atoms in the intralayer grooves (small open circles = Ce; large open circles = Cu; circles with octant shading = S; circles without octant shading = K) (96).

By changing the K_2S/Cu/Ce/S ratio to 4.2 : 1 : 0.5 : 8 and slightly lowering the reaction temperature to 260°C for 4 days, $K_2Cu_2CeS_4$ was isolated in approximately 60% yield (96). Pure synthesis of this compound is highly temperature dependent; increasing the reaction temperature above 260°C results in contamination by KCu_4S_3. The structure of $K_2Cu_2CeS_4$ is shown in Fig. 51, and although layered like the previous material, is strikingly different. Here the Ce atoms are octahedrally coordinated by monosulfides, and these [CeS_6] units share equatorial edges forming chains in 1D. The chains of [CeS_6] octahedra are linked into layers via [CuS_4] tetrahedra by sharing two edges with the oc-

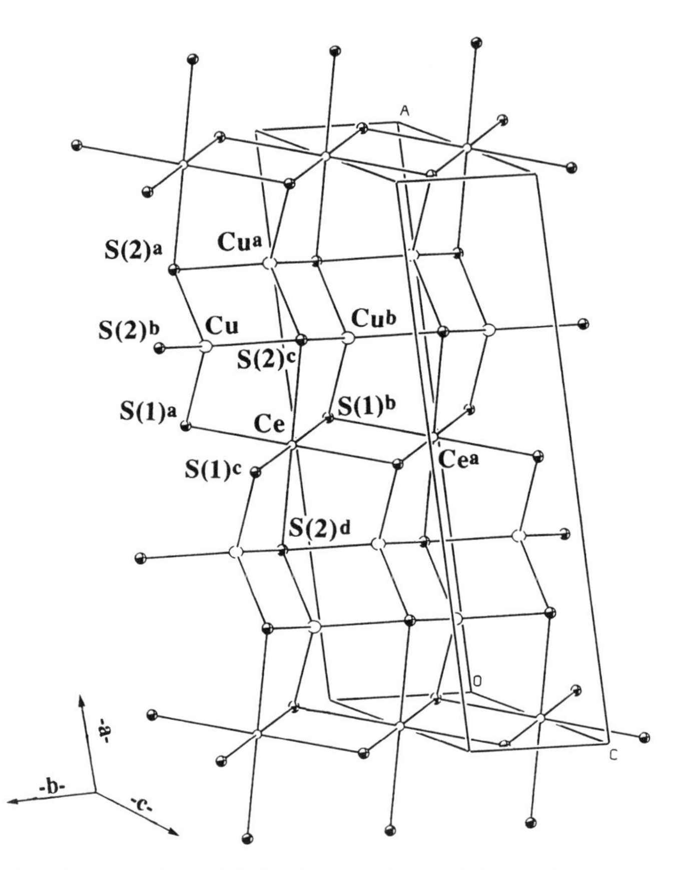

Figure 51. A single anionic layer of $K_2Cu_2CeS_4$ (small open circles = Ce; large open circles = Cu; circles with octant shading = S) (96).

tahedra of one chain and one axial corner with an octahedra in the chain opposite. The $[CuS_4]$ tetrahedra also edge share among themselves, forming a fragment of anti-PbO structure type. The formal oxidation states of $K_2Cu_2CeS_4$ do not balance as readily as they did in $KCuCe_2S_6$, and if all monosulfides are considered as 2−, then the charges on the metals reduce to either $K_2(Cu^{1+})_2(Ce^{4+})S_4$ or $K_2(Cu^{1+})(Cu^{2+})(Ce^{3+})S_4$. Both formalisms, however, possess cations so oxidizing (Cu^{2+} and Ce^{4+}) relative to S^{2-} that neither has been observed in exclusively monosulfide environments. Such quandaries are reminiscent of various copper sulfides of which KCu_4S_3 is a prominent example (55). Materials like this, which are formally Cu mixed-valent, have instead been found to possess S^{2-}/S^{1-} mixed valence manifesting itself as holes in the valence band that is mainly S 3p in character. Such a formalism has already been applied to several compounds in this chapter, as well. By analogy, the formalism $K_2(Cu^{1+})_2(Ce^{3+})(S^{2-})_3(S^{1-})$ would be reasonable. Magnetic studies indicate the presence of Ce^{3+} alone, which corresponds to the S^{2-}/S^{1-} formalism if the hole in the sulfur band is delocalized. Thermopower studies have shown *p*-type metallic behavior for the material, also agreeing with the S^{2-}/S^{1-} formalism, although the magnitude of the thermopower values are unusually high for a metal, perhaps indicating some other conducting mechanism.

The third structure type to be discussed here is found in the compound $KCuUSe_3$ (97), which is isostructural to the phases $KCuZrQ_3$ (Q = S, Se, or Te) synthesized from polychalcogenide fluxes at high temperatures (Section III). In $KCuUSe_3$, $[USe_6]$ octahedra form corrugated layers by edge sharing in 1D and corner sharing in 2D. The Cu^{1+} ion then resides in tetrahedral sites within the corrugated folds (see Fig. 10). The compound $KCuUSe_3$ is highly stable, forming in fluxes of K_2Se/Se ranging from 0.5:8 through 3:8, but is accompanied by Se impurity under such conditions. Pure phase was formed from the reaction of a 1:1:1:6 molar ratio of $K_2Se/Cu/U/Se$ heated at 320°C for 5 days. The material forms as black needles in upwards of 85% yield.

B. The A/M/Sn/S System

In considering which systems would likely lead to new quaternary materials, lessons may be taken from known metal sulfur chemistry. The tetrathiometallates, $(MS_4)^{2-}$ (M = Mo or W), provide a relevant example of metal containing anions that have shown utility as ligands towards other metals. Many discrete molecular clusters have resulted from the solution chemistry of such reactions (98) as well as two polymeric solid state materials: $(NH_4)CuMoS_4$ (99) and Cu_2WS_4 (100). These two particular tetrathiometallates, synthesized from solution, have limited thermal stability, but they do serve to illustrate what would be possible if analogous species could survive the more thermally robust conditions of a molten A_2Q_x flux. Such discrete anionic species would be beneficial in that being stable in the A_2Q_x solution they would be able to diffuse through it readily without crystallizing, and hence could react more readily with a second element in the flux. As seen previously in the ternary chemistry of Sn (Section IV.G), the species $(SnS_4)^{4-}$ and $(Sn_2S_6)^{4-}$ have already demonstrated an appreciable stability under molten salt conditions and so may act as ligands in A_2Q_x solutions in the same way $(MoS_4)^{2-}$ and $(WS_4)^{2-}$ do in conventional solutions. Investigations into such chemistry have lead to several new heterometallic quaternary sulfides that will now be discussed.

We begin with the new structure type $ACuSnQ_3$ (101), which has been seen in high yields of pure phase for A = K or Rb and Q = S or Se but always in mixtures with known binaries and ternaries for A = Na or Cs and Q = S. Structural refinement on single-crystal data for $RbCuSnS_3$ was poor due to severe twinning of the crystals, but an X-ray powder diffraction pattern calculated from the structural model of that study was found to match exactly with that of the material. Although fine details are not available, the structure (Fig. 52) is related to the ternary parent compound Cu_2SnS_3, which possess the 3D zinc-blende (adamantine) structure type (102). Inclusion of Rb^+ has the effect of breaking up the parent structure into a layered offspring. The layers are composed of $[CuS_4]$ tetrahedra sandwiched between two layers of $[SnS_4]$ tetrahedra. In the Cu layer the tetrahedra form a corner-sharing net in two directions but

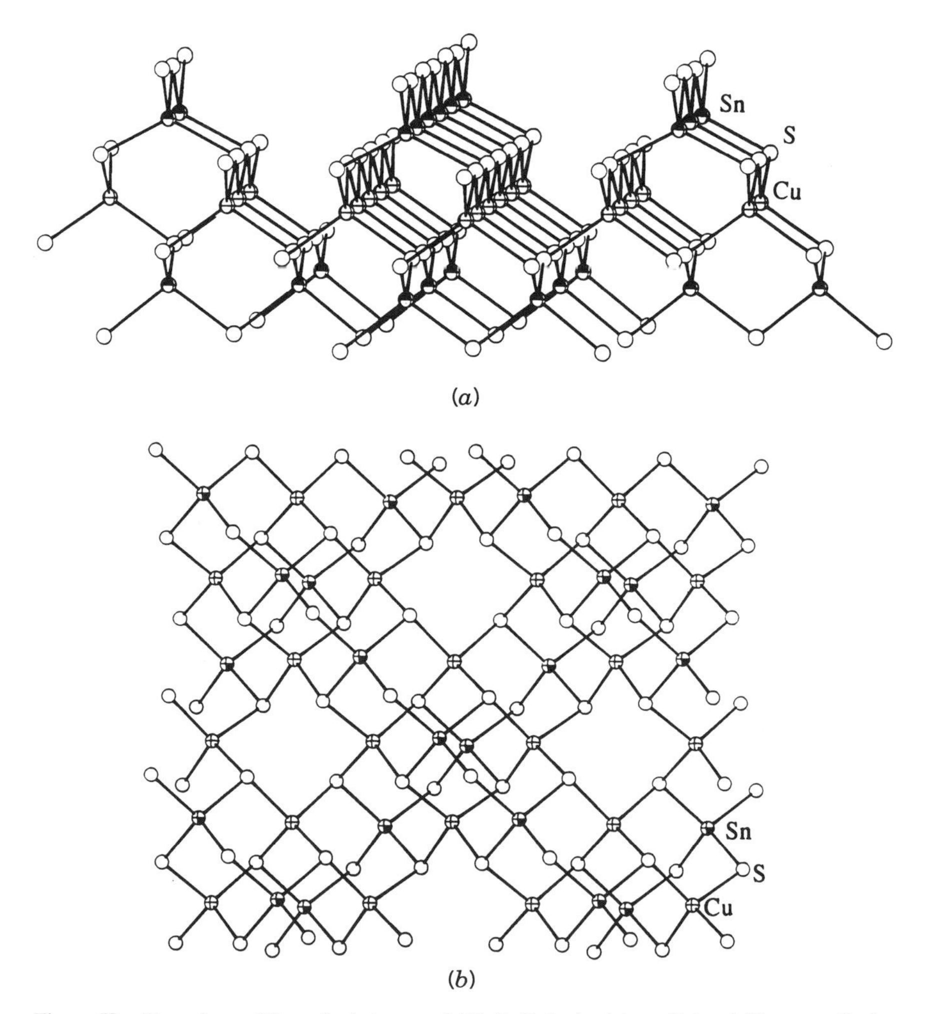

Figure 52. Two views of the anionic layers of $ACuSnQ_3$ in the (*a*) parallel and (*b*) perpendicular directions (101).

with every other Cu site empty resulting in a checkerboard pattern. In the sandwiching Sn layers, chains of edge-sharing [SnS_4] tetrahedra are formed, which run such that the direction of the chains in the top layer is perpendicular to that of the bottom layer. This leads to grooves above and below the Cu containing layer. The $RbCuSnS_3$ compound was formed from a reaction of Rb_2S/Cu/Sn/S in a molar ratio of 4:1–2:1:16 heated at 400°C for 4 days, and similar conditions were employed in the syntheses of the related compounds. The material crystallizes as thin black plates and has a band gap of 1.47 eV.

By simply changing the Cu/Sn ratio of the previous reaction to 4:1, the compound $Rb_2Cu_2SnS_4$ resulted (101). The material has the morphology of orange platelike crystals and a measured band gap of 2.08 eV. The structure is again layered and built from [CuS_4] and [SnS_4] tetrahedra, but now the parent structure type is anti-PbO (edge-sharing tetrahedra) rather than zincblende (corner-sharing tetrahedra). In the $[Cu_2SnS_4]_n^{2n-}$ layers, one half of the tetrahedral sites are occupied by Cu, one quarter by Sn, and one quarter remain empty. The arrangement is such that [CuS_4] units form chains of edge-sharing tetrahedra linked together by sulfur-bound Sn^{4+} ions at every other site (Fig. 53). This structure is related to the known ternary compounds $A_2M_3S_4$ (A = Rb or Cs and M = Mn, Co, or Zn) (103), and to KCu_2NbS_4 (38) (see Fig. 7), where analogous linking of $[CuNbS_4]_n^{2n-}$ chains by [CuS_4] tetrahedra leads to corrugated layers rather than the flat layers of $Rb_2Cu_2SnS_4$.

In these systems, Au has shown a preference for the 1+ oxidation state and subsequent linear coordination, as seen in $K_2Au_2SnS_4$ (101). The material is 1D in nature and composed of [SnS_4] tetrahedra linked along the *b* axis by linear [AuS_2] units (Fig. 54). Conceptually, the chain is merely that of edge-sharing tetrahedra with Au and S atoms inserted into the linkages. The conformation of the chains is such that [AuS_2] fragments between the same two [SnS_4] lie parallel to each other. This results in a zigzag motif between the tetrahedra and [$Sn(SAuS)_2Sn$] rings within the chain. The material was synthesized from a 4:2:1:16 molar ratio of K_2S/Au/Sn/S heated at 350°C for 4 days. It forms as yellow parallelepiped crystals in approximately 66% yield and is air stable but water soluble. A fairly high band gap of 2.78 eV was measured. This compound is structurally related to $K_2Hg_3S_4$.

Reaction of Au and Sn in a less basic flux (one with longer polysulfides) made from a 2:1.5:1:16 ratio of K_2S/Au/Sn/S resulted in the formation of $K_2Au_2Sn_2S_6$ after 4 days of heating at 350°C (101). In this structure the [SnS_4] tetrahedra of $K_2Au_2SnS_4$ have been replaced with the dimer of edge-sharing tetrahedra, [Sn_2S_6], which are subsequently linked into 1D chains via [AuS_2] as before (Fig. 55). The chain is now in a fully extended conformation, rather than zigzagging. This material is stable in air and fully insoluble in water and common organic solvents.

The $[SnS_4]^{2-}$ ligand has also shown an affinity towards Hg. The compound

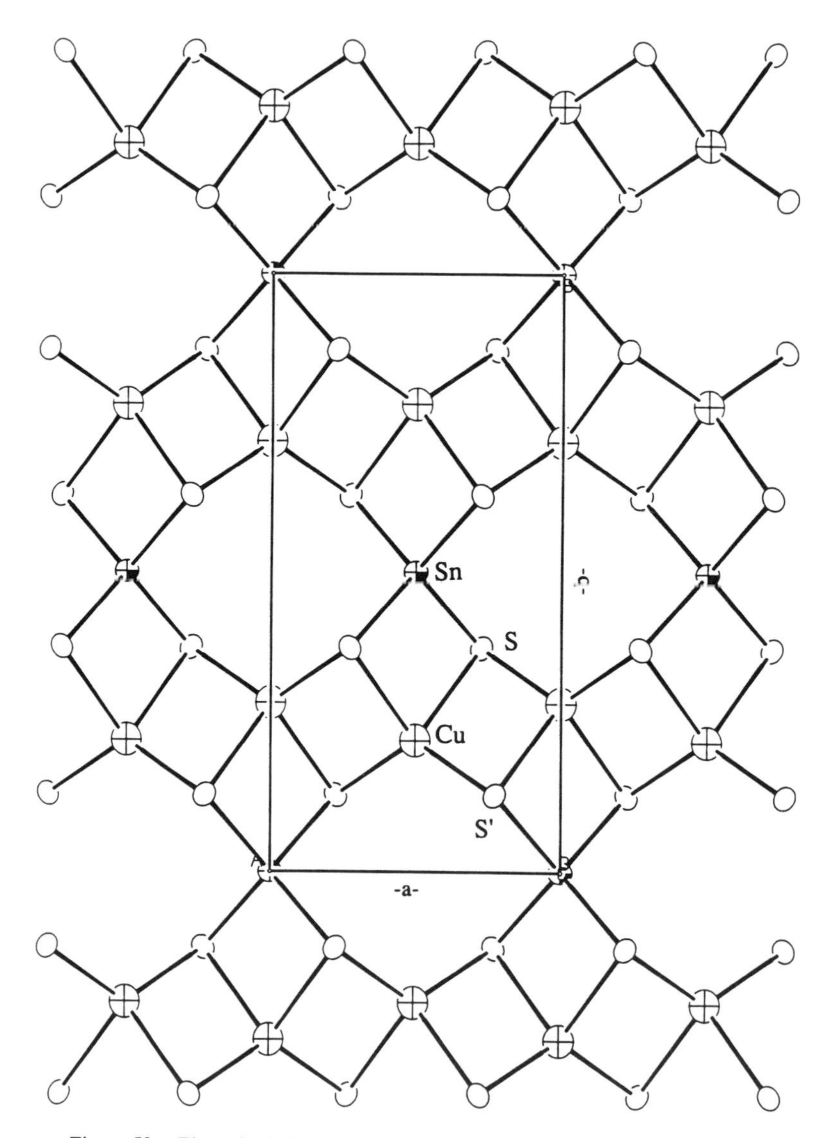

Figure 53. The anionic layers of $Rb_2Cu_2SnS_4$ as seen down the *c* axis (101).

$K_2HgSn_2S_6$ was found in an inhomogeneous mixture of products from the reaction of $K_2S/HgS/Sn/S$ in a 2 : 2 : 1 : 8 molar ratio heated at 350°C for 4 days (104). The structure is layered and shown in Fig. 56. One-dimensional chains of corner-sharing [SnS_4] tetrahedra run along the *a* axis, possessing a corrugated conformation that repeats itself every third tetrahedron analogous to that of the

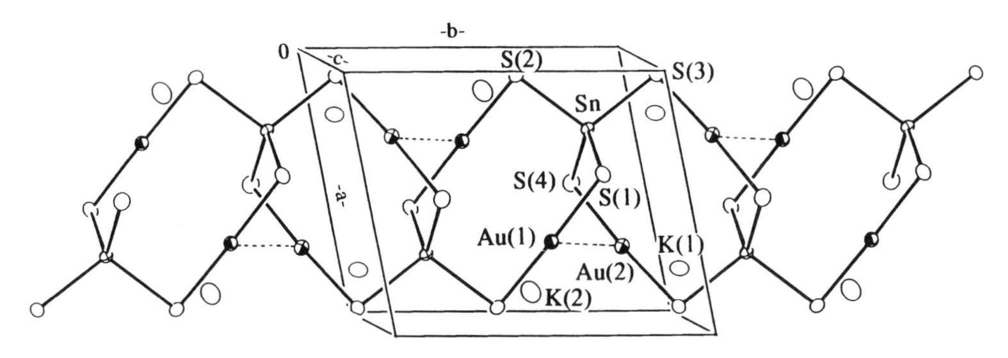

Figure 54. A fragment of an anionic chain from $K_2Au_2SnS_4$ (101).

chains in $K_2SnS_3 \cdot H_2O$ (105). Linking these chains into layers are [HgS_4] tetrahedra that corner share between non-Sn bridging S atoms in opposite chains. The Hg atoms occupy only alternating tetrahedral sites, leading to the formation of intralayer 12-member rings.

Reactions with Mn as the second metal were investigated in order to produce new materials with possibly novel magnetic properties. From a 4 : 1 : 1 : 16 molar mixture of K_2S/Mn/Sn/S heated at 400°C for 4 days, K_2MnSnS_4 resulted (104). The structure is 2D and isostructural to $KInS_2$ with Sn and Mn disordered equally over the In sites. The layers (Fig. 57) are made of four [MS_4] tetrahedra that form a tetrameric cluster with the geometry of P_4O_{10}, and hence is a fragment of the adamantine structure. The [M_4S_{10}] clusters then link via corner sharing into layers. The material proves to be insoluble in water and common organic solvents. Magnetic measurements show paramagnetism down to 150 K where a broad antiferromagnetic transition occurs.

The chemistry of Mn^{2+} in this system overlaps with that of another 2+ metal, Zn. Both have been found to form the phase $A_2MSn_2S_6$, where A = Rb for M = Zn and A = Cs for M = Mn (104). The Zn analogue results from a 2 : 1 : 2 : 16 molar ratio of Rb_2S/Zn/Sn/S heated at 400°C for 4 days, while a somewhat more basic flux (Cs_2S/Mn/Sn/S = 4.14 : 1 : 2.1 : 16.59) is required to form the Cs/Mn containing compound. Within the lattice are units built from one [MS_4] and two [SnS_4] bonded in a corner-sharing fashion such that a central [MSn_2S_3] ring is created. These rings are linked together into layers by sharing four of the six terminal S atoms coming off the ring. The remaining two terminal S atoms connect the layers into a 3D array, one bonded to the layer above and one to the layer below. The resulting tunnels that house the A^+ cations are narrow ($\approx$ 3.2 Å) and run in both the [100] and [001] directions (Fig. 58). Despite disorder problems between the metals in the previous Mn compound, structure refinement showed no metal disorder in either of the $A_2MSn_2S_6$ phases.

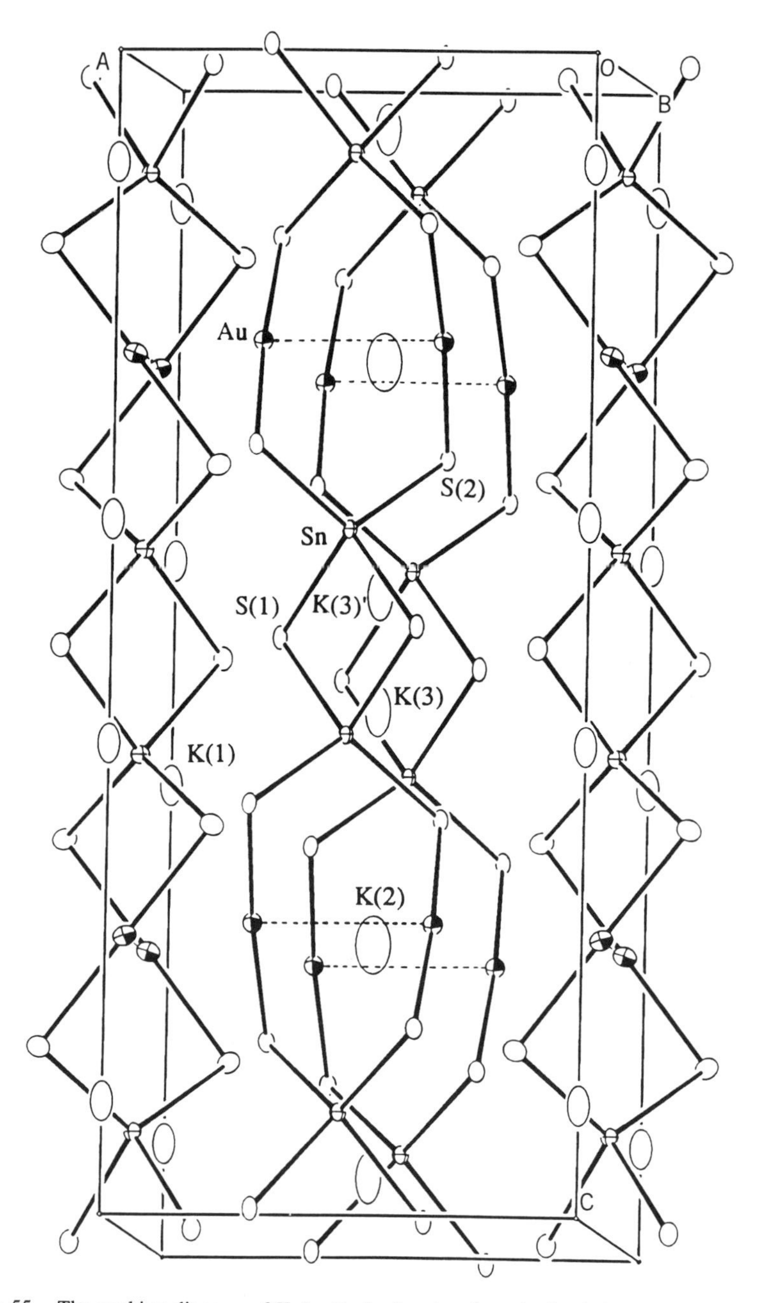

Figure 55. The packing diagram of $K_2Au_2Sn_2S_6$ showing the anionic chains running parallel to the c axis (101).

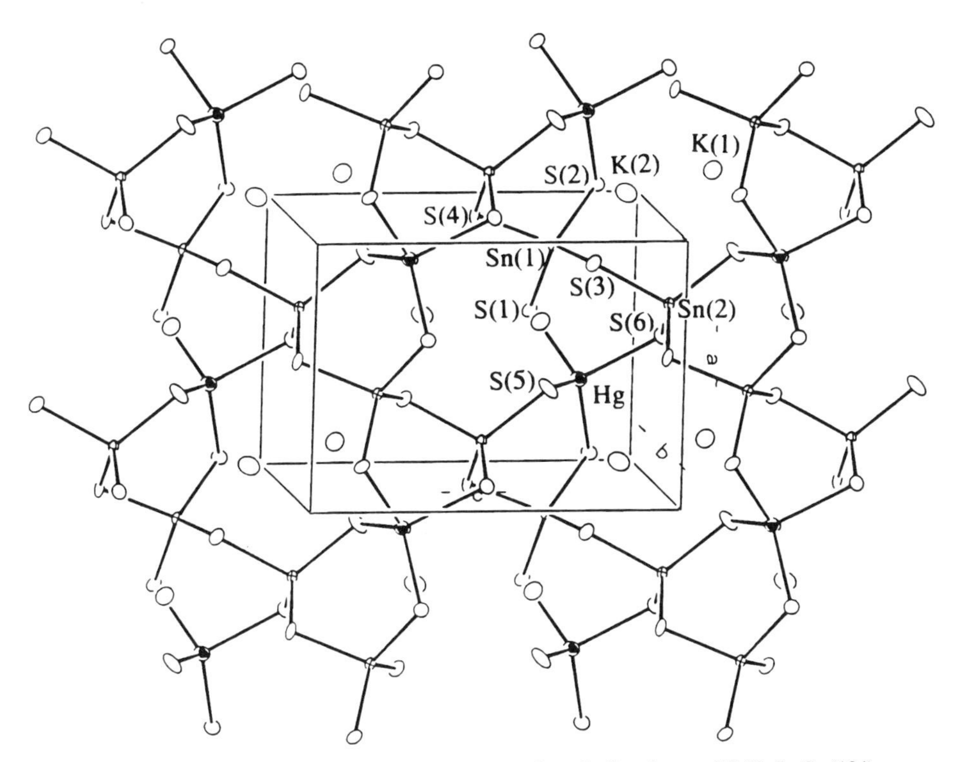

Figure 56. A single anionic layer, with accompanying alkali cations, of $K_2HgSn_2S_6$ (104).

C. Metal Reactivity in $A_xP_yQ_z$ Fluxes

The successes in the A/M/Sn/S system provided an important foundation for further exploratory synthesis by proving that species other that polychalcogenides could be formed *in situ* and then used for further chemistry in the flux. In this context the molten polychalcogenides become less like a low-temperature "heat and beat" method and more like a bonafide solvent in which, under appropriate conditions, various species can be stabilized and allowed to react preferentially in the flux. It now becomes conceptually possible to form any chalcogenide-containing ligand *in situ* with the only limitation being that such ligands have the modest (in solid state terms) thermal stability necessary to remain viable within the flux. One such class of ligands would be the chalcophosphates, $(P_yQ_z)^{n-}$. Only thiophosphate compounds have been seen to any extent in the literature; these are predominately the MPS_3 compounds (106), which contain the ethane-like $(P_2S_6)^{4-}$ ligand, but various other examples of

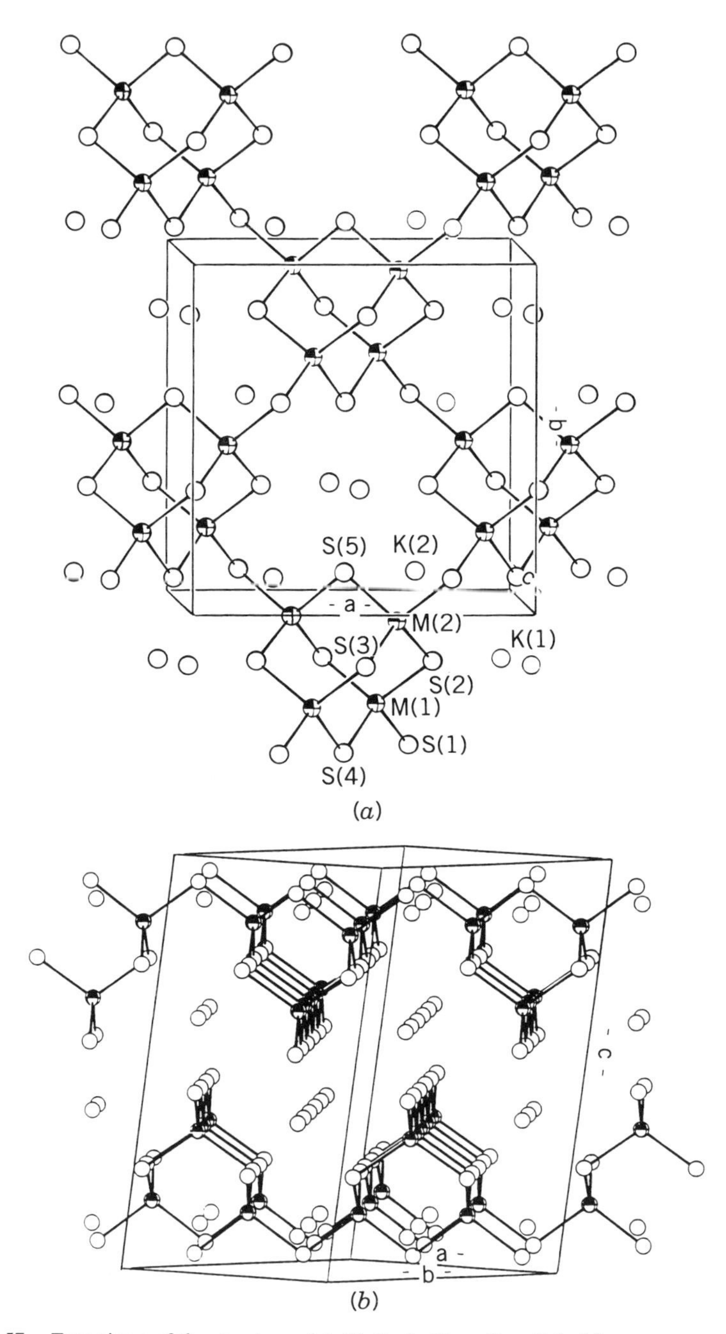

Figure 57. Two views of the structure of $A_2MnSn_2S_6$ (A = K or Cs): (*a*) as seen perpendicular to a single anionic layer and (*b*) as a packing diagram oriented down the [110] direction (104).

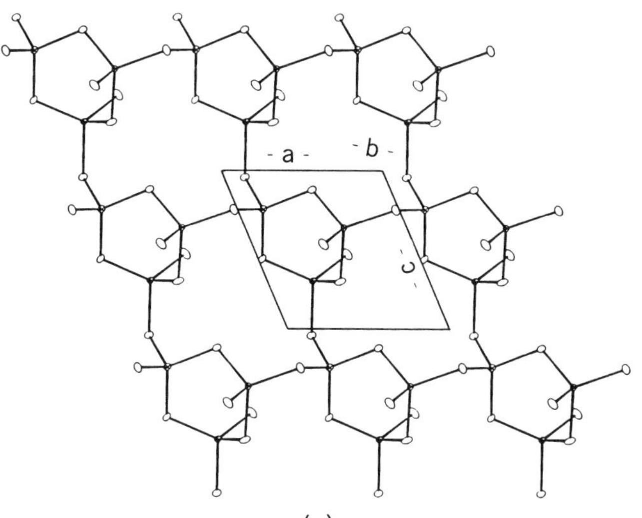

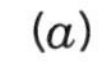

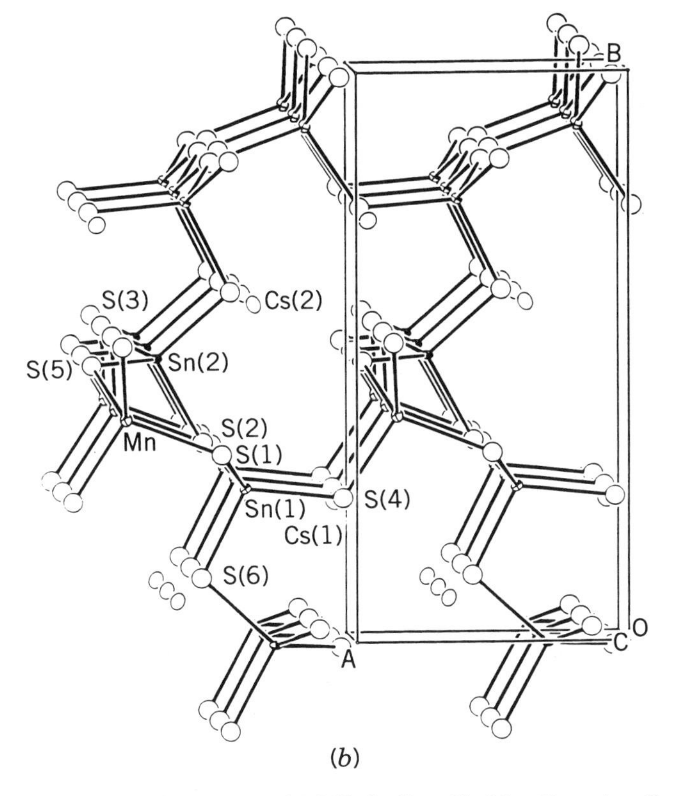

Figure 58. Two views of the structure of $A_2MSn_2S_6$ (A = Rb; M = Zn or A = Cs; M = Mn): (*a*) a single anionic ''sheet'' as seen down the *b* axis and (*b*) a packing diagram showing the full 3D anionic network and the resulting tunnels parallel to the *c* axis (104).

$(P_yS_z)^{n-}$ containing species also exist (107–109). Typically, they are synthesized by direct combination of the elements at temperatures between 500–800°C, and under these conditions, complex acid–base equilibria have been observed between species, such as $(P_2S_6)^{4-}$, $(P_2S_6)^{2-}$, $(P_2S_7)^{4-}$, and $(PS_4)^{3-}$ (110). The parallels between A_2Q_x fluxes are clear, and so investigations into alkali metal polychalcophosphate fluxes, $A_xP_yQ_z$, were engaged. These fluxes are formed by simple *in situ* fusion of $A_2Q/P_2Q_5/Q$ in the same manner used for the A_2Q_x fluxes. Although the fluxes are conceptually $A_xP_yQ_z$, a more accurate description is probably that of the $(P_yQ_z)^{n-}$ species solubilized is excess polychalcogenide flux, which also continues to serve as the oxidant to solubilize metallic elements into the flux for reaction. The inclusion of P into the fluxes renders them somewhat more basic than their all chalcogenide counterparts, but the melting points of $A_xP_yQ_z$ are also in the intermediate temperature range (minimum melting points: 300–400°C) making possible the synthesis of metastable phases. In addition, the $(P_yS_z)^{n-}$ species should act as effective mineralizers, promoting large crystal growth. Also being highly soluble in water and many organic solvents, reactions of metals in $A_xP_yQ_z$ fluxes appeared poised to be just as synthetically rich as those in polychalcogenide salts.

The first compound reported from a $A_xP_yS_z$ flux reaction at intermediate temperature was $ABiP_2S_7$ (A = K or Rb) (111). It was synthesized from a 2 : 1 : 3 : 4 molar ratio of $K_2S/Bi/P_2S_5/S$ heated at 400°C for 4 days. After the excess flux was dissolved with DMF, red crystals of $KBiP_2S_7$ remained in 68% yield. A similar ratio was used to form the Rb analogue, but a higher reaction temperature (420°C) was employed. The structure is 2D and composed of Bi atoms coordinated by $(P_2S_7)^{4-}$ ligands (Fig. 59). The $(P_2S_7)^{4-}$ ligand, "pyrothiophosphate," is actually two $[PS_4]$ tetrahedra joined by a corner-sharing connection. It acts as a multidentate ligand, bridging four different Bi^{3+} in four different modes as shown below

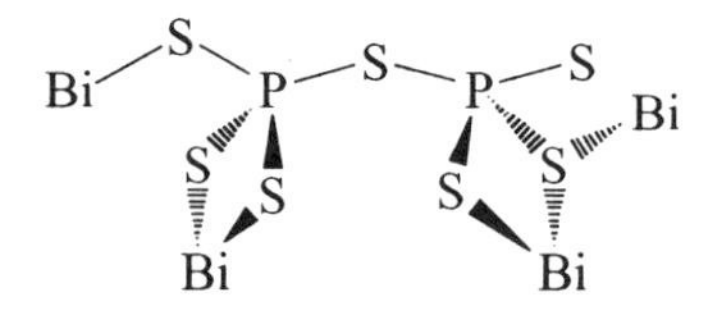

This leads to a complicated connectivity within the layer, the net effect being that $[BiSPS]_2$ lined intralayer rings are formed that stack in registry between the layers. Each Bi atom resides in a seven-coordinate capped trigonal prism that is distorted by a stereochemically active lone pair bending the S atoms away from the idealized rectangular faces. The $(P_2S_7)^{4-}$ species has been seen previously in $Ag_7(PS_4)(P_2S_7)$ (108b), $RbVP_2S_7$ (112), $Ag_4P_2S_7$ (113), $Hg_2P_2S_7$ (114), and $As_2P_2S_7$ (115).

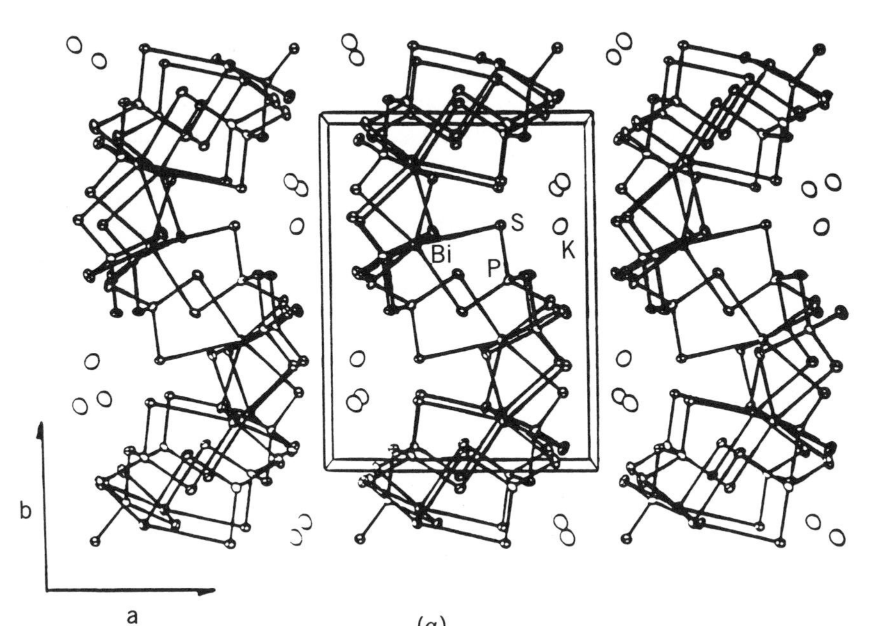

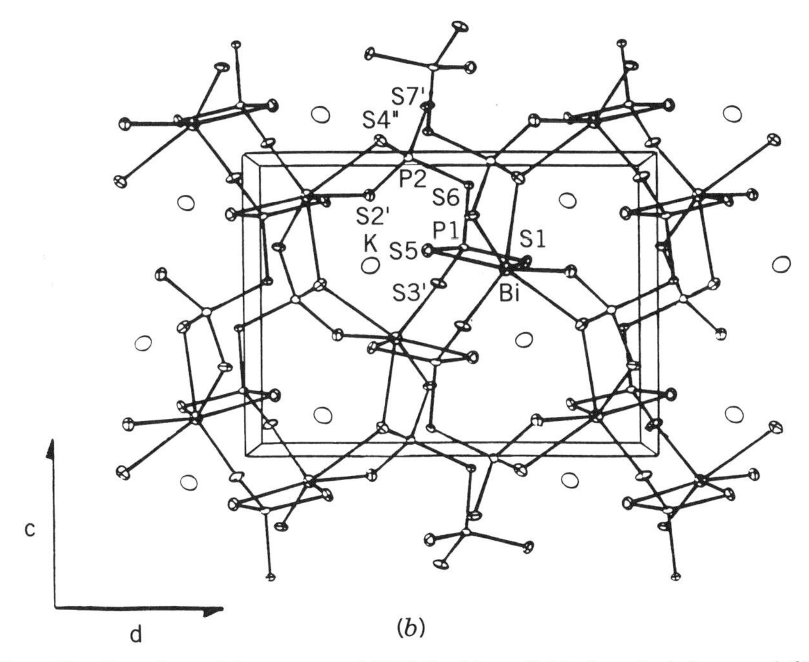

Figure 59. Two views of the structure of $KBiP_2S_7$: (*a*) parallel to the anionic layers and (*b*) perpendicular to a single anionic layer (111).

As in the A_2Q_x fluxes, modest changes in the composition of the $A_xP_yQ_z$ flux can alter the equilibria between $(P_yQ_z)^{n-}$ species, changing which ligand is incorporated into the final product. The compound $K_3Bi(PS_4)_2$ was isolated (116) from a reaction of $K_2S/Bi/P_2S_5/S$ in a 4 : 1 : 2 : 12 molar ratio heated at 400–450°C for 4 days. The increased sulfur content of this flux (relative to that in the $KBiP_2S_7$ synthesis) was sufficient to favor the most sulfur rich $(P_yS_z)^{n-}$ ligand possible. In the structure of $K_3Bi(PS_4)_2$ (Fig. 60), $[BiS_5]$ square pyramids

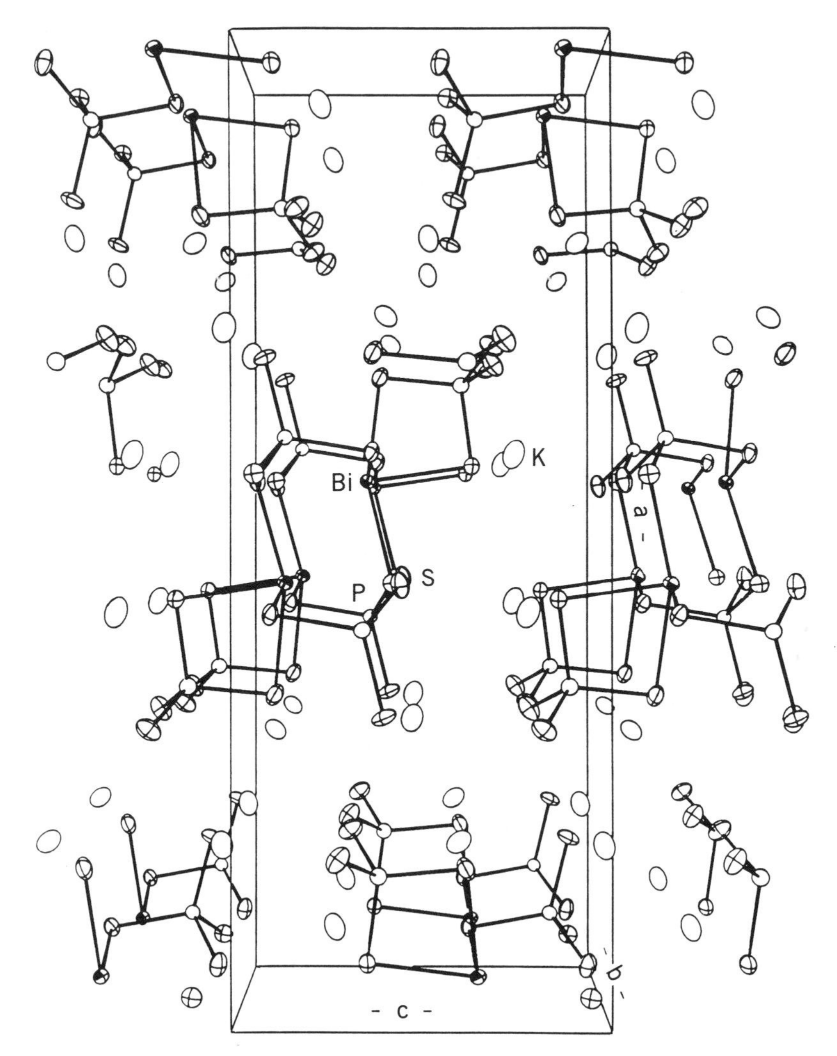

Figure 60. The packing diagram of $K_3Bi(PS_4)_2$ as seen parallel to the *b* axis (116).

are linked into helical chains by $[PS_4]$ tetrahedra, with the lone electron pair of Bi^{3+} presumably active at the base of the pyramid. The two $(PS_4)^{3-}$ ligands each have a different bonding mode to the Bi atoms. One ligand possesses two terminal S atoms and two S atoms that chelate to a single Bi atom, and the second ligand also chelates to a Bi atom but has one S atom bonding to a neighboring Bi atom while the remaining S atom is terminal. The analogous compounds have been made with A = K, Rb, or Cs and M = Sb or Bi, indicating that the phase has an appreciable stability (116).

The reaction of Ag in $K_xP_yS_z$ fluxes has also lead to a new phase for this system, although it is isostructural to a previously known material. The compound KAg_2PS_4 was isolated (117) from a $K_2S/Ag/P_2S_5/S$ flux in a 4 : 1 : 2 : 16 molar ratio heated at 400°C for 6 days. Its structure is shown in Fig. 61. Tetrahedra of $[AgS_4]$ form layers via corner sharing in the fashion of HgI_2, and these layers are linked by $[PS_4]$ tetrahedra, which edge share to $[AgS_4]$ in opposite layers. This 3D network is isostructural to that seen in $BaAg_2GeS_4$ (118), and the presence of $[GeS_4]$ units prompted an early examination of whether other main group chalcogenide ligands can undergo analogous chemistry under

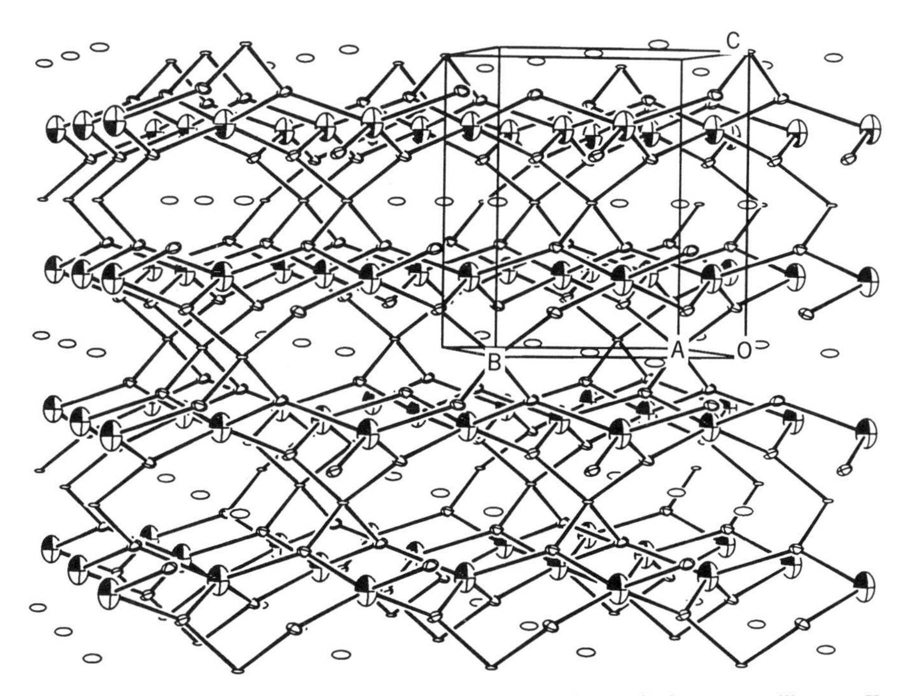

Figure 61. The packing diagram of KAg_2PS_4 as seen down the *a* axis (large open ellipses = K; ellipses with octant shading = Ag; ellipses without octant shading = S; small open ellipses = P) (117).

flux conditions. Although not yet studied extensively, the isostructural phase KAg_2AsS_4 has been isolated from a reaction of $K_2S/Ag/As_2S_3/S$ in a 4:1:2:12 ratio heated at 450°C for 4 days.

Modifying a A_2Q_x flux with P is not a technique limited to sulfide fluxes. Selenides have proven amenable to this approach as well. The compounds KMP_2Se_6 (M = Sb or Bi) have been synthesized from mixtures of $K_2Se/M/P_2Se_5/Se$ in 2:1:3:8 ratios heated at 410°C for 4 days when M = Bi and 450°C for 6 days when M = Sb (119). The structure for the Bi analogue is shown in Fig. 62, and the Sb compound was shown to be isostructural by X-ray powder diffraction. As can be inferred from the formula, the selenophosphate ligand manifested here is $(P_2Se_6)^{4-}$, in which two $[PSe_3]$ trigonal pyramids are connected through a central P—P bond, resulting in an ethane style species. The ligand exhibits a varied multidenticity. One Se binds to one Bi atom; one Se atom is μ^2 to two Bi atoms, and three Se atoms (two from one side of the P—P bond and one from the other) are involved in a tridentate chelation to a Bi atom. The Bi atoms are octahedrally coordinated, and chains are formed by a corner sharing of the central $[BiSe_6]$ along the *c* axis. These chains are further linked into layers by a Se—Bi bond between chains. The

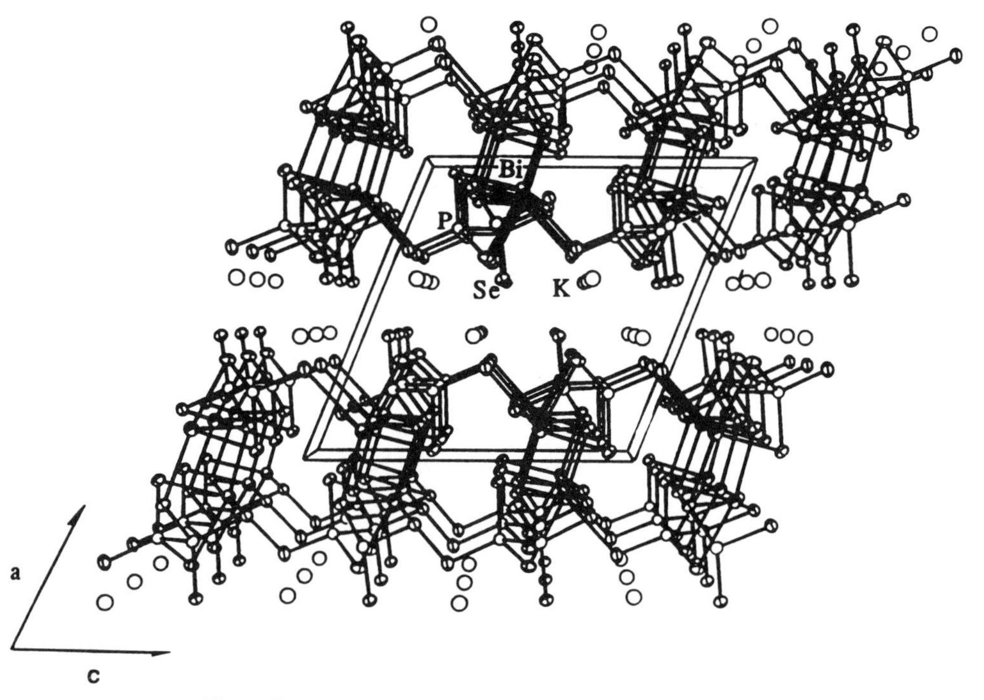

Figure 62. The view parallel to the *b* axis of $KBiP_2Se_6$ (119).

KMP_2Se_6 compounds have high melting points for both metals (554°C for M = Bi; 479°C for M = Sb), but the electronic properties are significantly different. The Bi compound is dark red colored and has a band gap of 1.61 eV, while the light red Sb analogue has a wider band gap of 2.00 eV.

The $(P_2Se_6)^{4-}$ ligand again manifests itself in reactions involving Mn and Fe. The compounds $A_2MP_2Se_6$ were found as the K^+ and Cs^+ salts of both metals (116). The molar ratios of $A_2Se/M/P_2Se_5/Se$ were 2:1:2:10 for A = Cs, M = Mn and A = K, M = Fe but were 2:1.2:2:10 for A = K, M = Mn, and 2:1:3:10 for A = Cs and M = Fe. All ratios were heated at 450°C for 6 days. The materials, all isostructural, possess 1D anionic chains in which the M^{2+} are octahedrally coordinated and bridged by $(P_2Se_6)^{4-}$ making two tridentate chelations to two separate M atoms (Fig. 63). If the center of the P—P bond in the ligand is taken as an atomic position, then the structure can simply be considered as a chain of face-sharing octahedra (TiI_3 structure type).

In these few examples, $(P_2Se_6)^{4-}$ ligands already demonstrated a remarkable variety of binding modes to the metal centers. Such adaptability runs wild in the astonishing compound $Cs_8M_4(P_2Se_6)_5$ (M = Sb or Bi) (116). This material possesses three distinct $(P_2Se_6)^{4-}$ ligands, each with a different binding mode. These modes are shown in Fig. 64 and consist of (a) three Se involved in one tridentate chelation and three terminal Se; (b) three Se involved in one tridentate chelation, two Se each bonding to one M atom and one terminal Se; and (c) four Se each bonding to one M atom and two terminal Se. The basic repeating unit of this structure is shown in Fig. 65; it features two $[MSe_5]$ square pyramids held by Se—P—Se bridges with eclipsed pyramidal bases facing each other. The repeating units are linked in 1D by $(P_2Se_6)^{4-}$ in the type-*b* binding mode [Fig. 66(*a*)]; the ligand is tridentate to one unit with its two monodentate Se atoms forming one of the Se—P—Se bridges between the square bases of the neighboring unit. The repeating units extend to form a zigzag chain as shown in Fig. 66(*b*). The chains are further linked into layers via ligands in the type-*c* mode; these function as the Se—P—Se square base bridges in two neighboring chains. Finally, type-*a* ligands are simply terminal tridentate chelaters, and

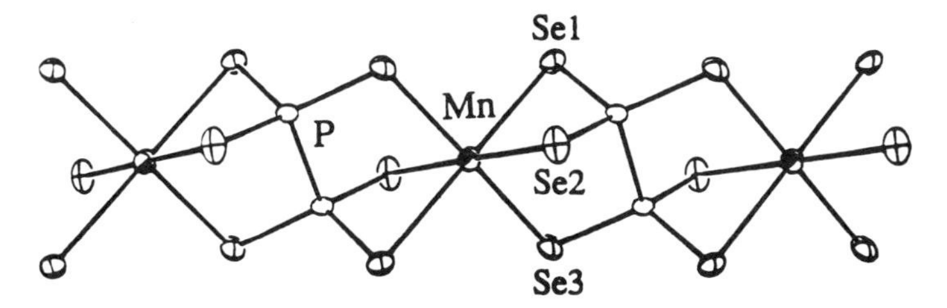

Figure 63. A fragment of the anionic chains of $A_2MnP_2Se_6$ (A = K or Cs) (116).

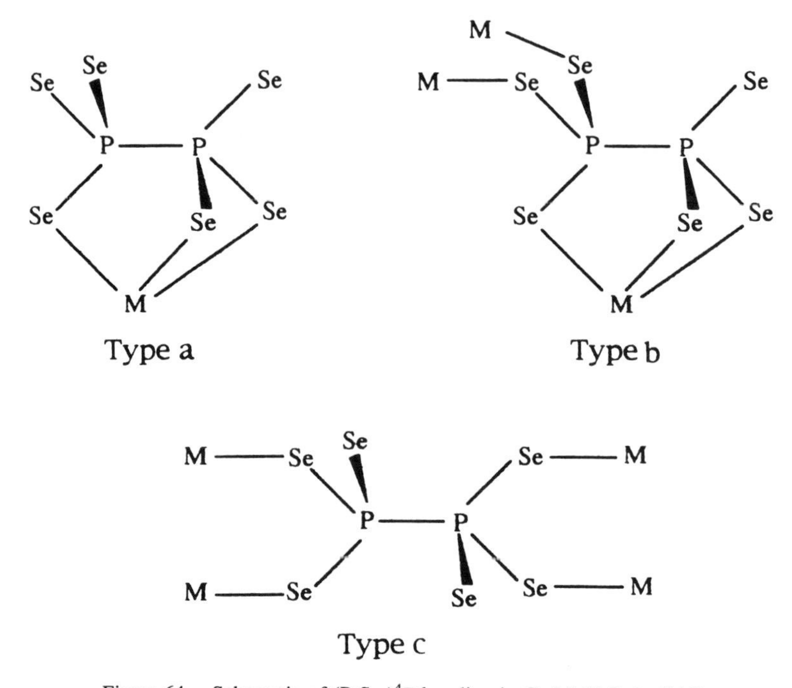

Figure 64. Schematic of $(P_2Se_6)^{4-}$ bonding in $Cs_8M_4(P_2Se_6)_5$ (116).

so the anionic framework of the entire monster is layered (Fig. 67). The M—M distances across the bases of the square pyramids are 3.534(2) Å for M = Bi and 3.441(2) Å for M = Sb, fairly long in both cases although the positioning of the lone pairs at those sites is suggestive of some manner of weak interaction. Both materials are semiconductors with a 1.44-eV band gap for M = Bi and 1.58-eV band gap for M = Sb. Each was made from similar reaction conditions as well ($Cs_2Se/M/P_2Se_5/Se$ = 2:1:3:4 heated at 460°C for 6 days).

The diverse chemistry of this system prompts us to close this section with a summary of three features that surely foreshadow the great wealth of $A_xP_yQ_z$ flux chemistry yet to come: (1) the internal flux equilibria that leads to a variety of ligands; (2) the adaptability of the binding modes of these ligands, even within the same structure; (3) clear indications that analogous chemistry may occur with many different types of chalcogenide containing ligands.

D. Metal Reactivity in $A_2Te_xS_y$ Fluxes

The electronegativity differences among the chalcogenides are one of the prime factors resulting in the diverse chemistry each displays under molten salt

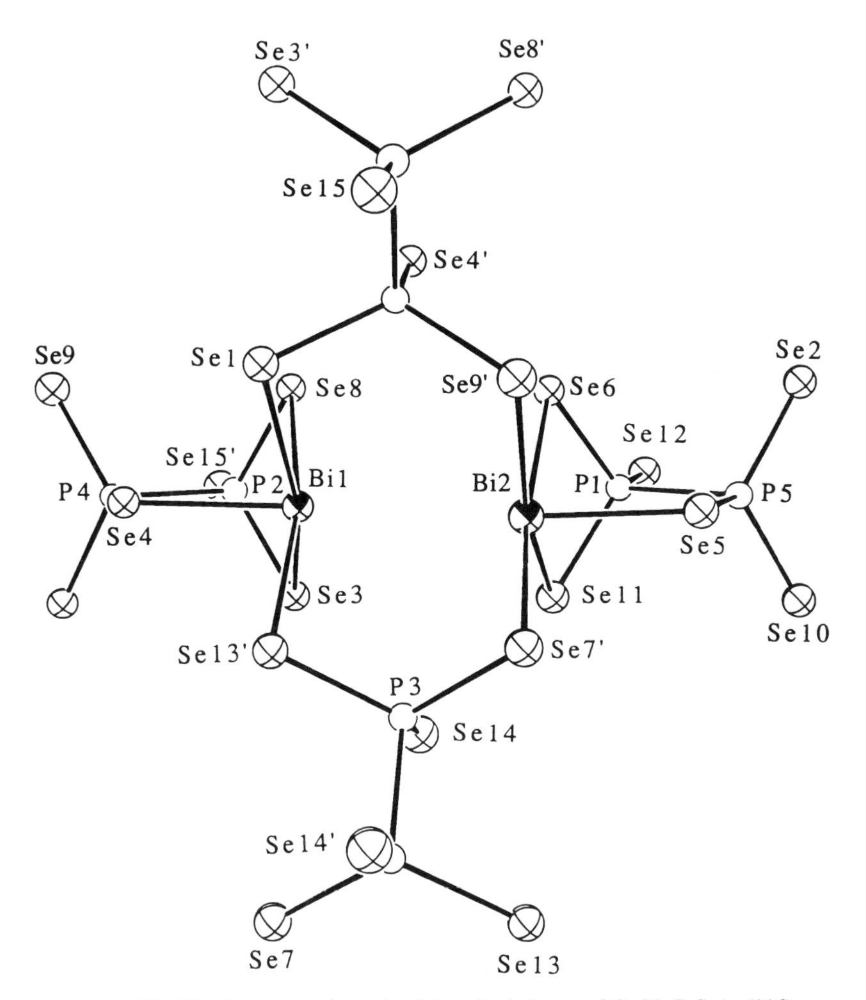

Figure 65. The basic repeating unit of the anionic layers of $Cs_8M_4(P_2Se_6)_5$ (116).

conditions. As with other quaternary systems, the questions of structure–property relationships resulting from mixtures of chemically dissimilar elements prompted investigations into mixed-chalcogenide systems. Initial experiments in S/Se systems did not result in well-characterized materials due to the high disorder, which the crystallographically similar S and Se atoms are prone to. Attention then focused on the S/Te system, where the sizes of the atoms are sufficiently different to allow for better structure determination of the resulting phases.

Although some compounds have been isolated with separate S^{2-} and Te^{2-}

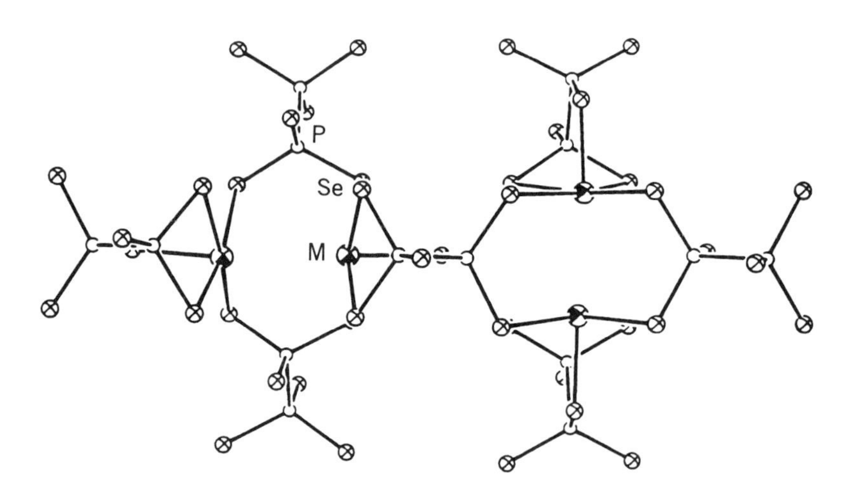

(a)

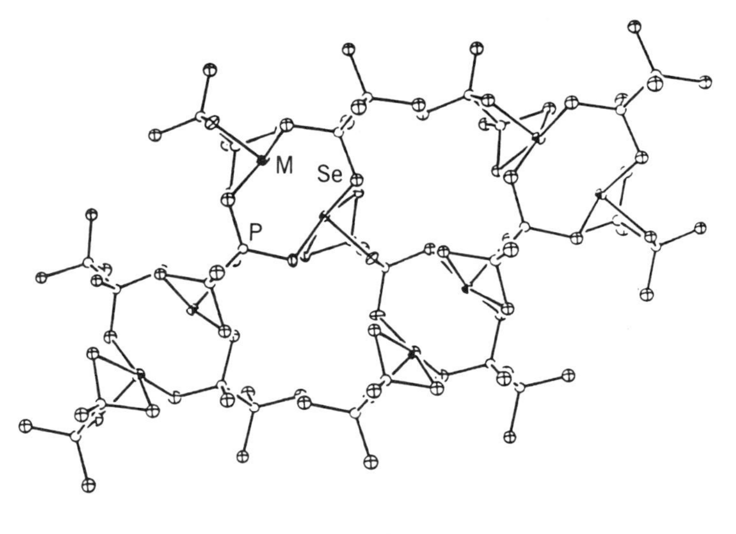

(b)

Figure 66. Building the extended structure of $Cs_8M_4(P_2Se_6)_5$ by starting with (a) two repeating units linking into a dimer then (b) further connection of the dimers into a 1D zigzagging ribbon (116).

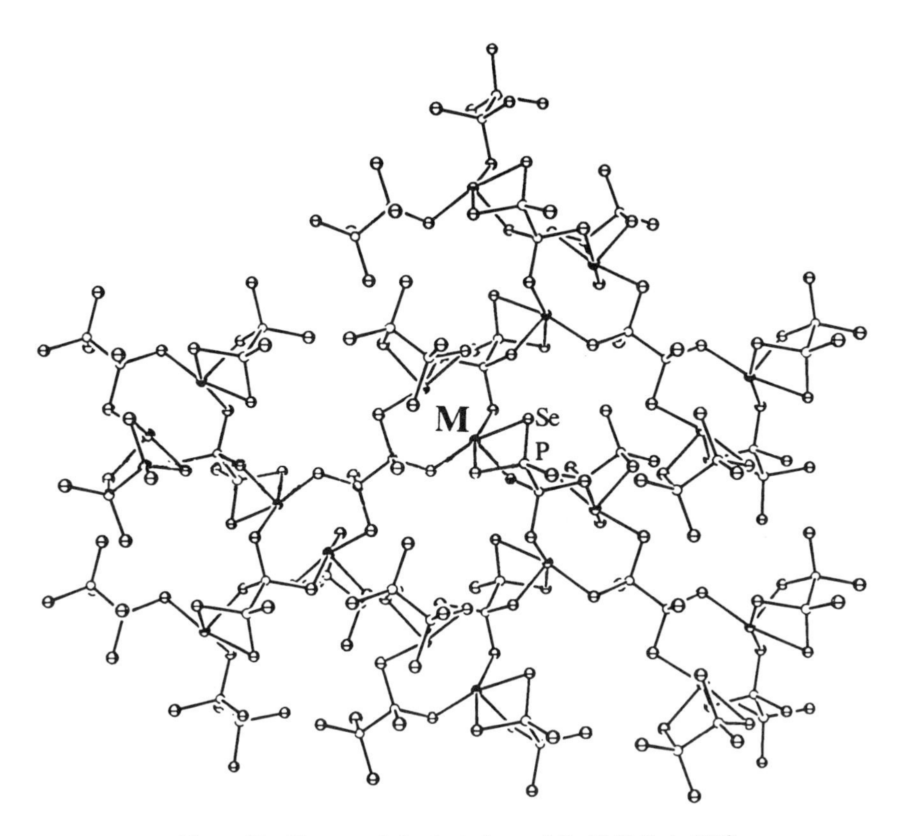

Figure 67. The extended anionic layer of $Cs_8M_4(P_2Se_6)_5$ (116).

species, a recurring ligand observed in products from mixed S/Te fluxes has been $(TeS_3)^{2-}$, which has proven to be another example of a stable ligand formed *in situ* in polychalcogenide fluxes. Generally occurring in fluxes rich in $(S_x)^{2-}$, its formation is rooted in the large electronegativity difference between S and Te. Even though chainlike polymers analogous to $(Q_x)^{2-}$ are conceptually possible with S/Te mixtures, the Te atoms are in reality oxidized by excess polysulfide just as is any metallic element. The $(TeS_3)^{2-}$ anion is isoelectronic to $(SO_3)^{2-}$ but coordinates to metals only through the S atoms, as the large electronegativity difference between S and Te is much more restricting to the lone pair of Te^{4+} than is the difference between S and O to the S^{4+} lone pair in $(SO_3)^{2-}$. Even without Te involved in bonding, $(TeS_3)^{2-}$ is still multidentate, which has lead to structural diversity in a number of systems.

A new 2D structure type containing $(TeS_3)^{2-}$ was found in $RbCuTeS_3$ (120).

The material was isolated from a 2 : 1 : 4 molar ratio of $Rb_2Te/Cu/S$ heated at 300°C for 4 days in 61% yield and was also made, although in lower yield (39%), from a more sulfur rich flux ($Rb_2S/Cu/Te/S$ = 1 : 1 : 1 : 8 and 260°C for 4 days). The lattice type is fairly stable in that by using variations of the later synthetic conditions, the K^+ (β-form), Rb^+, and Cs^+ salts of the Ag analogue, have all been synthesized as well (see Table V). The structure of $RbCuTeS_3$ consists of chains of corner-sharing $[CuS_4]$ tetrahedra that are linked into layers via $(TeS_3)^{2-}$ ligands (Fig. 68). The layers are relatively flat, and the $(TeS_3)^{2-}$ trigonal pyramid is oriented such that the Te^{4+} lone pair is directed towards the center of the layers. No unusually short contacts between the Te^{4+} and atoms other than S are evident. The band gap of the $RbCuTeS_3$ analogue, which forms as red needles, was measured at 1.95 eV, while that of all of the Ag salts was 2.4 eV. The compound $RbCuTeS_3$ was also found to melt incongruently at 343°C, decomposing to $Cu_{17.6}Te_8S_{26}$, a phase that will be discussed later in this section.

The anionic layers of β-$KAgTeS_3$ find themselves undergoing distortions in the α form of the compound. The compound α-$KAgTeS_3$ is formed upon increasing both the Te content of the flux ($K_2S/Ag/Te/S$ = 1 : 1 : 1 : 8) and the reaction temperature (350°C for 4 days) (120). Again the central feature is chains of corner-sharing $[AgS_4]$ tetrahedra linked into layers by $(TeS_3)^{2-}$ trigonal pyramids, but in the α form, the $(TeS_3)^{2-}$ is now oriented with the Te^{4+} lone pair directed out of the anionic layers (Fig. 69). This causes the layers to become corrugated and severely distorts the tetrahedron about the Ag atoms [S(1)—Ag—S(2) = 155.8(1)°; S(3)—Ag—S(3)′ = 131.3(2)°]. Again, no short contacts are evident between the Te^{4+} and any non-S atoms. The distortion about the Ag atoms is apparently sufficient to effect a lowering of the band gap from that of the β form to 2.2 eV. Differential scanning calorimetry does show a phase change between the two forms but only from the β to the α. The α form congruently melts at 334°C, but the β form possesses two endotherms, one at 329°C and the other at 334°C. Presumably the first represents a phase change of β to α, and the second, melting of the α. Upon subsequent heating cycles of

TABLE V
Synthetic Conditions for the Formation of $AAgTeS_3$ Phases

Compound	$A_2S/Ag/Te/S$ Ratio	Heating Program	Cooling Rate
α-$KAgTeS_3$	1/1/1/8	350°C (4 days)	2°C h^{-1}
β-$KAgTeS_3$	1/1/0.5/8	270°C (4 days)	2°C h^{-1}
$RbAgTeS_3$	1/1/1/6	350°C (4 days)	2°C h^{-1}
$CsAgTeS_3$	1/1/1/6	300°C (4 days)	2°C h^{-1}

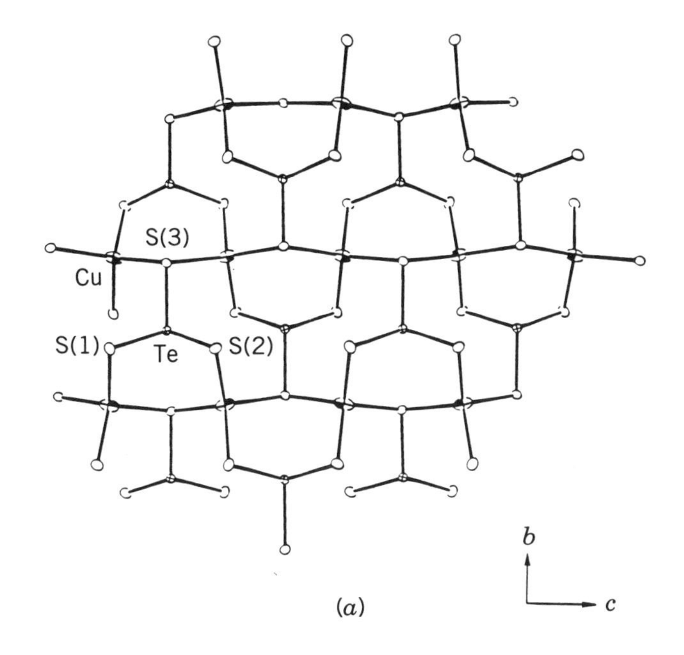

(a)

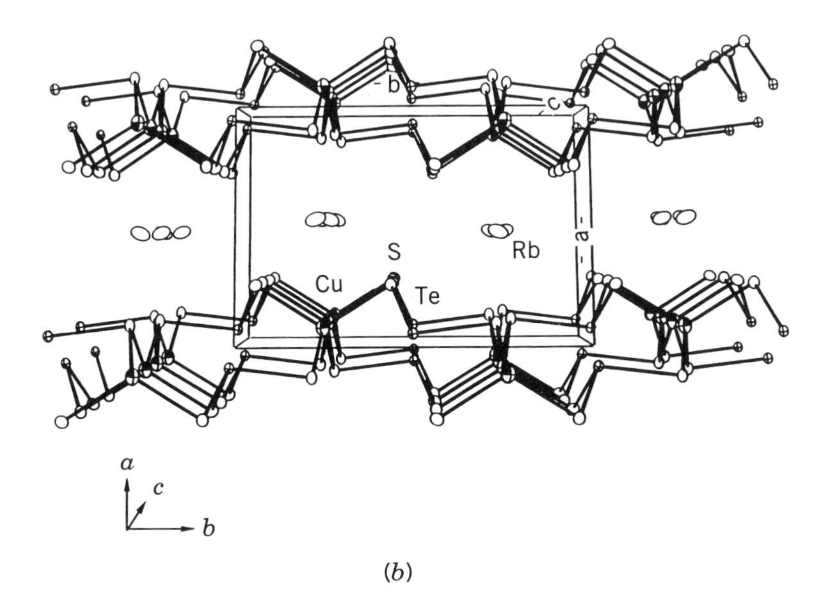

(b)

Figure 68. Two views of the structure of $RbCuTeS_3$: (a) perpendicular and (b) parallel to the anionic layers (120).

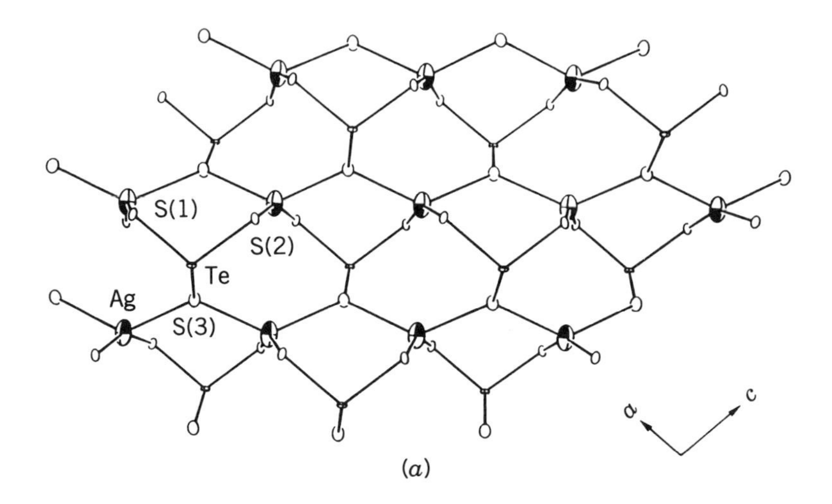

(*a*)

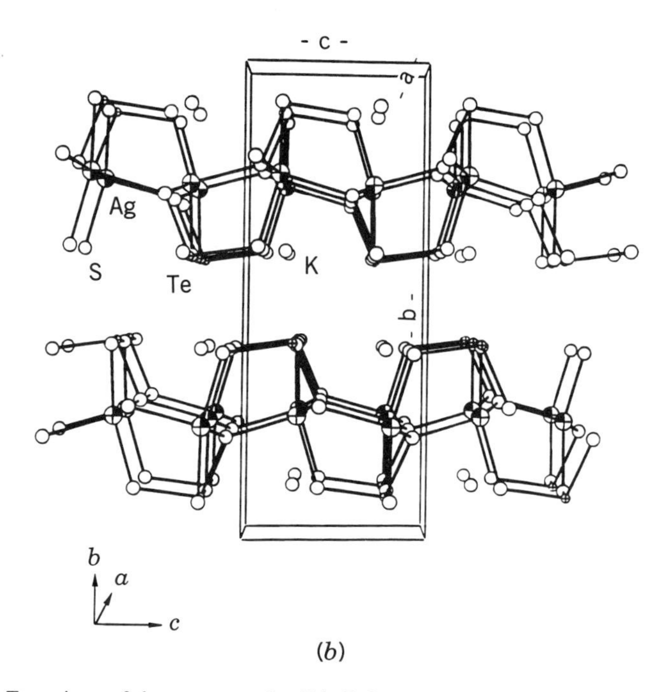

(*b*)

Figure 69. Two views of the structure of α-$KAgTeS_3$: (*a*) perpendicular and (*b*) parallel to the anionic layers (120).

the original β form, only a single peak at 334°C is seen, confirming that the phase change to α is irreversible. Interestingly, the Rb^+ and Cs^+ salts of the β form have no phase change themselves, melting congruently at 354 and 293°C, respectively.

Usually upon moving to a larger alkali ion, structure types are either retained or are changed to one of lower dimensionality. The compound $CsCuTeS_3$ illustrates an exception, forming a 3D lattice, albeit one with less overall metal–sulfur bonding (120). The phase was formed from a reaction of Cs_2S/Cu/Te/S in a 2 : 1 : 1 : 8 molar ratio heated at 260°C for 4 days and was contaminated by trace amounts of $Cu_{17.6}Te_8S_{26}$ (see below). The structure possesses cubic symmetry, and a view is shown in Fig. 70. Copper is now in a nearly ideal trigonal

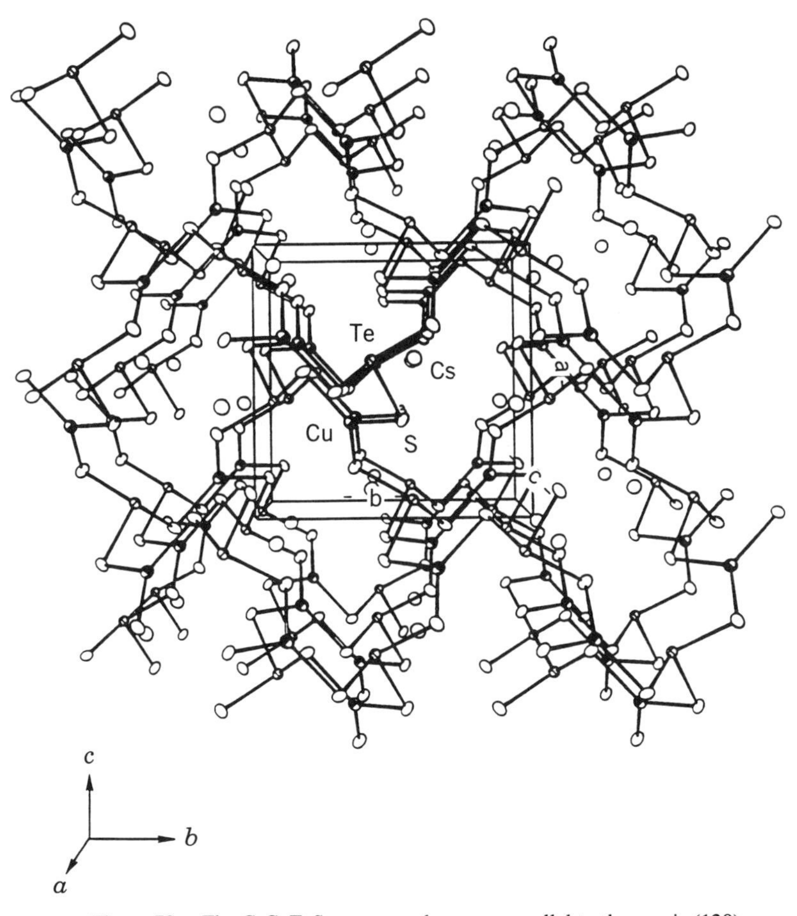

Figure 70. The $CsCuTeS_3$ compound as seen parallel to the a axis (120).

planar coordination [S—Cu—S = 119.6(1)°]. The [CuS_3] units do not connect directly to each other but rather are bridged by $(TeS_3)^{2-}$ ligands that bond to three different [CuS_3]. The Cs^+ ions reside along the walls of the lattice channels, having close contacts with S atoms rather that the Te^{4+} lone pair. Due to the lesser amount of Cu—S bonding present in the Cs^+ salt relative to the Rb^+ and K^+ analogues, a larger band gap than that of $RbCuTeS_3$ is achieved (2.15 vs. 1.95 eV).

Manganese has also shown reactivity towards fluxes with $(TeS_3)^{2-}$ ligands. The compound $A_2Mn(TeS_3)_2$ (A = Rb or Cs) has been synthesized from A_2S/Mn/Te/S in a 2:1:2:4 molar ratio for A = Rb and a 2:1:2:16 molar ratio for A = Cs, both reactions heated at 270°C for 4 days (121). The material possesses 2D anionic layers composed of [MnS_6] octahedra linked via the $(TeS_3)^{2-}$ ligands (Fig. 71). The [MnS_6] octahedra have no direct linkages with each other. All connectivity is through $(TeS_3)^{2-}$, which bridges three separate [MnS_6] units, and the $(TeS_3)^{2-}$ is oriented with the Te^{4+} lone pair directed towards the interlayer gallery. If the $(TeS_3)^{2-}$ ligand were reduced to a single point, the anionic layers would simply be a net of corner-sharing octahedra of the CdI_2 structure type (122). As is, this structure is related to a sulfite compound, $MnSO_3 \cdot 3H_2O$ (123), which possesses 1D ribbons containing fragments resembling $[Mn(TeS_3)_2]_n^{n-}$.

Despite having excess polysulfide in the flux, all of the compounds thus far described have exhibited amazing selectivity in possessing exclusively $(TeS_3)^{2-}$ ligands. There is only one example to date of a compound that has both $(TeS_3)^{2-}$ and polysulfide ligands: $Cs_6Cu_2(TeS_3)_2(S_6)_2$ (53). The material was synthesized from a flux of analogous composition to that used in previous reactions (Cs_2S/Cu/Te/S = 1.5:1:1:8), but the reaction temperature was somewhat lower (230°C for 4 days), which may be playing a role in stabilizing the resulting $(S_6)^{2-}$ ligand. The material forms as dark red crystals in 69% yield. Single-crystal studies revealed that $[Cu_2(TeS_3)_2(S_6)_2]^{6-}$ is a discrete anionic complex, and so under proper conditions, large Cs^+ does break up the previously discussed $(MTeS_3)^{1-}$ layers, with a little help from $(S_6)^{2-}$. Figure 72 shows two views of the anion. Its core possesses two Cu atoms bridged by two $(TeS_3)^{2-}$ ligands, each of which is functioning as a monodentate ligand. The tetrahedral coordination of each Cu atom is completed by chelating $(S_6)^{2-}$ ligands. Although it represented the first reported transition metal complex containing $(TeS_3)^{2-}$ ligands, $(S_6)^{2-}$ ligands had previously been reported in several examples including $[M(S_6)S_5]^{2-}$ (124), $[Hg(S_6)_2]^{2-}$ (125), $CsCuS_6$ (53), $[Bi_2(S_7)(S_6)]^{4-}$ (126), $[Sn(S_6)(S_4)]^{2-}$ (74), and $[In(S_4)(S_6)_2Br]^{2-}$ (127).

The phase $Cu_{17.6}Te_8S_{26}$ (52) has been mentioned previously as a side product in several reactions. It is in fact a new phase possessing $(TeS_3)^{2-}$ ligands and no charge balancing alkali cations. The material was successfully synthesized as black cubic crystals along with a small amount of yellow needles of K_2TeS_3,

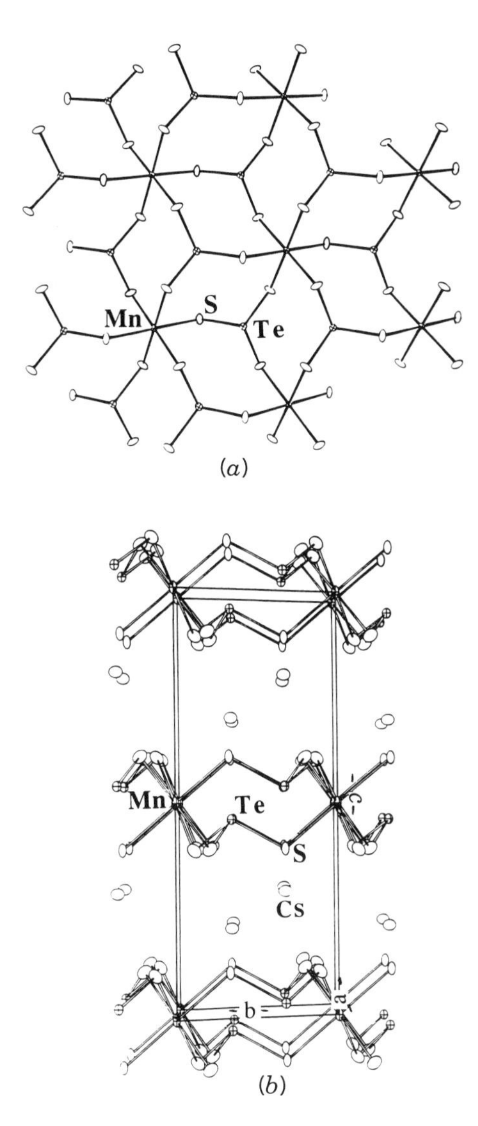

Figure 71. Two views of the structure of $A_2Mn(TeS_3)_2$: (*a*) perpendicular and (*b*) parallel to the anionic layers (121).

which was soluble in water and subsequently removed during the isolation. The flux used was a $K_2Te/Cu/S$ mixture in a 1 : 1 : 8 molar ratio heated at 350°C for 4 days. The compound $Cu_{17.6}Te_8S_{26}$ was found to be related to the tetrahedrite structure type, $M_{12}X_{14}S_{13}$ (M = Cu, Ag, Zn, Cd, Fe, or Hg; X = Sb, As, or Bi) (128), and both are derivatives of the ZnS (sphalerite) structure. The structure is shown in Fig. 73. Both Cu(1) and Te atoms occupy Zn sites in a sphal-

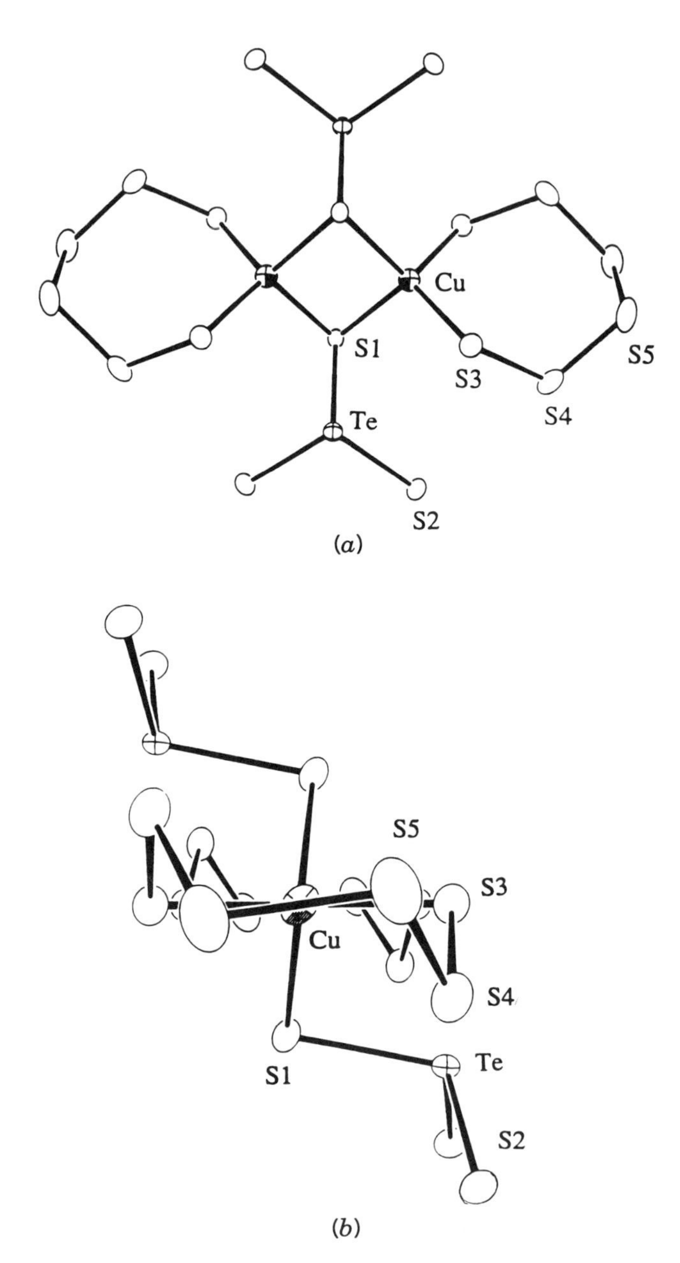

Figure 72. Two views of the anionic cluster: $[Cu_2(TeS_3)_2(S_6)_2]^{6-}$ (53).

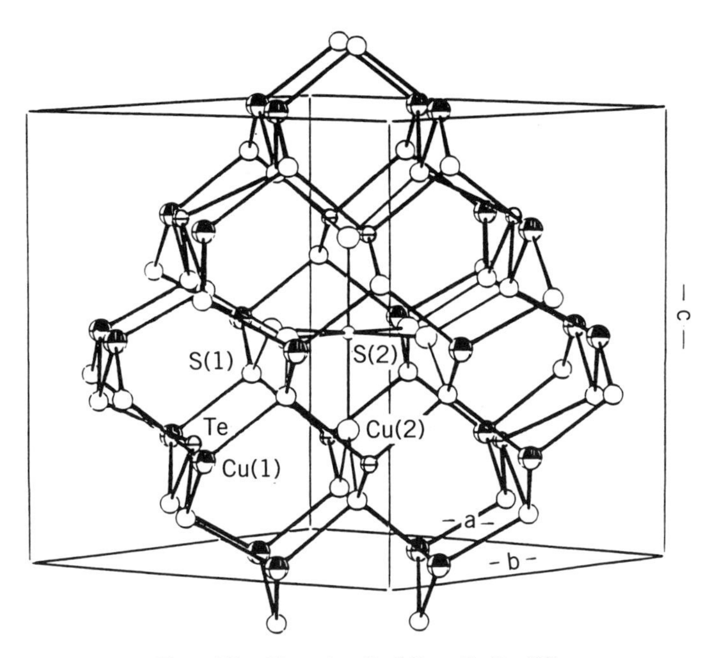

Figure 73. The unit cell of $Cu_{17.6}Te_8S_{26}$ (52).

erite lattice, but while Cu(1) maintains a distorted tetrahedral geometry, Te is part of the trigonal pyramidal $(TeS_3)^{2-}$ unit. By preventing coordination at that site, the lone pair on the Te^{4+} actually causes a void space to be opened in the sphalerite lattice. The resulting $[Cu_{12}(TeS_3)_8]$ cage is filled with an octahedral $[Cu_6S]$ fragment with each Cu(2) of the fragment having trigonal planar coordination. The Cu defects in this material are located on these positions; refinement of their occupancy revealed that 2.8 Cu atoms are disordered equally over those six sites.

In general the $(TeS_3)^{2-}$ ligand had been stabilized in fluxes that have possessed excess polysulfide. By reversing this and making Te the major chalcogenide component in the flux, the $(TeS_3)^{2-}$ ligand does not form, and compounds are isolated in which S^{2-} and Te^{2-} each possess discrete positions within the lattice.

The reaction of K_2S/Cu/Te in a 3 : 1.5 : 8 molar ratio at 450°C for 4 days yielded such a compound: KCu_4S_2Te (52). The material formed as black rectangular crystals in an 83.6% yield. As may be inferred from its formula, the phase is isostructural with KCu_4S_3 (129) in which two layers of anti-PbO type are fused together into a bilayer with K^+ between the layers (Fig. 74). The Te atoms are very selective in their choice of sites, occupying solely the eight-coordinate square prismatic positions within the layer while S is in the square

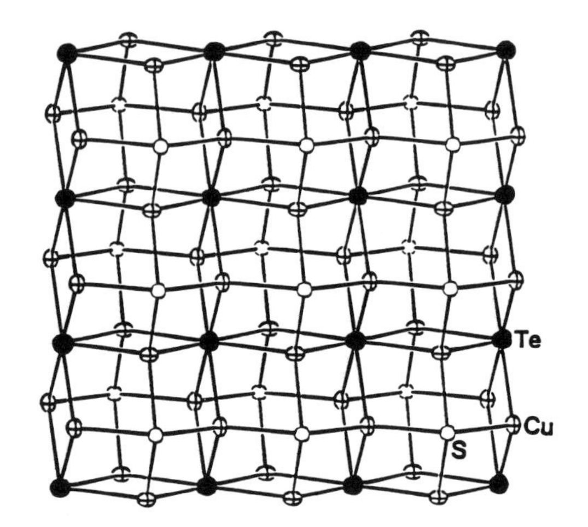

Figure 74. The view perpendicular to a single anionic layer of KCu_4S_2Te (52).

pyramidal sites above and below the layer. The hole formalism used to balance charges is the parent compound, $K(Cu^{1+})_4(S^{2-})_2(S^{1-})$, is just as applicable in KCu_4S_2Te. The lower electronegativity of Te suggests that it would make a more adequate receptacle for the 1− charge ($K(Cu^{1+})_4(S^{2-})_2(Te^{1-})$), although some degree of mixing between S and Te orbitals cannot entirely be ruled out. Physical properties are also comparable to the all sulfide phase. Resistivity increases with increasing temperature, and KCu_4S_2Te has a room temperature resistivity of 1.2×10^{-4} Ω cm (KCu_4S_3 room temperature resistivity: 2.5×10^{-4} Ω cm), indicating that Te substitution has little effect on the overall band structure. Thermopower studies revealed KCu_4S_2Te to be a *p*-type conductor, and magnetic studies showed Pauli-type paramagnetism at greater than 50 K with increasing interference from magnetic impurities at less than 50 K.

A second compound that also has partially oxidized monotellurides was found in $K_3Cu_8S_4Te_2$ (52). This species was synthesized in a K_2S/Cu/Te flux in a 3 : 1 : 5 molar ratio heated at 450°C for 4 days and forms as black needles in 71% yield. This material is isostructural with another mixed S^{2-}/S^{1-} copper sulfide, $K_3Cu_8S_6$ (130), suggesting that such structure types may have a tendency to form with the less electronegative Te present to act as a more easily oxidized receptacle for the required holes. Formally, the oxidation states become $K_3(Cu^{1+})_8(S^{2-})_4(Te^{1.5-})_2$ if the hole is considered delocalized over both Te atoms. The structure is shown in Fig. 75. The material's anionic layers are composed of $[Cu_4Q_4]_n^{3n-}$ columns (analogous to those of $Na_3Cu_4Se_4$) linked together by anti-PbO type fragments. The structure has four distinct Cu atoms. In the $[Cu_4Q_4]$ columns, where all Cu atoms are in distorted trigonal planar geometry, Cu(1) bonds to two S(2) atoms and one Te while Cu(4) bonds to one

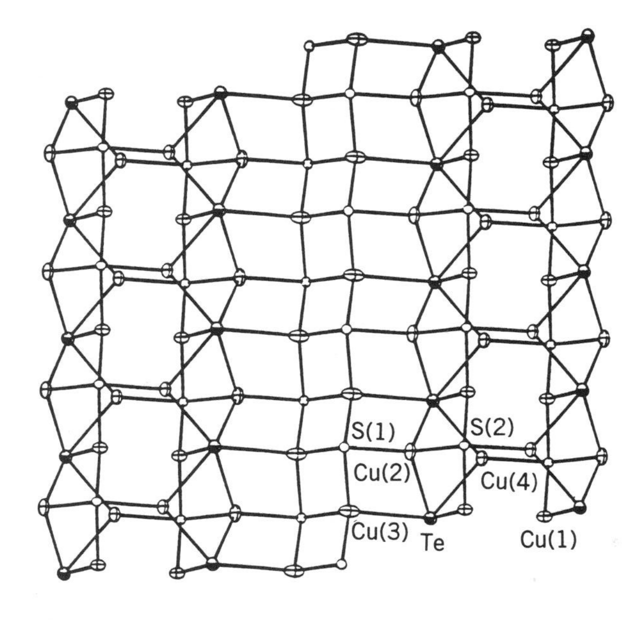

(*a*)

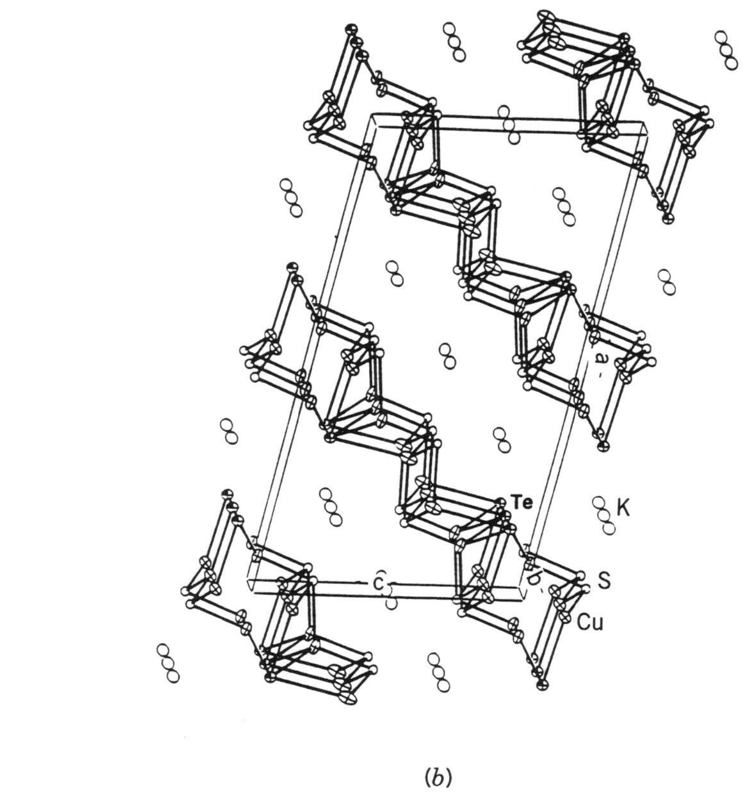

(*b*)

Figure 75. Two views of the structure of $K_3Cu_8S_4Te_2$: (*a*) perpendicular and (*b*) parallel to the anionic layers (52).

S(2) and two Te. The anti-PbO fragment also has Cu atoms that differ in the number of chalcogenides bonded to them: Cu(2) has two S bonds and two Te bonds; and Cu(3), three S bonds and one Te bond. The Cu(3)—Te bond at 2.914(3) Å is longer than typical, but this bond occupies the same position as a long Cu—S bond in $K_3Cu_8S_6$ (2.84 Å) and so appears to be a structurally based phenomenon. This particular site in the lattice has ramifications on the charge-transport properties of the two phases. In the all sulfide analogue, the material is metallic at high temperature but goes through a metal-to-semiconductor transition as the temperature drops. This change is thought to occur as the S atom with long bonds to Cu becomes localized in a site that is essentially too big for it, resulting in a structural distortion at low temperatures (130b and c). The larger Te atom occupying that site in $K_3Cu_8S_4Te_2$ is simply a snugger fit in the cavity. As such, no structural distortion occurs with falling temperature, and the material remains metallic all the way to 4 K. Magnetic measurements on the mixed chalcogenide show Pauli paramagnetism above 50 K and paramagnetic impurity below 50 K.

E. The A/Ba/Cu/Te Systems

One of the recurring themes in this chapter has been the influence that the size of the alkali metal cation has upon the final structure of the anionic framework. In general, if a larger cation does not stabilize the same framework as a smaller cation, then the change in packing requirements often results in materials with reduced dimensionality as the larger cation "pries apart" the parent anionic framework. Homoleptic systems have demonstrated this effect amply; however, the question of how sensitive the effect is has yet to be addressed. Specifically, if a heteroleptic mixture of countercations were employed, could the resulting anionic structures be fine tuned as a function of the composition of that mixture. A preliminary look at this question was taken in the reaction of Cu in polytelluride flux with an eye towards the structural diversity of the $[Cu_8(Te_2)_6]$ cage cluster, and the resulting compound, $K_{0.9}Cs_{2.1}Cu_8Te_{10}$, has already been discussed. Size is, of course, not the only property of the countercation that influences the structure; the alkali metals are all monovalent, and the anionic structure naturally responds to that as well. If alkaline earth cations could be included in the flux reactions, then the structure–property relationships as a function of changing cation charge could be gauged as well.

Preliminary investigations into various A/Ba/Cu/Te systems have lead to successful isolations of mixed alkali metal–barium phases. This system was initially chosen because it had already demonstrated stability towards mixed cations. The phases isolated, $A_2BaCu_8Te_{10}$, where A = K, Rb, or Cs, are all isostructural with $A_3Cu_8Te_{10}$ (see Section IV.C), and their synthetic conditions are given in Table VI (60). The smaller alkali cations (K and Rb) form with the

TABLE VI
Synthetic Conditions for the Formation of $A_2BaCu_8Te_{10}$ Phases

Compound	$A_2Te/BaTe/Cu/Te$ Ratio	Heating Program	Cooling Rate
$K_2BaCu_8Te_{10}$	4/1/2/12	370°C (4 days)	2°C h^{-1}
$Rb_2BaCu_8Te_{10}$	2/1/2/8	480°C (4 days)	2°C h^{-1}
$Cs_2BaCu_8Te_{10}$	2/2/1/6	420°C (4 days)	2°C h^{-1}

anionic layers in the monoclinic space group; the larger Cs cations, the orthorhombic space group. The alkali cations are found exclusively between the layers, while Ba^{2+} is consistently encapsulated inside the $[Cu_8(Te_2)_6]$ clusters, indicating that the larger positive charge on the alkali earth cation exerts a greater stabilization on the cage. Although structural effects are minimal in this case, inclusion of Ba^{2+} should radically alter the physical properties of these materials. Where $A_3Cu_8Te_{10}$ possessed two monotellurides with formal charges of -1.5 and had *p*-type metallic conductivity, $A_2BaCu_8Te_{10}$ is now valence precise and exhibits semiconducting properties.

In the first example, a Ba^{2+} containing flux has lead to a previously seen structure type; however, a radically different material has been found in the family $ABa_6Cu_3Te_{14}$ (60). This material was synthesized from a 3:1:1:7 ratio of $Na_2Te/BaTe/Cu/Te$ heated at 400°C for 4 days and from a 2:1:1:5 ratio of $A_2Te/BaTe/Cu/Te$ (A = K or Rb) at 420°C for 4 days. The Na^+ and K^+ analogues were well-characterized crystallographically, but severe atomic disorder apparently caused an unsatisfactory refinement on the Rb^+ phase, although it appears structurally similar by X-ray powder diffraction. All products form as black chunky crystals with small amounts of powdered impurity, and the phases are stable in air, water, and common organic solvents. In Fig. 76 the basic repeating unit of $ABa_6Cu_3Te_{14}$ is shown: a $[Cu_3Te_3(Te_3)_3]^{9-}$ anionic cluster. The anion has a paddle wheel shape with a central $[Cu_3Te_3]$ ring as the core and three $(Te_3)^{2-}$ ligands perpendicular to the plane of the ring, one chelating to each of the Cu atoms. The geometry about the Cu atoms is distorted tetrahedral with angles ranging from 96.4(1)° to 112.6(1)°. Figure 77 shows the compound packing diagram. The clusters stack in a staggered fashion along the *c* axis with a lone A^+ cation sandwiched between them and two columns of free Te^{2-} ions present in the lattice. That the clusters pack in such close proximity despite their 9− charge is intriguing and cannot be fully accounted for by the sandwiched alkali cation. Close examination of the structure reveals a possible source of stability. Each Te(1) atom (within the ring) has short contacts to the Te(2) atoms (in the paddle) on the neighboring chain. These distances are 3.262(2) Å in the Na^+ compound and 3.292 Å in the K^+ and are both much shorter than the sum of the van der Waals radii (4.4 Å). If considered as

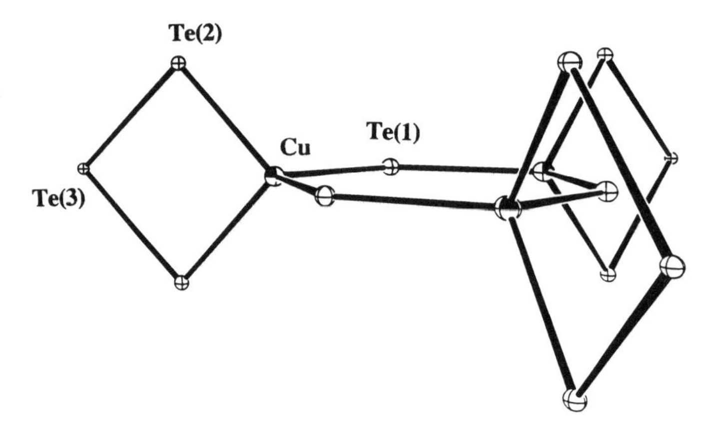

Figure 76. The anionic cluster $[Cu_3(Te)_3(Te_3)_3]^{9-}$ (60).

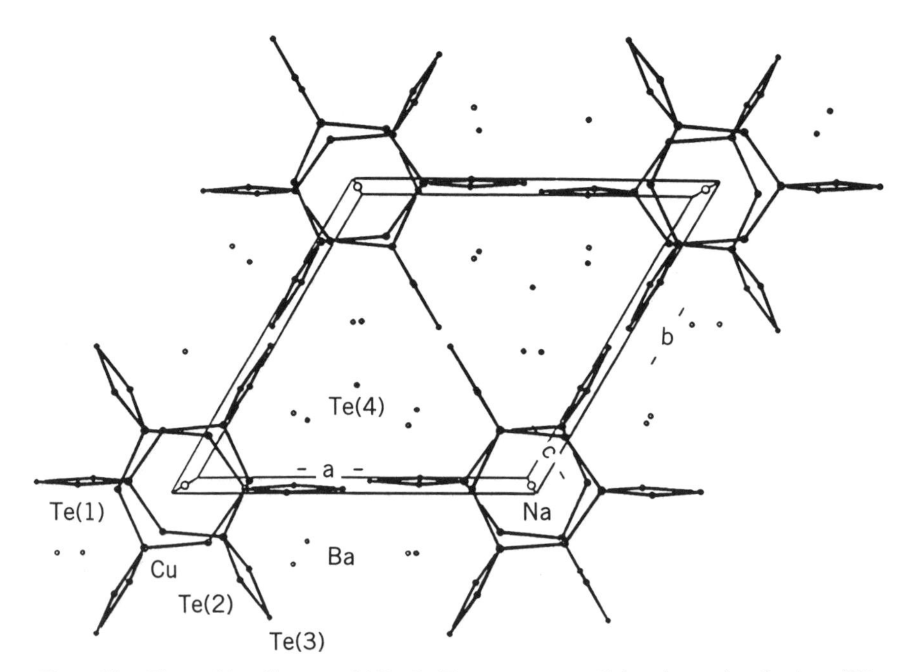

Figure 77. The packing diagram of $ABa_6Cu_3Te_{14}$, as seen parallel to the stacks of anions (60).

at least a partial bonding interaction, this would lead to the formation of intercluster Te chains as shown in Fig. 78. The Te(1) atom has a distorted tetrahedral geometry, and Te(3) has a simple bent configuration; both are common for sp^3 hybridization. The Te(2) atom, however, is nearly T shaped, suggesting an sp^3d hybridization. These intercluster interactions may, in fact, be providing the little extra stability needed to form this phase.

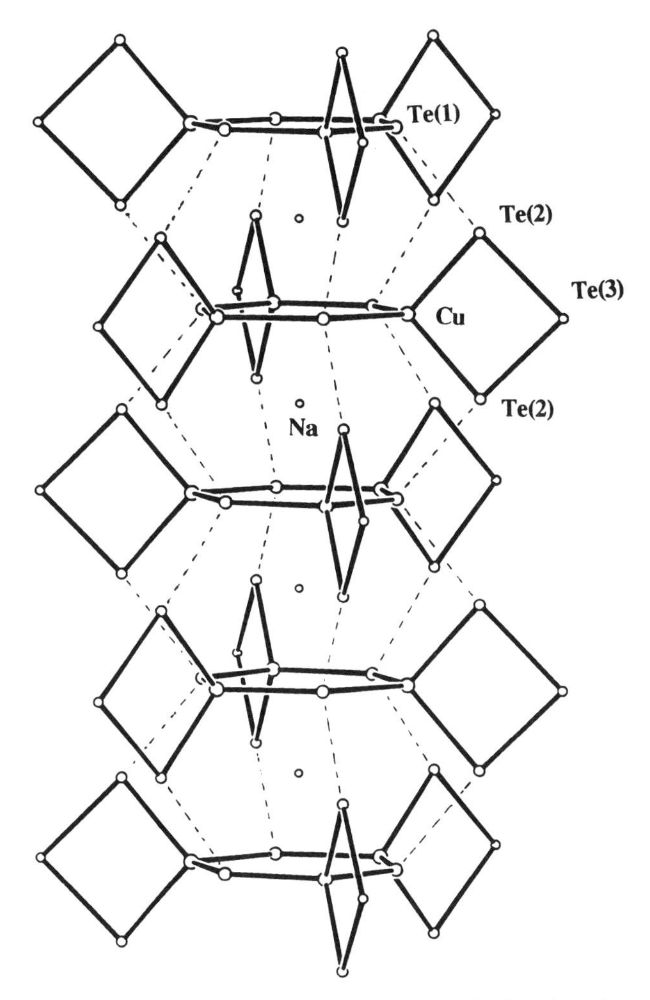

Figure 78. A view of the stacks of anions from $ABa_6Cu_3Te_{14}$. The long intercluster Te—Te interactions are shown as dashed lines. Taking these interactions into account: Te(1) is tetrahedrally coordinated; Te(2) has a T-shaped coordination; and Te(3) retains a bent coordination (60).

The success of these reactions in incorporating Ba^{2+} into the final structure has implications for further chemistry using all the alkaline earth cations. Although it is just as feasible to use alkaline earth/polychalcogenide fluxes (BQ_x) in nearly the same manner as their alkali metal counterparts, there is one glaring difference: the melting points of BQ_x fluxes are much higher than those of the A_2Q_x fluxes. Except for BaS_x [e.g., BaS_3 has a melting point of 554°C (131)], all other alkaline earth/polychalcogenides have melting points higher than the intermediate temperature window that we were originally interested in investigating. The syntheses of $A_2BaCu_8Te_{10}$ and $ABa_6Cu_3Te_{14}$, however, demonstrate that phases potentially very rich in alkaline earth cations can be synthesized using low melting A_2Q_x flux. This now grants the same benefits to alkaline earth chemistry that have already proven so fruitful with the alkali metals.

VII. REACTIVITY IN ORGANIC CATION/POLYCHALCOGENIDE SALTS

The quaternary systems of the previous section demonstrated that polychalcogenide fluxes have the potential to go beyond their original base as a low-temperature route to ternary compounds. As long as a species is stable in the flux, it has the potential to be used as a ligand or building block in these intermediate temperature reactions. Further expansion of the technique now becomes simply a matter of determining the conditions under which species will remain viable in the flux. Such investigations do not have to be limited to those containing alkali or alkaline earth chalcogenides. Particularly intriguing would be the use of organic compounds in molten $(Q_x)^{2-}$ reactions. The first steps in investigating mixed organic–polychalcogenide fluxes have already been taken in our laboratory by employing polychalcogenide salts in which the alkali metals have been substituted by large organic cations. These mixed salts are formed simply by precipitating them out of solutions containing the organic cation and polychalcogenide. Because of the limited thermal stability of the organic cations, these salts have a narrower temperature window than their alkali brethren, but as long as the reactions are held within that window ($\leq$ 200°C), chemistry analogous to the fully inorganic fluxes is possible. The much larger sizes of the polyatomic organic cations have shown a dramatic effect on anionic structure and pave the way for more extensive studies into both counterion size (and shape) dependence and mixed organic–polychalcogenide systems.

The first compounds synthesized from mixed organic–polychalcogenide fluxes employed polyselenides with tetraphenylphosphonium ($[PPh_4]^+$) as the organic cation. These phases were of the formula $[PPh_4][M(Se_6)_2]$, where M = Ga, In, or Tl (132). The exact synthetic conditions for each phase are given in Table VII. The reactions are similar in that each involved very Se rich fluxes ($[PPh_4]^+/Se \geq 2:21$) and isotherm temperatures of only 200°C. Yields are

TABLE VII
Synthetic Conditions for the Formation of $[PPh_4][M(Se_6)_2]$ Phases

Compound	$[PPh_4]_2Se_5$/M/Se Ratio	Heating Program
$[PPh_4][Ga(Se_6)_2]$	1/2.16/16	200°C (2 days)
$[PPh_4][In(Se_6)_2]$	1/2/16	200°C (2 days)
$[PPh_4][Tl(Se_6)_2]$	1/2/24	200°C (2 days)

highest for the In^{3+} analogue (95%); the Ga^{3+} containing compound occurs in 60% yield, and the Tl^{3+} analogue forms with Se contamination. The structure of $[PPh_4][M(Se_6)_2]$ is shown in Fig. 79. The M^{3+} cations are tetrahedrally coordinated and are joined into a 2D anionic lattice solely by bridging $(Se_6)^{2-}$ chains. Each metal is coordinated to the terminal atom of four different $(Se_6)^{2-}$ ligands, and each ligand is bonded to only two metal atoms, one at each ter-

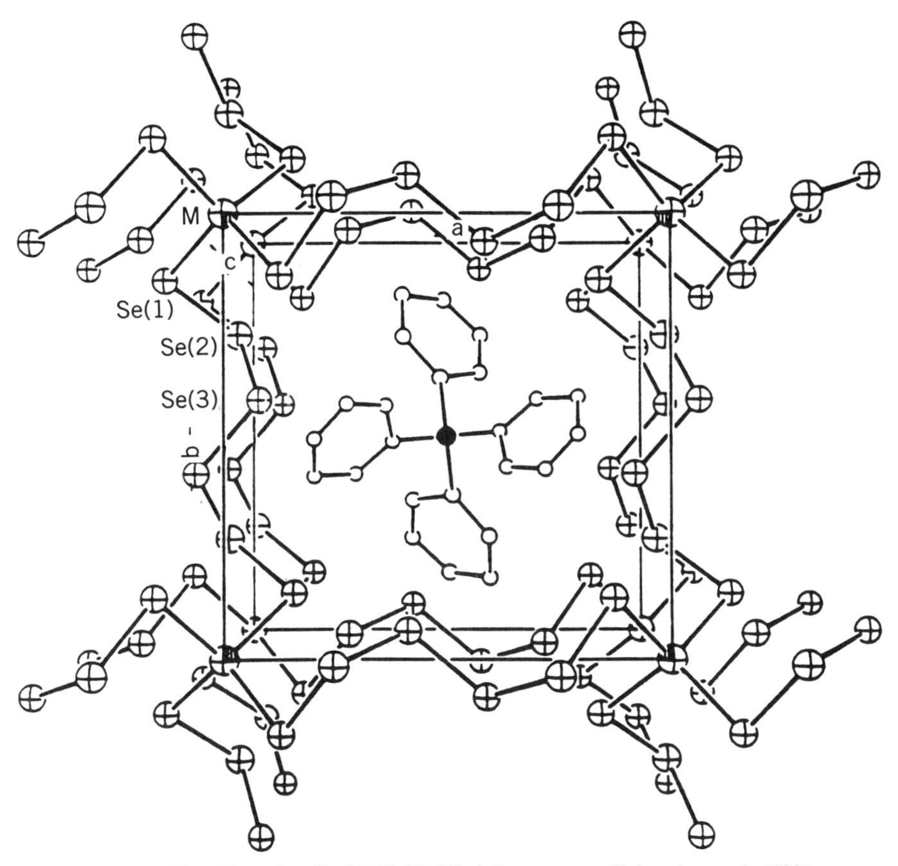

Figure 79. The unit cell of $[PPh_4][M(Se_6)_2]$ as seen parallel to the *c* axis (132).

minal Se atom. Conceptually, this 2D layer is related to the layers of corner-sharing tetrahedra of HgI_2 if the $(Se_6)^{2-}$ ligands are treated as the I^- equivalents. The organic cations reside within the anionic layers themselves, being surrounded by a 28-member $[M_4(Se_6)_4]$ ring (Fig. 80). This leaves the interlayer galleries empty. These rings also are lined up from layer to layer, forming tunnels parallel to the *c* axis. That the cations should be within the anionic layers, rather than between them, suggests that it performs a templating function in the compound's formation. This role appears to be crucial as the analogous reactions with various $[R_4N]^+$ containing fluxes (R = methyl, ethyl, propyl, and butyl) did not result in analogous compounds. Although the Se rich flux and the low reaction temperatures make the stabilization of the $(Se_6)^{2-}$ ligands possible, it is the large $[PPh_4]^+$ cations that enforce a reduced connectivity between metal and ligand, driving the structure to such an open framework.

The $[PPh_4][M(Se_6)_2]$ compounds also exhibit unusual thermal properties, as revealed by differential scanning calorimetry (DSC) experiments. This behavior was most extensively studied for the In phase. Its melting point was found to be 242°C; however, the material does not recrystallize upon cooling, as evidenced by a lack of any thermal event. Rather, the compound reverts to a glassy state that is metastable with respect to temperature. Reheating of this glassy state leads to a broad exotherm that peaks for the In phase at 165°C, with an

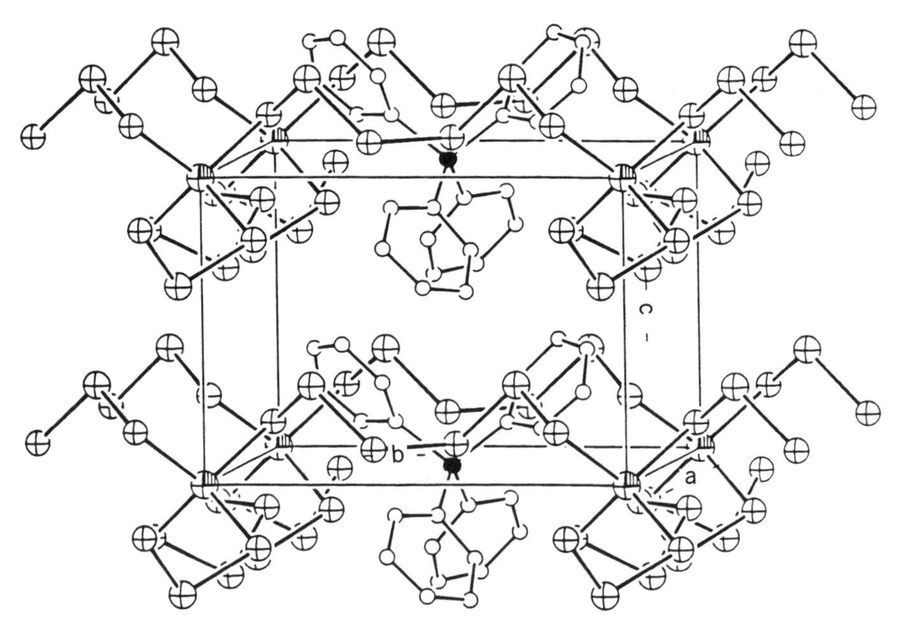

Figure 80. The view parallel to the layers in $[PPh_4][M(Se_6)_2]$ (132).

onset at 135 °C. After this reheating, powder X-ray diffraction confirms that the original crystalline phase is regenerated. Similar behavior was seen for the Ga and Tl analogues, which have melting points of 272 and 213 °C, respectively.

Further studies of In reactivity in $[PPh_4]_2[Se_x]$ fluxes have resulted in another new phase. When much less Se-rich conditions are used ($[PPh_4]^+/Se = 1:5$) an inhomogeneous mixture results. One component is a microcrystalline phase that has not yet been structurally characterized, but the second one, forming as red needles, was found from single-crystal studies to be $[PPh_4]_2[In_4Se_9]$ (133). A view of the 1D anionic chains of this compound is shown in Fig. 81. The In^{3+} cations are again all tetrahedrally coordinated, but the reduced Se content in the flux has lead to the incorporation of only $(Se)^{2-}$ and $(Se_2)^{2-}$ ligands in the chain. The main features within the structure are a series of fused In/Se rings: five-membered rings of $[In(Se^{2-})In\ (Se_2)^{2-}]$ and two different six-membered rings, one with two In^{3+} and two $(Se_2)^{2-}$ and the other with three In^{3+}, two $(Se)^{2-}$ and one Se from an $(Se_2)^{2-}$ ligand. This particular compound is highly metastable. It has yet to be made in pure form, and both longer isotherm times and slower cooling rates nearly eliminate this phase from the reaction products.

Reactivity towards $[PPh_4]_2[Se_x]$ fluxes has also been seen with Sn, and two new phases have been observed. The first, $[PPh_4]_4[Sn_6Se_{21}]$, has been synthesized in pure form from the reaction of $[PPh_4]_2Se_5/Sn/Se$ in a 1:2:4 molar ratio heated at 200 °C for 2 days (133). Its structure is shown in Fig. 82. The Sn atoms have either tetrahedral or highly distorted trigonal bipyramidal geometry and are coordinated by a mixture of bridging Se^{2-}, $(Se_2)^{2-}$, and $(Se_3)^{2-}$. A second new phase, $[PPh_4]_2[Sn_3Se_8]$, is observed when Sn is allowed to react with equimolar amounts of $[PPh_4]_2[Se_5]$, but the first phase occurs as a coprod-

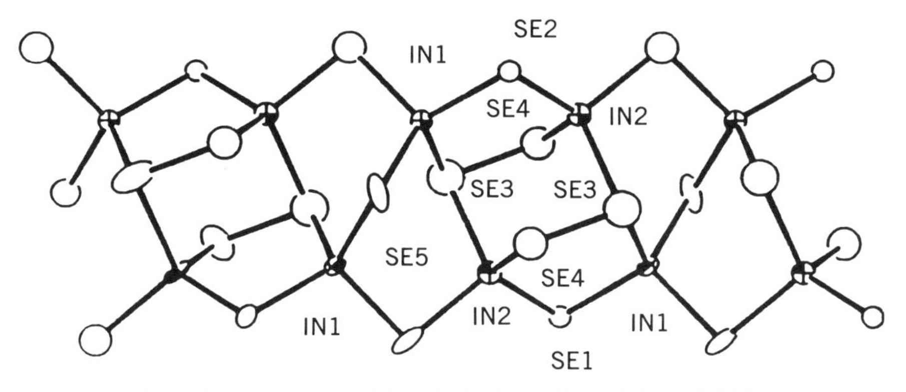

Figure 81. A fragment of the anionic chains of $[PPh_4]_2[In_4Se_9]$ (133).

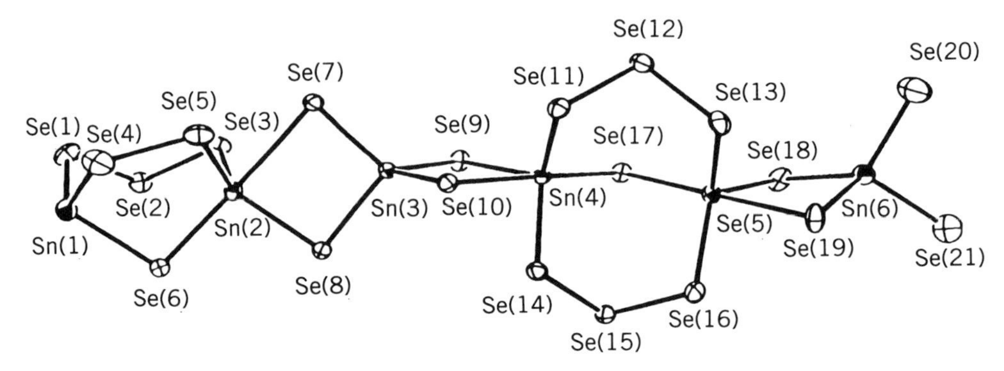

Figure 82. A fragment of the anionic chains of $[PPh_4]_4[Sn_6Se_{21}]$ (133).

uct. The second material is also composed of 1D anionic chains (Fig. 83) in which trigonal bipyramidal $[SnSe_5]$ share axial–equatorial edges in the chain direction and $[SnSe_4]$ tetrahedra form a bridge between every two $[SnSe_5]$ units. All of the Se atoms are simply monoselenides except for one $(Se_2)^{2-}$ formed between a Se atom in an $[SnSe_4]$ tetrahedron and an axial Se atom in a $[SnSe_5]$ units.

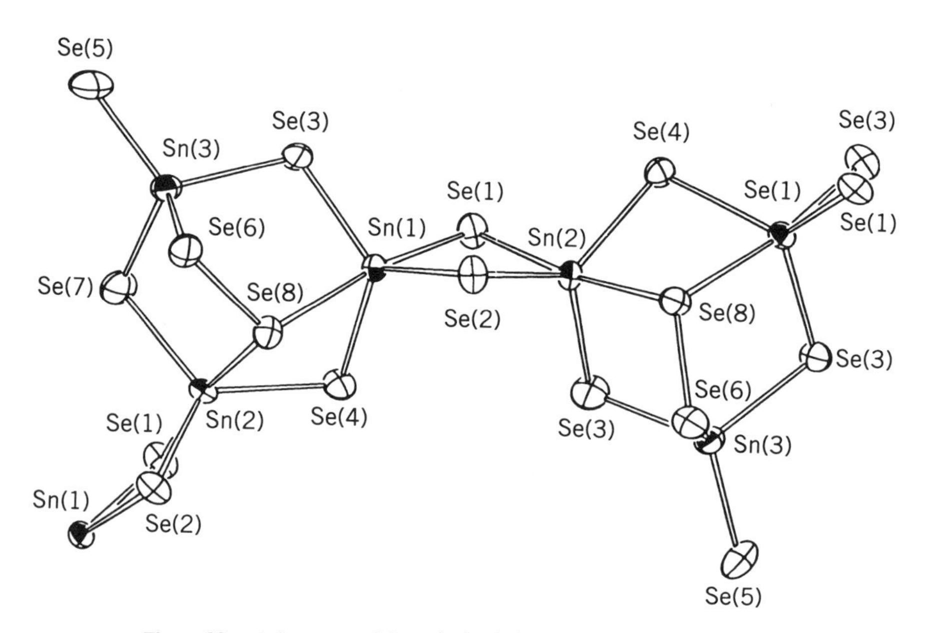

Figure 83. A fragment of the anionic chains of $[PPh_4]_2[Sn_3Se_8]$ (133).

VIII. CONCLUSIONS AND OUTLOOK

The original impetus for using molten A_2Q_x fluxes as a reaction media was to access a temperature regime much lower than what is normally used in solid state synthesis and thereby attempt to gain some measure of kinetic control in forming metastable solid state compounds. Now that the current progress of this approach has been reviewed, it is time to reexamine that original thesis to see where the potential of this method has been met, exceeded, or may yet lie.

We have shown numerous phases synthesized from polychalcogenide fluxes that have been "metastable" in the classic definition; that is, the phases change irreversibly when heated. Such a change can involve a transition to a different structure type of the same stoichiometry as the original, but, as is more commonly seen in this work, the change is simply a decomposition to more thermally stable binary compounds. One example worth mentioning again is that of α-$KCuS_4$, which when heated to 250°C decomposes to CuS (observed by powder X-ray diffraction) and, presumably, amorphous K_2S_x. Perhaps "metastable" is not entirely appropriate in the strict definition of the word. "Kinetically stable" may be better, but since "metastable" is used quite loosely by many investigators working in widely varying fields it is not always clear what is meant by it. Having read this chapter, we hope the reader understands in what context we use the term. We note, however, that "metastable" does not mean unstable, and that there are many stable compounds or materials that can be isolated and handled that are in fact metastable. One example is graphite (with respect to diamond) another is glass.

There are many other new phases, however, which although cannot be described with the strict definition of "metastable," are nonetheless so closely tied to the synthetic conditions of the flux that a broader consideration of the term is in order. Again, recall the case of α- and β-$KCuS_4$. Both phases are structurally so similar that a simple thermal conversion between the two would be expected. None, in fact, exists, and the α phase would prefer to decompose rather than undergo what would appear to be a simple thermal cosmetic change. This is even more surprising when remembering that the formation of the β phase is favored over the α simply by increasing the reaction temperature by 35°C. Another equally temperamental example is that of K_3AuSe_{13} and $KAuSe_5$. Both were formed from reactions run at 250°C for 99 h, but the ratio of the K_2Se/Au/Se starting materials was 1.8 : 1 : 8 for K_3AuSe_{13} and 2 : 1 : 8 for $KAuSe_5$. Does a 0.2 equivalent of K_2Se really change the chemistry of the flux all that much? Apparently, just enough.

The real key to understanding such sensitive formation conditions lies in a place that remains beyond our ability to fully comprehend. That place is the wonderfully mysterious entity that we have continually referred to as "the flux." To shed some light on the nature of these fluxes, detailed studies with a wide variety of experimental techniques and over a wide range of temperatures are needed. Some interesting research activity into the physicochemical characterization of molten alkali polychalcogenides has been recently reported in the literature (134). Examples like those in the previous paragraph point to a complexity of equilibria in the reaction media that we can only guess at. Apparently, these equilibria are such that many different intermediates can be stabilized in the flux depending on the conditions of the reaction, and then the stabilized intermediates give rise to specific products. The solubility, or better yet the insolubility of the products also plays a significant role in determining which one will deposit from these "solutions." This is why α- and β-$KCuS_4$ cannot be interconverted directly; without the flux to form the necessary intermediates, there is no pathway by which such an interconversion can take place. This shifts a considerable burden of stability away from the final products and onto the intermediates. As such a new phase may have any measure of stability, but the intermediates necessary to form that phase may only exist under specific conditions of temperature, time, and flux composition. Metastable or just kinetically stable intermediates can still result in highly stable products. The question of intermediates has always been a consideration in molecular chemistry, but in classical solid state chemistry, intermediates have been a largely moot point due to the lack of kinetic control. Unfortunately, our ability to investigate the intermediates in the polychalcogenide flux directly are also limited. The conditions of the flux are simply too extreme for most tools such as spectroscopy and electrochemistry, requiring that special methodologies and apparatus be developed. Although such physicochemical studies would be highly desirable, for now the only recourse is the systematic investigation of reaction conditions.

In addition to the "metastability" of the flux intermediates, another important factor governing the products of these reactions has been the choice of countercation. The initial investigations performed using alkali metals saw many examples in which the size of the countercation played a role in the dimensionality of the resulting phase. The forming anionic network can apparently respond to the changes in packing forces that a change in cation size represents. Hence, larger cations can help to drive a lattice to reduced dimensionality. Although not a consistent principle, the effect is common enough that logical extrapolations could be made. For instance, if cation size is a factor, what about cation charge? A divalent alkaline earth represents an effective decrease in cation volume over a monovalent alkali, and so studies into mixed-cation fluxes were begun. Another question was whether Cs^+ was the largest species we could hope to use or could some larger polyatomic species find utility in $(Q_x)^{2-}$ flux chemistry? This question led to the work with the $[PPh_4]_2[Se_x]$ fluxes. Both of these offshoots are still in their early stages, and the asking of creative questions can always lead to other "subdisciplines." One such "subdiscipline" may be microporous chalcogenides. For example, instead of opting to form a low-dimensional structure when faced with large countercations, a material may adopt an open framework structure templated by the large counterions. This is the case with $[PPh_4][In(Se_6)_2]$. If the factors that influence structure are better understood it could lead to a new class of microporous materials which, in addition to catalytic properties, would feature electronic properties as well.

Originally, this method was used to examine the simplest of systems: alkali metal/transition metal/chalcogenides. The complexity of the reactions has steadily increased from that base. The first extrapolations were made into quaternary systems; that is, what if two metals were added in the same flux? Or two chalcogenides? From there we have come to realize that specific polyatomic building blocks can be stabilized in the flux and used as ligands. To date, these species are the thiostannates ($[SnS_4]^{4-}$ and $[Sn_2S_6]^{2-}$), the $(TeS_3)^{2-}$ ligand, the various chalcophosphates, and the large organic cation, $[PPh_4]^+$. All have in common that they are already fully oxidized and so can remain intact under the corrosive conditions of the $(Q_x)^{2-}$ flux. The stabilization of discrete species is the first step in changing these reactions from a pot of equilibria and unknown intermediates into more rational chemistry where some of the intermediates can be known beforehand. The future potential of molten polychalcogenide fluxes now is only limited to what ligands, both organic and inorganic, can be manipulated in the media.

This chapter has described a few important steps that have been taken in one small, but elegant, area of synthetic inorganic chemistry. While this chapter was in preparation another shorter account focusing on a narrower area of polychalcogenide flux synthesis has appeared (135). Similar steps are being taken in other systems as solid state chemists attempt to increase the rationality of their work by employing gentler synthetic techniques that allow for the use of such molecular chemistry concepts as kinetic control and polyatomic building blocks. Simultaneously, molecular chemists have become increasingly interested in the designed aggregation of discrete fragments into larger and larger assemblies. These studies have come under the umbrella of buzzwords such as "supramolecular," "quantum superlattice," and "self-assembly." Chemists of all disciplines have become increasingly aware of the value of understanding structure as a necessary prerequisite to synthesizing materials with specific bulk properties. Hence, molecular and solid state chemists have been "borrowing" from each other in order to reach the same goal: exploitation of structure–property relationships. Both groups are indeed two sides of the same coin, and as solid state chemists become more like molecular chemists, and vice versa, the eventual meeting in the middle will result in whole new disciplines, which will make traditional synthetic boundaries anachronistic. Such fields will certainly be fertile ground for those with creative interdisciplinary approaches.

Although techniques will become more sophisticated, 100% synthetic predictability is not likely to be achieved in any chemical discipline. Thank God.

We will always have the unknown to frustrate, motivate, and inspire us. However, our desire for rationality and control is as strong as our love of the unknown, because in rationality lies our ability to impact our lives. Random forces will surely leave their mark, but if our intellect is to have any value, then the marks we make must be by design, and that design, in turn, must come from our ability to be imaginative and to make important connections between different fields of intellectual discourse.

ACKNOWLEDGMENTS

Of course, none of our contributions to the field presented in this chapter would have been possible were it not for several outstanding and extraordinarily dedicated students and postdocs. Namely, Younbong Park, Ju-Hsiou Liao, Tim J. McCarthy, Xiang Zhang, Sandeep Dhingra, Andy Axtell, and Deby Chakrabarty. We thank the National Science Foundation for generous funding of this work. Specifically, we gratefully acknowledge a Presidential Young Investigator Award (1989–1994) and a grant (DMR-92-02428). We also thank the Center for Fundamental Materials Research at Michigan State University and the Beckman Foundation for financial support. ACS thanks NASA for a graduate fellowship 1991–1994. MGK has a A. P. Sloan Foundation Fellowship and a Camille and Henry Dreyfus Teacher–Scholar award 1993–1995.

REFERENCES

1. For a recent review see M. L. Steigerwald, in *Inorganometallic Chemistry*, T. P. Fehlner, Ed., Plenum, New York, 1992, pp. 333–356.
2. (a) C. Niu and C. M. Lieber, *J. Am. Chem. Soc.*, *114*, 3570 (1992). (b) T. Novet and D. Johnson, *J. Am. Chem. Soc.*, *113*, 3398 (1991).
3. (a) C. J. Warren, S. S. Dhingra, R. C. Haushalter, and A. B. Bocarsley, *J. Solid State Chem.*, *112*, 340 (1994) (b) C. J. Warren, D. M. Ho, R. C. Haushalter, and A. B. Bocarsley, *J. Chem. Soc. Chem. Commun.*, 361 (1994).
4. A. Rabenau, *Angew. Chem. Int. Ed. Engl.*, *24*, 1026 (1985).
5. R. M. Barrer, *Hydrothermal Chemistry of Zeolites*, Academic, New York, 1982.
6. R. C. Haushalter and L. A. Mundi, *Chem. Mater.*, *4*, 31 (1992).
7. (a) R. E. Morris, A. P. Wilkinson, and A. K. Cheetham, *Inorg. Chem.*, *31*, 4774 (1992). (b) W. T. A. Harrison, G. D. Stucky, R. E. Morris, and A. K. Cheetham, *Acta Crystallogr. Ser. C*, *48*, 1365 (1992). (c) R. E. Morris, W. T. A. Harrison, G. D. Stucky, and A. K. Cheetham, *J. Solid State Chem.*, *94*, 227 (1991). (d) W. T. A. Harrison, R. E. Morris, and A. K. Cheetham, *Acta Crystallogr. Ser. C*, *48*, 1182 (1992). (e) R. E. Morris, J. A. Hriljac, and A. K. Cheetham, *Acta Crystallogr. Ser. C*, *46*, 2013 (1990). (f) R. E. Morris and A. K. Cheetham, *Chem. Mater.*, *6*, 67 (1994). (g) W. T. A. Harrison, G. D. Stucky, and A. K. Cheetham, *Eur. J. Solid State Inorg. Chem.*, *30*, 347 (1993).

8. (a) M. I. Khan, Q. Chen, J. Zubieta, and D. P. Goshorn, *Inorg. Chem.*, *31*, 1556 (1992). (b) M. I. Khan, Q. Chen, and J. Zubieta, *J. Chem. Soc. Chem. Commun.*, 305 (1992). (c) M. I. Khan, Q. Chen, D. P. Goshorn, H. Hope, S. Parkin, and J. Zubieta, *J. Am. Chem. Soc.*, *114*, 3341 (1992). (d) M. I. Khan and J. Zubieta, *J. Am. Chem. Soc.*, *114*, 10058 (1992). (e) M. I. Khan, Q. Chen, D. P. Goshorn, and J. Zubieta, *Inorg. Chem.*, *33*, 672 (1993).
9. (a) J.-H. Liao and M. G. Kanatzidis, *Inorg. Chem.*, *31*, 431 (1992). (b) K.-W. Kim and M. G. Kanatzidis, *J. Am. Chem. Soc.*, *114*, 4878 (1992).
10. P. T. Wood, W. T. Pennington, and J. W. Kolis, *J. Am. Chem. Soc.*, *114*, 9233 (1992).
11. D. Elwell and H. J. Scheel, *Crystal Growth from High-Temperature Solutions*, Academic, New York, 1975.
12. (a) R. R. Chianelli, T. A. Pecoraro, T. R. Halbert, W.-H. Pan, and E. I. Stiefel, *J. Catal.*, *86*, 226 (1984). (b) T. A. Pecoraro and R. R. Chianelli, *J. Catal.*, *67*, 430 (1981). (c) S. Harris and R. R. Chianelli, *J. Catal.*, *86*, 400 (1984).
13. (a) H. Eckert, *Angew. Chem. Int. Ed. Engl.*, *28*, 1723 (1989). (b) R. Zallen, *Physics of Amorphous Solids*, Wiley, New York, 1983. (c) D. Strand and D. Adler, *Proc. SPIE Int. Soc. Opt. Eng.*, *420*, 200 (1983). (d) N. Yamada, N. Ohno, N. Akahira, K. Nishiuchi, K. Nagata, and M. Takeo, *Proc. Int. Symp. Opt. Memory 1987, J. Appl. Phys.*, *26 (Suppl. 26-4)*, 61 (1987).
14. (a) M. S. Whittingham, in *Solid State Ionic Devices July 18–23 1988 Singapore*, B. V. R. Chowdari and S. Radhakrishna, Eds., World Scientific, Singapore, 1988, pp. 55–74. (b) W. L. Bowen, L. H. Burnette, and D. L. DeMuth, *J. Electrochem. Soc.*, *136*, 1614 (1989). (c) D. W. Murphy and F. A. Trumbore, *J. Electrochem. Soc.*, *123*, 960 (1976).
15. (a) A. P. Brown, *J. Electrochem. Soc.*, *134*, 2506 (1987). (b) J. L. Sudworth and A. R. Tilley, in *The Sulfur Electrode, Fused Salts and Solid Electrolytes*, R. P. Tischer, Ed., Academic, New York, 1983, p. 235. (c) A. P. Brown, *J. Electrochem. Soc.*, *134*, 1921 (1987). (d) M. Liu and L. C. DeJonghe, *J. Electrochem. Soc.*, *135*, 741 (1988). (e) H. Yamin, A. Gorenshtein, J. Penciner, Y. Sternberg, and E. Peled, *J. Electrochem. Soc.*, *135*, 1045 (1988). (f) R. Knödler, *J. App. Electrochem.*, *18*, 653 (1988). (g) K. E. Heusler, A. Grzegorzewski, and R. Knödler, *J. Electrochem. Soc.*, *140*, 426 (1993).
16. (a) H. J. Möller, *Semiconductors for Solar Cells*, Artech House, Boston, MA, 1993. (b) K. Zweibel, *Basic Photovoltaic Principles and Methods*, Van Nostrand-Reinhold, New York, 1984. (c) J. H. Armstrong, C. O. Pistole, M. S. Misra, V. K. Kapur, and B. M. Basol, in *Space Photovoltaics Research and Technology 1991*, NASA Conference Publication 3121, NASA, Washington DC, 1991, pp. 19-1–19-8.
17. (a) D. M. Rowe and C. M. Bhandari, *Modern Thermoelectrics*, Holt, Rinehart, and Winston, London, UK, 1983, p. 103. (b) K. Borkowski and J. Pyzyluski, *J. Mater. Res. Bull.*, *22*, 381 (1987).
18. H. J. Scheel, *J. Cryst. Growth*, *24/25*, 669 (1974).
19. (a) R. Sanjines, H. Berger, and F. Levy, *Mater. Res. Bull.*, *23*, 549 (1988). (b) R. W. Garner and W. B. White, *J. Cryst. Growth*, *7*, 343 (1970).

20. S. A. Sunshine, D. Kang, and J. A. Ibers, *J. Am. Chem. Soc.*, *109*, 6202 (1987).
21. (a) *The Sodium–Sulfur Battery*, J. L. Sudworth and A. R. Tilley, Eds., Chapman & Hall, New York, 1985. (b) W. Fischer, *Mater. Res. Soc. Symp. Proc.*, *135*, 541 (1989). (c) R. W. Powers and B. R. Karas, *J. Electrochem. Soc.*, *136*, 2787 (1989).
22. (a) T. G. Pearson and P. L. Robinson, *J. Chem. Soc.*, 1304 (1931). (b) G. H. Mathewson, *J. Am. Chem. Soc.*, *29*, 867 (1907). (c) W. Klemm, H. Sodomann, and P. Langmesser, *Z. Anorg. Allg. Chem.*, *241*, 281 (1939). (d) *Gmelin's Handbuch der Anorganischen Chemie*, Verlag Chemie, Weiheim/Brgstr, FRG, 1966; Soduim, Suppl. Part 3, pp. 1202–1205 and references cited therein.
23. (a) R. J. H. Clark and D. G. Cobbold, *Inorg. Chem.*, *17*, 3169 (1978). (b) P. Dubois, J. P. Lelieur, and G. Lepourte, *Inorg. Chem.*, *27*, 73 (1988). (c) K. W. Sharp and W. H. Koeheler, *Inorg. Chem.*, *16*, 2258 (1977). (d) L. Schultz and W. H. Koeheler, *Inorg. Chem.*, *26*, 1989 (1987).
24. (a) R. W. M. Wardle, S. Bhaduri, C. N. Chau, and J. A. Ibers, *Inorg. Chem.*, *27*, 1747 (1988). (b) W. A. Herrmann, J. Rohrmann, E. Herdtweck, C. Hecht, M. L. Ziegler, and O. Serhadi, *J. Organomet. Chem.*, *314*, 295 (1986). (c) L. Y. Coh, C. Wei, and E. Sinn, *J. Chem. Soc. Chem. Commun.*, 462 (1985). (d) A. L. Rheingold, C. M. Bolinger, and T. B. Rauchfuss, *Acta Crystallogr. Ser. C*, *42*, 1878 (1986). (e) D. M. Giolando, M. Papavassiliou, J. Pickardt, and T. B. Rauchfuss, *Inorg. Chem.*, *27*, 2596 (1988). (f) B. W. Eichhorn, R. C. Haushalter, F. A. Cotton, and B. Wilson, *Inorg. Chem.*, *27*, 4084 (1988). (g) R. C. Haushalter, *Angew. Chem. Int. Ed. Engl.*, *24*, 433 (1985). (h) P. M. Fritz, W. Beck, U. Nagel, K. Polborn, W. A. Hermann, C. Hecht, and J. Rohmann, *Z. Naturforsch.*, *43B*, 665 (1988). (i) O. Scheidsteger, G. Huttner, K. Dehnicke, and J. Pebler, *Angew. Chem. Int. Ed. Engl.*, *24*, 428 (1985). (j) M. G. Kanatzidis and S.-P. Huang, *Inorg. Chem.*, *30*, 3572 (1991). (k) M. G. Kanatzidis and S. Dhingra, *Inorg. Chem.*, *28*, 2024 (1989). (l) M. G. Kanatzidis and S.-P. Huang, *Inorg. Chem.*, *28*, 4667 (1989). (m) H. Strasdeit, B. Krebs, and G. Henkel, *Inorg. Chim. Acta*, *89*, L11 (1984). (o) M. G. Kanatzidis and S. Dhingra, unpublished results. (o) M. G. Kanatzidis and S.-P. Huang, *Angew. Chem. Int. Ed. Engl.*, *28*, 1513 (1989). (p) M. G. Kanatzidis and S.-P. Huang, *Inorg. Chem.*, *30*, 1455 (1991).
25. F. Feher, in *Handbuch der Praparativen Anorganischen Chemie*, G. Brauer, Ed., Ferdinand Enke, Stuttgart, Germany, 1954, pp. 280–281.
26. S. Schreiner, L. E. Aleandri, D. Kang, and J. A. Ibers, *Inorg. Chem.*, *28*, 392 (1989).
27. D. Coucouvanis and A. Hadjikyriacou, *Inorg. Chem.*, *26*, 1 (1987).
28. J. M. Manoli, C. Potvin, and F. Secheresse, *Inorg. Chem.*, *26*, 340 (1987).
29. K. O. Klepp, *Z. Naturforsch.*, *47b*, 937 (1992).
30. P. M. Keane and J. A. Ibers, *Acta Crystallogr. Ser. C*, *48*, 1301 (1992).
31. P. M. Keane and J. A. Ibers, *Inorg. Chem.*, *30*, 1327 (1991).
32. P. Böttcher, *Angew. Chem. Int. Ed. Engl.*, *27*, 759 (1988).
33. L. Brattas and A. Kjekshus, *Acta Chem. Scand.*, *26*, 1 (1972).
34. P. Wu, Y.-J. Lu, and J. A. Ibers, *J. Solid State Chem.*, *97*, 383 (1992).

35. Y.-J. Lu and J. A. Ibers, *Inorg. Chem.*, *30*, 3317 (1991).
36. Y.-J. Lu and J. A. Ibers, *J. Solid State Chem.*, *94*, 381 (1991).
37. Y.-J. Lu and J. A. Ibers, *J. Solid State Chem.*, *107*, 58 (1993).
38. Y.-J. Lu and J. A. Ibers, *J. Solid State Chem.*, *98*, 312 (1992).
39. F. Jellinek, *Ark. Kemi.*, *20*, 447 (1963).
40. F. Jellinek, *Int. Rev. Sci. Inorg. Chem. Ser. One*, *5*, 351 (1972).
41. (a) M. Latroche and J. A. Ibers, *Inorg. Chem.*, *29*, 1503 (1990). (b) H. Yun, C. R. Randall, and J. A. Ibers, *J. Solid State Chem.*, *76*, 109 (1988). L. E. Rendon-Diazmiron, C. F. Campana, and H. Steinfink, *J. Solid State Chem.*, *47*, 322 (1983).
42. M. F. Mansuetto, P. M. Keane, and J. A. Ibers, *J. Solid State Chem.*, *101*, 257 (1992).
43. M. F. Mansuetto, P. M. Keane, and J. A. Ibers, *J. Solid State Chem.*, *105*, 580 (1993).
44. P. Matkovic and K. Schubert, *J. Less-Common Met.*, *52*, 217 (1977).
45. R. Hoffman, *Angew. Chem. Int. Ed. Engl.*, *26*, 846 (1987).
46. E. A. Axtell and M. G. Kanatzidis, manuscript in preparation.
47. P. Wu and J. A. Ibers, *J. Solid State Chem.*, *107*, 347 (1993).
48. (a) S. Furuseth, L. Brattas, and A. Kjekshus, *Acta Chem. Scand.*, *A29*, 623 (1975). (b) H. Hahn and B. Harder, *Z. Anorg. Allg. Chem.*, *288*, 257 (1956). (c) J. Huster, *Z. Naturforsh., B Anorg. Chem., Org. Chem.*, *35B*, 775 (1980). (d) H. Hahn and B. Harder, *Z. Anorg. Allg. Chem.*, *288*, 241 (1956).
49. D. Kang and J. A. Ibers, *Inorg. Chem.*, *27*, 549 (1988).
50. Y. Park and M. G. Kanatzidis, *J. Am. Chem. Soc.*, *111*, 3767 (1989).
51. (a) G. Gattow and O. Rosenberg, *Z. Anorg. Allg. Chem.*, *332*, 269 (1964). (b) C. Burschka, *Z. Naturforsch.*, *35b*, 1511 (1980).
52. Y. Park, Ph.D. Thesis, Michigan State University, East Lansing, MI, 1992.
53. T. M. McCarthy, X. Zhang, and M. G. Kanatzidis, *Inorg. Chem.*, *32*, 2944 (1993).
54. (a) C. Burschka, *Z. Naturforsch.*, *34b*, 396 (1979). (b) Z. Peplinski, D. B. Brown, T. Watt, W. E. Hatfield, and P. Day, *Inorg. Chem.*, *21*, 1752 (1982). (c) M. H. Whangbo and E. Canadell, *Inorg. Chem.*, *29*, 1387 (1990).
55. J. C. W. Folmer and F. Jellinek, *J. Less-Common Met.*, *76*, 153 (1980).
56. Y. Park, D. C. Degroot, J. Shindler, C. Kannewurf, and M. G. Kanatzidis, *Chem. Mater.*, *5*, 8 (1993).
57. R. Berger and L. Eriksson, *J. Less-Common Met.*, *161*, 101 (1990).
58. Y. Park, D. C. Degroot, J. Shindler, C. R. Kannewurf, and M. G. Kanatzidis, *Angew. Chem. Int. Ed. Engl.*, *30*, 1325 (1991).
59. Y. Park and M. G. Kanatzidis, *Chem. Mater.*, *3*, 781 (1991).
60. X. Zhang, J. L. Schindler, T. Hogan, J. Albritton-Thomas, C. R. Kannewurf, and Mercouri G. Kanatzidis *Angew. Chem.*, *34*, 68, (1995).
61. M. G. Kanatzidis, *Chem. Mater.*, *2*, 353 (1990).

62. Y. Park and M. G. Kanatzidis, *Angew. Chem. Int. Ed. Engl.*, *29*, 914 (1990).
63. C.-N. Chau, W. M. Wardle, and J. A. Ibers, *Inorg. Chem.*, *26*, 2740 (1987).
64. D. Fenske, J. Adel, and K. Dehnicke, *Z. Naturforsch B*, *42*, 931 (1987).
65. A. F. Wells, *Structural Inorganic Chemistry*, Oxford University Press, New York, 1987, p. 414.
66. W. Bronger and O. Gönter, *J. Less-Common Met.*, *27*, 73 (1972).
67. E. A. Axtell, Z. Pikramenou, J.-H. Liao, and M. G. Kanatzidis, submitted for publication.
68. E. A. Axtell, J.-H. Liao, Z. Pikramenou, Y. Park, and M. G. Kanatzidis, *J. Am. Chem. Soc.*, *115*, 12191 (1993).
69. M. A. Ansari and J. A. Ibers, *J. Solid State Chem.*, *103*, 293 (1993).
70. Y. Park and M. G. Kanatzidis, *Chem. Mater.*, *2*, 99 (1990).
71. (a) F. G. Donika, G. A. Kiosse, S. I. Radautsan, S. A. Semiletov, and V. F. Zhitar, *Sov. Phys. Crystallogr.*, *12*, 745 (1968). (b) R. S. Boorman and J. K. Sutherland, *J. Mater. Sci.*, *4(8)*, 658 (1969). (c) D. E. Barnett, R. S. Boorman, and J. K. Sutherland, *Phys. Status Solidi A*, *4(1)*, K49 (1971).
72. K. O. Klepp, *J. Alloys Comp.*, *182*, 281 (1992).
73. B. Krebs, S. Pohl, and W. Schiwy, *Z. Anorg. Allg. Chemie.*, *393*, 63 (1973).
74. J.-H. Liao, C. Varotsis, and M. G. Kanatzidis, *Inorg. Chem.*, *32*, 2453 (1993).
75. K. O. Klepp, *Z. Naturforsch.*, *42b*, 197 (1992).
76. G. Z. Eulenberger, *Z. Naturforsch.*, *36b*, 687 (1981).
77. A. Müller, *Chimia*, *39*, 25 (1985).
78. R. M. Banda, J. Cusick, M. L. Scudder, D. C. Craig, and I. G. Dance, *Polyhedron*, 8, 1999 (1985).
79. S. Dhingra, S.-P. Huang, and M. G. Kanatzidis, *Polyhedron*, *11*, 1389 (1990).
80. T. M. McCarthy and M. G. Kanatzidis, *Inorg. Chem.*, *33*, 1205 (1994).
81. T. M. McCarthy, T. Tanzer, and M. G. Kanatzidis, *J. Am. Chem. Soc.*, *117*, 1294 (1995).
82. T. J. McCarthy, S.-P. Ngeyi, J. H. Liao, D. C. Degroot, T. Hogan, C. R. Kannewurf, and M. G. Kanatzidis, *Chem. Mater.*, *5*, 331 (1993).
83. A. S. Kanishcheva, Y. N. Mikhailov, and V. G. Batog, *Dokl. Akad. Nauk. SSSR*, *251*, 603 (1980).
84. Y. V. Vorisholov, E. Y. Peresh, and M. I. Golovei, *Inorg. Mater.*, *8*, 677 (1972).
85. P. Villars and L. D. Calvert, *Pearson's Handbook of Crystallographic Data for Intermetallic Phases*, American Society for Metals, Metals Park, OH, 1985, p. 1494.
86. S. Scavnicar, *Z. Kristallogr.*, *114*, 85 (1960).
87. R. S. Mitchel, *Z. Kristallogr.*, *108*, 296 (1956).
88. G. Cordier, H. Shafer, and C. Schwidetzky, *Rev. Chim. Miner.*, *22*, 676 (1985).
89. G. Cordier, H. Shafer, and C. Schwidetzky, *Rev. Chim. Miner.*, *22*, 631 (1985).
90. A. C. Sutorik and M. G. Kanatzidis, manuscript in preparation.

91. A. C. Sutorik and M. G. Kanatzidis, *Angew. Chem. Int. Ed., Engl.*, *31*, 1594 (1992).
92. A. C. Sutorik and M. G. Kanatzidis, *J. Am. Chem. Soc.*, *113*, 7754 (1991).
93. A. C. Sutorik and M. G. Kanatzidis, manuscript in preparation.
94. W. Kroniert and K. Plieth, *Z. Anorg. Allg. Chem.*, *336*, 207 (1965).
95. P. M. Keane and J. A. Ibers, *J. Solid State Chem.*, *93*, 291 (1991).
96. A. C. Sutorik and M. G. Kanatzidis, *J. Am. Chem. Soc.*, *116*, 7706 (1994).
97. A. C. Sutorik and M. G. Kanatzidis, to be published.
98. (a) D. Coucouvanis, *Acc. Chem. Res.*, *14*, 201 (1981). (b) R. H. Holm, *Chem. Soc. Rev.*, *10*, 455 (1981). (c) A. Müller, E. Diemann, R. Jostes, and H. Bögge, *Angew. Chem., Int. Ed., Engl.*, *20*, 934 (1981).
99. W. P. Binnie, M. J. Redman, and W. Mallio, *Inorg. Chem.*, *9*, 1449 (1970).
100. E. A. Pruss and A. M. Stacy, *203rd ACS National Meeting 1992*, Washington DC.
101. J.-H. Liao and M. G. Kanatzidis, *Chem. Mater.*, *5*, 1561 (1993).
102. G. K. Averkieva, A. A. Vaipolin, and N. A. Goryunova, in *Some Ternary Compounds of the $A_2(I)B(IV)C_3(VI)$ Type and Solid Solutions Based on Them*, Soviet Research in New Semiconductor Materials, D. N. Nasledov, Ed., Consultants Bureau, New York, 1965, pp. 26–34.
103. (a) W. Bronger and P. Böttcher, *Z. Anorg. Alleg. Chem.*, *390*, 1 (1972). (b) W. Bronger and U. Hendriks, *Rev. Chim. Miner.*, *17*, 555 (1980). (c) W. Bronger and P. Müller, *J. Less-Common Met.*, *100*, 241 (1984).
104. J.-H. Liao, Ph.D. Thesis, Michigan State University, East Lansing, MI, 1993.
105. W. Schiwy, Chr. Blutau, D. Gäthje, and B. Krebs, *Z. Anorg. Allg. Chem.*, *412*, 1 (1975).
106. (a) H. Hahn and W. Klingen, *Naturwissenschaften*, *52*, 494 (1965). (b) H. Hahn, R. Ott, and W. Klingen, *Z. Anorg. Allg. Chem.*, *396*, 271 (1973).
107. (a) R. Diehl and C.-D. Carpentier, *Acta Crystallogr. Ser. B*, *34*, 1097 (1978). (b) P. Buck and C.-D. Carpentier, *Acta Crystallogr. Ser. B*, 29, 1864 (1973). (c) H. Zimmerman, C.-D. Carpentier, and R. Nitsche, *Acta Crystallogr. Ser. B*, *31*, 2003 (1975). (d) R. Becker, W. Brockner, and B. Eisenmann, *Z. Naturforsch.*, *42a*, 1309 (1987). (e) A. Ferrari and L. Cavalca, *Gazz. Chim. Ital.* *78*, 283 (1948). (f) R. Diehl and C.-D. Carpentier, *Acta Crystallogr. Ser. B*, *33*, 1399 (1977). (g) A. Simon, K. Peters, E.-M. Peters, and H. Hahn, *Z. Naturforsch.*, *38b*, 426 (1983). (h) M. Jansen and U. Henseler, *J. Solid State Chem.*, *99*, 110 (1992).
108. (a) P. Toffoli, J. C. Rouland, P. Khodadad, and N. Rodier, *Acta Crystallogr. Ser. C*, *41*, 645 (1985). (b) P. Toffoli, P. Khodadad, and N. Rodier, *Acta Crystallogr. Ser. B*, *38*, 2374 (1982). (c) P. Toffoli, P. Khodadad, and N. Rodier, *Bull. Soc. Chim. Fr.*, *11*, 429 (1981).
109. (a) R. Mercier, J.-P. Malugani, B. Fahys, and G. Robert, *Acta Crystallogr. Ser. B*, *38*, 1887 (1982). (b) S. Fiechter, W. F. Kuhs, and R. Nitsche, *Acta Crystallogr., Ser. B*, *36*, 2217 (1980). (c) H. Schafer, G. Schafer, and A. Weiss, *Z. Naturforsch.*, *20b*, 811 (1965). (d) R. Brec, M. Evain, P. Grenouilleau, and J.

Rouxel, *Rev. Chim. Min.*, *20*, 283 (1983). (e) R. Brec, M. Evain, P. Grenouilleau, and J. Rouxel, *Rev. Chim. Min.*, *20*, 283 (1983). (e) R. Brec, P. Grenouilleau, M. Evain, and J. Rouxel, *Rev. Chim. Min.*, *20*, 295 (1983). (f) M. Evain, S. Lee, M. Queignec, and R. Brec, *J. Solid State Chem.*, *71*, 139 (1987). (g) M. Z. Jandali, G. Eulenberger, and H. Hahn, *Z. Anorg. Allg. Chem.*, *530*, 144 (1985). (h) A. Weiss and H. Schafer, *Z. Naturforsch*, *18b*, 81 (1963).

110. F. Menzel, L. Ohse, and W. Brockner, *Heteroatom. Chem.*, *1(5)*, 357 (1990).
111. T. M. McCarthy and M. G. Kanatzidis, *Chem. Mater.*, *5*, 1061 (1993).
112. E. Durand, M. Evain, and R. Brec, *J. Solid State Chem.*, *102*, 146 (1992).
113. P. Toffoli, P. Khodadad, and N. Rodier, *Acta Crystallogr. Ser. B*, *33*, 1492 (1977).
114. M. Z. Jandali, G. Eulenberger, and H. Hahn, *Z. Anorg. Allg. Chem.*, *445*, 184 (1978).
115. W. Honle, C. Wibbelmann, and W. Brockner, *Z. Naturforusch.*, *39b*, 1088 (1984).
116. (a) T. M. McCarthy, Ph.D. Thesis, (b) T. McCarthy and M. G. Kanatzidis, submitted for publication. (c) Timothy J. McCarthy and Mercouri G. Kanatzidis, *Inorg. Chem.*, *34*, 1257 (1995). (d) T. J. McCarthy, T. Hogan, C. R. Kannewurf, and M. G. Kanatzidis, *Chem. Mater.*, *6*, 1072 (1994).
117. T. M. McCarthy and M. G. Kanatzidis, unpublished results.
118. C. L. Teske, *Z. Naturforsch.*, *34B*, 544 (1979).
119. T. M. McCarthy and M. G. Kanatzidis, *J. Chem. Soc. Chem. Commun.*, 1089 (1994).
120. X. Zhang and M. G. Kanatzidis, *J. Am. Chem. Soc.*, *116*, 1890 (1994).
121. X. Zhang and M. G. Kanatzidis, *Inorg. Chem.*, *33*, 1238 (1994).
122. A. F. Wells, *Structural Inorganic Chemistry*, 5th ed., Clarendon Press, Oxford, UK, 1984, pp. 258–260.
123. L. Gmelin, *Handbuch der Anorganischen Chemie*, Springer-Verlag, Berlin, System, Number 56, Teil C6, 1976, pp. 77–78.
124. D. Coucouvanis, P. B. Patil, M. G. Kanatzidis, B. Detering, and N. C. Baenziger, *Inorg. Chem.*, *24*, 24 (1985).
125. A. Müller, M. Zimmerman, and H. Bögge, *Angew. Chem. Int. Ed. Engl.*, *25*, 273 (1986).
126. A. Müller, J. Schimanski, and U. Schimanski, *Angew. Chem. Int. Ed. Engl.*, *23*, 159 (1984).
127. S. Dhingra and M. G. Kanatzidis, *Polyhedron*, *10*, 1069 (1991).
128. (a) J. M. Charnock, C. D. Garner, R. A. Pattrick, and D. J. Vaughan, *J. Solid State Chem.*, *82*, 279 (1989). (b) B. J. Wuensch, *Z. Kristallogr.*, *119*, 437 (1964). (c) L. Pauling and E. W. Newmann, *Z. Kristallogr.*, *88*, 54 (1934).
129. (a) W. Rüdorff, H. G. Scharz, and M. Walter, *Z. Anorg. Allg. Chem.*, *269*, 141 (1952). (b) D. B. Brown, J. A. Zubieta, P. A. Vella, T. Wrobleski, T. Watt, W. E. Hatfield, and P. Day, *Inorg. Chem.*, *19*, 1945 (1980). (c) K. O. Klepp, H. Boller, and H. Völlenkle, *Monatsh. Chem.*, *111*, 727 (1980). C. Burschka, *Z. Anorg. Allg. Chem.*, *463*, 65 (1980).

130. (a) C. Burschka, *Z. Naturforsch.*, *34b*, 675 (1979). (b) L. W. Huar, F. J. DiSalvo, H. E. Bair, R. M. Fleming, J. V. Waszczak, and W. E. Hatfield, *Phys. Rev. B*, *35(4)*, 1932 (1987). (c) R. M. Flemming, L. W. Haar, and F. J. DiSalvo, *Phys. Rev. B*, *35(10)*, 5388 (1987).
131. G. J. Janz, E. Roduner, J. W. Coutts, and J. R. Downey, Jr., *Inorg. Chem.*, *15(8)*, 1751 (1976).
132. S. Dhingra and M. G. Kanatzidis, *Science*, *258*, 1769 (1992).
133. D. Chakrabarty and M. G. Kanatzidis, to be published.
134. J. Fortner, M.-L. Saboungi, and J. E. Enderby, *Phys. Rev. Lett.* *69*, 1415 (1992).
135. For another recent review, see J. A. Cody, M. F. Mansuetto, S. Chien, and J. A. Ibers, *Mat. Sci. Forum*, *152–153*, 35 (1994).

Mechanistic and Kinetic Aspects of Transition Metal Oxygen Chemistry

ANDREJA BAKAC

Ames Laboratory
Iowa State University
Ames, IA

CONTENTS

Progress in Inorganic Chemistry, Vol. 43, Edited by Kenneth D. Karlin.
ISBN 0-471-12336-6

I. INTRODUCTION

The oxidation of organic materials by molecular oxygen is thermodynamically favored, but kinetically slow. The reason is the spin mismatch between the ground states of O_2 (triplet) and organic reductants (singlet). The problem can be overcome in several ways, including the excitation of either $^3O_2$ to $^1O_2$ or ^{1}(organic molecule) to ^{3}(organic molecule). Much more common, however, is the activation of O_2 (1–12) by partial reduction (to yield $HO_2/O_2^{\bullet -}$, H_2O_2, or $HO^\bullet$) or by coordination to transition metals. In the latter case, the spin–orbit coupling in MO_2 (M = transition metal) relaxes the spin restriction.

A number of biological and industrial processes, including oxygen utilization by living organisms, production of artificial blood, catalytic oxygenations, and functioning of fuel cells, at some point involve the interaction of dioxygen with metal complexes. It therefore comes as no surprise that this interaction in both synthetic and naturally occurring compounds has become a subject of intense investigations, resulting in an enormous number of published papers. Some of the work on structural, spectroscopic, and chemical aspects of metal–oxygen chemistry has been summarized and discussed in several books and reviews (1–15), and most recently a complete issue of *Chemical Reviews* has been devoted to the subject. The aim of this chapter is to use a number of examples in the recent literature to highlight the kinetic and mechanistic behavior of different types of species formed when dioxygen reacts with transition metal complexes. For the sake of completeness and to provide the necessary background, some older work will be discussed as well. Only the metal-containing species will be addressed. The chemistry of nonmetallic intermediates $O_2^{\bullet -}/HO_2$, H_2O_2, and $HO^\bullet$ has been covered elsewhere (13, 16, 17).

Many of the metal-containing intermediates are too short lived to be detected under ordinary conditions, yet others are so persistent as to allow isolation and crystal structure determination. Given such a scale of lifetimes, a number of spectroscopic and kinetic methods have been used to detect, characterize, and study the chemistry of superoxo, peroxo, and high-valent metal–oxo species in a variety of environments. The readers are referred to the original literature for the description of experimental techniques. Unless specified otherwise, the kinetic data are given at 25°C.

Whenever possible, the formulas of metal complexes show all the ligands, including molecules of solvent. The notation (aq)M is used for aqua complexes and their hydrolyzed forms when the exact composition of the coordination sphere has not been established and cannot be reasonably estimated [e.g., $(aq)Fe^{V}$ or $(aq)Fe^{IV}$] or when several hydrolytic forms coexist under reaction conditions [e.g., $(aq)Cr^{VI}$ in acidic solutions]. The formula MO_2 represents a metal–dioxygen complex irrespective of the oxidation state (superoxo and peroxo) and geometry (end-on and side-on) of bound O_2. Similarly, hydroperoxo and μ-peroxo complexes are denoted MO_2H and MO_2M, respectively. To emphasize the nature of bonding in certain cases, the formulas MOO, MOOH, and MOOM are also used. Other abbreviations are defined at the end of the chapter.

The order in which different metals are discussed is arbitrary. This order is based on the degree of completeness of the overall mechanistic picture, the number and types of different kinds of intermediates observed and characterized, the level of mechanistic and kinetic information available, the historical role these metals have played in the development of the field, and the availability of recent reviews.

Figure 1 shows a simplified general scheme for the reaction of O_2 with a mononuclear metal complex M (ligands not shown), allowing only three oxidation states for the metal and neglecting any nonmetallic intermediates. Even in the absence of any additional reagents, several intermediates are involved

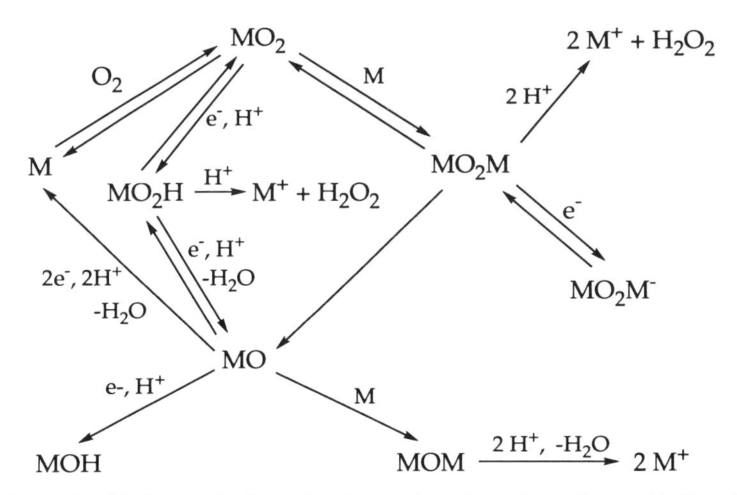

Figure 1. A simplified general scheme for the reaction of metal complexes with O_2, with only three oxidation states allowed for the metal.

(MO_2, MO_2M, MO, or MOM). Their lifetimes in solution depend on the metal, ligands, and experimental conditions, as discussed in detail later.

Chemical and electrochemical reduction produces additional reactive species (hydroperoxo and dinuclear superoxo) with a rich chemistry of their own. As shown in the scheme, these species sometimes provide additional routes for the interconversion of various "primary" intermediates and for the regeneration of the starting complex M. Examples of all the species in Fig. 1 have been observed experimentally, although rarely has it been possible to identify and characterize more than two or three such species for the same M.

A. Superoxometal Complexes, $M^{n+1}(O_2^-)$

Coordinatively unsaturated and/or substitutionally labile reducing metal complexes often react with O_2 to produce observable superoxometal ions, as shown in Eq. 1 for a mononuclear complex M^n (ligands not shown). The same species is sometimes produced in the reaction of $O_2^{\cdot -}$ with M^{n+1}. Spectroscopic (ESR, IR, and resonance Raman) and crystallographic evidence exists for both η^1 (end-on) and η^2 (side-on) coordination modes, with the former being much more common. Only two examples of side-on superoxocobalt complexes, $[HB(3\text{-}t\text{-}Bu\text{-}5\text{-}Mepz)_3]Co^{II}O_2$ and $[HB(3\text{-}i\text{-}Pr\text{-}5\text{-}Mepz)_3]Co^{II}O_2$, have been reported so far (18a and b), and none are known for cobalt in the oxidation state 3+. A superoxocopper(II) complex $[HB(3\text{-}t\text{-}Bu\text{-}5\text{-}i\text{-}Prpz)_3]CuO_2$ also has side-on geometry (18c), suggesting that tris(pyrazolyl)borate ligands favor side-on coordination. A superoxo copper complex of tris[(6-(pivaloylamino)-2-pyridyl)methyl] ammine, exhibits end-on geometry (18d). In the case of iron, both geometries have been observed at low temperatures (19–22). However, the lower O—O stretching frequencies for the side-on adducts indicate a substantial contribution of the peroxo structure.

$$M^n + O_2 \rightleftharpoons [M^{n+1}(O_2^-)] \qquad k_{on}, k_{off}, K_{on} \tag{1}$$

The assignment of discrete oxidation states to the metal and oxygen in MO_2 has an electron-counting purpose only, and is not meant to imply complete electron transfer. In fact, based on an analysis of anisotropic cobalt hyperfine coupling for several macrocyclic cobalt–dioxygen complexes, it has been estimated that the electron transfer from cobalt to dioxygen is only 10–80% complete, the individual values depending strongly on the macrocycle (23–25). As shown in Table I, the vibrational frequencies (ν_{O-O}) and the available O—O bond lengths of superoxo complexes also cover a range of values (1, 15, 26–30), from those coinciding with the values for free superoxide to those indicating a stronger and shorter bond, as would be expected for O_2 bearing a charge

TABLE I
Selected Structural and Spectroscopic Data for O_2, $O_2^{\cdot -}$, and End-On MO_2 Complexes[a]

Compound	O—O Distance (Å)	ν_{O-O} (cm^{-1})
O_2	1.207	1555
$O_2^{\cdot -}$(KO_2)	1.28	1108
Co(t-Bu-salen)(py)(O_2)	1.350(11)	
[Co(saltmen)(BzIm)(O_2)] · thf	1.277(3)	
(Co-myoglobin)O_2	1.26(8)	1134
Co(bzacen)(py)(O_2)	1.26(4)	1128
$(NEt_4)_3[Co(CN)_5O_2]$ · $5H_2O$	1.240(17)	1138
Co(acacen)(B)(O_2)[b]		~1130
Co(3-MeO-salen)(py)(O_2)		1140
Co(TPP)(MeIm)(O_2)		1144
Co(salen)(O_2)		1235
Co(OEP)O_2		1275
Co(TPP)O_2		1278
Myoglobin-O_2	1.21(10)	1103
Fe(OEP)O_2		1190[c]
Fe(TPP)O_2		1195[d]
Fe(Pc)O_2		1207
$L^1(H_2O)CrO_2^{2+}$ [e]		1140
Cr(TPP)(py)(O_2)		1142
$(H_2O)_5CrO_2^{2+}$		1165

[a]Data from (1, p. 5) and (15, 27–30).
[b]B = py, Me—py, NH_2—py, 4—CN—py.
[c]ν_{O-O} = 1104 cm^{-1} for the side-on form.
[d]ν_{O-O} = 1106 cm^{-1} for the side-on-form.
[e]L^1 = [14]aneN_4.

between 0 and −1. It has been pointed out, however, that X-ray and IR data cannot be used reliably to estimate the extent of electron transfer (24).

With the proper choice of metal and/or experimental conditions, the reaction between M and O_2 stops at the MO_2 level, and the role of M is that of a dioxygen carrier, the most familiar examples of which are the respiratory proteins hemoglobin, hemerythrin, and hemocyanin. The stability constants K_{on}, defined in Eq. 1, have been determined for a number of such MO_2 complexes (29). As expected, K_{on} decreases with the ligand bulk and increases with increasing electron density (29, 31–34) on the metal, as measured by the reduction potential of the unoxygenated pair M^+/M, pK_a of the ligands, and Hammett substituent constants. The increase in electron density stabilizes the higher oxidation state of the metal and promotes electron transfer to O_2. Aprotic solvents and nitrogenous bases coordinated trans to O_2 greatly increase the stability of MO_2 (32).

Some synthetic MO_2 complexes are sufficiently stable to be used as oxygen-storing materials for medicinal and industrial purposes. The release of dioxygen can be induced by chemical or physical stimuli, such as a change in pH, temperature, pressure or solvent, chemical and electrochemical oxidation of MO_2, and UV–vis photoexcitation, as discussed in Section III.A.

However, all dioxygen carriers, including the biological ones, eventually lose their activity because irreversible oxidation of either the metal (35, 36) or ligands (37–39) takes place. Also, the dissociation of MO_2 in the reverse of Reaction 1 is a built-in source of reducing equivalents in the form of M. A common and often unavoidable route for the loss of MO_2 is thus the reduction by M, often yielding dimetallic μ-peroxo complexes, MO_2M, and their decomposition products.

$$MO_2 + M \rightleftharpoons MO_2M \qquad k_{2,on}, k_{2,off}, K_{2,on} \tag{2}$$

$$MO_2M \longrightarrow \text{products} \tag{3}$$

Consistent with the reaction scheme of Eqs. 1–3, the loss of MO_2 is minimized at high concentrations of O_2, and it is under these conditions that the detection, characterization, and possible isolation of such complexes is typically carried out. For the same reason, kinetic studies of Reaction 1 are also preferentially carried out in the presence of a large excess of O_2.

B. Peroxometal Complexes, $M^{n+2}(O_2^{2-})$

Not all mononuclear metal complexes that react with O_2 form superoxo species. A two-electron initial interaction is in fact quite common for metals with favorable reduction potentials for M^{n+2}/M^n couples. Complexes of Mn^{II} are typical examples, yielding routinely peroxo manganese(IV) intermediates, $Mn^{IV}(O_2^{2-})$. Similarly, the second- and third-row elements often yield side-on peroxo species.

Again, the assignment of discrete oxidation states may be a matter of semantics and should not be taken literally (40). From a practical point of view, however, it is useful to be able to categorize the MO_2 adducts into $M^{n+1}(O_2^-)$ and $M^{n+2}(O_2^{2-})$, even if the picture is oversimplified and incomplete. The former class is characterized by end-on bonding (with rare exceptions) and O—O stretching frequencies of greater than 1075 cm^{-1}. The $M^{n+2}(O_2^{2-})$ class has the peroxide bound side-on and O—O stretching frequencies of less than 1000 cm^{-1}. [The end-on binding of peroxide in oxyhemerythrin (41, 42) is exceptional and related to hydrogen-bonding interactions.]

The structures of intermediates discussed in this chapter are shown in Fig. 2.

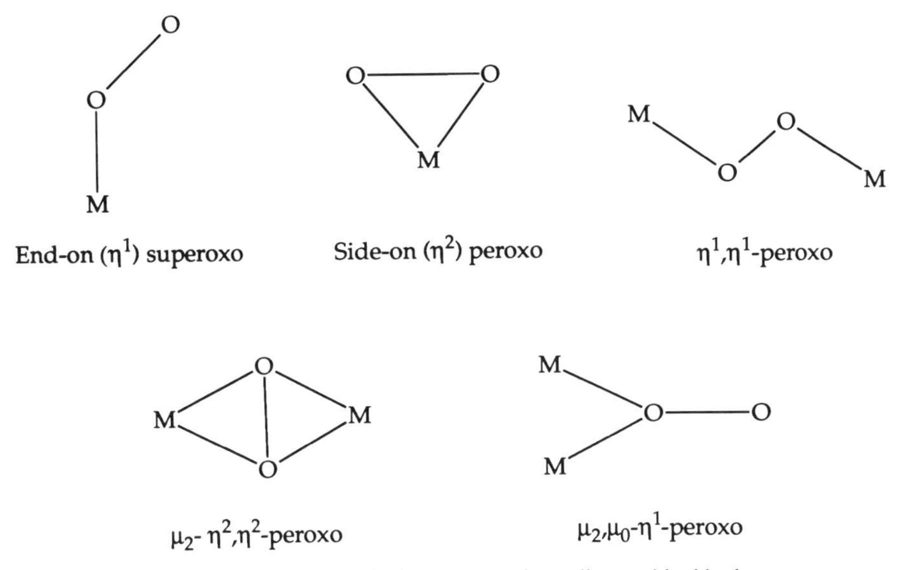

Figure 2. Structures of metal–dioxygen complexes discussed in this chapter.

II. INTERACTIONS OF DIOXYGEN WITH TRANSITION METAL COMPLEXES

A. Cobalt

Cobalt is among the most frequently studied metals in the realm of transition metal–oxygen chemistry. In fact, many of the early clues on the nature of metal–dioxygen interactions came in the area of cobalt chemistry (27, 43–45). This is mainly due to the large number of Co^{II} complexes available, and to the relatively long lifetimes of the (formally) Co^{III} intermediates, a number of which have been isolated and characterized spectroscopically, chemically, structurally, and photochemically. Electron spin resonance is an especially useful spectroscopic technique for the study of $LCoO_2$ complexes (46) in (frozen) solution, owing to the presence of a single unpaired electron and the ^{59}Co nuclear spin of $\frac{7}{2}$. The g values, ^{59}Co-induced hyperfine splitting and nitrogen base induced superhyperfine splitting, have been used in numerous cases (27, 45, 47–50) to deduce the coordination number and geometry and, as mentioned above, even the extent of electron transfer for a number of complexes (23, 24).

A vast number of equilibrium constants for the formation of $LCoO_2$ (L = ligand system) and $LCoO_2CoL$ have been determined by spectrophotometric, potentiometric, polarographic, and manometric techniques (29, 33, 51). The range of values observed covers over 10 orders of magnitude, such that some complexes react quantitatively with even minute concentrations of O_2, while others require low temperatures and high concentrations and pressures to produce measurable amounts of products (1, 29).

It has been suggested (43) that the kinetics of O_2 binding are mainly determined by the substitution rates on the cobalt. As these vary only moderately in the absence of drastic structural changes, it would follow that the spread in the dissociation rate constants k_{off} and $k_{2,off}$ is the main factor responsible for the large observed differences in equilibrium binding constants (52). However, the kinetic determination of the individual rate constants has been limited by the complexity of the kinetic scheme and the lack of appropriate experimental techniques for direct measurement of large reaction rates. In the earlier studies (27, 44, 53, 54) most of the kinetic data were obtained under conditions where both Reactions 4 and 5 contribute and the rate law of Eq. 6 applies. (In these equations L denotes the complete ligand system, which usually differs for the reactants and products by a molecule of coordinated solvent.) This approach has rarely allowed the evaluation of all four individual rate constants for a given complex, but it was possible to bring out different terms of Eq. 6 by careful variation of reaction conditions. By this method the rate constants k_{on}, $k_{2,off}$ and the ratio $k_{2,on}/k_{off}$ have been obtained for several complexes (1). Initially, the range of values for $k_{2,off}$ was indeed significantly greater than that for k_{on}, as predicted. With time, however, more kinetic data became available (52, 55–62), extending the ranges of all four rate constants as more structural and electronic variety was built into the cobalt complexes. Some of the typical data are shown in Table II.

$$\mathrm{LCo} + \mathrm{O_2} \underset{k_{off}}{\overset{k_{on}}{\rightleftharpoons}} \mathrm{LCoO_2} \qquad K_{on} \tag{4}$$

$$\mathrm{LCoO_2} + \mathrm{LCo} \underset{k_{2,off}}{\overset{k_{2,on}}{\rightleftharpoons}} \mathrm{LCoO_2CoL} \qquad K_{2,on} \tag{5}$$

$$k_{obs} = \frac{k_{on}k_{2,on}[\mathrm{LCo}]^2[\mathrm{O_2}]}{k_{off} + k_{2,on}[\mathrm{LCo}]} \tag{6}$$

The reaction of O_2 with $L^1(H_2O)_2Co^{2+}$ (Fig. 3) in aqueous solution was originally studied by a stopped-flow method (58, 59) to become the first case where all four rate constants have been determined.

Later, a laser flash photolysis study (55) utilized photohomolysis of the superoxo complex $L^1(H_2O)CoO_2^{2+}$, Eq. 7, to generate small concentrations of $L^1(H_2O)_2Co^{2+}$ in the presence of a large excess of O_2. Reaction 5 was kinetically unimportant under these conditions, and subsequent thermal equilibration yielded directly $k_{on} = 1.2 \times 10^7\ M^{-1}\ s^{-1}$, a value that is some 20 times greater than that reported earlier (58).

TABLE II
Effect of Ligands on Oxygenation Kinetics of Some Cobalt Complexes in Aqueous Solution[a,b]

Co^{II} Complex	k_{on} ($M^{-1}s^{-1}$)	k_{off} (s^{-1})	$k_{2,on}$ ($M^{-1}s^{-1}$)	$k_{2,off}$ (s^{-1})
$Co(papd)^{2+}$	1.4×10^7	0.65	1.2×10^7	2.2×10^{-6}
$L^1Co(H_2O)_2^{2+}$	1.18×10^7	63	4.9×10^5	0.6
$L^2Co(H_2O)_2^{2+}$	5.0×10^6	2.06×10^4	*c*	*c*
$Co(dmgH)_2^{2+}$			1.9×10^6	16
$Co(tetren)(H_2O)^{2+}$	10^5			10^{-5}
$Co(tetren)(NO_2)^+$	1.0×10^5 6.8×10^4[d]			
$Co(trien)(H_2O)_2^{2+}$	2.5×10^4			
$Co(NH_3)_5H_2O^{2+}$	2.5×10^4			56
$Co(his)_2(H_2O)_2^{2+}$	1.8×10^4			0.014
$Co(dien)_2^{2+}$	1.2×10^3			0.016
$Co(trien)(NO_2)^+$	3.2×10^2 2.7×10^5[d]			

[a]Rate constants defined in Eqs. 4 and 5. Data from (1, 52–58, and 61).
[b]L^1 = [14]aneN_4, L^2 = Me_6-[14]aneN_4.
[c]Reaction not observed.
[d]In MeCN (56).

$$L^1(H_2O)CoO_2^{2+} \underset{}{\overset{h\nu,\,H_2O}{\rightleftharpoons}} L^1(H_2O)_2Co^{2+} + O_2 \tag{7}$$

It is now clear that the reaction is too fast for k_{on} to be determined accurately by the stopped-flow technique. The results of an independent equilibrium study (60) are in qualitative agreement with the new (55), larger value of K_{on}. The data shown in Table II are the best combined values from the two studies (55, 58).

A related cobalt(II) macrocycle, $L^2(H_2O)_2Co^{2+}$, has $k_{on} = 5.0 \times 10^6$ and $k_{off} = 2.17 \times 10^4\ s^{-1}$ (55, 61). The 500-fold decrease in K_{on} upon methylation

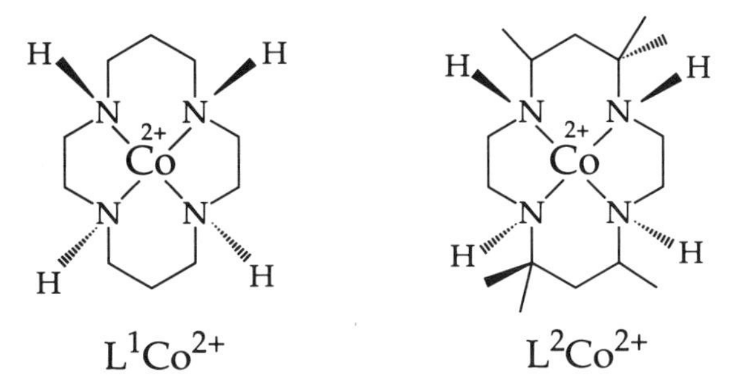

Figure 3. Structures of complexes L^1Co^{2+} and L^2Co^{2+} (L^1 = [14]aneN_4, L^2 = Me_6-[14]aneN_4, axial water molecules not shown).

of the macrocycle is almost entirely due to the large increase in k_{off}, the forward rate constants for the L^1 and L^2 complexes differing only by a factor of about 2. Formation of the dicobalt μ-peroxo complex was not observed for this sterically crowded ion.

Axially substituted complexes $(X)L^2(H_2O)Co^+$ (X = Cl, SCN, or OH) all bind O_2 more tightly than does the parent diaqua ion, Fig. 4 and Table III (61, 62). However, k_{on} is almost unchanged throughout the series, and the stabilization of the superoxocobalt(III) product by anions is again reflected mostly in the k_{off} term. Thus for a series of related cobalt complexes the dissociation rates dominate the thermodynamics of both Reactions 4 and 5, as suggested earlier (52).

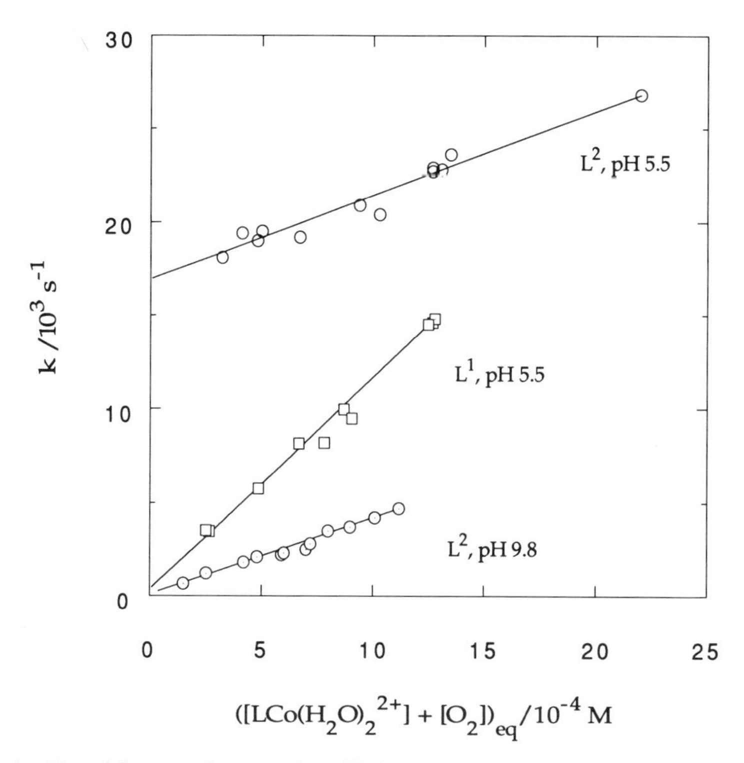

Figure 4. Plot of k versus the sum of equilibrium concentrations for the reaction of O_2 with $LCo(H_2O)_2^{2+}$ (L^1 = [14]aneN$_4$, L^2 = Me$_6$-[14]aneN$_4$). The slopes of the lines represent the binding constants k_{on}, and the intercepts the dissociation constants k_{off}, as defined in Eq. 4. The values of k_{on} differ by less than a factor of 3 [$L^1(H_2O)_2Co^{2+}$, $k_{on} = 1.2 \times 10^7\ M^{-1}\ s^{-1}$ at pH 1-7; $L^2(H_2O)_2Co^{2+}$, $k_{on} = 5 \times 10^6\ M^{-1}\ s^{-1}$ at pH 1-10], but k_{off} increases from 63 s^{-1} [$(H_2O)L^1CoO_2^{2+}$] to $2.06 \times 10^4\ s^{-1}$ [$(H_2O)L^2CoO_2^{2+}$]. At pH 9.8 the hydroxo species, $(OH)L^2CoO_2^+$, becomes the predominant oxygenated form of the L^2 complex and $k_{off} < 10\ s^{-1}$. At pH > 12 (data not shown), $k_{off} \sim 0.02\ s^{-1}$ for the L^2 derivative. Data from (55, 61, and 62).

TABLE III
Kinetics and Equilibrium Data for Dioxygen Binding by Macrocyclic Cobalt Complexes $(X)(L)(H_2O)Co^{n+}$ in Aqueous Solution at 25°C[a]

Complex	k_{on} ($M^{-1}s^{-1}$)	k_{off}	$K_{on} = k_{on}/k_{off}$
$(H_2O)_2L^2Co^{2+}$	5.00×10^6	2.06×10^4	2.3×10^2
$(Cl)L^2(H_2O)Co^+$	1.80×10^6	3.21×10^3	5.6×10^2
$(SCN)L^2(H_2O)Co^+$	7.29×10^6	1.77×10^1	4.1×10^5
$(HO)L^2(H_2O)Co^+$	8.9×10^5	2.1×10^{-2}	4.2×10^7

[a]Data from (55, 61, and 62). L^1 = [14]aneN_4; L^2 = Me_6-[14]aneN_4.

Heterolytic dissociation of $L(H_2O)CoO_2^{2+}$ complexes to $O_2^{\bullet -}$ and LCo-$(H_2O)_2^{3+}$ has not been observed for L = L^1 and L^2, but a superoxo complex of a synthetic analogue of cobalt bleomycin has recently been shown to release $O_2^{\bullet -}$ in DMF solution (63).

The equilibrium constant for heterolysis of $L(H_2O)CoO_2^{2+}$ can be calculated from the value of K_{on} and appropriate reduction potentials (55, 58, 64), as shown in Scheme 1 for $L^1(H_2O)CoO_2^{2+}$.

$$O_2 + e^- \rightleftharpoons O_2^{\bullet -} \qquad E^0 = -0.16\ V$$

$$L^1Co(H_2O)_2^{2+} \rightleftharpoons L^1Co(H_2O)_2^{3+} + e^- \qquad E^0 = -0.42\ V$$

$$L^1(H_2O)CoO_2^{2+} + H_2O \rightleftharpoons L^1Co(H_2O)_2^{2+} + O_2$$

$$K = 5 \times 10^{-6}\ M^{-1}$$

$$\text{Net:} \quad L^1(H_2O)CoO_2^{2+} + H_2O \rightleftharpoons L^1Co(H_2O)_2^{3+} + O_2^{\bullet -}$$

$$K_{heter} = 8 \times 10^{-16}\ M$$

Scheme 1

The binding constant K_{bind} (= $1/K_{heter}$) (55, 64) has an unusually large value of $1 \times 10^{15}\ M^{-1}$. Thus the affinity of $L^1(H_2O)_2Co^{3+}$ is greater for $O_2^{\bullet -}$ than for OH^- ($K_{bind,OH} = 2 \times 10^{11}\ M^{-1}$) by four orders of magnitude, despite the 10^9-fold lower basicity of $O_2^{\bullet -}$ [pK_w = 14, pK_a ($HO_2^{\bullet}/O_2^{\bullet -}$) = 4.7] (65).

Results of similar calculations for $L^2(H_2O)CoO_2^{2+}$ are equally impressive, Table IV. Using E^0 = 0.59 V for $L^2Co(H_2O)_2^{3+/2+}$, one calculates the binding constant of $9 \times 10^{-16}\ M^{-1}$, almost identical with that for the L^1 complex. The last two entries in Table IV express the same (unusual) point somewhat differently: The dissociation of OH^- from (hydroxo)superoxocobalt(III) is more favorable than dissociation of $O_2^{\bullet -}$ from the same molecule. These calculations

TABLE IV
Thermodynamic Data for the Dissociation of $O^{\cdot -}$ and OH^- from Cobalt Macrocycles[a,b]

Reaction	K
$(H_2O)L^1CoO_2^{2+} + H_2O \rightleftarrows (H_2O)_2L^1Co^{3+} + O_2^{\cdot -}$	8×10^{-16}
$(H_2O)L^2CoO_2^{2+} + H_2O \rightleftarrows (H_2O)_2L^2Co^{3+} + O_2^{\cdot -}$	9×10^{-16}
$(HO)L^2CoO_2^{+} + H_2O \rightleftarrows (HO)L^2Co(H_2O)^{2+} + O_2^{\cdot -}$	4×10^{-12}
$(HO)L^2CoO_2^{+} + H_2O \rightleftarrows (H_2O)L^2CoO_2^{2+} + OH^-$	2×10^{-8}

[a] L^1 = [14]aneN_4 and L^2 = Me_6-[14]aneN_4.
[b] Calculated from data in (55) and (61).

use pK_a = 2.7 for $L^2Co(H_2O)_2^{3+}$ (66), 11.7 for $L^2Co(H_2O)_2^{2+}$ (61a), and 6.4 for $L^2(H_2O)CoO_2^{2+}$ (61a).

It has been suggested (64) that the $Co^{IV}O_2^{2-}$ form contributes significantly to the overall electronic structure of $(H_2O)LCoO_2^{2+}$ complexes, resulting in unusually tight binding of dioxygen to cobalt. Similar arguments may be applicable to the chromium complexes $Cr(H_2O)_5O_2^{2+}$ and $L^1(H_2O)CrO_2^{2+}$, which also exhibit unusually large binding constants (67, 68).

Hydrogen bonding between the coordinated $O_2^{\cdot -}$ and other ligands, Fig. 5, may also be responsible for some of the stability of these superoxometal ions. Kinetic and thermodynamic evidence provides strong support for the stabilization of dioxygen adducts of metalloproteins and porphyrin complexes by distal polar interactions and hydrogen bonding (8, 15, 69–75). In the case of superstructured "hanging base" iron porphyrin complexes (71), the rate constant for the binding of O_2 is unaffected, but the dissociation is approximately an order of magnitude slower when hydrogen bonding is available. Important new insights may be gained by exploring the possibility of intramolecular hydrogen

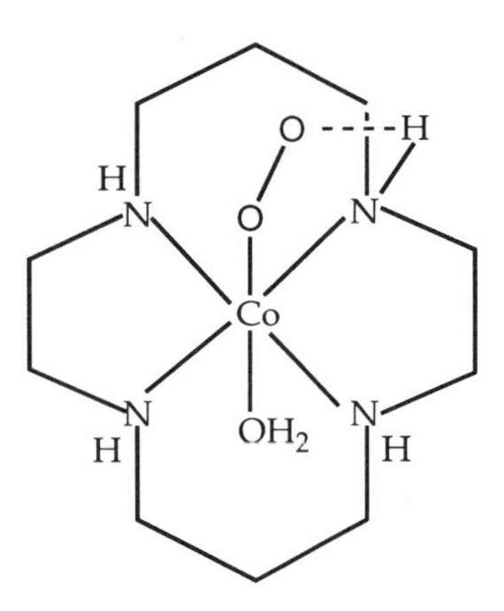

Figure 5. Representation of possible hydrogen bonding between the macrocyclic ligand and coordinated O_2 in $(H_2O)([14]aneN_4)CoO_2^{2+}$.

bonding for complexes such as $L^1(H_2O)CoO_2^{2+}$ and $(H_2O)_5CrO_2^{2+}$, although it is highly improbable that this alone will account for the thermodynamic stability of these complexes.

A number of intermediates have been observed in the reaction of $Co(CN)_5^{3-}$ with O_2 (76–78). As with most other complexes, the mononuclear, end-on superoxo (79), $(CN)_5CoO_2^{3-}$, and binuclear peroxo species, $(CN)_5$-$CoO_2Co(CN)_5^{6-}$, are formed first. Depending on conditions, the μ-peroxo complex reacts either with O_2 to give the μ-superoxo ion, or with protons to yield the hydroperoxide $(CN)_5Co(O_2H)^{3-}$.

$$(CN)_5CoO_2Co(CN)_5^{6-} \xrightarrow{O_2} (CN)_5CoO_2Co(CN)_5^{5-} \quad (8)$$

$$(CN)_5CoO_2Co(CN)_5^{6-} + H^+ \rightleftharpoons (CN)_5Co(O_2H)Co(CN)_5^{5-} \quad (9)$$

$$(CN)_5Co(O_2H)Co(CN)_5^{5-} \xrightarrow{H_2O} (CN)_5Co(O_2H)^{3-} + (CN)_5CoOH^{3-} + H^+ \quad (10)$$

The hydroperoxo complex, which also can be prepared from $(CN)_5CoH^{3-}$ and O_2, or by chemical reduction of $(CN)_5CoO_2^{3-}$, decomposes to H_2O_2 and $(CN)_5Co(H_2O)^{2-}$ in a pH-dependent reaction.

$$(CN)_5Co(O_2H)^{3-} + H^+ \xrightarrow{H_2O} (CN)_5Co(H_2O)^{2-} + H_2O_2 \quad (11)$$

The reaction of $[HB(3\text{-}t\text{-Bu-5-Mepz})_3]Co^I$ with O_2 produces an unusual, side-on superoxocobalt(II) complex (18a). The activation of O_2 by the closely related $[HB(3\text{-}i\text{-Pr-5-Mepz})_3]Co^I$ results in the oxidation of the ligand isopropyl groups (18b). The reaction produces a μ-peroxo intermediate, which then decays to the final products in a first-order process, $k = 2.27 \times 10^{-3}\ s^{-1}$ in CH_2Cl_2 at 281 K, $\Delta H^\ddagger = 68.6\ kJ\ mol^{-1}$, $\Delta S^\ddagger = -50\ J\ mol^{-1}\ K^{-1}$. The large kinetic isotope effect for isopropyl hydrogen atoms, $k_H/k_D = 22$, was taken as evidence for C—H bond breaking and the involvement of tunneling in the rate-determining step. In the proposed mechanism, the cleavage of the O—O bond is accompanied by the formation of two O—H bonds. The hydrogen atoms are supplied by the ligand isopropyl groups, as shown in Eq. 12, where L—$CHMe_2$ stands for $HB(3\text{-}i\text{-Pr-5-Mepz})_3$. Finally, the alkyl radical disproportionation yields the observed products.

$$(L-CHMe_2)CoOOCo(L-CHMe_2) \longrightarrow 2(L-C^{\bullet}Me_2)CoOH \longrightarrow$$
$$(L-CHMe_2)CoOH + (L-CMeCH_2)CoOH \quad (12)$$

B. Iron

The presence of iron at the active site of a large number of biologically important molecules has sparked intense research efforts in the area of synthetic and mechanistic iron–oxygen chemistry. Accordingly, most of the work has been done with either naturally occurring compounds or the synthetic models of such molecules.

Iron complexes of small saturated macrocycles undergo facile ligand-based oxidation (38, 80, 81), which severely limits their use as dioxygen carriers and dioxygen-activating reagents. For example, excess O_2 converts L^1Fe^{2+} to a strongly colored dimetallic complex (81), Fig. 6, analogous to the one formed by H_2O_2 oxidation of L^1Ni^{2+} (82). The difference in the chemistry of iron and cobalt macrocycles is undoubtedly the result of different reduction potentials for the two metals in the same coordination environment.

On the other hand, complexes of some unsaturated macrocycles, such as L^5 (tim) (83) and L^6 (84), Eq. 13 and Fig. 7, yield stable di-μ-peroxo species, much like macrocyclic cobalt complexes do, but unlike iron porphyrins, which typically yield μ-oxo dimers, (P)FeOFe(P) (P = porphyrin), as discussed later.

$$2L^6Fe^{2+} + O_2 \longrightarrow L^6FeO_2FeL^6 \qquad (13)$$

With L^6Fe^{2+} present in a large excess over O_2, Reaction 13 obeys a mixed second-order rate law, $k = k_{on}[O_2][L^6Fe^{2+}]$. The pH-dependent k_{on} has a value of $1.4 \times 10^2\ M^{-1}\ s^{-1}$ at pH 8 (84). The first step, formation of $L^6FeO_2^{2+}$, is apparently rate determining in this system.

The autoxidation of the hexaaqua ion, $Fe(H_2O)_6^{2+}$, is slow in the absence of coordinating anions. The stoichiometry and rate law in aqueous $HClO_4$ (85) are

$$4Fe(H_2O)_6^{2+} + O_2 + 4H^+ \longrightarrow 4Fe(H_2O)_6^{3+} + 2H_2O \qquad (14)$$

$$-d[Fe(H_2O)_6^{2+}]/dt = k_{Fe}\ [Fe(H_2O)_6^{2+}]^2\ [O_2] \qquad (15)$$

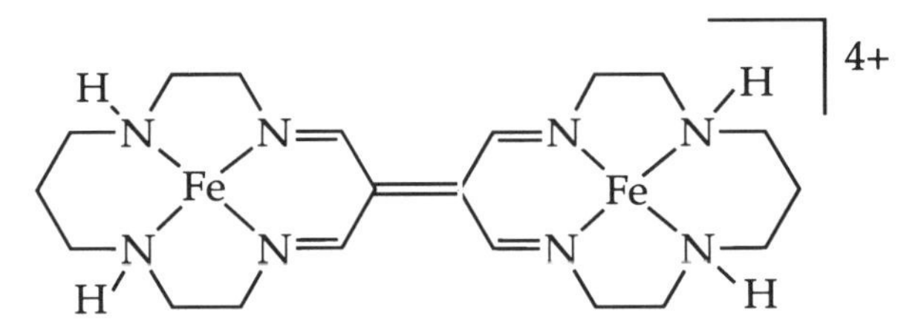

Figure 6. Product of autoxidation of $Fe([14]aneN_4)^{2+}$. [Reproduced with permission from H. S. Mountford, L. O. Spreer, J. W. Otvos, M. Calvin, K. J. Brewer, M. Richter, and B. Scott, *Inorg. Chem.*, *31*, 717 (1992). Copyright © 1992 American Chemical Society.]

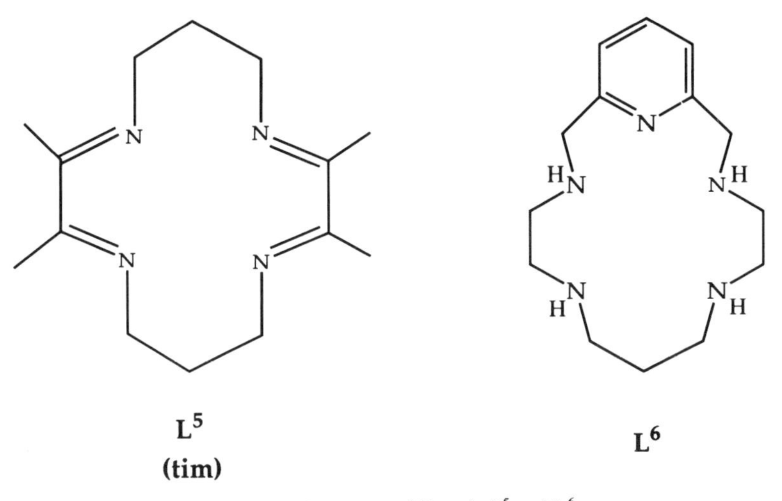

Figure 7. Structures of ligands L^5 and L^6.

The rate constant k_{Fe} is a function of $[H^+]$ and ionic strength. In 1 *M* $HClO_4$ the value is 2.9×10^{-4} M^{-2} s^{-1} [calculated from the reported (85) value of 2.9×10^{-7} M^{-1} $atm^{-1}s^{-1}$, and taking the solubility of O_2 as 1.0 m*M* atm^{-1} at this ionic strength].

A mechanism consistent with the observations has the reaction of $Fe(H_2O)_6^{2+}$ with $(H_2O)_5FeO_2^{2+}$ as the rate-determining step, followed by the rapid oxidation of two additional moles of $Fe(H_2O)_6^{2+}$ by H_2O_2, to give $k_{Fe} = 4\,K_{16}k_{17}$.

$$Fe(H_2O)_6^{2+} + O_2 \rightleftharpoons (H_2O)_5FeO_2^{2+} + H_2O \qquad (16)$$

$$(H_2O)_5FeO_2^{2+} + Fe(H_2O)_6^{2+} \xrightarrow{2H^+,\ H_2O} 2Fe(H_2O)_6^{3+} + H_2O_2 \qquad (17)$$

$$2Fe(H_2O)_6^{2+} + H_2O_2 \xrightarrow{2H^+} 2Fe(H_2O)_6^{3+} + 2H_2O \qquad (18)$$

Reaction 17 is almost certainly not an elementary step. Possible intermediates (4, 85) are $(H_2O)_5FeOOFe(H_2O)_5^{4+}$ and $(H_2O)_5FeOOH^{2+}$.

Coordinating anions, such as Cl^- (86), accelerate the reaction by providing an additional pathway that is first order in $[Fe^{II}]$. For $X^- = Cl^-$, the chemistry is probably best described by Eqs. 19, 16a, and 17a. For illustration purposes, only a single chloride ion is shown in the reaction scheme, but the more highly substituted complexes may be involved at any stage of the reaction. At high $[Cl^-]$ the catalytic term dominates and the rate law of Eq. 20 applies, where $k_{Fe,X}$ is $[H^+]$ dependent.

$$Fe(H_2O)_6^{2+} + X^- \rightleftharpoons (H_2O)_5FeX^+ + H_2O \qquad K_X \tag{19}$$

$$(H_2O)_5FeX^+ + O_2 \rightleftharpoons (H_2O)_4(X)FeO_2^+ + H_2O \tag{16a}$$

$$(H_2O)_4(X)FeO_2^+ + Fe(H_2O)_6^{2+} \xrightarrow{2H^+,\ H_2O} (H_2O)_5FeX^{2+} + Fe(H_2O)_6^{3+} + H_2O_2 \tag{17a}$$

$$-d[Fe^{II}]/dt = k_{Fe,X}\,[Fe^{II}][O_2][X^-] \tag{20}$$

The catalysis is generally accepted to arise from the stabilizing effect of X^- on Fe^{III}. Provided the proposed mechanism holds, the first-order dependence on $[Fe^{II}]$ indicates that Reaction 16a is rate determining, and $k_{Fe,X} = K_X k_{16a}$, implying that $k_{-16a} << k_{-16}$ and $k_{17a} >> k_{17}$. Both inequalities appear reasonable. Recent examples (62) in cobalt chemistry show that the dissociation of O_2 is indeed slower for $(X)LCoO_2^+$ than for $(H_2O)LCoO_2^{2+}$, and the stabilization of Fe^{III} in $(H_2O)_4(X)FeO_2^{2+}$ may impart more superoxide character to coordinated dioxygen, resulting in $k_{17a} >> k_{17}$. Another possibility, which is highly unlikely, has Reaction 17a rate determining and the equilibrium Reactions 19 and 16a shifted completely to the right.

The dissociation of one molecule of imidazole from $Fe(dmgH)_2(Im)_2$ precedes the reaction with O_2 (87), Scheme 2. It was proposed that the aquated species captures O_2 and yields the superoxo complex, which then reacts rapidly with $Fe(dmgH)_2(Im)_2$ to give H_2O_2 and $Fe(dmgH)_2(Im)_2^+$.

$$Fe(dmgH)_2(Im)_2 + H_2O \underset{170\ M^{-1}s^{-1}}{\overset{0.36\ s^{-1}}{\rightleftharpoons}} Fe(dmgH)_2(Im)(H_2O) + Im$$

$$Fe(dmgH)_2(Im)(H_2O) + O_2 \overset{20\ M^{-1}s^{-1}}{\rightleftharpoons} Fe(dmgH)_2(Im)O_2 + H_2O$$

$$Fe(dmgH)_2(Im)O_2 + Fe(dmgH)_2(Im)_2 \xrightarrow[\text{fast}]{H^+,\ Im} H_2O_2 + 2\ Fe(dmgH)_2(Im)_2^+$$

Scheme 2

The reaction is strongly catalyzed by Cu^{II} owing to the rapid reduction of $Cu(Im)_4^{2+}$ by $Fe(dmgH)_2(Im)_2$ ($k = 1.2 \times 10^6\ M^{-1}s^{-1}$, $K = 0.33$). The reaction of O_2 with $Cu(Im)_4^+$ then forms the superoxocopper(II) ion, $(Im)_nCuO_2^+$, which reacts rapidly with $Fe(dmgH)_2(Im)_2$.

Reactions of synthetic iron complexes with O_2 have been studied in some detail (88). Perhaps the most successful dioxygen carriers are the cyclidenes (89–94), Fig. 8, whose O_2 affinity can be controlled by structural variations, especially those that affect cavity size at the O_2-binding site.

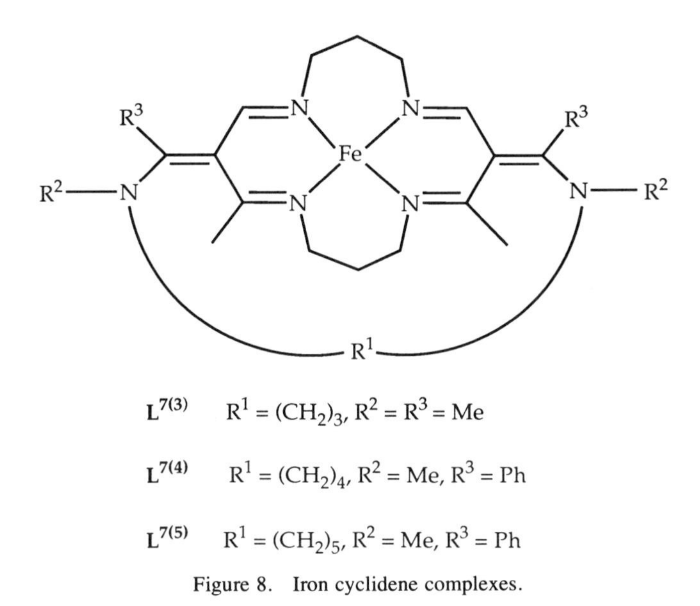

Figure 8. Iron cyclidene complexes.

Also, as shown in a recent study (89), the mechanism of autoxidation of an iron cyclidene complex, $L^{7(5)}Fe^{2+}$, parallels that of hemoglobin and myoglobin. In all three cases the rate of autoxidation increases with $[O_2]$ at low O_2 pressures, reaches a maximum, and then decreases at high $[O_2]$, Fig. 9. By comparison, the kinetics of autoxidation of the cyclidene complex $L^{7(3)}Fe^{2+}$ are first order in O_2 throughout the range investigated, whereas $L^{7(4)}Fe^{2+}$ exhibits saturation behavior at high $[O_2]$. The main difference between the three L^7 complexes is the size of the cavity, which is too small to allow O_2 binding for $L^{7(3)}Fe^{2+}$, is just large enough to allow O_2 binding for $L^{7(4)}Fe^{2+}$, and is sufficiently large to allow not only O_2, but also the solvent to enter the cavity of $L^{7(5)}Fe^{2+}$.

A common mechanism for autoxidation of $L^{7(5)}Fe^{2+}$, Hb, and Mb, shown in Eqs. 21–24, takes into account all of the observations (89). According to this mechanism, the autoxidation takes place by outer-sphere electron transfer in parallel with O_2 binding, as opposed to an inner-sphere process, whereby electron transfer would occur within the $LFeO_2$ adduct. Also, the solvento species, LFe(S), is introduced in the new mechanism, in addition to $LFeO_2$ and the pentacoordinated LFe^{II}, to rationalize the complex kinetic behavior of $L^{7(5)}Fe^{2+}$, Hb, and Mb. The rate law of Eq. 25 treats LFe(S) as a steady-state intermediate, and correctly predicts the change in the rate dependence on $[O_2]$ with changes in O_2 concentration.

$$LFe^{II} + O_2 \rightleftharpoons LFeO_2 \qquad K_{on} \tag{21}$$

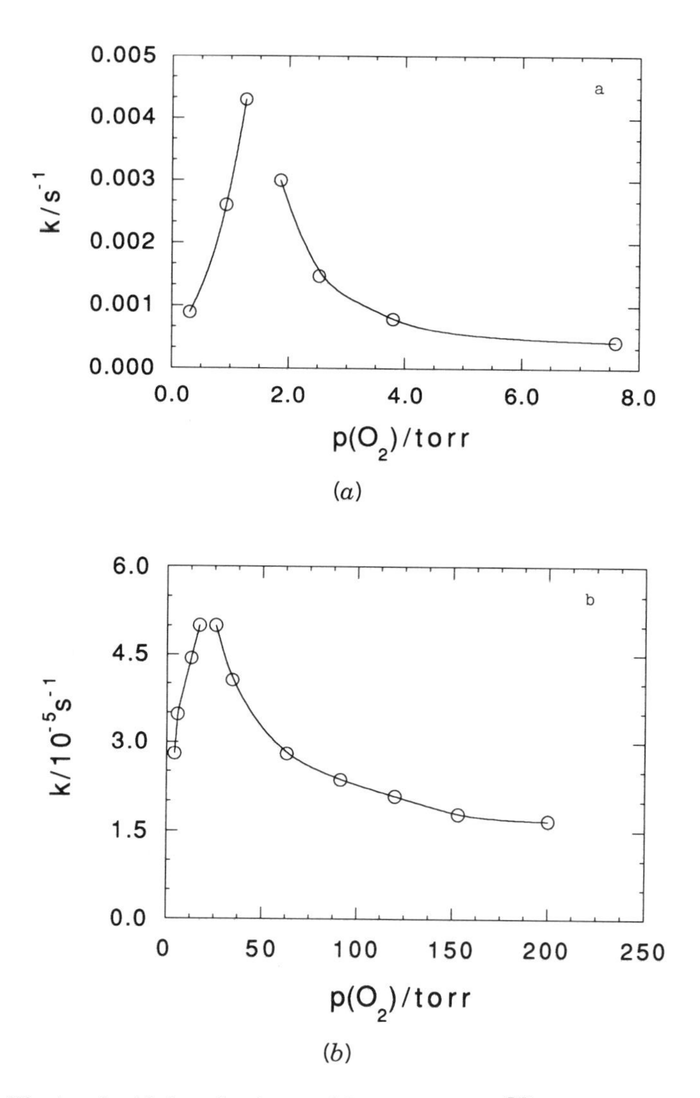

Figure 9. Kinetics of oxidation of an iron cyclidene complex $FeL^{7(5)}$ [(*a*) R = $(CH_2)_5$, at 0°C) and of hemoglobin (*b*), at 30°C) as a function of O_2 pressure]. [Adapted with permission from L. D. Dickerson, A. Sauer-Masarwa, N. Herron, C. N. Fendrick, and D. H. Busca, *J. Am. Chem. Soc.*, *115*, 3623 (1993). Copyright © 1993 American Chemical Society.] The rate constants for the hemoglobin reaction in Fig. 1 of (89), and in the original paper [J. Brooks, *Proc. R. Soc. Chem.*, *118*, 560 (1935)] appear to be in error by a constant factor of 10^4. The values given here have been corrected by this factor.

$$LFe^{II} + O_2 \longrightarrow LFe^{III} + O_2^{\bullet -} \qquad (22)$$

$$LFe^{II} + (S) \rightleftharpoons LFe(S) \qquad (23)$$

$$LFe(S) + O_2 \longrightarrow LFe^{III}(S) + O_2^{\bullet -} \qquad (24)$$

$$\frac{-d[LFe^{II}]}{dt} = \left\{k_{22} + \frac{k_{23}k_{24}[S]}{(k_{-23} + k_{24}[O_2])}\right\} \frac{[LFe^{II}]_T[O_2]}{1 + K_{on}[O_2]} \qquad (25)$$

Dioxygen-carrying properties of biological molecules containing heme iron are not duplicated by iron porphyrins except when special steric features are built into the porphyrin. In the absence of such steric effects, the reaction occurs as in Eq. 26 (95). The first intermediate, superoxoiron, is rapidly captured by additional (P)Fe to yield the unstable (P)FeOOFe(P), which homolyzes to (P)FeO. The rapid reaction of (P)FeO by (P)Fe then yields the μ-oxo-dimer. Even at low temperatures, that are standard in work with iron porphyrins, the superoxo and μ-peroxo intermediates are short lived and difficult to observe (22, 96, 97).

$$(P)Fe \xrightarrow{O_2} (P)FeO_2 \xrightarrow{(P)Fe} (P)FeOOFe(P) \longrightarrow$$
$$2(P)Fe{=}O \xrightarrow{(P)Fe} (P)FeOFe(P) \qquad (26)$$

To prevent the formation of diiron intermediates, and thus inhibit the autoxidation process of Eq. 26, a large number of modified porphyrins have been prepared (98–103). In addition to the steric bulk and/or other architectural features that prevent dimerization, the successful synthetic molecules have an N base (imidazole or pyridine) coordinated to an iron(II) porphyrin in a hydrophobic environment.

By use of laser flash photolysis, the kinetics of O_2 binding and release by iron porphyrins, as well as by the natural compounds, can now be determined almost routinely (104–106). For this purpose, iron(II) porphyrins are usually produced by photolysis of (P)Fe(CO) in the presence of O_2. The reequilibration then takes place according to Scheme 3.

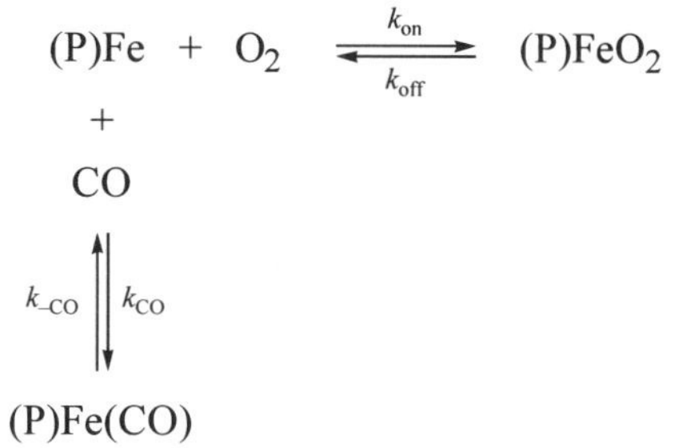

$$\begin{array}{ccc} (P)Fe + O_2 & \underset{k_{off}}{\overset{k_{on}}{\rightleftharpoons}} & (P)FeO_2 \\ + & & \\ CO & & \\ k_{-CO} \uparrow\downarrow k_{CO} & & \\ (P)Fe(CO) & & \end{array}$$

Scheme 3

The rate constants k_{on} for O_2 binding by (P)Fe are typically more than 10 times greater than those for binding of CO, but the thermodynamic stability of (P)Fe(CO) is significantly greater than that of (P)FeO$_2$. As a result, only the reaction of (P)Fe with O_2 is kinetically important immediately after the flash, and the rate law of Eq. 27 applies.

$$k_{obs} = k_{on}[O_2] + k_{off} \tag{27}$$

The subsequent slower equilibration restores (P)Fe(CO) according to the rate law

$$k_{obs} = \frac{k_{off}k_{CO}[CO]}{k_{CO}[CO] + k_{on}[O_2] + k_{off}} \tag{28}$$

Values of k_{on} are affected greatly by the changes in ligand structure. Extensive data compilations in (1) and (105–107) list values that are at one end comparable to those for hemoglobin (108, 109), Table V, and at the other some three to four orders of magnitude lower. The range of k_{off} values also covers several orders of magnitude.

Direct photolysis of (P)(B)FeO$_2$ (B = nitrogenous base) in O_2-containing solutions to generate (P)(B)Fe^{II} for kinetic purposes works only in exceptional cases (104) owing to the rapid autoxidation of most iron(II) porphyrins under such conditions.

Another success of synthetic iron heme chemistry involves the preparation of model compounds that exhibit cooperative O_2 binding (75, 110) similar to that of hemoglobin itself. Cooperativity in model compounds has been achieved through ligand-induced dimerization and conformational changes.

Cytochrome P450 enzymes, cytochrome *c* oxidases, and their model compounds have also received increasing attention in recent years. The nature of

TABLE V
Kinetic Data for the Interaction of Hemoglobin with O_2[a,b]

k_{on}, R state	$5.9 \times 10^7\ M^{-1}\ s^{-1}$ (α chain)[b]
	$5.9 \times 10^7\ M^{-1}\ s^{-1}$ (β chain)[b]
k_{off}, R state	$12\ s^{-1}$ (α chain)[b]
	$21\ s^{-1}$ (β chain)[b]
k_{on}, T state	$2.9 \times 10^6\ M^{-1}\ s^{-1}$ (α chain)[b]
	$1.2 \times 10^7\ M^{-1}\ s^{-1}$ (β chain)[b]
k_{off}, T state	$183\ s^{-1}$ (α chain)[b]
	$2480\ s^{-1}$ (β chain)[b]

[a]Data from (108) and (109).
[b]The α and β chains are the subunits of the tetrameric homoglobin molecule, which has an $\alpha_2\beta_2$ structure.

the proximal axial ligands in cytochromes P450 (thiolate from a cysteine residue) and O_2-carrying hemoproteins (imidazole from a histidine residue) is the main chemical difference between the two classes of enzymes. The cytochrome P450 enzymes play their detoxifying roles by catalyzing the oxidation of various molecules by O_2 according to the generally accepted mechanism in Fig. 10 (111).

Cytochrome *c* oxidases are membrane bound, multimetallic enzymes, that catalyze the four-electron reduction of O_2 to H_2O. The binuclear center at the active site contains iron and copper ions. Recently, the first example of an O_2 adduct of a model compound has been reported (112). The 1 : 1 ([Cu,Fe] : O_2) adduct forms rapidly and irreversibly and requires four equivalents of cobaltocene for complete reduction.

The chemistry of non-porphyrin diiron metalloproteins hemerythrin (Hr) (73, 113), methane monooxygenase (MMO) (114–117), ribonucleotide reductase (RR) (118–122) and several other oxygenases and oxidases, and of biomimetic non-porphyrin iron complexes, has been reviewed recently (3–5).

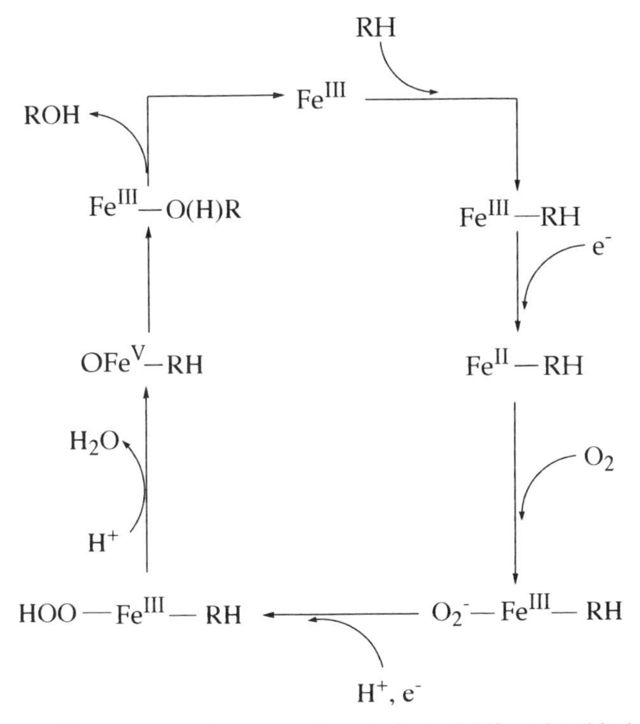

Figure 10. Generally accepted mechanism for the cytochrome P450-catalyzed hydroxylation of organic substrates by O_2.

The structures of the diiron centers in Hr, MMO, and RR are closely related, but the nature of the terminal ligands is different and considered responsible for the different functions of the enzymes: The N-rich environment in Hr promotes O_2 binding, and O-rich ligation in MMO and RR promotes O_2 activation (118). In the reaction with Hr, O_2 is reduced to a coordinated hydroperoxide and is stabilized by hydrogen bonding to the Fe—O—Fe oxo bridge. The equilibrium constant for the binding is about $10^6\ M^{-1}$ (4).

Methane monooxygenase, which is a mixed-function oxidase, hydroxylates methane and a number of other substrates. The initial interaction of O_2 with the reduced enzyme probably produces a diiron(III) peroxide intermediate. The details of subsequent chemistry, which incorporates one of the oxygen atoms into substrate, have been a subject of intense research (114–117). A number of observations led to proposed mechanisms that are similar to that for hydroxylation by cytochrome P450, featuring high-valent ferryl intermediates and substrate derived radicals and radical cations as intermediates. A mechanism involving a cysteine-derived S-centered radical and an iron-bound oxyl has also been proposed (4). According to this mechanism, the reaction with methane yields the cysteine thiol and a coordinated alcohol in a concerted process that bypasses the formation of substrate-derived radicals. This proposal is consistent with the lack of rearranged products in the radical clock study of MMO from *Methylococcus capsulatus* (115b).

The kinetics of the reactions of O_2 with a series of three synthetic dinuclear, carboxylato-bridged, pentacoordinated iron(II) complexes in propionitrile were studied as a function of temperature and the nature of the ligands (123). All three reactions yield the corresponding μ-peroxodiiron(III) complexes. The compounds with sterically less demanding ligands obey a mixed second-order rate law and exhibit enthalpies of activation that are comparable to that for hemerythrin ($\Delta H^{\ddagger} = 16.8$ kJ mol^{-1}). For a complex with bulky ligands, there is a change in the rate law, and a substantial increase in $\Delta H^{\ddagger}$, demonstrating the importance of steric crowding and suggesting that a structural rearrangement, such as an elongation of iron–ligand bond(s) or ligand dissociation, precedes the reaction with O_2.

Pentacoordinated, monomeric carboxylate ferrous complexes of HB(3,5-*i*-$Pr_2pz)_3$ have recently been synthesized and found to react with O_2 at $< -20°C$ to yield μ-peroxo dinuclear ferric complexes (124). Above $-20°C$, the reaction with O_2 causes irreversible oxidation and yields trimeric ferric products.

C. Chromium

The reaction of $Cr(H_2O)_6^{2+}$ with O_2 yields dimeric Cr^{III} as a major product (125) under all conditions. Some Cr^{VI} is also produced when O_2 is used in large excess. These data and others presented below are consistent with the mechanism of Eqs. 29–33. Under conditions of excess Cr^{2+}, Reaction 33 is unimportant and a 4 : 1 $[Cr^{2+}]/[O_2]$ stoichiometry is observed.

$$Cr(H_2O)_6^{2+} + O_2 \underset{2.5 \times 10^{-4} s^{-1}}{\overset{1.6 \times 10^8 M^{-1}s^{-1}}{\rightleftarrows}} (H_2O)_5CrO_2^{2+} + H_2O$$

$$k_{on}, k_{off}, K_{on} \qquad (29)$$

$$(H_2O)_5CrO_2^{2+} + Cr(H_2O)_6^{2+} \xrightarrow{8 \times 10^8 M^{-1}s^{-1}} [(H_2O)_5CrOOCr(H_2O)_5^{4+}] \quad k_{2,on} \qquad (30)$$

$$[(H_2O)_5CrOOCr(H_2O)_5^{4+}] + Cr(H_2O)_6^{2+} \longrightarrow [(H_2O)_5CrOCr(H_2O)_5^{4+}] + (H_2O)_5CrO^{2+} + H_2O \qquad (31)$$

$$(H_2O)_5CrO^{2+} + Cr(H_2O)_6^{2+} \longrightarrow [(H_2O)_5CrOCr(H_2O)_5^{4+}] + H_2O \longrightarrow (H_2O)_5Cr(OH)_2Cr(H_2O)_5^{4+} \qquad (32)$$

$$2(H_2O)_5CrO_2^{2+} \xrightarrow{6.0 M^{-1}s^{-1}} (aq)Cr^{VI} + nH^+ + nH_2O \quad k_{bi} \qquad (33)$$

The UV–vis (67, 126), Fig. 11, and resonance Raman (30) data convincingly support an end-on superoxo structure for the first observable intermediate, $(H_2O)_5CrO_2^{2+}$.

The rapid formation ($k_{on} = 1.6 \times 10^8\ M^{-1}s^{-1}$) (126, 127) and slow homolysis ($k_{off} = 2.5 \times 10^{-4} s^{-1}$) (67) make $(H_2O)_5CrO_2^{2+}$ one of the most stable and long-lived mononuclear superoxometal complexes in aqueous solution. The strongly reducing potential of the $Cr(H_2O)_6^{3+/2+}$ couple ($E^0 = -0.41$ V) and substitutional inertness of Cr^{III} are undoubtedly responsible for this stability. A detailed analysis has also shown that the equilibrium constant for Reaction 34 is unusually high ($K_{bind} = 3 \times 10^7\ M^{-1}$) (67), but less so than for the macrocyclic cobalt complexes discussed earlier (64, 67). In fact, the driving force for heterolysis ($K_{heter} = 1/K_{bind} = 3 \times 10^{-8}\ M$) of $(H_2O)_5CrO_2^{2+}$ is greater than for homolysis ($K_{hom} = 1/K_{on} = 2 \times 10^{-12}\ M$), but only homolysis has been observed experimentally. This point will be discussed in more detail in Section III.A.6.

$$Cr(H_2O)_6^{3+} + O_2^{\cdot-} \rightleftarrows (H_2O)_5CrO_2^{2+} + H_2O \quad K_{bind} = 3 \times 10^7\ M^{-1} \qquad (34)$$

The complex $(H_2O)_5CrO_2^{2+}$ exhibits a unique combination of stability and reactivity, and thus has a great advantage over most other mononuclear superoxometal complexes for kinetic and mechanistic studies. The slowness of O_2

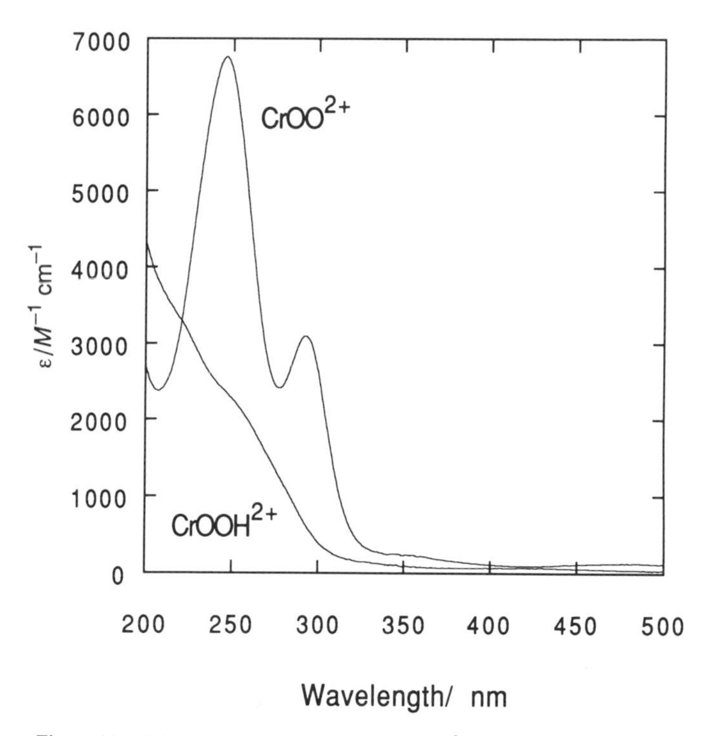

Figure 11. Ultraviolet spectra of $(H_2O)_5CrOO^{2+}$ and $(H_2O)_5CrOOH^{2+}$.

dissociation makes it possible to remove excess O_2 after the preparation and to conduct studies with air-sensitive materials under straightforward conditions and without interference from the reversible binding and dissociation of O_2. At the same time the reactivity of $(H_2O)_5CrO_2^{2+}$ toward a variety of reagents matches or exceeds that of many less persistent metal superoxides, as discussed in Section III.

In addition to homolysis, $(H_2O)_5CrO_2^{2+}$ decomposes in a second-order pathway of Eq. 33. The complete rate law, Eq. 35, for the loss of $(H_2O)_5CrO_2^{2+}$ in Reactions 29–33 under conditions of excess $[O_2]$ treats $Cr(H_2O)_6^{2+}$ as a steady-state intermediate.

$$\frac{-d[(H_2O)_5CrO_2^{2+}]}{dt} = \frac{2k_{off}k_{2,on}[(H_2O)_5CrO_2^{2+}]^2}{k_{on}[O_2] + k_{2,on}[(H_2O)_5CrO_2^{2+}]} + 2k_{bi}[(H_2O)_5CrO_2^{2+}]^2 \tag{35}$$

The Cr^{VI}-producing bimolecular reaction of Eq. 33 is an intriguing one that appears to have no precedent in the chemistry of superoxometal complexes.

The reaction may be proceeding through a cyclic transition state, $[\overline{CrOOCrOO}]^{4+}$ (67), that yields (aq)Cr^{VI} directly. Another possibility, shown in Eq. 36, is the formation of a head-to-head dimer, $CrOOOOCr^{4+}$, that eliminates O_2 and yields $(H_2O)_5CrO^{2+}$, which has been independently shown to yield (aq)Cr^{VI} on standing (128, 129). Reaction 36 is written in analogy to the well-established stepwise processes in the self-reactions of organic peroxyl radicals (130–133). Experiments with labeled O_2 should either confirm or rule out this possibility.

$$2(H_2O)_5CrO_2^{2+} \longrightarrow (H_2O)_5CrOOOOCr(H_2O)_5^{4+} \longrightarrow O_2 + 2(H_2O)_5CrO^{2+} \tag{36}$$

Until recently, the disproportionation of $(H_2O)_5CrO_2^{2+}$, Eq. 37, was considered too unfavorable thermodynamically to provide a path for the bimolecular decomposition route. The plausibility of this mechanism grew, however, after the reduction potentials (134, 135) for the couples $(H_2O)_5\text{-}CrO_2^{2+}/(H_2O)_5CrO_2H^{2+}$ (1.03 V in 1 M H^+) and $(H_2O)_5CrO_2^{3+}/(H_2O)_5CrO_2^{2+}$ (0.97–1.4 V) became available. In this scenario, the observed (aq)Cr^{VI} is the decomposition product of $(H_2O)_5CrO_2H^{2+}$ (136).

$$2(H_2O)_5CrO_2^{2+} \xrightleftharpoons{H^+} (H_2O)_5CrO_2^{3+} + (H_2O)_5CrO_2H^{2+} \quad (10^1 > K > 10^{-6}) \tag{37}$$

Of all the remaining intermediates in the autoxidation of $(H_2O)_6Cr^{2+}$, Eqs. 29–33, only $(H_2O)_5CrO^{2+}$ has been observed directly (128, 129). The compounds $(H_2O)_5CrOOCr(H_2O)_5^{4+}$ and $(H_2O)_5CrOCr(H_2O)_5^{4+}$ are either not formed, or are too short lived to be observed under experimental conditions. A long-lived (several minutes at room temperature) compound that analyzes for $(H_2O)_5CrOOCr(H_2O)_5^{4+}$ has been prepared (137) independently from (aq)Cr^{VI} and H_2O_2, and shown (138) not to yield $(H_2O)_5CrO^{2+}$ spontaneously. The $Cr(H_2O)_6^{2+}$ reduction of $(H_2O)_5CrOOCr(H_2O)_5^{4+}$ does, however, produce $(H_2O)_5CrO^{2+}$ in a synthetically useful, but mechanistically unsolved reaction. The chemistry in Eqs. 30 and 31 was written under the assumption that the peroxo complex $(H_2O)_5CrOOCr(H_2O)_5^{4+}$ is chemically similar (identical?) to the known (137) compound.

The μ-oxo dimer, $(H_2O)_5CrOCr(H_2O)_5^{4+}$, is an unknown species (139) and may not be an intermediate in the autoxidation, although it is difficult to see how the final product, $(H_2O)_5Cr(OH)_2Cr(H_2O)_5^{4+}$, could be formed directly from $(H_2O)_5CrOOCr(H_2O)_5^{4+}$. The absence of an observable μ-oxo dimer in the aquachromium family is surprising, given that the pentaammine (140), porphyrin (141), and other (142) chromium complexes yield stable μ-oxo dimers.

In the absence of alcohols, pure solutions of $(H_2O)_5CrO_2^{2+}$ can be prepared only when O_2 is used in ≥20-fold excess over $Cr(H_2O)_6^{2+}$. For smaller ratios the reaction yields a mixture of $(H_2O)_5CrO_2^{2+}$ and $(H_2O)_5CrO^{2+}$. Addition of an alcohol (such as methanol or 2-propanol) to such mixtures, or the presence of an alcohol during the preparation of $(H_2O)_5CrO_2^{2+}$, increases the yield and lifetime of $(H_2O)_5CrO_2^{2+}$. These observations, which initially led to the discovery of $(H_2O)_5CrO^{2+}$, are now understood in terms of Reaction 38, followed by the rapid capture of $Cr(H_2O)_6^{2+}$ by O_2 to regenerate $(H_2O)_5CrO_2^{2+}$.

$$(H_2O)_5CrO^{2+} + MeOH \longrightarrow Cr(H_2O)_6^{2+} + CH_2O \qquad (38)$$

The $(H_2O)_5CrO^{2+}$, produced as an intermediate in Eq. 31 (and possibly Eq. 33) during the decomposition of $(H_2O)_5CrO_2^{2+}$, is reconverted to $(H_2O)_5CrO_2^{2+}$ in Reaction 38, hence the increase in lifetime of $(H_2O)_5CrO_2^{2+}$ in the presence of alcohols. The detailed mechanism of Reaction 38 will be discussed in Section III.B.

The macrocyclic chromium complex $L^1Cr(H_2O)_2^{2+}$ (L^1 = [14]ane N_4) (143) reacts with O_2 (k_{on} = 1×10^8 $M^{-1}s^{-1}$) (144) to yield a long-lived $L^1(H_2O)CrO_2^{2+}$ in solution. The UV–vis and Raman data (30), Table I, are similar to those for $(H_2O)_5CrO_2^{2+}$ and consistent with the end-on superoxo formulation. The monomeric $L^1Cr(H_2O)_2^{3+}$ is a major final product of the reaction of $L^1Cr(H_2O)_2^{2+}$ with O_2 (143). The steric bulk of the macrocycle apparently disfavors the formation of dinuclear Cr^{III} products, which appear only as a minor fraction.

D. Copper

Superoxocopper(II) complexes are typically short lived and solid kinetic evidence has been obtained for only a few such species.

The $[H^+]$-dependence in the autoxidation of $(aq)Cu^+$ in aqueous acetonitrile

$$-d[O_2]/dt = \frac{k_a[(aq)Cu^+][O_2](1 + k_b[H^+])}{1 + k_c[H^+]} \qquad (39)$$

can be reasonably explained by the initial formation of $(aq)CuO_2^+$ (145), followed by H^+-assisted and H^+-independent displacement of superoxide, Eqs. 40–43. The values of equilibrium constants for the formation of various Cu^I–acetonitrile complexes are available, which allows one to calculate the concentration of the uncomplexed $(aq)Cu^+$. According to this mechanism, $k_a = k_{on}k_{42}/(k_{off} + k_{42}) = 3.5 \times 10^4\ M^{-1}\ s^{-1}$, $k_b = k_{41}/k_{42} = 410\ M^{-1}$, $k_c = k_{41}/(k_{off} + k_{42}) = 15\ M^{-1}$, $k_{on} = k_ak_b/k_c = 9.5 \times 10^5\ M^{-1}\ s^{-1}$, and $k_{on}k_{41}/k_{off} = 1.5 \times 10^7\ M^{-2}\ s^{-1}$, all at 20°C and 0.2 M ionic strength.

$$\text{(aq)Cu}^+ + \text{O}_2 \rightleftharpoons \text{(aq)CuO}_2^+ \qquad k_{on}(9.5 \times 10^5\ M^{-1}\ \text{s}^{-1}),\ k_{off} \tag{40}$$

$$\text{(aq)CuO}_2^+ + \text{H}^+ \longrightarrow \text{(aq)Cu}^{2+} + \text{HO}_2^{\bullet} \tag{41}$$

$$\text{(aq)CuO}_2^+ \longrightarrow \text{(aq)Cu}^{2+} + \text{O}_2^{\bullet-} \tag{42}$$

$$\text{(aq)Cu}^+ + \text{HO}_2^{\bullet} + \text{H}^+ \xrightarrow{\text{fast}} \text{(aq)Cu}^{2+} + \text{H}_2\text{O}_2 \tag{43}$$

In the presence of ligands such as MeCN (146) and imidazoles (147, 148), the order in Cu^I increases from one to two as the concentration of added Cu^{II} increases. This observation provides strong evidence for a competition between the unimolecular and bimolecular reactions of CuO_2^+, as depicted in Eqs. 44 and 45, where Cu^+ and Cu^{2+} represent all the complexes of a given oxidation state present in such solutions.

$$\text{CuO}_2^+ \rightleftharpoons \text{Cu}^{2+} + \text{O}_2^{\bullet-} \tag{44}$$

$$\text{CuO}_2^+ + \text{Cu}^+ \xrightarrow{2\text{H}^+} 2\text{Cu}^{2+} + \text{H}_2\text{O}_2 \tag{45}$$

The $(aq)CuO_2^+$ complex was also invoked in the reactions of $Cu(alkene)^+$ complexes with O_2 in aqueous solution (149). A stable superoxo copper complex, $[HB(Me_2pz)_3]CuO_2$ was produced in the reaction of O_2 with $[HB(Me_2pz)_3]Cu(C_2H_4)$ (150).

The oxidation of $Cu(phen)_2^+$ by O_2 in aqueous solution at pH 7 is second order in $[Cu(phen)_2^+]$, but changes to first order in the presence of superoxide dismutase, a fact that was taken as evidence for the involvement of free $O_2^{\bullet-}$, according to the mechanism in Eqs. 46–49 (151).

$$\text{Cu(phen)}_2^+ + \text{O}_2 \underset{1.9 \times 10^9}{\overset{5 \times 10^4}{\rightleftharpoons}} \text{Cu(phen)}_2^{2+} + \text{O}_2^{\bullet-} \tag{46}$$

$$\text{Cu(phen)}_2^+ + \text{O}_2^{\bullet-} + 2\text{H}^+ \xrightarrow{3 \times 10^8} \text{Cu(phen)}_2^{2+} + \text{H}_2\text{O}_2 \tag{47}$$

$$\text{Cu(phen)}_2^+ + \text{O}_2 \rightleftharpoons \text{Cu(phen)}_2\text{O}_2^+ \tag{48}$$

$$\text{Cu(phen)}_2\text{O}_2^+ + \text{Cu(phen)}_2^+ + 2\text{H}^+ \longrightarrow 2\text{Cu(phen)}_2^{2+} + \text{H}_2\text{O}_2 \tag{49}$$

The rate law in Eq. 50 was obtained by applying the steady-state approximation to the concentrations of $O_2^{\bullet-}$ and $Cu(phen)_2O_2^+$, and assuming that $k_{48} > k_{49}[Cu(phen)_2^+]$ and $k_{46} > k_{47}$. Reactions 46 and 47 provide an efficient pathway for the $Cu(phen)_2^{2+}$-catalyzed dismutation of superoxide.

$$-d[\text{Cu(phen)}_2^+/dt = 2\left(\frac{k_{46}k_{47}}{k_{-46}[\text{Cu(phen)}_2^{2+}]} + \frac{k_{48}k_{49}}{k_{-48}}\right)[\text{Cu(phen)}_2^+]^2[\text{O}_2] \tag{50}$$

Major advances in the synthesis of biomimetic copper complexes have been recorded in recent years (3, 9, 152–155). Probably the most celebrated case is the preparation of a peroxodicopper(II) complex, $[Cu[HB(3,5\text{-}i\text{-}Pr_2pz)_3]_2(O_2)$ (153), which has a planar μ_2-$\eta^2\eta^2$ structure (Fig. 2). This is the first synthetic compound whose magnetic and spectral properties match those of oxy-hemocyanin, and thus it has been proposed that the latter also has the unusual μ_2-$\eta^2\eta^2$ structure. Recently, this has been confirmed for oxyhemocyanin from *Limulus polyphemus* (156).

A complete kinetic and thermodynamic description is available for only a handful of biomimetic copper–dioxygen systems. A dinuclear Cu^I complex $L^8Cu_2^{2+}$, containing two bridged tripodal tridentate ligand units, produces a μ-peroxo dicopper species, $L^8Cu_2(O_2)^{2+}$ as the only observable intermediate (157), Fig. 12. Spectral and reactivity data suggest a bent μ_2-η^2,η^2 side-on structure for this peroxo complex. Subsequent electrophilic attack of the bound peroxide on the aromatic ring yields the final hydroxylated product. No experimental evidence was obtained for the intermediacy of the monodentate superoxo species. The kinetic data for the parent complex (R = H) and several ring-substituted derivatives (157) are summarized in Table VI.

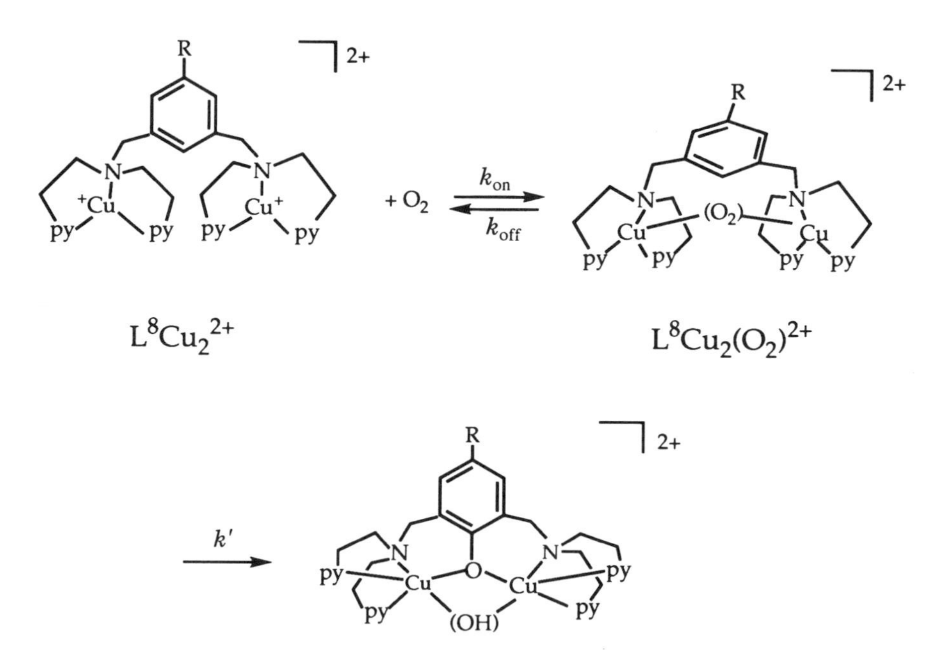

Figure 12. Reaction of a copper monooxygenase model complex with O_2. [Adapted with permission from R. W. Cruse, S. Kaderli, K. D. Karlin, and A. D. Zuberbühler, *J. Am. Chem. Soc.*, *110*, 6882 (1988). Copyright © 1988 American Chemical Society.]

TABLE VI
Kinetic (298 K) and Thermodynamic Data for the Reaction of $L^8Cu_2^{2+}$ with O_2 in CH_2Cl_2[a]

Parameter (Units)	R = H	*t*-Bu	F	NO_2
k_{on} ($M^{-1}s^{-1}$)	5.1×10^3	8.1×10^3	2.2×10^4	890
k_{off} (s^{-1})	1.3×10^3	1.2×10^4	4×10^3	99
k' (s^{-1})	172	360	50	31
K_{eq} (M^{-1}) = k_{on}/k_{off}	3.9	0.7	5.8	9.0
$\Delta H^{\ddagger}_{on}$ (kJ mol^{-1})	8.2	9.1	29	6.4
$\Delta S^{\ddagger}_{on}$ (J mol^{-1} K^{-1})	−146	−140	−66	−167
$\Delta H^{\ddagger}_{off}$ (kJ mol^{-1})	70	83	81	59
$\Delta S^{\ddagger}_{off}$ (J mol^{-1} K^{-1})	50	110	90	−8

[a]Reproduced in part, with permission, from K. D. Karlin, M. S. Nasir, B. I. Cohen, R. W. Cruse, S. Kaderli, and A. D. Zuberbühler, *J. Am. Chem. Soc.*, *116*, 1329 (1994).

A ligand modification (157) that decreases the Cu—Cu distance results in an increase in both the rate and equilibrium constants for O_2 binding, bringing these parameters closer to those for hemocyanin. The geometric constraint imposed by the protein thus seems to play an important role in the natural system.

The reaction of a macrocyclic dicopper complex with O_2 also results in ligand hydroxylation, Fig. 13. The oxygen inserted into the ligand is derived from O_2, as shown by $^{18}O_2$ labeling experiments (158). The likely intermediate, dicopper peroxo complex, was not observed for this macrocyclic tyrosinase model complex.

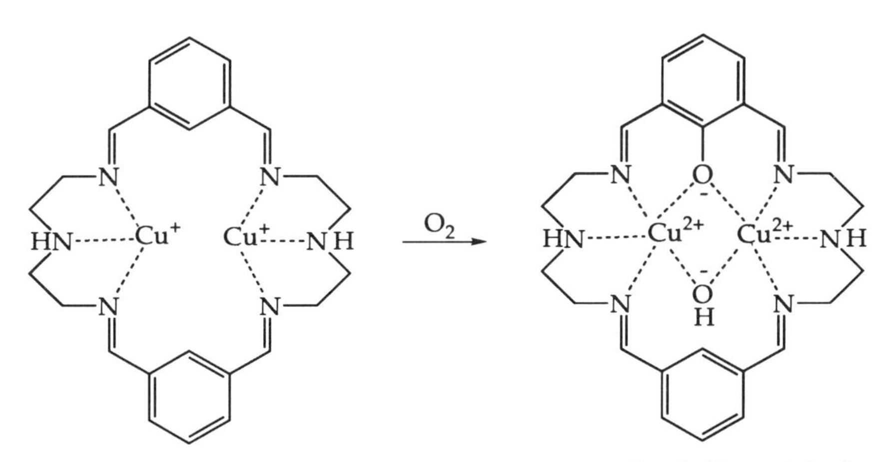

Figure 13. Reaction of a macrocyclic copper complex with O_2. [Adapted with permission from R. Menif, A. E. Martell, P. J. Squattrito, and A. Clearfield, *Inorg. Chem.*, *29*, 4723 (1990). Copyright © 1990 American Chemical Society.]

The complex $L(MeCN)Cu^+$ ($L = (1,4,7\text{-}i\text{-Pr})_3$-tacn) reacts with O_2 at $-78°C$ to give a μ-$\eta^2\eta^2$ peroxo complex $(LCu)_2O_2^{2+}$, which decomposes to a hydroxo-bridged compound $(LCu)_2(OH)_2^{2+}$ at greater than $-70°C$ (159). Isotopic labeling has shown that the hydroxyl hydrogen atoms are derived from the macrocycle isopropyl groups. The conversion of $(LCu)_2O_2^{2+}$ to $(LCu)_2(OH)_2^{2+}$ takes place with first-order kinetics and exhibits a large kinetic isotope effect, $k_H/k_D = 18$ ($\Delta H^\ddagger_H = 56.5$ kJ mol^{-1}, $\Delta S^\ddagger_H = -50$ J mol^{-1} K^{-1}, $\Delta H^\ddagger_D = 58.6$ kJ mol^{-1}, $\Delta S^\ddagger_D = -67$ J mol^{-1} K^{-1}). The rate-determining step in the decomposition of the peroxo species thus involves cleavage of a C—H(D) bond of the isopropyl groups. This appears to be the first case where an aliphatic C—H bond has been activated in a synthetic copper–peroxo complex. In the proposed mechanism, two intramolecular hydrogen-atom abstractions take place simultaneously with the peroxo O—O bond cleavage. The trapping of the resulting alkyl radicals by solvent yields the observed products. This mechanism differs from that proposed earlier in a related reaction of $[(HB(3\text{-}i\text{-Pr-5-Mepz})_3)Co^{II}]_2(O_2)$ (18b) in that tunnelling does not appear to be a factor in the reaction of the copper complex.

Reversible formation of 1:1 Cu—O_2 adducts in propionitrile was reported recently (160) for the mononuclear Cu^I complexes of ligands L^9, L^{10}, and L^{11}, Fig. 14. Equations 1 and 2 ($M = LCu^+$) fully describe the overall kinetic

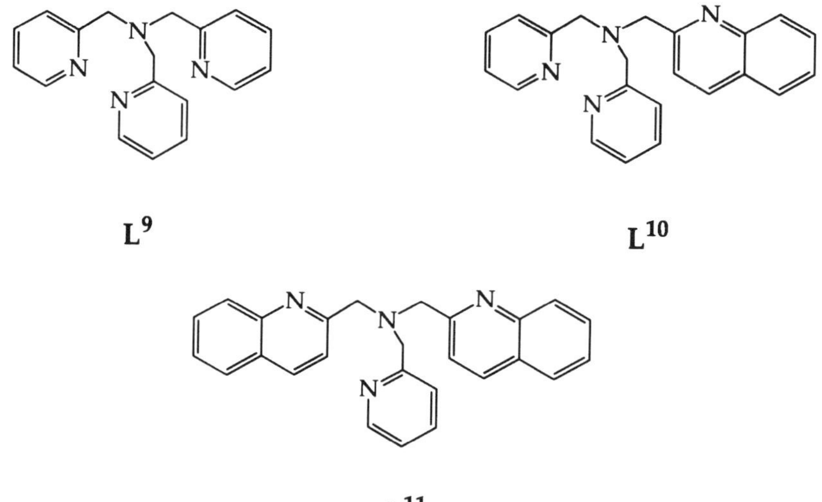

Figure 14. Structures of ligands L^9, L^{10}, and L^{11} that form mononuclear complexes with Cu^I (160).

behavior, although only in the case of L^8Cu^+ has it been possible to obtain all four individual rate constants. The values at 25°C are $k_{on} = 8 \times 10^7\ M^{-1}\ s^{-1}$, $k_{off} = 2 \times 10^8\ s^{-1}$, $k_{2,on} = 1.8 \times 10^6\ M^{-1}\ s^{-1}$, $k_{2,off} = 1.2 \times 10^3\ s^{-1}$. The unfavorable equilibrium constants K_{on} and $K_{2,on}$ for copper complexes at room temperature (for L^9Cu^+, $K_{on} = 0.34\ M^{-1}$, $K_{on}K_{2,on} = 500\ M^{-2}$) have frustrated previous efforts to detect 1:1 $Cu{-}O_2$ adducts. It has now become clear (160) that large negative reaction entropies are responsible (L^9Cu^+, $\Delta S_1^0 = -123$ J $mol^{-1}\ K^{-1}$, $\Delta S_2^0 = -97$ J $mol^{-1}\ K^{-1}$). Both 1:1 and 2:1 copper–dioxygen adducts are stabilized at low temperatures because of the favorable enthalpic terms (L^9Cu^+, $\Delta H_1^0 = -34$ kJ mol^{-1}, $\Delta H_2^0 = -47$ kJ mol^{-1}).

The final product of the reaction of $L^{11}Cu^+$ with O_2 is a 1:1 adduct, although a 2:1 complex was observed as a transient in the early stages of an experiment that consisted of bubbling O_2 through a solution of $L^{11}Cu^+$. The equilibrium constants are apparently such that $(L^{11}Cu)_2O_2^{2+}$ is produced at low $[O_2]/[L^{11}Cu^+]$ ratios, but reverts to $L^{11}CuO_2^+$ as this ratio increases. These results (161) illustrate the role of steric effects in determining the overall stoichiometry of O_2 binding to copper.

Dimeric and tetrameric (halo)(pyridine)copper(I) complexes react with O_2 (162, 163) in nonaqueous solvents to yield tetrameric copper(II) oxo products according to the stoichiometry of Eqs. 51 and 52. Tetrameric complexes are oxidized with a second-order rate law, $k_{tetra}[Cu_4L_4X_4][O_2]$ (L = py), consistent with the insertion of O_2 into the halo core of the reactant. The reaction of dimeric complexes obeys a mixed third-order rate law, $k_{di}[Cu_2L_4X_2]^2[O_2]$, which suggests that minor concentrations of the intermediate $Cu_2L_4X_2O_2$ (not observed) are reduced by $Cu_2L_4X_2$ in the rate-determining step. The distribution of the oxo groups in the product is ligand dependent, as shown in Fig. 15.

$$2Cu_2L_4X_2 + O_2 \longrightarrow Cu_4L_4X_4O_2 + 4L \tag{51}$$

$$Cu_4L_4X_4 + O_2 \longrightarrow Cu_4L_4X_4O_2 \tag{52}$$

Dimeric, colorless $Cu_2(R_4\text{-diamine})_2X_2$ (R = Me, Et, Pr, or amyl) complexes yield dark greenish-brown oxo dimers (164) in a reaction that again obeys a mixed third-order rate law, Eqs. 53 and 54. Reaction rates are strongly affected by changes in diamine, X^-, and solvent. The reactivity order for X^- is $k_{Cl} > k_{Br}$, and for R, $k_{Me} > k_{Et} > k_{Pr}$. The latter is a result of steric restrictions, caused by interaction between the alkyl substituents of the two dimers in the activated complex for oxidation.

$$2Cu_2(\text{diamine})_2X_2 + O_2 \longrightarrow 2Cu_2(\text{diamine})_2X_2O \tag{53}$$

$$d[Cu_2(\text{diamine})_2X_2O]/dt = k[Cu_2(\text{diamine})_2X_2]^2[O_2] \tag{54}$$

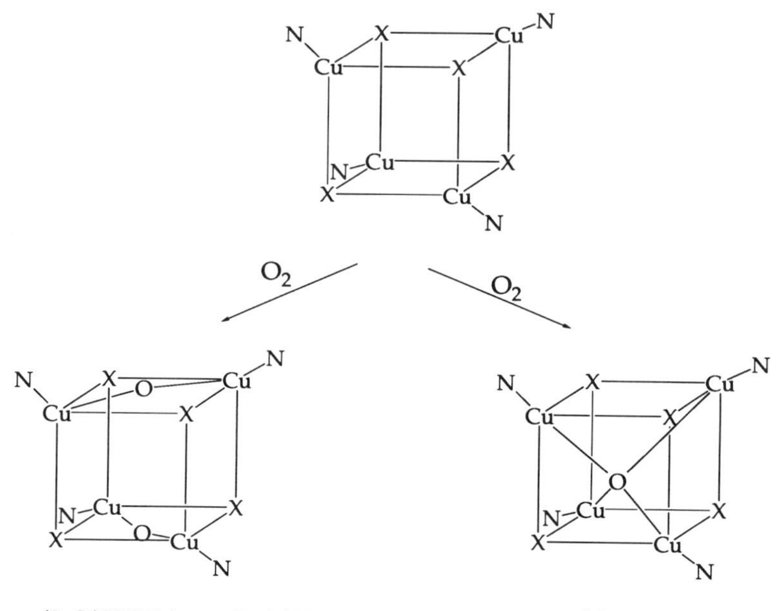

Figure 15. Reaction of halo(pyridine)copper complexes with O_2. [Reproduced with permission from G. Davies and M. A. El-Sayed, *Inorg. Chem.*, *22*, 1257 (1983). Copyright © 1983 American Chemical Society.]

E. Manganese

Manganese is present in superoxide dismutase, in several redox enzymes, and in the oxygen-evolving complex of photosystem II. In the laboratory, manganese complexes have proved to be good epoxidation catalysts. As a result, there is a keen interest in manganese–oxygen chemistry, which has been reviewed extensively in the past few years (165–169). Therefore, only a brief summary of the mechanistic features of $Mn{-}O_2$ chemistry is given here.

The interaction of O_2 with mononuclear manganese(II) complexes typically yields η^2-peroxomanganese(IV) species as initial products (166, 170–172). This appears to be the case with Mn(Pc) as well (173), despite the earlier belief that an end-on $PcMn^{III}(O_2^-)$ is formed (174). Reversible release of O_2 from $LMn^{IV}(O_2^{2-})$ complexes is rare, and has been observed only in a handful of cases (27, 169–171). The chemistry of $LMn^{IV}(O_2^{2-})$ is a function of ligands, solvent, and reaction conditions, and yields a variety of binuclear and polynuclear products containing manganese in oxidation states 2+, 3+, and 4+ (175). Thus, unlike the cases of cobalt and iron, there has been no general mechanism proposed to date for O_2 activation by manganese (166).

Some Mn^{II}–Schiff base complexes act as electrocatalysts, but lose their activity after a limited number of cycles. This has been explained by the formation of an inactive dimer, η^2,η^2-$[LMn^{IV}(\mu\text{-}O)]_2$, according to Eqs. 55 and 56 (176). Studies utilizing cyclic voltammetry have established first-order dependence on LMn^{II} for the dimer formation, suggesting that Reaction 55 is rate determining. The success in isolating solid $LMn^{IV}(O)_2Mn^{IV}L$ complexes for Schiff bases, but not for porphyrins, was attributed to the conformational mobility of the Schiff base ligands (166).

$$LMn^{II} + O_2 \longrightarrow LMn^{IV}\langle O_2 \rangle \quad (55)$$

$$LMn^{IV}\langle O_2 \rangle + LMn^{II} \longrightarrow LMn^{IV}(\mu\text{-}O)_2Mn^{IV}L \quad (56)$$

An interesting aspect of the chemistry of $LMn^{IV}(O)_2Mn^{IV}L$ is its demonstrated ability to participate in intermetal oxygen-atom transfer reactions. Both complete and incomplete oxygen transfers have been observed (Eqs. 57 and 58), where L and L′ are Schiff base ligands of the salprn and salen type (166, 176).

$$LMn^{IV}(O)_2Mn^{IV}L + 4L'Fe^{II} \longrightarrow 2LMn^{II} + 2L'Fe^{III}OFe^{III}L' \quad (57)$$

$$LMn^{IV}(O)_2Mn^{IV}L + 2LMn^{II} \longrightarrow [LMn^{III}OMn^{III}]_2 \quad (58)$$

An unusual, pentacoordinated Mn^{III} dimer, $Li_2[Mn(PHAB)]_2$, Figure 16, reacts with O_2 in THF in the presence of agents that complex lithium ions

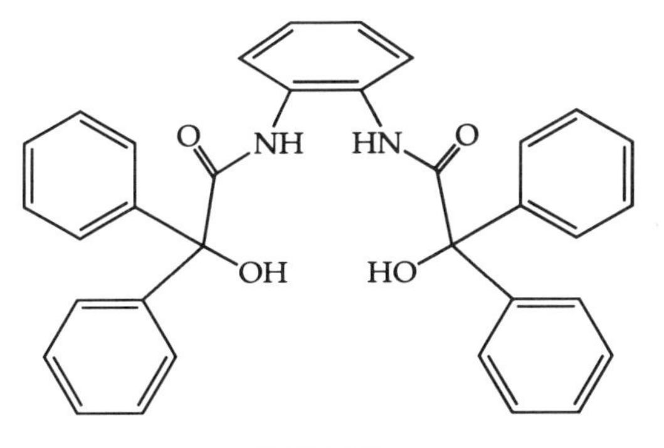

Figure 16. Structure of the ligand H_4PHAB.

(177a). The reaction shows signs of autocatalysis and yields 2-hydroxy-thf and the square pyramidal $Mn^{V}(O)(PHAB)^{-}$, which is structurally related to other known oxomanganese(V) complexes (177b–d). The kinetic behavior was rationalized by proposing the formation of thf-hydroperoxide as an intermediate that reacts rapidly with the starting material.

A side-on peroxo manganese(III) complex, (3,5-*i*-Pr_2pzH)[HB(3,5-*i*-Pr_2pz$)_3$]MnO_2, exhibits thermochromism (178). The color changes reversibly from brown at −20°C to dark blue at −78°C. The IR and structural data for two isolated species provide strong evidence that the thermochromism is caused by the formation of a hydrogen bond between the pyrazole proton and the peroxide at low temperatures, and the loss of hydrogen bonding at higher temperatures.

F. Other Transition Metal Complexes

1. *Titanium*

Dioxygen does not react with $Ti(H_2O)_6^{3+}$, but oxidizes $(H_2O)_5TiOH^{2+}$ to $(H_2O)_5TiO^{2+}$, $k = 4.25\ M^{-1}\ s^{-1}$. The reaction proceeds by an outer-sphere mechanism and does not yield $(H_2O)_5TiO_2^{3+}$ (179). The protonated superoxotitanium(IV) ion, $(H_2O)_4TiO(HO_2)^{2+}$, is produced however, by Ce^{IV} oxidation of a stable peroxotitanium(IV) complex, $(H_2O)_5Ti(O_2)^{2+}$ (179, 180).

$$(H_2O)_5Ti(O_2)^{2+} + Ce^{IV} \longrightarrow (H_2O)_4TiO(HO_2)^{2+} + Ce^{III} + H^+ \quad (59)$$

The decomposition of $(H_2O)_4TiO(HO_2)^{2+}$ regenerates $(H_2O)_5Ti(O_2)^{2+}$ according to the rate law of Eqs. 60 and 61 (179). The general mechanism of Eqs. 1–3 cannot account for the inhibition by $(H_2O)_5TiO^{2+}$. It was proposed that the process begins with the reversible heterolytic dissociation of $(H_2O)_4TiO(HO_2)^{2+}$ to $HO_2^{\bullet}$ and $(H_2O)_5TiO^{2+}$, Eq. 62, followed by rapid reduction of the second mole of $(H_2O)_4TiO(HO_2)^{2+}$ by $HO_2^{\bullet}$, Eq. 63. The latter reaction is a special case of hydroperoxyl disproportionation, with one of the partners coordinated to Ti^{IV}.

$$-d[(H_2O)_4TiO(HO_2)^{2+}]/dt = 2k_{obs}[(H_2O)_4TiO(HO_2)^{2+}] \quad (60)$$

$$k_{obs} = \frac{k_{62}}{1 + \dfrac{k_{-62}}{k_{63}} \dfrac{[(H_2O)_5TiO^{2+}]}{[(H_2O)_4TiO(HO_2)^{2+}]}} \quad (61)$$

$$(H_2O)_4TiO(HO_2)^{2+} + H_2O \rightleftharpoons (H_2O)_5TiO^{2+} + HO_2^{\bullet} \quad (62)$$

$$(H_2O)_4TiO(HO_2)^{2+} + HO_2^{\bullet} \longrightarrow (H_2O)_5Ti(O_2)^{2+} + O_2 \quad (63)$$

At 0.6 M H^+ ($\mu = 1\ M$), the kinetic parameters have values $k_{62} = 0.0985\ s^{-1}$ and $k_{-62}/k_{63} = 3.21 \times 10^{-3}$. The dissociation of $HO_2^{\bullet}$ from

$(H_2O)_4TiO(HO_2)^{2+}$ occurs in two parallel reactions, one of which is H^+ catalyzed, $k_{62} = k_{H_2O} + k_H[H^+]$. The activation parameters associated with the two terms have values $\Delta H^{\ddagger}_{H_2O} = 62$ kJ mol^{-1}, $\Delta S^{\ddagger}_{H_2O} = -68$ J K^{-1} mol^{-1}, $\Delta H^{\ddagger}_{H} = 81$ kJ mol^{-1}, $\Delta S^{\ddagger}_{H} = 7.6$ J K^{-1} mol^{-1}. The ionic strength dependence of k_H is consistent with a 2+ charge on the molecule, providing some evidence that the coordinated superoxide is indeed protonated in strongly acidic solutions (179).

Perhydroxytitanyl ions, $(H_2O)_4TiO(HO_2)^{2+}$, can be oxidized further by Ce^{IV} to give $(H_2O)_5TiO^{2+}$ and O_2, $k \sim 400\ M^{-1}\ s^{-1}$ in 1 M $HClO_4$ (180).

2. *Vanadium*

The kinetically complex autoxidation of $V(H_2O)_6^{2+}$ yields $V(H_2O)_6^{3+}$ and $(H_2O)_5VO^{2+}$ in different ratios depending on the initial concentration of $V(H_2O)_6^{2+}$ (181). At $[V(H_2O)_6^{2+}] < 5$ mM, 2 mol of $V(H_2O)_6^{3+}$ and 1 mol of $(H_2O)_5VO^{2+}$ are produced per mole of O_2. As $[V(H_2O)_6^{2+}]$ increases, the yield of $(H_2O)_5VO^{2+}$ increases at the expense of $V(H_2O)_6^{3+}$, until in the theoretical limit the reaction yields 2 mol of $(H_2O)_5VO^{2+}$ and no $V(H_2O)_6^{3+}$. Because the formation of $(H_2O)_5VO^{2+}$ bypasses $V(H_2O)_6^{3+}$, an outer-sphere reaction between O_2 and $V(H_2O)_6^{2+}$ is ruled out. A substitution-controlled inner-sphere reaction must also be discounted, because the rate constant for the first step ($k = 2.0 \times 10^3\ M^{-1}\ s^{-1}$) exceeds the substitution rate at $V(H_2O)_6^{2+}$ (182, 183). The kinetics and products are consistent with the initial formation of a seven-coordinated superoxo complex $(H_2O)_6VO_2^{2+}$, which either decomposes to $(H_2O)_5VO^{2+}$ and H_2O_2 [yielding 2 mol of $V(H_2O)_6^{3+}$], or reacts reversibly with $V(H_2O)_6^{2+}$ to give $(H_2O)_5VOOV(H_2O)_5^{4+}$, whose homolytic cleavage yields 2 mol of $(H_2O)_5VO^{2+}$, Scheme 4. The seven-coordinated $(H_2O)_6VO_2^{2+}$ is analogous to the intermediates proposed in the reaction of $V(H_2O)_6^{2+}$ with halogens (184).

$$V(H_2O)_6^{2+} + O_2 \xrightarrow{2 \times 10^3\ M^{-1}\ s^{-1}} (H_2O)_6VO_2^{2+}$$

$$(H_2O)_6VO_2^{2+} + V(H_2O)_6^{2+} \underset{20\ s^{-1}}{\overset{3.7 \times 10^3 M^{-1} s^{-1}}{\rightleftarrows}} (H_2O)_5VOOV(H_2O)_5^{4+} + 2H_2O$$

$$(H_2O)_6VO_2^{2+} \overset{\sim 100\ s^{-1}}{\rightleftarrows} (H_2O)_5VO^{2+} + H_2O_2$$

$$2V(H_2O)_6^{2+} + H_2O_2 + 2H^+ \xrightarrow{17.2\ M^{-1}\ s^{-1}} 2V(H_2O)_6^{3+} + 2H_2O$$

$$(H_2O)_5VOOV(H_2O)_5^{4+} \xrightarrow{35\ s^{-1}} 2(H_2O)_5VO^{2+}$$

Scheme 4

Both the $[H^+]$ independent ($k = 140\ M^{-1}\ s^{-1}$) and $[H^+]$ assisted ($k = 2.08 \times 10^8\ M^{-2}\ s^{-1}$) pathways for the autoxidation of a binuclear, oxo-bridged vanadium(III) complex, $V_2O(ttha)^{2-}$, produce the mixed-valence superoxo intermediate $[V^{III}V^{IV}{-}O_2^-]$, Eq. 64. Catalysis by H^+ arises from ligand protonation that opens up a coordination site for O_2 in $V_2O(ttha)^{2-}$. In the presence of O_2 the intermediate is converted to the final product, $[V^{IV}V^{IV}]$ (185). Neither $O_2^{\bullet -}$ nor H_2O_2 could be detected at any stage of the reaction by enzyme or chemical tests.

$$[V^{III}V^{III}] + O_2 \longrightarrow [V^{III}V^{IV}{-}O_2^-] \tag{64}$$

When solutions of the superoxo complex were quickly purged with argon, up to 50% of the starting $[V^{III}V^{III}]$ complex was regenerated. It was proposed that dimeric V^{III} rapidly reduces the superoxo complex $[V^{III}V^{IV}{-}O_2^-]$ to the mixed-valence state, $[V^{III}V^{IV}]$, which either reacts with O_2 or undergoes comproportionation in the absence of O_2, Eqs. 65–67. The kinetics of this last step were studied by use of independently prepared samples of the mixed-valence compound and yielded $k_{comp}/M^{-1}\ s^{-1} = 0.0508 + 6.35 \times 10^5\ [H^+]$. Again the source of $[H^+]$ catalysis is the protonation of either a carboxylate or an amine trans to the bridging oxo group. Both would facilitate the structural change accompanying the oxidation of vanadium(III) to vanadyl(IV). Equation 67 explains the observed approximate 50% recovery of $[V^{III}V^{III}]$ upon removal of excess O_2.

$$[V^{III}V^{IV}{-}O_2^-] + 3[V^{III}V^{III}] \xrightarrow{\text{fast}} 4[V^{III}V^{IV}] + 2H_2O \tag{65}$$

$$[V^{III}V^{IV}] \xrightarrow{O_2} [V^{IV}V^{IV}] \tag{66}$$

$$[V^{III}V^{IV}] \underset{k_{disp}}{\overset{k_{comp}}{\rightleftharpoons}} [V^{III}V^{III}] + [V^{IV}V^{IV}] \qquad K = 6.7 \tag{67}$$

An alternate route to the mixed-valence species may involve the dismutation of the superoxo complex via a peroxy intermediate.

$$2[V^{III}V^{IV}{-}O_2^-] \longrightarrow [V^{III}V^{IV}] + [V^{III}V^{IV}{-}O_2^{2-}] + O_2 \tag{68}$$

$$2[V^{III}V^{IV}{-}O_2^{2-}] \xrightarrow{\text{fast}} 2[V^{III}V^{IV}] + O_2 + 2H_2O \tag{69}$$

The reaction of the binuclear V^{III} complex with O_2 is several orders of magnitude faster than the corresponding reaction of the open-chain form and that of the closely related, but monomeric, $V(Hedta)H_2O$ (185). This result suggests a superoxo-bridged structure for $[V^{III}V^{IV}{-}O_2^-]$, a form that is not readily available to the other two complexes.

3. Molybdenum

The superoxomolybdenum(IV) ion, $(H_2O)_5MoO_2^{3+}$, forms reversibly in aqueous HPTS ($k_{on} = 180\ M^{-1}\ s^{-1}$, $k_{off} = 0.47\ s^{-1}$) as a short-lived intermediate in the autoxidation of $Mo(H_2O)_6^{3+}$ (186). In the presence of excess $Mo(H_2O)_6^{3+}$, the rest of the reaction scheme parallels that given for a general case in Eqs. 2 and 3, $k_{2,on} = 42\ M^{-1}\ s^{-1}$, $k_{2,off} = 0.08\ s^{-1}$. The intensely yellow $(H_2O)_5MoOOMo(H_2O)_5^{6+}$ decomposes to the final product, $Mo_2O_4(H_2O)_6^{2+}$ (186, 187) in a reaction having the rate proportional to $[H^+]^{-1}$ (k = $0.23\ s^{-1}$ in 2.0 M HPTS). At least some oxygen in $Mo_2O_4(H_2O)_6^{2+}$ is derived from molecular oxygen, as demonstrated by use of $^{18}O_2$.

Experiments using excess O_2 showed no evidence for $(H_2O)_5MoOOMo(H_2O)_5^{6+}$. Under these conditions the rate law of Eq. 70 applies, where k_{dec} $(0.037 + 0.012\ [H^+]^{-1})$ represents a first-order pathway for the decomposition of $(H_2O)_5MoO_2^{3+}$.

$$d[Mo_2O_4(H_2O)_6^{2+}]/dt = (k_{on}/k_{off})k_{dec}[(Mo(H_2O)_6^{3+}][O_2] \tag{70}$$

4. Rhodium

The autoxidation of dinuclear aquarhodium(II) ions (188–191) yields different products depending on the concentration of O_2. Slow diffusion of O_2 into solutions of $(aq)Rh_2^{4+}$ produces yellow dimeric or polymeric Rh^{III} cations. The violet dimer of the proposed formula $(H_2O)_nRh(O_2)(OH)_2Rh(H_2O)_n^{3+}$ is obtained in strongly acidic solutions of $(aq)Rh_2^{4+}$ by vigorous bubbling of O_2. This μ-superoxo complex was characterized spectroscopically, chemically, and electrochemically, but no kinetic data for the oxygenation reaction are available. It has been shown recently (192–194) that chemical oxidation of Rh^{III} in water also generates the dimeric superoxo complex, and not Rh complexes in oxidation states 4–6, as previously thought.

Ultraviolet photolysis of *cis*- and *trans*-$Rh^{III}(en)_2(NO_2)_2^+$ in the presence of O_2 and Cl^- yields the superoxo complexes $Rh(en)_2(Cl)(O_2)^+$, Eqs. 71 and 72, which have been characterized by ESR, Raman, and vis spectroscopies (195). The reaction also produces superoxo bridged dimeric Rh^{III} products. Both monomeric and dimeric species act as one-electron oxidants toward I^- and Fe^{2+}.

$$Rh(en)_2(NO_2)_2^+ \xrightarrow{h\nu} Rh(en)_2^{2+} + NO_2^- + NO_2^{\bullet} \tag{71}$$

$$Rh(en)_2^{2+} + O_2 \xrightarrow{Cl^-} Rh(en)_2(Cl)(O_2)^+ \tag{72}$$

Another mononuclear rhodium(II) complex, $Rh(NH_3)_4(H_2O)_n^{2+}$, was generated by the reduction of Rh^{III} complexes with hydrated electron (196). A pulse radiolytic study of the reaction of $Rh(NH_3)_4(H_2O)_n^{2+}$ with O_2 yielded the rate constant $k_{on} = 3.1 \times 10^8\ M^{-1}\ s^{-1}$ for the formation of (presumably trans) $(NH_3)_4(H_2O)RhO_2^{2+}$.

III. REACTIVITY

A. Superoxo, Hydroperoxo, and Peroxo Complexes

1. Superoxo Complexes

In reactions with various substrates, superoxometal complexes typically behave as oxidants. Inner- and outer-sphere electron transfers are common, and hydrogen-atom abstraction, demonstrating the free radical nature of bound O_2, was proposed in a number of reactions (197–199). In some of these cases, however, alternative mechanistic assignments have been advanced. For example, the oxidation of phenols (ArOH) by superoxocobalt(III) complexes was proposed to take place by an initial S_N2 displacement (200) of peroxide by phenol, Eqs. 73a or b, rather than by hydrogen-atom abstraction of Eq. 74 (25, 199, 201).

$$LCoO_2 + ArOH \longrightarrow LCoOAr + HO_2^{\bullet} \tag{73a}$$

$$LCoOOCoL + ArOH \longrightarrow LCoOAr + LCoO_2H \tag{73b}$$

$$LCoO_2 + ArOH \longrightarrow LCoO_2H + ArO^{\bullet} \tag{74}$$

The superoxo complexes $(H_2O)_5CrO_2^{2+}$ (202, 203) and $(H_2O)([14]\text{-aneN}_4)CoO_2^{2+}$ (204, 205) react rapidly with outer-sphere reductants, such as $Ru(NH_3)_6^{2+}$, $V(H_2O)_6^{2+}$, and $Co(sep)^{2+}$, as shown in Eqs. 75 and 76, and summarized in Table VII.

$$(H_2O)([14]aneN_4)CoO_2^{2+} \xrightarrow{Ru(NH_3)_6^{2+},\ H^+} (H_2O)([14]aneN_4)CoO_2H^{2+} \tag{75}$$

$$(H_2O)_5CrO_2^{2+} \xrightarrow{Ru(NH_3)_6^{2+},\ H^+} (H_2O)_5CrO_2H^{2+} \tag{76}$$

The identification of the products as end-on hydroperoxides is based in part on the fact that at least one of them, $(H_2O)_5CrO_2H^{2+}$, can be reconverted to the end-on superoxide by oxidation with Ce^{IV}. This reaction confirms that no loss of peroxide from the metal has occurred during or after the reduction by $Ru(NH_3)_6^{2+}$.

$$(H_2O)_5CrO_2H^{2+} \xrightarrow{Ce^{IV},\ -H^+} (H_2O)_5CrO_2^{2+} \tag{77}$$

TABLE VII
Kinetic Data (M^{-1} s^{-1}) for the Reactions of Superoxo Complexes of Chromium and Cobalt with One-Electron Reductants[a,b]

Reductant	$(H_2O)_5CrO_2^{2+}$	$(H_2O)([14]aneN_4)CoO_2^{2+}$	Mechanism[c]
$Cr(H_2O)_6^{2+}$	8×10^8		is
$CuCl_3^{2-}$		1.1×10^6 [d]	is
$(H_2O)_2Co([14]aneN_4)^{2+}$	7×10^6	4.9×10^5	is
$Cu(C_2H_4)(H_2O)_3^+$		2.4×10^5	is
$(H_2O)_2Co([15]aneN_4)^{2+}$	6.2×10^5	3.7×10^4	is
$Fe(H_2O)_6^{2+}$	4.5×10^3	1.1×10^3	is
$Cu(trpy)(H_2O)^+$		$\sim 5 \times 10^5$ [e]	os
$Ru(NH_3)_6^{2+}$	9.2×10^5	2.3×10^5	os
$Co(sep)^{2+}$	8.5×10^5	10^6	os
$V(H_2O)_6^{2+}$	2.3×10^5	1.8×10^5	os
$ABTS^{2-}$	1.36×10^3		os
$N_2H_5^+$	58.1[f]		os

[a] Data from (129) and (202–205).
[b] Acidic aqueous solutions, $\mu = 0.1\ M$.
[c] Inner-sphere = is, outer-sphere = os.
[d] $\mu = 0.4\ M$.
[e] $\mu = 0.01\ M$.
[f] Units M^{-2} s^{-1} (rate = $+k[N_2H_5^+][H^+][(H_2O)_5CrO_2^{2+}]$).

Inner-sphere reductants eventually yield stable Co^{III} and Cr^{III} products via observable, but short-lived intermediates, believed to be μ-peroxo complexes.

$$(H_2O)_5CrO_2^{2+} + M^m \longrightarrow (H_2O)_5CrOOM^{2+m} \longrightarrow \longrightarrow \longrightarrow Cr^{III} \quad (78a)$$

$$L^1(H_2O)CoO_2^{2+} + M^m \longrightarrow L^1(H_2O)CoOOM^{2+m} \longrightarrow \text{products} \quad (78b)$$

This proposal is supported by the fact that a number of kinetically stable μ-peroxides, including $[L^1(H_2O)Co]_2O_2^{4+}$, are known independently (206).

The oxidation of an organochromium complex, $(H_2O)_5CrCH_2OH^{2+}$, by $(H_2O)_5CrO_2^{2+}$ is rapid, and provides a pathway for the catalyzed autoxidation of $(H_2O)_5CrCH_2OH^{2+}$ (207). The rate law of Eq. 79 ($k_{Cr} = 137\ M^{-1}\ s^{-1}$) and the mechanism of Scheme 5 have been established. The crucial factors respon-

$$(H_2O)_5CrO_2^{2+} + (H_2O)_5CrCH_2OH^{2+} \xrightarrow{k_{Cr},\ H_2O} (H_2O)_5CrO_2H^{2+} + CH_2O + Cr(H_2O)_6^{2+}$$

$$Cr(H_2O)_6^{2+} + O_2 \longrightarrow (H_2O)_5CrO_2^{2+} + H_2O$$

Scheme 5

sible for the catalysis are the formation of $Cr(H_2O)_6^{2+}$ in the first oxidation step, and the presence of chromium in both complexes. In the absence of added $(H_2O)_5CrO_2^{2+}$, the reaction of $(H_2O)_5CrCH_2OH^{2+}$ with O_2 serves as its own source of $(H_2O)_5CrO_2^{2+}$ and shows signs of autocatalysis.

$$-d[(H_2O)_5CrCH_2OH^{2+}]/dt = k_{Cr}[(H_2O)_5CrCH_2OH^{2+}][(H_2O)_5CrO_2^{2+}] \quad (79)$$

A dinuclear superoxorhodium complex, $(aq)Rh^{III}(O_2)(OH)_2Rh^{III}(aq)^{3+}$, undergoes clean one-electron reduction by a number of metallic and nonmetallic reductants to yield the peroxo complex $(aq)Rh^{III}(O_2)(OH)_2Rh^{III}(aq)^{2+}$ (191, 208). These reactions utilize parallel $[H^+]$-independent and $[H^+]^{-1}$-proportional paths, and in two cases, $Ti(H_2O)_6^{3+}$ and $V(H_2O)_6^{3+}$, the rate law contains a term proportional to $[H^+]^{-2}$. Only the reactions with I^- and $IrCl_6^{3-}$ are mildly catalyzed by acid. This acid dependence has been rationalized by invoking the protonation of the oxidant, and a $pK_a < -1$ was calculated for $(aq)Rh^{III}(O_2)(OH)Rh^{III}(aq)^{4+}$ from the kinetic data. All of the other reductants have dissociable protons, causing the prominent $1/[H^+]$ terms to mask the minor term proportional to $[H^+]$. The values of the reported second-order rate constants in 1 *M* $HClO_4$ are summarized in Table VIII.

At longer times the peroxo complex disappears in a process that regenerates up to 67% of the starting superoxide, similar to the disproportionation of an (ammine)rhodium peroxide reported earlier (209).

$$3Rh^{III}(O_2^{2-})(OH)_2Rh^{III} + 4H^+ \longrightarrow 2Rh^{III}(O_2^-)Rh^{III} + 2Rh^{III} \quad (80)$$

TABLE VIII
Kinetic Data for One-Electron Reduction of $(H_2O)_nRh(O_2)(OH)_2Rh(H_2O)_n^{3+}$ in 1 *M* $HClO_4^a$

Reductant	k (M^{-1} s^{-1})
Ascorbic acid	1.50×10^3
I^-	1.07×10^3
Hydroquinone	6.5×10^2
L-Cysteine	0.47
$V(H_2O)_6^{2+}$	1.4×10^6
$(aq)Mo_3^{III}$	1.2×10^6
$IrCl_6^{3-}$	3.01×10^5
$(aq)Eu^{2+}$	2.8×10^5
$Ti(H_2O)_6^{3+}$	3.94×10^2
$Fe(H_2O)_6^{2+}$	3.51×10^2
$V(H_2O)_6^{3+}$	0.162

[a]Calculated from data in (191) and (208).

Interestingly, the kinetics of this complicated reaction are first order, $k_{obs} = 2.1 \times 10^{-3}\ s^{-1}$, independent of $[H^+]$ and of the reductant used to generate the peroxide. The proposed mechanism consits of the homolytic O—O bond cleavage, followed by rapid reactions of the (formally) Rh^{IV} fragments, $Rh^{IV}O^{\bullet}$, to give the observed products.

$$Rh^{III}(O_2^{2-})Rh^{III} \longrightarrow 2Rh^{IV}O^{\bullet} \tag{81}$$

$$Rh^{IV}O^{\bullet} + Rh^{III}(O_2^{2-})Rh^{III} \xrightarrow{\text{fast}} Rh^{III}O^- + Rh^{III}(O_2^-)Rh^{III} \tag{82}$$

$$2Rh^{III}O^- + 2H^+ \longrightarrow Rh(OH)_2Rh \tag{83}$$

It does not appear that the reported (191) rate constant, $2.1 \times 10^{-3}\ s^{-1}$, has been corrected for the stoichiometric factor of 3. Thus the value of k_{81} is probably $7 \times 10^{-4}\ s^{-1}$.

The ESR parameters of electrochemically generated $(AP)_4Rh_2(O_2)^-$ in CH_2Cl_2 are consistent with a superoxodirhodium(II) structure having the O_2^- unit axially coordinated to a single Rh^{II} center (210). The complex reacts with the solvent CH_2Cl_2 ($k = 0.2\ s^{-1}$) to yield CH_2O, Cl^-, and the starting dirhodium complex $(AP)_4Rh_2$.

2. Hydroperoxo Complexes

Reactive hydroperoxo and peroxo complexes figure prominently in mechanisms for a number of O_2-activating systems, such as cytochrome P450, methane monooxygenase, hemerythrin, and others. A detailed discussion of such systems has been given elsewhere (3–5, 113b, 211, 212).

The lifetimes of $(H_2O)_5CrO_2H^{2+}$ (213) and $(H_2O)L^1CoO_2H^{2+}$ (59, 214) are sufficiently long for detailed mechanistic studies of their reactions to be carried out. The reduction of the two hydroperoxides by transition metal ions was observed only in cases where an inner-sphere mechanism is feasible. Thus the need for H_2O_2 to bind to the reductant in order for electron transfer to take place is a feature that perpetuates itself even when peroxide is coordinated to Cr^{III} and Co^{III}. Figures 11 and 17 show the UV–vis spectra of the two hydroperoxides.

The reduction of $(H_2O)_5CrO_2H^{2+}$ by metal complexes (136, 213) takes place with a 2:1 ($[M^n]$: $[(H_2O)_5CrO_2H^{2+}]$) stoichiometry. A Fenton-type mechanism of Eqs. 84 and 85

$$(H_2O)_5CrO_2H^{2+} + M^n \longrightarrow M^{n+1}OH + (H_2O)_5CrO^{2+} \tag{84}$$

$$(H_2O)_5CrO^{2+} + M^n + 2H^+ \longrightarrow Cr(H_2O)_6^{3+} + M^{n+1} \tag{85}$$

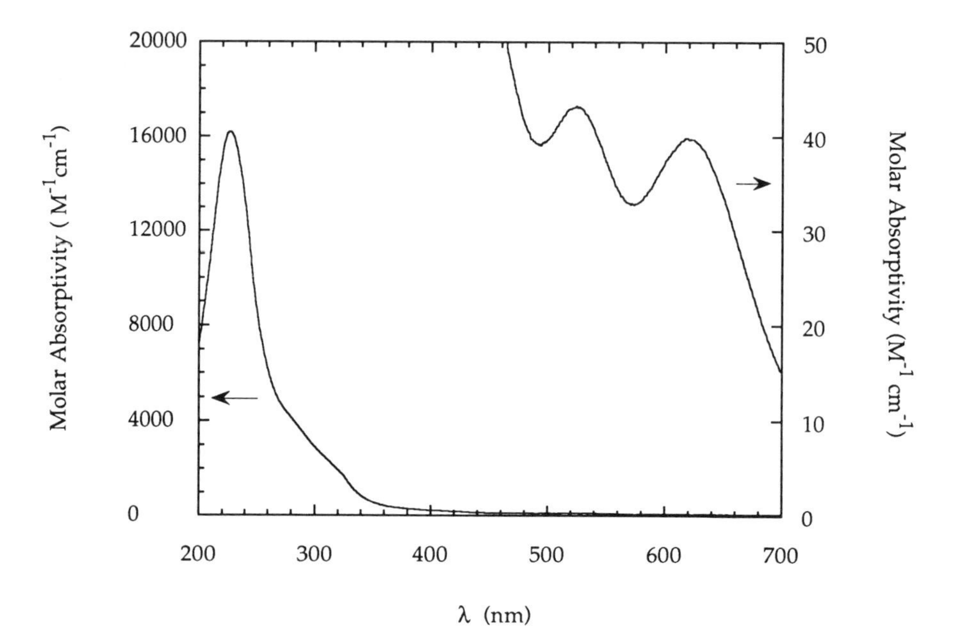

Figure 17. Ultraviolet–vis spectrum of $(H_2O)([14]aneN_4)CoOOH^{2+}$. [Reproduced from W.-D. Wang, A. Bakac, and J. H. Espenson, submitted for publication in *Inorg. Chem.* Unpublished work copyright © 1995 American Chemical Society.]

is supported by both the chemical trapping of oxochromium(IV) ion (formally an analogue of OH radicals) and the linearity of an isokinetic plot of $\Delta H^{\ddagger}$ versus $\Delta S^{\ddagger}$ for the reduction of H_2O_2 and $(H_2O)_5CrO_2H^{2+}$ by a number of metal complexes.

On the other hand, chemical tests have failed to detect M^{IV} oxo species in the reduction of $(H_2O)L^1CoO_2H^{2+}$ by $Fe(H_2O)_6^{2+}$ and by other metal ion reductants (215). These observations are consistent with the large free energy change expected for the formation of $(H_2O)L^1CoO^{2+}$ or $(H_2O)_5FeO^{2+}$. The overall reaction again takes place with a 2:1 stoichiometry and presumably yields dimetallic peroxo intermediates, which are then rapidly reduced by the second mole of reductant, Eqs. 86 and 87 and Table IX. For strongly reducing M^n ions, the scheme is completed by Reaction 88, in which case an overall 3:1 stoichiometry is observed.

$$(H_2O)L^1CoO_2H^{2+} + M^n \longrightarrow [(H_2O)L^1CoO_2(H)M]^{n+2} \quad (86)$$

$$[(H_2O)L^1CoO_2(H)M]^{n+2} + M^n + 3H^+ \longrightarrow L^1Co(H_2O)_2^{3+} + 2M^{n+1} + H_2O \quad (87)$$

$$L^1Co(H_2O)_2^{3+} + M^n \longrightarrow L^1Co(H_2O)_2^{2+} + M^{n+1} \quad (88)$$

TABLE IX
Kinetic Data (M^{-1} s^{-1}) for the Reactions of Metal Hydroperoxides and H_2O_2 with Transition Metal Reductants and Nucleophiles[a]

	Electron-Transfer Reactions[b]		
Reactant	$(H_2O)_5CrO_2H^{2+}$	$(H_2O)L^1CoO_2H^{2+}$	H_2O_2
$Cr(H_2O)_6^{2+}$		8400	70000
$CoL^1(H_2O)_2^{2+}$	1530		3970
Cu_{aq}^+	700		4100
$Fe(H_2O)_6^{2+}$	48.4	4.86	58
$Ti(H_2O)_6^{3+}$	28	1.8[c]	920
$CoL^2(H_2O)_2^{2+}$	24		265[d]
$(H_2O)_5VO^{2+}$	6	1.43	5.8
$V(H_2O)_6^{2+}$	2.5	5.5	17
$Ru(NH_3)_6^{2+}$	No reaction	$\ll 1$	<0.01
	Nucleophilic substitutions		
Br^-	0.063		0.00023
$CoSR^{2+}$, [e]	20.5	4.8	1.36
PPh_3	75	No reaction	3

[a]Data from (136) and (215), where L^1 = [14]aneN_4 and L^2 = Me_6-[14]aneN_4. The medium is 0.1 M aqueous $HClO_4$.
[b]$(H_2O)_5CrO_2H^{2+}$ and H_2O_2 react by Fenton-type mechanisms and $(H_2O)L^1CoO_2H^{2+}$ reacts by formation of dimetallic peroxo intermediates.
[c]39.6°C, $\mu = 0.5$ M.
[d]Non-Fenton mechanism.
[e]the compound is $(en)_2\overline{CoSCH_2CH_2NH_2}^{2+}$.

The similarity in the rate constants for one-electron oxidations by H_2O_2 and $(H_2O)_5CrO_2H^{2+}$ is contrasted sharply by the differences between the two in nucleophilic substitutions reactions with $(en)_2\overline{CoSCH_2CH_2NH_2}^{2+}$, PPh_3, and Br^-. As shown in Table IX, $(H_2O)_5CrO_2H^{2+}$ reacts more rapidly. The kinetics are $[H^+]$-dependent, and the data in Table IX are given at 0.1 M H^+.

Several other mechanisms also have been proposed for the oxidation of phosphines by metal hydroperoxides and dioxygen complexes. For example, isotopic labeling by ^{18}O and ^2H has established that the conversion of Ph_3P to Ph_3PO in Eq. 89 is an intramolecular process (216).

$$(acac)(PPh_3)_2Rh(Cl)(O_2H) \xrightarrow{PPh_3,\ CHCl_3} (acac)(PPh_3)_2Rh(Cl)(OH) + Ph_3PO \tag{89}$$

In another study, the oxidation of $PMePh_2$ and PMe_2Ph to the corresponding oxides by $Pt{-}O_2$ complexes was shown to involve a number of steps and intermediates as shown in Scheme 6 (217). Two of the intermediates, HO_2^- and $Pt(PR_3)_4^{2+}$, were demonstrated to be the active oxidants for PR_3.

$$Pt(PR_3)_2O_2 + 2PR_3 \xrightarrow{R'OH} Pt(PR_3)_4^{2+} + HO_2^- + R'O^-$$

$$HO_2^- + PR_3 \longrightarrow R_3PO + OH^-$$

$$Pt(PR_3)_4^{2+} + R_3P \xrightarrow{OH^-, R'O^-} Pt(PR_3)_4 + R_3PO + R'OH$$

$$Pt(PR_3)_4 + O_2 \longrightarrow Pt(PR_3)_2O_2 + 2PR_3$$

Scheme 6

Iron-bleomycin model complexes $Fe^{II}(PMA)^+$ and $Fe^{III}(PMA)^{2+}$, Fig. 18, react with O_2 and H_2O_2, respectively, to yield the low-spin hydroperoxo complex $(PMA)Fe^{III}O_2H^+$ (218).

$$2Fe(PMA)^+ + O_2 \xrightarrow{MeOH, H^+} (PMA)FeO_2H^+ + Fe(PMA)^{2+} \qquad (90)$$

$$Fe(PMA)^{2+} + H_2O_2 \xrightarrow{DMSO, -H^+} (PMA)FeO_2H^+ \qquad (91)$$

When carried out in the presence of DNA, Reactions 90 and 91 inflict the strand scission with the same sequence specificity as Fe-bleomycins, implicating iron(III) hydroperoxides as active species in the reactions of both Fe-bleo-

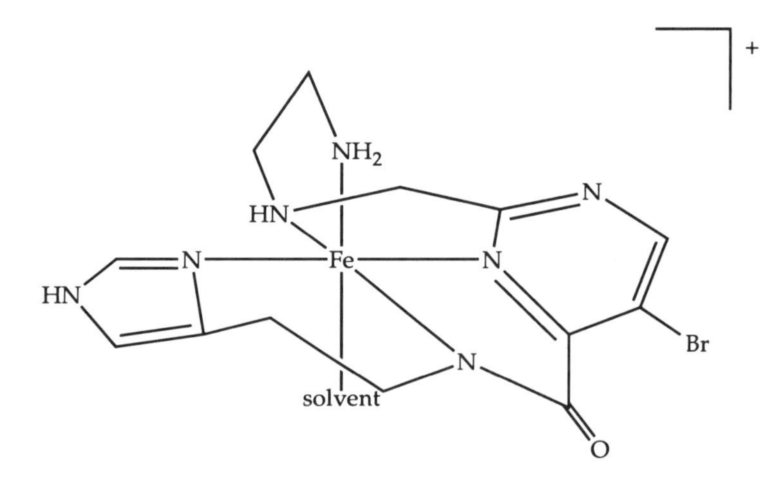

Figure 18. Structure of $Fe^{II}(PMA)^+$.

mycin and the model compounds. The Fe(PMA) complexes are also comparable to Fe-bleomycin in their ability to promote stereospecific oxo transfer to alkenes.

The replacement of H by Me at the secondary amine group of $Fe(PMA)^{2+}$ has no effect on the quantity of the hydroperoxide produced in the equivalent of Eq. 90, or on the subsequent chemistry of the hydroperoxide (219). This finding rules out the involvement of intramolecular hydrogen bonding that was recently proposed (220) to stabilize $(PMA)FeO_2H^+$.

A rare case of O—O bond formation at iron was observed in the reaction of $Fe(PMA)^{2+}$ with PhIO in basic solutions (218).

$$(PMA)Fe^{2+} + PhIO + OH^- \longrightarrow (PMA)FeO_2H^+ + PhI \qquad (92)$$

3. *Dinuclear μ-Peroxo Complexes*

The preparation and chemistry of dinuclear μ-peroxo cobalt complexes (221–225) have been reviewed (52). Most of the work on dicobalt complexes has centered around the oxidation of bound peroxide and the reduction of μ-superoxide by metallic (52, 226, 227) and nonmetallic (227–229) reductants. The reduction of $(NC)_5Co(O_2)Co(NC)_5^{5-}$ by thiols is strongly catalyzed by copper ions (229) according to the simplified scheme of Eqs. 93 and 94, followed by hydrolysis of $(NC)_5Co(O_2)Co(NC)_5^{6-}$, Eqs. 9–11. Copper catalysis has been observed in several other cases involving redox chemistry of metal–oxygen species (87, 228, 230, 231).

$$Cu^{II} + RS^- \longrightarrow Cu^I + 0.5R_2S_2 \qquad (93)$$

$$Cu^I + (NC)_5Co(O_2)Co(NC)_5^{5-} \longrightarrow Cu^{II} + (NC)_5Co(O_2)Co(NC)_5^{6-} \qquad (94)$$

The pivotal role of binuclear μ-peroxo complexes as intermediates in many metal-catalyzed oxygenations has been discussed (10, 52, 200). Recent kinetic results on metal-catalyzed autoxidations of $Co(tim)^{2+}$ show convincingly that the substitutional lability and redox potentials of all the species involved, including intermediates, play a major role in determining the mechanism of such reactions. These cases are especially interesting because a clean four-electron reduction of O_2 to H_2O takes place (232, 233).

The catalysis by $Fe(H_2O)_6^{2+}$ in acidic aqueous solutions is unaffected by acidity and the presence of halide ions, $k_{Fe} = 940\ M^{-2}\ s^{-1}$ at $\mu = 0.2\ M$ (232).

$$4Co(tim)^{2+} + O_2 + 4H^+ \xrightarrow{Fe(H_2O)_6^{2+}} 4Co(tim)^{3+} + 2H_2O \qquad (95)$$

$$-d[Co(tim)^{2+}]/dt = k_{Fe}[Co(tim)^{2+}][Fe(H_2O)_6^{2+}][O_2] \qquad (96)$$

A Co^{II} macrocycle, $Co(dmgBF_2)_2$ (abbreviated as Co_B), is also a catalyst, but the reaction is strongly influenced by H^+ and bromide ions, $k_{Co} = 6.5 \times 10^{11}$ M^{-5} s^{-1} at $\mu = 0.1$ M (233).

$$4Co(tim)^{2+} + O_2 + 4H^+ \xrightarrow{Co_B, Br^-, H^+} 4Co(tim)^{3+} + 2H_2O \quad (97)$$

$$-d[Co(tim)^{2+}]/dt = k_{Co}[Co(tim)^{2+}][Co_B]^2[O_2][H^+][Br^-] \quad (98)$$

The difference in the behavior of the two systems was traced to the difference in kinetic lability of the oxidized forms of the two catalysts. In both cases the reaction initially produces a μ-peroxo intermediate, (tim)CoOO(Cat). For Cat = $Fe(H_2O)_6^{2+}$, the reaction is completed by the dissociation of $Fe(H_2O)_6^{3+}$ and rapid reduction of $(tim)CoOO^+$ by additional equivalents of $Co(tim)^{2+}$, Scheme 7.

This route is not available to the intermediate produced in the Co_B-catalyzed reaction because the dissociation of either Co^{III} complex from (tim)-$CoOOCo_B^{2+}$ is slow. An overall sixth-order rate law of Eq. 98 testifies to the fact that different and more complicated pathways need to be utilized for the reaction to proceed. The reduction of $(tim)CoOOCo_B^{2+}$ by Co_B provides such a route, but only when an efficient electron-transfer bridge is available—thus the catalysis by Br^-, Scheme 7. The ease of bromide substitution at cobalt(III) implies kinetic lability of the positions trans to the peroxo bridge, a conclusion that has been reached earlier for some other peroxocobalt complexes (52).

$$Co(tim)^{2+} + O_2 + Cat^n \rightleftharpoons (tim)CoOOCat^{n+2} \qquad (n = 0 \text{ or } 2)$$

$$Cat = Fe(H_2O)_6^{2+}$$

$$(tim)CoOOFe(H_2O)_5^{4+} \xrightarrow{H_2O} (tim)CoOO^+ + Fe(H_2O)_6^{3+}$$

$$(tim)CoOO^+ + 2Co(tim)^{2+} \xrightarrow{4H^+, fast} 3Co(tim)^{3+} + 2H_2O$$

$$Co(tim)^{2+} + Fe(H_2O)_6^{3+} \longrightarrow Co(tim)^{3+} + Fe(H_2O)_6^{2+}$$

$$Cat = Co_B$$

$$(tim)CoOOCo_B^{2+} + Co_B + Br^- + H^+ \longrightarrow [Co_BOO(H)Co(tim)BrCo_B^{2+}]$$

$$Co_BOO(H)Co(tim)BrCo_B^{2+} + 3Co(tim)^{2+} \xrightarrow{3H^+, fast}$$

$$4Co(tim)^{3+} + 2Co_B + 2H_2O + Br^-$$

Scheme 7

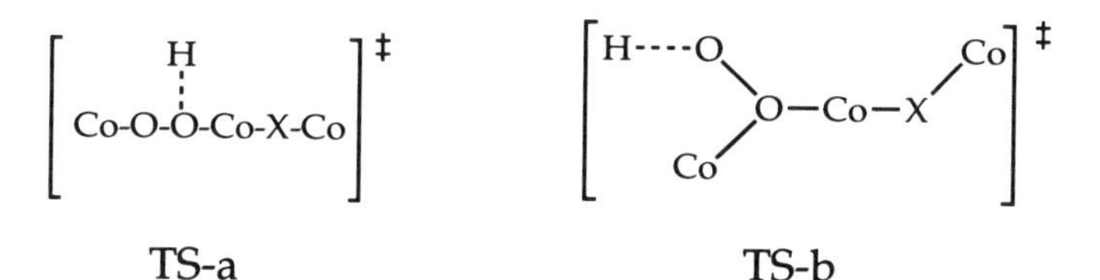

Figure 19. Proposed transition states for the $Co(dmgBF_2)_2$-catalyzed autoxidation of $Co(tim)^{2+}$, (233). [Reproduced with permission from A. Marchaj, A. Bakac, and J. H. Espenson, *Inorg. Chem.*, *31*, 4860 (1992). Copyright © 1992 American Chemical Society.]

The simplest structure of the transition state for the Co_B-catalyzed reaction is that labeled TS-a in Fig. 19, although the catalysis by H^+ may indicate a less common peroxide bridging as in TS-b.

A peroxo-bridged dicobalt(III) complex of a binucleating macrocycle undergoes intramolecular oxo transfer to a coordinated phosphite, Fig. 20. The product phosphate contains one oxygen atom derived from molecular oxygen, as established by ^{18}O labeling (234). A similar ligand hydroxylation in a related dicopper complex proceeds without an observable peroxo intermediate (158).

4. *Mononuclear Peroxo and Oxo Peroxo Complexes*

High-valent early transition metals react with H_2O_2 to form peroxo and oxo peroxo complexes, many of which are good oxo-transfer reagents. Such metal-catalyzed oxidations by H_2O_2 (235) have received considerable attention in the context of "green" (environmentally friendly) chemistry and as functional models for various peroxidases. Complexes of molybdenum and tungsten (236–238), rhenium (239–243), vanadium (244–246), and chromium (247–249) have received most attention. The discussion of the chemistry involved is beyond the scope of this chapter, and the reader is referred to the original literature and recent reviews for more details.

Some peroxo complexes can also act as weak *reductants*, either directly or indirectly. Hydrogen peroxide is released from $(H_2O)_5Ti(O_2)^{2+}$ with the rate constant $k_r = 0.019\ s^{-1}$ in 1 *M* $HClO_4$, providing the only route for the reduction of Cl_2 by $(H_2O)_5Ti(O_2)^{2+}$ (180).

$$(H_2O)_5TiO^{2+} + H_2O_2 \underset{k_r}{\overset{k_f}{\rightleftharpoons}} (H_2O)_5Ti(O_2)^{2+} + H_2O$$

$$K = 8.7 \times 10^3\ M^{-1} \qquad (99)$$

The reduction of Ce^{IV} by $(H_2O)_5Ti(O_2)^{2+}$, Eq. 59 (179, 180), is a direct bimolecular process. This conclusion is supported by the large value of the rate

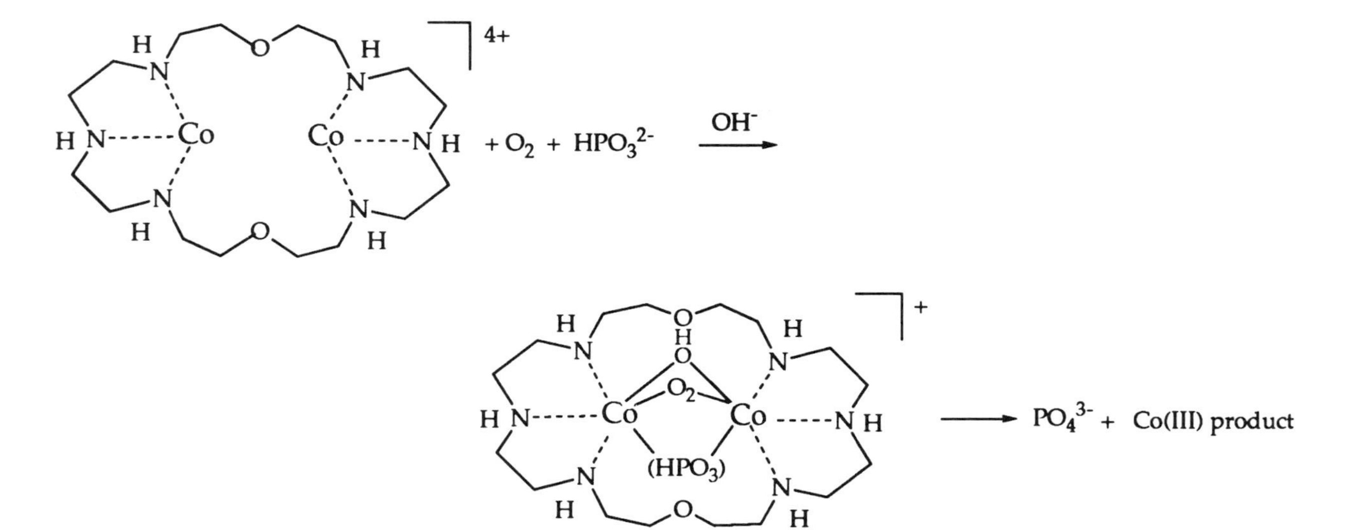

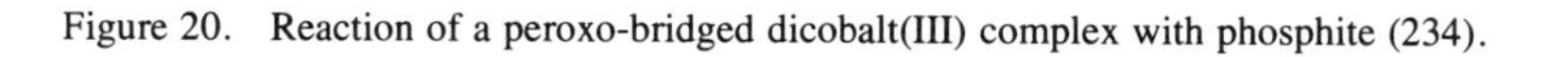
Figure 20. Reaction of a peroxo-bridged dicobalt(III) complex with phosphite (234).

constant (k_{59} = 1.1 × 10^5 $M^{-1}s^{-1}$ in 1 M $HClO_4$), which rules out significant involvement of Reaction 99.

HSO_5^- oxidizes $(H_2O)_4V^V(O)(O_2)^+$ in a reaction that is strongly catalyzed by $(H_2O)_5VO^{2+}$,

$$HSO_5^- + (H_2O)_4V(O)(O_2)^+ \xrightarrow{(H_2O)_5VO^{2+}} HSO_4^- + (H_2O)_4V(O)_2^+ + O_2 \tag{100}$$

according to the mechanism of Eqs. 101–103, where k_{101} = 13.3 M^{-1} s^{-1} in 1 M $HClO_4$ and the ratio k_{102a}/k_{102b} = 38 (250).

$$HSO_5^- + (H_2O)_5VO^{2+} \longrightarrow SO_4^{\bullet -} + (H_2O)_4VO_2^+ + H^+ + H_2O \tag{101}$$

$$SO_4^{\bullet -} \xrightarrow{(H_2O)_4V(O)(O_2)^+,\ H^+} HSO_4^- + (H_2O)_4V(O)(O_2)^{2+} \tag{102a}$$

$$SO_4^{\bullet -} \xrightarrow{(H_2O)_5VO^{2+}} HSO_4^- + (H_2O)_4VO_2^+ + H^+ \tag{102b}$$

$$(H_2O)_4V(O)(O_2)^{2+} + H_2O \longrightarrow (H_2O)_5VO^{2+} + O_2 \tag{103}$$

The complex $Co(H_2O)_6^{3+}$ oxidizes $(H_2O)_4V(O)(O_2)^+$ directly (k = 18.2 M^{-1} s^{-1} in 1 M $HClO_4$) (250) to the (formally) superoxo complex $(H_2O)_4V(O)(O_2)^{2+}$, which again undergoes rapid intramolecular electron transfer as in Eq. 103.

5. *Electron-Transfer Induced Dioxygen Release from Superoxo and Peroxo Complexes*

In addition to their typical role as oxidants, superoxo complexes have been shown to be efficient reductants as well. This reactivity mode may provide a potentially useful mechanism for the controlled release of O_2 from dioxygen carriers.

The oxidation of $(H_2O)_5CrO_2^{2+}$ by Ru^{III} and Fe^{III} complexes yields $Cr(H_2O)_6^{3+}$ and O_2. The kinetics (251) [M^{III} = $Ru(bpy)_3^{3+}$, k = 2.63 × 10^3 M^{-1} s^{-1}, $Ru(5,6\text{-}Me_2\text{-}phen)_3^{3+}$, 1.06 × 10^3, $Fe(phen)_3^{3+}$, 81.9] are consistent with outer-sphere electron transfer to yield $(H_2O)_5CrO_2^{3+}$, which rapidly (> 10^6 s^{-1}) dissociates O_2, Eqs. 104 and 105. The dissociation constant K_{105} for $(H_2O)_5CrO_2^{3+}$ has been estimated (134) from electrochemical data to lie in the range 10^9–10^{16} M, which represents an amazing greater than 10^{20}-fold increase compared to the reduced form, $(H_2O)_5CrO_2^{2+}$ ($K = 2 \times 10^{-12}$ M).

$$(H_2O)_5CrO_2^{2+} + M^{III} \longrightarrow (H_2O)_5CrO_2^{3+} + M^{II} \tag{104}$$

$$(H_2O)_5CrO_2^{3+} \overset{fast}{\rightleftharpoons} Cr(H_2O)_6^{3+} + O_2 \tag{105}$$

Electrochemical oxidation in DMSO of a O_2 adduct of an iron(II) capped porphyrin, $(P)(MeIm)FeO_2$, results in the production of O_2 and Fe^{III} (252). Spectroelectrochemical experiments detected an intermediate, proposed to be the intact, one-electron oxidized complex, $(P)(MeIm)FeO_2^+$. The dissociation of O_2 from this intermediate is apparently slow and takes place in competition with electroreduction to $(P)(MeIm)FeO_2$, Eq. 106. On the other hand, the two-electron electrochemical oxidation of a μ-peroxo cobalt complex, $[(dmso)(salen)Co]_2O_2$ in DMSO yields $Co(salen)(dmso)_2^+$ and O_2 without detectable intermediates (253, 254).

$$(P)(MeIm)FeO_2 \underset{+e^-}{\overset{-e^-}{\rightleftharpoons}} (P)(MeIm)FeO_2^+ \longrightarrow (P)(MeIm)Fe^+ + O_2 \tag{106}$$

Photochemically induced dissociation of O_2 from a superoxocobalt porphyrin having an axially bound azaferrocene, Fig. 21, takes place in a kinetically observable process at 180 K in toluene (255–257). The reaction appears to be initiated by intramolecular electron transfer from Fe^{II} to Co^{III}, and thus represents a case of reductive homolysis.

In a broad sense, all of the photolytic cleavage reactions of L_nMO_2 complexes to L_nM and O_2 (55, 258) could be classified as intramolecular reductive homolyses, given that the oxidation states of the metal and O_2 formally change as in Eq. 107. The photolysis of some peroxo complexes yields singlet O_2 (259–262).

$$L_nM^n(O_2^-) \xrightarrow{h\nu} M^{n-1} + O_2 \tag{107}$$

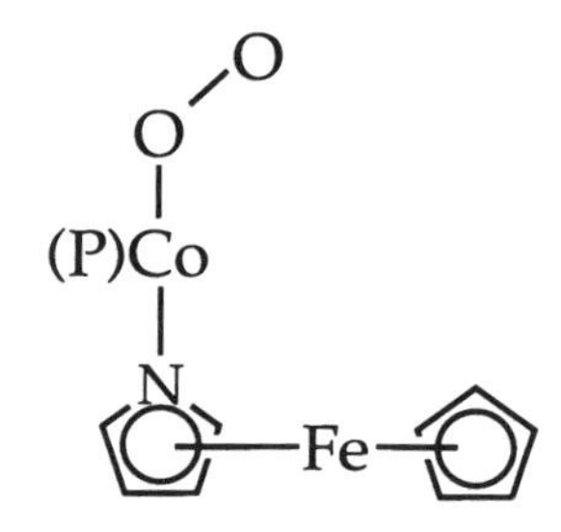

Figure 21. Dioxygen adduct of azaferrocenecobalt(tetra-*p*-tolylporphyrin) (256).

TABLE X
Kinetic Data for the Oxidation of $(H_2O)L^1CoO_2H^{2+}$ by Transition Metal Ions[a,b]

Oxidant	k (M^{-1} s^{-1})
Outer-Sphere Reactions	
$Ru(bpy)_3^{3+}$	$1510 + 790[H^+]^{-1}$
$IrCl_6^{2-}$	$190 + 9.2[H^+]^{-1}$
$Fe(phen)_3^{3+}$	191[c]
$Ru(NH_3)_4(phen)^{3+}$	$200 + 0.09[H^+]^{-1}$[d]
Inner-Sphere Reactions	
$Fe(H_2O)_6^{3+}$	$0.63 + 0.26[H^+]^{-1}$
$(H_2O)_4VO_2^+$	254[c], 21[e]

[a] L^1 = [14]aneN_4. Data from (204) and (215).
[b] Ionic strength = 1.0 M.
[c] $[H^+]$ = 1.0 M.
[d] μ = 0.10 M.
[e] $[H^+]$ = 0.10 M.

The oxidation (204, 215) of the hydroperoxo complex $L^1(H_2O)CoO_2H^{2+}$ by $IrCl_6^{2-}$, $Fe(phen)_3^{3+}$, $Ru(bpy)_3^{3+}$, and $Ru(NH_3)_4(phen)^{3+}$, Table X, proceeds as in Eq. 108. The rate law of Eq. 109 indicates that both $L^1(H_2O)CoO_2H^{2+}$ (k_{CoOOH}) and the conjugate base form, $L^1(H_2O)CoO_2^+$ (k_{CoOO}), are reactive, the respective rate constants being $k_0 = k_{CoOOH}$ and $k_{-1} = K_a k_{CoOO}$. The pK_a of $L^1(H_2O)CoO_2H^{2+}$ has been estimated as about 8 (204).

$$L^1(H_2O)CoO_2H^{2+} + 2M^n \xrightarrow{H_2O} L^1(H_2O)_2Co^{3+} + O_2 + H^+ + 2M^{n-1} \tag{108}$$

$$-d[(L^1(H_2O)CoO_2H^{2+}]/dt = (k_0 + k_{-1}[H^+]^{-1})[L^1(H_2O)CoO_2H^{2+}][M^n] \tag{109}$$

The initial outer-sphere electron transfer to the hydroperoxo complex is believed to be followed by the rapid loss of O_2 from $L^1(H_2O)CoO_2^{2+}$ and oxidation of $L^1(H_2O)_2Co^{2+}$ by an additional mole of oxidant, as shown for $IrCl_6^{2-}$ in Eqs. 110–112. It is, however, also possible that some of the oxidants in Table X react with $L^1(H_2O)CoO_2^{2+}$ directly.

$$L^1(H_2O)CoO_2H^{2+} + IrCl_6^{2-} \xrightarrow[-H^+]{k_0, k-1} L^1(H_2O)CoO_2^{2+} + IrCl_6^{3-} \tag{110}$$

$$L^1(H_2O)CoO_2^{2+} + H_2O \rightleftharpoons L^1(H_2O)_2Co^{2+} + O_2 \tag{111}$$

$$L^1(H_2O)_2Co^{2+} + IrCl_6^{2-} \xrightarrow{fast} L^1(H_2O)_2Co^{3+} + IrCl_6^{3-} \tag{112}$$

The two remaining oxidants, $Fe(H_2O)_6^{3+}$ and $(H_2O)_4VO_2^+$, react by inner-sphere mechanisms. In the case of $(H_2O)_4VO_2^+$, the rates are proportional to $[H^+]$, as expected for the conversion of $(H_2O)_4VO_2^+$ to $(H_2O)_5VO^{2+}$.

6. *Common Features in the Chemistry of Superoxo and Alkyl Complexes*

Oxidative homolysis is a feature shared by metal superoxides and metal alkyls. The chemistry analogous to that in Eqs. 104 and 105, and a large decrease in homolytic stability upon oxidation, have been demonstrated for a number of transition and main group organometallic compounds (263–268).

There are many other common threads and remarkable similarities in the chemistry of the two classes of compounds. In retrospect, this might have been anticipated, given that both are produced from reduced metal complexes and radical species ($R^{\cdot}$ or $^3O_2$), Eq. 113, and in neither case is the electron transfer from metal complete. This point has already been discussed for MO_2 complexes, and similar arguments are applicable to the compounds containing metal–carbon bonds. For example, it has been noted (269) that the lability of the positions trans to the alkyl group in organocobalt(III) complexes is intermediate between those of cobalt(II) and cobalt(III).

$$M + R^{\cdot} \quad (\text{or } ^3O_2) \underset{k_r}{\overset{k_f}{\rightleftharpoons}} M{-}R \quad (\text{or } M{-}O_2) \tag{113}$$

The less-than-quantitative electron-transfer manifests itself in the occurrence of the reverse reactions, homolytic cleavages of metal–alkyl and metal–dioxygen bonds. Such homolyses can also be induced photochemically under conditions that are not too different for the two types of complexes of a given metal. For example, both alkyl and superoxo complexes of $(H_2O)LCo^{2+}$ ($L = L^1$ and L^2) have weak, but photoactive bands in the visible spectrum, and irradiation with light of $\lambda \leq 550$ nm results in clean homolysis of both (55, 270, 271). The chromium complexes, on the other hand, require UV excitation for photohomolysis to take place. The lower wavelengths required for chromium are qualitatively consistent with the greater bond dissociation energies and smaller rates of thermal homolysis compared to those of cobalt, Table XI. These factors are, however, insufficient to account quantitatively for the large difference in the photochemical threshold energies for the two metals.

Kinetic and thermodynamic data for the formation of alkyl and superoxo complexes of the same metal and ligand system are limited. A detailed analysis of the data in Table XI (55, 58, 61, 67, 126, 127, 272–280) would thus seem unjustified at the present time, but it is worth nothing that complexes with R = primary alkyl are typically much more stable than metal superoxides. In fact, many complexes that do not react with O_2 do react with alkyl radicals, but the opposite is not true.

TABLE XI
Kinetic and Equilibrium Data for the Formation and Homolysis of Some Superoxo and Alkyl Complexes of Cobalt and Chromium[a]

Complex	k_f (M^{-1} s^{-1})	k_r (s^{-1})	$K = k_f/k_r$
$(H_2O)_5CrEt^{2+}$	1.9×10^8 [b]	$<2 \times 10^{-5}$ [c]	$>10^{13}$
$(H_2O)_5CrO_2^{2+}$	1.6×10^8 [d]	2.5×10^{-4} [e]	6.4×10^{11}
$(H_2O)_5CrCHMe_2^{2+}$	$\sim 1 \times 10^8$ [f]	1.7×10^{-4} [g]	$\sim 3 \times 10^{11}$
$(H_2O)_5CrCH_2Ph^{2+}$	8.5×10^7 [h]	2.6×10^{-3} [g]	3.3×10^{10}
$(H_2O)_5CrCMe_2OH^{2+}$	5×10^7 [i]	0.127[j]	3.9×10^8
$(H_2O)L^1CoCH_2Ph^{2+}$	$\sim 10^7$ [k]	0.094[l]	1×10^8
$(H_2O)L^1CoCHMe_2^{2+}$	$\sim 10^7$ [k]	$\gg 0.1$ [m]	$\ll 10^8$
$(H_2O)L^1CoO_2^{2+}$	1.2×10^7 [n]	63[o]	1.9×10^5
$(H_2O)L^2CoEt^{2+}$	2.9×10^7 [p]	6.7×10^{-4} [q]	4.3×10^{10}
$(H_2O)L^2CoCHMe_2^{2+}$	$\sim 10^7$ [k]	$\gg 1 \times 10^{-3}$ [r]	$\ll 10^{10}$
$(H_2O)L^2CoO_2^{2+}$	5.0×10^6 [n]	2.06×10^4 [s]	2.5×10^2

[a]In aqueous solution, where L^1 = [14]aneN$_4$, L^2 = Me$_6$-[14]aneN$_4$; k_f and k_r are defined in Eq. 113.
[b]Reference (272).
[c]Assuming k_r < 10% of total decomposition rate (273).
[d]References (126) and (127).
[e]Reference (67).
[f]Estimated from data in (272).
[g]Reference (273).
[h]Reference (274).
[i]Reference (275).
[j]Reference (276).
[k]Estimated from data in (277).
[l]Reference (278).
[m]Complex homolyzes more rapidly than $(H_2O)L^1CoCH_2Ph^{2+}$ (279).
[n]Reference (55).
[o]Reference (58).
[p]Reference (277).
[q]Reference (280).
[r]Complex homolyzes much more rapidly than $(H_2O)L^2CoEt^{2+}$ (280).
[s]Reference (61).

For both (incomplete) series available [M = $Cr(H_2O)_5^{2+}$ and $(H_2O)L^1Co^{2+}$], the data for superoxo complexes lie close to those for R = 2-propyl and benzyl. The similarity is surprising because the steric bulk and the reduction potentials for the two series [$Cr(H_2O)_6^{3+/2+}$, $E^0 = -0.41$ V, $(H_2O)_2L^1Co^{3+/2+}$, $E^0 = 0.42$ V] (58) differ greatly. The third group, M = $(H_2O)L^2Co^{2+}$, qualitatively fits the picture as well in that neither 2-propyl nor benzyl complex could be prepared at room temperature owing to their facile homolysis.

Earlier it was shown that the kinetics of homolytic bond cleavage reactions depend on both steric and electronic factors. The data in Table XI imply that the combination of the two factors is comparable for $CHMe_2$ and end-on $O_2^{\bullet -}$ when complexed to $Cr(H_2O)_5^{2+}$ and $(H_2O)L^1Co^{2+}$. Additional data are needed to establish whether this pattern extends to other series as well.

The acidolysis of Eq. 114 is a major decomposition route for organochromium(III) complexes, but the corresponding reaction has not been observed for $(H_2O)_5CrO_2^{2+}$.

$$(H_2O)_5CrR^{2+} + H^+ + H_2O \longrightarrow Cr(H_2O)_6^{3+} + RH \qquad (114)$$

As commented in Section II, the unfavorable thermodynamics of heterolytic dissociation of $(H_2O)_5CrO_2^{2+}$ and $(H_2O)LCoO_2^{2+}$ are puzzling. However, even the favorable thermodynamics are no guarantee that a reaction will proceed rapidly, as is clearly illustrated by the observations for organochromium complexes, Reaction 114. From a thermodynamic cycle in Scheme 8, similar to that used earlier for superoxo complexes, one calculates $K_{heter} \sim 1 \times 10^{26}$ for heterolysis of $(H_2O)_5CrCHMe_2^{2+}$. In these calculations, E^0 for the couple $Me_2CH^{\bullet}/Me_2CH_2$ was estimated from $E^0(CH_3^{\bullet}/CH_4) = 2.25$ V (281) and the difference in bond dissociation energies between $CH_3{-}H$ (438.9 kJ mol^{-1}) and $Me_2CH{-}H$ (397.5 kJ mol^{-1}).

$$Cr(H_2O)_6^{2+} \rightleftharpoons Cr(H_2O)_6^{3+} + e^- \qquad E = 0.41\text{ V}$$

$$(H_2O)_5CrCHMe_2^{2+} + H_2O \rightleftharpoons Cr(H_2O)_6^{2+} + Me_2CH^{\bullet} \qquad K = 2 \times 10^{-12}\,M^{-1}$$

$$Me_2CH^{\bullet} + H^+ + e^- \rightleftharpoons Me_2CH_2 \qquad E = 1.82\text{ V}$$

$$(H_2O)_5CrCHMe_2^{2+} + H^+ + H_2O \rightleftharpoons Cr(H_2O)_6^{3+} + Me_2CH_2 \qquad K_{heter} = 1 \times 10^{26}$$

Scheme 8

The reaction with H_2O, Eq. 115, has $K = 1 \times 10^{22}\,M^{-1}$. Despite the favorable thermodynamics, both Reactions 114 and 115 are very slow; for $(H_2O)_5CrCHMe_2^{2+}$, k_{114} is too small to observe ($< 1 \times 10^{-5}\,M^{-1}\,s^{-1}$) and $k_{115} = 1.05 \times 10^{-4}\,s^{-1}$ (282).

$$(H_2O)_5CrCHMe_2^{2+} + H_2O \rightleftharpoons (H_2O)_5CrOH^{2+} + Me_2CH_2 \qquad (115)$$

The OH^- and anion-induced hydrolysis, both of which have been well established for organochromium complexes (283–285), Eqs. 116 and 117, appear to operate for $(H_2O)_5CrO_2^{2+}$ and $(H_2O)L^1CrO_2^{2+}$, as well, Eq. 118 (144). These reactions are similar, but not necessarily identical to the previously proposed H^+-promoted nucleophilic displacement of superoxide in the autoxidation of hemoglobin (286).

$$MeCOO^- + (H_2O)_5CrR^{2+} \xrightarrow{H_2O, H^+} (H_2O)_5CrOC(O)Me^{2+} + RH \quad (116)$$

$$([15]aneN_4)(H_2O)CrR^{2+} + H_2O \xrightarrow{OH^-} ([15]aneN_4)(H_2O)CrOH^{2+} + RH \quad (117)$$

$$LCrO_2^{2+} + OH^- \longrightarrow LCr(OH)^{2+} + O_2^{\bullet -} \quad (118)$$

Bimolecular homolytic (S_H2) displacement reactions, such as those in Eqs. 119 and 120, are also well documented in organometallic chemistry (287, 288), but no related examples have been reported for R = O_2. It is unlikely that O_2 transfer from MO_2 to a reducing metal M′ (equivalent of Eq. 119) could be observed, because direct redox reactions between MO_2 and M′ are usually rapid, as already discussed. The displacement of O_2 from MO_2 by an alkyl radical, Eq. 121a, should be facile for R = primary alkyl because of favorable thermodynamics (Table XI) and the expected kinetic lability of the position trans to coordinated superoxide. A site for attack by $R^{\bullet}$ thus appears available. There are also other possibilities, such as attack at coordinated superoxide, Eq. 121b. These intriguing possibilities may be difficult to demonstrate experimentally because of the unavoidable side reactions, including those between carbon-centered radicals and O_2 and MO_2.

$$RM + M' \rightleftharpoons RM' + M \quad (119)$$

$$RM + R' \rightleftharpoons R'M + R \quad (120)$$

$$MO_2 + R^{\bullet} \longrightarrow RM + O_2 \quad (121a)$$

$$MO_2 + R^{\bullet} \longrightarrow M + ROO^{\bullet} \text{ or } M^+ + ROO^- \quad (121b)$$

B. Metal–Oxo Complexes, MO and $M(O)_2$

1. *Chromium(IV)–(V)*

Ligands such as porphyrins (141, 289–291), macrocyclic tetraamides (L^{12}) (292), salen derivatives (L^{13}) (293, 294), chelating carboxylates (L^{14}) (295, 296), and crown ethers (297) stabilize chromium as an oxo ion in the unusual 4+ and 5+ oxidation states, Fig. 22. Such complexes have been prepared by oxidation of Cr^{II} and Cr^{III} with a number of oxygen-atom donors, and by O_2 itself (141, 291, 298).

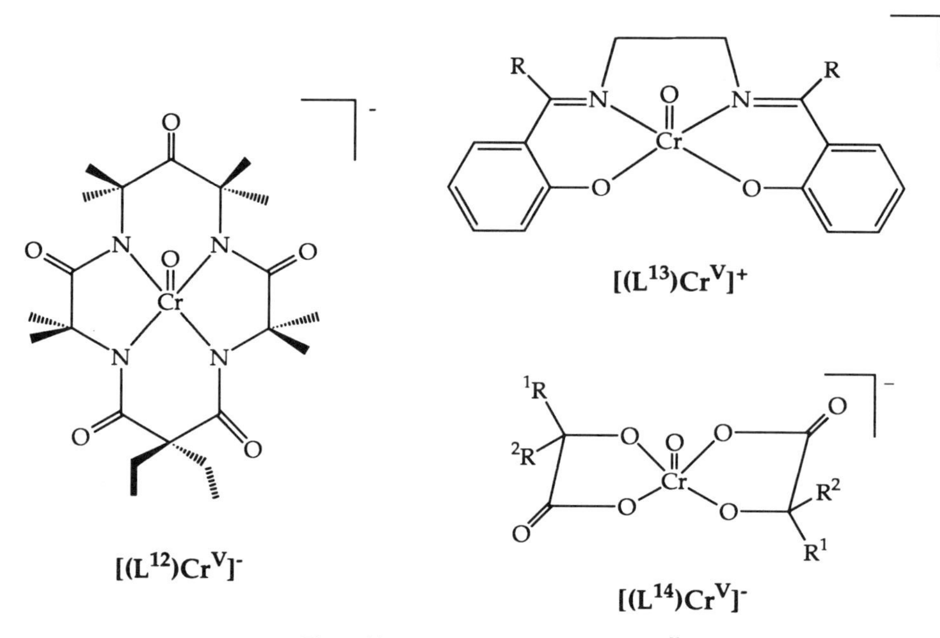

Figure 22. Some stable complexes of Cr^V.

In aqueous solution and in the absence of such stabilizing ligands, both oxidation states were believed to have only fleeting existence. In the case of Cr^{IV}, evidence to the contrary was obtained recently, when it was shown that $(H_2O)_5CrO^{2+}$ has a respectable lifetime of about 1 min at room temperature. This species can be prepared either in the reaction of $Cr(H_2O)_6^{2+}$ with O_2, as described in Section II.C, or by oxidation of $Cr(H_2O)_6^{2+}$ with two-electron oxidants, such as Tl^{III}, under anaerobic conditions (128, 129).

Kinetic studies, which sometimes necessitated the use of a colored probe $ABTS^{2-}$, Fig. 23 (299), identified several mechanistic pathways for oxidations

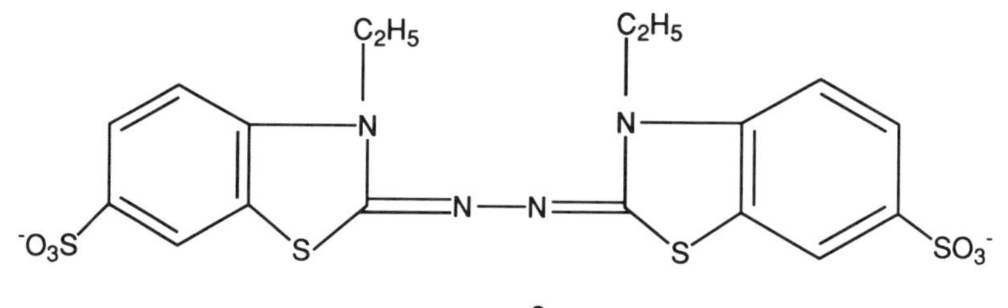

Figure 23. Structure of the kinetic probe $ABTS^{2-}$.

by $(H_2O)_5CrO^{2+}$. The PPh_3 reacts by oxygen-atom transfer (129), alcohols by hydride transfer, phenols by hydrogen-atom transfer, and one-electron reductants Fe^{2+} and $ABTS^{2-}$ by electron transfer. Kinetic data are summarized in Table XII and discussed below.

The oxidation of alcohols in the presence of O_2, Eq. 122, produces $(H_2O)_5CrO_2^{2+}$, which can easily be identified by its UV spectrum, Fig. 11.

$$CR_2HOH + (H_2O)_5CrO^{2+} \xrightarrow{O_2} CR_2O + (H_2O)_5CrO_2^{2+} + H_2O \quad (122)$$

The reaction takes place by hydride transfer yielding $Cr(H_2O)_6^{2+}$ and an aldehyde or ketone. The rapid capture of $Cr(H_2O)_6^{2+}$ by O_2 completes the process. This mechanism is supported by the formation of $(H_2O)_5CrO_2^{2+}$ as a product in the presence of O_2, by substantial kinetic isotope effects (k_H/k_D), and by the narrow range of rate constants observed, Table XII. The one-electron, hydrogen-atom transfer path of Eq. 123, was not observed for acyclic alcohols, and was shown to be thermodynamically less favorable than hydride transfer (129). Cyclobutanol, on the other hand, reacts by hydrogen-atom transfer and yields acyclic products.

$$CR_2HOH + (H_2O)_5CrO^{2+} \not\longrightarrow (H_2O)_5CrOH^{2+} + {}^{\bullet}CR_2OH \quad (123)$$

Aldehydes, ethers, carboxylic acids, and diols (129, 300) react similarly by two-electron pathways, with an occasional exception when stabilized radicals are produced by hydrogen-atom transfer.

The oxidation of phenol (301) initially produces phenoxyl radicals in a reaction exhibiting a large kinetic isotope effect for the phenolic hydrogen ($k_H/k_D = 14.7$) and none for the ring hydrogen atoms ($k_H/k_D = 1$). The proposed mechanism consists of hydrogen-atom abstraction from the phenolic OH group, followed by the known chemistry of phenoxyl radicals to yield benzoquinone as a major product. A very different mechanism, electrophilic attack on the aromatic ring, has been proposed in the oxidation of phenols by $Ru^{IV}(bpy)_2(py)O^{2+}$ (302).

Hydrogen peroxide reacts with $(H_2O)_5CrO^{2+}$ to yield $(H_2O)_5CrO_2^{2+}$ even in the absence of O_2, clearly ruling out the involvement of free $Cr(H_2O)_6^{2+}$(138).

$$(H_2O)_5CrO^{2+} + H_2O_2 \longrightarrow (H_2O)_5CrO_2^{2+} + H_2O \quad (124)$$

The stoichiometry alone does not provide a clue as to the nature of the reaction, which can be defined either as a two-electron oxidation of $(H_2O)_5CrO^{2+}$ (formally oxygen-atom transfer from H_2O_2), or a one-electron reduction of $(H_2O)_5CrO^{2+}$ [formally $Cr^{IV} + H_2O_2 \longrightarrow Cr^{III}(O_2^-)$]. Mechanistic considerations support the second alternative. Experiments conducted in H_2O and D_2O

TABLE XII

Kinetic Data for Representative Reactions of $(H_2O)_5CrO^{2+}$ in Acidic Aqueous Solution[a,b]

Substrate	k (M^{-1} s^{-1})	k_H/k_D	$\Delta H^{\ddagger}$ (kJ mol^{-1})	$\Delta S^{\ddagger}$ (J mol^{-1} K^{-1})
		Hydride Abstraction		
MeOH	52.2		34	−99
CD_3OH	15.1	3.5	38	−95
EtOH	88.4			
C_2D_5OH	41.5	2.1		
Me_2CHOH	12.0		33	112
$(CD_3)_2CDOH$	4.6	4.8		
$PhCH_2OH$	56.0			
PhCH(OH)Me	29.6			
Ph_2CHOH	10.5			
Cyclopentanol	44.1			
$HOCH_2CH_2OH$	33			
$HOCD_2CD_2OH$	8.9	3.7		
$DOCH_2CH_2OD$	28	1.2		
cis-1,2-$(OH)_2$-c-C_6H_{10}	30		41	−78
cis-1,2-$(OD)_2$-c-C_6H_{10}	26	1.2		
trans-1,2-$(OH)_2$-c-C_6H_{10}	57		49	−47
$[Me_2C(OH)]_2$	3.7		60	−31
CH_2O	91.7			
Et_2O	4.45			
HCOOH	11.6			
$HCOO^-$	6,680			
		Hydrogen Atom Abstraction		
$(H_2O)_5CrOOH^{2+}$[c]	1,340			
$(D_2O)_5CrOOD^{2+}$[c,d]	266	5.0		
PhOH[c]	471		15.2	−144
d_5-PhOH[c]	465	1		
PhOD[c,d]	32	14.7		
H_2O_2[c,e]	190		25	−116
D_2O_2[c–e]	53	3.6		
Cyclobutanol	44.1		46	−61
Me_3CCHO	37.1			
		Oxygen-Atom Transfer		
PPh_3[f]	2,100			
		Electron Transfer		
$ABTS^{2-}$[c]	79,000			
$Fe(H_2O)_6^{2+}$[c]	3,800		28.3	−81

[a]Source: (129), (136), (138), (211), (281), and (282).

[b]$[H^+] = 0.1\ M$, $\mu = 1.0\ M$.

[c]$\mu = 0.1\ M$.

[d]In D_2O.

[e]Hydride or hydrogen-atom transfer.

[f]85% $MeCN/H_2O$.

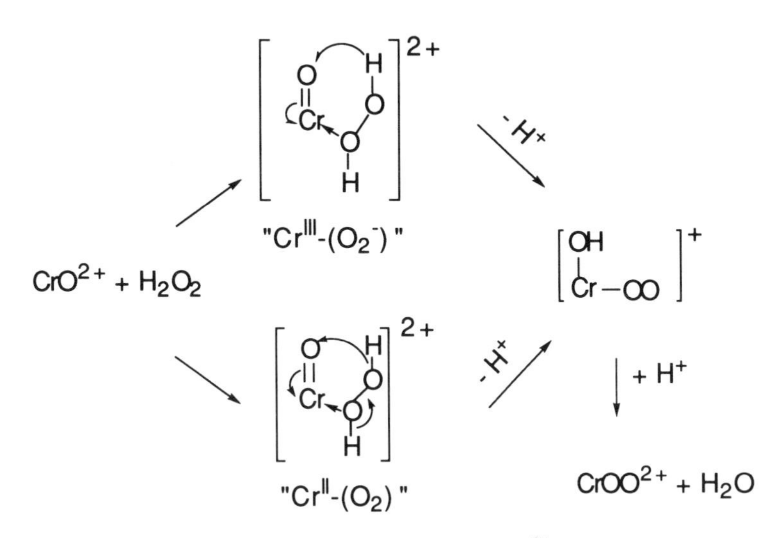

Figure 24. Mechanistic scheme for the reaction of $(H_2O)_5CrO^{2+}$ with H_2O_2. [Modified with permission from A. M. Al-Ajlouni, J. H. Espenson, and A. Bakac, *Inorg. Chem.*, *32*, 3162 (1993). Copyright © 1993 American Chemical Society.]

yielded a kinetic isotope effect $k_H/k_D = 3.6$, consistent with both hydrogen atom and hydride transfers, as shown in Fig. 24. The reaction of another oxo ion, $(bpy)(py)RuO^{2+}$, with H_2O_2 also takes place by hydrogen-atom transfer (or proton-coupled electron transfer) (303), although in this case the precoordination of peroxide to the metal does not take place. Some other high-valent metal ions oxidize H_2O_2 to O_2 or MO_2 via hydroperoxo intermediates (304, 305).

The use of Reaction 122 as a mechanistic fingerprint for $(H_2O)_5CrO^{2+}$ has been demonstrated in the reduction of Cr^{VI} by alcohols, a reaction for which several different mechanisms had been suggested over the years (306–308). It has now been shown (129) that the reaction of 2-propanol with Cr^{VI} in O_2-saturated acidic aqueous solutions yields $(H_2O)_5CrO_2^{2+}$, as expected if Cr^{IV} is produced and subsequently consumed in Reaction 122. The involvement of Cr^{IV} and $Cr(H_2O)_6^{2+}$ according to the mechanism in Scheme 9 has thus been unequivocally demonstrated. The data also suggest, but do not prove, that the Cr^{IV} produced in this reaction is the same chromyl ion as that in Eq. 122. The Cr^{VI} in these equations is presumably $HCrO_4^-/CrO_4^{2-}/Cr_2O_7^{2-}$, although this may have to be reevaluated when more information becomes available on exact nature of Cr^{VI} species in acidic aqueous solutions (309, 310).

Recent work on the reduction of $(aq)Cr^{VI}$ by alcohols, aldehydes, and carboxylic acids (311, 312) has provided additional support for the two-electron nature of $(aq)Cr^{IV}$ reactions with these substrates. Also, the relative rate constants obtained in this work for the $(aq)Cr^{IV}$ reactions (312) in most cases agree

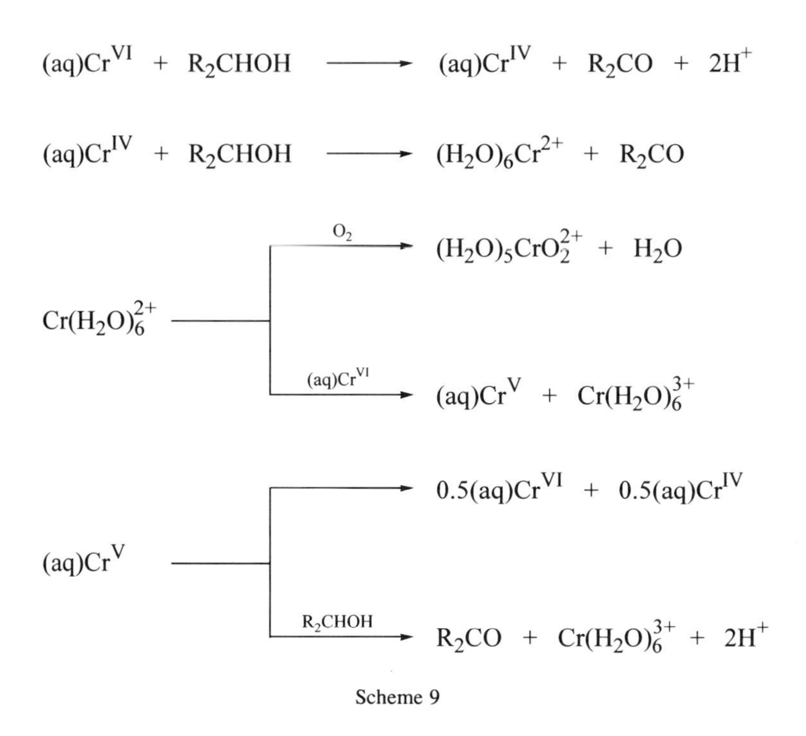

Scheme 9

well with the data for $(H_2O)_5CrO^{2+}$ (129), again supporting the notion that the reduction of chromate and oxidation of $Cr(H_2O)_6^{2+}$ yield the same $(aq)Cr^{IV}$ species. From the inhibiting effect of $Mn(H_2O)_6^{2+}$ and $(aq)Ce^{III}$ on the reactions of chromate with organic substrates, the rate constants for the reduction of $(aq)Cr^{IV}$ were estimated at about 10^6 M^{-1} s^{-1} for both $Mn(H_2O)_6^{2+}$ and $(aq)Ce^{III}$.

The kinetic and thermodynamic data that have been presented in preceding paragraphs and sections on chromium–O_2 interactions can be combined with other information to assess the thermodynamic effect of Cr^{III} coordination on every step of the four-electron reduction of O_2 to H_2O. The results (135) are shown in Fig. 25.

It is noteworthy that $HO_2^{\bullet}$ and $HO^{\bullet}$ are much more strongly oxidizing than their Cr-substituted analogues, $(H_2O)_5CrO_2^{2+}$ and $(H_2O)_5CrO^{2+}$. On the other hand, $(H_2O)_5CrO_2H^{2+}$ is a more potent oxidant than H_2O_2. Taking into account the substitutional inertness and low redox reactivity associated with Cr^{III}, the effect of replacement of H^+ by $(H_2O)_5Cr^{3+}$ is almost spectacular. This appears to be the first system for which the role of the metal in step-by-step thermodynamics of the O_2/H_2O has been evaluated quantitatively.

The finding that the 5+ oxidation state of chromium is responsible for car-

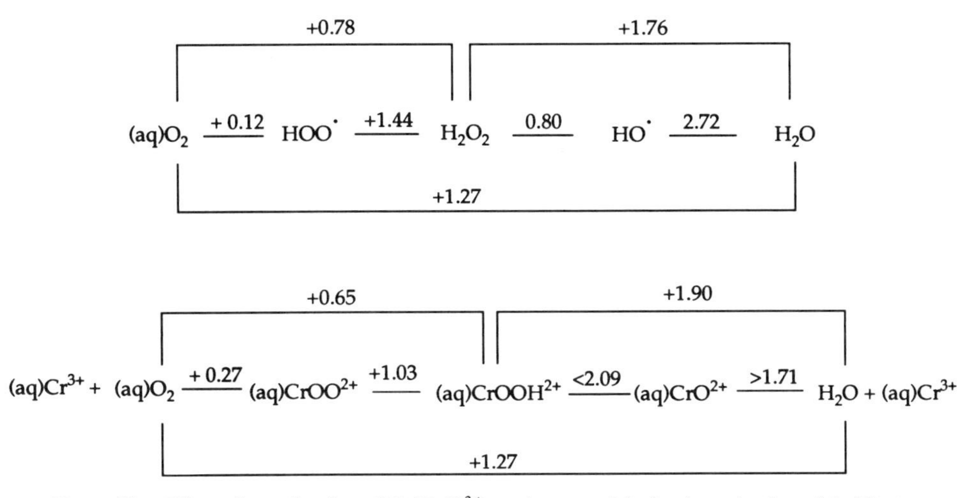

Figure 25. Effect of coordination of $Cr(H_2O)_5^{3+}$ on the potentials for the reduction of (aq)O_2 to H_2O at pH 0. The reference state is 1 *M* O_2(aq). In view of the latest electrochemical results (134) the values listed for the couples $(H_2O)_5CrOO^{2+}/(H_2O)_5CrOOH^{2+}$ and $(H_2O)_5CrOOH^{2+}/(H_2O)_5CO^{2+}$ are slightly different from those in the original paper. [Modified with permission from J. H. Espenson, A. Bakac, and J. Janni, *J. Am. Chem. Soc.*, *116*, 3436 (1994). Copyright © 1994 American Chemical Society.]

cinogenicity of chromate has sparked intense interest in the chemistry of Cr^V (313–318). Another aspect is concerned with chromium(V) complexes as epoxidation agents in both stoichiometric and catalytic systems (294, 319–321).

The coordination number of five and the presence of at least one oxo group are typical for Cr^V, as established by crystal structure determinations of several stable complexes (292, 294, 296, 322, 323). However, in the absence of large steric effects, hexacoordinated Cr^V can be formed if suitable ligands are available. Electrochemistry and ESR and UV–vis spectroscopies have been used most successfully in identifying Cr^V species as intermediates in the oxidation of organic substrates by Cr^{VI}.

In the absence of stabilizing ligands, aqueous solutions of Cr^V are stable only at high pH (324). In neutral and acidic solutions the complex disproportionates rapidly to yield Cr^{III} and Cr^{VI}. No kinetic or mechanistic data are available for these reactions.

$$2(aq)Cr^V \rightleftharpoons (aq)Cr^{IV} + (aq)Cr^{VI} \tag{125}$$

$$(aq)Cr^{IV} + (aq)Cr^V \longrightarrow (aq)Cr^{III} + (aq)Cr^{VI} \tag{126}$$

Studies of redox chemistry of Cr^V compounds with labile coordination spheres are complicated by disproportionation, ligand substitution, and possibly dimerization (325). Despite these difficulties, a number of reactions of the 2-ethyl-2-hydroxy butanoate complex $L^{14}Cr^{V}O^{-}$ with organic and inorganic substrates have been studied and found to take place by both inner-sphere and outer-sphere electron transfer (295, 326) and occasionally feature autocatalytic behavior (327). Some of the earlier mechanistic assignments have been disputed recently (325). The chromium(IV) complex, which can be prepared independently (328, 329), has been observed as an intermediate in some of the reactions. It has been suggested that this Cr^{IV} species may not be "free," but rather complexed to Cr^{V} (325). Several reviews of Cr^{V} chemistry are available (295, 324, 325, 330).

Strongly coordinating tetradentate ligands render Cr^{V} much less reactive, but complexes such as $(salen)CrO^{+}$ and chromium(V) porphyrins do epoxidize alkenes by oxygen-atom transfer (294, 319). For example, $(salen)CrO^{+}$ epoxidizes norbornene in MeCN, $k = 0.028\ M^{-1}\ s^{-1}$ at 21°C. The $(salen)CrO^{+}$ molecule also oxidizes alkynes, but these reactions are rather slow (331). The oxidation of diphenylacetylene takes place according to the stoichiometry of Eq. 127 and the rate law of Eq. 128. These observations are consistent with the formation of an intermediate, probably metallaoxetane, in the rate-determining step, followed by the rapid oxygen-atom transfer from the second molecule of $(salen)CrO^{+}$ to the intermediate.

$$\mathrm{PhCCPh} + 2(\mathrm{salen})\mathrm{CrO}^{+} \longrightarrow \mathrm{PhC(O)C(O)Ph} + \mathrm{Cr(salen)}^{+} \qquad (127)$$

$$-d[(\mathrm{salen})\mathrm{CrO}^{+}]/dt = 4.7 \times 10^{-5}\,[(\mathrm{salen})\mathrm{CrO}^{+}][\mathrm{PhCCPh}] \qquad (128)$$

The first examples of oxo-alkyl complexes of Cr^{V} and Cr^{VI}, Fig. 26, have been prepared recently in reactions of Eqs. 129 and 130 (332). Both complexes are surprisingly stable in benzene solutions at room temperature. The Cr—C bond in $Cp^{*}Cr^{VI}(O)_2Me$ is not cleaved by MeOH, and neither $Cp^{*}Cr^{V}(O)Me_2$ nor $Cp^{*}Cr^{VI}(O)_2Me$ react with C_2H_4 at room temperature and ambient pressure. Catalytic properties of these complexes have not been examined, and it is not known how they might compare with the (formally) related $MeRe(O)_3$, which is an efficient oxidation catalyst (239–243).

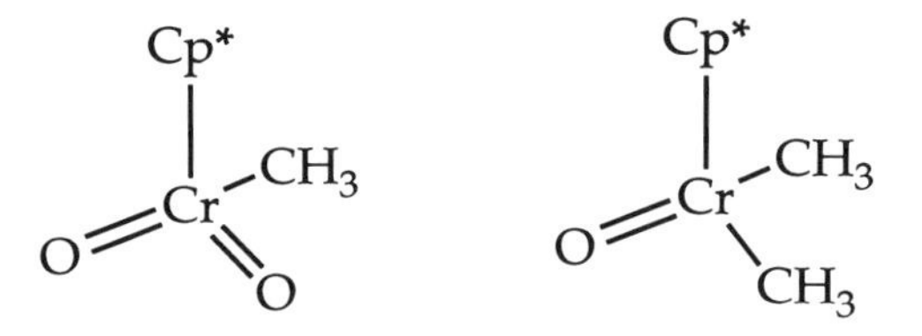

Figure 26. Chromium oxo–alkyl complexes $Cp^{*}Cr(O)_2Me$ and $Cp^{*}Cr(O)(Me)_2$.

$$Cp^*Cr^{III}(py)Me_2 + 0.5O_2 \longrightarrow Cp^*Cr^{V}(O)Me_2 + py \quad (129)$$

$$Cp^*Cr^{III}(py)Me_2 \xrightarrow{O_2(excess)} Cp^*Cr^{VI}(O)_2Me \quad (130)$$

Oxo-transfer abilities of heteropoly anions (333–335), including those of Cr^V, are receiving increasing attention (336). For example, complexes $(R_4N)_n[XW_{11}O_{39}CrO]$ [R = alkyl, X = P (n = 4), or Si (n = 5)] oxidize PPh_3, alkanes, alkenes, and alcohols at ≥50°C. Based on kinetic and detailed product analysis studies, it was proposed (336) that the oxidation of alkenes takes place by radical-like addition of Cr^V to the double bond.

$$Cr^V{=}O + \text{(alkene)} \rightleftharpoons Cr^{IV}{-}O{-}C{-}\dot{C} \longrightarrow Cr^{III} + \text{(epoxide)} \quad (131)$$

The formation of other oxidation products, in addition to epoxides, was taken as evidence against concerted oxo transfer, and the observed solvent effect ($k_{PhH} > k_{MeCN}$) rules out the involvement of radical cations.

An electrochemical study of the oxidation of benzyl alcohol to benzaldehyde by $PW_{11}O_{39}Cr^VO^{4-}$ and $P_2W_{17}O_{61}Cr^VO^{7-}$ reports a rate constant $k = 1.5 \times 10^{-3}\ M^{-1}\ s^{-1}$ for both reactions (337). The rate is too low for these Cr^V polyoxometalates to be useful as electrocatalysts.

2. *Iron(IV)–(VI)*

Interest in high-valent iron derives from its probable involvement in iron-catalyzed oxidations, both biological and industrial, and in Fenton reactions in neutral solutions, as shown below (338).

$$Fe^{II} + H_2O_2 \longrightarrow Fe^{III}OH + HO^{\bullet} \quad (132)$$

$$HO^{\bullet} + H_2O_2 \rightleftharpoons O_2^{\bullet-}/HO_2^{\bullet} + H_2O \quad (133)$$

$$HO^{\bullet}/HO_2^{\bullet}/H_2O_2 + Fe^{II}/Fe^{III} \longrightarrow \text{oxidizing intermediates} \quad (134)$$

Intermediates involved in these reactions are believed to be either hypervalent iron or peroxoiron species. The presence of known traps for $HO^{\bullet}$, H_2O_2, and $O_2^{\bullet-}/HO_2^{\bullet}$ has little effect on chemical or biological activity of these intermediates (339).

The generally accepted mechanism of Reaction 132 has been challenged recently (340, 341) because a number of substrates yielded oxidation products that are inconsistent with the involvement of $HO^{\bullet}$. This discrepancy has been addressed by another mechanistic proposal (342), according to which the reaction produces two or more reactive intermediates, as shown in Eq. 135.

$$Fe^{II} + H_2O_2 \longrightarrow Fe^{II}(H_2O_2) \begin{cases} \nearrow Fe^{III}OH + HO^{\bullet} \\ \searrow Fe^{IV}? \end{cases} \tag{135}$$

Depending on the nature and concentration of the substrate, the active oxidant may be $(H_2O)_5Fe(H_2O_2)^{2+}$ or $HO^{\bullet}$ or Fe^{IV}. Such a scheme, where the actual oxidizing intermediate may be different for different substrates, explains the large number of conflicting reports on the involvement of hydroxyl radicals in Reaction 132.

Aqueous solutions of iron in oxidation states above 3+ have a considerable lifetime only at high pH, much like the Cr^{V} complexes discussed earlier. The ions $Fe^{VI}O_4^{2-}$, $Fe^{V}O_4^{3-}$, $(HO)_nFe^{IV}O^{2-n}$, and a number of associated hydrolytic forms decompose by first- and second-order pathways yielding Fe^{III} and O_2/H_2O_2 (343).

Kinetic studies of these reactions were carried out by pulse radiolysis. The species $(aq)Fe^{IV}$ and $(aq)Fe^{V}$ were produced by radical oxidation and reduction of iron in stable oxidation states (344).

$$(aq)Fe^{III} + HO^{\bullet} \longrightarrow (aq)Fe^{IV} \tag{136}$$

$$(aq)Fe^{VI} + CO_2^{\bullet -} \longrightarrow (aq)Fe^{V} + CO_2 \tag{137}$$

As illustrated by the data in Table XIII, both $(aq)Fe^{VI}$ and $(aq)Fe^{V}$ react with a variety of reductants. As expected, $(aq)Fe^{V}$ is much more reactive and the rate constants increase with $[H^+]$. The oxidation of α-OH and α-NH_2-carboxylic acids takes place in a one-step two-electron process yielding $(aq)Fe^{III}$ and the oxidized product (345). Thermodynamically, this path is almost certainly more favorable than the one-electron process yielding $(aq)Fe^{IV}$ and the organic radical.

$$RCH(OH)COO^- + (aq)Fe^{V} \longrightarrow RC(O)COO^- + (aq)Fe^{III} \tag{138}$$

TABLE XIII
Kinetic Data (M^{-1} s^{-1}) for the Reactions of (aq)Fe^{V} and (aq)Fe^{VI} with Reductants in Aqueous Solution[a]

Reductant	(aq)Fe^{VI}	(aq)Fe^{V}	pH
$^{+}NH_3CH^{\cdot}COO^{-}$	1.4×10^{9}		12.4
$CO_2^{\cdot -}$	3.5×10^{8}		9.5–10.5
$O_2^{\cdot -}$	5.71×10^{3}		9.5–10.5
$HSCH_2CH(NH_3^{+})COO^{-}$	7.6×10^{2}	4.0×10^{9}	12.4
$^{+}NH_3CH_2COO^{-}$	97	8.4×10^{6}	12.4
NADH	77.0		11.0
$^{-}OOCCH_2CH(NH_3^{+})COO^{-}$		2.6×10^{6}	12.4
$^{-}COOCH_2CHOHCOO^{-}$		1.7×10^{2}	12.4
$^{-}OOCCH_2CH_2COO^{-}$		2.0×10^{1}	12.4
$HCOO^{-}$	2.33×10^{-2}	2.48×10^{3}	10.5
	4.0×10^{-1}		8.0
EtOH	1.47×10^{-3}		10.5
	8×10^{-2}		8.0

[a]Data from (338) and (343–345).

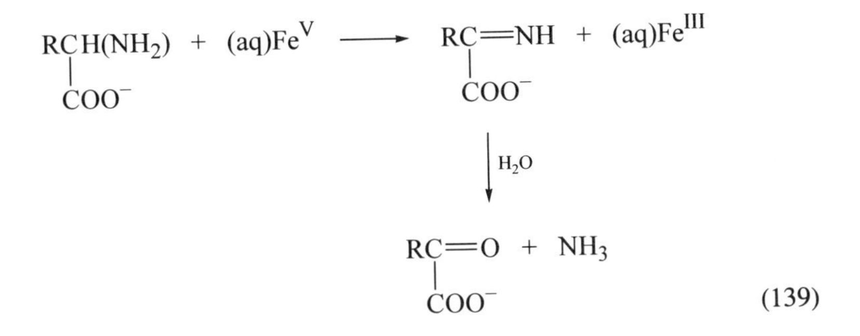

(139)

A pyrophosphate complex of Fe^{IV} oxidizes divalent pyrophosphate complexes of Mn^{II}, Fe^{II}, and Co^{II}, $k \sim 10^{6}\ M^{-1}\ s^{-1}$, as well as H_2O_2 ($k = 3.6 \times 10^{5}\ M^{-1}\ s^{-1}$). The disproportionation to Fe^{III} and Fe^{V} takes place with $k_{disp} = 1.0 \times 10^{6}\ M^{-1}\ s^{-1}$ at pH 10 (343).

The chemistry of stable Fe^{IV} complexes (346, 347) of ligand L^{12} (Fig. 22) has been explored only marginally.

The oxidation of phenols by the kinetically stable, pentacoordinated OFe^{IV}–(T2MPyP), Table XIV, has been examined (348) in aqueous solution as a model for the corresponding reaction of horseradish peroxidase compound II (HP II). The pH dependence of the OFe^{IV}(T2MPyP) reaction in the pH range 7.7–8.6 establishes the un-ionized ArOH as the reactive form. The reaction of 4-fluorophenol has a solvent kinetic isotope effect of 1.32. The results of several

TABLE XIV
Kinetic Data for the Oxidation of Phenols $X{-}C_6H_4OH$ by OFe^{IV}(T2MPyP) in Aqueous Solutions at pH 7.7[a]

X	$10^{-4}\,k\ (M^{-1}\,s^{-1})$
3-CN	0.424
3-F	1.57
H	1.78
4-F	1.92
4-Cl	2.44
4-Me	4.05
4-OMe	19.6

[a]Reproduced with permission from N. Colclough and J. R. L. Smith, *J. Chem. Soc. Perkin Trans. 2*, 1139 (1994).]

linear free energy correlations were taken as evidence that the reactions of OFe^{IV}-(T2MPyP) take place by hydrogen-atom abstraction, and those of HP II by electron transfer followed by proton loss. The latter proposal differs from that advanced earlier (349) according to which HP II reacts by simultaneous loss of H^+ and e^-.

3. *Ruthenium(IV)–(VI)*

The oxidation of alcohols (350, 351) and aromatic hydrocarbons (352) by the polypyridine oxo complexes $Ru(trpy)(bpy)O^{2+}$ and $Ru(bpy)_2(py)O^{2+}$ takes place by a concerted hydride transfer from α-C—H bonds to the Ru=O group. Strikingly, the hydrocarbons are more reactive than alcohols. All of the reactions are completed by rapid comproportionation (353) and, in the reactions with alcohols, subsequent slow reduction of Ru^{III}.

$$Ru^{IV}O + R_2CHOH \longrightarrow Ru^{II}(H_2O) + R_2CO \qquad (140)$$

$$Ru^{II}(H_2O) + Ru^{IV}O \longrightarrow 2Ru^{III}(OH) \qquad (141)$$

$$2Ru^{III}(OH) + R_2CHOH \longrightarrow 2Ru^{II}(H_2O) + R_2CO \qquad (142)$$

The hydride-transfer step is supported by enormous kinetic isotope effects, ranging from 9 (methanol) to 50 (benzyl alcohol) for oxidations with $Ru(bpy)_2(py)O^{2+}$. The reaction with HCO_2^- is sufficiently rapid to allow the observation of Ru^{II} (354) before the comproportionation step can take place, thus confirming a direct two-electron process. Consistent with hydride transfer, the reaction of $CHMe_2OH$ with $Ru(bpy)_2(py)^{18}O^{2+}$ failed to produce $Me_2C^{18}O^{2+}$ (350).

The compound $Ru(bpy)_2(py)O^{2+}$, unlike $(H_2O)_5CrO^{2+}$, exhibits clear reactivity preferences in its reactions with alcohols. The rate constants in acidic aqueous solutions have values 3.8×10^{-4} M^{-1} s^{-1} (MeOH), 10^{-3}–10^{-2} (primary alcohols), 10^{-2}–10^{-1} (secondary alcohols), 10^{0} (allyl-OH), 10^{0} (benzyl-OH).

The transfer of oxygen from $Ru(bpy)_2(py)O^{2+}$ to PPh_3 in MeCN is quantitative, as shown by O^{18} labeling experiments (355). The reaction proceeds through an observable intermediate, $(bpy)_2(py)RuOPPh_3^{2+}$, which forms rapidly (1.75×10^5 M^{-1} s^{-1}) and decomposes slowly (1.15×10^{-4} s^{-1}) to the final products, $Ru(bpy)_2(py)(MeCN)^{2+}$ and $OPPh_3$.

A closely related Ru^{IV} complex, $Ru(bpy)_2(O)(PPh_3)^{2+}$, and substituted analogues oxidize benzyl alcohol and its derivatives in reactions that also exhibit large primary kinetic isotope effects. For the reactions of unsubstituted $PhCH_2OH$ with $Ru(bpy)_2(O)(P(p\text{-}C_6H_4F)_3)^{2+}$ ($k_H = 0.85$ M^{-1} s^{-1}, in H_2O at pH 2) and $Ru(bpy)_2(O)(P[(C_6H_5)(p\text{-}C_6H_4CF_3)_2])^{2+}$ ($k_H = 1.80$ M^{-1} s^{-1}), the values of k_H/k_D are 17 and 13, respectively (356). The effects of substituents on both reaction partners were examined systematically. On the basis of free energy correlations, it was concluded that these reactions take place by hydrogen-atom abstraction from the benzylic carbon. The primary products, Ru^{III} and C-centered radicals, then react rapidly with each other to yield Ru^{II} and benzaldehyde.

The oxidation of alcohols by $LRu^{V}{=}O^{2+}$ (357), (L = [2-hydroxy-2-(2-pyridyl)ethyl]bis[2-(2-pyridyl)ethyl]amine) takes place according to Eq. 143. In aqueous 0.1 M $HClO_4$, the rate constants are 0.15 M^{-1} s^{-1} (MeOH), 1.7 (EtOH), 13.5 (2-PrOH), 107 (c-$C_6H_{11}OH$), and 117 ($PhCH_2OH$). This trend parallels that observed in the reaction of alcohols with $Ru(bpy)_2(py)O^{2+}$.

$$LRu^{V}(O)^{2+} + R_2CHOH \longrightarrow LRu^{III}(H_2O)^{2+} + R_2CO \qquad (143)$$

The same Ru^{V} complex also oxidizes cyclohexane and thf ($k_{thf} = 31$ M^{-1} s^{-1}).

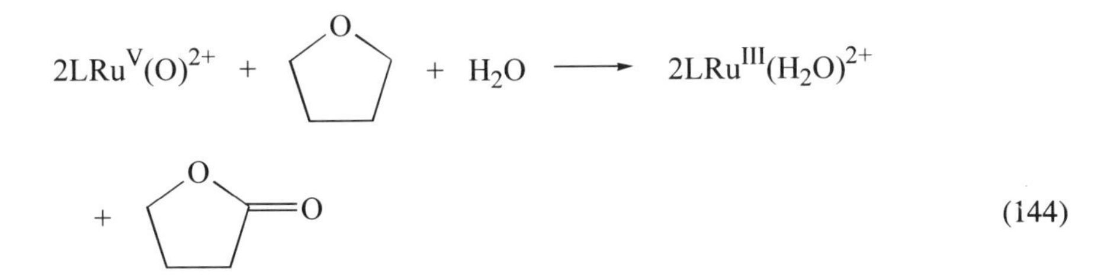

$$2LRu^{V}(O)^{2+} + \text{cyclohexane} \longrightarrow 2LRu^{III}(H_2O) + \text{cyclohexanone} \qquad (145)$$

The kinetic isotope effects k_H/k_D have values 5.3–5.9 for all the reactions examined (357). No evidence was obtained for the involvement of radical intermediates, and cyclobutanol was oxidized cleanly to cyclobutanone. The two proposed mechanistic possibilities include (a) hydride transfer and (b) hydrogen-atom transfer followed by rapid in-cage transformation of $\{Ru^{IV}{-}OH^{\bullet}, R^{\bullet}\}$ to $\{Ru^{III} + ROH\}$.

Ruthenium(V) complexes of tmc are reasonably potent oxidants. The unsubstituted form is unstable and disproportionates in acidic aqueous solutions (358)

$$(tmc)Ru^V(O)_2^+ + H^+ \overset{K_p}{\rightleftharpoons} (tmc)Ru^V(O)(OH)^{2+} \qquad K_p = 615\ M^{-1} \tag{146}$$

$$(tmc)Ru^V(O)(OH)^{2+} + (tmc)Ru^V(O)_2^+ \longrightarrow (tmc)Ru^{VI}(O)_2^{2+} + (tmc)Ru^{IV}(O)(OH)^+ \qquad k_{disp} = 2.72 \times 10^6\ M^{-1}\ s^{-1} \tag{147}$$

$$-d[Ru^V]/dt = \frac{2k_{disp}K_p[H^+][Ru(V)]^2}{(1 + K_p[H^+])^2} \tag{148}$$

but the halide and pseudohalide complexes readily oxidize $PhCH_2OH$ to PhCHO in MeCN. For example, $k = 210\ M^{-1}\ s^{-1}$ for $(tmc)Ru^V(O)(Cl)^{2+}$ and $140\ M^{-1}\ s^{-1}$ for $(tmc)Ru^V(O)(NCO)^{2+}$. The two complexes also act as electrocatalysts for the oxidation of $PhCH_2OH$ (359). This chemistry has been reviewed recently (360).

The Ru^{VI} complex *trans*-$(tmc)Ru(O)_2^{2+}$ oxidizes PPh_3 to $OPPh_3$, $k = 128\ M^{-1}\ s^{-1}$ in MeCN (361). The oxidation of alcohols is much slower ($PhCH_2OH$, $k = 2 \times 10^{-4}\ M^{-1}\ s^{-1}$) (362), in keeping with the low two-electron $Ru^{VI/IV}$ potential, 0.66 V versus SCE in MeCN at pH 1 (363). The more strongly oxidizing *trans*-$Ru(bpy)_2(O)_2^{2+}$ (E = 1.01 V vs. SCE) reacts more rapidly, $k = 21\ M^{-1}\ s^{-1}$ ($PhCH_2OH$) and 2.0 (2-PrOH) (362).

The stereochemistry around the nitrogen atoms for Ru(tmc) complexes in solution has not been established, but crystal structure determinations for several $(X)(tmc)Ru^{IV}(O)^{n+}$ ions (X = Cl, NCO, or MeCN) have shown that a number of isomers exist, including R, R, S, S (two up, two down), R, S, R, S (four up), and R, S, R, R (three up, one down), Fig. 27 (363–365). It would appear desirable to establish the reactivity of different isomers in oxidation reactions.

4. *Other Metal–Oxo Species*

Oxygen atom transfer reactions have been summarized recently (366) for complexes of molybdenum (367–374), rhenium (375, 376), tungsten (377), vanadium (378), and others. Both intermetallic oxygen transfer and that be-

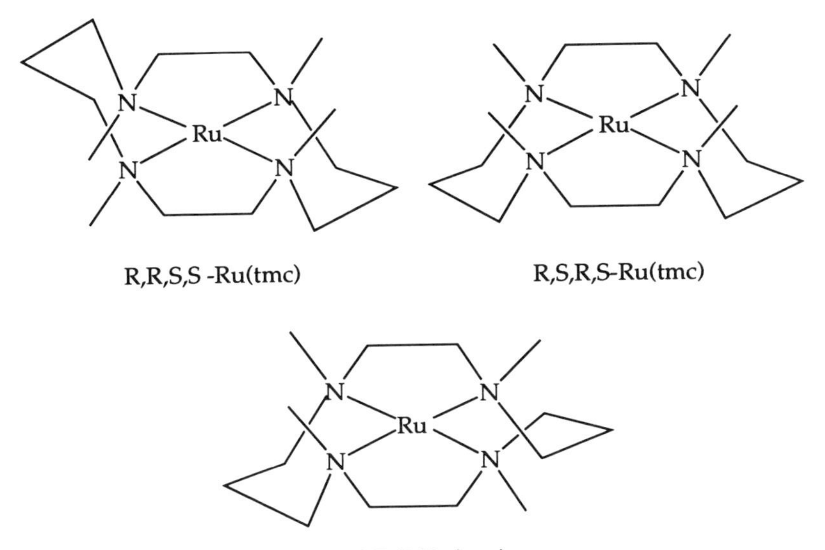

Figure 27. Schematic representation of the three structurally characterized forms of Rn(tmc) complexes.

tween metal complexes and nonmetallic substrates [X/XO = Me_2S/Me_2SO, $(Ar)_3P/(Ar)_3PO$, etc.], Eqs. 149 and 150, have been examined.

$$LMO + X \rightleftharpoons LM + XO \tag{149}$$

$$LM(O)_2 + L'M' \rightleftharpoons LM(O) + L'M'O \tag{150}$$

Most frequently, the complexes that have been used in these studies are synthetic models for enzymes transferases, which are involved in a number of biological oxygen-transfer reactions.

IV. CONCLUSIONS

The last two decades have seen tremendous advances in our understanding of metal–oxygen chemistry. However, many challenges remain. The mechanistic picture of O_2 activation by most naturally occurring systems is far from complete. In addition to intellectual satisfaction, such an understanding would be invaluable in the design of synthetic compounds with desired O_2-carrying and O_2-activating properties.

The mechanistic role of the metal and the chemical environment at various stages of O_2 activation will have to be uncovered in studies of thermodynamic, kinetic, and structural properties of intermediates involved in such reactions. Independent generation of short-lived intermediates and synthesis of their long-lived analogues with different metals and ligand systems will undoubtedly continue to be at the heart of future research efforts directed at unraveling the many remaining mysteries in this area.

From the more practical standpoint, the design of novel compounds that will reproduce or exceed the O_2-activating properties of compounds already existing in nature is a goal worth pursuing. The environmental and energy considerations clearly point to O_2 as one of the most desirable future industrial oxidants, requiring that efficient new metal-based catalysts be developed. Both theoretical and practical aspects of such an endeavor present challenges that almost guarantee strong interaction between scientists in various fields.

ABBREVIATIONS

$ABTS^{2-}$	2,2′-Azinobis(3-ethylbenzothiazoline-6-sulfonate) ion
acac	Acetylacetonate
acacen	*N,N′*-Ethylenebis(acetylacetoneiminato) ion (2−)
[14]aneN_4	1,4,8,11-Tetraazacyclotetradecane
AP	2-Anilinopyridinato anion
(aq)Mn^{n+}	Metal ion with all the accompanying aqua, hydroxo, and oxo ligands in aqueous solution
Ar	Aryl
bpy	2,2′-Bipyridine
bzacen	*N,N′*-Ethylenebis(benzoylacetoneiminato) ion (2−)
cat	Catalyst
Cp*	η^5-Pentamethylcyclopentadienyl ion (−)
DMF	*N,N*-Dimethylformamide
dien	Diethylenetriamine
dmgH_2	Dimethylglyoxime
dmso	Dimethyl sulfoxide (ligand)
DMSO	Dimethyl sulfoxide (solvent)
edta	Ethylenediaminetetraacetate ion (4−)
en	Ethylenediamine = 1,2-diaminoethane
ESR	Electron spin resonance
his	Histidine
HB$(pz)_3$	Hydridotris(pyrazolyl)borate
HPTS	*p*-Toluene sulfonic acid
HR	Hemerythrin
im	Imidazole
IR	Infrared

Me_6-[14]aneN_4	*meso*-1,4,8,11-Tetraaza-5,7,7,12,14,14,-hexamethyl-cyclotetradecane
MeIm	1-Methylimidazole
MMO	Methane monooxygenase
NADH	Nicotinamide adenine dinucleotide
OEP	Octaethylporphyrinato ion (2−)
papd	1,5,8,11,15-Pentaazapentadecane
Pc	Phthalocyanine
phen	1,10-Phenanthroline
PhIO	Iodosobenzene
py	Pyridine
pzH	Pyrazol
RR	Rubonucleotide reductase
salen	*N,N′*-Ethylenebis(salicylideneiminato) ion (2−)
salprn	*N,N′*-propylenebis(salicylideneiminato) ion (2−)
saltmen	*N,N′*-(1,1,2,2-tetramethylethylene)bis(salicylideniminato) ion (2−)
SCE	Standard calomel electrode
sep	Sepulchrate = 1,3,6,8,10,13,16,19-octaazabicyclo-[6.6.6]eicosane
tacn	1,4,7-Triazacyclononane
T2MPyP	Tetra(2-methylpyridyl)porphyrinato ion (2−)
tetren	1,4,7,10,13-Pentaazatridecane (tetraethylenepentamine)
thf	Tetrahydrofuran (ligand or reactant)
THF	Tetrahydrofuran (solvent)
tim	1,4,8,11-Tetraaza-2,3,9,10-tetramethylcyclotetradeca-1,3,8,10-tetraene
tmc	Tetramethylcyclam = 1,4,8,11-tetramethyl-1,4,8,11-tetraazacyclotetradecane
TpivPP	*meso*-Tetra($\alpha,\alpha,\alpha,\alpha$-*o*-pivalamidophenyl)porphyrinato ion (2−)
TPP	*meso*-Tetraphenylporphyrinato ion (2−)
trien	Triethylenetetramine
trpy	2,2′,2″-Terpyridine
ttha	Triethylenetetraminehexaacetate ion (6−)
UV	Ultraviolet
vis	visible

ACKNOWLEDGMENTS

Our group's work described in this chapter was supported by the US Department of Energy, Office of Basic Energy Research, Chemical Sciences Division, under Contract

W-7405-Eng-82, and by the National Science Foundation, grants CHE-9007283 and CHE-9303388. I am grateful to all my co-workers for their hard work and enthusiasm for this project, to A. Feig, S. Lippard, and F. Anson for preprints of their papers, and to the reviewers for many useful suggestions and comments.

REFERENCES

1. L. I. Simandi, *Catalytic Activation of Dioxygen by Metal Complexes*, Kluwer Academic Publishers, Dordrecht/Boston/London, 1992.
2. A. Sobkowiak, H.-C. Tung, and D. T. Sawyer, *Progress in Inorganic Chemistry*, Wiley-Interscience, New York, 1992, Vol. 40, pp. 291–352.
3. K. D. Karlin, *Science*, *261*, 701 (1993).
4. A. L. Feig and S. J. Lippard, *Chem. Rev.*, *94*, 759 (1994).
5. D. H. R. Barton, A. E. Martell, and D. T. Sawyer, Eds., *The Activation of Dioxygen and Homogeneous Catalytic Oxidation*, Plenum, New York, 1993.
6. A. E. Martell and D. T. Sawyer, Eds., *Oxygen Complexes and Oxygen Activation by Transition Metals*, Plenum, New York, 1988.
7. K. Shikama, *Coord. Chem. Rev.*, *83*, 73 (1988).
8. P. C. Wilkins and R. G. Wilkins, *Coord. Chem. Rev.*, *79*, 195 (1987).
9. K. Karlin, Z. Tyeklar, and A. D. Zuberbühler, in *Bioinorganic Catalysis*, J. Reedijk, Ed., Marcel-Dekker, New York, 1993, pp. 261–315.
10. R. A. Sheldon and J. K. Kochi, *Metal-Catalyzed Oxidations of Organic Compounds*, Academic, New York, 1981.
11. G. McLendon and A. E. Martell, *Coord. Chem. Rev.*, *19*, 1 (1976).
12. L. I. Simandi, Ed., *Dioxygen Activation and Homogeneous Catalytic Oxidation*, Elsevier, The Netherlands, 1991.
13. R. S. Drago, *Coord. Chem. Rev.*, *117*, 185 (1992).
14. I. P. Skibida, *Russ. Chem. Rev.*, *54*, 875 (1985).
15. K. Nakamoto, *Coord. Chem. Rev.*, *100*, 363 (1990).
16. G. V. Buxton, C. L. Greenstock, W. P. Helman, and A. B. Ross, *J. Phys. Chem. Ref. Data*, *17*, 513 (1988).
17. B. H. J. Bielski, D. E. Cabelli, R. L. Arudi, and A. B. Ross, *J. Phys. Chem. Ref. Data*, *14*, 1041 (1985).
18. (a) J. W. Egan, Jr., B. S. Haggerty, A. L. Rheingold, S. C. Sendlinger, and K. H. Theopold, *J. Am. Chem. Soc.*, *112*, 2445 (1990). (b) O. M. Reinaud and K. H. Theopold, *J. Am. Chem. Soc.*, *116*, 6979 (1994); (c) K. Fujisawa, M. Tanaka, Y. Moro-oka, and N. Kitajima, *J. Am. Chem. Soc.*, *116*, 12079 (1994); (d) M. Harate, K. Jitsukawa, H. Masuda, and H. Einaga, *J. Am. Chem. Soc.*, *116*, 10817 (1994).
19. L. M. Proniewicz, T. Isobe, and K. Nakamoto, *Inorg. Chim. Acta*, *155*, 91 (1989).
20. T. Watanabe, T. Ama, and K. Nakamoto, *J. Phys. Chem.*, *88*, 440 (1984).

21. S. Chang, G. Blyholder, and J. Fernandez, *Inorg. Chem.*, *20*, 2813 (1981).
22. K. Nakamoto, T. Watanabe, T. Ama, and M. W. Urban, *J. Am. Chem. Soc.*, *104*, 3744 (1982).
23. B. S. Tovrog, D. J. Kitko, and R. S. Drago, *J. Am. Chem. Soc.*, *98*, 5144 (1976).
24. R. S. Drago and B. B. Corden, *Acc. Chem. Res.*, *13*, 353 (1980).
25. C. L. Bailey and R. S. Drago, *Coord. Chem. Rev.*, *79*, 321 (1987).
26. L. Vaska, *Acc. Chem. Res.*, *9*, 175 (1976).
27. R. D. Jones, D. A. Summerville, and F. Basolo, *Chem. Rev.*, *79*, 139 (1979).
28. J. P. Collman, J. I. Brauman, T. R. Halbert, and K. S. Suslick, *Proc. Natl. Acad. Sci. USA*, *73*, 3333 (1976).
29. E. C. Niederhofer, J. H. Timmons, and A. E. Martell, *Chem. Rev.*, *84*, 137 (1984).
30. A. Bakac, S. L. Scott, J. H. Espenson, and K. L. Rodgers, submitted.
31. M. J. Carter, D. P. Rillema, and F. Basolo, *J. Am. Chem. Soc.*, *96*, 392 (1974).
32. F. Basolo, B. M. Hoffman, and J. A. Ibers, *Acc. Chem. Res.*, *8*, 384 (1975).
33. A. E. Martell, *Acc. Chem. Res.*, *15*, 155 (1982).
34. P. Zanello, R. Cini, A. Cinquantini, and P. L. Orioli, *J. Chem. Soc. Dalton Trans.*, 2159 (1983).
35. D. H. Busch, in *Oxygen Complexes and Oxygen Activation by Transition Metals*, A. E. Martell and D. T. Sawyer, Eds., Plenum, New York and London, 1988, pp. 61–85.
36. B. K. Coltrain, N. Herron, and D. H. Busch, in *The Activation of Dioxygen and Homogeneous Catalytic Oxidation*, D. H. R. Barton, A. E. Martell, and D. T. Sawyer, Eds., Plenum, New York and London, 1988, pp. 359–380.
37. A. E. Martell, A. K. Basak, and C. J. Raleigh, *Pure Appl. Chem.*, *60*, 1325 (1988).
38. W. Nam, R. Ho, and J. S. Valentine, *J. Am. Chem. Soc.*, *113*, 7052 (1991).
39. A. Böttcher, H. Elias, E.-G. Jäger, H. Langfelderova, M. Mazur, L. Müller, H. Paulus, P. Pelikan, M. Rudolph, and M. Valko, *Inorg. Chem.*, *32*, 4131 (1993).
40. M. B. Hall, in *Oxygen Complexes and Oxygen Activation by Transition Metals*, A. E. Martell and D. T. Sawyer, Eds., Plenum, New York and London, 1988, pp. 3–16.
41. D. M. Kurtz, Jr., D. F. Shriver, and I. M. Klotz, *J. Am. Chem. Soc.*, *98*, 5033 (1976).
42. R. E. Stenkamp, *Chem. Rev.*, *94*, 715 (1994).
43. R. G. Wilkins, *Adv. Chem. Ser.*, *100*, 111 (1971).
44. J. Simplicio and R. G. Wilkins, *J. Am. Chem. Soc.*, *91*, 1325 (1969).
45. V. L. Goedken, N. K. Kildahl, and D. H. Busch, *J. Coord. Chem.*, *7*, 89 (1977).
46. T. D. Smith and J. R. Pilbrow, *Coord. Chem. Rev.*, *39*, 295 (1981).
47. W. Lubitz, C. J. Winscom, H. Diegruber, and R. Moseler, *Z. Naturforsch.*, *42a*, 970 (1987).

48. S. Kawanishi and S. Sano, *J. Chem. Soc. Chem. Commun.*, 1628 (1984).
49. E. Kimura, M. Shionoya, T. Mita, and Y. Iitaka, *J. Chem. Soc. Chem. Commun.*, 1712 (1987).
50. F. A. Walker, *J. Am. Chem. Soc.*, *92*, 4235 (1970).
51. A. S. Abusamleh, P. J. Chmielewski, P. R. Warburton, L. Morales, N. A. Stephenson, and D. H. Busch, *J. Coord. Chem.*, *23*, 91 (1991).
52. S. Falab and P. R. Mitchell, *Adv. Inorg. Bioinorg. Mech.*, *3*, 311 (1984).
53. L. I. Simandi, C. R. Savage, Z. A. Schelly, and S. Nemeth, *Inorg. Chem.*, *21*, 2765 (1982).
54. F. Miller, J. Simplicio, and R. G. Wilkins, *J. Am. Chem. Soc.*, *91*, 1962 (1969).
55. A. Bakac and J. H. Espenson, *J. Am. Chem. Soc.*, *112*, 2273 (1990).
56. T. Dhanasekaran and P. Natarajan, *J. Am. Chem. Soc.*, *114*, 4621 (1992).
57. M. Maeder and H. R. Macke, *Inorg. Chem.*, *33*, 3135 (1994).
58. (a) C.-L. Wong, J. A. Switzer, K. P. Balakrishnan, and J. F. Endicott, *J. Am. Chem. Soc.*, *102*, 5511 (1980); (b) C.-L. Wong, and J. F. Endicott, *Inorg. Chem.*, *20*, 2233 (1981).
59. J. F. Endicott and K. Kumar, *ACS Symp. Ser.*, *198*, 425 (1982).
60. N. Shinohara, K. Ishii, and M. Hirose, *J. Chem. Soc. Chem. Commun.*, 700 (1990).
61. (a) A. Marchaj, A. Bakac, and J. H. Espenson, *Inorg. Chem.*, *31*, 4164 (1992); (b) M. Zhang, R. van Eldik, J. H. Espenson, and A. Bakac, *Inorg. Chem.*, *33*, 130 (1994).
62. A. Bakac and J. H. Espenson, *Inorg. Chem.*, *29*, 2062 (1990).
63. E. Farinas, N. Baidya, and P. K. Mascharak, *Inorg. Chem.*, *33*, 5970 (1994).
64. H. Taube, *Progress in Inorganic Chemistry*, Wiley-Interscience, New York, 1986, Vol. 34, pp. 607–625.
65. B. H. J. Bielski, *Photochem. Photobiol.*, *28*, 645 (1978).
66. M. P. Liteplo and J. F. Endicott, *Inorg. Chem.*, *10*, 1420 (1971).
67. M. E. Brynildson, A. Bakac, and J. H. Espenson, *J. Am. Chem. Soc.*, *109*, 4579 (1987).
68. A. Bakac and J. H. Espenson, *Acc. Chem. Res.*, *26*, 519 (1993).
69. C. K. Chang and M. P. Kondylis, *J. Chem. Soc. Chem. Commun.*, 316 (1986).
70. R. S. Drago, J. P. Cannady, and K. A. Leslie, *J. Am. Chem. Soc.*, *102*, 6014 (1980).
71. M. Momenteau and D. Lavalette, *J. Chem. Soc. Chem. Commun.*, 341 (1982).
72. I. P. Gerothanasis, M. Momenteau, and B. Loock, *J. Am. Chem. Soc.*, *111*, 7006 (1989).
73. D. M. Kurtz, Jr., *Chem. Rev.*, *90*, 585 (1990).
74. T. M. Loehr, in *Oxygen Complexes and Oxygen Activation by Transition Metals*, A. E. Martell and D. T. Sawyer, Eds., Plenum, New York and London, 1988, pp. 17–32.

75. M. Momenteau and C. A. Reed, *Chem. Rev.*, *94*, 659 (1994).
76. M. H. Gubelmann, S. Ruttimann, B. Bocquet, and A. F. Williams, *Helv. Chim. Acta*, *73*, 1219 (1990).
77. J. H. Bayston and M. E. Winfield, *J. Catal.*, *3*, 123 (1964).
78. J. H. Bayston, R. N. Beale, N. K. King, and M. E. Winfield, *Aust. J. Chem.*, *16*, 954 (1963).
79. L. D. Brown and K. N. Raymond, *Inorg. Chem.*, *14*, 2595 (1975).
80. J. C. Dabrowiak and D. H. Busch, *Inorg. Chem.*, *14*, 1881 (1975).
81. H. S. Mountford, L. O. Spreer, J. W. Otvos, M. Calvin, K. J. Brewer, M. Richter, and B. Scott, *Inorg. Chem.*, *31*, 717 (1992).
82. A. McAuley and C. Xu, *Inorg. Chem.*, *31*, 5549 (1992).
83. N. M. Levy, E. Stadler, A. S. Mangrich, M. A. C. Melo, and V. Drago, *Trans. Met. Chem.*, *18*, 613 (1993).
84. E. Kimura, M. Kodama, R. Machida, and K. Ishizu, *Inorg. Chem.*, *21*, 595 (1982).
85. P. George, *J. Chem. Soc.*, 4349 (1954).
86. A. M. Posner, *Trans. Faraday Soc.*, *49*, 382 (1953).
87. H. E. Toma and A. C. C. Silva, *Can. J. Chem.*, *64*, 1280 (1986).
88. E. R. Brown and J. D. Mazzarella, *J. Electroanal. Chem.*, *222*, 173 (1987).
89. L. D. Dickerson, A. Sauer-Masarwa, N. Herron, C. M. Fendrick, and D. H. Busch, *J. Am. Chem. Soc.*, *115*, 3623 (1993).
90. N. Herron and D. H. Busch, *J. Am. Chem. Soc.*, *103*, 1236 (1981).
91. N. Herron, L. L. Zimmer, J. J. Grzybowski, D. J. Olszanski, S. C. Jackels, R. W. Callahan, J. H. Cameron, G. G. Christoph, and D. H. Busch, *J. Am. Chem. Soc.*, *105*, 6585 (1983).
92. J. H. Cameron, M. Kojima, B. Korybut-Daszkiewicz, B. K. Coltrain, T. J. Meade, N. W. Alcock, and D. H. Busch, *Inorg. Chem.*, *26*, 427 (1987).
93. A. Sauer-Masarwa, N. Herron, C. M. Fendrick, and D. H. Busch, *Inorg. Chem.*, *32*, 1086 (1993).
94. D. H. Busch and N. W. Alcock, *Chem. Rev.*, *94*, 585 (1994).
95. J. P. Collman, *Acc. Chem. Res.*, *10*, 265 (1977).
96. D.-H. Chin, J. D. Gaudio, G. N. LaMar, and A. L. Balch, *J. Am. Chem. Soc.*, *99*, 5486 (1977).
97. I. R. Paeng, H. Shiwaku, and K. Nakamoto, *J. Am. Chem. Soc.*, *110*, 1995 (1988).
98. J. P. Collman, R. R. Gagne, C. A. Reed, T. R. Halbert, G. Lang, and W. T. Robinson, *J. Am. Chem. Soc.*, *97*, 1427 (1975).
99. E. Tsuchida and H. Nishide, *Top. Curr. Chem.*, *132*, 63 (1986).
100. J. Almong, J. E. Baldwin, M. J. Crossley, J. F. Debernadis, R. L. Dyer, J. R. Huff, and M. K. Peters, *Tetrahedron*, *37*, 3589 (1981).
101. J. E. Baldwin and P. Perlmutter, *Top. Curr. Chem.*, *121*, 181 (1984).
102. M. Momenteau, *Pure Appl. Chem.*, *58*, 1493 (1986).

103. J. P. Collman, J. I. Brauman, T. J. Collins, B. L. Iverson, G. Lang, R. B. Pettman, J. L. Sessler, and M. A. Walters, *J. Am. Chem. Soc.*, *105*, 3038 (1983).

104. D. El-Kasmi, C. Tetreau, D. Lavalette, and M. Momenteau, *J. Chem. Soc. Perkin Trans. 2*, 1799 (1993).

105. S. David, B. R. James, D. Dolphin, T. G. Traylor, and M. A. Lopez, *J. Am. Chem. Soc.*, *116*, 6 (1994).

106. J. P. Collman, J. I. Brauman, B. L. Iverson, J. L. Sessler, R. M. Morris, and Q. H. Gibson, *J. Am. Chem. Soc.*, *105*, 3052 (1983).

107. E. Rose, B. Boitrel, M. Quelquejeu, and A. Kossanyi, *Tetrahedron Lett.*, *34*, 7267 (1993).

108. C. A. Savicki and Q. H. Gibson, *J. Biol. Chem.*, *252*, 7538 (1977).

109. G. Ilgenfritz and T. M. Schuster, *J. Biol. Chem.*, *249*, 2959 (1974).

110. (a) E. Tsuchida, E. Hasegawa, and K. Honda, *Biochem. Biophys. Res. Commun.*, *67*, 864 (1975); (b) E. Bayer and G. Holzbach, *Angew. Chem., Int. Ed. Engl.*, *16*, 117 (1977). (c) T. G. Traylor, M. J. Mitchell, J. P. Ciccone, and S. Nelson, *J. Am. Chem. Soc.*, *104*, 4986 (1982); (d) J. Rebek, Jr., R. V. Wattley, T. Costello, R. Gadwood, and L. Marshall, *Angew. Chem. Int. Ed. Engl.*, *20*, 605 (1981).

111. F. P. Guengerich, *Am. Sci.*, *81*, 440 (1993).

112. J. P. Collman, P. C. Herrmann, B. Boitrel, X. Zhang, T. A. Eberspacher, and L. Fu, *J. Am. Chem. Soc.*, *116*, 9783 (1994).

113. (a) W. Micklitz, S. G. Bott, J. G. Bentsen, and S. J. Lippard, *J. Am. Chem. Soc.*, *111*, 372 (1989); (b) L. Que, Jr., and A. E. True, *Progress in Inorganic Chemistry*, Wiley Interscience, New York, 1990, Vol. 38, pp. 97–200; (c) B. Mauerer, J. Crane, J. Schuler, K. Wieghardt, and B. Nuber, *Angew. Chem. Int. Ed. Engl.*, *32*, 289 (1993).

114. A. C. Rosenzweig, C. A. Frederick, S. J. Lippard, and P. Nordlund, *Nature (London)*, *366*, 537 (1993).

115. (a) W. B. Tolman, S. Liu, J. G. Bentsen, and S. J. Lippard, *J. Am. Chem. Soc.*, *113*, 152 (1991); (b) K. E. Liu, C. C. Johnson, M. Newcomb, and S. J. Lippard, *J. Am. Chem. Soc.*, *115*, 939 (1993).

116. B. G. Fox, W. A. Froland, J. E. Dege, and J. D. Lipscomb, *J. Biol. Chem.*, *264*, 10023 (1989).

117. J. Green and H. J. Dalton, *J. Biol. Chem.*, *264*, 17698 (1989).

118. L. Que, Jr., *Science*, *253*, 273 (1991).

119. J. M. Bollinger, Jr., D. E. Edmondson, B.-H. Huynh, J. Filley, J. R. Norton, and J. Stubbe, *Science*, *253*, 292 (1991).

120. J. M. Bollinger, Jr., J. Stubbe, B.-H. Huynh, and D. E. Edmondson, *J. Am. Chem. Soc.*, *113*, 6289 (1991).

121. D. P. Goldberg, S. P. Watton, A. Masschelein, L. Wimmer, and S. J. Lippard, *J. Am. Chem. Soc.*, *115*, 5346 (1993).

122. Y. Dong, S. Menage, B. A. Brennan, T. E. Elgren, H. G. Jang, L. L. Pearce, and L. Que, Jr., *J. Am. Chem. Soc.*, *115*, 1851 (1993).

123. A. L. Feig and S. J. Lippard, *J. Am. Chem. Soc.*, *116*, 8410 (1994).

124. N. Kitajima, N. Tamura, H. Amagai, H. Fukui, Y. Moro-oka, Y. Mizutani, T. Kitagawa, R. Mathur, K. Heerwegh, C. A. Reed, C. R. Randall, L. Que, Jr., and K. Tatsumi, *J. Am. Chem. Soc.*, *116*, 9071 (1994).
125. (a) J. Piccard, *Berichte*, *46*, 2477 (1913); (b) M. Ardon and R. A. Plane, *J. Am. Chem. Soc.*, *81*, 3197 (1959); (c) R. W. Kolaczkowski and R. A. Plane, *Inorg. Chem.*, *3*, 322 (1964).
126. Y. A. Ilan, G. Czapski, and M. Ardon, *Isr. J. Chem.*, *13*, 15 (1975).
127. R. M. Sellers and M. G. Simic, *J. Am. Chem. Soc.*, *98*, 6145 (1976).
128. S. L. Scott, A. Bakac, and J. H. Espenson, *J. Am. Chem. Soc.*, *113*, 7787 (1991).
129. S. L. Scott, A. Bakac, and J. H. Espenson, *J. Am. Chem. Soc.*, *114*, 4205 (1992).
130. L. Batt, *Int. Rev. Phys. Chem.*, *6*, 53 (1987).
131. C. von Sonntag and H.-P. Schuchmann, *Angew. Chem. Int. Ed. Engl.*, *30*, 1229 (1991).
132. Q. Niu and G. D. Mendenhall, *J. Am. Chem. Soc.*, *112*, 1656 (1990).
133. Q. J. Niu and G. D. Mendenhall, *J. Am. Chem. Soc.*, *114*, 165 (1992).
134. C. Kang and F. C. Anson, *Inorg. Chem.*, *33*, 2624 (1994).
135. J. H. Espenson, A. Bakac, and J. Janni, *J. Am. Chem. Soc.*, *116*, 3436 (1994).
136. W.-D. Wang, A. Bakac, and J. H. Espenson, *Inorg. Chem.*, *32*, 5034 (1993).
137. A. C. Adams, J. R. Crook, F. Bockhoff, and E. L. King, *J. Am. Chem. Soc.*, *90*, 5761 (1968).
138. A. Al-Ajlouni, J. H. Espenson, and A. Bakac, *Inorg. Chem.*, *32*, 3162 (1993).
139. S. L. Scott, A. Bakac, and J. H. Espenson, *J. Am. Chem. Soc.*, *114*, 4605 (1992).
140. E. Pedersen, *Acta Chem. Scand.*, *26*, 333 (1972).
141. D. J. Liston and B. O. West, *Inorg. Chem.*, *24*, 1568 (1985).
142. M. Di Vaira and F. Mani, *Inorg. Chem.*, *23*, 409 (1984).
143. A. Bakac and J. H. Espenson, *Inorg. Chem.*, *31*, 1108 (1992).
144. A. Bakac, unpublished observations.
145. A. D. Zuberbühler, *Helv. Chim. Acta*, *53*, 473 (1970).
146. L. Mi and A. D. Zuberbühler, *Helv. Chim. Acta*, *74*, 1679 (1991).
147. A. D. Zuberbühler, *Helv. Chim. Acta*, *59*, 1448 (1976).
148. M. Guntensperger and A. D. Zuberbühler, *Helv. Chim. Acta*, *60*, 2584 (1977).
149. G. V. Buxton, J. C. Green, and R. M. Sellers, *J. Chem. Soc. Dalton Trans.*, 2160 (1976).
150. J. S. Thompson, *J. Am. Chem. Soc.*, *106*, 4057 (1984).
151. S. Goldstein and G. Czapski, *J. Am. Chem. Soc.*, *105*, 7276 (1983).
152. A. Nanthakumar, S. Fox, S. M. Nasir, N. Ravi, B.-H. Huynh, R. D. Orosz, E. P. Day, K. S. Hagen, and K. D. Karlin, in *The Activation of Dioxygen and Homogeneous Catalytic Oxidation*, D. H. R. Barton, A. E. Martell, and D. T. Sawyer, Eds., Plenum, New York and London, 1988, pp. 381–394.
153. N. Kitajima, K. Fujisawa, C. Fujimoto, Y. Moro-oka, S. Hashimoto, T. Kitagawa, K. Toriumi, K. Tatsumi, and A. Nakamura, *J. Am. Chem. Soc.*, *114*, 1277 (1992).

154. Z. Tyeklar and K. D. Karlin, *Acc. Chem. Res.*, *22*, 241 (1989).

155. N. Kitajima, *Adv. Inorg. Chem.*, *39*, 1 (1992); (b) N. Kitajima and Y. Moro-oka, *Chem. Rev.*, *94*, 737 (1994).

156. K. A. Magnus, H. Ton-That, and J. E. Carpenter, in *Bioinorganic Chemistry of Copper*; K. D. Karlin and Z. Tyeklar, Eds.; Chapman & Hall, New York, 1993, pp. 143–150.

157. (a) R. W. Cruse, S. Kaderli, K. D. Karlin, and A. D. Zuberbühler, *J. Am. Chem. Soc.*, *110*, 6882 (1988); (b) K. D. Karlin, M. S. Nasir, B. I. Cohen, R. W. Cruse, S. Kaderli, and A. D. Zuberbühler, *J. Am. Chem. Soc.*, *116*, 1324 (1994).

158. R. Menif, A. E. Martell, P. J. Squattrito, and A. Clearfield, *Inorg. Chem.*, *29*, 4723 (1990).

159. S. Mahapatra, J. A. Halfen, E. C. Wilkinson, L. Que, Jr., and W. B. Tolman, *J. Am. Chem. Soc.*, *116*, 9785 (1994).

160. (a) K. D. Karlin, N. Wei, B. Jung, S. Kaderli, P. Niklaus, and A. D. Zuberbühler, *J. Am. Chem. Soc.*, *115*, 9506 (1993); (b) K. D. Karlin, N. Wei, B. Jung, S. Kaderli, and A. D. Zuberbühler, *J. Am. Chem. Soc.*, *113*, 5868 (1991).

161. N. Wei, N. N. Murthy, Q. Chen, J. Zubieta, and K. D. Karlin, *Inorg. Chem.*, *33*, 1953 (1994).

162. G. Davies and M. A. El-Sayed, *Inorg. Chem.*, *22*, 1257 (1983).

163. M. A. El-Sayed, A. Abu-Raqabah, G. Davies, and A. El-Toukhy, *Inorg. Chem.*, *28*, 1909 (1989).

164. M. A. El-Sayed, A. El-Toukhy, and G. Davies, *Inorg. Chem.*, *24*, 3387 (1985).

165. (a) H. H. Thorp and G. W. Brudvig, *New J. Chem.*, *15*, 479 (1991); (b) G. W. Brudvig and R. H. Crabtree, *Progress in Inorganic Chemistry*, Wiley-Interscience, New York, 1989, Vol. 37, pp. 99–142.

166. C. P. Horwitz and G. C. Dailey, *Comm. Inorg. Chem.*, *14*, 283 (1993).

167. G. Christou, *Acc. Chem. Res.*, *22*, 328 (1989).

168. V. McKee, *Adv. Inorg. Chem.*, *40*, 323 (1994).

169. V. L. Pecoraro, M. J. Baldwin, and A. Gelasco, *Chem. Rev.*, *94*, 807 (1994).

170. C. A. McAuliffe, H. F. Al-Khateeb, D. S. Barratt, J. C. Briggs, A. Challita, A. Hosseiny, M. G. Little, A. G. Mackie, and K. Minten, *J. Chem. Soc. Dalton Trans.* 2147 (1983).

171. B. M. Hoffman, T. Szymanski, T. G. Brown, and F. Basolo, *J. Am. Chem. Soc.*, *100*, 7253 (1978).

172. M. W. Urban, K. Nakamoto, and F. Basolo, *Inorg. Chem.*, *21*, 3406 (1982).

173. T. Watanabe, T. Ama, and K. Nakamoto, *Inorg. Chem.*, *22*, 2470 (1983).

174. A. B. P. Lever, J. P. Wilshire, and S. K. Quan, *J. Am. Chem. Soc.*, *101*, 3668 (1979).

175. M. Perree-Fauvet, A. Gaudemer, J. Bonvoisin, J. J. Girerd, C. Boucly-Goester, and P. Boucly, *Inorg. Chem.*, *28*, 3533 (1989).

176. (a) C. P. Horwitz, Y. Ciringh, C. Liu, and S. Park, *Inorg. Chem.*, *32*, 5951 (1993); (b) C. P. Horwitz, P. J. Winslow, J. T. Warden, and C. A. Lisek, *Inorg. Chem.*, *32*, 82 (1993).

177. (a) F. M. MacDonnell, N. L. P. Fackler, C. Stern, and T. V. O'Halloran, *J. Am. Chem. Soc.*, *116*, 7431 (1994); (b) T. J. Collins, R. D. Powell, C. Slebodnick, and E. S. Uffelman, *J. Am. Chem. Soc.*, *112*, 899 (1990); (c) J. M. Workman, R. D. Powell, A. D. Procyk, T. J. Collins, and D. F. Bocian, *Inorg. Chem.*, *31*, 1548 (1992); (d) T. J. Collins and S. W. Gordon-Wylie, *J. Am. Chem. Soc.*, *111*, 4511 (1989).
178. N. Kitajima, H. Komatsuzaki, S. Hikichi, M. Osawa, and Y. Moro-oka, *J. Am. Chem. Soc.*, *116*, 11596 (1994).
179. F. P. Rotzinger and M. Grätzel, *Inorg. Chem.*, *26*, 3704 (1987).
180. R. C. Thompson, *Inorg. Chem.*, *23*, 1794 (1984).
181. J. D. Rush and B. H. J. Bielski, *Inorg. Chem.*, *24*, 4282 (1985).
182. W. Kruse and D. Thusius, *Inorg. Chem.*, *7*, 464 (1968).
183. J. M. Malin and J. H. Swinehart, *Inorg. Chem.*, *7*, 250 (1968).
184. J. M. Malin and J. H. Swinehart, *Inorg. Chem.*, *8*, 1407 (1969).
185. T. K. Myser and R. E. Shepherd, *Inorg. Chem.*, *26*, 1544 (1987).
186. (a) Y. Sasaki and A. G. Sykes, *J. Chem. Soc. Chem. Commun.*, 767 (1973); (b) E. F. Hills, P. R. Norman, T. Ramasami, D. T. Richens, and A. G. Sykes, *J. Chem. Soc. Dalton Trans.*, 157 (1986); (c) E. F. Hills and A. G. Sykes, *Polyhedron*, *5*, 511 (1986).
187. Y. Sasaki and A. G. Sykes, *J. Chem. Soc. Dalton Trans.*, 1048 (1975).
188. M. Moszner and J. J. Ziolkowski, *Inorg. Chim. Acta*, *145*, 299 (1988).
189. M. Moszner, M. Wilgocki, and J. J. Ziolkowski, *J. Coord. Chem.*, *20*, 219 (1989).
190. M. Moszner and J. J. Ziolkowski, *Trans. Met. Chem.*, *18*, 248 (1993).
191. S. P. Ghosh, E. Gelerinter, G. Pyrka, and E. S. Gould, *Inorg. Chem.*, *32*, 4780 (1993).
192. I. J. Ellison, R. D. Gillard, M. Moszner, M. Wilgocki, and J. J. Ziolkowski, *J. Chem. Soc. Dalton Trans.*, 2531 (1994).
193. I. J. Ellison and R. D. Gillard, *J. Chem. Soc. Chem. Commun.*, 851 (1992).
194. (a) A. N. Buckley, J. A. Busby, I. J. Ellison, and R. D. Gillard, *Polyhedron*, 12, 247 (1993); (b) N. S. A. Edwards, I. J. Ellison, R. D. Gillard, B. Mile, and J. Maher, *Polyhedron*, *12*, 371 (1993).
195. R. D. Gillard and J. D. P. de Jesus, *J. Chem. Soc. Dalton Trans.*, 1895 (1984).
196. J. Lilie, M. G. Simic, and J. F. Endicott, *Inorg. Chem.*, *14*, 2129 (1975).
197. E. W. Abel, J. M. Pratt, R. Whelan, and P. J. Wilkinson, *J. Am. Chem. Soc.*, *96*, 7119 (1974).
198. A. Nishinaga, H. Yamato, T. Abe, K. Maruyama, and T. Matsuura, *Tetrahedron Lett.*, *29*, 6309 (1988).
199. A. Nishinaga and H. Tomita, *J. Mol. Catal.*, *7*, 179 (1980).
200. R. A. Sheldon, in *The Activation of Dioxygen and Homogeneous Catalytic Oxidation*, D. H. R. Barton, A. E. Martell, and D. T. Sawyer, Eds., Plenum, New York, and London, 1993, pp. 9–30.
201. A. Nishinaga, K. Nishizawa, H. Tomita, and T. Matsuura, *J. Am. Chem. Soc.*, *99*, 1287 (1977).

202. M. E. Brynildson, A. Bakac, and J. H. Espenson, *Inorg. Chem.*, *27*, 2592 (1988).
203. S. L. Bruhn, A. Bakac, and J. H. Espenson, *Inorg. Chem.*, *25*, 535 (1986).
204. K. Kumar and J. F. Endicott, *Inorg. Chem.*, *23*, 2447 (1984).
205. M. Munakata and J. F. Endicott, *Inorg. Chem.*, *23*, 3693 (1984).
206. B. Bosnich, C. K. Poon, and M. L. Tobe, *Inorg. Chem.*, *5*, 1514 (1966).
207. S. Scott, A. Bakac, and J. H. Espenson, *Inorg. Chem.*, *30*, 4112 (1991).
208. S. P. Ghosh, M. C. Ghosh, and E. S. Gould, *Int. J. Chem. Kin.*, *26*, 665 (1994).
209. J. Springborg and M. Zehnder, *Acta Chem. Scand. A*, *A41*, 484 (1987).
210. J. L. Bear, C.-L. Yao, F. J. Capdevielle, and K. M. Kadish, *Inorg. Chem.*, *27*, 3782 (1988).
211. F. P. Guengerich and T. L. Macdonald, *Acc. Chem. Res.*, *17*, 9 (1984).
212. J. H. Dawson, *Science*, *240*, 433 (1988).
213. W.-D. Wang, A. Bakac, and J. H. Espenson, *Inorg. Chem.*, *32*, 2005 (1993).
214. T. Geiger and F. C. Anson, *J. Am. Chem. Soc.*, *103*, 7489 (1981).
215. W.-D. Wang, A. Bakac, and J. H. Espenson, *Inorg. Chem.*, submitted for publication.
216. H. Suzuki, S. Matsuura, Y. Moro-Oka, and T. Ikawa, *J. Organomet. Chem.*, *286*, 247 (1985).
217. A. Sen and J. Halpern, *J. Am. Chem. Soc.*, *99*, 8337 (1977).
218. R. J. Guajardo, S. E. Hudson, S. J. Brown, and P. K. Mascharak, *J. Am. Chem. Soc.*, *115*, 7971 (1993).
219. R. J. Guajardo, J. D. Tan, and P. K. Mascharak, *Inorg. Chem.*, *33*, 2838 (1994).
220. Y.-D. Wu, K. N. Houk, J. S. Valentine, and W. Nam, *Inorg. Chem.*, *31*, 718 (1992).
221. G. McLendon and A. E. Martell, *Inorg. Chem.*, *15*, 2662 (1976).
222. A. B. Hoffman and H. Taube, *Inorg. Chem.*, *7*, 1971 (1968).
223. K. M. Davies and A. G. Sykes, *J. Chem. Soc. A*, 2831 (1968).
224. K. B. Yatsimirskii, Y. I. Bratushko, and A. B. Kondratjuk, *J. Coord. Chem.*, *23*, 335 (1991).
225. G. McLendon and W. F. Mooney, *Inorg. Chem.*, *19*, 12 (1980).
226. M. R. Hyde and A. G. Sykes, *J. Chem. Soc. Dalton Trans.*, 1550 (1974).
227. J. D. Edwards, C.-H. Yang, and A. G. Sykes, *J. Chem. Soc. Dalton Trans.*, 1561 (1974).
228. S. K. Saha, M. C. Ghosh, and E. S. Gould, *Inorg. Chem.*, *31*, 5439 (1992).
229. S. K. Ghosh, S. K. Saha, M. C. Ghosh, R. N. Bose, J. W. Reed, and E. S. Gould, *Inorg. Chem.*, *31*, 3358 (1992).
230. S. K. Ghosh and E. S. Gould, *Inorg. Chem.*, *28*, 3651 (1989).
231. S. K. Ghosh and E. S. Gould, *Inorg. Chem.*, *28*, 1948 (1989).
232. A. Marchaj, A. Bakac, and J. H. Espenson, *Inorg. Chem.*, 32, 2399 (1993).
233. A. Marchaj, A. Bakac, and J. H. Espenson, *Inorg. Chem.*, *31*, 4860 (1992).

234. R. J. Motekaitis and A. E. Martell, *Inorg. Chem.*, *33*, 1032 (1994).
235. G. Strukul, *Catalytic Oxidations with Hydrogen Peroxide as Oxidant*, Kluwer Academic Publishers, Dordrecht, The Netherlands, 1992.
236. G. E. Meister and A. Butler, *Inorg. Chem.*, *33*, 3269 (1994).
237. T.-J. Won, B. M. Sudam, and R. C. Thompson, *Inorg. Chem.*, *33*, 3804 (1994).
238. A. F. Ghiron and R. C. Thompson, *Inorg. Chem.*, *28*, 3647 (1989).
239. W. A. Herrmann, R. W. Fischer, W. Scherer, and M. U. Rauch, *Angew. Chem. Int. Ed. Engl.* *32*, 1157 (1993).
240. W. A. Herrmann, W. Wagner, U. N. Flessner, U. Volkhardt, and H. Komber *Angew. Chem. Int. Ed. Engl.*, *30*, 1636 (1991).
241. W. A. Herrmann, R. W. Fischer, and D. W. Marz, *Angew. Chem. Int. Ed. Engl.* *30*, 1638 (1991).
242. P. Huston, J. H. Espenson, and A. Bakac, *Inorg. Chem.*, *32*, 4517 (1993).
243. J. H. Espenson, O. Pestovsky, P. Huston, and S. Staudt, *J. Am. Chem. Soc.*, *116*, 2869 (1994).
244. A. F. Ghiron and R. C. Thompson, *Inorg. Chem.*, *29*, 4457 (1990).
245. C. Djordjevic, *Chem. Br.*, 554 (1982).
246. A. Butler, M. J. Clague, and G. E. Meister, *Chem. Rev.*, *94*, 625 (1994).
247. M. H. Dickman and M. T. Pope, *Chem. Rev.*, *94*, 569 (1994).
248. (a) S. K. Ghosh and E. S. Gould, *Inorg. Chem.*, *28*, 1948 (1989); (b) S. K. Ghosh and E. S. Gould, *Inorg. Chem.*, *28*, 3651 (1989).
249. M. T. H. Tarafder, P. Bhattacharjee, and A. K. Sarkar, *Polyhedron*, *11*, 795 (1992).
250. R. C. Thompson, *Inorg. Chem.*, *21*, 859 (1982).
251. A. Bakac, J. H. Espenson, and J. A. Janni, *J. Chem. Soc. Chem. Commun.*, 315 (1994).
252. J. H. Cameron and S. C. Turner, *J. Chem. Soc. Dalton Trans.*, 1941 (1993).
253. J. H. Cameron, P. C. Morgan, and S. C. Turner, *J. Chem. Soc. Chem. Commun.*, 1617 (1990).
254. J. H. Cameron and S. C. Turner, *J. Chem. Soc. Dalton Trans.*, 3285 (1992).
255. J. Zakrzewski and C. Giannotti, *J. Chem. Soc. Chem. Commun.*, 662 (1992).
256. J. Zakrzewski and C. Giannotti, *J. Chem. Soc. Chem. Commun.*, 743 (1990).
257. J. Zakrzewski and C. Giannotti, *J. Photochem. Photobiol. A*, *57*, 479 (1991).
258. M. Hoshino, *Chem. Phys. Lett.*, *120*, 50 (1985).
259. M. Seip and H.-D. Brauer, *J. Photochem. Photobiol. A: Chem.*, *79*, 19 (1994).
260. A. Vogler and H. Kunkely, *J. Am. Chem. Soc.*, *103*, 6222 (1981).
261. P. Bergamini, S. Sostero, O. Traverso, P. Deplano, and L. J. Wilson, *J. Chem. Soc. Dalton Trans.*, 2311 (1986).
262. M. Seip and H.-D. Brauer, *J. Photochem. Photobiol. A: Chem.*, *76*, 1 (1993).
263. J. K. Kochi, *Angew. Chem. Int. Ed. Engl.*, *27*, 1227 (1988).
264. J. Halpern, *Ann. N. Y. Acad. Sci.*, *239*, 2 (1974).

265. W. H. Tamblyn, R. J. Klingler, W. S. Hwang, and J. K. Kochi, *J. Am. Chem. Soc.*, *103*, 3161 (1981).
266. J. Halpern, M. S. Chan, T. S. Roche, and G. M. Tom, *Acta Chem. Scand. A*, *33A*, 141 (1979).
267. T. Katsuyama, A. Bakac, and J. H. Espenson, *Inorg. Chem.*, *28*, 339 (1989).
268. A. Bakac and J. H. Espenson, *J. Am. Chem. Soc.*, *110*, 3453 (1988).
269. J. F. Endicott, K. Kumar, C. L. Schwarz, M. W. Perkovic, and W.-K. Lin, *J. Am. Chem. Soc.*, *111*, 7411 (1989).
270. C. Y. Mok and J. F. Endicott, *J. Am. Chem. Soc.*, *99*, 1276 (1977).
271. C. Y. Mok and J. F. Endicott, *J. Am. Chem. Soc.*, *100*, 123 (1978).
272. A. Bakac and J. H. Espenson, *Inorg. Chem.*, *28*, 3901 (1989).
273. J. H. Espenson, *Adv. Inorg. Bioinorg. Mech.*, *1*, 1 (1982).
274. R. J. Blau, J. H. Espenson, and A. Bakac, *Inorg. Chem.*, *23*, 3526 (1984).
275. H. Cohen and D. Meyerstein, *Inorg. Chem.*, *13*, 2434 (1974).
276. G. W. Kirker, A. Bakac, and J. H. Espenson, *J. Am. Chem. Soc.*, *104*, 1249 (1982).
277. A. Bakac and J. H. Espenson, *Inorg. Chem.*, *28*, 4319 (1989).
278. A. Bakac and J. H. Espenson, *Inorg. Chem.*, *26*, 4305 (1987).
279. A Bakac and J. H. Espenson, *Inorg. Chem.*, *29*, 4353 (1987).
280. S. Lee, J. H. Espenson, and A. Bakac, *Inorg. Chem.*, *29*, 3442 (1990).
281. J. F. Endicott, in *Concepts of Inorganic Photochemistry*, A. W. Adamson and P. D. Fleischauer, Eds., Wiley, New York, 1975, p. 88.
282. D. A. Ryan and J. H. Espenson, *J. Am. Chem. Soc.*, *104*, 704 (1982).
283. S. Shi, J. H. Espenson, and A. Bakac, *Inorg. Chem.*, *29*, 4318 (1990).
284. H. Ogino, M. Shimura, and N. Tanaka, *J. Chem. Soc. Chem. Commun.*, 1063 (1983).
285. H. Cohen, W. Gaede, A. Gerhard, D. Meyerstein, and R. van Eldik, *Inorg. Chem.*, *31*, 3805 (1992).
286. W. J. Wallace, J. C. Maxwell, and W. S. Caughey, *Biochem. Biophys. Res. Commun.*, *57*, 1104 (1974).
287. A. Bakac and J. H. Espenson, *J. Am. Chem. Soc.*, *106*, 5197 (1984).
288. M. D. Johnson, *Acc. Chem. Res.*, *16*, 343 (1983).
289. J. W. Buchler, K. L. Lay, L. Castle, and V. Ullrich, *Inorg. Chem.*, *21*, 842 (1982).
290. L.-C. Yuan and T. C. Bruice, *J. Am. Chem. Soc.*, *107*, 512 (1985).
291. J. R. Budge, B. K. Gatehouse, M. C. Nesbit, and B. West, *J. Chem. Soc. Chem. Commun.*, 370 (1981).
292. T. J. Collins, C. Slebodnick, and E. S. Uffelman, *Inorg. Chem.*, *29*, 3433 (1990).
293. T. L. Siddall, N. Miyaura, J. C. Huffman, and J. K. Kochi, *J. Chem. Soc. Chem. Commun.*, 1185 (1983).
294. K. Srinivasan and J. K. Kochi, *Inorg. Chem.*, *24*, 4671 (1985).

295. E. S. Gould, *Acc. Chem. Res.*, *19*, 66 (1986).
296. M. Krumpolc and J. Rocek, *J. Am. Chem. Soc.*, *101*, 3206 (1979).
297. M. Mitewa, P. Russev, P. R. Bontchev, K. Kabassanov, and A. Malinovski, *Inorg. Chem. Acta*, *70*, 179 (1983).
298. S. K. Cheung, C. J. Grimes, J. Wong, and C. A. Reed, *J. Am. Chem. Soc.*, *98*, 5028 (1976).
299. A. Bakac, *Croat. Chem. Acta*, *66*, 435 (1993).
300. A. Al-Ajlouni, A. Bakac, and J. H. Espenson, *Inorg. Chem.*, *33*, 1011 (1994).
301. A. Al-Ajlouni, A. Bakac, and J. H. Espenson, *Inorg. Chem.*, *32*, 5792 (1993).
302. W. K. Seok and T. J. Meyer, *J. Am. Chem. Soc.*, *110*, 7358 (1988).
303. J. Gilbert, L. Roecker, and T. J. Meyer, *Inorg. Chem.*, *26*, 1126 (1987).
304. C. F. Wells and D. Fox, *J. Chem. Soc. Dalton Trans.*, 1498 (1977).
305. H. Sigel, C. Flierl, and R. Griesser, *J. Am. Chem. Soc.*, *91*, 1061 (1969).
306. F. H. Westheimer, *Chem. Rev.*, *45*, 419 (1949).
307. J. Rocek and A. E. Radkowsky, *J. Am. Chem. Soc.*, *90*, 2968 (1968).
308. K. B. Wiberg and H. Schafer, *J. Am. Chem. Soc.*, *91*, 933 (1969).
309. G. Michel and R. Machiroux, *J. Raman Spectrosc.*, *14*, 22 (1983).
310. G. Michel and R. Cahay, *J. Raman Spectrosc.*, *17*, 79 (1986).
311. J. F. Perez-Benito, C. Arias, and D. Lamrhari, *J. Chem. Soc. Chem. Commun.*, 472 (1992).
312. J. F. Perez-Benito and C. Arias, *Can. J. Chem.*, *71*, 649 (1993).
313. P. O'Brien and G. Wang, *J. Chem. Soc. Chem. Commun.*, 690 (1992).
314. X. Shi, N. S. Dalal, and V. Vallyathan, *Arch. Biochem. Biophys.*, *90*, 381 (1991).
315. D. M. L. Goodgame and A. M. Joy, *J. Inorg. Biochem.*, *26*, 219 (1986).
316. D. Rai, L. E. Eary, and J. M. Zachara, *Sci. Total Environ.*, *86*, 15 (1989).
317. S. I. Shupack, *Environ. Health Persp.*, *92*, 7 (1991).
318. S. L. Boyko and D. M. L. Goodgame, *Inorg. Chim. Acta*, *123*, 189 (1986).
319. G.-X. He, R. D. Arasasingham, G.-H. Zhang, and T. C. Bruice, *J. Am. Chem. Soc.*, *113*, 9828 (1991).
320. J. Muzart, *Chem. Rev.*, *92*, 113 (1992).
321. K. A. Jorgensen, *Chem. Rev.*, *89*, 431 (1989).
322. R. J. Judd, T. W. Hambley, and P. A. Lay, *J. Chem. Soc. Dalton Trans.*, 2205 (1989).
323. B. Gahan, D. C. Garner, L. H. Hill, F. E. Mabbs, K. D. Hargrave, and A. T. McPhail, *J. Chem. Soc. Dalton Trans.*, 1726 (1977).
324. M. Mitewa and P. R. Bontchev, *Coord. Chem. Rev.*, *61*, 241 (1985).
325. R. P. Farrell and P. A. Lay, *Comments Inorg. Chem.*, *13*, 133 (1992).
326. M. C. Ghosh and E. S. Gould, *J. Am. Chem. Soc.*, *115*, 3167 (1993).
327. S. K. Ghosh, R. N. Bose, and E. S. Gould, *Inorg. Chem.*, *26*, 3722 (1987).
328. M. C. Ghosh and E. S. Gould, *Inorg. Chem.*, *30*, 491 (1991).

329. M. C. Ghosh and E. S. Gould, *Inorg. Chem.*, *29*, 4258 (1990).
330. K. Nag and S. N. Bose, *Struct. Bonding (Berlin)*, *63*, 153 (1985).
331. B. Rihter, S. SriHari, S. Hunter, and J. Masnovi, *J. Am. Chem. Soc.*, *115*, 3918 (1993).
332. S.-K. Noh, R. A. Heintz, B. S. Haggerty, A. L. Rheingold, and K. H. Theopold, *J. Am. Chem. Soc.*, *114*, 1892 (1992).
333. K. Picpgrass and M. T. Pope, *J. Am. Chem. Soc.*, *111*, 753 (1989).
334. B. S. Jaynes and C. L. Hill, *J. Am. Chem. Soc.*, *115*, 12212 (1993).
335. Y. Hou and C. L. Hill, *J. Am. Chem. Soc.*, *115*, 11823 (1993).
336. A. M. Khenkin and C. L. Hill, *J. Am. Chem. Soc.*, *115*, 8178 (1993).
337. C. Rong and F. C. Anson, *Inorg. Chem.*, *33*, 1064 (1994).
338. V. K. Sharma and B. H. J. Bielski, *Inorg. Chem.*, *30*, 4306 (1991).
339. S. Rahhal and H. W. Richter, *J. Am. Chem. Soc.*, *110*, 3126 (1988).
340. D. T. Sawyer, C. Kang, A. Llobet, and C. Redman, *J. Am. Chem. Soc.*, *115*, 5817 (1993).
341. D. A. Wink, R. W. Nims, J. E. Saavedra, W. E. Utermahlen, Jr., and P. C. Ford, *Proc. Natl. Acad. Sci. USA*, *91*, 6604 (1994).
342. S. Goldstein, D. Meyerstein, and G. Czapski, *Free Rad. Biol. Med.*, *15*, 435 (1993).
343. B. H. J. Bielski, *Free Rad. Res. Commun.*, *12–13*, 469 (1991).
344. B. H. J. Bielski and M. J. Thomas, *J. Am. Chem. Soc.*, *109*, 7761 (1987).
345. B. H. J. Bielski, V. K. Sharma, and G. Czapski, *Radiat. Phys. Chem.*, *44*, 479 (1994).
346. T. J. Collins, B. G. Fox, Z. G. Hu, K. L. Kostka, E. Münck, C. E. F. Rickard, and L. J. Wright, *J. Am. Chem. Soc.*, *114*, 8724 (1992).
347. T. J. Collins, K. L. Kostka, E. Münck, and E. S. Uffelman, *J. Am. Chem. Soc.*, *112*, 5637 (1990).
348. N. Colclough and J. R. L. Smith, *J. Chem. Soc. Perkin Trans.*, *2*, 1139 (1994).
349. H. B. Dunford and A. J. Adeniran, *Arch. Biochem. Biophys.*, *251*, 536 (1986).
350. M. S. Thompson and T. J. Meyer, *J. Am. Chem. Soc.*, *104*, 4106 (1982).
351. L. Roecker and T. J. Meyer, *J. Am. Chem. Soc.*, *109*, 746 (1987).
352. M. S. Thompson and T. J. Meyer, *J. Am. Chem. Soc.*, *104*, 5070 (1982).
353. R. A. Binstead and T. J. Meyer, *J. Am. Chem. Soc.*, *109*, 3287 (1987).
354. L. Roecker and T. J. Meyer, *J. Am. Chem. Soc.*, *108*, 4066 (1986).
355. B. A. Moyer, B. K. Sipe, and T. J. Meyer, *Inorg. Chem.*, *20*, 1475 (1981).
356. J. G. Muller, J. H. Acquaye, and K. J. Takeuchi, *Inorg. Chem.*, *31*, 4552 (1992).
357. C.-M. Che, C. Ho, and T.-C. Lau, *J. Chem. Soc. Dalton Trans.*, 1259 (1991).
358. C.-M. Che, K. Lau, T.-C. Lau, and C.-K. Poon, *J. Am. Chem. Soc.*, *112*, 5176 (1990).
359. K.-Y. Wong, C.-M. Che, and F. C. Anson, *Inorg. Chem.*, *26*, 737 (1987).
360. C.-M. Che and V. W.-W. Yam, *Adv. Inorg. Chem.*, *39*, 233 (1992).

361. C.-M. Che and K.-Y. Wong, *J. Chem. Soc. Dalton Trans.*, 2065 (1989).
362. C.-M. Che, W.-T. Tang, W.-O. Lee, K.-Y. Wong, and T.-C. Lau, *J. Chem. Soc. Dalton Trans.*, 1551 (1992).
363. C.-M. Che, T.-F. Lai, and K.-Y. Wong, *Inorg. Chem.*, *26*, 2289 (1987).
364. C.-M. Che, K.-Y. Wong, and T. C. W. Mak, *J. Chem. Soc. Chem. Commun.*, 988 (1985).
365. C.-M. Che, K.-Y. Wong, and T. C. W. Mak, *J. Chem. Soc. Chem. Commun.*, 546 (1985).
366. R. H. Holm and J. P. Donahue, *Polyhedron*, *12*, 571 (1993).
367. (a) E. W. Harlan, J. M. Berg, and R. H. Holm, *J. Am. Chem. Soc.*, *108*, 6992 (1986); (b) E. W. Harlan and R. H. Holm, *J. Am. Chem. Soc.*, *112*, 186 (1990).
368. S. K. Das, P. K. Chaudhury, D. Biswas, and S. Sarkar, *J. Am. Chem. Soc.*, *116*, 9061 (1994).
369. B. E. Schultz and R. H. Holm, *Inorg. Chem.*, *32*, 4244 (1993).
370. B. E. Schultz, S. F. Gheller, M. C. Muetterties, M. J. Scott, and R. H. Holm, *J. Am. Chem. Soc.*, *115*, 2714 (1993).
371. S. F. Gheller, B. E. Shultz, M. J. Scott, and R. H. Holm, *J. Am. Chem. Soc.*, *114*, 6934 (1992).
372. Z. Xiao, C. G. Young, J. H. Enemark, and A. G. Wedd, *J. Am. Chem. Soc.*, *114*, 9194 (1992).
373. H. Oku, N. Ueyama, M. Kondo, and A. Nakamura, *Inorg. Chem.*, *33*, 209 (1994).
374. H. Arzoumanian, R. Lopez, and G. Agrifoglio, *Inorg. Chem.*, *33*, 3177 (1994).
375. R. H. Holm, *Chem. Rev.*, *87*, 1401 (1987).
376. J. K. Felixberger, J. G. Kuchler, E. Herdtweck, R. A. Paciello, and W. A. Herrmann, *Angew. Chem. Int. Ed. Engl.*, *27*, 946 (1988).
377. D. E. Over, S. C. Critchlow, and J. M. Mayer, *Inorg. Chem.*, *33*, 4643 (1992).
378. Y. Zhang and R. H. Holm, *Inorg. Chem.*, *29*, 911 (1990).

The Chemistry of Metal Complexes with Selenolate and Tellurolate Ligands

JOHN ARNOLD

Department of Chemistry
University of California
Berkeley, CA

CONTENTS

Progress in Inorganic Chemistry, Vol. 43, Edited by Kenneth D. Karlin.
ISBN 0-471-12336-6

I. INTRODUCTION

The chemistry of metal complexes containing alkoxide and thiolate ligands ($L_nM{-}OR$ and $-SR$) has been developed extensively over the last several decades as evidenced by the information contained in a number of reviews and books on the subject (1–5). When one considers homologs involving the heavier elements, selenium and tellurium, however, it is clear that our understanding of related selenolate and tellurolate derivatives ($L_nM{-}SeR$ and $-TeR$) is still relatively poor (6, 7). In the last major review on the subject, Gysling (8) suggested one major reason for this lack of development was the general assumption that organo derivatives of Se and Te are difficult to prepare and handle as a result of their instability, toxicity, and foul smell (Fig. 1). In recent years, these misconceptions have been overcome to the point where selenolate and tellurolate chemistry is now being explored across a wide front (9). In addition to basic curiosity regarding the structures and reactivities of compounds of this

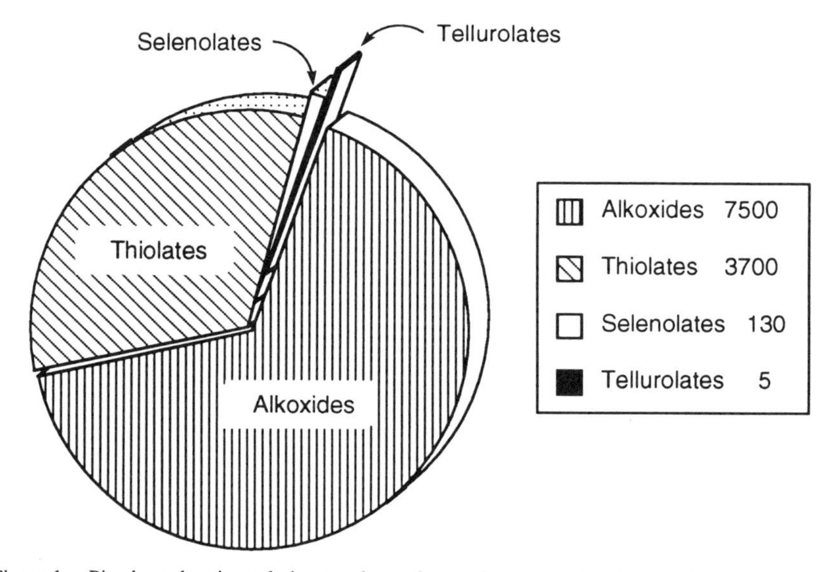

Figure 1. Pie chart showing relative numbers of crystallographically characterized metal chalcogenolates. Data from Cambridge Structural Database, July 1992.

type, a recent added impetus to the study of selenolate and tellurolate complexes stems from the discovery that these materials may be used as precusors to semiconducting metal selenides and tellurides (10, 11).

The aim of this chapter is to comprehensively review metal selenolate and tellurolate literature with a particular emphasis on some of the more novel developments since the subject was last covered. For a thorough overview of the more general aspects of metal complexes with ligands incorporating Se and Te, the reader is referred to Gysling's most recent article (8). The chemistry of selenoethers and telluroethers is quite separate from that of the chalcogenolates and is considered to be outside the scope of the present review (12).

The last few years have seen a tremendous growth in chalcogenolate chemistry involving sterically demanding ligands. These developments have resulted in the synthesis of stable, well-characterized crystalline derivatives with low molecularities. Their solubility in nonreactive hydrocarbon solvents has allowed for detailed reactivity studies; X-ray diffraction studies have led to significant advances in our understanding of metal–chalcogen interactions.

Work in this area began to appear in the mid-1980s with two reports describing the alkyl (13) and aryl (14) ligands shown below.

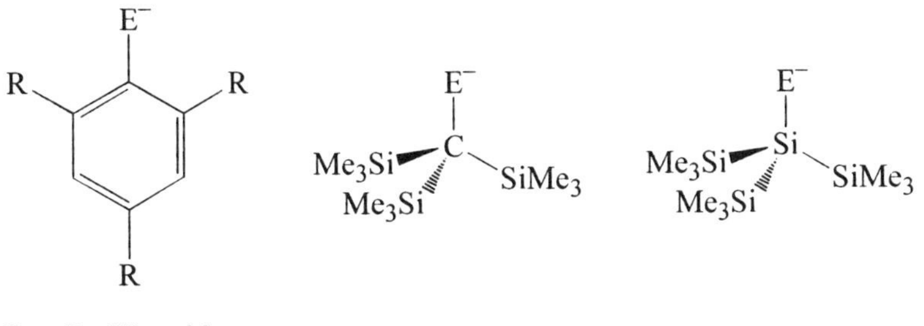

Since the first silyl tellurolate derivative was published in 1991 (15), work on this particular ligand has shown it to be useful for the preparation of a wide range of metal and nonmetal complexes, as described in detail in Section III.

II. SURVEY OF SYNTHETIC REAGENTS AND ROUTES

A. Chemistry Involving Alkali or Alkaline Earth Reagents

The majority of preparative routes involve, at least at some stage, insertion of elemental selenium or tellurium into a reactive M—R bond. Even if the metal–chalcogen bond is not prepared directly by this route, it is usually involved in making the chalcogenolate starting material. Chemistry involving alkali metal salts has been known for many years, although most of these reactions generated reagents that were used *in situ* (16); these anions were then used directly in metathesis reactions of general formula (Eq. 1):

$$L_nMX_m + mER^- \longrightarrow L_nM(ER)_m + mX^- \qquad (1)$$

More recently, selenolate and tellurolate salts have been isolated and fully characterized by elemental analysis, NMR and, in some cases, by X-ray structural methods (14, 15, 17–21). As an example, in polar solvents such as THF both tellurium and selenium insert rapidly into Li—aryl bonds, such as in the mesityl derivative in Eq. 2 (21, 22), where E = Se or Te.

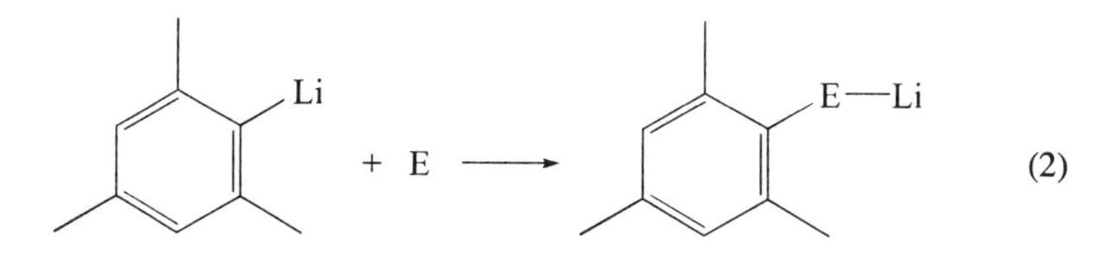

Sodium and potassium derivatives are best prepared by reduction of the air-stable ditellurides, using the reagents in Eqs. 3 and 4, for example.

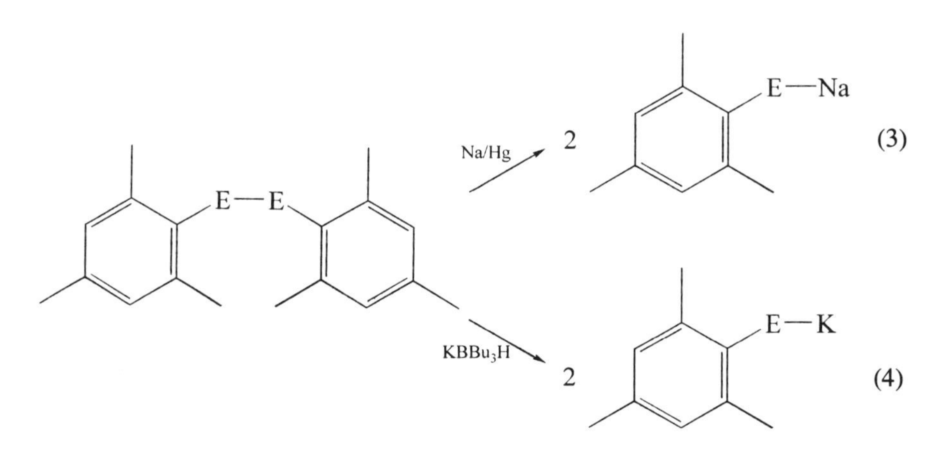

In general they are less soluble than their lithium counterparts and while they have been used as reagents in a few instances, they appear to offer few real advantages over the latter. The most notable exception occurs in lanthanide chemistry where it has been shown that the potassium salts are often more useful in metathesis-type reactions with metal halides, a finding that is presumably related to the relative ease of removal of the potassium halide byproduct from the metal chalcogenolate (23–25). All alkali metal salts are rapidly oxidized to the corresponding dichalcogenides on exposure to oxygen and hydrolysis yields the corresponding tellurol, at least transiently (see below).

A number of alkali metal tellurolates and selenolates have now been crystallographically characterized. Most of these, for example, $(thf)_3LiE(t\text{-}Bu_3C_6H_2)$ [E = Se (26, 27) or (21, 22)], $(18\text{-}C\text{-}6)KTe(i\text{-}Pr_3C_6H_2)$ (22), $(tmeda)_2Na$-Te(mesityl) (22), and the intramolecularly stabilized $(dme)LiTe(o\text{-}C_6H_4CH_2\text{-}NMe_2)$ (21) and $(dme)Li\{CpFe[C_5H_3(CH_2NMe_2)Te]\}$ (28) are monomeric. The propensity of tellurium to form bridging interactions is demonstrated in the case of $(thf)_{1.33}KTe(i\text{-}Pr_3C_6H_2)$, however, which crystallizes in an infinite ladder-type structure with alternating K—Te "rungs" (21).

The reactions shown in Eqs. 2–4 appear to be quite general for most alkali metal aryls and alkyls (20, 29–31), and have also been used to prepare sterically hindered silyl tellurolate derivatives, as shown in Eq. 5 (15, 17, 32, 33).

$$2(thf)_3LiSi(SiMe_3)_3 + 2Te \longrightarrow [(thf)_2LiTeSi(SiMe_3)_3]_2 \qquad (5)$$

Scheme 1 shows some simple reaction chemistry of the silyl tellurolate anion (17). Interestingly, selenium and sulfur analogues could not be prepared cleanly using a similar route to that in Eq. 5, but were made by displacement of tellurium at low temperature (−50°C) in DME (33) (Eq. 6).

$$[(thf)_2LiTeSi(SiMe_3)_3]_2 + 2Se \longrightarrow [(thf)_2LiSeSi(SiMe_3)_3]_2 + 2Te \qquad (6)$$

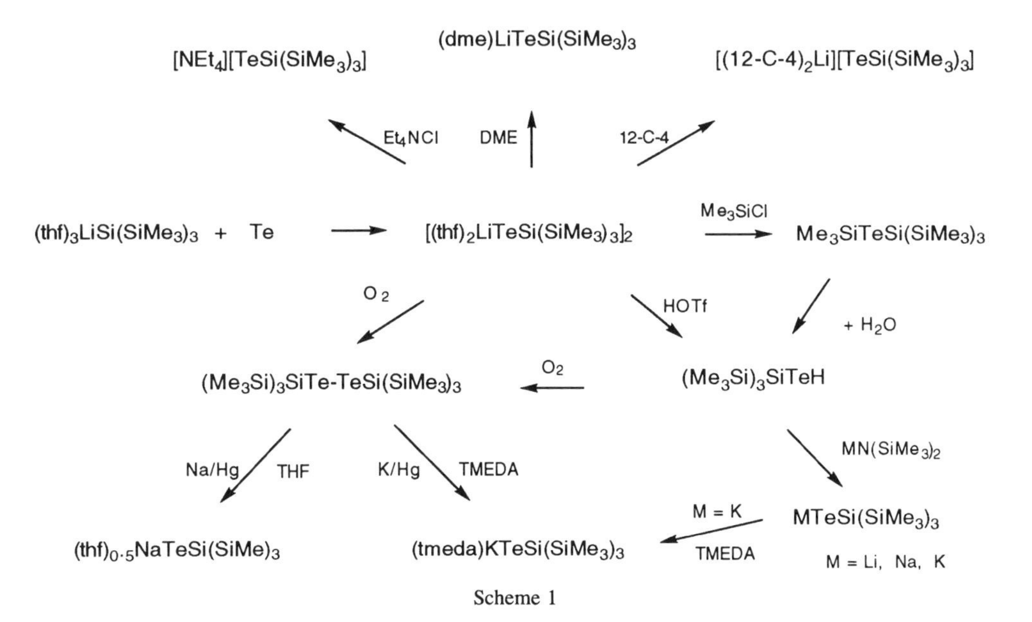

Scheme 1

X-ray structures of four different solvates of the silyl tellurolate have been determined. The first to be described was the dimeric bis-thf salt $[(thf)_2LiTeSi(SiMe_3)_3]_2$ (17) shown in Fig. 2. The dme complex is quite similar (34) being virtually identical to the selenolate (35), but these differ from the mono-thf species $[(thf)LiTeSi(SiMe_3)_3]_2$ (33), which features three-coordinate lithium in a related Li_2Te_2 core. In the crown derivative $[(12\text{-}C\text{-}4)_2Li]$-$[TeSi(SiMe_3)_3]$ (Fig. 3) the tellurolate anion is noncoordinated (17), as is also the case in the thienyl species $[Ph_4P][TeC_4H_3STe]$ (36).

Alkyl (13, 37) and germyl analogues of the lithium silyl tellurolates described above were made by direct insertion routes; in these cases chalcogen metathesis was not required (33). Indeed, it was shown that insertion of Se or S into the Li—Te bond in $LiTeC(SiMe_3)_3$ leads to tellurenyl chalcogenolates $LiETeC(SiMe_3)_3$ rather than metathesis (Eq. 7).

$$Li{-}TeC(SiMe_3)_3 + E \longrightarrow Li{-}E{-}TeC(SiMe_3)_3 \qquad (E = S \text{ or } Se) \qquad (7)$$

This result confirmed earlier findings in studies of chalcogen insertions into tellurolates with lower molecular weight alkyl and aryl substituents (13, 38). Both studies found that only chalcogens with electronegativities higher than that already present in the Li—E bond will insert. Thus, S and Se insert into Li—Te

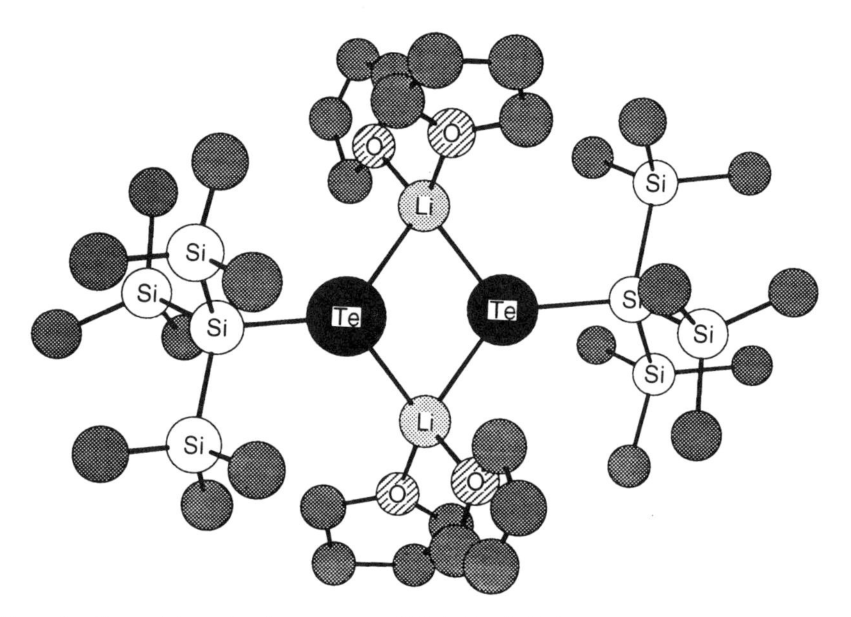

Figure 2. View of the molecular structure of $[(thf)_2LiTeSi(SiMe_3)_3]_2$ (methyl groups omitted for clarity) (17).

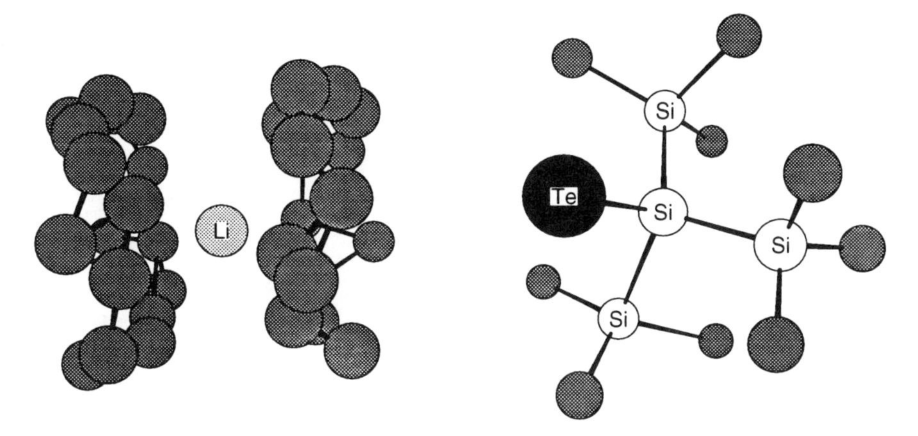

Figure 3. View of the molecular structure of $[Li(12\text{-crown-}4)_2][TeSi(SiMe_3)_3]$ (17).

bonds, but only S inserts into Li—Se. In all cases, the chalcogenyl chalcogenates were stable only in solution and multinuclear NMR spectroscopy was the sole characterization technique. It was suggested that in the silyl derivatives, where chalcogen metathesis is rapid and no evidence for Li—E—Te—SiR_3 species were detected, hypervalent silicon plays a role in providing a low-energy pathway for the elimination of elemental tellurium (33).

Insertion of chalcogens into Mg—C bonds in Grignard reagents has also been known for many years (16, 39, 40), although it was not until very recently that well-characterized alkaline earth complexes were isolated and characterized. A series of derivatives were prepared according to the reaction shown in Eq. 8 (41, 42).

$$M[N(SiMe_3)_2]_2 + 2HESi(SiMe_3)_3 \longrightarrow M[ESi(SiMe_3)_3]_2 + 2HN(SiMe_3)_2 \quad (8)$$

In general, the compounds were isolated as adducts $M[ESi(SiMe_3)_3]_2L_n$ with L = thf, pyr, or tmeda and n varying from 2 to 5 as the ionic radius of the metal (M = Mg, Ca, Sr, or Ba) increases. A number of these derivatives have been structurally characterized, for example, the Mg and Ba complexes shown in Figs. 4 and 5. The tendency of Mg to adopt four coordinate structures is not altered on reaction of the homoleptic selenolate with the potentially tridentate phosphine trmpsi. Although a 1 : 1 complex was isolated, the solid state structure, determined by X-ray crystallography, shows that only two arms of the ligand coordinate to Mg. In solution, it was found by ^1H and ^{31}P NMR spectroscopy that the molecule is fluxional at room temperature in toluene with only one signal being observed for the PMe_2 groups. On cooling to −80°C, how-

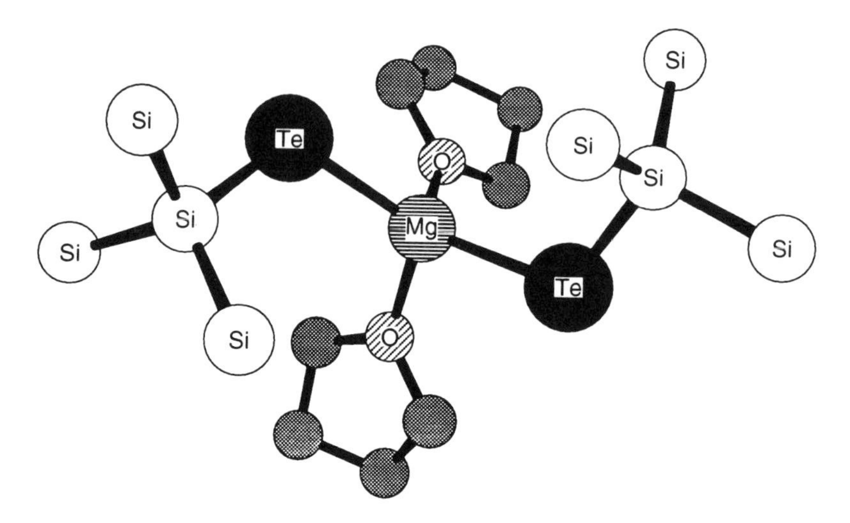

Figure 4. View of the molecular structure of $Mg[TeSi(SiMe_3)_3]_2(thf)_2$ (methyl groups omitted for clarity) (41, 42).

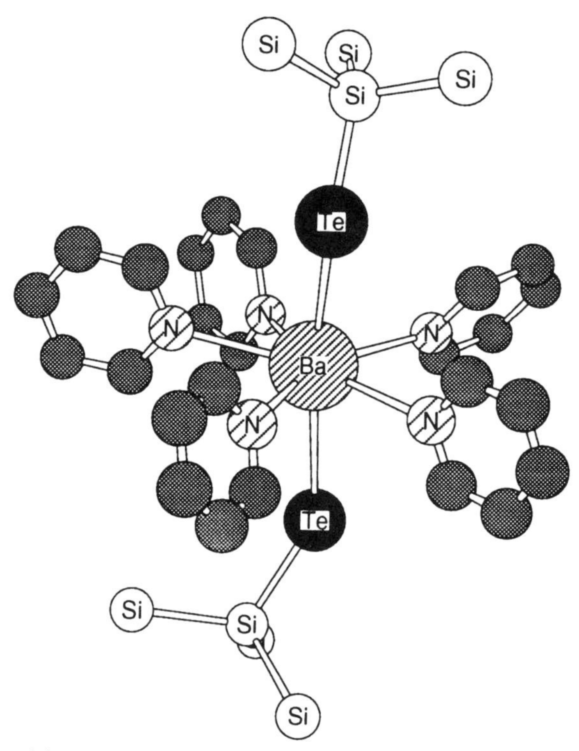

Figure 5. View of the molecular structure of $Ba[TeSi(SiMe_3)_3]_2(pyr)_5$ (methyl groups omitted for clarity) (42).

ever, these groups become inequivalent and two peaks in the ratio 2:1 were seen, the most intense of which was flanked by ^{77}Se satellites (42).

Treatment of the magnesium tellurolate with the crown ether 18-C-6 proceeds as shown in Eq. 9 (41).

$$Mg[TeSi(SiMe_3)_3]_2(thf)_2 + 2(18\text{-}C\text{-}6) \longrightarrow [Mg(18\text{-}C\text{-}6)_2][TeSi(SiMe_3)_3]_2 \quad (9)$$

Conductivity measurements confirmed the 2:1 stoichiometry of this salt; ^{1}H NMR spectroscopy showed a singlet for the two free tellurolates and a complex AA′BB′ pattern for the ethylene groups in the coordinated crown ethers.

The magnesium selenolate was isolated as a crystalline, base-free, homoleptic species. It is fluxional in hydrocarbon solution, with only one signal for the rapidly equilibrating selenolate ligands by ^{1}H and ^{77}Se NMR spectroscopy at room temperature. On cooling, however, these signals split into two peaks, consistent with bridging and terminal selenolates in a dimeric structure.

B. Chemistry Involving Chalcogenolysis

It is generally the case that whenever possible, chalcogenolysis reactions of the type shown in Eq. 10 are more useful than metathesis chemistry, where E = Se or Te; R = allyl, aryl, or silyl.

$$L_nM{-}X + HER \longrightarrow L_nM{-}ER + HX \quad (10)$$

Good examples of X leaving groups are any highly basic species, such as alkyl, aryl, or amido ligands. Particular advantages of this route relative to metathesis include: (a) reactions can be carried out in nonpolar solvents, (b) conditions are generally milder, (c) product isolation and purification is often much easier, and (d) yields are usually higher, approaching quantitative in many instances.

Selenolysis has been used as a synthetic route to M—SeR bonds for many years (6, 8). A wide range of selenol reagents is known and some of the simple ones are commercially available. With the recent drive to prepare low-coordinate metal chalcogenolates, recent work in this area has focused on preparing novel selenols with sterically hindered substituents. These include very bulky ligands such as (2,4,6-t-BuC_6H_2 or *supermesityl*) (43), [2,4,6-$(CF_3)_3C_6H_2$] (44), and $C(SiMe_3)_3$ (33, 37), and more recently, the silyl analogue $Si(SiMe_3)_3$ (33). The selenols are nearly always prepared by simple protonolysis of alkali metal selenolates. The more volatile derivatives are malodorous and all are air-sensitive to varying degrees.

In contrast to the large body of literature relating to selenols, there are very few reports of the corresponding tellurium chemistry, although given the fact that nearly all examples of tellurols are thermally unstable and light sensitive, this disparity is easily reconciled (31, 45). Methane tellurol (46–48) and phenyl tellurol (49) are reported to be thermally sensitive liquids; accordingly, they have seen little use in synthesis. Bulkier aryl tellurols such as mesityl and supermesityl tellurol, 2,4,6-t-$Bu_3C_6H_2TeH$, were made by protonolysis of the corresponding anions (29). Again, these materials decompose rapidly at ambient temperature and are oxidized to the ditelluride on exposure to air. When techniques are available to deal with these problems, however, the compounds may be used as reagents as shown by Bochmann et al. (29) in the preparation of Group 12 (IIB) tellurolates.

A remarkably stable tellurol is formed on protonolysis of the bulky silyl tellurolate anion as shown in Eq. 11 (15, 17).

$$LiTeSi(SiMe_3)_3 + HX \longrightarrow HTeSi(SiMe_3)_3 + LiX \qquad (11)$$

The tellurol was isolated and purified by sublimation on scales up to about 50 g. It is a colorless, waxy solid that decomposes on exposure to oxygen, but is otherwise quite stable, melting at 128°C without decomposition. A strong ν_{TeH} occurs at 2021 cm^{-1} in the IR and a doublet was seen in the ^{125}Te NMR spectrum (δ −955, $|J_{TeH}|$ = 74 Hz).

Subsequently, further examples of stable, or relatively stable, tellurols have appeared. Similar methodology to that described above was used to prepare the alkyl and germyl tellurol analogues, $HTeC(SiMe_3)_3$ and $HTeGe(SiMe_3)_3$ (33). A feature common to all tellurols is the high-field shifted Te-H resonance in the ^1H NMR spectrum. This ranges from as high as −9.02 in $HTeGe(SiMe_3)_3$ down to −1.25 ppm in the aryl tellurol 2,4,6-t-$Bu_3C_6H_2TeH$. In all cases, these signals are flanked by ^{125}Te satellites with $|J_{TeH}|$ in the range 27–101 Hz. Nonaqueous titrations in methyl(isobuyl)ketone were used to measure the pK_a values of a series of related tellurols and selenols. These data showed that for a given chalcogen, the silyl derivatives were the most acidic, with $HTeSi(SiMe_3)_3$ having the lowest pK_a (7.3). The tellurols were the strongest acids followed by the selenols, then thiols. Clearly, the trends in pK_a values are related to the ability of the chalcogenolate group to stabilize the negative charge on the chalcogenolate anion.

The first crystallographically characterized tellurol and selenol were reported recently. The solid state structure of $HTeSiPh_3$, prepared by protonolysis of the lithium salt, is shown in Fig. 6 (50). There are no close contacts between monomers and the Te—Te distance [4.6412(6) Å] is outside the sum of van der Waals radii (4.40 Å). The absence of hydrogen bonding in the tellurol is presumably due to the small charge/radius ratio and low electronegativity of tellurium. A description of the structure of the selenol [2,4,6-$(CF_3)_3C_6H_2SeH$]

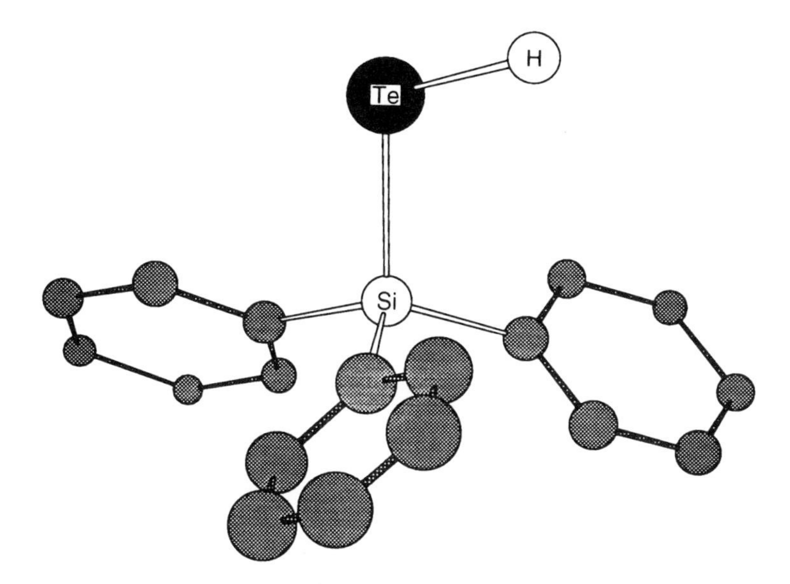

Figure 6. View of the molecular structure of $HTeSiPh_3$ (the tellurolic hydrogen was not located in the X-ray determination) (50).

shows that this is hydrogen bonded in the solid state (44). Here, the selenol monomers are associated in a continuous zigzag pattern with a Se—Se contact 0.418 Å less than the sum of van der Waals radii.

The availability of synthetically useful tellurol reagents has transformed metal tellurolate chemistry, since they have permitted a much more general and flexible approach to these compounds. In addition, the steric bulk and hydrocarbon solubility imparted by $-TeSi(SiMe_3)_3$ and $-TeC(SiMe_3)_3$ ligands to their resulting complexes has allowed for the isolation and characterization of structural types that are still unprecedented in related aryl tellurolate systems (see Section III).

C. Miscellaneous Routes

There are a few reports describing direct insertion of chalcogens into transition metal–carbon bonds (Eq. 12).

$$L_nM{-}R + E \longrightarrow L_nM{-}ER \tag{12}$$

From what little is known, it appears that this route is best suited to early metal, coordinatively unsaturated systems. Since the chalcogens are insoluble in common solvents, elevated temperatures are often required. Alternatively, an increasingly common practice is to add strongly basic phosphines, which generates soluble R_3PE (16, 51) species. Presumably this approach can only be used when the phosphine does not completely block coordination sites at the metal center and/or when incorporation of the phosphine in the final product is not undesirable (52). Piers and co-workers (53, 54) recently prepared scandium tellurolates by direct insertion routes and some years earlier, Klapötke and co-workers (55, 56) isolated the metallocene species described in Section III. Manganese tellurolates have also been prepared by insertion of Te into a Mn—C(alkyl) bonds (57, 58).

Two-electron oxidative addition reactions of the type outlined in Eq. 13 were used to prepare some of the first selenolate and tellurolate complexes (6, 8), and they are still popular as evidenced by more recent literature (59–61).

$$L_nM + RE{-}ER \longrightarrow L_nM(ER)_2 \qquad (13)$$

There are clear limitations to this route, however. The metal starting materials must be readily oxidized to a stable higher oxidation state, and the dichalcogenide cannot be too bulky otherwise the reaction may be too slow to be useful synthetically.

There is a single report describing use of aluminum reagents to prepare metal selenolates: metathesis reactions using the tris selenolates $Al(SeR)_3$ (R = Ph or naphthyl) are described in more detail in Section III (62).

Finally, the use of silyl reagents should be mentioned. Again by analogy with alkoxide and thiolate chemistry, several groups have made use of $-SiMe_3$ as a leaving group in reactions such as that shown below (Eq. 14) (17, 63–66).

$$L_nM{-}X + RE{-}SiMe_3 \longrightarrow L_nM{-}ER + X{-}SiMe_3 \qquad (14)$$

Full details of these miscellaneous reactions are given in the appropriate sections below.

III. TRANSITION METAL DERIVATIVES

A. Titanium, Zirconium, Hafnium, Vanadium, Niobium, and Tantalum

Metallocene species were among the first tellurolate complexes to be prepared for the early transition metals. Sato and Yoshida (67) made $Cp_2M(TePh)_2$ for Ti and Zr by metathesis from lithium phenyltellurolate and the respective metal chloride salts. The corresponding selenolates are also known, the first example being prepared by reaction of Cp_2TiCl_2 with HSePh in the presence of NEt_3 (68). Sometime later, Köpf and Klapötke (55, 56) prepared a novel tellurium chatecholate analogue $Cp_2Ti(o\text{-}Te_2C_6H_4)$ by two routes (Eqs. 15 and 16).

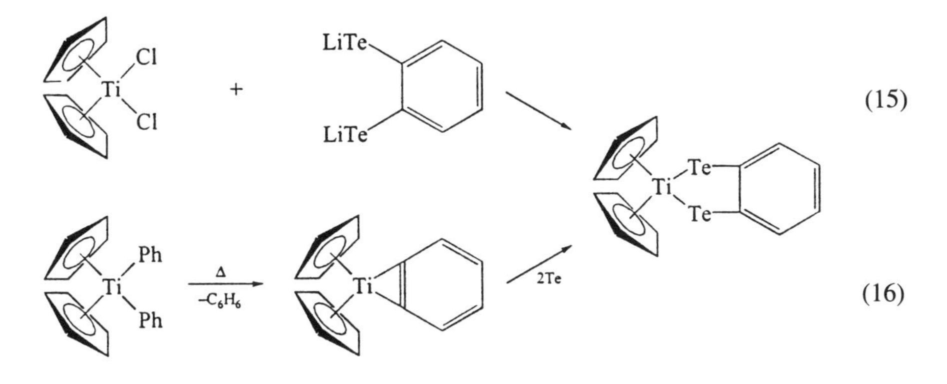

The latter reaction required heating to 99 °C in heptane, under which conditions formation of the metal benzyne complex is likely. Insertion of two equivalents of Te into the Ti—C bonds of the metalloaryne would then give the product. In common with homologous S and Se derivatives, the tellurolene ring undergoes inversion on the NMR time scale, with an activation barrier at the coalescence temperature (−38 °C) of 51 kJ mol^{-1}. Surprisingly similar values were obtained for the thiolate (53 kJ mol^{-1}) and selenolate (57 kJ mol^{-1}). Further evidence for insertion of selenium into M—C bonds was described by Gautheron and co-workers (69, 70). Reaction of Zr and Hf metallocene dimethyls with Se in boiling heptane gave the bis methyl selenolates, and in reactions similar to those above for Ti, thermolysis of $(t\text{-}BuC_5H_4)_2MPh_2$ in the presence of Se afforded $(t\text{-}BuC_5H_4)_2M(o\text{-}Se_2C_6H_4)$, again by a mechanism assumed to involve metalloaryne intermediates.

Andra (62) reported the only use to date of Al reagents in the synthesis of M—Se bonds. Although little characterization data was given, the complexes $M^{IV}(SePh)_4$ (M = Ti, Zr, or W), $M^{III}(SePh)_3$ (M = Ti, Nb, Ta, or Cr), and $M^{II}(SePh)_2$ (M = Ni or Co), and some naphthyl analogues were reported to be formed by treatment of the metal halides with $Al(SeR)_3$ (R = Ph or Nap).

A complete series of Group 4 (IVB) metallocene derivatives based on the bulky tris(trimethylsilyl)silyl tellurolate ligand has been investigated (15, 41, 50, 71). The compounds were readily prepared by metathesis or tellurolysis routes. In addition to the more common alkali metal compounds used to deliver tellurolate anions in preparations of this type, a novel magnesium bis-tellurolate was also used as a reagent for the first time. Mixed-ligand species such as $Cp_2Zr[TeSi(SiMe_3)_3]X$ where X = Cl, or Me, were characterized. The methyl complex reacted with CO to form the 18-electron (18 e^-) complex $Cp_2Zr[TeSi(SiMe_3)_3](\eta^2\text{-}COMe)$. The X-ray crystal structure showed a Zr—Te bond length of 2.990(1) Å, which is substantially longer than the same parameter in the 16-electron (16 e^-) species $Cp_2Zr[TeSi(SiMe_3)_3]_2$ of 2.866(1) Å (71). There are two possible explanations for the difference in Zr—Te bond lengths. The first, based on electronic arguments, invokes the idea of Te → Zr π-bonding in the 16 e^- complex, a feature that is bound to be absent in the 18 e^- derivative. Alternatively, differences in coordination number (i.e., steric factors) may be argued since the η^2-acyl ligand being a bidentate ligand results in a nine-coordinate structure versus eight for the 16 e^- bis-tellurolate.

Lewis bases, such as phosphines and isocyanides, reduced the Ti^{IV} tellurolate to Ti^{III} and ditelluride (Eq. 17).

$$Cp_2Ti[TeSi(SiMe_3)_3]_2 + L \longrightarrow Cp_2Ti[TeSi(SiMe_3)_3]L + \tfrac{1}{2}[TeSi(SiMe_3)_3]_2 \quad (17)$$

The Ti^{III} tellurolates were characterized by ESR and magnetic measurements and the X-ray crystal structure of the PMe_3 derivative was reported. As expected, reduction of the metal results in a lengthening of the Ti—Te bond, from 2.788(4) Å in $Cp_2Ti[TeSi(SiMe_3)_3]_2$ to 2.912(3) Å in the reduced species, as determined by X-ray crystallography (Fig. 7). Treatment of $Cp_2Ti[TeSi(SiMe_3)_3]_2$ with CO (3 atm) resulted in reduction to Ti^{II} and the formation of $Cp_2Ti(CO)_2$ and an equivalent of $[TeSi(SiMe_3)_3]_2$. The latter product was also

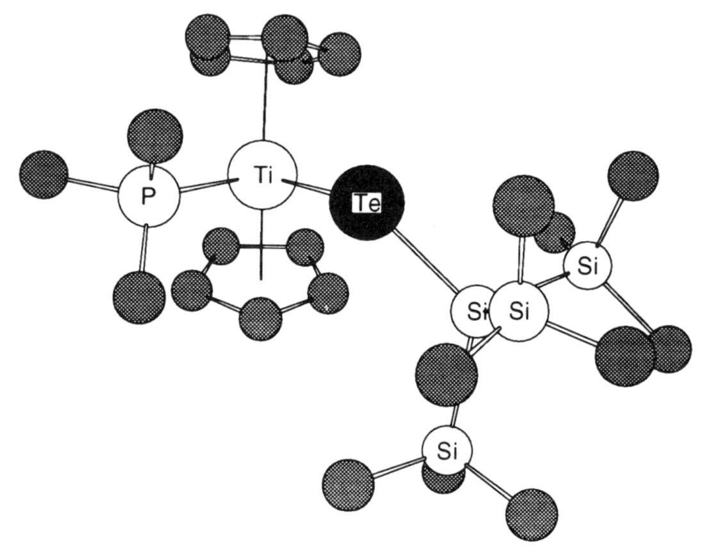

Figure 7. View of the molecular structure of $Cp_2Ti[TeSi(SiMe_3)_3]PMe_3$ (71).

obtained in reactions of $Cp_2Ti[TeSi(SiMe_3)_3]_2$ with CO_2 or CS_2; in these cases the metal containing products were somewhat more complex, although both had been previously characterized (Eqs. 18 and 19) (72, 73).

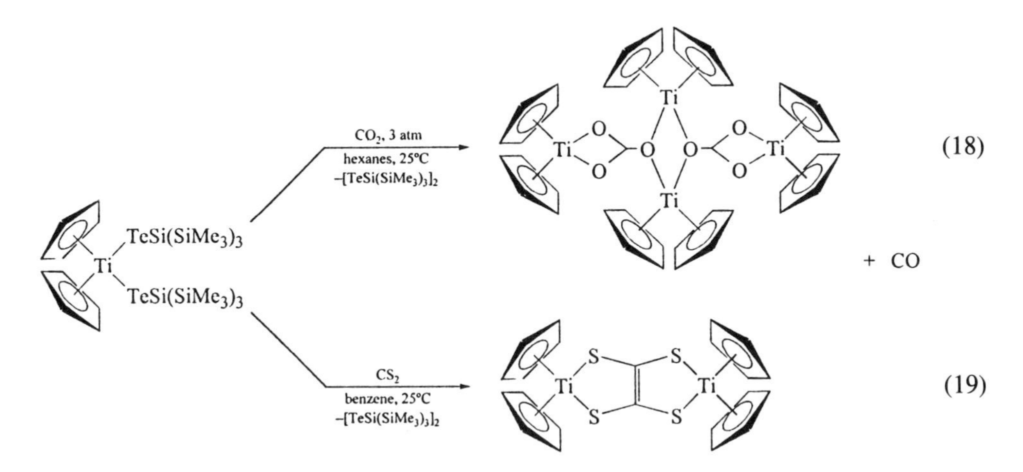

Under the same conditions the Zr and Hf derivatives were unreactive, results that may be rationalized on the basis of the correspondingly higher redox potentials of these metals relative to Ti.

More recent results in systems based on $-TeSiPh_3$ ligands show these compounds to be substantially more reactive than their $-TeSi(SiMe_3)_3$ counterparts (50). In the Ti system, $Cp_2Ti(TeSiPh_3)_2$ was found to react with pyridine to yield complex mixtures of other products, but no evidence for simple redox chemistry was detected. The only well-characterized metal-containing product was the cluster $(CpTi)_6(\mu_3\text{-Te})_6(\mu_3\text{-O})_2$, which was isolated in very low yield along with the disilyltelluride $Te(SiPh_3)_2$. The X-ray crystal structure of the cluster was determined; the source of oxygen was ascribed to trace water impurities in the solvent. In the methyl substituted case, $Cp'_2Ti(TeSiPh_3)_2$ afforded a low yield of the known bridging *ditelluride* $[Cp'_2TiTe_2]_2$ (74).

Reactions with the Zr complexes were much cleaner and more easily understood. As shown in Eq. 20, a smooth elimination reaction takes place that serves as a model for related processes in which a wide range of metal bis-tellurolates eliminate diorganotellurides to form metal tellurides (see Section VII) (50).

$$Cp_2Zr(TeSiPh_3)_2 + py \longrightarrow \tfrac{1}{2}\,[Cp_2ZrTe]_2 + Te(SiPh_3)_2 + pyr \quad (20)$$

In the Cp case, the bridging telluride precipitated during the course of the reaction, however, use of the *t*-Bu substituted complex ensured that all components were soluble throughout the reaction so that the kinetics of the reaction could be followed by NMR spectroscopy. These data showed that the reaction was first order in added base (in this case *tert*-butylpyridine) and first order in metal complex, that is, they show that elimination of $Te(SiPh_3)_2$ from the bis-tellurolate is an *intramolecular* process. The following reaction was proposed to account for the results (Eq. 21).

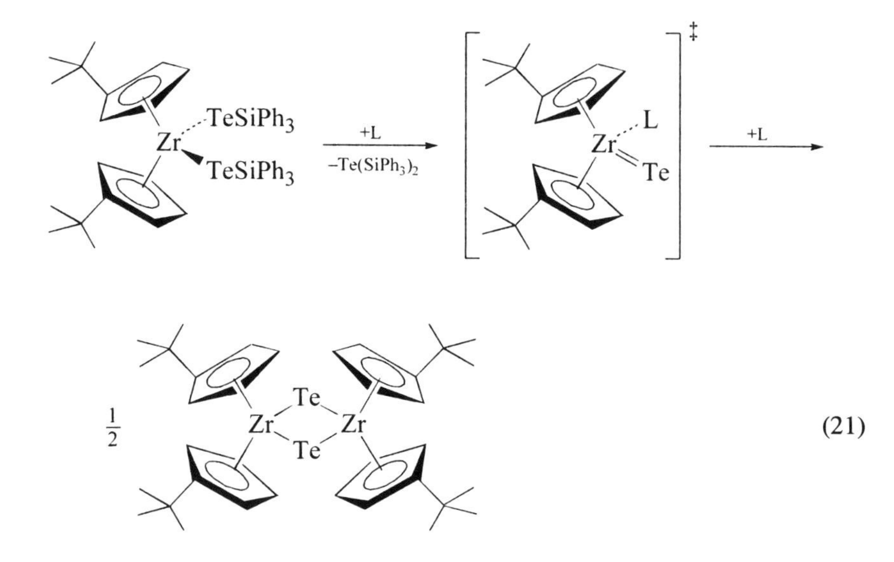

(21)

Loss of pyridine and dimerization of the terminal telluride species to form the bis μ-telluride complex is presumably favored when the Cp ligand is only partly substituted. The X-ray crystal structure of $Cp_2Zr(TeSiPh_3)_2$ showed it to be almost identical to the $-Si(SiMe_3)_3$ complex (Zr—Te = 2.876 Å).

In general, the metallocene bis-tellurolates are highly colored, despite their d^0 electronic configuration, with the Ti complexes being very dark red or black, Zr violet or red, and Hf orange or yellow. An interesting ^{125}Te NMR correlation has been made, which shows a linear dependence of the lowest energy excited state (λ_{max}) with ^{125}Te chemical shifts (see Section VI).

Attempts to prepare related bulky *alkyl* analogues to the silyl complexes discussed above highlighted a remarkable difference in stabilities between the two. The bis-alkyltellurolate $Cp_2Zr[TeC(SiMe_3)_3]_2$, formed at $-60°C$ in a metathesis reaction analogous to the silyl cases, decomposed above $-20°C$ as monitored by ^1H NMR spectroscopy [Scheme 2 (71)]. This contrasts with the very high thermal stability of the corresponding silyl tellurolate that melts without decomposition at 204°C!

A complete series of homoleptic Group 4 (IVB) tellurolates with the general formula $M[TeSi(SiMe_3)_3]_4$ was synthesized by either metathesis or tellurolysis

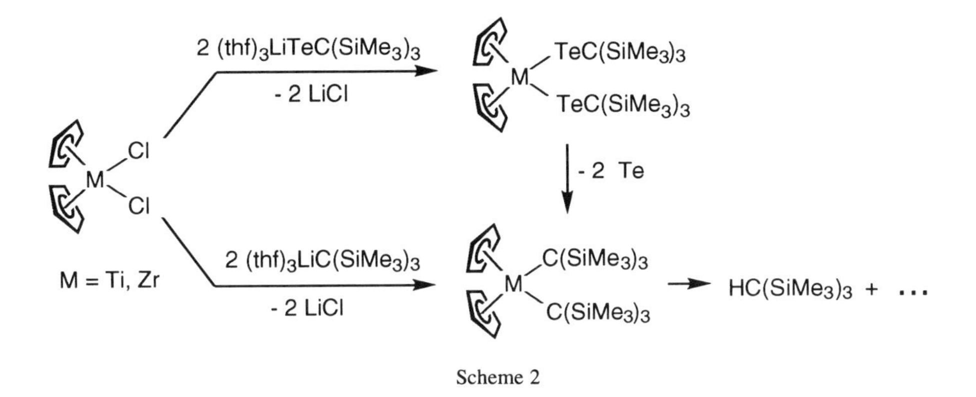

Scheme 2

(75). Interestingly, the Ti complex could only be obtained by metathesis between the magnesium bis-tellurolate and $TiCl_4(pyr)_2$. The complexes were highly colored (Ti, black; Zr, green; Hf, red) and were characterized by numerous methods, including ^{125}Te NMR and X-ray crystallography for the Zr and Hf compounds. The M—Te bond lengths in the two compounds are almost identical, falling in the range 2.724–2.751 Å (Fig. 8). These values are substantially shorter than those in the eight- and nine-coordinate metallocene species described above, and the same steric and electronic arguments discussed above apply here also.

In common with most early transition metal tellurolates, the homoleptic compounds react with water and oxygen to yield tellurol and ditelluride in addition to metal oxides–hydroxides. Despite the steric bulk of the four tellurolates, the Zr and Hf derivatives still manifest their electronically unsaturated character by coordinating Lewis bases. For example, two equivalents of alkyl or aryl isocyanide react to form isolable *trans*-$M[TeSi(SiMe_3)_3]_4(RNC)_2$ complexes. In the case of chelating phosphines, such as dmpe, the related six-coordinate adduct was also isolated (Eq. 22); treatment with a second equivalent of base resulted in a clean elimination reaction to form the novel terminal telluride–tellurolate species $M(Te)[TeSi(SiMe_3)_3]_2(dmpe)_2$ shown in Fig. 9.

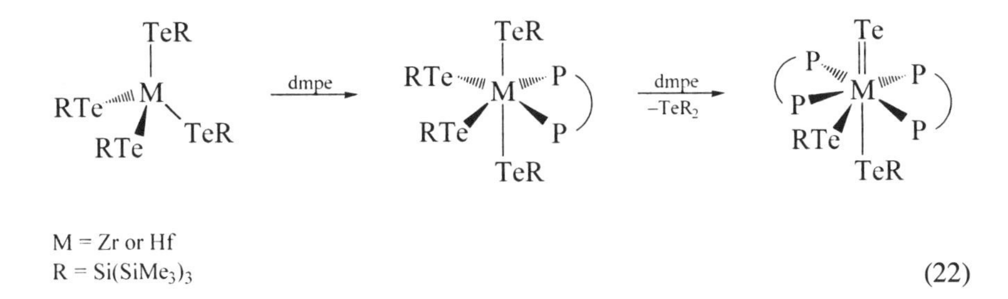

(22)

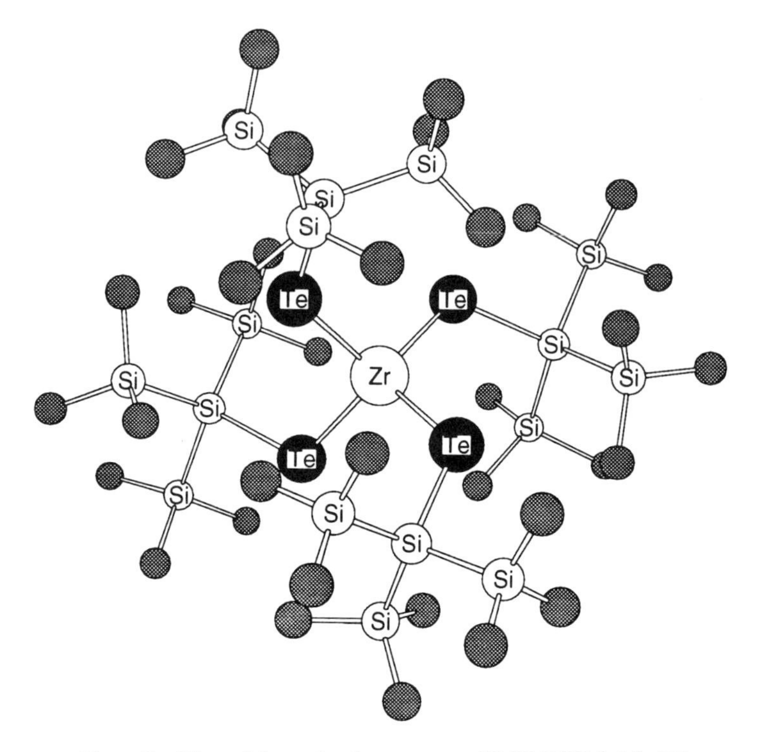

Figure 8. View of the molecular structure of $Zr[TeSi(SiMe_3)_3]_4$ (75).

Three signals were detected in the ^{125}Te NMR spectra of the compounds, one at low field (δ −706), assigned to the terminal telluride, and two at higher field (δ −1173 (t, $|J_{TeP}|$ 166 Hz), −1197 for Zr), assigned to the tellurolate ligands cis and trans, respectively, to the telluride. X-ray crystal structure determinations for both complexes allowed comparison of metal–tellurium double and single bonds in the same complex for the first time. The Zr=Te bond [2.650(1) Å] is considerably shorter than either of the two different Zr—Te single bonds, which differ slightly depending on their relationship to the terminal telluride [cis 3.028(1) Å; trans. 2.939(1) Å].

Vanadocene and niobocene phenyl tellurolates and selenolates were reported in 1975. They were prepared by metathesis routes from LiTePh and the metallocene dichlorides (76). Since this study, only one other paper involving these metals has appeared. In 1990 Herberhold et al. (77) used the interaction of dilithio-1-1′-diselenoferrocene with $CpV(O)Cl_2$ to prepare $[Fe(C_5H_4Se)_2]$-V(O)Cp (77). The Cp* analogue was also made and structurally characterized (Fig. 10).

Tantalum tellurolates and selenolates were implicated in the reactions shown in Eq. 23, although they are short lived, even at −78°C (78).

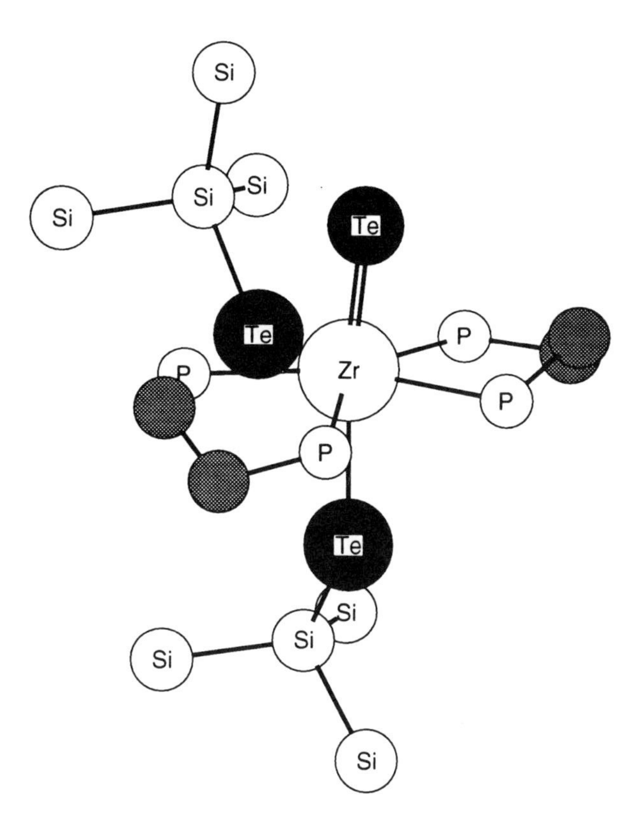

Figure 9. View of the molecular structure of Zr=Te[TeSi(SiMe$_3$)$_3$]$_2$(dmpe)$_2$ (methyl groups omitted for clarity) (75).

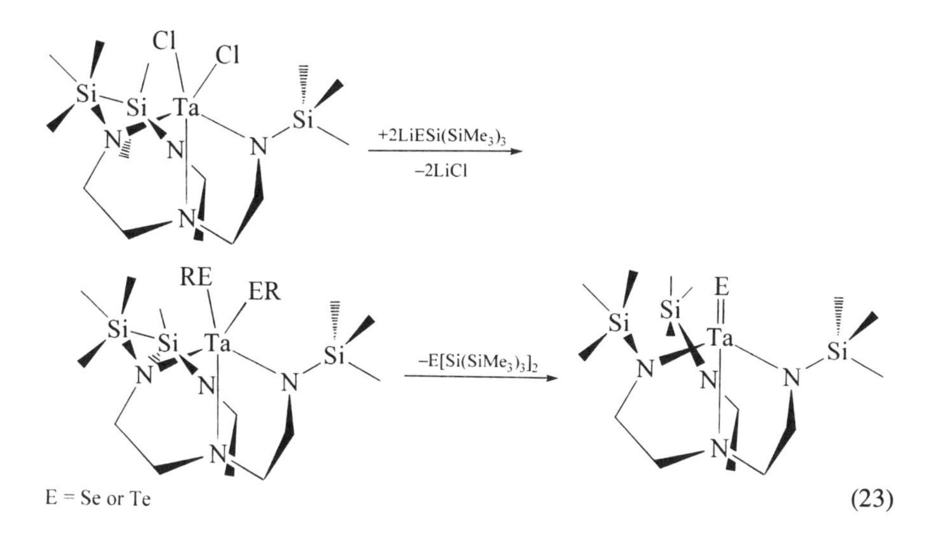

(23)

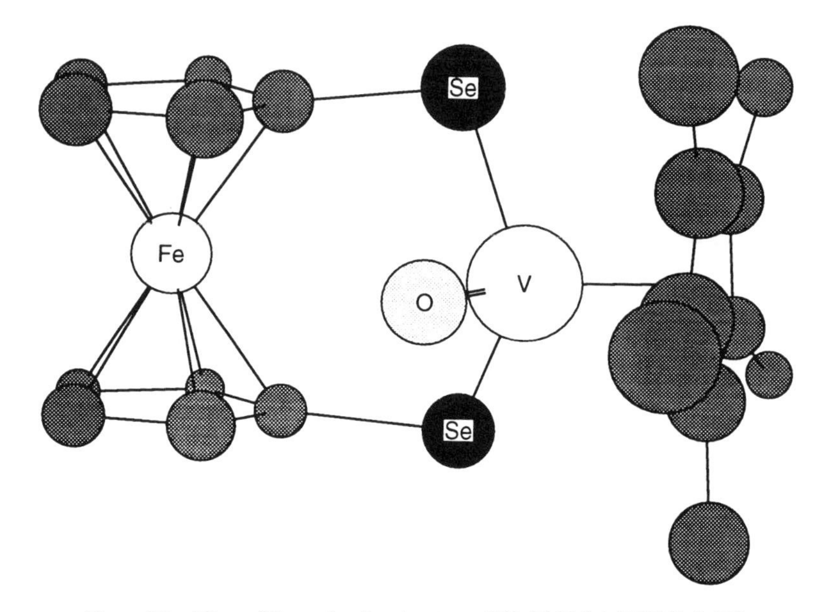

Figure 10. View of the molecular structure of $[Fe(C_5H_4Se)_2]V(O)Cp^*$ (77).

These reactions are clearly related to the ligand induced eliminations discussed above for zirconium and hafnium. In Reaction 23, however, the bulky silylated–tren ligand acts as a template promoting the elimination process under very mild conditions.

B. Chromium, Molybdenum, Tungsten, Manganese, Technetium, and Rhenium

Some of the earliest work with these metals used *in situ* generated magnesium reagents to prepare molybdenocene and tungstenocene methyl, phenyl, and *p*-tolyl tellurolates and selenolates (Eq. 24) (76), where E = Se or Te and R = Me or Ph.

$$Cp_2MCl_2 + 2REMgBr \longrightarrow Cp_2M(ER)_2 + 2MgBrCl \qquad (24)$$

All these derivatives reacted with MeI to form Cp_2MI_2 with the exception of $Cp_2Mo(SeMe)_2$, which produced the salt $[Cp_2Mo(SeMe_2)I]I$.

Cycloheptatrienyl Mo phenyl selenolates and tellurolates were prepared and structurally characterized in 1981 (79). The structure of the first example of a triply bridging selenolate $(\eta^7\text{-}C_7H_7)Mo(\mu\text{-}SePh)_3Mo(CO)_3$ was described along with $(\eta^7\text{-}C_7H_7)Mo(SePh)(CO)_3$ and $(\eta^7\text{-}C_7H_7)Mo(TePh)(CO)_3$. The structure of a related complex $[Mo(\mu\text{-}TePh)(CO)_4]_2$ prepared in low yield by oxidative addition of Ph_2Te_2 to $Mo(CO)_6$, has also been reported (80).

There are several recent accounts describing chromium tellurolates and selenolates. The work of Andra (62), who prepared $W(SePh)_4$ and $Cr(SePh)_3$, was already mentioned. Gindelberger and Arnold (81) isolated the orange, paramagnetic Cr^{II} bis-tellurolate $Cr[TeSi(SiMe_3)_3]_2(dmpe)_2$ (μ = 2.7 BM) in high yield by reaction of $CrCl_3(dmpe)_2$ with $LiTeSi(SiMe_3)_3$ in hexanes. Goh and co-workers (82, 83) prepared $CpCr(SePh)(CO)_3$ by e^- oxidation of $Cp_2Cr_2(CO)_6$ with Ph_2Se_2 and structurally characterized closely related species such as $[CpCr(SePh)]_2Se$ and $[CpCr(SePh)]_2Se[Cr(CO)_5]$. Chalcogenolate-bridged species are quite common, especially for selenolates. Zeigler prepared and structurally characterized $Cp_2Cr(NO)_2(\mu\text{-}SePh)_2$ and $Cp_2Cr(NO)_2(\mu\text{-}n\text{-}BuSe)(\mu\text{-}OH)$ along with related alkyl and aryl selenolates and tellurolates (84). Fischer and co-workers (85) fully characterized the related $Cr_2(CO)_8(\mu\text{-}SePh)_2$, and in an interesting migration reaction the aryl selenolates *trans*-$Cr(SeAr)(CO)_4(CNEt_2)$ were formed by transfer of selenolate groups from the carbene in $Cr(CO)_5[C(SeAr)NEt_2]$ to the metal. A first-order mechanism is involved in which the nature of the Ar group has little influence on the reaction. The structure of one of the products (Ar = $p\text{-}FC_6H_4$) was determined by X-ray crystallography.

A number of new molybdenum and tungsten tellurolates and selenolates have been prepared in the last few years. Electronic and redox properties of oxomolybdenum and tungsten phenylselenolates were reported by Wedd and coworkers (86, 87). Specific examples include $(Et_4N)[M(O)(SePh)_4]$ and the dinuclear complex $(Et_4N)[Mo_2O_2(SePh)_6(OMe)]$. Boorman and co-workers (20, 66, 88–91) investigated a range of selenolate derivatives with both Mo and W. Examples include the oxo–selenolate triphos complex $MoO[PH_2P(O)CH_2CH_2P(PH)CH_2CH_2PPH_2][Se(mesityl)]_2$ (89), the bimetallic $Cu(PPh_3)(\mu\text{-}p\text{-}tolylSe)_3W(\mu\text{-}p\text{-}tolylSe)_3Cu(PPh_3)$ (91), and the homoleptic derivative $W[Se(mesityl)]_4$ (90). The latter reacts with *t*-BuNC to form the interesting homobimetallic complex $(t\text{-}BuNC)[Se(mesityl)]_2W(\mu\text{-}Se)_2W[Se(mesityl)]_2(t\text{-}BuNC)$. The selenide–selenolate dimer $W_2Se_2(PMePh_2)(\mu\text{-}SePh)_4$ was prepared in low yield by reacting $WCl_4(PMePh_2)$ with $Me_3SiSePh$ (88). The X-ray structure shows the complex to contain two bridging selenides with four terminal selenolates.

Dinuclear Mo and W selenolates containing M—M bonds were recently isolated and characterized by Chisholm and co-workers (92, 93). These include $M_2(OR)_2[Se(mesityl)]_4$ and $M_2[Se(mesityl)]_6$ (R = *t*-Bu, CH_2 *t*-Bu, or *i*-Pr), which were all derived from reactions of M_2X_6 (X = NMe_2 or OR) or closely related species with mesityl selenol. The $M_2[Se(mesityl)]_6$ compounds are isostructural as determined by X-ray diffraction with staggered confirmations (Fig. 11).

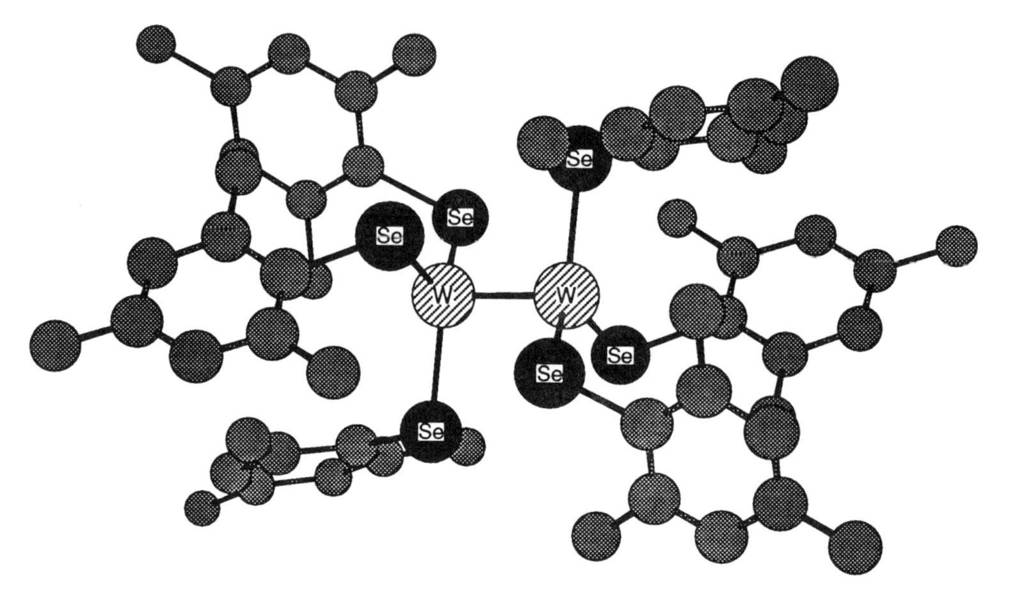

Figure 11. View of the molecular structure of $W_2[Se(mesityl)]_6$ (92).

Four-coordinate $Mn[TeSi(SiMe_3)_3]_2(dmpe)$ and $Mn[TeSi(SiMe_3)_3]_2(4\text{-}t\text{-}BuC_5H_4N)_2$ were prepared by tellurolysis of $Mn[N(SiMe_3)_2]_2(thf)_2$ with $HTeSi(SiMe_3)_3$ in the presence of excess dmpe or $4\text{-}t\text{-}BuC_5H_4N$, respectively (81). Both compounds are paramagnetic, with magnetic moments of 5.9 BM in accord with high-spin, d^5 electronic configurations. In the dmpe complex, the Mn—Te bond [2.690(2) Å] is shorter than in either $[Mn(TePh)_4]^{2-}$ [2.722–2.760 Å (94)] or $Mn(TeCH_2Ph)(CO)_3(PEt_3)_2$ [2.705 Å (57)]. Related aryl selenolates were described by Bochmann et al. (95) a year later. Selenolysis of $Mn[N(SiMe_3)_2]_2(thf)$ with $HSe(2,4,6\text{-}i\text{-}PrC_6H_2)$ gave $Mn[Se(2,4,6\text{-}i\text{-}PrC_6H_2)]_2(thf)$ by way of the isolated intermediate $Mn[N(SiMe_3)_2][Se(2,4,6\text{-}i\text{-}PrC_6H_2)(thf)$, as shown by X-ray crystallography (95). The homoleptic derivatives $Mn(EPh)_4^{2-}$ [E = Se (96) or Te] are isostructural with slightly distorted tetrahedral Mn^{II} centers (94).

A number of Mn complexes involving carbonyl ligands have been prepared. Exploiting the electrophilicity of the telluride ligand in $(\mu_3\text{-}Te)[Mn(CO)_3Cp]_3$, Hermann et al. (97) were able to prepare a triply bridging methyl tellurolate anion $(\mu_3\text{-}TeMe)[Mn(CO)_3Cp]_3^-$. Determination of the crystal structures of both complexes showed that the planar Mn_3Te core in the precursor was significantly pyramidalized by methylation at Te. Huttner et al. (98) found that the Mn—TePh bond in $Cp_2(CO)_4Mn_2(\mu\text{-}TePh)_2$ (99) is stable to oxidation and reduction by $AgPF_6$ and $NaBH_4$, respectively. A report from the same group investigated the oxidation of $(Cp')Mn(CO)_2[E(aryl)]$ (E = Se or Te), which

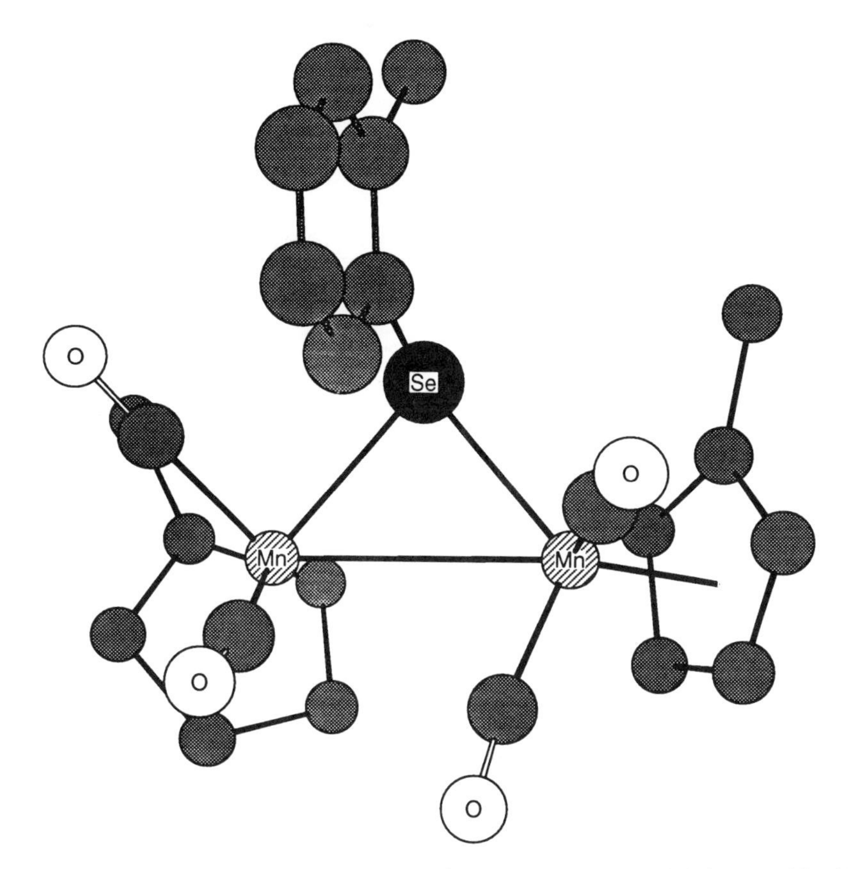

Figure 12. View of the structure of the anion in $\{[(Cp'Mn(CO)_2]_2[Se(o\text{-tolyl})]\}[PF_6]$ (100, 101).

gave ''inidene'' complexes as the salts $\{[Cp'Mn(CO)_2]_2[E(aryl)]\}^+$ (Fig. 12) (100, 101).

Insertion of Te into Mn—C bonds is a useful route to complexes such as $Mn(TeR)(CO)_3(PR'_3)_2$ (R = Me or CH_2Ph; R′ = Me or Et) (58). The same authors also used metathesis chemistry to prepare $[Mn(CO)_4(\mu\text{-}TeR)]_2$ dimers, with R = Me, Et, *i*-Pr, CH_2SiMe_3, and $Si(SiMe_3)_3$.

Not surprisingly, there are no Tc selenolates or tellurolates and very few examples for Re. Cotton and Dunbar (59) reduced the triple bond in $Re_2Cl_4(\mu\text{-}dppm)_2$ by oxidative addition of Ph_2Se_2 to yield the doubly bonded product $Re_2Cl_4(\mu\text{-}SePh)_2(\mu\text{-}dppm)_2$. X-ray crystallography revealed average Re—Se bond lengths of 2.462(1) Å and a metal–metal bond length of 2.656(1) Å. Trimethylstannyl selenolates and tellurolates of Re and Mn of the type $M(ESnMe_3)(CO)_4$, prepared from the reactions of $M(CO)_5X$ (X = Cl or Br) with $E(SnMe_3)_2$ (Eq. 25), rearrange on heating to the tetramers $[M(ESnMe_3)(CO)_3]_4$ (102).

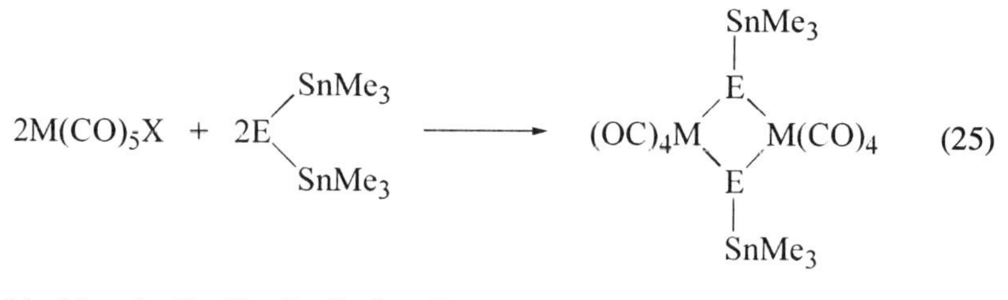

M = Mn or Re; X = Cl or Br; E = Se or Te

C. Iron, Ruthenium, Osmium, Cobalt, Rhodium, and Iridium

There is now a fairly substantial body of information on iron tellurolates and selenolates and while most of these involve bridging chalcogenolates, there are a few complexes containing terminal interactions. McConnachie and Ibers (96) structurally characterized the homoleptic derivative $[Fe(SePh)_4]^{2-}$ as its NEt_4^+ salt as part of a wider study of $M(SePh)_4^{2-}$ complexes. It shows a slightly distorted tetrahedral geometry with Se—Fe—Se angles ranging from 103.6 to 114.9(1)°. The corresponding tellurolate anion reacts with Na_2Te to form the tellurolate–telluride cluster $[Fe_4Te_4(TePh)_4]^{3-}$ (Fig. 13), the structure of which is related to the more well studied sulfur analogues (103).

Treatment of $FeCl_2(dmpe)_2$ with one or two equivalents of $LiTeSi(SiMe_3)_3$ yielded the diamagnetic iron(II) tellurolates $Fe[TeSi(SiMe_3)_3]Cl(dmpe)_2$ and $Fe[TeSi(SiMe_3)_3]_2(dmpe)_2$ as orange and dark green crystals, respectively (81). One tellurolate ligand in $Fe[TeSi(SiMe_3)_3]_2(dmpe)_2$ was displaced by MeCN to form the salt $[FeTeSi(SiMe_3)_3(dmpe)_2(MeCN)][TeSi(SiMe_3)_3]$. In a related reaction that demonstrated the lability of tellurolate ligands in this system, it was shown that redistribution occurred between $FeCl_2(dmpe)_2$ and $Fe[TeSi(SiMe_3)_3]_2(dmpe)_2$ to form two equivalents of the mixed tellurolate–chloride complex.

Low-valent Fe^0 complexes oxidatively add diselenides and ditellurides to form a range of Fe^{II} derivatives as demonstrated recently by Liaw and co-workers (104–109). Examples include $[Na(thf)_3][Fe(CO)_3(SePh)_3]$, and *fac*-$[PPN][Fe(CO)_3(TePh)_n(SePh)_{3-n}]$ (n = 1 or 2). Huttner and co-workers (110) prepared and crystallographically characterized the neutral trinuclear cluster $Fe_3(CO)_9(\mu_3\text{-SCy})(\mu_2\text{-SePh})$ (Fig. 14). Alkenyl substituted cluster species related to the latter complex were also reported (111). Oxidative addition reactions of the phosphido-bridged cluster $Fe_3(CO)_{10}(\mu_3\text{-P})$ with Ph_2Se_2 and (mes-

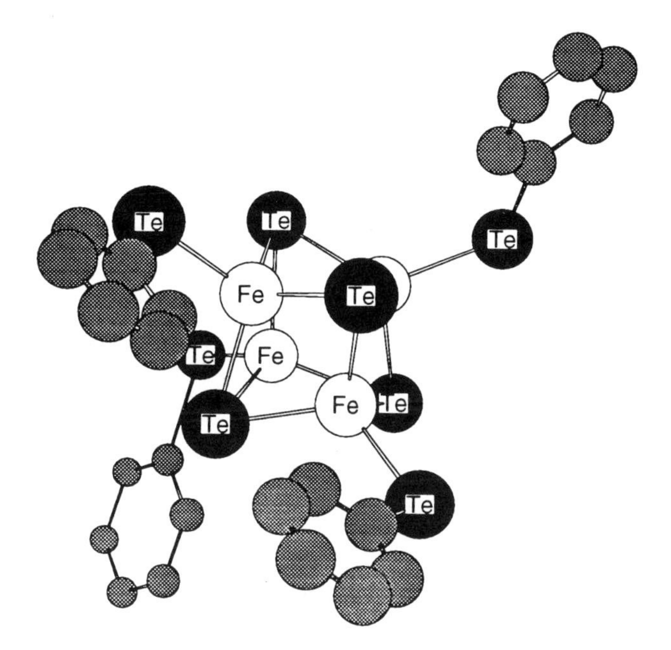

Figure 13. View of the structure of the anion in $[NEt_4]_3[Fe_4Te_4(TePh)_4]$ (103).

ityl)$_2$Te$_2$ led to mixtures of selenolate and tellurolate complexes, including $Fe_3(CO)_9(\mu_3\text{-P})(\mu_3\text{-ER})$, $Fe_2(CO)_6(\mu_2\text{-PE})(\mu_2\text{-RPER})$, and $Fe_3(CO)_8(\mu_3\text{-PR})(\mu_2\text{-ER})_2$ (112). An unusual doubly bridging bidentate tellurolate ligand was found to span two iron carbonyl fragments in the complex $[(\mu\text{-TeMe})Fe_2(CO)_6]_2(\mu\text{-}Te_2CH_2CH_2Te\text{-}\mu)$, as shown by X-ray crystallography (113). The possibility that two bridging chalcogenolates could give rise to different geometric isomers under suitable conditions was confirmed by structures of *cis*- and *trans*-$\{CpFe(CO)[Te(p\text{-}EtOC_6H_4)]\}_2$ (114).

A five coordinate Ru^{IV} selenolate $Ru[Se(2,3,5,6\text{-}Me_4C_6H)]_4(MeCN)$ was prepared and its structure determined for comparison to the analogous thiolate (Fig. 15) (115). As expected, the Ru—Se bonds were slightly longer than in the thiolate [$Ru—Se_{eq}$ av = 2.322(6) Å; $Ru—Se_{ax}$ = 2.495(3) Å]. Selenolate ligands were proposed to enhance the metal's π basicity as evidenced by the relatively shorter Ru—N distance and lower CN stretching frequency in the selenolate.

Compounds with bridging chalcogenolates are much more common. Andreu et al. (61) followed up earlier chemistry of Schermer and Baddley (116) with a report on the chemistry and structure of $Ru_2(CO)_6(\mu\text{-SePh})_2$. These results confirmed that the complex with an anti configuration of the phenyl selenolate bridges was preferred in solution and the solid state. More complex clusters are

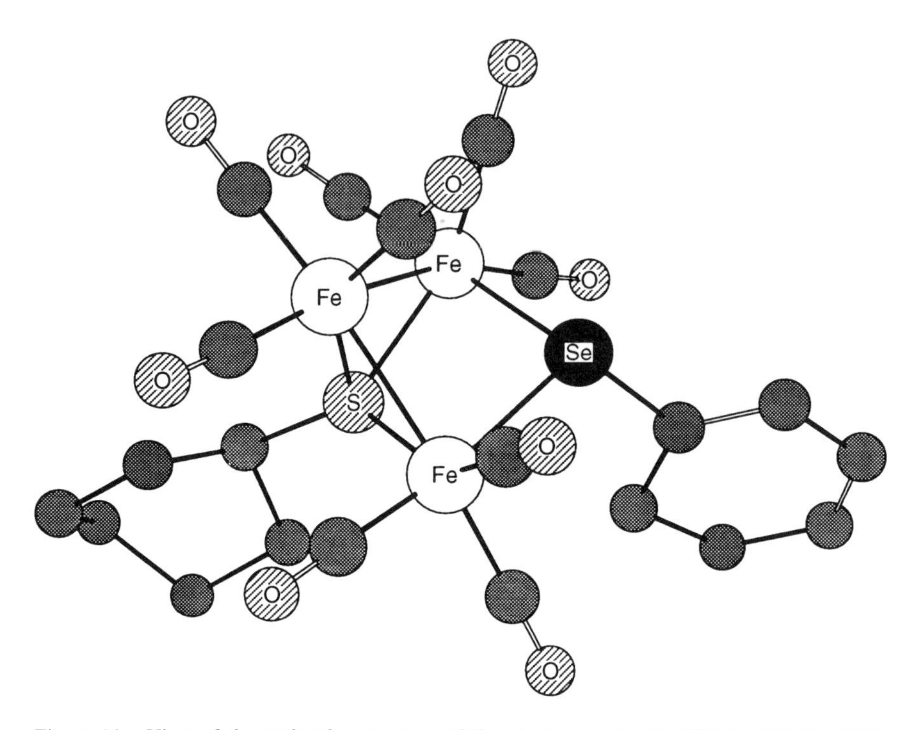

Figure 14. View of the molecular structure of the cluster species $Fe_3(CO)_9(\mu_3\text{-SCy})(\mu_2\text{-SePh})$ (110).

also known, for example, the carbide species $[Ru_6C(CO)_{16}]^{2-}$ reacts with PhSeCl to form the anion $[Ru_6C(CO)_{15}(SePh)]^-$, which was characterized structurally as its PPN salt (117). Reactions of triosmium clusters with diselenides originally reported in 1982 (118) were the subject of a reinvestigation by Deeming and co-workers (119). Two different isomers of $Os_3(CO)_{10}(\mu\text{-SeR})_2$ (R = Ph or Me) were identified as the major products following chromatography of the reaction mixtures. The kinetic isomer was shown to rearrange by an edge-to-edge shift of SeR groups from a species in which they are on different edges of the cluster.

More recent studies used reactions in Eqs. 26 and 27 to prepare tellurolates and selenolates with the ubiquitous Cp*Ru fragment; both complexes were structurally characterized by X-ray diffraction (120).

The insoluble Co^{II} homoleptic species $Co(SePh)_2$ was claimed many years ago, but nothing is known about its (presumably polymeric) structure (62). The four coordinate $[Co(SePh)_4]^{2-}$ is well characterized, however, having been prepared by Ibers and co-worker (96) from the reaction of $CoCl_2$ with LiSePh. The

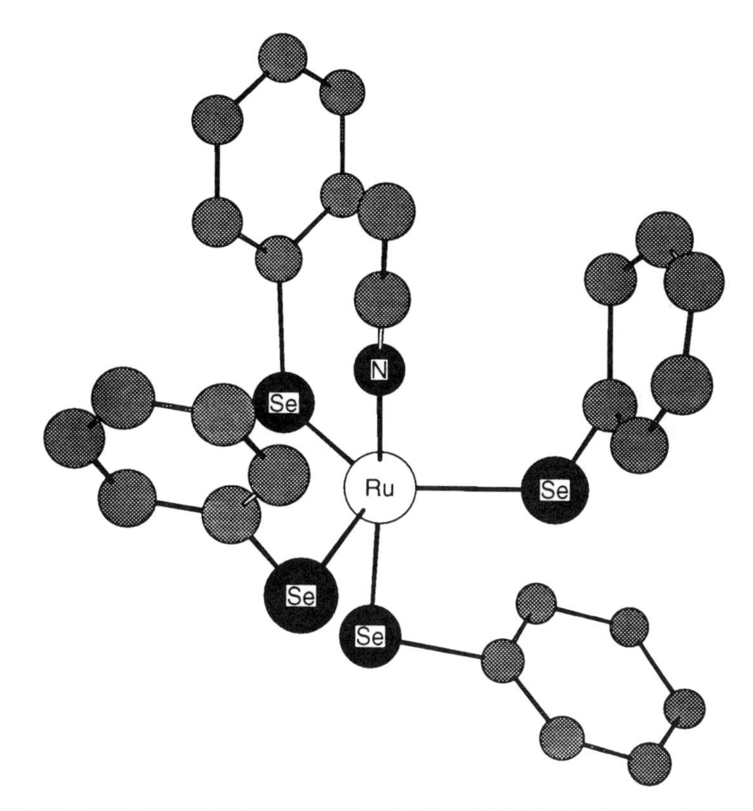

Figure 15. View of the molecular structure of $Ru[Se(2,3,5,6\text{-}Me_4C_6H)]_4(MeCN)$ (Me groups on phenyl rings omitted for clarity) (115).

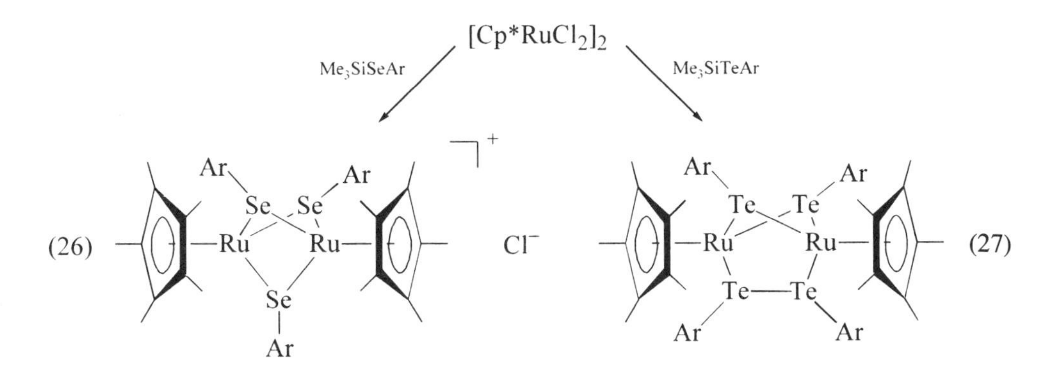

Ar = Ph or *p*-tolyl

reducing power of tellurolate ligands was again demonstrated in attempts to prepare Co^{II} derivatives (Eq. 28) (81).

$$Co^{II}Br_2(PMe_3)_3 + LiTeSi(SiMe_3)_3 \longrightarrow$$
$$Co^{I}[TeSi(SiMe_3)_3](PMe_3)_3 + \tfrac{1}{2}[TeSi(SiMe_3)_3]_2 \quad (28)$$

Magnetic measurements showed the dark orange Co tellurolate to be low-spin d^8, with two unpaired electrons (μ = 3.8 BM); X-ray crystallography confirmed it to be four coordinate in the solid state with a Co—Te bond length of 2.543(1) Å (Fig. 16). Reaction of this 16 e^- species with carbon monoxide resulted in displacement of one phosphine and formation of the five-coordinate 18 e^- complex $Co[TeSi(SiMe_3)_3](CO)_2(PMe_3)_2$.

Tellurolates or selenolates of Rh are extremely few in number. Collman et al. (121) isolated and structurally characterized the six-coordinate Rh^{III} complex shown below by oxidative addition of Ph_2Se_2 to a Rh^{I} precursor (Eq. 29).

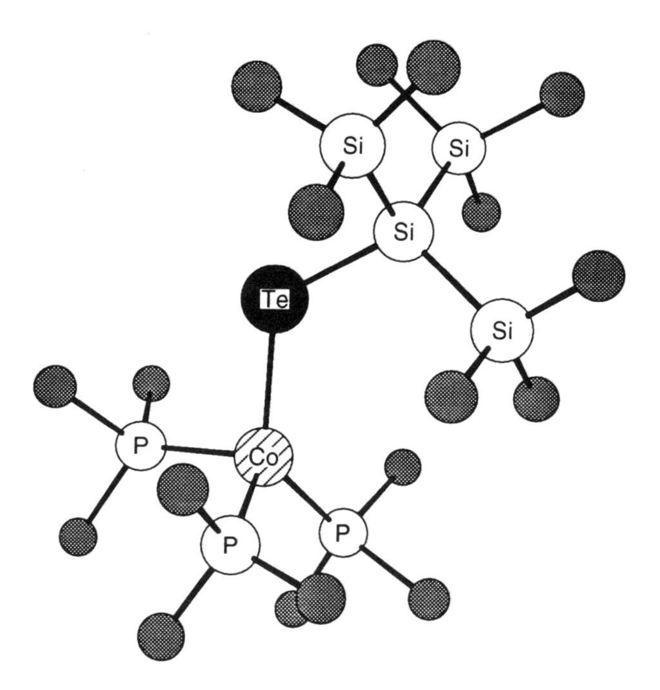

Figure 16. View of the molecular structure of $Co[TeSi(SiMe_3)_3](PMe_3)_3$ (81).

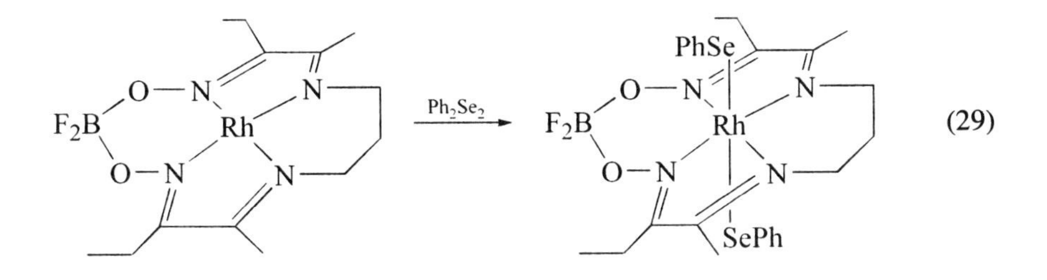

A Rh^{III} methyl selenolate $\{Rh[P(CH_2CH_2PPh_2)_3](H)SeMe\}BPh_4$ was crystallographically characterized by Di Vaira et al. (122). The only tellurolates of Rh are the ill-defined phenyl derivatives $Rh(TePh)_3$ and $Rh(CO)Cl_2(TePh)(TePh_2)_2$ (123).

An unusual cis-Vaska's-type complex has been prepared using a bulky aryl tellurolate ligand as shown in Eq. 30 (21).

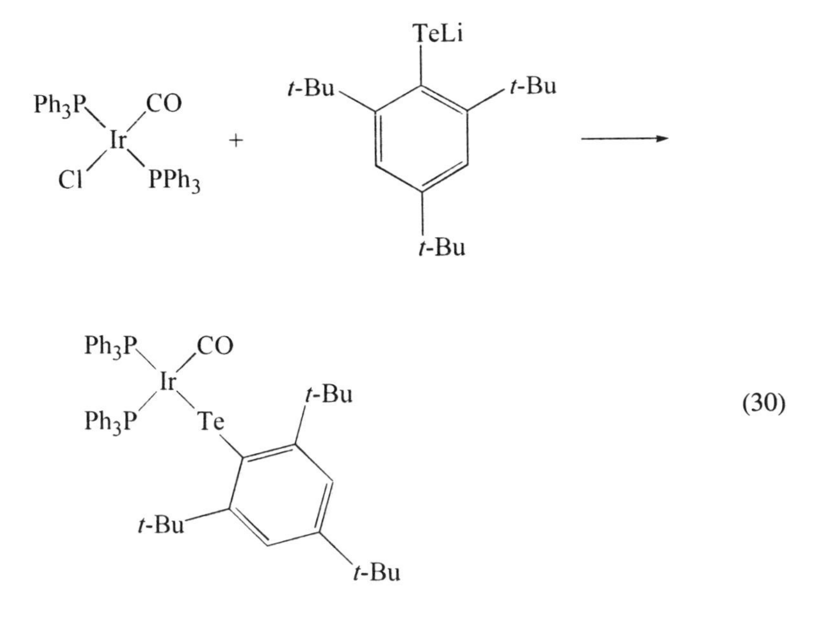

No evidence was obtained for the intermediacy of the more usual trans isomer. The structure of the complex, determined by X-ray crystallography, showed the geometry about Ir is far from square planar and may be viewed as having a slight tetragonal distortion. The aryl ring in the tellurolate ligand is bent into a saddle-shaped geometry with the ipso carbon 0.26 Å out of the plane of the other five carbon atoms.

D. Nickel, Palladium, and Platinum

An impetus for much of the research in chalcogenolate chemistry involving these metals relates to the novel spectroscopic, electronic, and magnetic properties of analogous dithiolene-type complexes (124). Complexes with Ni—Se bonds, such as the structurally characterized $Ni(SeC_6H_4PPh_2)_2$ (125) and $CpNi(SePh)(Pn\text{-}Bu_3)$ (126) have been in the literature for many years (8). In more recent work, Sandman et al. (127) prepared and fully characterized Ni, Co, and Cu complexes of general formula $M(o\text{-}Se_2C_6H_4)_2$, and shortly afterwards Hoffman and co-workers (128) described their studies of ferromagnetic interactions in Cp_2^*Fe salts of the Ni derivative.

Biological aspects of nickel selenolate chemistry have recently been addressed by Mascharak and co-workers (129, 130). Their work was motivated by findings that some nickel-containing enzymes also involve close interactions to selenium. For example, in its first coordination sphere, the active site of the enzyme *Desulfovibrio baculatus* has a Ni—Se contact of 2.44 Å and in *Methanococcus vannielii* it has been suggested that there are two close Ni—Se interactions. Model complexes with the general formula $Ni(SeAr)_2(L)$ (Ar = Ph or mesityl; L = tripyridine or dimethylphenanthrolene) and anions with four Ni—Se bonds $[Ni(SeCH_2CH_2Se)_2]^{2-}$ were isolated and structurally characterized.

There are many examples of Pt^{II} tellurolates and selenolates, most of which are prepared either by metathesis reactions involving aryl tellurolate ligands or by oxidative additions of dichalcogenides to Pt^0 compounds (6, 8, 131, 132). One of the first to be structurally characterized, the ditellurolene $Pt(o\text{-}Te_2C_6H_4)(PPh_3)_2$, was prepared by Rauchfuss and co-workers (Fig. 17) (133).

Bonasia and Arnold (21) isolated complexes of the type *cis*-$Pt[Te(2,4,6\text{-}R_3C_6H_2)]_2(PPh_3)_2$ (R = *t*-Bu or *i*-Pr) and structurally characterized the *t*-Bu derivative. An early report of $Ni(SePh)_2$ did not address the structure of the complex, noting only that it was thought to be polymeric (62). Trifluoromethyl substituted compounds of general formula $\{M[Se_2C_2(CF_3)_2]_2\}^{-z}$ (M = Pt, z = 0, 1, or 2; M = Au, z = 1) have been prepared and fully characterized by a range of spectroscopic and crystallographic methods (134).

Two groups of workers in India have studied a broad range of Pt and Pd complexes with terminal and bridging selenolate and tellurolate ligands and have characterized numerous examples by X-ray diffraction (135–140).

E. Copper, Silver, and Gold

Interest in Se and Te derivatives of these metals (particularly Cu) has increased recently with the realization that they may serve as precursors to electronic materials (see Section VII). While the clinical use of gold(I) thiolates for

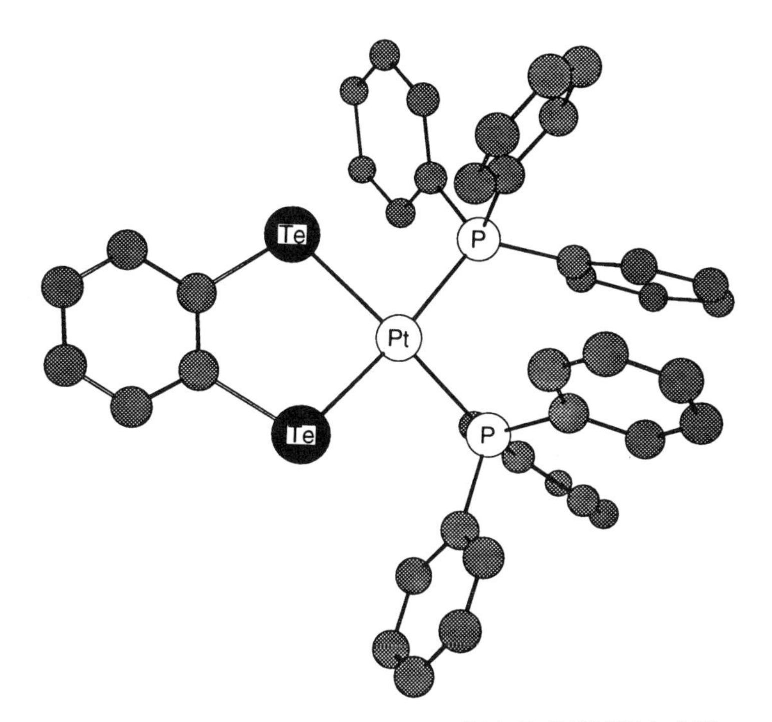

Figure 17. View of the molecular structure of $Pt(o\text{-}Te_2C_6H_4)(PPh_3)_2$ (133).

the treatment of rheumatoid arthritis (141, 142) has unquestionably provided an impetus to study the chemistry of gold thiolates, related chemistry involving selenolates and tellurolates has received much less attention. In addition to further potential chemotherapeutic uses in the treatment of arthritis and cancer (143, 144), gold chalcogenolates are beginning to find applications in the sensitization of photographic materials (145), electron microscopy (141), and the preparation of conducting, semiconducting, and laser recording devices (141). Polymeric CuTeR (R = alkyl or aryl) complexes, made from trialkyl tin reagents and CuCl, were first reported in 1976 (146, 147). The structural and electronic properties of $[NMe_4][Cu(Se_5C_3)]$ were described by Matsubayashi and Yokozawa (148) who found that in contrast to related dithiolenes and other diselenolenes, the geometry was not square planar, showing a dihedral angle of 53.7° between the ligand planes. Work on a novel electrochemical synthesis of $(Ph_3P)Cu(\mu\text{-}SePh)_2Cu(PPh_3)$ involved electrooxidation of a copper anode in MeCN containing PPh_3 and Ph_3Se_2 (149, 150). The copper in the Cu_2Se_2 core is pseudotetrahedral with bridging selenolates that are unsymmetrically disposed such that Cu—Se distances range from 2.406(1) to 2.617(1) Å (149).

Efforts to produce a single-source ternary precursor to the solar cell material $CuInSe_2$ used the bridging tendency of selenolate ligands to good effect in the following synthesis (Eq. 31) (151).

$$[In(SeR)_4]^- + [Cu(MeCN)_2(PPh_3)_2]^+ \longrightarrow (RSe)_2In(\mu\text{-}SeR)_2Cu(PPh_3)_2 \qquad (31)$$

The structure of the ethyl complex shows a planar $CuIn(SeEt)_2$ core similar to the thiolato analogue.

Two recent reports have expanded the tellurolate chemistry of the Group 11 (IB) metals considerably. Homoleptic derivatives with the bulky alkyl or silyl tellurolate ligands $TeX(SiMe_3)_3$ (X = C or Si) were unstable for Cu and Ag, and metathetical reactions of the type shown in Eq. 32 led to redox reactions (152).

$$MCl + LiTeSi(SiMe_3)_3 \longrightarrow \tfrac{1}{2}[TeSi(SiMe_3)_3]_2 + LiCl \qquad (M = Cu \text{ or } Ag) \qquad (32)$$

In the case of Cu, this reaction is the best way to prepare useful quantities of the ditellurides synthetically. Addition of phosphines such as PCy_3 to the above reactions yielded the yellow, mildly air-sensitive derivatives $M[EC(SiMe_3)_3]$-PCy_3 (M = Cu or Ag; E = Se or Te). The structure of the copper selenolate revealed a dimeric Cu_2Se_2 core that is severely folded or "butterflied."

The commonly observed higher stability of selenolates relative to tellurolates was again demonstrated by the ability of Bonasia et al. (152) to isolate the homoleptic selenolate $Cu_4[SeC(SiMe_3)_3]_4$ from a simple metathesis reaction using CuCl. In the solid state, the complex crystallizes as a tetramer with Cu—Se bond lengths ranging from 2.473(2) to 2.488(2) Å. These are shorter than in the PCy_3 complex where the Cu—Se bonds ranged in length from 2.396(2) to 2.519(2) Å. Silver derivatives were prepared by analogous means and the structure of the homoleptic selenolate is shown in Fig. 18.

The first examples of gold tellurolates are the homoleptic derivatives isolated from metathesis reactions using AuCl(tht) as the metal starting material (153). It was shown, using X-ray crystallography, that the alkyl tellurolate exists as a tetramer $Au_4[TeC(SiMe_3)_3]_4$ in the solid state with a Te_4Au_4 core similar to the Cu selenolate above. The corresponding selenolate and thiolate were also re-

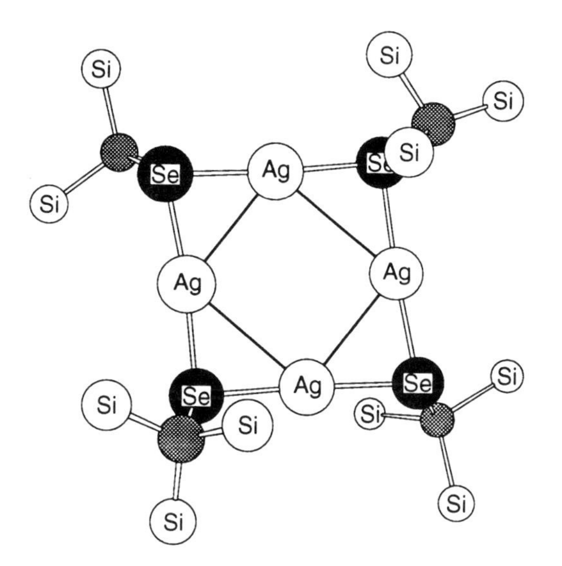

Figure 18. View of the molecular structure of $Ag_4[SeC(SiMe_3)_3]_4$ (methyl groups omitted for clarity (152).

ported. Interestingly, for the Group 11 (IB) metals the silyl and germyl analogues were much less stable than the alkyls, an effect opposite to that seen for all other metal complexes of these ligands reported to date.

Phosphine-substituted Au derivatives are readily prepared by either metathesis or chalcogenolysis with a range of metal starting materials (152, 153). The complexes $Au[EX(SiMe_3)_3]PPh_3$ (E = S, Se, or Te; X = C, Si, or Ge) were isolated and the X-ray crystal structure of the alkyl tellurolate showed it to be a loose dimer with long Au—Te intermolecular contacts [3.740(1) Å]. The intramolecular Au—Te bond length [2.566(1) Å] is effectively identical to that in the homoleptic complex described above (av 2.56(3) Å] (153).

The first silver selenolate to be structurally characterized used the novel thienyl selenolate anion as its PPh_4^+ salt (36); interaction of this with $AgNO_3$ in DMF produced the tetrameric cluster anion $[Ag_4(TeR)_6]^{2-}$ (Fig. 19). The only silver or gold tellurolates are those described above, but there are a few examples of gold selenolates; those that have been structurally characterized include $[Au(CH_2)_2PPh_2]_2(Cl)(SePh)$ (154), $Ph_3PAuSePh$ (64), the cation $[(Ph_3PAu)_2(SePh)]^+$, and its benzyl and C_6F_5 analogues (64, 155, 156), and $[(Ph_3PAu)_2(SeC_{10}H_7)]SbF_6[(Ph_3PAu)_2(SeC_{10}H_7)]$ (156). All the PPh_3 derivatives were prepared by reactions of $AuCl(PPh_3)$ with the appropriate diselenide and $AgSbF_6$. A feature common to all is at least some degree of Au· · ·Au interaction.

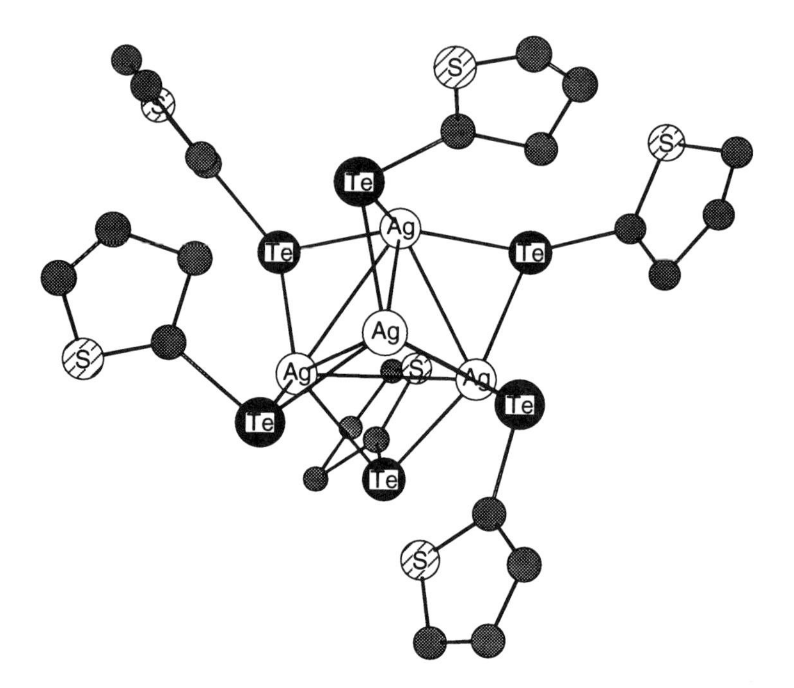

Figure 19. View of the molecular structure of the anion in $[PPh_3]_2[Ag_4(TeR)_6]$ (36).

IV. MAIN GROUP METAL DERIVATIVES

By far the most interest in main group tellurolates and selenolates has focused on zinc, cadmium, and mercury, for reasons explained in more detail in the following sections. Other than these, there are relatively few derivatives of the remainder of the main group metals, although several very recent reports have expanded the range considerably.

A. Zinc, Cadmium, and Mercury

Selenolate and tellurolate derivatives of the Group 12 (IIB) elements are among the oldest and most common metal complexes with these ligands. Recent work using sterically hindered ligands has led to further advances in Group 12 (IIB) chalcogenolate chemistry, including the first detailed structural data and the use of these complexes as single-source precursors to II–VI semiconductors (Section VII). In addition to these motivations, compounds of this type have been studied for their relevance to metal–sulfur metalloproteins, such as rubredoxin and horse liver alcohol dehydrogenase.

There are two main classes of compound in the literature, one involving aryl chalcogenolates, the other bulky silyl- or alkyl-substituted ligands. Although metathetical reactions are useful in certain instances, it is generally more convenient to employ chalcogenolysis, since this route is usually cleaner, the yields are higher, and a wide range of metal starting materials is readily available.

The three-coordinate anion $[Hg(TePh)_3]^-$ was one of the first metal tellurolates to be structurally characterized by X-ray methods (157); Kolis and coworkers (158) described the analogous thienyl complex in a recent article. Selenolate derivatives are much more common, the simplest being $Hg(SeMe)_2$, which exists as a polymer with pseudotetrahedral Hg atoms being bridged by μ-SeMe groups in infinite one-dimensional chains (159). Tetrameric pyridine complexes $[HgCl(pyr)(SeEt)]_4$ and $[HgCl(pyr)_{0.5}(Se\textit{t}\text{-}Bu)]_4$ were characterized in the same study. Four coordinate $[M(SePh)_4]^{2-}$ anions have been prepared and characterized by numerous techniques, including crystallography (160). The series of complexes of general formula $M(ESiEt_3)_2$ (M = Zn or Cd and E = Se; M = Hg and E = Te) were made by treatment of the metal dialkyls with the respective selenol or tellurol (161).

Early reports of the chemistry of $Hg(TePh)_2$ (162, 163) were reinvestigated by Steigerwald and Sprinkle (10) whose results are summarized in Eq. 33.

$$R_2Te_2 + Hg \rightleftharpoons Hg(TeR)_2 \longleftarrow HgCl_2 + 2LiTeR \qquad (33)$$

$$Hg(TeR)_2 \xrightarrow{\Delta} HgTe + TeR_2$$

R = Ph or *p*-tolyl

Due to their limited solubility in all but coordinating media, the structures of $Hg(TePh)_2$ and its cadmium analogue are unknown, but are assumed to be polymeric. Formation of the bulk metal tellurides occurred at relatively low temperature (Hg, 120°C; Cd, 200°C) suggesting for the first time that tellurolate complexes of this type may be useful precursors to solid state compounds (see Section VII). In subsequent work, bidentate phosphines were employed to prepare $[Cd(SePh)_2]_2(depe)$, which shows Cd_2Se_2 cores with planar $(CdSePh)_2$ units bridged by phosphines, and the interesting dimer $[Hg(TePh)_2(depe)]_2$, which contains terminal tellurolates and a 10-membered $(HgPCH_2CH_2P)_2$ ring (164, 165).

Bochmann et al. (166) made use of sterically hindered aryl ligands to prepare low-coordinate Group 12 (IIB) chalcogenolates. Homoleptic chalcogenolate-bridged dimers $[M(ER)_2]_2$ are formed when very large ligands such as 2,4,6-*t*-$Bu_3C_6H_2E$ are used; coordination polymers (e.g., $\{Cd[Te(mes)]_2\}_n$, Fig. 20)

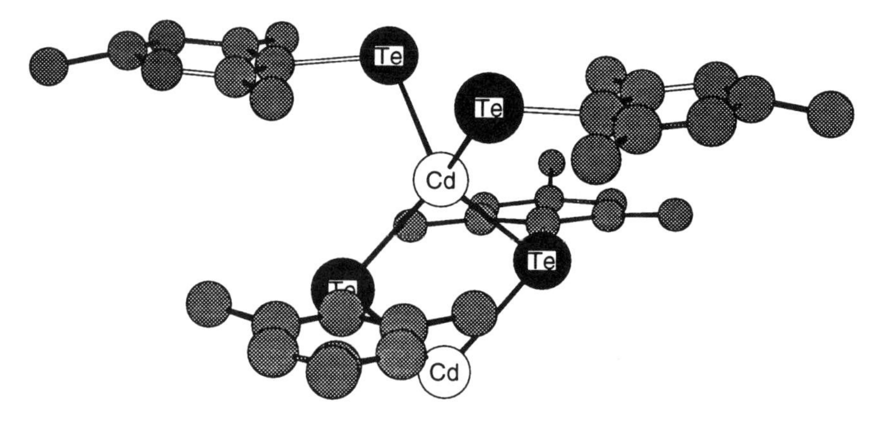

Figure 20. View of the molecular structure of the repeating unit in the polymer $\{Cd[Te(mes)]_2\}_n$ (29).

result when the less bulky mesityl analogue is prepared (29). Two equivalents of *tert*-butyl isocyanide react with the Lewis acidic Zn center in the homoleptics to produce four-coordinate $Zn[Se(2,4,6\text{-}t\text{-}Bu_3C_6H_2)]_2(t\text{-}BuNC)_2$ as shown by a range of spectroscopic and crystallographic techniques (167). Related *dimeric* aldehyde and ketone complexes were prepared, one of which, $\{Zn[Se(2,4,6\text{-}t\text{-}Bu_3C_6H_2)]_2[O{=}CH(p\text{-}MeOC_6H_4)]\}_2$, was structurally characterized (168).

Mono selenolate derivatives M(SeAr)R (M = Zn, Cd, or Hg; Ar = 2,4,6-i-$Pr_3C_6H_2$; R = Me, Et, *i*-Pr, or *n*-Pr) were made by reactions of MR_2 (M = Zn or Cd) or $HgMe(NO_3)$ with sterically hindered selenols and the structure of Hg(SeAr)Me was determined (169).

The use of silyl tellurides in the preparation of Zn and Cd chalcogenolates (E = S, Se, or Te) via dealkylsilylations such as that shown in Eq. 34 has also been explored (63).

$$CdMe_2 + 2PhESiMe_3 \longrightarrow Cd(EPh)_2 + 2SiMe_4 \qquad (34)$$

A complete series of compounds of general formula $M[EX(SiMe_3)_3]_2$ (M = Zn, Cd, or Hg; E = S, Se, or Te; X = C, Si, or Ge) has been prepared (170, 171). X-ray structural data is available for two of these compounds: $\{Zn[TeSi(SiMe_3)_3]_2\}_2$ (171) and $\{Cd[SeC(SiMe_3)_3]_2\}_2$ (172). Both are dimeric in the solid state, with relatively long M—E interactions involving quite symmetrically disposed bridging chalcogenolate ligands. The M_2E_2 cores are moderately butterflied with pyramidalized chalcogens and trigonal metal centers. In the gas phase the compounds are monomeric as determined by electron impact–mass spectrometry (EI–MS) and in hydrocarbon solution, multinuclear NMR data suggest a monomer–dimer equilibrium is present (see Fig. 21). In the ^{125}Te

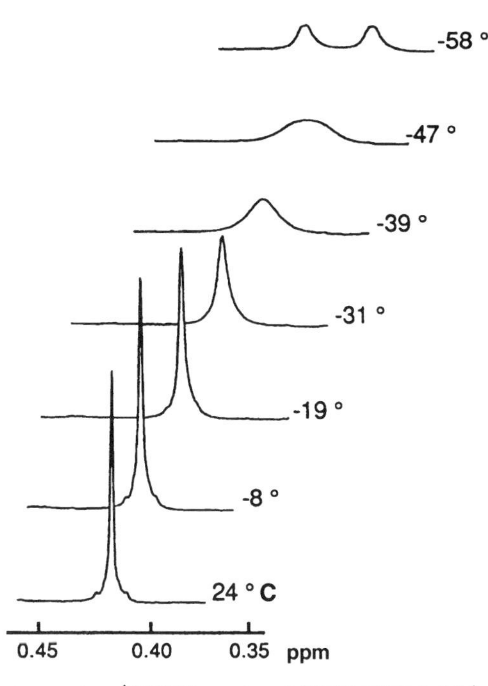

Figure 21. Variable temperature ^{1}H NMR spectra of $\{Cd[TeSi(SiMe_3)_3]_2\}_2$ [adapted from (171)].

NMR spectrum of $\{Cd[TeSi(SiMe_3)_3]_2\}_2$, for example, two peaks with different Cd satellites were observed. Based on considerations of isotopomeric distributions, the lower field of the two (-933 ppm; $|J_{TeCd}| = 1059$ Hz) was assigned to the bridging tellurolate and the higher field signal (-1338 ppm; $|J_{Te^{111}Cd}| =$ 1821 Hz, $|J_{Te^{113}Cd}| = 1897$ Hz) to the terminal interaction. (see Fig. 22) The ^{113}Cd NMR spectrum confirmed these assignments. For the tellurolates, vapor pressure molecular weight measurements also imply a slight degree of association in solution.

The metal centers in the homoleptic derivatives are electrophilic, displaying decreasing Lewis acidity as the group is descended from Zn to Hg. A variety of adducts with N and P donors been isolated and characterized for M = Zn and Hg. These include the structurally characterized $Zn[TeSi(SiMe_3)_3]_2(pyr)_2$ and $Cd[TeSi(SiMe_3)_3]_2(dmpe)$ (Fig. 23) (171). Similar four-coordinate structures are present in solution as shown by ^{125}Te and ^{113}Cd NMR spectroscopy (see Fig. 24).

The propensity of Zn, Cd, and Hg chalcogenolates to form cluster species has been well documented. For example, the tetramer $[M_4(SePh)_{10}]^{2-}$ (M = Zn or Cd) reacts with halogens (Cl_2, Br_2, or I_2) to form anions of the type

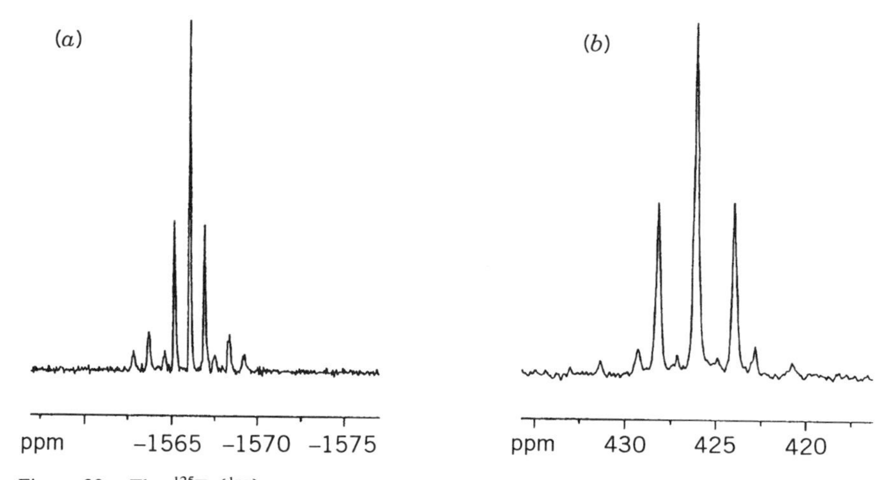

Figure 22. The $^{125}Te\{^{1}H\}$ NMR spectrum of $\{Cd[TeSi(SiMe_3)_3]_2\}_2$ at $-65°C$ [adapted from (171)].

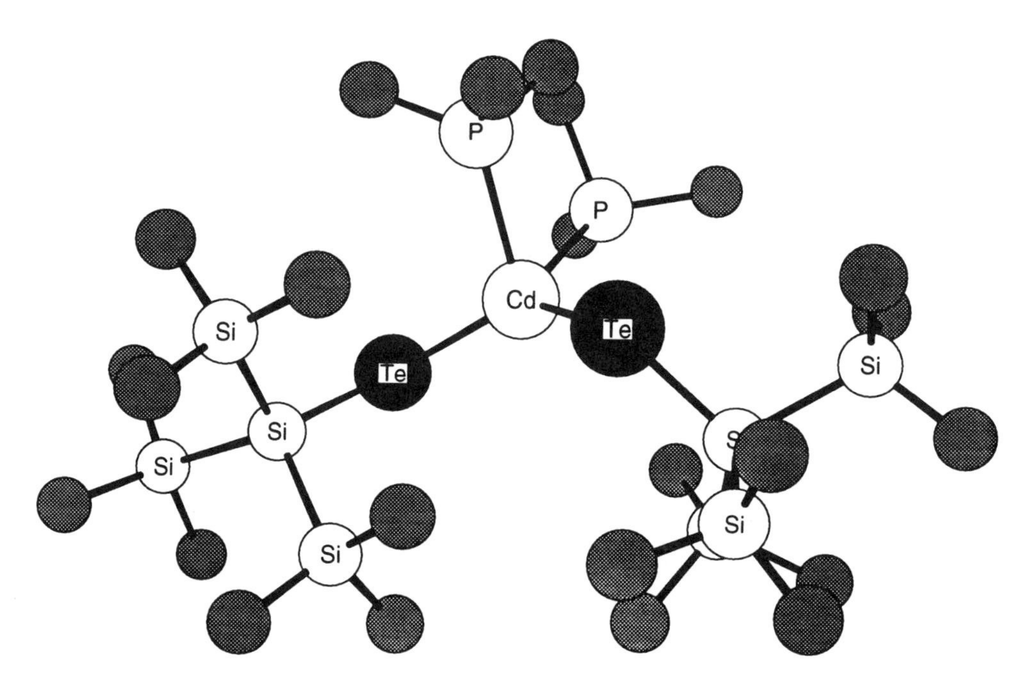

Figure 23. View of the molecular structure of $Cd[TeSi(SiMe_3)_3]_2(dmpe)$ (171).

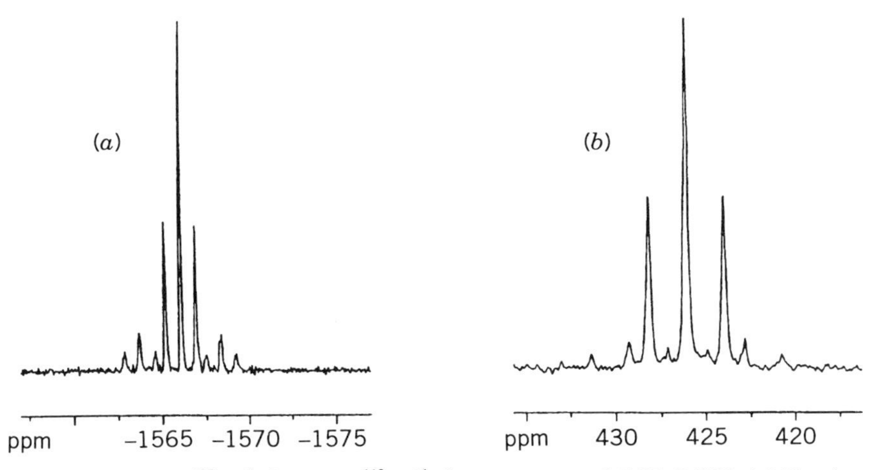

Figure 24. (*a*) The $^{125}Te\{^1H\}$ and (*b*) $^{113}Cd\{^1H\}$ NMR spectra of $Cd[TeSi(SiMe_3)_3]_2(dmpe)$ at −40 and −70°C, respectively [adapted from (171)].

$[(MX)_4(\mu\text{-SePh})_6]^{2-}$ (173). The structures of several compounds of this type are known, some of the most interesting of which are adamantane-like cluster cations such as $[M(SePh)_4(\mu\text{-SePH})_6]^{2-}$ (M = Zn or Cd) (174, 175); a view of the zinc derivative is given in Fig. 25. Still larger clusters are reported to be produced by self-assembly of a mixture of EPh^-, Na_2E (or NaEH), and CdI_2 (E = Se or Te) to form $[Cd_8(E)(EPh)_{16}]^{2-}$ anions (176).

B. Aluminum, Gallium, Indium, Thallium, Tin, Lead, Antimony, and Bismuth

The review by Krebs (177) covers general aspects of interactions between Se and Te and main group metals. The aluminum tellurolate $[Al\ t\text{-}Bu_2(\mu\text{-Te}\ t\text{-}Bu)]_2$ prepared from the reaction of Al t-Bu_3 with Te has been isolated and structurally characterized (178). Gysling and co-workers (30) described the synthesis and molecular structure of $[Ga(\mu\text{-TePh})(CH_2CMe_3)_2]_2$, prepared by the reaction of $Ga(CH_2CMe_3)_2Cl$ with LiTePh in Et_2O. The complex is dimeric in the solid state with bridging —TePh groups (Fig. 26).

A year later, Becker and co-workers (179) described the synthesis of a related complex, the *monomeric* bulky silyl tellurolate $Ga[TeSi(SiMe_3)_3][CH(SiMe_3)_2]_2$.

Three-coordinate Ga and In homoleptic derivatives have been isolated by two research groups using different ligand systems. Bulky aryl selenolates $M[Se(2,4,6\text{-}t\text{-}Bu_3C_6H_2)]_3$ [M = Ga (27); In (180)] were obtained by metathesis and selenolyis reactions, respectively. X-ray structures of both molecules showed essentially trigonal planar coordination geometries (Fig. 27).

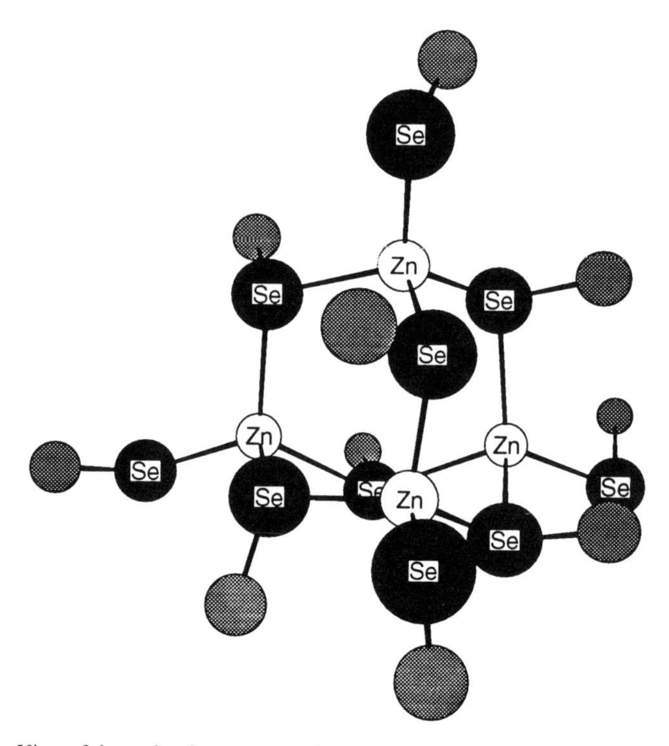

Figure 25. View of the molecular structure of the cation in $[Zn(SePh)_4(\mu\text{-}SePh)_6](ClO_4)_2$ (175).

Arnold and co-workers (181) described bulky alkyl and silyl derivatives $M[EX(SiMe_3)_3]_3$ (M = Ga or In; E = Se or Te; X = C or Si), and determined the crystal structures of $In[SeC(SiMe_3)_3]_3$ and $Ga[TeSi(SiMe_3)_3]_3$. The complexes react with various Lewis donors such as phosphorus and nitrogen donors, one of which, the μ-dmpe derivative $\{In[TeSi(SiMe_3)_3]_3\}_2(\mu\text{-dmpe})$, was structurally characterized (Fig. 28).

Divalent Sn^{II} and Pb^{II} compounds with bulky silyl chalcogenolate ligands were reported very recently by Seligson and Arnold (182) (Eq. 35).

$$M[N(SiMe_3)_2]_2 + 2HESi(SiMe_3)_3 \longrightarrow M[ESi(SiMe_3)_3]_2 + 2HN(SiMe_3)_2 \tag{35}$$

M = Sn or Pb; E = S, Se, or Te

The compounds are highly colored, thermally stable materials that oxidize rapidly on exposure to air. They decompose under fairly mild conditions to form

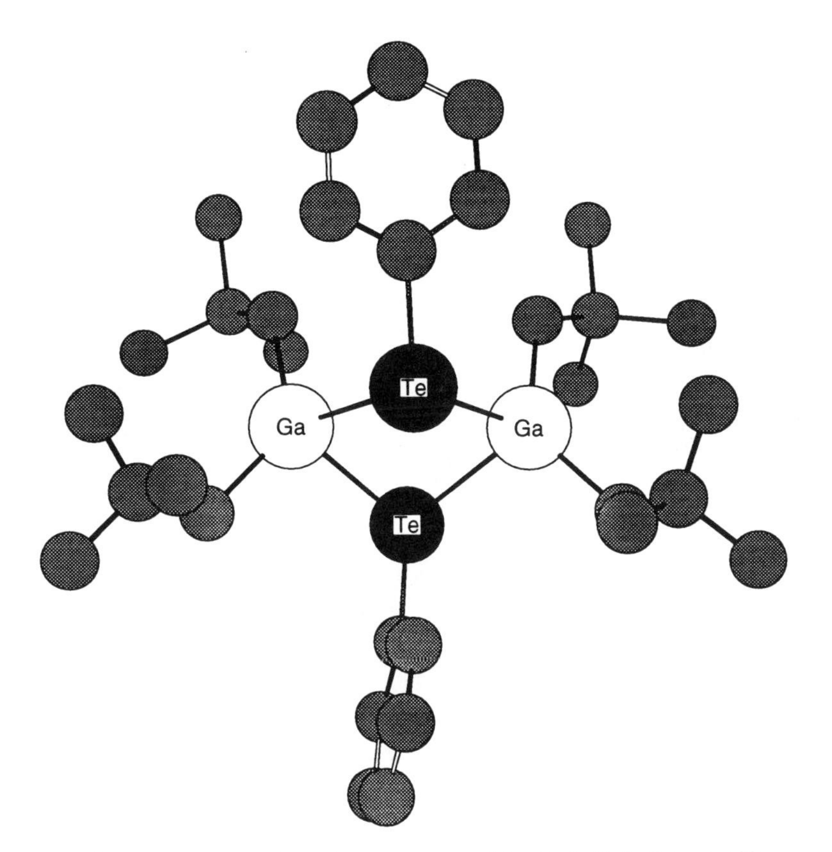

Figure 26. View of the molecular structure of $[Ga(TePh)(CH_2CMe_3)_2]_2$ (30).

[ME] and the respective disilylchalcogenide $E[Si(SiMe_3)_3]_2$, as described in more detail in Section VII. They all appear to be monomers in nondonor solvents, however, the metal center in the Sn derivatives manifests its Lewis acidity by reacting with strong donors to form adducts of general formula $M[ESi(SiMe_3)_3]L$ (L = PMe_3, PEt_3, or $\frac{1}{2}$ dmpe). Poorly basic phosphines do not form observable adducts and no reaction was seen with any of the more weakly acidic Pb^{II} homoleptic complexes. X-ray crystallographic studies of two of these compounds, $Sn[TeSi(SiMe_3)_3]PMe_3$ and $Sn[TeSi(SiMe_3)_3]_2$ were described (Figs. 29 and 30). The latter exists as a dimer with bridging tellurolates in a structure that is similar in appearance to the related Zn derivative, but which differs markedly in that the core is severely butterflied. As a result, the terminal tellurolates are cis to one another, whereas they are trans in the Zn case. The root cause of these differences can be ascribed to the large stereochemically

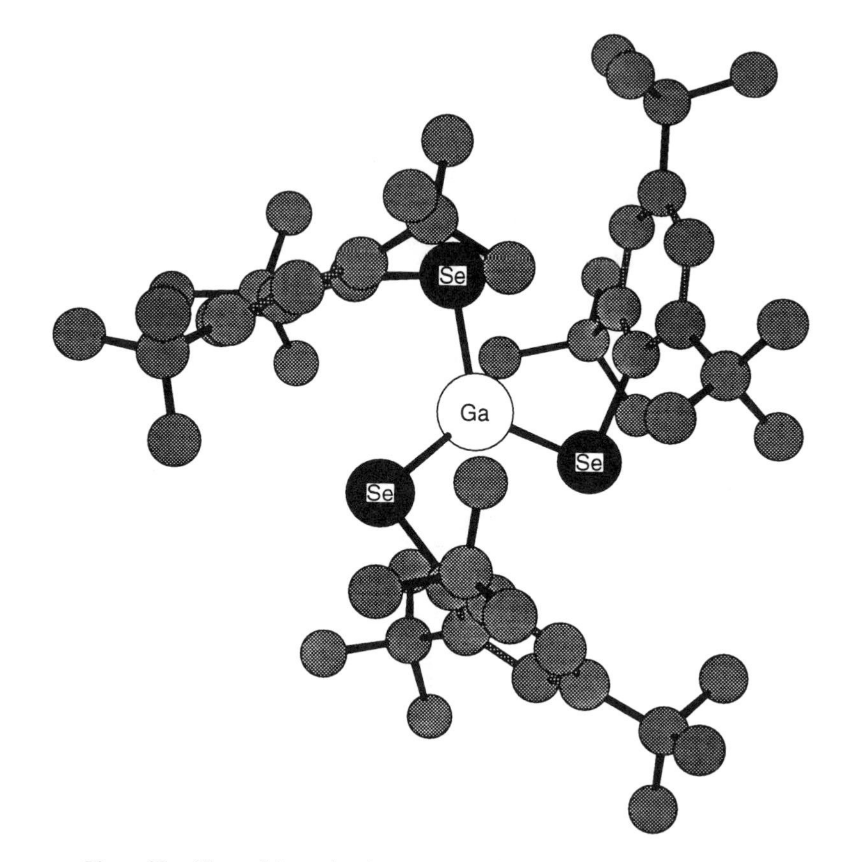

Figure 27. View of the molecular structure of $Ga[Se(2,4,6\text{-}t\text{-}Bu_3C_6H_2)_3]$ (27).

active lone pair on the Sn^{II} center. In the PMe_3 adduct, the Sn is trigonal pyramidal with the lone pair occupying the remaining coordination site. Attempts to induce the lone pair on Sn^{II} to react and form adducts with Lewis acids were unsuccessful.

Other reports of complexes for Sn and Pb include the three-coordinate trigonal pyramidal Pb^{II} selenolate $(AsPh_4)[Pb(SePh)_3]$, isolated and structurally characterized by Dean et al. (183), and four-coordinate $Sn(SePh)_4$ (184).

For antimony and bismuth, a series of silyl chalcogenolates of general formula $M[ESi(SiMe_3)_3]_3$ (M = Sb or Bi; E = Se or Te) has been prepared by chalcogenolysis of the metal bis(trimethylsilyl)amides (181). Apart from the bismuth tellurolate, which is deep purple and sensitive to heat and light, the remainder are orange, crystalline solids that can be handled under inert atmospheres with no special precautions.

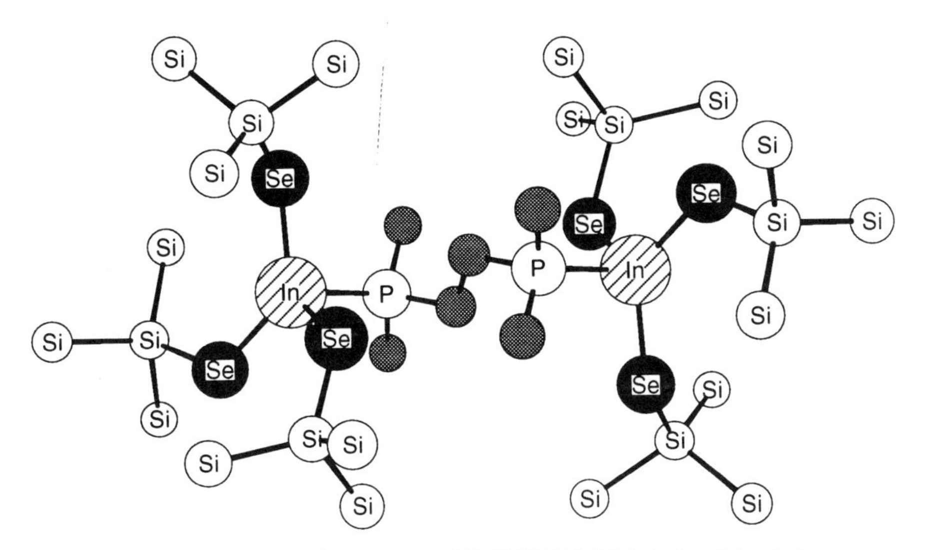

Figure 28. View of the molecular structure of $\{In[SeSi(SiMe_3)_3]_3\}_2(\mu\text{-dmpe})$ (methyl groups on Si omitted for clarity) (181).

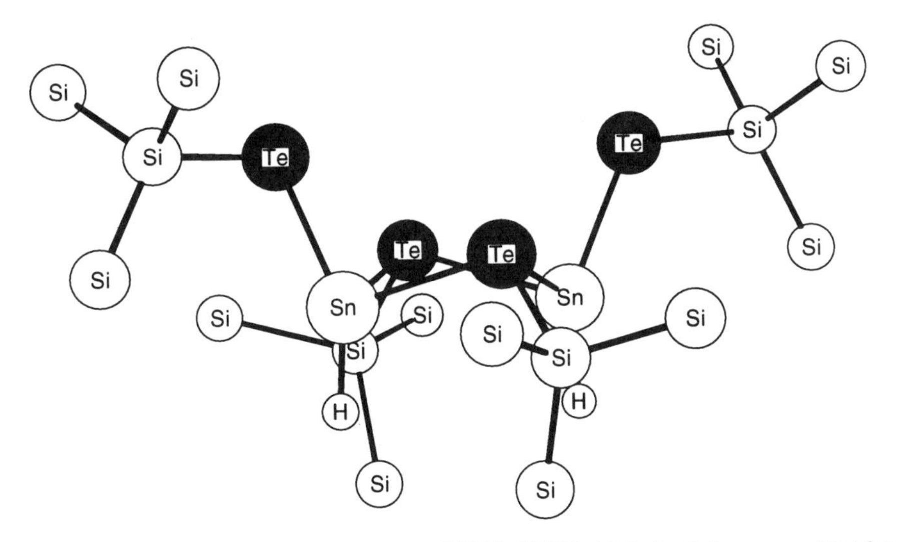

Figure 29. View of the molecular structure of $\{Sn[TeSi(SiMe_3)_3]_2\}_2$ (methyl groups omitted for clarity) (182).

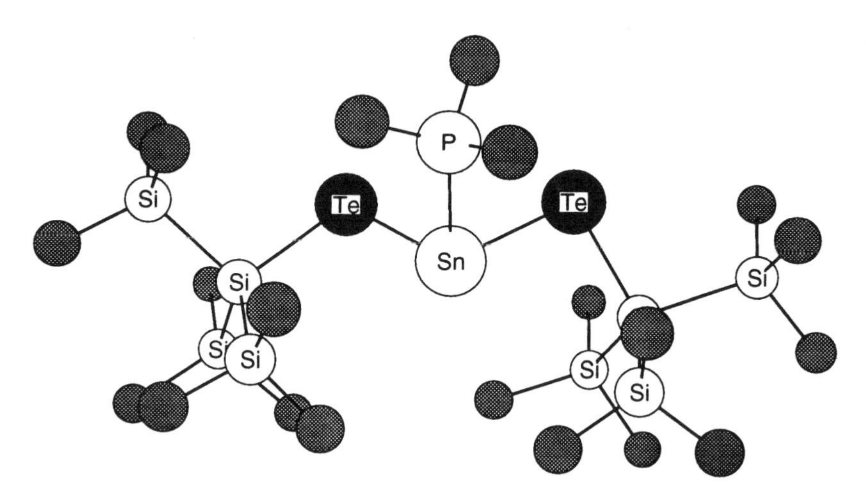

Figure 30. View of the molecular structure of $Sn[TeSi(SiMe_3)_3]PMe_3$ (182).

V. LANTHANIDE DERIVATIVES

The chemistry of lanthanide chalcogenolates has recently attracted much attention, presumably in part as a result of the interest in these molecular species as precursors to important solid state chalcogenides; more will be said on these driving forces in Section VII. The first trivalent lanthanide tellurolate was reported by Berg and Andersen (185) in 1988 and was prepared by a e^- oxidation of $Cp_2^*Yb(L)$ ($L = OEt_2$ or NH_3) by Ph_2Te_2. Related alkoxide, thiolate, and selenolates were also isolated and the X-ray crystal structure of $Cp_2^*Yb(TePh)(NH_3)$ was determined. Edelmann and co-workers (186) prepared the corresponding Sm derivatives by identical methodology and described the X-ray structures of $Cp_2^*Sm[Te(mesityl)](thf)$ and the perfluoro analogue $Cp_2^*Sm[Se(2,4,6\text{-}(CF_3)_3C_6H_2](thf)$. A comparison of the two M—Te bond lengths in the Yb and Sm complexes reveals a difference of only 0.048 Å [Yb—Te, 3.039(1) Å; Sm—Te 3.087(2) Å], roughly in line with differences in ionic radii between the two metals. In closely related reactions the benzamidinate complex $[PhCN(SiMe_3)_2]_2Yb(thf)_2$ reacted with R_2E_2 to yield Yb^{III} selenolates and tellurolates (187). In the mesityl selenolate, the Yb—Se bond length is 2.793(2) Å.

The first report of divalent lanthanide selenolates and tellurolates appeared in 1992 (25). Metal complexes of Yb, Sm, and Eu were prepared using metathesis reactions with solvated metal halides and potassium salts of bulky aryl chalcogenolates. In addition to characterization by standard analytical techniques, ^{125}Te and ^{171}Yb NMR data were obtained, although no coupling between the two nuclei was detected.

Cary and Arnold recently described trivalent (188) and divalent (23) tellurolates and selenolates. Tellurolysis of $M[N(SiMe_3)_2]_3$ (M = La or Ce) with three equivalents of $HTeSi(SiMe_3)_3$ in hexane gave unstable $M[TeSi(SiMe_3)_3]_3$ complexes that were characterized by NMR spectroscopy (for the La complex), by derivatization and by monitoring their decomposition reactions. Addition of excess DMPE to solutions of the homoleptic compounds gave seven-coordinate adducts $M[TeSi(SiMe_3)_3]_3(dmpe)_2$, one of which (M = La) was structurally characterized by X-ray crystallography (Fig. 31). Variable temperature ^1H NMR studies showed that the molecules are stereochemically nonrigid in solution. The room temperature spectrum, which shows three equivalent $-TeSi(SiMe_3)_3$ ligands and equivalent dmpe ligands, changes markedly on cooling to $-84\,°C$. At this temperature, two tellurolate signals are observed in the ratio 2:1 and the dmpe ligands give rise to four sets of signals, two for each inequivalent half of the ligand. These results confirm that the low-temperature solution and solid state structures are identical, with two *trans*-tellurolates across an equatorial plane containing the third tellurolate and the two dmpe ligands.

In the absence of a suitable trap, the homoleptic $M[TeSi(SiMe_3)_3]_3$ complexes decompose to give complex mixtures from which only two components have been characterized, one by X-ray crystallography. The structure of the cluster species $Ce_5[TeSi(SiMe_3)_3]_3[\mu\text{-}TeSi(SiMc_3)_3]_6[\mu\text{-}Te]_3$ isolated in moderate yields from hexane, is given in Fig. 32 (188). The La complex decom-

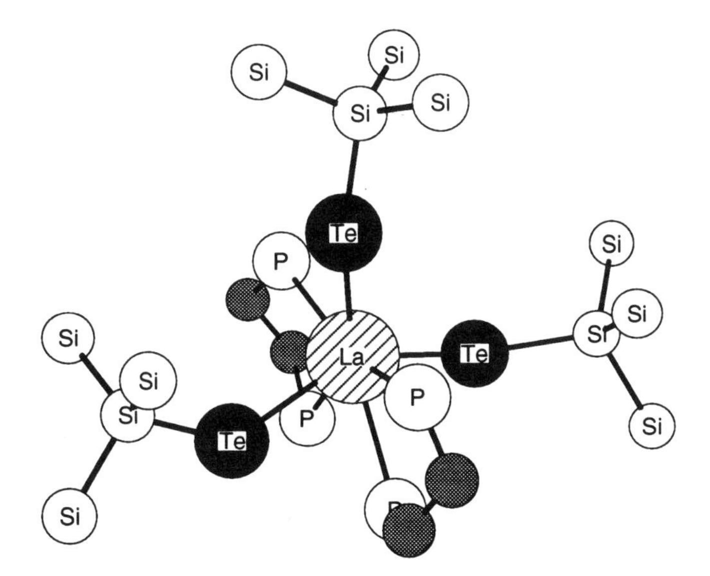

Figure 31. View of the molecular structure of $La[TeSi(SiMe_3)_3]_3(dmpe)_2$ (methyl groups omitted for clarity) (188).

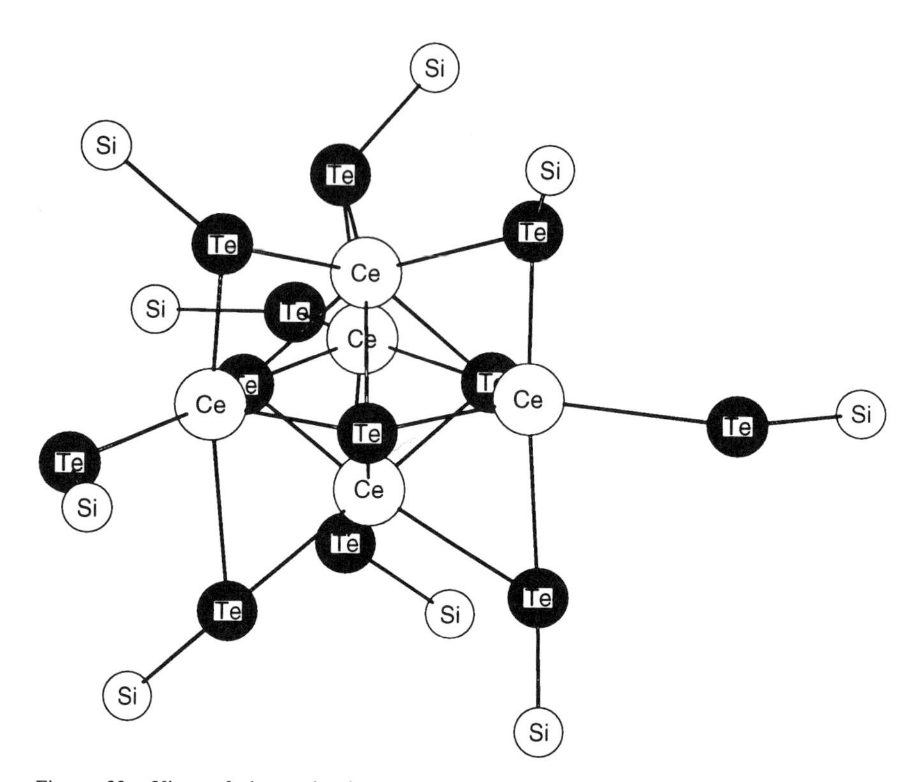

Figure 32. View of the molecular structure of the cluster species $Ce_5[TeSi(SiMe_3)_3]_3[\mu$-$TeSi(SiMe_3)_3]_6[\mu$-$Te]_3$ ($SiMe_3$ groups omitted for clarity) (188).

poses similarly. Formation of the other product, the disilyl telluride $Te[Si(SiMe_3)_3]_2$, is typically associated with elimination reactions that produce metal tellurides. Although the mechanism of this complex reaction remains to be determined, it is likely that these $M[TeSi(SiMe_3)_3]_3$ decompose via coordinatively unsaturated $M[TeSi(SiMe_3)_3](Te)$ type species, which then assemble into the observed clusters. On pyrolysis the clusters decompose to the metal sesquitellurides M_2Te_3.

A series of divalent selenolates and tellurolates of general formula $Ln[ESi(SiMe_3)_3]_2(tmeda)_2$ (Ln = Sm, Eu, or Yb; E = Se or Te) was prepared by chalcogenolysis of the divalent amides $Ln[N(SiMe_3)_2]_2(thf)_2$ (23). When dmpe was used instead of tmeda in the Eu reaction, a μ-dmpe species was formed in which each metal is seven coordinate, with two bidentate and one monodentate dmpe. Both $Yb[SeSi(SiMe_3)_3]_2(tmeda)_2$ and $\{Eu[ESi(SiMe_3)_3]_2(dmpe)_2\}_2$-($\mu$-dmpe) were structurally characterized by X-ray crystallography. Multinuclear NMR spectroscopy (^1H, ^{13}C, or ^{125}Te) was used to characterize the solution state structures of the Yb and Sm complexes. Attempts to prepare trivalent tellurolates for these metals resulted in reduction to the divalent species and formation of ditelluride.

Brennan and co-workers (24) isolated and structurally characterized the coordination polymers $[(L)_nEu^{II}(\mu\text{-SePh})_2]_\infty$ (L = pyr, n = 2; L = thf, n = 3) from metathesis reactions starting from NaSePh and $EuCl_3$. The complexes rapidly lose coordinated solvent on isolation, leading to amorphous materials containing a single equivalent of pyr or thf. In the case of Yb, reactiuon of the metal trichloride with LiSePh in THF, followed by treatment with pyridine gave a low yield of the crystalline complex $Yb^{III}(pyr)_2(SePh)_2(\mu\text{-SePh})_2Li(pyr)_2$, whose X-ray crystal structure was determined (189).

Scandium tellurolate chemistry was the subject of two recent communications by Piers et al. (53) and (54). In the first of these, the preparation of Cp_2^*ScTeR ($R = CH_2SiMe_3$, CH_2Ph, Me, or $CH_2CH_2CMe_3$) by insertion of Te into Sc–alkyl bonds was described, as shown in Eq. 36.

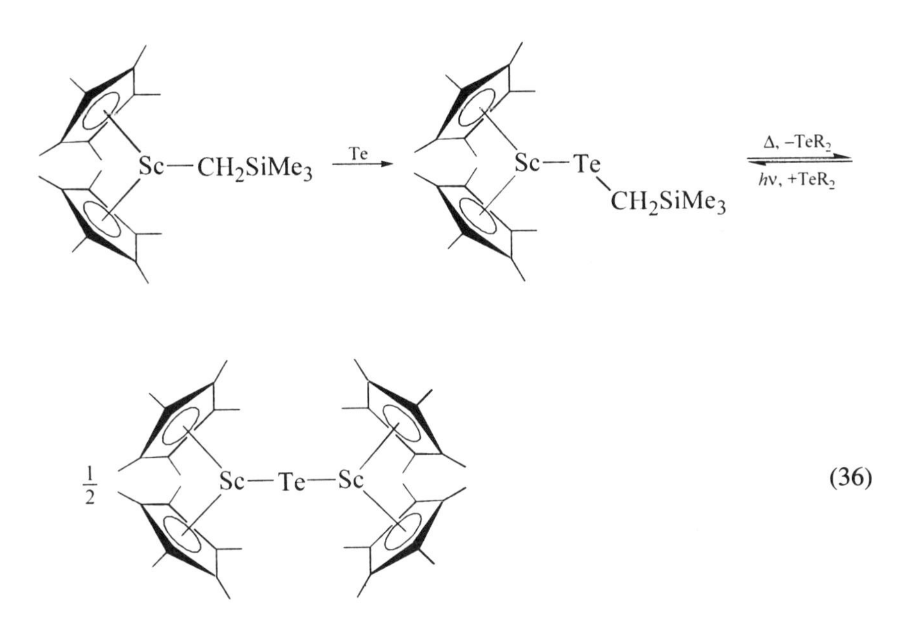

(36)

Thermal elimination of dialkyl telluride from these complexes is reversed by photolysis; the metal-containing product, $[Cp_2^*Sc]_2(\mu\text{-Te})$, was isolated and structurally characterized. As demonstrated in a subsequent note, insertion of Te and extrusion of TeR_2 proceed by concerted transition states (54). Thus, insertion of Te into the Sc–alkyl bond in *erythro*-$Cp_2^*Sc(CHDCHDCMe_3)$ proceeds with retention of the stereochemical probe (Eq. 37).

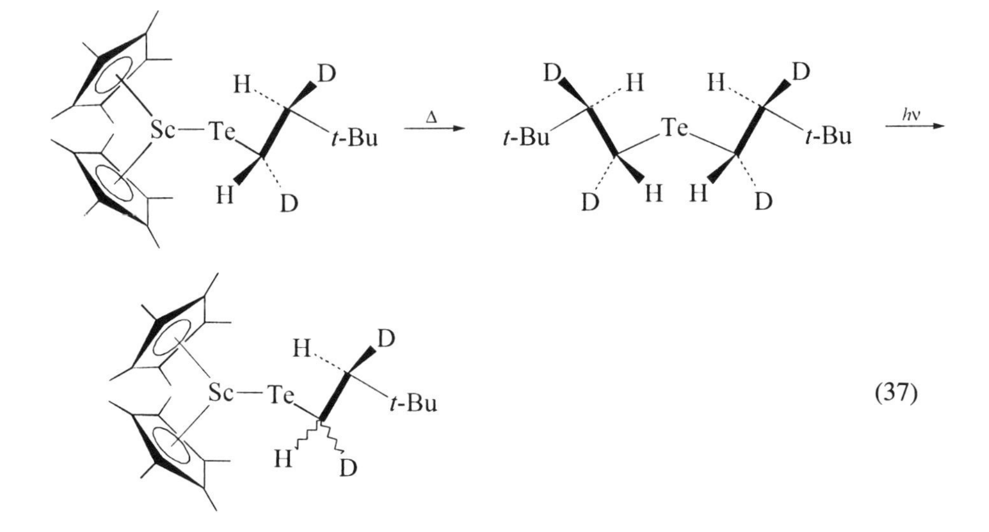

On thermolysis (90°C, 24 h), greater than 90% of *erythro, erythro*-Te-$(CHDCHDCMe_3)_2$ was detected, implying a concerted, bimolecular elimination reaction.

Elemental tellurium inserts into the Y—Me bond in $[(t\text{-BuCp})_2Y(\mu\text{-Me})]_2$ to form $[(t\text{-BuCp})_2Y(\mu\text{-TeMe})]_2$ in high yield (190). Related Lu derivatives were briefly mentioned in a preliminary account, but little characterization data was given for either compound. Beletskaya (190) also prepared Y and Lu selenolates by cleavage of metal–alkyl bonds in $[(t\text{-BuCp})_2M(\mu\text{-Me})]_2$ (M = Y or Lu) with R_2Se_2. The X-ray crystal structure of one of these derivatives, $[(t\text{-BuCp})_2Y(\mu\text{-SePh})]_2$ was also reported.

VI. ^{77}Se AND ^{125}Te NUCLEAR MAGNETIC RESONANCE SPECTROSCOPY

Instrumental advances in multinuclear NMR spectrometers over the last decade have made routine acquisition of ^{77}Se and ^{125}Te NMR data fairly routine (191). The nuclear characteristics of these elements result in extremely broad shift ranges (~2000 ppm for ^{77}Se; 4000 ppm for ^{125}Te!), making chemical shifts acutely sensitive to changes in electronic environments in the molecule. Important NMR properties of the elements are listed in Table I.

The high sensitivity of modern spectrometers and good relative receptivity of ^{77}Se and ^{125}Te allows spectra to be obtained on 20–50 mg of compound in a 5-mm NMR tube over periods of 20 min to several hours. Recent studies with sterically hindered, strongly hydrophobic ligands such as supermesityl or

TABLE I
Nuclear Magnetic Resonance Properties of ^{77}Se and ^{125}Te[a]

Nucleus	I	Natural Abundance (%)	R	$\gamma(10^7$ rad s^{-1} T$^{-1})$	Reference Std. (0 ppm)
^{77}Se	1/2	7.58	3.02	5.1214	SeM_2 (ℓ)
^{125}Te	1/2	6.99	0.906	−8.5087	$TeMe_2$ (ℓ)

[a]In this table I is the nuclear spin quantum number, R^C is the receptivity relative to ^{13}C, and γ is the magnetogyric ratio.

tris(trimethylsilyl)silyl have benefited from the high hydrocarbon solubility imparted by these groups. The NMR data on more insoluble compounds, such as phenyl chalcogenolates and cluster species, is much more difficult to obtain. A selection of currently available data is provided in Table II, which shows the enormous range in chemical shifts for these nuclei, even between homologous series of compounds.

Several recent attempts have been made to probe substituent effects on chemical shifts, for example, two separate studies of a wide range of lithium alkyl and aryl tellurolates by the groups of Sladky and co-workers (38) and Irgolic and co-workers (192). By using the relationship

$$\delta\ ^{125}\text{Te} = 1.81\ \delta\ ^{77}\text{Se} - 141.3$$

Irgolic was able to linearly correlate shifts of lithium alkyl tellurolates and selenolates (slope = 1.81; R = 0.998). The same paper discussed a useful dual-parameter correlation of ^{125}Te shifts with substituent parameters in para- and meta-substituted aryl tellurolates following an extensive investigation of their chemical shifts.

In the series of alkaline earth complexes $M[TeSi(SiMe_3)_3]_2(pyr)_n$, where d orbital influences can be safely ignored, there is only a small effect, such that $\delta\ ^{125}$Te shifts to marginally lower field as the group is descended. In contrast, for d^0 early transition metals (for which a significant amount of data is now available) there is a very large *upfield* shift on moving to heavier elements in a given group in the periodic table and some interesting correlations have been demonstrated between the lowest energy excited state (λ_{max}) and ^{125}Te chemical shifts in a series of metallocene tellurolates (Fig. 33) (50). This relationship holds for both the $-TeSiPh_3$ and $-TeSi(SiMe_3)_3$ derivatives and is due to the well-known influence of low-lying excited states on the paramagnetic term in the shielding equation for heavier nuclei (196). A similar relationship was found in the ^{77}Se NMR data of two series of selenolates $M[SeSi(SiMe_3)_3]_4$ and $M[SeSi(SiMe_3)_3]_3(CH_2Ph)$ (M = Ti, Zr, or Hf) (195).

TABLE II
^{77}Se and ^{125}Te NMR Data

Compound	δ ^{77}Se, ^{125}Te (ppm)	References
$HTeSi(SiMe_3)_3$	−955 ($|J_{TeH}|$74)	17
$HTeC(SiMe_3)_3$	−63 ($|J_{TeH}|$101)	33
$HTeGe(SiMe_3)_3$	−940 ($|J_{TeH}|$95)	33
$HTeSiPh_3$	−684 ($|J_{TeH}|$27)	50
$HSeSi(SiMe_3)_3$	−4.7 ($|J_{SeH}|$36)	33
$HSeC(SiMe_3)_3$	−1 ($|J_{SeH}|$25)	33
2,4,6-*t*-$Bu_3C_6H_2TeH$	154 ($|J_{TeH}|$52)	29
2,4,6-*i*-$Pr_3C_6H_2TeH$	−135($|J_{TeH}|$49)	29
2,4,6-$Me_3C_6H_2TeH$	−91.4 ($|J_{TeH}|$63)	29
2,4,6-*t*-$Bu_3C_6H_2SeH$	139.9 ($|J_{SeH}|$51.9)	43
2,4,6-*i*-$Pr_3C_6H_2SeH$	9.78 ($|J_{SeH}|$48.8)	43
2,4,6-$Me_3C_6H_2SeH$	24.77 ($|J_{SeH}|$53.7)	43
$[(thf)_2LiTeSi(SiMe_3)_3]_2$	−1622	17
$[(thf)LiTeSi(SiMe_3)_3]_2$	−1578	33
$(thf)_2LiTeC(SiMe_3)_3$	−287	33, 38
$(thf)_2LiTeGe(SiMe_3)_3$	−1515	33
$(thf)_2LiTeSiPh_3$	−1337	50
LiTeMe	−746	38
LiTeMe	−725	192
LiTeEt	−433	192
LiTePh	−127	38
LiTePh	−122.5	192
LiTePh	370	30
LiTe(*n*-Bu)	−520	38
LiTe(*n*-Bu)	−535	192
LiTe(*s*-Bu)	−256	38
LiTe(*s*-Bu)	−236	192
LiTe(*t*-Bu)	84	38
LiTe(*t*-Bu)	117	192
$(thf)_2LiSeSi(SiMe_3)_3$	−852	33
$(dme)LiSeC(SiMe_3)_3$	−15	33
$LiSeC(SiMe_3)_3$	188	13
$[TeSi(SiMe_3)_3]_2$	−678	75
$Te[Si(SiMe_3)_3]_2$	−1241	75
$[TeC(SiMe_3)_3]_2$	340	193
$Te(SiPh_3)_2$	−851	50
$Me_3SiTeSi(SiMe_3)_3$	−1081	50
$TePh_2$	688	30
$[SeC(SiMe_3)_3]_2$	395	194
$Mg[TeSi(SiMe_3)_3]_2(pyr)_2$	−1578	41
$Ca[TeSi(SiMe_3)_2(pyr)_4$	−1458	41
$Sr[TeSi(SiMe_3)_3]_2(pyr)_4$	−1482	41
$Ba[TeSi(SiMe_3)_3]_2(pyr)_5$	−1405	41

TABLE II. (*Continued*)

Compound	δ ^{77}Se, ^{125}Te (ppm)	References
$Cp_2Ti[TeSi(SiMe_3)_3]_2$	810	71
$Cp_2Zr[TeSi(SiMe_3)_3]_2$	−26	71
$Cp_2Hf[TeSi(SiMe_3)_3]_2$	−233	71
$Cp_2Zr[TeSi(SiMe_3)_3]Me$	−207	71
$Cp_2Zr[TeSi(SiMe_3)_3](\eta^2\text{-}COMe)$	−1146	71
$Cp'_2Ti[TeSi(SiMe_3)_3]_2$	783	71
$Cp_2It(TeSiPh_3)_2$	709	50
$Cp_2Zr(TeSiPh_3)_2$	15	50
$Cp_2Hf(TeSiPh_3)_2$	−170	50
$Cp'_2Ti(TeSiPh_3)_2$	659	50
$Cp^t_2Zr(TeSiPh_3)_2$	75	50
$Cp^t_2Hf(TeSiPh_3)_2$	−117	50
$Cp_2Ti(SePh)_2$	847	55
$Cp_2Ti(o\text{-}Se_2C_6H_4)$	982	55
$Zr[TeSi(SiMe_3)_3]_4$	459	75
$Hf[TeSi(SiMe_3)_3]_4$	193	75
$Zr[TeSi(SiMe_3)_3]_4[CN(mesityl)]_2$	203	75
$Hf[TeSi(SiMe_3)_3]_4[CN(mesityl)_2$	−14	75
$Zr[TeSi(SiMe_3)_3]_4(dmpe)$	151, −116	75
$Hf[TeSi(SiMe_3)_3]_4(dmpe)$	128, −349	75
$Zr[TeSi(SiMe_3)_3]_4(dmpe)$		75
$Zr{=}Te[TeSi(SiMe_3)_3]_2(dmpe)_2$	−706, −1173 (t, $\lvert J_{PTe}\rvert$166), −1197	75
$Hf{=}Te[TeSi(SiMe_3)_3]_2(dmpe)_2$	−701, −1212 (t, $\lvert J_{PTe}\rvert$ 190), −1250	75
$Ti[SeSi(SiMe_3)_3]_4$	865	195
$Zr[SeSi(SiMe_3)_3]_4$	356	195
$Hf[SeSi(SiMe_3)_3]_4$	211	195
$Ti[SeSi(SiMe_3)_3]_3(CH_2Ph)$	828	195
$Zr[SeSi(SiMe_3)_3(CH_2Ph)$	338	195
$Hf[SeSi(SiMe_3)_3]_3(CH_2Ph)$	206	195
$Hf[SeSi(SiMe_3)_3]_3(CH_2Ph)(dmpe)$	−28.7 (t, $\lvert J_{SeP}\rvert$28), −191 (t, $\lvert J_{SeP}\rvert$28)	195
$Ti[SeSi(SiMe_3)_3]_2(OEt)_2$	416	195
$Mn(TeCH_2Ph)(CO)_3(PMe_3)_2$	630	58
$Mn_2(\mu\text{-}TeMe)_2(CO)_8$	−964, −974	58
$Mn_2(\mu\text{-}TeCH_2SiMe_3)_2(CO)_8$	−927	58
$Mn_2[\mu\text{-}TeSi(SiMe_3)_3]_2(CO)_8$	−1996	58
$Fe[TeSi(SiMe_3)_3]_2(dmpe)_2$	−1189 (q, $\lvert J_{PTe}\rvert$80)	81
$Fe[TeSi(SiMe_3)_3]Cl(dmpe)_2$	−1350 (q, $\lvert J_{PTe}\rvert$78)	81
$Fe_3(CO)_8(PPr)[\mu_2\text{-}Te(mesityl)]_2$	250	112
$Co[TeSi(SiMe_3)_3](CO)_2(PMe_3)_2$	−1417	81
$Au_4[TeC(SiMe_3)_3]_4$	−302	153
$Au[TeC(SiMe_3)_3]PPh_3$	−148	153
$Au_4[TeSi(SiMe_3)_3]_4$	−1112	153
$Au_4[TeGe(SiMe_3)_3]_4$	−1093	153

TABLE II. (*Continued*)

Compound	δ ^{77}Se, ^{125}Te (ppm)	References
$Au_4[SeC(SiMe_3)_3]_4$	−124	153
$(AuSePh)_2(dppe)$	−181.5	156
$\{Cd[Te(mes)]_2\}_n$		
$Zn[TeSi(SiMe_3)_3]_2$	−783, −1215	171
$Cd[TeSi(SiMe_3)_3]_2$	−933 ($\|J_{TeCd}\|$ 1059), −1338 ($\|J_{TeCd}\|$ 1821)	171
$Hg[TeSi(SiMe_3)_3]_2$	−850	171
$Zn[TeSi(SiMe_3)_3]_2(pyr)_2$	−1469	171
$Cd[TeSi(SiMe_3)_3]_2(dmpe)_2$	−1565 ($\|J_{TeCd}\|$ 744; $\|J_{PTe}\|$ 137)	171
$[(HgPn\text{-}Bu_3)_4(\mu\text{-}TePh)_6](ClO_4)_2$	−263 ($\|J_{TeHg}\|$ 2720) (T = 214 K)	175
$[(HgPEt_3)_4(\mu\text{-}TePh)_6](ClO_4)_2$	−258 ($\|J_{TeHg}\|$ 2770) (T= 214 K)	175
$[(HgPPh_3)_4(\mu\text{-}TePh)_6](ClO_4)_2$	−253 ($\|J_{TeHg}\|$ 3050) (isomer A) (T = 181 K)	175
	−201 ($\|J_{TeHg}\|$ 3220) (isomer B)	
$[(HgP(p\text{-}MeC_6H_4)_3]_4(\mu\text{-}TePh)_6](ClO_4)_2$	−216 ($\|J_{TeHg}\|$ 2860) (T = 234 K)	175
$[(HgPPh_3)_4(\mu\text{-}TeMe)_6](ClO_4)_2$	−572 ($\|J_{TeHg}\|$ 3055) (T = 294 K)	175
$[(CdSe)_4(\mu\text{-}SePh)_6](ClO_4)_2$	18, −62	173
$[Ga(TePh)(CH_2CMe_3)_2]_2$	345	30
$\{Sn[TeSi(SiMe_3)_3]_2\}_2(dmpe)$	−1175 (d, $\|J_{PTe}\|$ 64)	182
$(Me_3P)Sn[TeSi(SiMe_3)_3]_2$	−1165 (d, $\|J_{PTe}\|$ 79; $\|J_{SnTe}\|$ 1327)	182
$Sb[TeSi(SiMe_3)_3]_3$	−584	181
$Bi[TeSi(SiMe_3)_3]_3$	−627	181
$Sb[SeSi(SiMe_3)_3]_3$	−207	181
$Bi[SeSi(SiMe_3)_3]_3$	−222	181
$La[TeSi(SiMe_3)_3]_3$	1018	188
$La[TeSi(SiMe_3)_3]_3(dmpe)_2$	−894, −1074	188
$Yb[TeSi(SiMe_3)_3]_2(tmeda)_2$	610	23
$Sm[TeSi(SiMe_3)_3]_2(tmeda)_2$	3243	23
$Yb[Te(mesityl)]_2(thf)_2$	−270	25

VII. APPLICATIONS

There have been sporadic reports describing applications of tellurolates and selenolates as reagents in organic synthesis (16) and as mentioned above, they have been studied as model systems for various biological systems. Generally, however, interest in these compounds has focused on their use as precursors to metal chalcogenides, many of which are important commercially due to their electronic and optoelectronic properties (11). The first workers to kindle activity in this area were Steigerwald and Sprinkle (10) who in 1987 showed that HgTe and CdTe could be synthesized at relatively low temperatures by pyrolysis of polymeric $[M(TePh)_2]_n$ precursors. Subsequent papers from this group have developed this chemistry extensively, and molecular complexes have since been used to prepare metal chalcogenide nanoclusters (52, 165) in addition to bulk

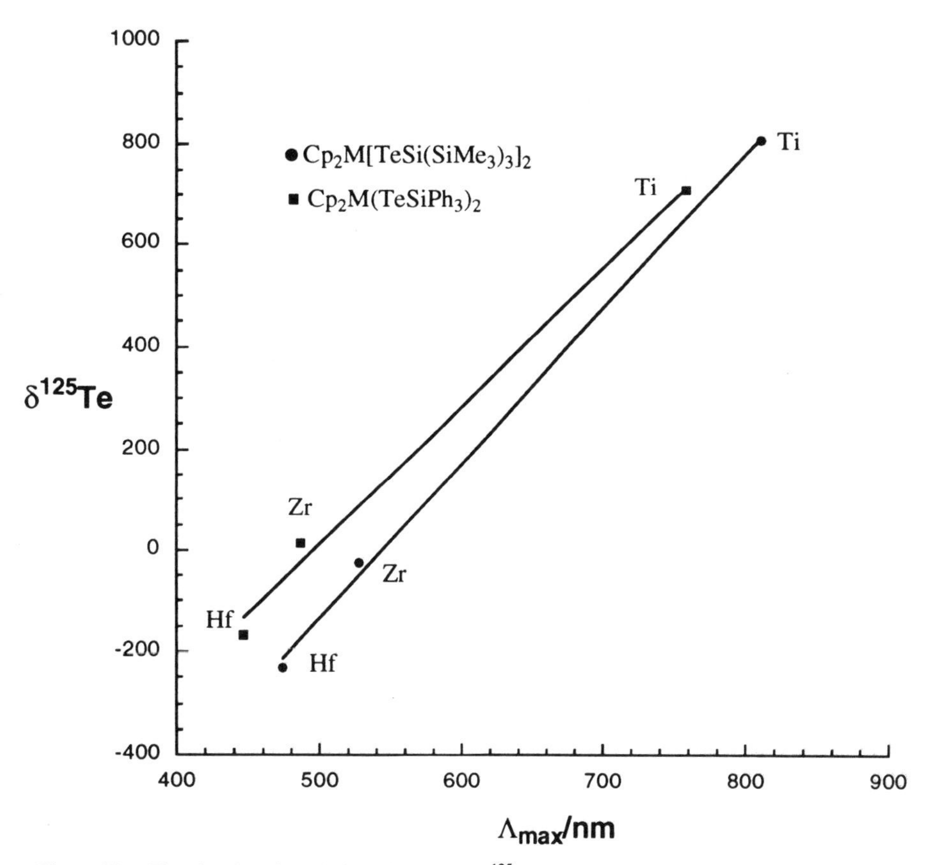

Figure 33. Plot showing the relationship between ^{125}Te NMR shift and lowest energy electronic absorption for metallocene bis-tellurolates [adapted from (50)].

solids. Four-coordinate $Cd[Te(\textit{o}\text{-}C_6H_4NMe_2)]_2$ also decomposes to CdTe and the diorganoditelluride on heating (21).

Significant developments have been made in the area of thin-film growth of II–VI (or 12–16) metal chalcogenides from single-source molecular precursors. Bochmann et al. (197) showed that Cd thiolates and selenolates based on the supermesityl ligand can be used as sources for the CVD of CdE (E = S or Se), however, the thermal instability of the tellurolates and of the Hg derivatives prevents their use as CVD sources. Aryl tellurolate ligands are therefore useful up to a point, but they are generally less stable than bulky alkyl and silyl derivatives based on $-X(SiMe_3)_3$ substituents (X = C or Si). For example, while $[Hg(TePh)_2]_n$ is completely nonvolatile and decomposes on heating to 120°C, $Hg[TeSi(SiMe_3)_3]_2$ sublimes cleanly at about 150–200°C and decomposes at around 400°C. Bourret and co-workers (190, 198) found that all Group 12 (IIB) alkyl and silyl chalcogenolates of general formula $M[EX(SiMe_3)_3]_2$ (M = Zn, Cd, or Hg; E = S, Se, or Te; X = C or Si) described above deposit ME films of varying quality under CVD conditions (199).

In addition to the II–VI (or 12–16) work, there are several recent accounts suggesting that chalcogenolate-to-chalcogen transformations may be easily applied to a wide range of metal systems. For example, McGregor and co-workers (57, 58) prepared films of MnTe by pyrolysis of $Mn(TeCH_2Ph)(CO)_3(PEt_3)_2$ at 300°C under a stream of hydrogen, the IV–VI (or 14–16) materials ME (M = Sn or Pb; E = S, Se, or Te) were products of the decomposition of $M[ESi(SiMe_3)_3]_2$ (182), and bulk lanthanide chalcogenides have been made from aryl and silyl-substituted precursors (23, 25) (Eqs. 38 and 39).

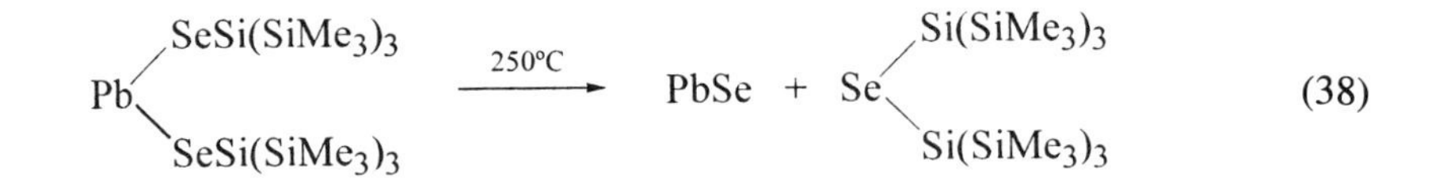

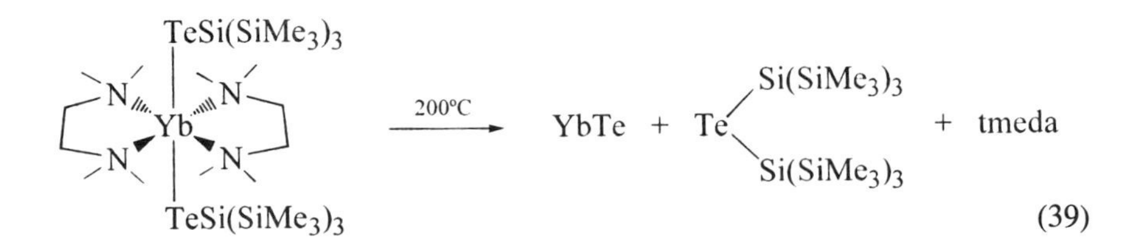

Figures 34 and 35 show characterization data for two of the cases described above: powder X-ray diffraction (XRD) data confirm that PbSe is produced in the reaction shown in Eq. 38 (Fig. 34) (182) and Rutherford back-scattering results show that Zn and Se are formed in a 1 : 1 ratio in the related gas-phase decomposition of $Zn[SeC(SiMe_3)_3]_2$ (Fig. 35) (170).

The general type of reaction discussed above [i.e., $M(ER)_2 \rightarrow ME + ER_2$] has been the subject of model studies aimed at (a) trapping M = E *molecular* species, and (b) understanding the mechanism of the transformation. As described in Sections III and V, early metal and lanthanide systems have provided valuable information in this regard (see Eqs. 21–23, 36, 37, and Figs. 9 and 32) (50, 53, 54, 71, 75, 78, 195).

The idea of using molecular chemistry to control stoichiometry of binary solid state materials has been taken a step further by work aimed at producing *ternary* materials, such as the high-efficiency solar cell material CIS. The work

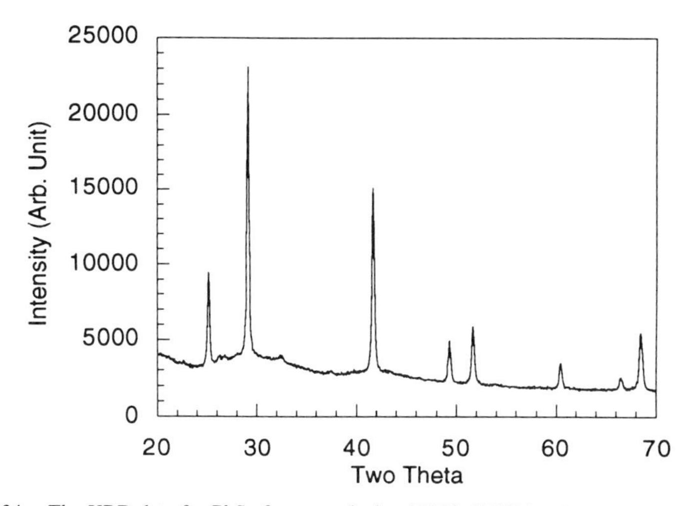

Figure 34. The XRD data for PbSe from pyrolysis of $Pb[SeSi(SiMe_3)_3]_2$ [adapted from (182)].

of Gysling et al. (200) toward this end used $Me_2InSePh$ and $In(SePh)_3$ in a spray CVD process to yield InSe and In_2Se_3 thin films, respectively. The first breakthrough in a single-source molecular approach to ternary compounds was reported in 1993 by Kanatzidis and co-workers (151) who prepared $(Ph_3P)_2CuIn(\mu\text{-}SeEt)_2(SeEt)_2$ and pyrolyzed it to form CIS (Eq. 40).

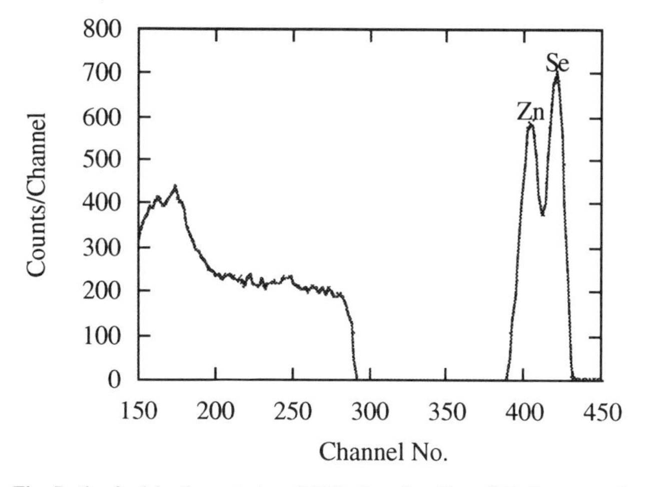

Figure 35. The Rutherford back-scattering (RBS) data for film of ZnSe grown by CVD using $Zn[SeC(SiMe_3)_3]_2$ as a single-source precursor [adapted from (170)].

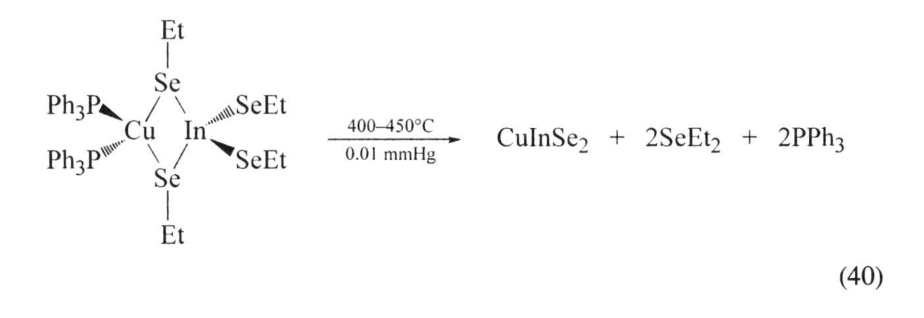

(40)

VIII. SUMMARY AND OUTLOOK

Since the last major review on the subject was published 9 years ago, metal selenolate and tellurolate chemistry has burgeoned into a well-defined field in its own right. A number of research groups have developed methodology that was originally used to synthesize related alkoxide and thiolate derivatives and successfully applied it to the heavier homologs. Selenolate and tellurolate chemistry differs considerably from the former, however, due to the weaker bond energies, larger sizes, and relative ease of oxidation of the heavier chalcogens. In general, these effects combine to produce less "innocent" ligands that bind to a broad range of metals to form labile complexes with a wide variety of structures and reactivities. Now that synthetic methodology for making M—E single bonds for these elements is on a much firmer footing, the prospects for continued growth and novel discoveries in this area are excellent.

NOTE ADDED IN PROOF

Work continues apace in this field and a number of relevant papers have appeared since this manuscript was submitted. Brennan and co-workers prepared a number of interesting chalcogenolates involving divalent lanthanides (201–203), and further example of cadmium and mercury selenolates have been isolated (204). Germanium(II) and (IV) aryl selenolates were described in a recent report by Kersting and Krebs (205). The nickel(II) complex CpNi(SePh)(PPh_3) was isolated and crystallographically characterized by Darkwa (206) and some related chromium tellurolate derivatives were the subject of a report by Goh et al. (207). Group 2 (IIA) selenolates incorporating the bulky aryl group 2,4,6-t-$Bu_3C_6H_2$— have been prepared by Ruhlandt-Senge et al. (208); specific examples include four-coordinate Mg[2,4,6-t-$Bu_3C_6H_2Se]_2(thf)_2$ and six-coordinate Sr[2,4,6-t-$Bu_3C_6H_2Se](thf)_4$.

ACKNOWLEDGMENTS

I thank my talented co-workers for their contributions to work presented in this chapter and the National Science Foundation for generous financial support.

ABBREVIATIONS

Cp	Cyclopentadienyl anion
Cp′	Methylcyclopentadienyl anion
Cp*	Pentamethylcyclopentadienyl anion
CVD	Chemical vapor deposition
Cy	Cyclohexyl
EI–MS	Electron impact–mass spectroscopy
ESR	Electron spin resonance
depe	1,2-Bis(diethylphosphino)ethane
DME	1,2-Dimethoxyethane (solvent)
dme	1,2-Dimethoxyethane (ligand)
dmpe	1,2-Bis(dimethylphosphino)ethane
dppe	1,2-Bis(diphenylphosphino)ethane
i-Bu	Isobutyl
mes	2,4,6-Trimethylphenyl anion
NMR	Nuclear magnetic resonance
OTf	Trifluoromethanesulfonate anion
pyr	Pyridine
RBS	Rutherford back-scattering
THF	Tetrahydrofuran (solvent)
thf	Tetrahydrofuran (ligand)
THT	Tetrahydrothiophene (solvent)
tht	Tetrahydrothiophene (ligand)
tmeda	*N*,*N*,*N*′,*N*′-Tetramethylethylenediamine
tren	2,2′,2″-Triaminotriethylamine
triphos	Tris(diphenylphosphinomethyl)ethane
trmpse	$MeSi(CH_2PMe_2)_3$
XRD	X-ray diffraction
12-C-4	12-crown-4
18-C-6	18-crown-6

REFERENCES

1. R. C. Mehrotra, A. Singh, and U. M. Tripathi, *Chem. Rev.*, *91*, 1287 (1991).
2. I. P. Rothwell, *Acc. Chem. Res.*, *21*, 153 (1988).

3. I. P. Rothwell and M. H. Chisholm, in *Comprehensive Coordination Chemistry*, G. Wilkinson, R. D. Gillard, and J. A. McCleverty, Eds., Pergamon, New York, 1987, Vol. 2, p. 335.
4. D. C. Bradley, R. C. Mehrotra, and D. P. Gaur, *Metal Alkoxides*, Academic, New York, 1978.
5. J. R. Dilworth and J. Hu, *Adv. Inorg. Chem.*, *40*, 411 (1993).
6. H. J. Gysling, *Coord. Chem. Rev.*, *42*, 133 (1982).
7. F. J. Berry, in *Comprehensive Coordination Chemistry*, G. Wilkinson, R. D. Gillard, and J. A. McCleverty, Eds., Pergamon, Oxford, 1987.
8. H. J. Gysling, in *The Chemistry of Organic Selenium and Tellurium Compounds*, S. Patai and Z. Rappoport, Eds., Wiley, New York, 1986, Vol. 1, p. 679.
9. U. Siemeling, *Angew. Chem. Int. Ed. Engl.*, *32*, 67 (1993).
10. M. L. Steigerwald and C. R. Sprinkle, *J. Am. Chem. Soc.*, *109*, 7200 (1987).
11. P. O'Brien, *Chemtronics*, *5*, 61 (1991).
12. E. G. Hope and W. Levason, *Coord. Chem. Rev.*, *122*, 109 (1993).
13. F. Sladky, B. Bildstein, C. Rieker, A. Gieren, H. Betz, and T. Hubner, *J. Chem. Soc. Chem. Commun.*, 1800 (1985).
14. L. Lange and W. W. du Mont, *J. Organomet. Chem.*, *286*, C1 (1985).
15. B. O. Dabbousi, P. J. Bonasia, and J. Arnold, *J. Am. Chem. Soc.*, *113*, 3186 (1991).
16. K. J. Irgolic, *The Organic Chemistry of Tellurium*, Gordon & Breach, New York, 1974.
17. P. J. Bonasia, D. E. Gindelberger, B. O. Dabbousi, and J. Arnold, *J. Am. Chem. Soc.*, *114*, 5209 (1992).
18. J. Liesk, P. Schulz, and G. Klar, *Z. Anorg. Allg. Chem.*, *435*, 98 (1977).
19. W. W. du Mont, S. Kubiniok, L. Lange, S. Pohl, W. Saak, and I. Wagner, *Chem. Ber.*, *124*, 1315 (1991).
20. J. M. Ball, P. M. Boorman, J. F. Fait, and A. S. Hinman, *Can. J. Chem.*, *67*, 751 (1988).
21. P. J. Bonasia and J. Arnold, *J. Organomet. Chem.*, *449*, 147 (1993).
22. P. J. Bonasia and J. Arnold, *J. Chem. Soc. Chem. Commun.*, 1299 (1990).
23. D. R. Cary and J. Arnold, *Inorg. Chem.*, *33*, 1791 (1994).
24. M. Berardini, T. Emge, and J. G. Brennan, *J. Am. Chem. Soc.*, *115*, 8501 (1993).
25. A. R. Strzelecki, P. A. Timinski, B. A. Helsel, and P. A. Bianconi, *J. Am. Chem. Soc.*, *114*, 3159 (1992).
26. W. W. du Mont, R. Hensel, and S. Kubiniok, *Phosphous Sulfur*, *38*, 85 (1988).
27. K. Ruhlandt-Senge and P. P. Power, *Inorg. Chem.*, *30*, 3683 (1991).
28. H. Gornitzka, S. Besser, R. Herbstirmer, U. Kilimann, and F. T. Edelmann, *Angew. Chem. Int. Ed. Engl.*, *31*, 1260 (1992).
29. M. Bochmann, A. P. Coleman, K. J. Webb, M. B. Hursthouse, and M. Mazid, *Angew. Chem. Int. Ed. Engl.*, *30*, 973 (1991).

30. M. A. Banks, O. T. Beachley, H. J. Gysling, and H. R. Luss, *Organometallics*, *9*, 1979 (1990).
31. M. Herberhold and M. W. Biersack, *J. Organomet. Chem.*, *443*, 1 (1993).
32. P. J. Bonasia and J. Arnold, *Inorg. Synth.*, *31*, in press.
33. P. J. Bonasia, V. Christou, and J. Arnold, *J. Am. Chem. Soc.*, *115*, 6777 (1993).
34. G. Becker, K. W. Klinkhammer, S. Lartiges, P. Bottcher, and W. Poll, *Z. Anorg. Allg. Chem.*, *613*, 7 (1992).
35. K. E. Flick, P. J. Bonasia, D. E. Gindelberger, J. E. B. Katari, and D. Schwartz, *Acta Crystallogr. Ser. C*, *50*, 674 (1994).
36. J. Zhao, D. Adcock, W. T. Pennington, and J. W. Kolis, *Inorg. Chem.*, *29*, 4358 (1990).
37. B. Bildstein, K. Giselbrecht, and F. Sladky, *Chem. Ber.*, *122*, 2279 (1989).
38. C. Köllemann, D. Obendorf, and F. Sladky, *Phosphous Sulfur*, *38*, 69 (1988).
39. N. Petragnani, *Chem. Ber.*, *96*, 247 (1963).
40. N. Petragnani and J. V. Comasseto, *Synthesis*, 1 (1986).
41. D. E. Gindelberger and J. Arnold, *J. Am. Chem. Soc.*, *114*, 6242 (1992).
42. D. E. Gindelberger and J. Arnold, *Inorg. Chem.*, *33*, 6293 (1994).
43. M. Bochmann, K. Webb, M. Harman, and M. B. Hursthouse, *Angew. Chem. Int. Ed. Engl.*, *29*, 638 (1990).
44. D. Labahn, M. Bohnen, R. Herbst-Irmer, E. Pohl, D. Stalke, and H. W. Roesky, *Z. Anorg. Allg. Chem.*, *620*, 41 (1994).
45. T. B. Rauchfuss, in *The Chemistry of Organic Selenium and Tellurium Compounds*, S. Patai, Ed., Wiley, New York, 1987, Vol. 2, p. 339.
46. C. W. Sink and A. B. Harvey, *J. Chem. Phys.*, *57*, 4434 (1972).
47. K. Hamada and H. Morishita, *Jpn. J. Appl. Phys.*, *15*, 748 (1976).
48. K. Hamada and H. Morishita, *Synth. React. Inorg. Met. Org. Chem.*, *7*, 355 (1977).
49. J. E. Drake and R. T. Hemmings, *Inorg. Chem.*, *19*, 1879 (1980).
50. D. E. Gindelberger and J. Arnold, *Organometallics*, *13*, 4462 (1994).
51. D. J. Berg, C. J. Burns, R. A. Andersen, and A. Zalkin, *Organometallics*, *8*, 1865 (1989).
52. M. L. Steigerwald and C. R. Sprinkle, *Organometallics*, *7*, 245 (1988).
53. W. E. Piers, L. R. Macgillivray, and M. Zaworotko, *Organometallics*, *12*, 4723 (1993).
54. W. E. Piers, *J. Chem. Soc. Chem. Commun.*, 309 (1994).
55. T. Klapötke, *Phosphorus Sulfur*, *41*, 105 (1989).
56. H. Kopf and T. Klapötke, *J. Chem. Soc. Chem. Commun.*, 1192 (1986).
57. K. McGregor, G. B. Deacon, R. S. Dickson, G. D. Fallon, R. S. Rowe, and B. O. West, *J. Chem. Soc. Chem. Commun.*, 1293 (1990).
58. A. P. Coleman, R. S. Dickson, G. B. Deacon, G. D. Fallon, M. Ke, K. McGregor, and B. O. West, *Polyhedron*, *13*, 1277 (1994).

59. F. A. Cotton and K. R. Dunbar, *Inorg. Chem.*, *26*, 1305 (1987).
60. J. A. M. Canich, F. A. Cotton, K. R. Dunbar, and L. R. Falvello, *Inorg. Chem.*, *27*, 804 (1988).
61. P. L. Andreau, J. A. Cabeza, D. Miguel, V. Riera, M. A. Villa, and S. Garciar-anda, *J. Chem. Soc. Dalton Trans*, 533 (1991).
62. K. Andra, *Z. Anorg. Allg. Chem.*, *373*, 209 (1970).
63. S. M. Stuczynski, J. G. Brennan, and M. L. Steigerwald, *Inorg. Chem.*, *28*, 4431 (1989).
64. P. G. Jones and C. Thone, *Chem. Ber.*, *123*, 1975 (1990).
65. J. E. Drake, B. M. Glavincevski, and R. T. Hemmings, *J. Inorg. Nucl. Chem.*, *41*, 457 (1979).
66. J. M. Ball, P. M. Boorman, J. F. Fait, H.-B. Kraatz, J. F. Richardson, D. Collison, and F. E. Mabbs, *Inorg. Chem.*, *29*, 3290 (1990).
67. M. Sato and T. Yoshida, *J. Organomet. Chem.*, *67*, 395 (1974).
68. H. Kopf, B. Block, and M. Schmidt, *Z. Naturforsch.*, *22b*, 1077 (1967).
69. B. Gautheron, G. Tainturier, and P. Meunier, *J. Organomet. Chem.*, *209*, C49 (1981).
70. B. Gautheron, G. Tainturier, S. Pouly, F. Theobald, H. Vivier, and A. Laarif, *Organometallics*, *3*, 1495 (1984).
71. V. Christou, S. P. Wuller, and J. Arnold, *J. Am. Chem. Soc.*, *115*, 10545 (1993).
72. G. Fachinetti, C. Floriani, A. Chiesi-Villa, and C. Guastini, *J. Am. Chem. Soc.*, *101*, 1767 (1979).
73. H. A. Harris, A. D. Rae, and L. F. Dahl, *J. Am. Chem. Soc.*, *109*, 4739 (1987).
74. D. Fenske and A. Grissinger, *Z. Naturforsch.*, *45b*, 1309 (1990).
75. V. Christou and J. Arnold, *J. Am. Chem. Soc.*, *114*, 6240 (1992).
76. M. Sato and T. Yoshida, *J. Organomet. Chem.*, *87*, 217 (1975).
77. M. Herberhold, M. Schrepfermann, and A. L. Rheingold, *J. Organomet. Chem.*, *394*, 113 (1990).
78. V. Christou and J. Arnold, *Angew. Chem. Int. Ed. Engl.*, *32*, 1450 (1993).
79. A. Rettenmeier, K. Weidenhammer, and M. L. Ziegler, *Z. Anorg. Allg. Chem.*, *473*, 91 (1981).
80. T. Vogt and J. Strähle, *Z. Naturforsch.*, *40b*, 1599 (1985).
81. D. E. Gindelberger and J. Arnold, *Inorg. Chem.*, *32*, 5813 (1993).
82. L. Y. Goh, M. S. Tay, Y. Y. Lim, W. Chen, Z.-Y. Zhou, and T. C. W. Mak, *J. Organomet. Chem.*, *441*, 51 (1992).
83. L. Y. Goh, Y. Y. Lim, M. S. Tay, T. C. W. Mak, and Z. Y. Zhou, *J. Chem. Soc. Dalton Trans.*, 1239 (1992).
84. J. Rott, E. Guggolz, A. Rettenmeier, and M. I. Ziegler, *Z. Naturforsch.*, *37b*, 13 (1982).
85. W. Röll, O. Fischer, D. Neugebauer, and U. Schubert, *Z. Naturforsch.*, *37b*, 1274 (1982).

86. G. R. Hanson, A. A. Brunette, A. C. McDonell, K. S. Murray, and A. G. Wedd, *J. Am. Chem. Soc.*, *104*, 1953 (1981).
87. J. R. Bradbury, A. F. Masters, A. C. McDonell, A. A. Brunette, A. M. Bond, and A. G. Wedd, *J. Am. Chem. Soc.*, *104*, 1959 (1981).
88. P. M. Boorman, H. B. Kraatz, and M. Parvez, *J. Chem. Soc. Dalton Trans.*, 3281 (1992).
89. P. M. Boorman, H. B. Kraatz, and M. Parvez, *Polyhedron*, *12*, 601 (1993).
90. H. B. Kraatz, P. M. Boorman, and M. Parvez, *Can. J. Chem.*, *71*, 1437 (1993).
91. P. M. Boorman, H. B. Kraatz, M. Parvez, and T. Ziegler, *J. Chem. Soc. Dalton Trans*, 433 (1993).
92. M. H. Chisholm, J. C. Huffman, I. P. Parkin, and W. Streib, *Polyhedron*, *9*, 2941 (1990).
93. M. H. Chisholm, I. P. Parkin, J. C. Huffamn, and W. B. Streib, *J. Chem. Soc. Chem. Commun.*, 920 (1990).
94. W. Tremel, B. Krebs, K. Greiwe, W. Simon, H. O. Stephan, and G. Henkel, *Z. Naturforsch.*, *47b*, 1580 (1992).
95. M. Bochmann, A. K. Powell, and X. J. Song, *Inorg. Chem.*, *33*, 400 (1994).
96. J. M. McConnachie and J. A. Ibers, *Inorg. Chem.*, *30*, 1770 (1991).
97. W. A. Herrmann, C. Hecht, and E. Herdtweck, *J. Organomet. Chem.*, *331*, 309 (1987).
98. G. Huttner, S. Schuler, L. Zsolnai, M. Gottlieb, H. Braunwarth, and M. Minelli, *J. Organomet. Chem.*, *299*, C4 (1986).
99. P. Jaitner, W. Wohlgenannt, A. Gieren, H. Betz, and T. Hübner, *J. Organomet. Chem.*, *297*, 281 (1985).
100. P. Lau, G. Huttner, and L. Zsolnai, *J. Organomet. Chem.*, *440*, 41 (1992).
101. P. Lau, G. Huttner, and L. Zsolnai, *Z. Naturforsch.*, *46b*, 719 (1991).
102. V. Küllmer and H. Vahrenkamp, *Chem. Ber.*, *110*, 228 (1977).
103. W. Simon, A. Wilk, B. Krebs, and G. Henkel, *Angew. Chem. Int. Ed. Engl*, *26*, 1009 (1987).
104. W. F. Liaw, M. H. Chiang, C. J. Liu, P. J. Harn, and L. K. Liu, *Inorg. Chem.*, *32*, 1536 (1993).
105. W. F. Liaw, C. H. Lai, C. K. Lee, G. H. Lee, and S. M. Peng, *J. Chem. Soc. Dalton Trans.*, 2421 (1993).
106. W. F. Liaw, Y. C. Horng, D. S. Ou, G. H. Lee, and S. M. Peng, *J. Chinese Chem. Soc.*, *40*, 367 (1993).
107. W. F. Liaw, S. J. Chiou, W. Z. Lee, G. H. Lee, and S. M. Peng, *J. Chinese Chem. Soc.*, *40*, 361 (1993).
108. W. F. Liaw, C. H. Lai, M. H. Chiang, C. K. Hsieh, G. H. Lee, and S. M. Peng, *J. Chinese Chem. Soc.*, *40*, 437 (1993).
109. W. F. Liaw, D. S. Ou, Y. C. Horng, C. H. Lai, G. H. Lee, and S. M. Peng, *Inorg. Chem.*, *33*, 2495 (1994).
110. A. Winter, L. Zsolnai, and G. Huttner, *J. Organomet. Chem.*, *250*, 409 (1983).

111. T. Fässler, D. Buchholz, G. Huttner, and L. Zsolnai, *J. Organomet. Chem.*, *369*, 297 (1989).
112. D. Buchholz, G. Huttner, L. Zsolnai, and W. Imhof, *J. Organomet. Chem.*, *377*, 25 (1989).
113. P. Mathur, D. V. Reddy, K. Das, and U. C. Sinha, *J. Organomet. Chem.*, *409*, 255 (1991).
114. R. E. Cobbledick, N. S. Dance, F. W. B. Einstein, C. H. W. Jones, and T. Jones, *Inorg. Chem.*, *20*, 4356 (1981).
115. M. M. Millar, T. O'Sullivan, N. de Vries, and S. A. Koch, *J. Am. Chem. Soc.*, *107*, 3714 (1985).
116. E. D. Schermer and W. H. Baddley, *J. Organomet. Chem.*, *30*, 67 (1971).
117. T. Chihara and H. Yamazaki, *J. Organomet. Chem.*, *428*, 169 (1992).
118. P. V. Broadhurst, B. F. G. Johnson, and J. Lewis, *J. Chem. Soc. Dalton Trans.*, 1881 (1982).
119. A. J. Arce, P. Arrojo, Y. Desanctis, A. J. Deeming, and D. J. West, *Polyhedron*, *11*, 1013 (1992).
120. H. Matsuzaka, T. Ogino, M. Nishio, Y. A. Nishibayashi, S. Uemura, and M. Hidai, *J. Chem. Soc. Chem. Commun.*, 223 (1994).
121. J. P. Collman, R. K. Rothrock, J. P. Sen, T. D. Tullius, and K. O. Hodgson, *Inorg. Chem.*, *15*, 2947 (1976).
122. M. Di Vaira, D. Rovai, P. Stoppioni, and M. Peruzzini, *J. Organomet. Chem.*, *420*, 135 (1991).
123. S. A. Gardner, *J. Organomet. Chem.*, *190*, 289 (1980).
124. R.-M. Olk, B. Olk, J. Rohloff, J. Reinhold, J. Sieler, K. Trubenbach, R. Kirmse, and E. Hoyer, *Z. Anorg. Allg. Chem.*, *609*, 103 (1992).
125. R. Curran, J. A. Cunningham, and R. Eisenberg, *Inorg. Chem.*, *9*, 2749 (1970).
126. M. Sato and T. Yoshida, *J. Organomet. Chem.*, *51*, 231 (1973).
127. D. J. Sandman, G. W. Allen, L. A. Acampora, J. C. Stark, S. Jansen, M. T. Jones, G. J. Ashwell, and B. M. Foxman, *Inorg. Chem.*, *26*, 1664 (1987).
128. W. E. Broderick, J. A. Thompson, M. R. Godfrey, M. Sabat, and B. M. Hoffman, *J. Am. Chem. Soc.*, *111*, 7656 (1989).
129. C. A. Marganian, N. Baidya, M. M. Olmstead, and P. K. Mascharak, *Inorg. Chem.*, *31*, 2992 (1992).
130. N. Baidya, B. C. Noll, M. M. Olmstead, and P. K. Mascharak, *Inorg. Chem.*, *31*, 2999 (1992).
131. C. F. Xu, J. W. Siria, and G. K. Anderson, *Inorg. Chem. Acta*, *206*, 123 (1993).
132. L.-Y. Chia and W. R. McWhinnie, *J. Organomet. Chem.*, *148*, 165 (1978).
133. D. M. Giolondo, T. B. Rauchfuss, and A. L. Rheingold, *Inorg. Chem.*, *26*, 1636 (1987).
134. W. B. Heuer, A. E. True, P. N. Swepston, and B. M. Hoffman, *Inorg. Chem.*, *27*, 1474 (1988).
135. V. K. Jain, S. Kannan, and R. Bohra, *Polyhedron*, *11*, 1551 (1992).

136. V. K. Jain, S. Kannan, R. J. Butcher, and J. P. Jasinski, *J. Chem. Soc. Dalton Trans*, 1509 (1993).
137. B. L. Khandelwal and S. K. Gupta, *Inorg. Chim. Acta*, *166*, 199 (1989).
138. V. K. Jain, S. Kannan, R. J. Butcher, and J. P. Jasinski, *J. Organomet. Chem.*, *468*, 285 (1994).
139. B. L. Khandelwal, K. Kundu, and S. K. Gupta, *Inorg. Chim. Acta*, *154*, 183 (1988).
140. B. L. Khandelwal, K. Kundu, and S. K. Gupta, *Inorg. Chim. Acta*, *148*, 255 (1988).
141. R. G. Raptis, H. H. Murray III, and J. P. Fackler Jr., *J. Chem. Soc. Chem. Commun.*, 737 (1987).
142. F. A. Cotton and G. Wilkinson, *Advanced Inorganic Chemistry*, 5th ed., Wiley-Interscience, New York, 1988.
143. G. Gafner and R. P. King, *Gold 100*, SAIMM Publications: Johannesburg, South Africa, 1986; Vol. 3.
144. S. J. Lippard, Ed., *Platinum, Gold and Other Metal Chemotherapeutic Agents: Chemistry and Biochemistry ACS Symposium Series, No. 209*, American Chemical Society, Washington, DC, 1983.
145. G. Marbach and J. Strähle, *Angew. Chem. Int. Ed. Engl.*, *23*, 715 (1984).
146. I. Davies and W. R. McWhinnie, *Inorg. Nucl. Chem. Lett.*, *12*, 763 (1976).
147. I. Davies, W. R. McWhinnie, N. S. Dance, and C. H. W. Jones, *Inorg. Chim. Acta*, *29*, L217 (1978).
148. G. E. Matsubayashi and A. Yokozawa, *Chem. Lett.*, 355 (1990).
149. J. Kampf, R. Kumar and J. P. Oliver, *Inorg. Chem.*, *31*, 3626 (1992).
150. R. Kumar and D. G. Tuck, *Can. J. Chem.*, *67*, 127 (1989).
151. W. Hirpo, S. Dhingra, A. C. Sutorik, and M. G. Kanatzidis, *J. Am. Chem. Soc.*, *115*, 1597 (1993).
152. P. J. Bonasia, G. P. Mitchell, F. J. Hollander, and J. Arnold, *Inorg. Chem.*, *33*, 1797 (1994).
153. P. J. Bonasia, D. E. Gindelberger, and J. Arnold, *Inorg. Chem.*, *32*, 5126 (1993).
154. L. C. Porter and J. P. Fackler, *Acta Crystallogr. C*, *43*, 29 (1987).
155. P. G. Jones and C. Thone, *Z. Naturforsch.*, *47b*, 600 (1992).
156. W. Eikens, C. Kienitz, P. G. Jones, and C. Thone, *J. Chem. Soc. Dalton Trans.*, 83 (1994).
157. U. Behrens, K. Hoffmann, and G. Klar, *Chem. Ber.*, *110*, 3672 (1977).
158. J. Zhao, J. W. Kolis, and W. T. Pennington, *Acta Crystallogr. Ser. C*, *49*, 1753 (1993).
159. A. P. Arnold, A. J. Canty, B. W. Skelton, and A. H. White, *J. Chem. Soc. Dalton Trans*, 607 (1982).
160. N. Ueyama, T. Sugawara, K. Sasaki, A. Nakamura, S. Yamashita, Y. Wakatsuki, H. Yamazaki, and N. Yasuoka, *Inorg. Chem.*, *27*, 741 (1988).
161. M. N. Bochkarev, *Russ. Chem. Rev.*, *49*, 800 (1980).

162. Y. Okamota and T. Yano, *J. Organomet. Chem.*, *29*, 99 (1971).
163. N. S. Dance and C. H. W. Jones, *J. Organomet. Chem.*, *152*, 175 (1978).
164. J. G. Brennan, T. Siegrist, P. J. Carroll, S. M. Stuczynski, L. E. Brus, and M. L. Steigerwald, *J. Am. Chem. Soc.*, *111*, 4141 (1989).
165. J. G. Brennan, T. Siegrist, P. J. Carroll, S. M. Stuczynski, P. Reynders, L. E. Brus, and M. L. Steigerwald, *Chem. Mater.*, *2*, 403 (1990).
166. M. Bochmann, K. J. Webb, M. B. Hursthouse, and M. Mazid, *J. Chem. Soc. Dalton Trans.*, 2317 (1991).
167. M. Bochmann, G. C. Bwembya, and A. K. Powell, *Polyhedron*, *12*, 2929 (1993).
168. M. Bochmann, K. J. Webb, H.-B. Hursthouse, and M. Mazid, *J. Chem. Soc. Chem. Commun.*, 1735 (1991).
169. M. Bochmann, A. P. Coleman, and A. K. Powell, *Polyhedron*, *11*, 507 (1992).
170. A. L. Seligson, P. J. Bonasia, J. Arnold, K.-M. Yu, J. M. Walker, and E. D. Bourret, *Proc. Mater. Res. Soc.*, *282*, 665 (1993).
171. P. J. Bonasia and J. Arnold, *Inorg. Chem.*, *31*, 2508 (1992).
172. P. J. Bonasia, Ph.D. Thesis, "The Synthesis and Reactivity of Sterically Hindered Chalcogenolate Ligands," 1993 University of California, Berkeley, CA.
173. P. A. W. Dean, J. J. Vittal, and N. C. Payne, *Inorg. Chem.*, *26*, 1683 (1987).
174. J. J. Vittal, D. A. W. Dean, and N. C. Payne, *Can. J. Chem.*, *70*, 792 (1992).
175. P. A. W. Dean, V. Manivannan, and J. J. Vittal, *Inorg. Chem.*, *28*, 2360 (1989).
176. G. S. H. Lee, K. J. Fisher, C. Craig, M. L. Scudder, and I. G. Dance, *J. Am. Chem. Soc.*, *112*, 6435 (1990).
177. B. Krebs, *Angew. Chem. Int. Ed. Engl.*, *22*, 113 (1983).
178. A. H. Cowley, R. A. Jones, P. R. Harris, D. A. Atwood, L. Contreras, and C. J. Burek, *Angew. Chem. Int. Ed. Engl.*, *30*, 1143 (1991).
179. W. Uhl, M. Layh, G. Becker, K. W. Klinkhammer, and T. Hildenbrand, *Chem. Ber.*, *125*, 1547 (1992).
180. K. Ruhlandt-Senge and P. P. Power, *Inorg. Chem.*, *32*, 3478 (1993).
181. S. P. Wuller, A. L. Seligson, G. P. Mitchell, and J. Arnold, *Inorg. Chem.*, submitted for publication.
182. A. L. Seligson and J. Arnold, *J. Am. Chem. Soc.*, *115*, 8214 (1993).
183. P. A. W. Dean, J. J. Vittal, and N. C. Payne, *Inorg. Chem.*, *23*, 4232 (1984).
184. D. H. R. Barton, H. Dadoun, and A. Gourdon, *Nouv. J. Chem.*, *6*, 53 (1982).
185. D. J. Berg, R. A. Andersen, and A. Zalkin, *Organometallics*, 7, 1858 (1988).
186. A. Recknagel, M. Noltemeyer, D. Stalke, U. Pieper, H.-G. Schmidt, and F. T. Edelmann, *J. Organomet. Chem.*, *411*, 347 (1991).
187. M. Wedler, A. Recknagel, J. W. Gilje, M. Noltemeyer, and F. T. Edelmann, *J. Organomet. Chem.*, *426*, 295 (1992).
188. D. R. Cary and J. Arnold, *J. Am. Chem. Soc.*, *115*, 2520 (1993).
189. M. Berardini, T. J. Emge, and J. G. Brennan, *J. Chem. Soc. Chem. Commun.*, 1537 (1993).

190. I. P. Beletskaya, A. Z. Voskoboynikov, A. K. Shestakova, and H. Schumann, *J. Organomet. Chem.*, *463*, C1 (1993).
191. H. C. E. McFarlane and W. McFarlane, in *Multinuclear NMR*, J. Mason, Ed., Plenum, New York, 1987, p. 417.
192. B. Bildstein, K. J. Irgolic, and D. H. O'Brien, *Phosphorus Sulfur*, *38*, 245 (1988).
193. K. Giselbrecht, B. Bildstein, and F. Sladky, *Chem. Ber.*, *122*, 1255 (1989).
194. W. W. du Mont and I. Wagner, *Chem. Ber.*, *121*, 2109 (1988).
195. C. Gerlach, V. Christou, and J. Arnold, *Inorg. Chem.*, submitted for publication.
196. M. Karplus and J. A. Pople, *J. Chem. Phys.*, *38*, 2803 (1963).
197. M. Bochmann, K. J. Webb, J. E. Hails, and D. Wolverson, *Eur. J. Solid State Inorg. Chem.*, *29*, 155 (1992).
198. J. Arnold, J. M. Walker, K. M. Yu, P. J. Bonasia, A. L. Seligson, and E. D. Bourret, *J. Crys. Growth*, *124*, 647 (1992).
199. J. Arnold and P. J. Bonasia, *United States Patent*, *5 157 136* (1992).
200. H. J. Gysling, A. A. Wernberg, and T. N. Blanton, *Chem. Mater.*, *4*, 900 (1992).
201. M. Brewer, D. Khasnis, M. Buretea, M. Berardini, T. J. Emge, and J. G. Brennan, *Inorg. Chem.*, *33*, 2743 (1994).
202. M. Berardini, T. Emge, and J. G. Brennan, *J. Am. Chem. Soc.*, *116*, 6941 (1994).
203. D. V. Khasnis, M. Brewer, J. S. Lee, T. J. Emge, and J. G. Brennan, *J. Am. Chem. Soc.*, *116*, 7129 (1994).
204. Y. F. Cheng, T. J. Emge, and J. G. Brennan, *Inorg. Chem.*, *33*, 3711 (1994).
205. B. Kersting and B. Krebs, *Inorg. Chem.*, *33*, 3886 (1994).
206. J. Darkwa, *Organometallics*, *13*, 3743 (1994).
207. L. Y. Goh, M. S. Tay, and C. Wei, *Organometallics*, *13*, 1813 (1994).
208. K. Ruhlandt-Senge, personal communication.

Coordination Chemistry with Sterically Hindered Hydrotris(pyrazolyl)borate Ligands: Organometallic and Bioinorganic Perspectives

NOBUMASA KITAJIMA†

Research Laboratory of Resources Utilizaиon
Tokyo Institute of Technology
Midori-ku, Yokohama 227, Japan

WILLIAM B. TOLMAN

Department of Chemistry
University of Minnesota
Minneapolis, MN

CONTENTS

†Deceased January 8, 1995.

Progress in Inorganic Chemistry, Vol. 43, Edited by Kenneth D. Karlin.
ISBN 0-471-12336-6

I. INTRODUCTION

A well-recognized synthetic strategy in inorganic and organometallic chemistry is to control the nuclearity, coordination number, geometry, physical properties, and reactivity of metal complexes through the use of supporting ligands that contain sterically bulky substituents. Examples abound of the use of such ligands to inhibit oligomerization reactions, access compounds of low-coordination number, increase the kinetic and thermodynamic stability of reactive complexes, and model the protected cavities within which metal ions in proteins reside. Alkyl- and aryl-substituted cyclopentadienyl (Cp) ligands (1), amides (2, 3), alkoxides (4), thiolates (5), porphyrins (6), and phosphines (7) represent just a few of the most popular auxiliaries employed for the steric and electronic tuning of the properties of metal-containing compounds. Recently, hydrotris(pyrazolyl)borate (Tp) ligands possessing various sterically bulky pyrazolyl ring substituents (Fig. 1) have been prepared and demonstrated to be advantageous supporting chelates in organometallic and bioinorganic studies. Steric effects on metal complex properties are particularly profound in these systems due to the orientation of the R^3 substituents, which enclose the bound metal ion in a highly protected pocket. The first of these ligands, $Tp^{t\text{-Bu}}$, as well as the key concept of steric control via substitution in the pyrazolyl 3 position, were introduced in 1986–1987 by Trofimenko and co-workers (8, 9). Several reviews that focus on the chemistry of the parent Tp and hydrotris(3,5-dimethylpyrazolyl)borate (Tp^{Me2}) ligands have been published (10–13), and in a recent ar-

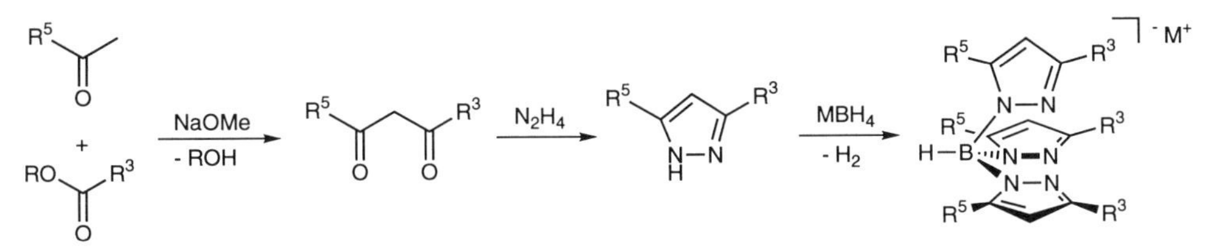

Figure 1. Standard synthesis of Tp ligands.

ticle the metal complexes that have been prepared with hindered Tp ligands were comprehensively cataloged (14). In this chapter, we update this review and delve deeper into the relationship between hindered Tp ligand structures and the structures, spectroscopic features, and reactivities of derived complexes. The organometallic and bioinorganic chemistry of complexes of Tp ligands with 3-pyrazolyl substituents larger than a methyl group will be emphasized, with discussion of the chemistry of complexes of the parent Tp and the Tp^{Me2} "scorpionates" to be limited to issues of particular relevance to those presented for the more sterically hindered ligands. In view of the all-inclusive nature of the previous review (14), no effort to present a comprehensive account will be made. Instead, specific cases will be chosen for discussion on the basis of their significance to the organometallic and bioinorganic subfields and/or because they exemplify important concepts concerning the relationship of ligand structure to metal complex properties.

II. SYNTHESIS AND PROPERTIES OF STERICALLY HINDERED HYDROTRIS(PYRAZOLYL)BORATE LIGANDS

A. Synthesis of Achiral Ligands

The now standard procedure (15) for the synthesis of poly(pyrazolyl)borate ligands has generally been found to be applicable to the preparation of sterically hindered variants, which are listed in Table I. The method involves thermolysis of a mixture of the suitably substituted pyrazole and a borohydride salt, MBH_4 (M = K or Na), either as a melt or in a high-boiling solvent, with evolution of H_2 (Fig. 1). The pyrazoles are typically prepared from the appropriate 1,3-dicarbonyl compound via a high-yield cyclocondensation with hydrazine. Substituents are introduced at the 3 and 5 positions during the preparation of the dicarbonyl component, most often by a Claisen condensation between a ketone and an ester. Alternative procedures have been reported [cf. (16)]. Introduction of an R^4 group, such as Br, is performed via electrophilic substitution on the isolated pyrazole.

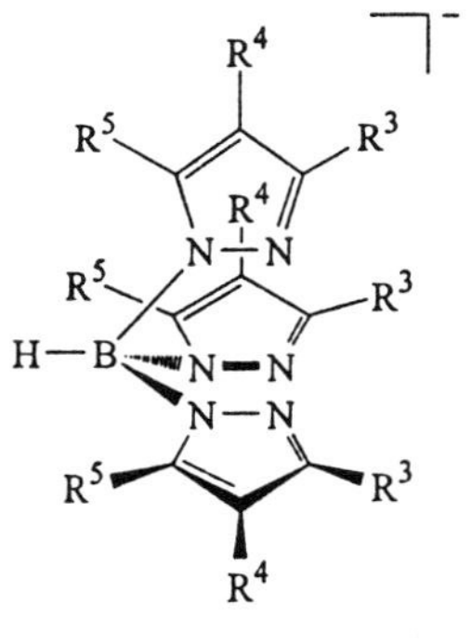

TABLE I
Sterically Hindered Tris(pyrazolyl)hydroborate Ligands

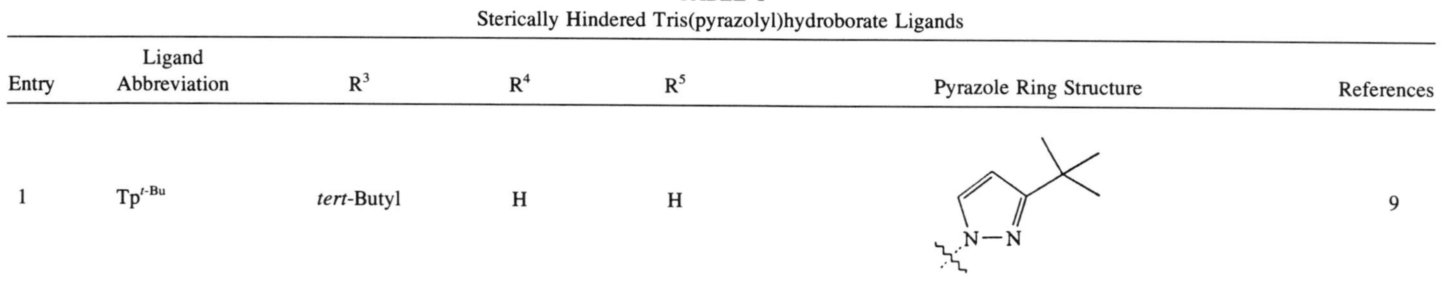

Entry	Ligand Abbreviation	R^3	R^4	R^5	Pyrazole Ring Structure	References
1	$Tp^{t\text{-Bu}}$	*tert*-Butyl	H	H		9

2	$Tp^{t\text{-Bu},4Br}$	*tert*-Butyl	Br	H	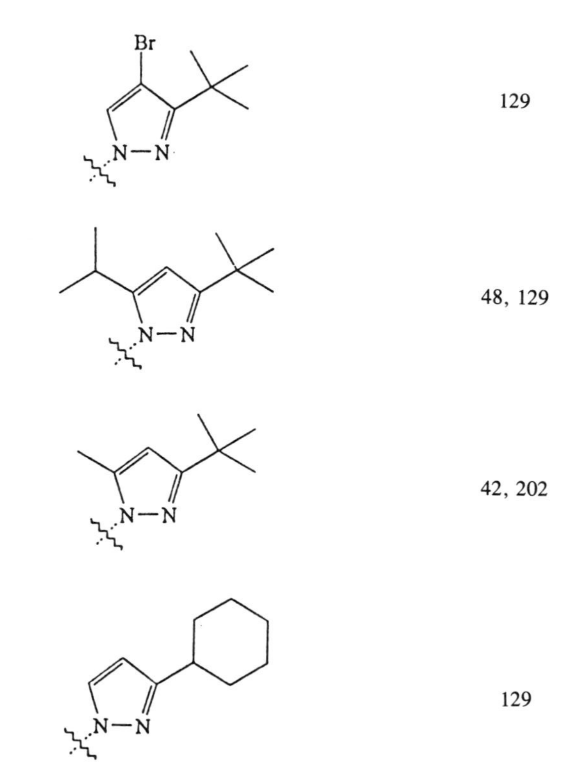	129
3	$Tp^{t\text{-Bu},i\text{-Pr}}$	*tert*-Butyl	H	Isopropyl		48, 129
4	$Tp^{t\text{-Bu},Me}$	*tert*-Butyl	H	Me		42, 202
5	Tp^{Cy}	Cyclohexyl	H	H		129

TABLE I. (*Continued*)

Entry	Ligand Abbreviation	R^3	R^4	R^5	Pyrazole Ring Structure	References
6	$Tp^{i\text{-}Pr}$	Isopropyl	H	H	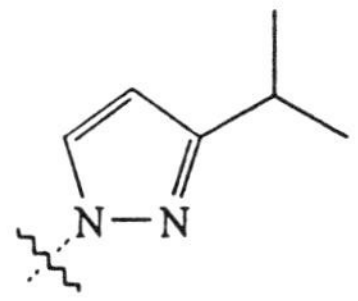	17
7	$Tp^{i\text{-}Pr,4Br}$	Isopropyl	Br	H	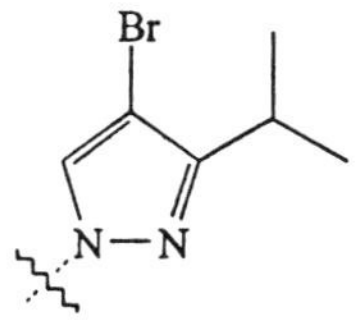	17
8	$Tp^{i\text{-}Pr,Me}$	Isopropyl	H	Me	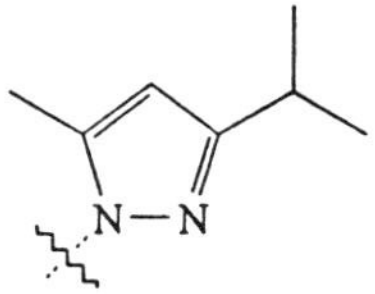	18
9	$Tp^{i\text{-}Pr2}$	Isopropyl	H	Isopropyl	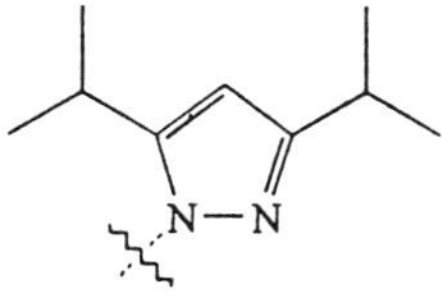	167

10	Tp^{Np}	Neopentyl	H	H		52
11	Tp^{Me2}	Me	H	Me		*a*
12	$Tp^{Me2,4Bz}$	Me	Benzyl	Me		*b*
13	Tp^{Tn}	Thienyl	H	H		39

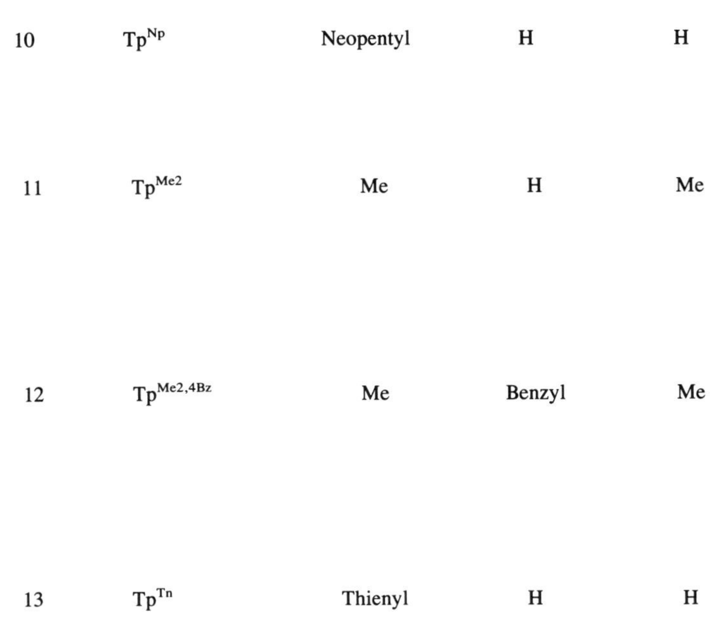

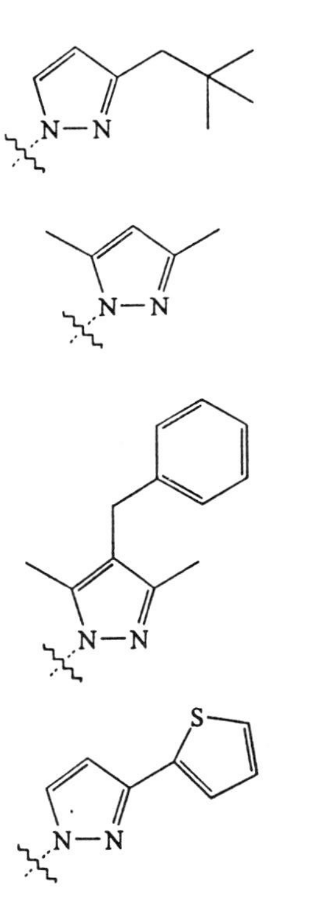

TABLE I. *(Continued)*

Entry	Ligand Abbreviation	R^3	R^4	R^5	Pyrazole Ring Structure	References
14	Tp^{CF_32}	CF_3	H	CF_3	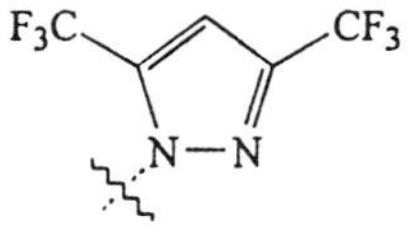	24
15	$Tp^{CF_3,Me}$	CF_3	H	Me	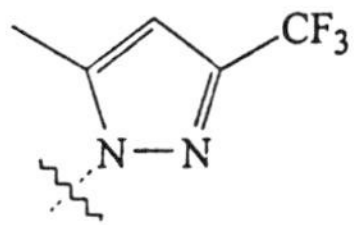	24
16	$Tp^{CF_3,Tn}$	CF_3	H	Thienyl	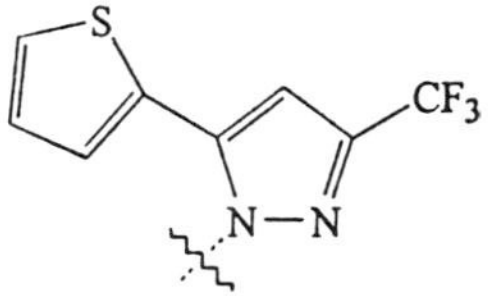	19, 129
17	Tp^{Ph}	Ph	H	H	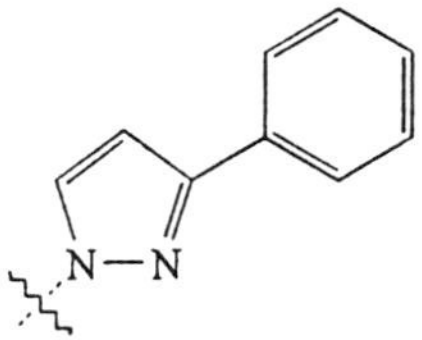	9

18	$Tp^{Ph,Me}$	Ph	H	Me		40
19	Tp^{Ph2}	Ph	H	Ph		167
20	$Tp^{p\text{-}Tol}$	*p*-Tolyl	H	H		58
21	$Tp^{p\text{-}An}$	*p*-Anisyl	H	H		42

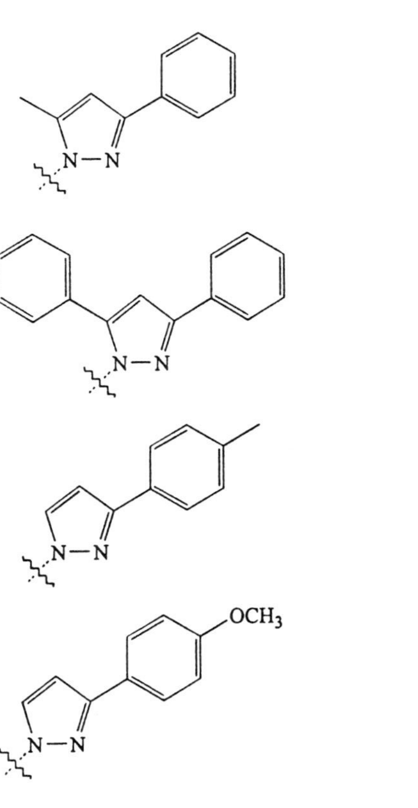

TABLE I. *(Continued)*

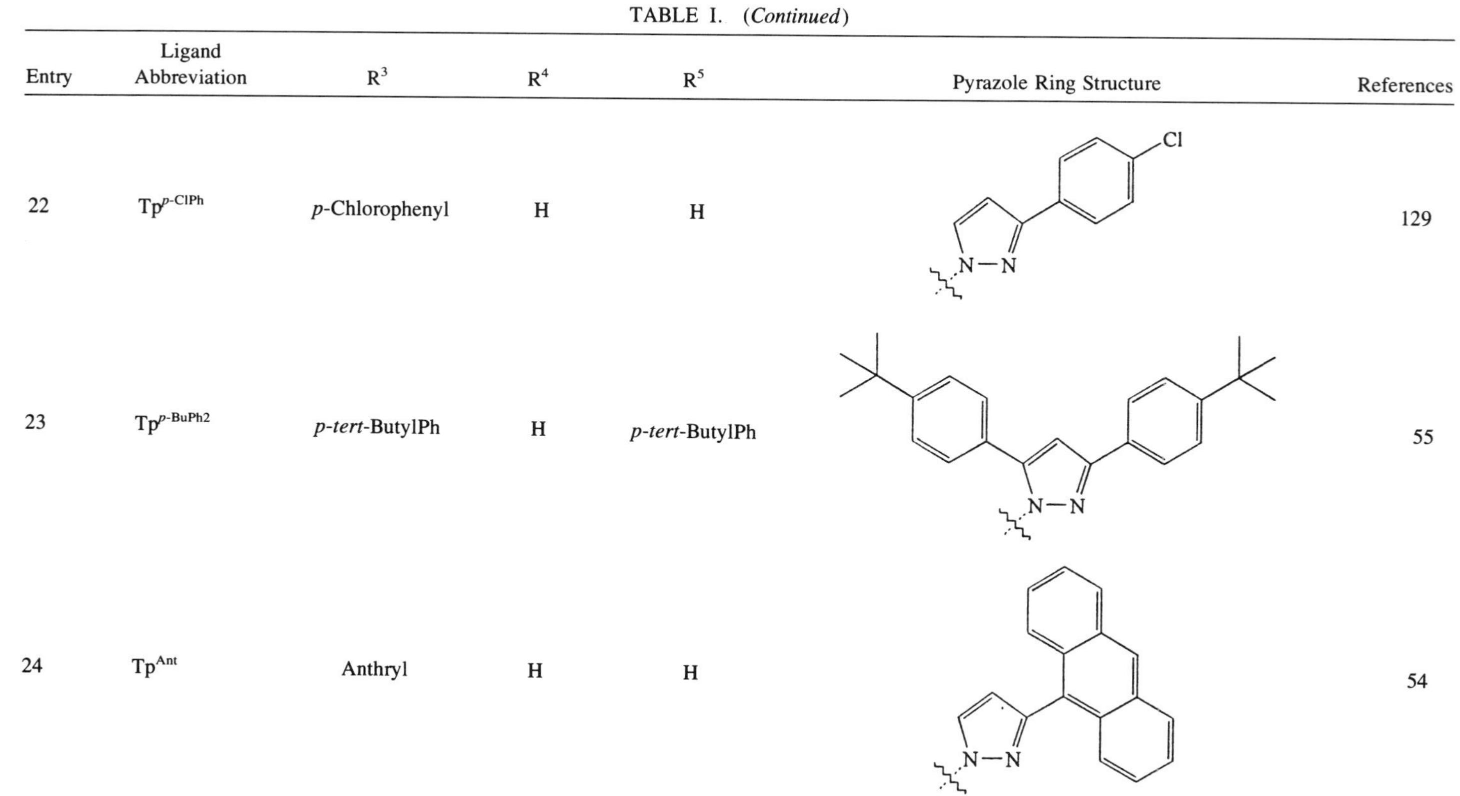

Entry	Ligand Abbreviation	R^3	R^4	R^5	Pyrazole Ring Structure	References
22	$Tp^{p\text{-ClPh}}$	*p*-Chlorophenyl	H	H		129
23	$Tp^{p\text{-BuPh2}}$	*p*-*tert*-ButylPh	H	*p*-*tert*-ButylPh		55
24	Tp^{Ant}	Anthryl	H	H		54

No.	Ligand	R^3	R^4	R^5	Structure	Ref.
25	Tp^{Ms}	Mesityl	H	H	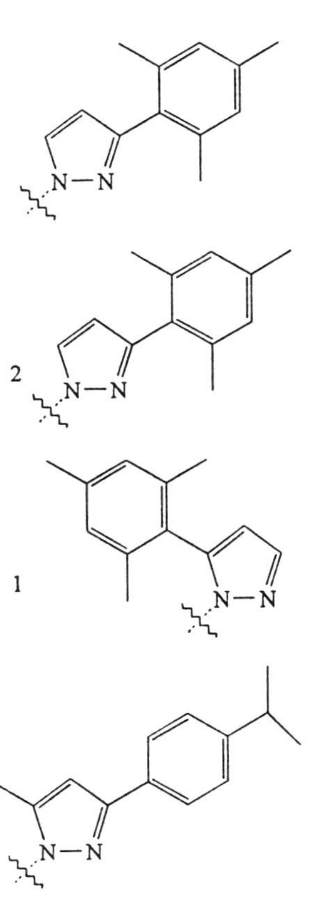	25
26	Tp^{Ms*}	Mesityl (2 rings)	H	H	2	25
		H (1 ring)	H	Mesityl (1 ring)	1 	
27	Tp^{Cum}	Cumyl	H	Me		25

TABLE I. (*Continued*)

Entry	Ligand Abbreviation	R^3	R^4	R^5	Pyrazole Ring Structure	References
28	Tp^{a}			H	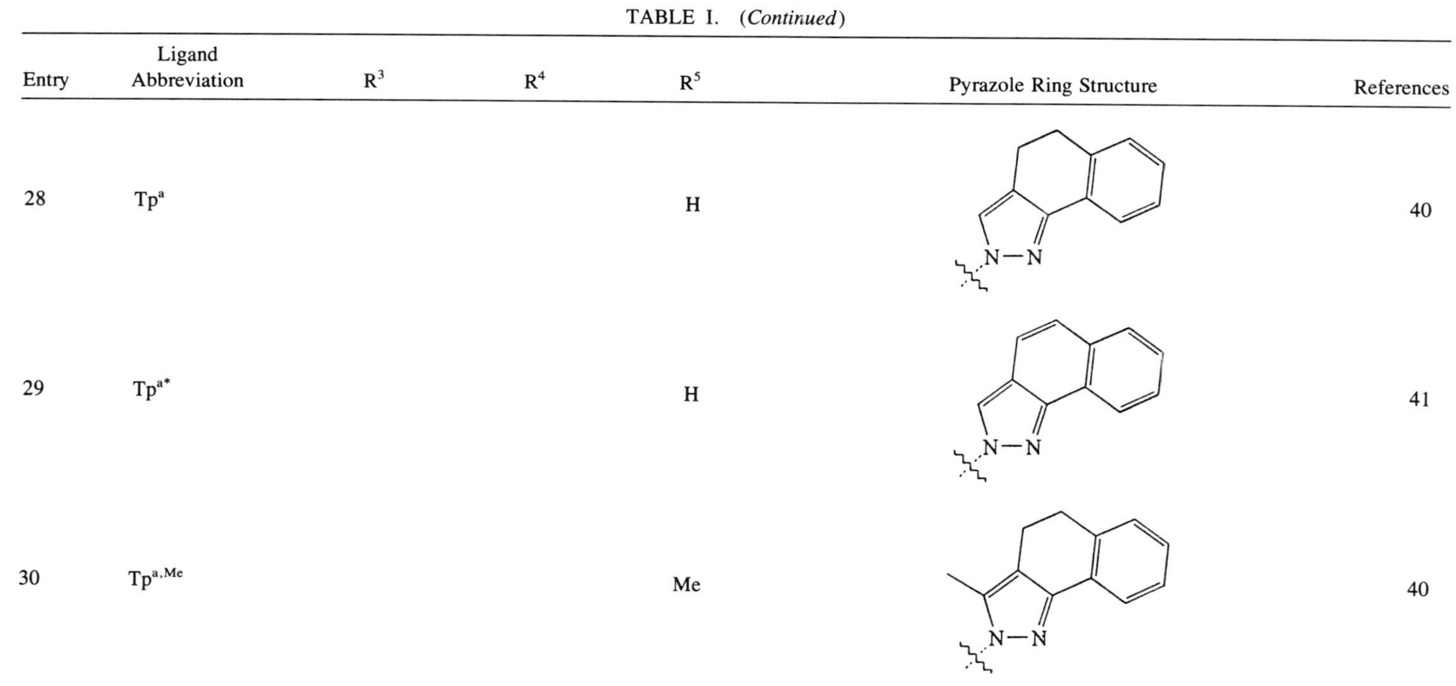	40
29	Tp^{a*}			H		41
30	$Tp^{a,Me}$			Me		40

31	Tp^{b}			H		40
32	Tp^{Menth}			H		32, 33
33	Tp^{Menth*}			H (2 rings)		21
		H (1 ring)	H			

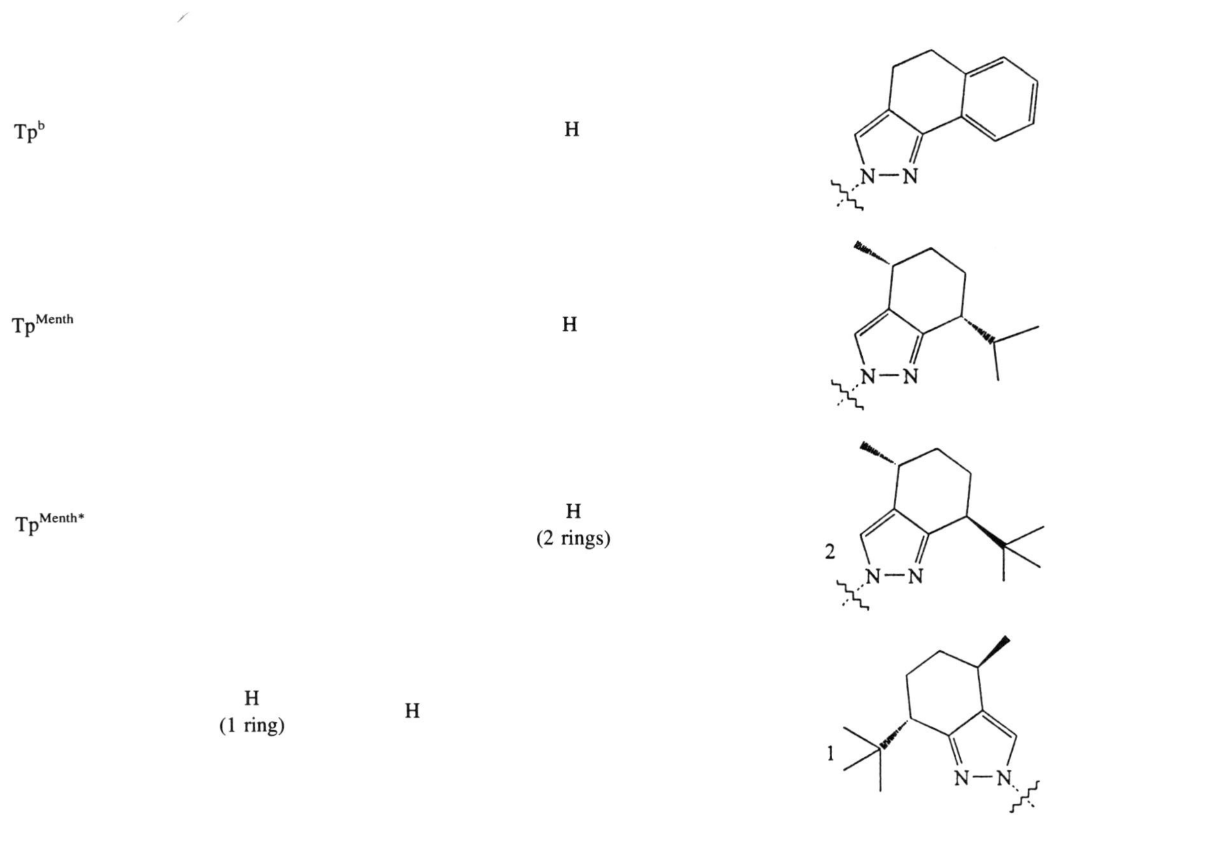

TABLE I. (*Continued*)

Entry	Ligand Abbreviation	R^3	R^4	R^5	Pyrazole Ring Structure	References
34	$Tp^{Mementh}$			H	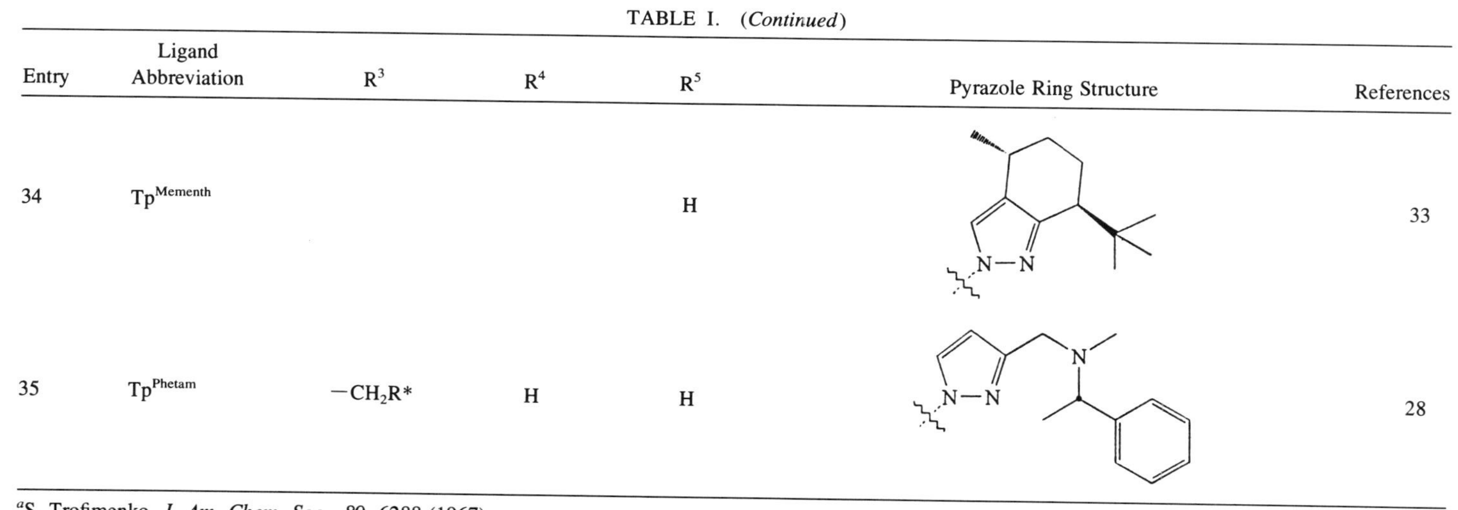	33
35	Tp^{Phetam}	$-CH_2R^*$	H	H		28

[a]S. Trofimenko, *J. Am. Chem. Soc.*, *89*, 6288 (1967).

[b]J. C. Jeffrey, S. S. Kurek, J. A. McCleverty, E. Psillakis, R. M. Richardson, M. D. Ward, and A. Wlodarczyk, *J. Chem. Soc. Dalton Trans.*, 2559 (1994).

As with pyrazole itself, the reactions affording the hindered Tp ligands proceed via the dihydrobis(pyrazolyl)borate anion, which in some instances has been isolated (9). If the temperature is too high in the case of the parent pyrazole ($R^3 = R^4 = R^5 = H$), the tetra(pyrazolyl)borate anion is isolated instead of the Tp ligand (15). With large R^3 groups, however, the formation of the more highly boron-substituted product is relatively disfavored, facilitating the preparation of the desired Tp in most cases. Isolation of pure samples of the sodium or potassium salts of the hindered $Tp^{RR'}$ ligands is often problematic, and despite the fact that metal complexes can be prepared by using crude solutions of these salts, conversion to crystalline thallium complexes by treatment with aqueous $TlNO_3$ or methanolic TlOAc is often performed in order to obtain the ligands in pure form. The thallium salts are particularly suitable reagents in metathesis reactions with metal halides and oxyanions due to the thermodynamic driving force in favor of the formation of insoluble TlX (X = halide or oxyanion). For R^3 = isopropyl, bromination at R^4 was found to be a convenient way of rendering both the K and Tl salts of this ligand crystalline (17).

A critical issue in the synthesis of hindered $Tp^{RR'}$ ligands that contain asymmetrically substituted pyrazolyl rings ($R^3 \neq R^5$) is the regiochemistry of B—N bond formation. As expected, larger substituents usually end up in the pyrazolyl ring 3 position, relatively distant from the site of the B—N bond (but see Note Added in Proof). This tendency is most pronounced when the steric differences between R^3 and R^5 are large, as in the regiospecific synthesis of $Tp^{t\text{-Bu}}$, $Tp^{i\text{-Pr}}$, and Tp^{Ph} (i.e., when R^5 = H). When these steric differences are less substantial a mixture of isomers is often formed, one of which is the anticipated C_{3v}-symmetric ligand, the other of which contains two pyrazolyl rings with the larger substituent in the 3 position but the third with the bulkier group in the 5 position. For example, reaction of 3-isopropyl-5-methylpyrazole with KBH_4 resulted in the formation of an approximate 4:1 ratio of $Tp^{i\text{-Pr},Me}$ and its isomer **I** (18).

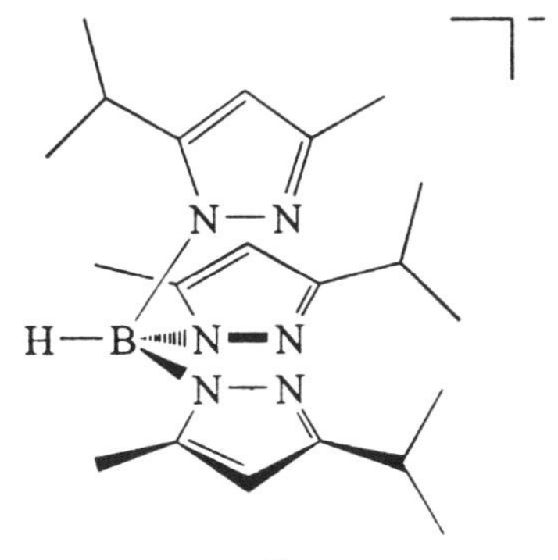

I

Similar isomeric mixtures were reported when 3-ethyl-5-methylpyrazole or 3-isobutyl-5-methylpyrazole were used as starting materials (19). In some instances, separation and isolation of the isomers are possible via fractional crystallization of metal complexes derived from the initial crude ligand mixture (18, 20, 21).

Electronic effects may also control the regiochemical course of the ligand syntheses, with electron-withdrawing groups preferring the 3 position. This tendency is most clearly manifested in $Tp^{CF_3,Me}$ and $Tp^{CF_3,Tn}$, in which electronically quite different substituents are present. The syntheses of these ligands are highly regioselective, with the *larger* CF_3 group residing exclusively in the 3 position. The preference of an electron-withdrawing group for the 3 position in Tp ligands can be explained via an argument analogous to that invoked to rationalize the equilibrium between the pyrazole tautomers **A** and **B** in Fig. 2. The inductive electron-withdrawing effect of the CF_3 group makes the distal nitrogen more basic, favoring tautomer **A** (22, 23). It has been suggested that the B—N bond preference is kinetic in origin, the less favored tautomer **B** leading to the more favorable transition state shown in Fig. 2 during the course of its reaction with BH_4^- (24). In view of the facility of borotropic rearrangements and the ease of pyrazole exchange processes under the thermolytic synthetic conditions used, however (see below), one must also consider that the isomeric preference derives from the greater thermodynamic stability of the system when the boron binds to the more basic nitrogen atom (i.e., like H in tautomer **A**).

The interplay of steric and electronic influences on the course of ligand synthetic reactions is evident from the results of $Tp^{RR'}$ preparation from 3-mesitylpyrazole (25). The asymmetric isomer Tp^{Ms*} (**II**) was the major product and it could be separated from the more symmetric Tp^{Ms} (minor component) after repeated crystallizations of the thallium complexes. Surprisingly, ligand **II** could be thermally isomerized to symmetric Tp^{Ms}, indicating that **II** is the favored product of the preparative reaction on kinetic, rather than thermodynamic, grounds. Thus, it would appear that the steric demands of the mesityl groups

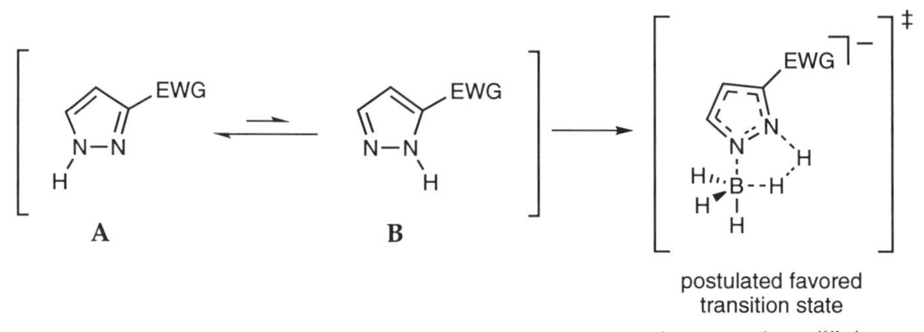

Figure 2. Effect of an electron-withdrawing group (EWG) on pyrazole tautomeric equilibrium.

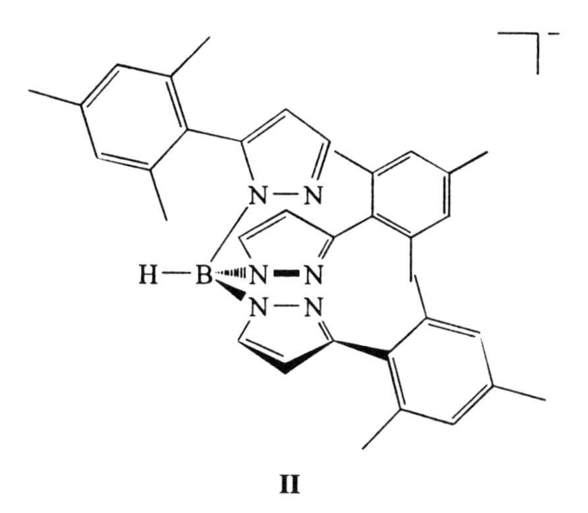

II

kinetically favor the less-hindered asymmetric isomer, but the electron-withdrawing properties of the same substituents render the ligand with all the mesityl groups in the 3 position more stable thermodynamically.

B. Synthesis of Chiral Ligands

The strong coordinating ability of Tp ligands to elements from throughout the periodic table coupled with the profound steric effects of substituted versions as detailed herein has led to an interest in preparing chiral Tp ligands for use in stereoselective metal centered molecular recognition and catalysis of organic chemical reactions. Efforts to construct optically active Tp ligands have focused on assembling C_3-symmetric molecules of generalized Structure **III**, which con-

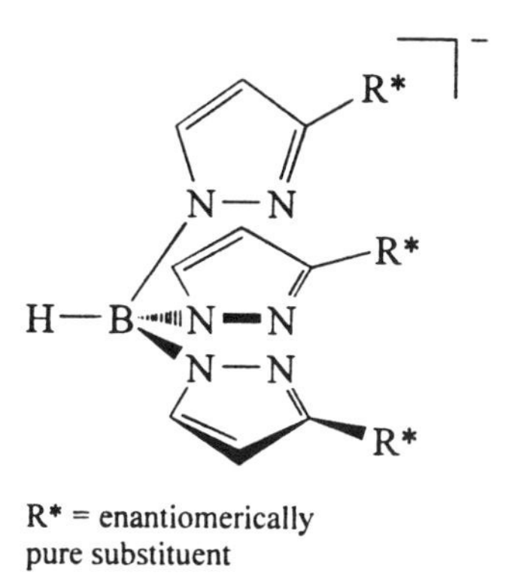

III

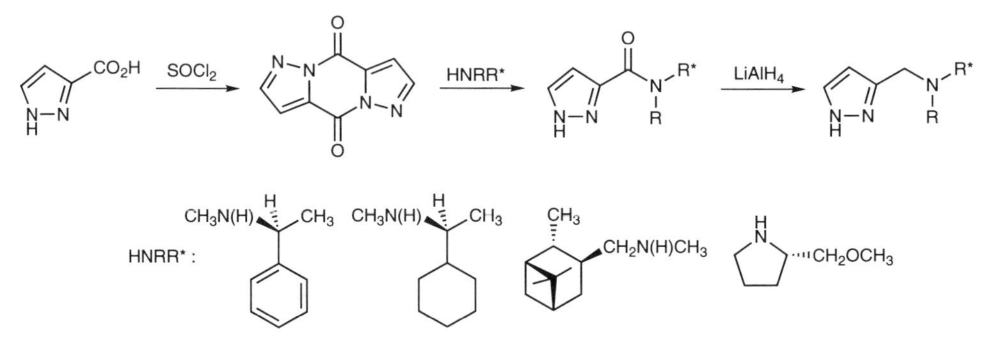

Figure 3. Synthesis of chiral pyrazoles from optically pure amines (27, 28).

tain three identical, enantiomerically pure pyrazoles [see (26) for another strategy]. In one approach to the synthesis of such pyrazoles, optically active amines were condensed with a dimer derived from pyrazole–carboxylic acid (Fig. 3) (27, 28). Thermolysis of one of these pyrazoles with KBH_4 was reported to yield the Tp^{Phetam} ligand (see Table I, Entry 35) according to mass spectrometric (MS) and ^1H NMR and IR spectroscopic measurements (28). While this route has the advantages of versatility due to the availability of a wide range of chiral amines and the possibility of potentially useful secondary binding interactions involving the amine nitrogen atom, the stereogenic center is necessarily well removed from the metal-coordinating pyrazole nitrogen and would thus be expected to have attenuated effects on the asymmetric reactivity of metal complexes.

An alternative synthesis of optically active pyrazoles involves the usual cyclocondensation between hydrazine and a 1,3-dicarbonyl compound, but with the latter having a stereogenic center α to one carbonyl unit. This center would then reside at the R^3 pyrazolyl ring position (proximate to a bound metal ion) in the final ligand. In this route, the problematic synthetic step is constructing the optically active 1,3-dicarbonyl without epimerizing the α carbon. In the synthesis of the pyrazole **IV** from (+)-camphor (Fig. 4), loss of α-carbon stereochemistry was circumvented because the critical stereogenic center is quaternary (29–31). Epimerization did occur during the preparation of the pyrazoles derived from (2*S*, 5*R*)-menthone (**V**) (32, 33), (2*S*, 5*R*)- and (2*R*, 5*R*)-methylmenthone (**VI**) (32, 33), and (2*S*, 5*R*)- and (2*R*, 5*R*)-phenylmenthone (**VII**) (Fig. 4) (34), but the presence of a second, fixed stereogenic center in each of these cases allowed the separation of the resulting diastereomers or their salts by fractional crystallization or chromatography.

Construction of Tp ligands from the highly hindered optically pure pyrazoles **IV–VII** has been achieved in some instances, but difficulties have arisen as a result of their unique steric properties. Although a tetra(pyrazolyl)borate **VIII**

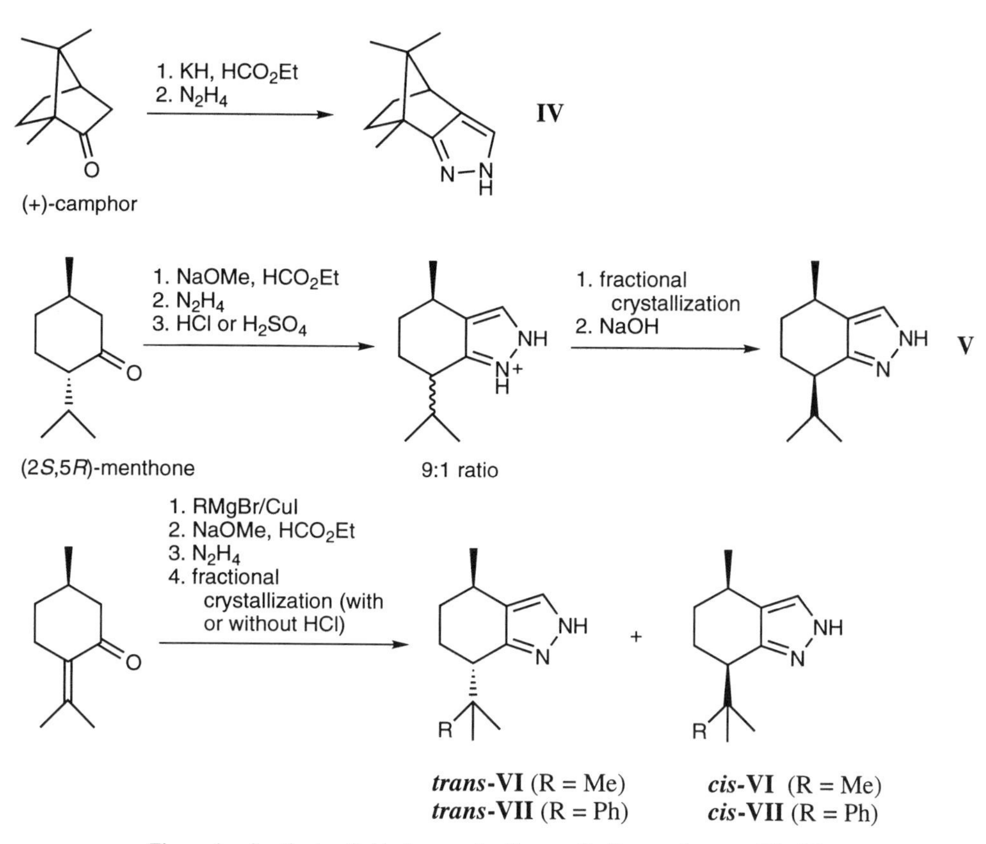

Figure 4. Synthesis of chiral pyrazoles from optically pure ketones (32, 33).

has been prepared from camphor-derived pyrazole **IV** (35), attempts to isolate a Tp anion using **IV** have not been successful to date (36). Poor regiospecificity in the linking of the N atoms of this pyrazole to other connecting fragments (e.g., $-CH_2-$) has been reported (29–31), suggesting that a similar problem may be the cause of the formation of intractable mixtures upon attempted Tp synthesis with this pyrazole. The accentuated steric differences between R^3 and R^5 in pyrazoles **V–VII** would be expected to give better regiospecificity in the B—N bond formation reaction. In fact, the C_3-symmetric Tp^{Menth} and $Tp^{Mementh}$ ligands (see Table I, Entries 32 and 34, respectively) were isolated, structurally characterized as thallium(I) salts, and used in metal complex synthesis (32, 33). Attempted synthesis of Tp from **VII** under conditions similar to those used to prepare **V** and **VI** has been unsuccessful to date, presumably because of the overwhelming size of the dimethylphenyl group attached to the R^3 stereogenic center in **VII**. Interestingly, while the major cis diastereomer of **V** can be elab-

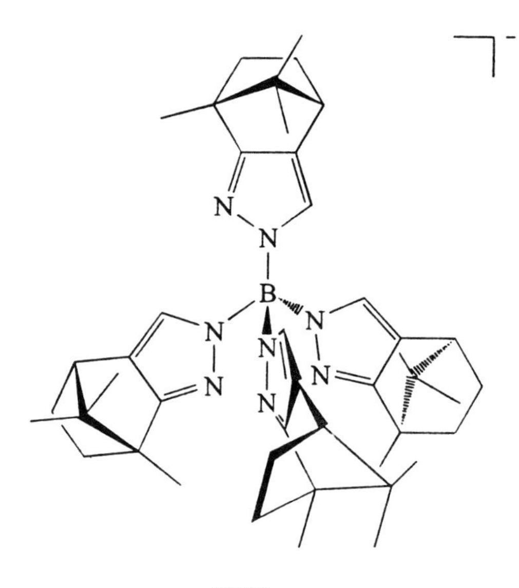

VIII

orated into ligand Tp^{Menth}, under similar conditions only *trans*-**VI** and not *cis*-**VI** will assemble into a $Tp^{Mementh}$. While the reasons for this are unclear, it would appear that steric interactions among substituents well removed from the "cage" comprised of the B and pyrazolyl N atoms can be important in ligand and, by inference, metal complex syntheses.

C. Ligand Properties

1. Steric Effects

Evaluation of the size of the variously substituted Tp ligands is critical for developing an understanding of the influence of pyrazolyl ring substituents on the chemistry of their metal complexes. Of the many conceivable ways by which ligand size can be defined, the "cone angle" has been used most often. In phosphine ligands, for which the concept was originally introduced, the cone angle was defined using Corey–Pauling–Koltun (CPK) models for symmetric cases (three identical substituents) as "the apex angle of a cylindrical cone, centered 2.28 Å from the center of the P atom, which just touches the van der Waals radii of the outermost atoms of the model" (Fig. 5) (7). As noted by C. A. Tolman and also critically discussed elsewhere (37, 38), the steric demand of phosphine ligands defined in this way, while qualitatively useful for understanding the behavior of a wide range of organometallic complexes, represents

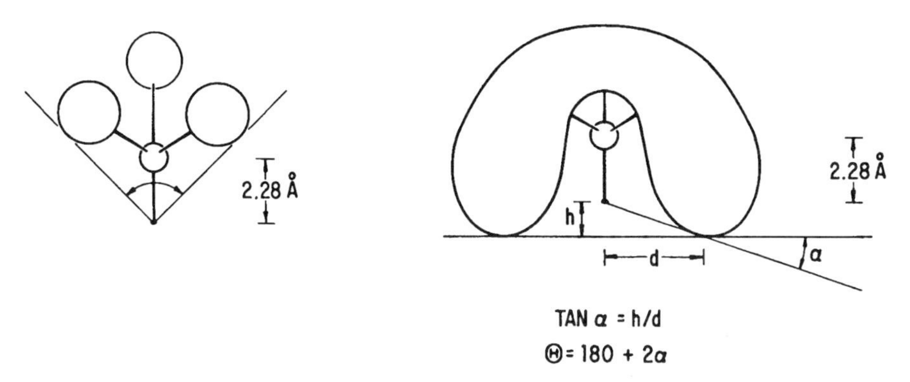

Figure 5. Definition of cone angles. [Reprinted with permission from C. A. Tolman, *Chem. Rev.*, *77*, 313 (1977). Copyright © 1977 American Chemical Society.]

a maximum value for the size of the ligands and does not always take into account possible ''intermeshing'' of ligand substituents and differences in orientations of substituents such as those resulting from rotations about M—P or P—C bond axes. Refinements to the original method of cone angle determination that attempt to take into account steric demand changes that accompany orientational variations have been introduced and have in some instances allowed rationalization of the structures of metal complexes (37).

As for the phosphines, application of the cone angle concept to Tp ligands does provide a general and qualitatively useful measure of their relative steric demand, but has drawbacks in some instances as a result of specific substituent shapes and orientations. Trofimenko and co-workers (15, 39–41) calculated cone angles (θ) for several $Tp^{RR'}$ ligands using the dimensions established by X-ray crystallography for selected complexes; these are listed in Table II along with values for other pertinent ligands for comparison. It is particularly noteworthy that all of the Tp ligands have $\theta > 180°$, the smallest Tp ($R^3 = R^5 = H$) being most similar to the larger Cp* ligand. The large cone angles for the Tp ligands reflect the enclosure of a bound metal ion by a ''protective pocket'' of R^3 substituents, a characteristic property of these chelates that differentiates them from the majority of Cp ligands (1). For nonplanar R^3 groups the expected trend in θ is seen: *t*-Bu > Me > H. As discussed below, the observed structures and chemical behavior of complexes of ligands with these substituents can be rationalized by their relative cone angles. The behavior of Tp^{Tn} ($\theta = 246°$) and Tp^{b} ($\theta = 242°$) is anomalous, however, in that they exhibit reactivity more closely resembling Tp than Tp^{Ph} or $Tp^{t\text{-Bu}}$, but have cone angles larger than the latter two ligands (39, 40). In addition, one might predict analogous coordinative behavior and reactivity for Tp^{Ph} and $Tp^{t\text{-Bu}}$ on the basis of their similar θ values, but in many (albeit not all) respects this is not observed (see below). It has thus been suggested that for planar substituents the cone angle concept is

TABLE II
Cone Angles (θ) and Wedge Angles (ω) for Selected Ligands

Ligand	θ (deg)	ω (deg)	Reference
Tp^{Ms}	281	7	40
Tp^{a*}	276	87	41
$Tp^{t\text{-}Bu}$	265	35	40
$Tp^{a,Me}$	263	44	40
Tp^{a}	262	44	40
$Tp^{i\text{-}Pr,4Br}$	262	36	40
$Tp^{Ph,Me}$	250[a]	32[a]	40
Tp^{Tn}	246[a]		39[b]
Tp^{b}	242	82	40
Tp^{Me2}	236	75	40
Tp^{Ph}	235[a]		39[b]
Tp	199	91	40
Cp* (C_5Me_5)	182[c]		*d*
Cp (C_5H_5)	150[c]		*d*

[a]Approximate values due to orientational differences among aryl rings.
[b]Wedge angles not determined, and van der Waals radii of H atoms of ligands apparently not taken into account in the calculation of these cone angles.
[c]Values calculated based on X-ray structures of Rh^{I} complexes, including the van der Waals radii for all hydrogen atoms (see *d*). Different numbers were cited in C. E. Davies, I. M. Gardiner, J. C. Green, M. L. H. Green, P. D. Grebenik, V. S. B. Mtetwa, and K. Prout, *J. Chem. Soc. Dalton Trans.*, 669 (1985): (Cp, θ = 110°; Cp*, θ = 142°) and in (7) (Cp, θ = 136°), reflecting the variation in cyclopentadienyl-metal bond distances used in the calculations.
[d]P. M. Maitlis, *Chem. Soc. Rev.*, 1 (1981).

not useful; rather that "what is meaningful is the size of the open wedgelike spaces between the pz rings, as defined by the size and orientation of the 3 substituents, through which a nucleophile ligand can approach the metal center" (39). Accurate definition of these "wedge angles" is somewhat problematic for the ligands with planar R^3 groups that adopt a range of orientations, but they have been calculated for a number of ligands (Table II) and correlations with properties have been uncovered (see below). While there is some indication of the steric properties of the chiral ligands from studies of their transition metal coordination behavior (see below), unambiguous definition of steric influences remains a difficult task in the absence of wide-ranging enantioselective reactivity data.

More conclusive and practical information on the steric demands of Tp ligands has come from experimental studies in which the structures, spectroscopic properties, and reactivity of a number of metal complexes were examined and in some cases directly compared. Here we summarize results of purely "inorganic" investigations, some of which have been discussed elsewhere (12); work involving complexes of particular organometallic or bioinorganic chemical interest will be discussed in later sections. The combined results from these studies can be summarized by the following series according to effective steric bulk at a complexed metal center (Eq. 1).

$$\begin{aligned} \mathrm{Tp}^{t\text{-Bu}} &> \mathrm{Tp}^{\mathrm{Menth}} = \mathrm{Tp}^{i\text{-Pr}} \approx \mathrm{Tp}^{\mathrm{Np}} \approx \mathrm{Tp}^{\mathrm{Ms}} \approx \mathrm{Tp}^{\mathrm{Ant}} \\ &> \mathrm{Tp}^{\mathrm{Ph}} \approx \mathrm{Tp}^{\mathrm{Tol}} \approx \mathrm{Tp}^{p\text{-An}} \approx \mathrm{Tp}^{p\text{-ClPh}} \approx \mathrm{Tp}^{\mathrm{a}} > \mathrm{Tp}^{\mathrm{CF_3,R'}} > \mathrm{Tp}^{\mathrm{Me}} \\ &> \mathrm{Tp}^{\mathrm{Tn}} \approx \mathrm{Tp}^{\mathrm{b}} > \mathrm{Tp} \end{aligned} \tag{1}$$

Ligands with identical R^3 groups but different R^4 or R^5 substituents should probably be placed in the same category as their analogues with $R^4 = R^5 = H$, although initial indications suggest that their effective size is slightly larger (see below). The remaining ligands listed in Table I are difficult to categorize due to a current paucity of information on their coordination behavior.

The most sterically demanding ligands in the series, $\mathrm{Tp}^{t\text{-Bu}}$, $\mathrm{Tp}^{t\text{-Bu,Me}}$, $\mathrm{Tp}^{t\text{-Bu,4Br}}$, and $\mathrm{Tp}^{t\text{-Bu},i\text{-Pr}}$, prohibit formation of Tp_2M and TpM(μ-L)MTp (L = bridging group) complexes and heavily favor four-coordinate compounds with C_{3v}-distorted tetrahedral geometries. A wide range of structurally characterized $\mathrm{Tp}^{t\text{-Bu}}$ complexes of divalent metal halides and pseudohalides exemplify these characteristic features, the structures of $\mathrm{Tp}^{t\text{-Bu,Me}}$Ni(NCS) (42) and $\mathrm{Tp}^{t\text{-Bu}}$CuCl (Fig. 6) (43) being representative cases. As is typical for the general class of TpMX complexes, in $\mathrm{Tp}^{t\text{-Bu,Me}}$Ni(NCS) the $N_{pz}-M-N_{pz}$ angles, where pz = pyrazolyl, (av. = 94.2°) are decreased and the $N_{pz}-M-X$ angles (av = 122.3°) are correspondingly increased from the tetrahedral value of 109.5°. The ubiquity of this topology in complexes of Tp ligands having R^3 = *t*-butyl has led to its characterization as a "tetrahedral enforcer" (17). As discussed in more detail below, the identification of bidentate coordination of small

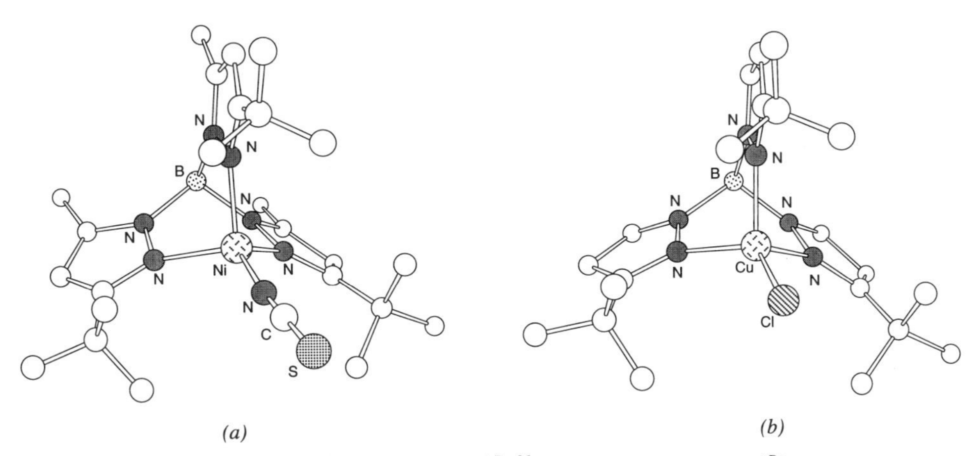

Figure 6. X-ray crystal structures of (*a*) $\mathrm{Tp}^{t\text{-Bu,Me}}$Ni(NCS) (42) and (*b*) $\mathrm{Tp}^{t\text{-Bu}}$CuCl (43).

coligands such as O_2^- (44), NO_2^- (45), and NO_3^- (43, 46) to afford essentially five-coordinate complexes, as well as monodentate binding of two or three coligands to yield five- or six-coordinate complexes, respectively, of second-row transition metals (15, 47) and lanthanides (see below), have shown that exceptions to the "tetrahedral" rule are possible.

The R^5 methyl and isopropyl groups in $Tp^{t\text{-Bu,Me}}$ and $Tp^{t\text{-Bu},i\text{-Pr}}$, respectively, serve to stabilize the B—N bonds of the ligands toward attack by external reagents and have thus facilitated the isolation of a number of unusual complexes (see below). In addition, nonbonded interactions involving these R^5 substituents appear to increase the ligands' "bite" (i.e., increase the cone angle). An illustrative example of this effect is evident from consideration of the divergent structures of $Tp^{t\text{-Bu}}Cu(NO_3)$ (43, 46) and $Tp^{t\text{-Bu},i\text{-Pr}}Cu(NO_3)$ (48) (Fig. 7). In the former complex, the pronounced tendency for Cu^{II} to adopt tetragonal structures results in symmetric, bidentate nitrate coordination, and a square pyramidal copper geometry. The increased ligand bite in the latter compound apparently counteracts this tendency, affording asymmetric bidentate nitrate coordination (ΔCu—O = 0.217 Å). While details of this structure, such as the orientation of the long Cu—O bond approximately along the B···Cu axis, as well as that of the highly distorted $Tp^{t\text{-Bu,Me}}CuCl$ (48, 49) are difficult to rationalize, it is evident that R^5 groups can appreciably influence metal coordination geometry.

The intermediate steric properties of the $Tp^{i\text{-Pr}}$, $Tp^{i\text{-Pr2}}$, $Tp^{i\text{-Pr,4Br}}$, $Tp^{i\text{-Pr,Me}}$, Tp^{Menth}, $Tp^{Mementh}$, and Tp^{Np} chelates are evidenced by the following charac-

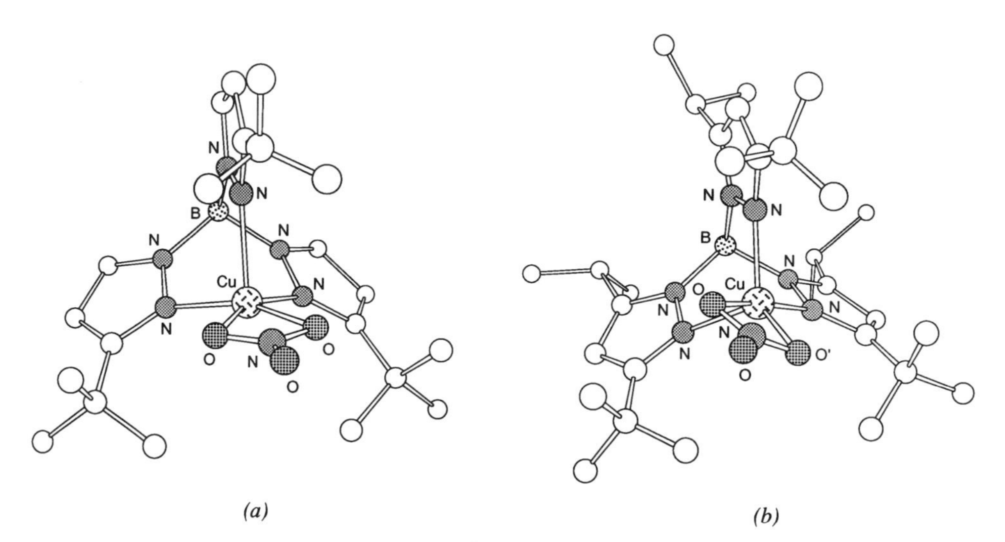

Figure 7. X-ray crystal structures of (*a*) $Tp^{t\text{-Bu}}Cu(NO_3)$ (43, 45) and (*b*) $Tp^{t\text{-Bu},i\text{-Pr}}Cu(NO_3)$ (48).

teristic coordination behavior. Like $Tp^{t\text{-Bu}}$, TpMX (M = divalent metal ion; X = halide or pseudohalide) complexes are accessible, but coordination of a solvent (sol) molecule or stronger binding of an additional TpM unit to form five-coordinate TpMX(sol) or $TpM(\mu\text{-L})_nMTp$ dimers, respectively, are possible for R^3 = isopropyl. The structures of $Tp^{i\text{-Pr2}}$CuCl(dmf), where dmf = *N,N*-dimethylformamide (ligand), (Fig. 8) (50) and $[(Tp^{i\text{-Pr},4\text{Br}}Ni)_2(\mu\text{-NCS})_2]$ (Fig. 9) (42) typify these motifs and can be compared to the four-coordinate $Tp^{t\text{-Bu},R}$ (R = H or Me) complexes shown in Fig. 6. The divergent monomeric versus dimeric structures, respectively, of the Cu^{II}-azide complexes of $Tp^{t\text{-Bu},Me}$ and $Tp^{i\text{-Pr2}}$ shown in Fig. 10 also illustrate the different effective sizes of these ligands (48). The 3-isopropyl substituent is large enough to prevent Tp_2M formation, although such complexes will form if a borotropic migration occurs that places an isopropyl group on one pyrazolyl ring in the 5 position {cf. $[HB(3\text{-}i\text{-Pr-4-Brpz})_2(5\text{-}i\text{-Pr-4-Brpz})]_2Co$ in Fig. 11} (17). Octahedral $Tp^{Np2}M$ (M = Ni or Co) complexes have been isolated and crystallographically characterized, but they are also thermodynamically unstable with respect to the formation of isomers resulting from a similar 1,2-borotropic shift (51). Unfavorable steric interactions among the isopropyl or neopentyl groups in the "equatorial belt" are apparently responsible for the lack of unrearranged examples of octahedral $Tp^{i\text{-Pr2}}M$ complexes and the tendency of $Tp^{Np2}M$ species to isomerize. The chiral Tp^{Menth} ligand, while able to stabilize pseudotetrahedral geometries like $Tp^{t\text{-Bu}}$,

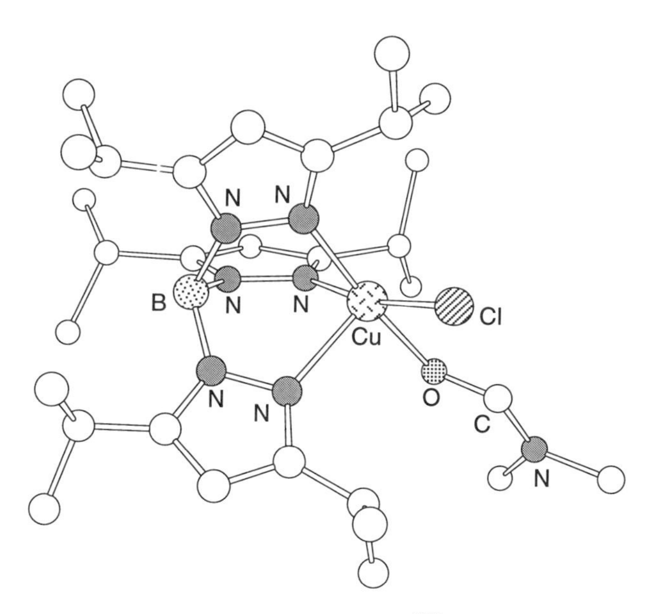

Figure 8. X-ray crystal structure of $Tp^{i\text{-Pr2}}$CuCl(dmf) (51).

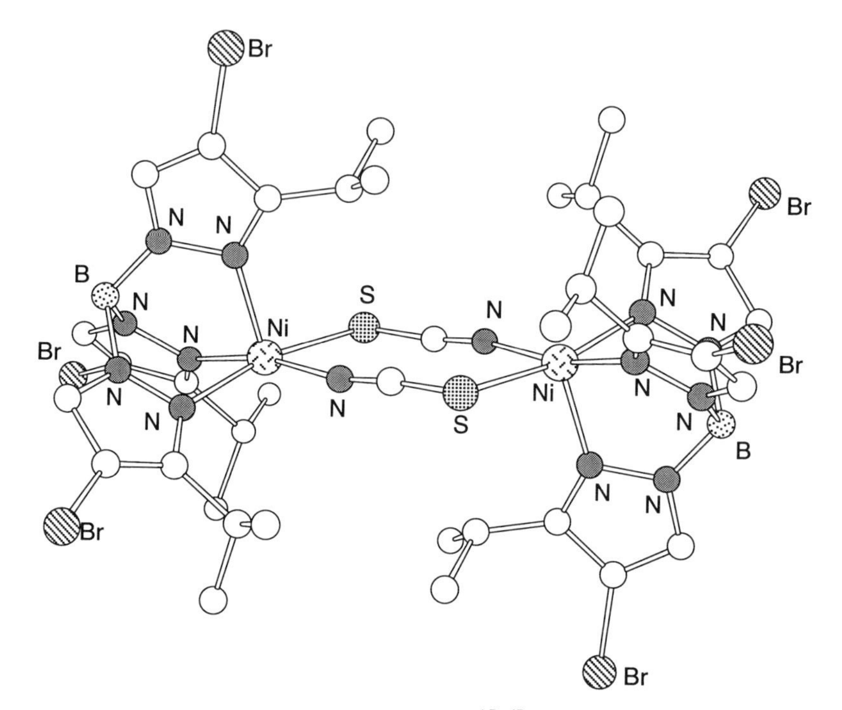

Figure 9. X-ray crystal structure of $[(Tp^{i\text{-Pr},4Br}Ni)_2(\mu\text{-NCS})_2]$ (42).

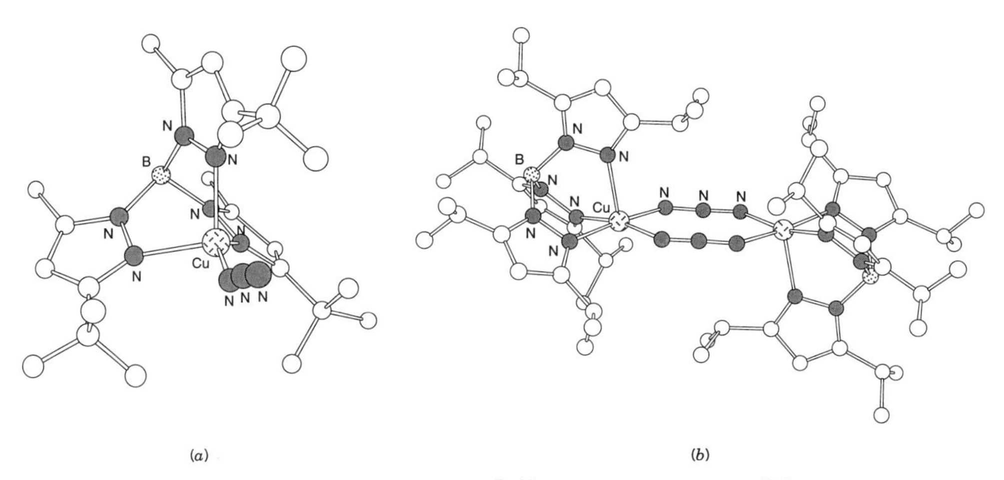

Figure 10. X-ray crystal structures of (*a*) $Tp^{t\text{-Bu},Me}Cu(N_3)$ (48) and (*b*) $[(Tp^{i\text{-Pr}2}Cu)_2(N_3)_2]$ (48).

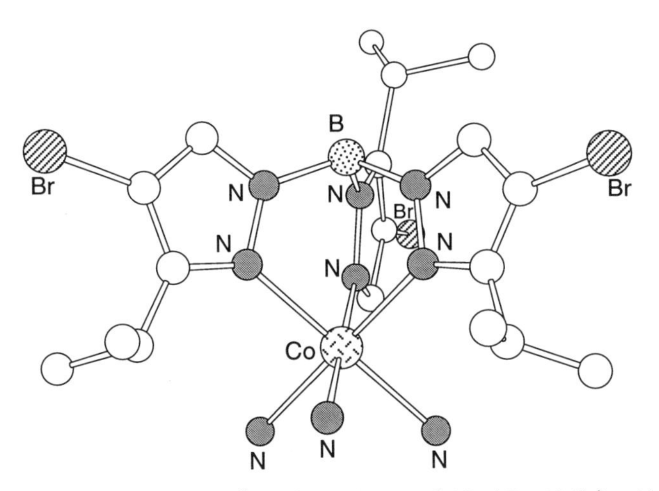

Figure 11. X-ray crystal structure of {[HB(3-*i*-Pr-4-Brpz)$_2$(5-*i*-Pr-4-Brpz)]$_2$Co}, with one ligand hidden for clarity (17).

has been placed in the same category as the 3-isopropyl-substituted ligands because five- and six-coordinate complexes (e.g., $Tp^{Menth}TiCl_3$) are accessible (21). Nonetheless, the observation of ready borotropic isomerization of $Tp^{Menth}TiCl_3$ to $Tp^{Menth*}TiCl_3$ (Fig. 12) illustrates the destabilizing effects of the large appendages of the Tp^{Menth} ligand (21).

A particularly interesting manifestation of the intermediate steric properties inherent to ligands containing 3-isopropyl groups appeared during attempts to isolate $Tp^{i\text{-Pr, Me}}CoI$ from crude mixtures containing other regioisomeric species with 5-isopropyl-3-methylpyrazolyl units (20). The desired symmetric isomer was separated pure by an unusual process that involved cooling of the crude reaction mixture to dissolve it in MeCN followed by *warming* to induce selective crystallization. The success of this procedure, termed "inverse recrystallization," was suggested to be due to the presence of equilibria between tetrahedral $Tp^{RR'}CoI$ and octahedral, solvated $[Tp^{RR'}Co(MeCN)_3]I$ (Fig. 13) that are characterized by a large negative ΔS [$-90(3)$ eu for the $Tp^{i\text{-Pr, Me}}$ complex] (20). While all of the regiosiomeric complexes participate in this equilibria, the affinity of the most hindered isomer, $Tp^{i\text{-Pr, Me}}CoI$, for MeCN is less than that of the others, so it more readily reverts to its insoluble, tetrahedral, unsolvated form upon warming. The fact that the 3-isopropyl groups in $Tp^{i\text{-Pr, Me}}$ allow binding of three acetonitrile molecules to its Co^{II} complex is indicative of the lesser size of these groups compared to *tert*-butyl substituents.

Finally, the special effects of the additional isopropyl group in the 5 position in $Tp^{i\text{-Pr}2}$ deserve noting. In addition to imparting greater solubility and crystallinity to complexes of $Tp^{i\text{-Pr}2}$, the 5-isopropyl group protects the B—N bonds

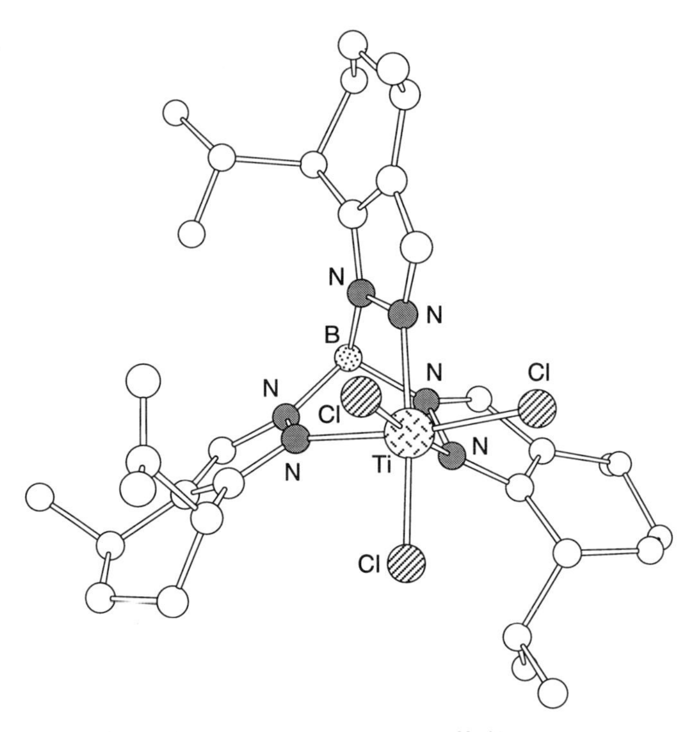

Figure 12. X-ray crystal structure of $Tp^{Menth*}TiCl_3$ (21).

from deleterious cleavage reactions and renders potentially complicating borotropic rearrangements degenerate. As discussed in more depth below, particular advantage has been taken of these properties for the isolation of a large number of important structural and functional models of metalloprotein active sites (52).

The flat shape and orientational flexibility of the aryl R^3 groups in Tp^{Ph}, Tp^{Ph2}, Tp^{Tol}, and Tp^{pAn} more readily allows for the formation of metal complexes with coordination numbers 5 or 6 compared to those in the category of

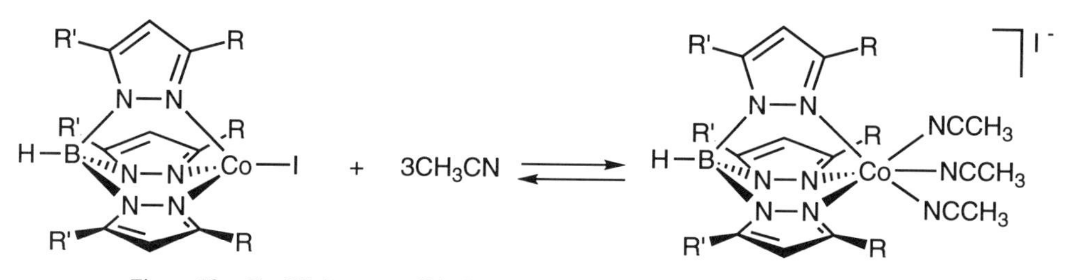

Figure 13. Equilibria responsible for ''inverse recrystallization'' phenomena (20).

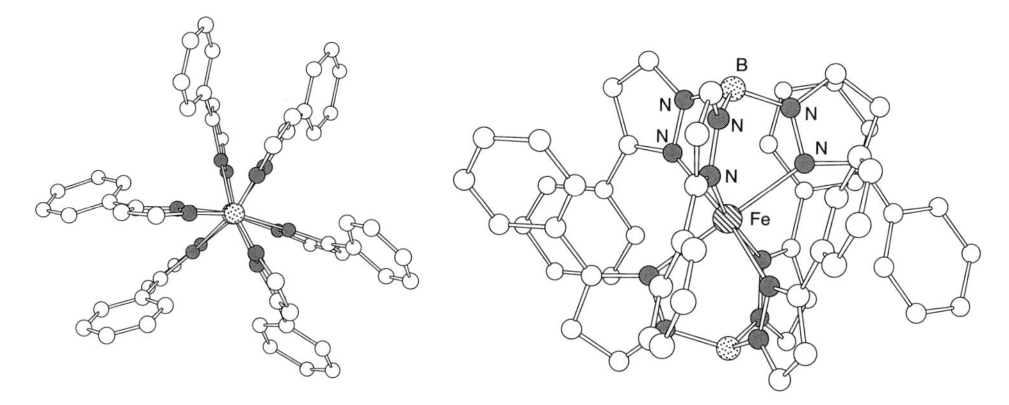

Figure 14. Two views of the X-ray crystal structure of $Tp_2^{Ph}Fe$ (16).

$Tp^{i\text{-}Pr}$ and Tp^{Np}. In addition to the presence of a larger substrate binding cavity in the aryl-substituted Tp scorpionates, the aromatic rings can interweave and π stack in species with multiple ligands. For example, consider the structure of $Tp_2^{Ph}Fe$ in Fig. 14, in which octahedral coordination is accessed by stacking of the phenyl groups from one ligand with those of the other (53). Despite the close approach of the two ligands that is facilitated by the phenyl groups' "intercalation," however, the large steric demand of the Tp^{Ph} chelates in this and an analogous Mn^{II} complex results in longer M—N bond lengths than in their respective $Tp_2^{Me2}M$ and Tp_2M congeners (Table III). Differences in physical properties can be traced to these bond length variations, the $E_{1/2}$ of +0.86 V versus SCE (CH_2Cl_2) for the reversible oxidation of $Tp_2^{Ph}Fe$ being particularly noteworthy. This large positive value approximately 0.6 V greater than that determined for Tp_2Fe has been suggested to be a consequence of the greater metal–ligand distances in the phenyl-substituted case that destabilize the Fe^{III} state (53). The electron withdrawing capability of the phenyl rings would also

TABLE III
Comparison of Structural Properties of $Tp_2^{RR'}M$ (M = Fe or Mn)[a]

Complex	$M—N_{pz}$ Distance (Å)
Tp_2Fe	1.973
$Tp_2^{Me2}Fe$	2.172
$Tp_2^{Ph}Fe$	2.246
$Tp_2^{Me2}Mn$	2.275
$Tp_2^{Ph2}Mn$	2.32

[a]All data from (53).

be expected to contribute to the positive potential shift (see discussion of electronic effects below).

Modification of the aryl rings in order to discourage "intercalation" has been reported, a primary objective being to inhibit Tp_2M formation while at the same time providing better access to external substrates than the more hindered $Tp^{t\text{-Bu}}$ and $Tp^{i\text{-Pr}}$ ligands. In one approach, each aryl group was designed so that it would be forced to adopt a conformation with its molecular plane oriented almost orthogonal to that of its connected pyrazolyl ring. This was achieved by adding *ortho*-methyl groups [Tp^{Ms}; cf. $Tp^{Ms}Mo(CO)_2$(methallyl) in Fig. 15] (25) or by annulating additional rings to the phenyls (Tp^{Ant}; Entry 24, Table I) (54). Note the extremely small wedge angle of 7.0° for Tp^{Ms} (Table II). Formation of Tp_2M complexes has also been anticipated to be prevented by using a ligand having *p-tert*-butylphenyl substituents ($Tp^{p\text{-BuPh2}}$) (55).

In complementary modifications of the aryl rings designed to decrease their orientational flexibility but *increase* the wedge angle, ligands having alkyl tethers between the pyrazole and phenyl rings were prepared (Tp^{a}, $Tp^{a,Me}$, and Tp^{b}; Table I) (40). The effective size of these ligands as shown by structural and reactivity studies is inversely correlated with their wedge angles (Table II),

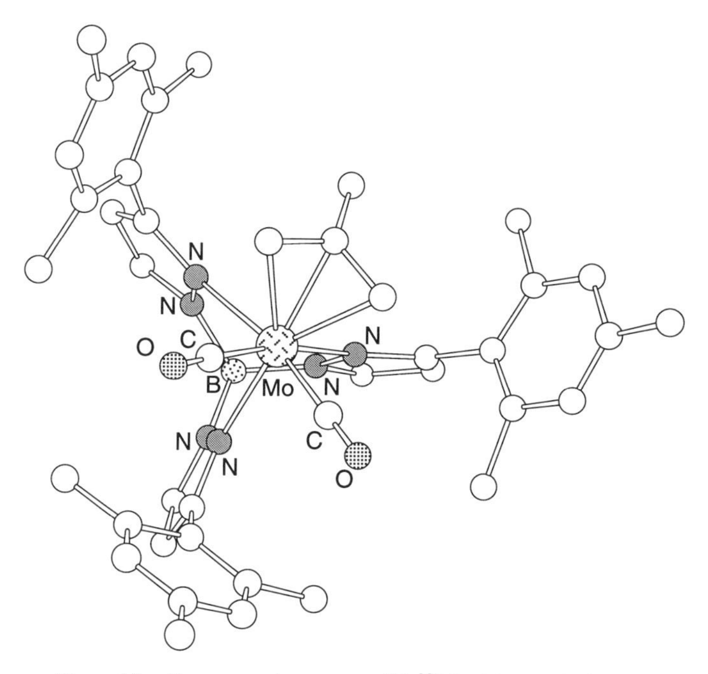

Figure 15. X-ray crystal structure of $Tp^{Ms}Mo(CO)_2$(methallyl) (25).

which are in turn controlled by the length of the alkyl tether. Thus, based on reactivity studies, Tp^{a} and $Tp^{a,Me}$ appear to be less hindered than $Tp^{Ph,Me}$, as reflected by their slightly larger wedge angles (44° vs. 32°). These wedge angles are significantly less than that of Tp^{b}, however, due to a "zigzag" orientation of the ethylene tethers in Tp^{a} and $Tp^{a,Me}$ that induces an approximate 21° deviation of the pyrazole and aryl rings from coplanarity. In Tp^{b}, the shorter methylene tether enforces pyrazole–aryl coplanarity to yield a large wedge angle of 82° and a smaller cone angle of 242°, thus resulting in a drastically decreased effective size akin to Tp^{Tn} and Tp (see below). More recently, coplanarity of the pyrazole and phenyl rings was enforced without shortening the tether, by changing the ethylene linker in Tp^{a} to ethyne (41). The resulting ligand Tp^{a*} based on 2*H*-benz[G]indazole thus has large wedge *and* cone angles of 87° and 276°, respectively (41).

The position of the next ligand on the size scale shown in Eq. 1, $Tp^{CF_3R'}$, must be viewed as tentative due to difficulties inherent in attempts to separate steric from electronic effects in the observed physical properties of complexes of this highly electron-withdrawing ligand. The interplay of these effects as revealed by the reactivity and spectroscopic properties of rhodium complexes will be discussed in more detail below; suffice it to say at this juncture that while there appears to be little doubt that a 3-CF_3 substituent exerts greater steric demands than a 3-CH_3 group, its effective size compared to 3-aryl and 3-isopropyl groups is not as clear.

The remaining ligands, Tp^{Me2}, Tp^{Tn}, Tp^{b}, and Tp, are characterized by a much greater tendency to form octahedral Tp_2M complexes with divalent metal ions than those listed higher on the size scale. The Tp^{Tn} and Tp^{b} ligands are of particular interest in the current context because they have large cone angles greater than that of Tp^{Ph} (Table II), yet both exhibit significantly decreased steric demand compared to Tp^{Ph} and actually are more similar to Tp than to Tp^{Me2} (39, 40). For example, only when M = Zn^{II} was it possible to prepare tetrahedral $Tp^{Tn}MX$ complexes from MX_2 starting materials, the rapidly formed and thermodynamically favored octahedral $Tp_2^{Tn}M$ compounds being isolated instead. In contrast, tetrahedral species are accessible from Tp^{Me2} and Tp^{Ph}. Additional evidence for the smaller effective size of Tp^{Tn} compared to Tp^{Ph} has come from comparison of $Tp^{RR'}$ reactivity with $Tp^{i\text{-Pr},4Br}CoCl$ (39, 56). Treatment of this complex with ligands having R^3 = H, Me, or thienyl produced the octahedral mixed-ligand $Tp^{i\text{-Pr},4Br}Tp^{RR'}Co$ complexes at rates paralleling the respective steric demands of the R^3 substituents (H ≈ thienyl > Me). With Tp^{Ph}, however, unfavorable steric interactions allowed only two 3-phenylpyrazolyl rings to be bonded to the Co^{II} ion and forced the third to dangle free, with the last coordination position on Co^{II} filled by an agostic B—H bond (Fig. 16). Finally, the small difference in average Co—N bond lengths in Tp_2Co and $Tp_2^{Tn}Co$ of 0.025 Å has been cited as evidence for the similar steric demand of

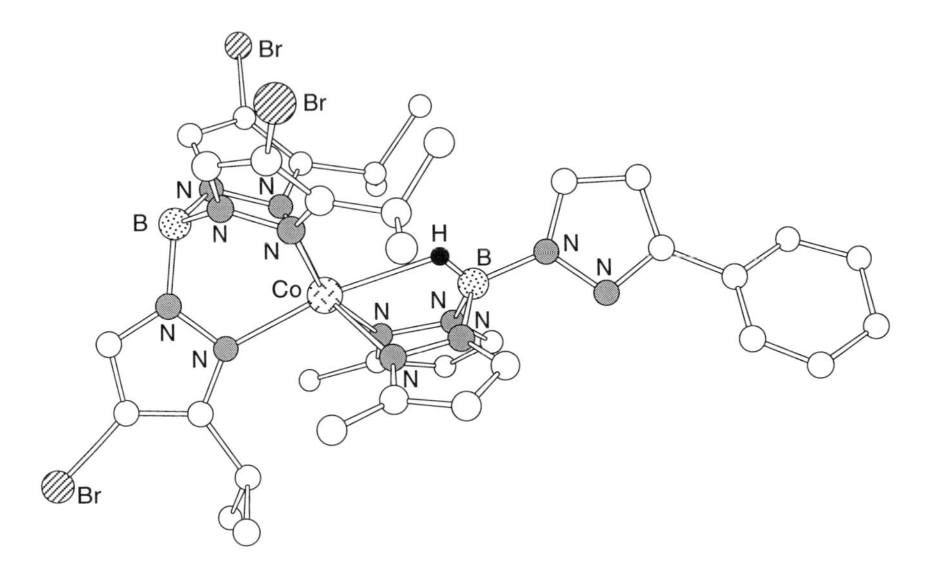

Figure 16. X-ray crystal structure of $Tp^{i\text{-}Pr,4Br}Tp^{Ph}Co$, with only ipso carbon atoms shown on two of the Tp^{Ph} pyrazolyl rings for clarity (39).

the two Tp ligands (39). Much larger differences in the Fe—N bond distances in Tp_2Fe, $Tp_2^{Me2}Fe$, and $Tp_2^{Ph}Fe$ (Table III) support this argument and attest to the more substantive steric differences between the 3-H, 3-Me, and 3-Ph substituents.

Other comparisons of the structures, physical properties, and reactivity of sets of similar compounds with homologous ligands provide additional insight into the steric demands of the variously substituted Tp ligands. For example, it is interesting to compare the X-ray crystal structures that have been determined for the series of isostructural Tl^I complexes (Figs. 17 and 18) (33, 40, 50, 55–58). The average Tl—N bond lengths for the complexes fall within the relatively narrow range 2.50–2.61 Å, values that have been suggested (55) to indicate that the bonding between Tl and the N donor atoms of the Tp ligands are composed of a combination of normal covalent and dative covalent interactions because they are greater than the sum (2.31 Å) of the covalent radii of Tl (1.57 Å) and N (0.74 Å) (59). All of the complexes consist of discrete monomers, except for $Tp^{tol}Tl$, which crystallizes with two independent molecules in the unit cell with their tolyl rings interwoven so as to allow a relatively close [3.8636(4) Å] Tl· · ·Tl contact (Fig. 17) (58). This result is yet another manifestation of the tendency of the simple 3-aryl-substituted ligands to adopt conformations that allow more facile approach and/or "intercalation" of other molecules. The observation of noninteracting monomers in the crystals of the Tl^I

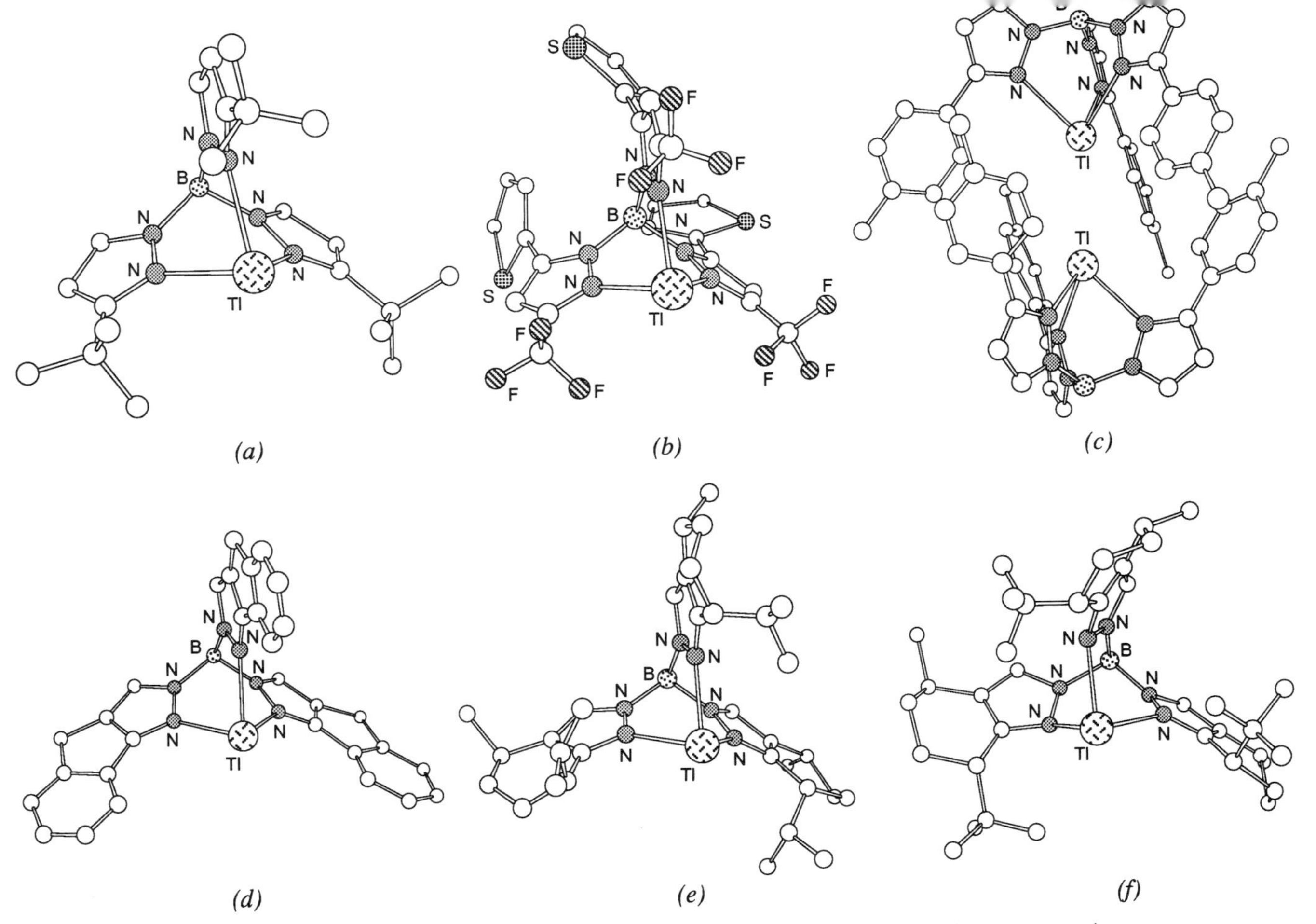

Figure 17. X-ray crystal structures of (*a*) $Tp^{t\text{-}Bu}Tl$ (56), (*b*) $Tp^{CF_3,Tn}Tl$ (49), (*c*) $Tp^{pTol}Tl$ (58), (*d*) $Tp^{b}Tl$ (40), (*e*) $Tp^{Menth}Tl$ (33), and (*f*) $Tp^{Mementh}Tl$ (33).

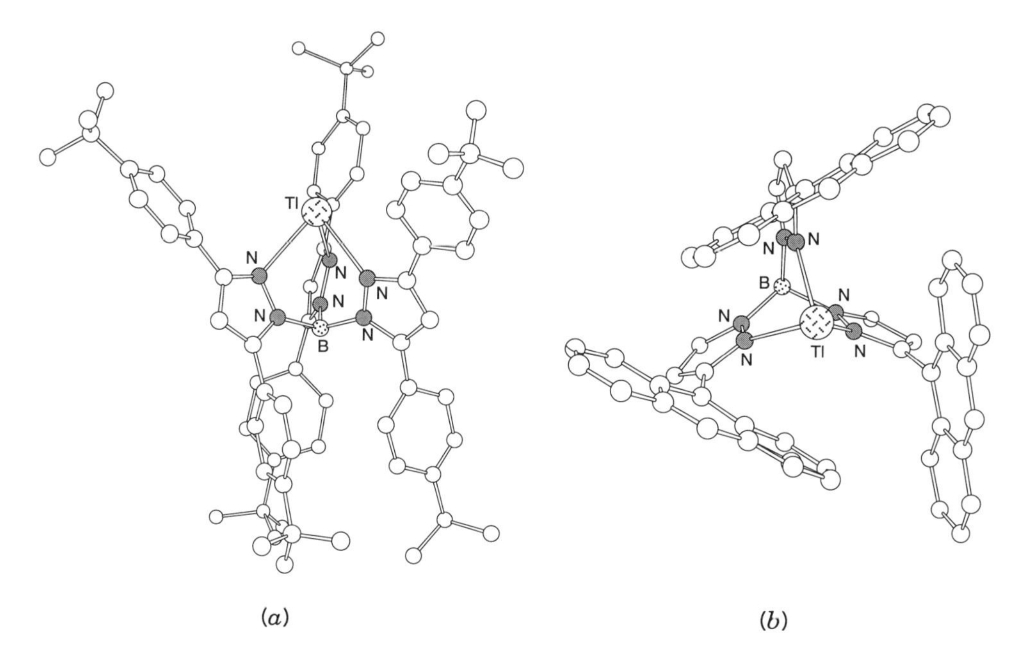

Figure 18. X-ray crystal structures of (*a*) $Tp^{p\text{-BuPh2}}Tl$ (55) and (*b*) $Tp^{Ant}Tl$ (54).

complexes of Tp^{Ant} (54) and $Tp^{p\text{-BuPh2}}$ (55) demonstrate how modifications of the aryl rings can either close the wedge angle or simply prevent aryl ring intercalation via nonbonded steric forces, respectively (Fig. 18). Finally, in $TlTp^{Mementh}$ [Fig. 17(*d*)] and $Tp^{p\text{-BuPh2}}$ [Fig. 18(*a*)] a significant "propellor-like" distortion is evident from the X-ray crystal structures that appears to arise from unusual steric demands of these ligands. This distortion is not evident in $Tp^{Mementh}ZnCl$ (21), however, which suggests that the larger size of the Tl^I ion is a contributing factor. An implication of the structural distortion is that there is a significant degree of ionic character in the bonding of $Tp^{Mementh}$ and $Tp^{p\text{-BuPh2}}$ to Tl^I, since the N_{pz} lone-pair orbitals are not directed along the Tl-N_{pz} bond vector.

Another set of complexes with analogous ligands with which to compare substituent steric effects are the dimers $[Tp^{RR'}Cu]_2$ ($Tp^{RR'} = Tp^{t\text{-Bu}}$, $Tp^{t\text{-Bu,Me}}$, Tp^{Ph2}, Tp^{Me2}, and Tp; Fig. 19) that have been found to be useful starting materials for bioinorganic modeling studies (49, 60, 61). The coordination numbers, geometries, and Cu—N distances within the set smoothly vary with the size of the ligand substituents. Thus, in $[Tp^{t\text{-Bu}}Cu]_2$ each Cu^I ion is two coordinate with a linear geometry (N—Cu—N angle = 177°) (60). The average Cu—N distance is 1.88 Å, with the $Tp^{t\text{-Bu}}$ ligand adopting an η^2 mode in order to reduce unfavorable nonbonded interactions involving the *tert*-butyl groups.

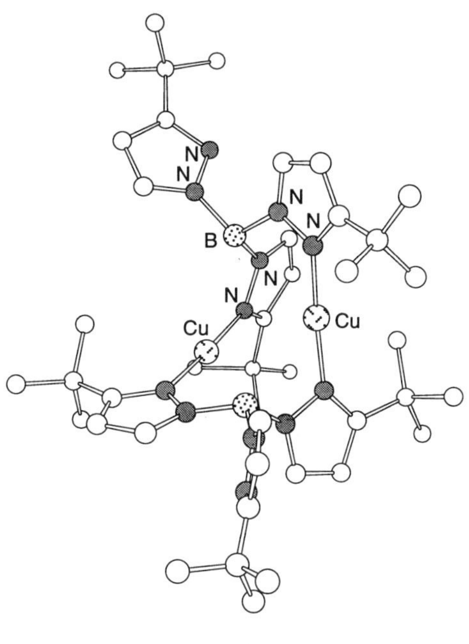

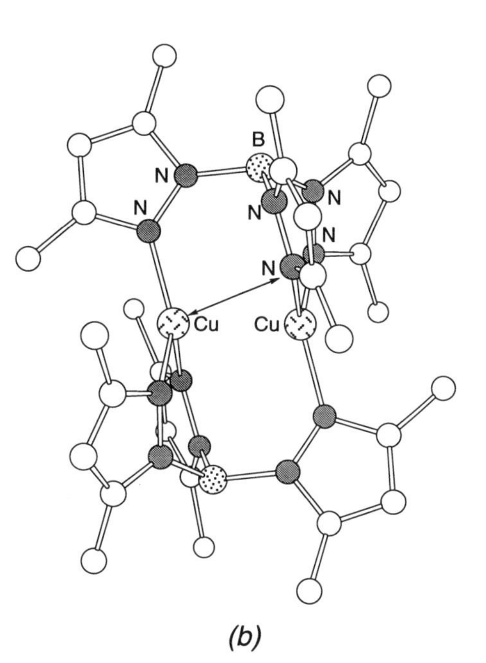

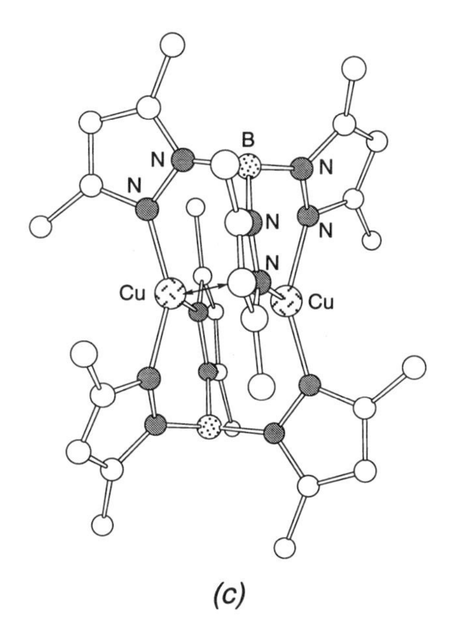

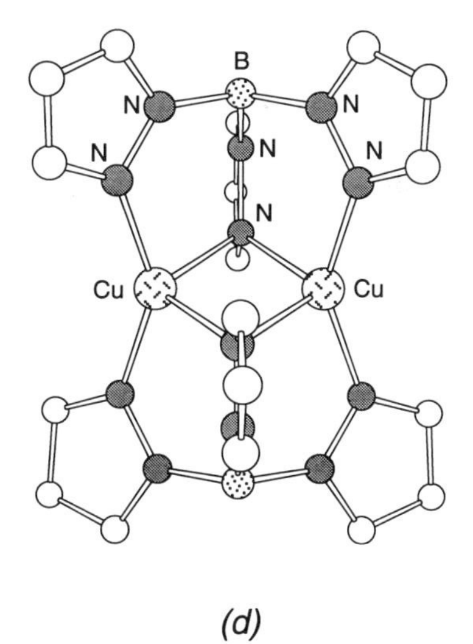

Figure 19. X-ray crystal structures of (*a*) $[Tp^{t\text{-Bu}}Cu]_2$ (60), (*b*) $[Tp^{Ph2}Cu]_2$ (all carbon atoms except ipso carbon atoms from pheny rings deleted for clarity) (60), (*c*) $[Tp^{Me2}Cu]_2$ (61), and (*d*) $[TpCu]_2$ (61). Arrows in (*b*) and (*c*) denote $Cu\cdots N_{pz}$ distances cited in text.

A similar structure is adopted by $[Tp^{t\text{-Bu,Me}}Cu]_2$ (49). All three pyrazolyl rings of each of the ligands are bonded to Cu^I in $[Tp^{Ph2}Cu]_2$, which contains three-coordinate Cu^I ions with distorted trigonal planar geometries and an elongated average Cu—N bond length of 1.97 Å (60). Again, interweaving of the phenyl substituents facilitates attainment of a higher metal coordination number. The structure of $[Tp^{Me2}Cu]_2$ is closely analogous to that of $[Tp^{Ph2}Cu]_2$, although the decreased steric influences of methyl groups in the former cause its core to be substantially distorted compared to the latter (61). A shorter Cu· · ·Cu distance [2.507(1) Å vs. 2.544(2) Å] and a "semibridging" interaction with one of the pyrazolyl rings indicated by a Cu—N distance (indicated by arrows in Fig. 19) of 2.777(4) Å (vs. 3.125 Å) comprise the major attributes of $[Tp^{Me2}Cu]_2$ that differ from those of $[Tp^{Ph2}Cu]_2$. Finally, in $[TpCu]_2$ one pyrazolyl ring adopts a true bridging binding mode, each Cu^I ion being four-coordinate with an average $Cu—N_{pz}$ distance that is further lengthened to 2.04 Å (61). Taken together, the structures of the $Tp^{RR'}$ complexes of the spherically symmetric d^{10} Cu^I ion directly demonstrate how R^3 substituent steric differences can control metal complex geometry in the absence of ligand field effects.

As is typical for Cu^I complexes, the $[Tp^{RR'}Cu]_2$ complexes undergo rapid fluxional processes in solution that render the pyrazolyl ring resonances equivalent in ^1H NMR spectra. The behavior of the more hindered $[Tp^{t\text{-Bu}}Cu]_2$ is distinct from that of the others with respect to both the apparent mechanism and rates of pyrazolyl group equilibration. Thus, solution molecular weight measurements and low-temperature ^1H NMR experiments indicated that dimer cleavage of $[Tp^{Me2}Cu]_2$ and $[TpCu]_2$ was rapid (61). Decomposition of $[Tp^{Ph2}Cu]_2$ to presumed monomeric species also occurred, but more slowly (60). In contrast, similarly obtained data for $[Tp^{t\text{-Bu}}Cu]_2$ showed that it exists predominantly as a dimer in solution and allowed identification by line shape analysis of a pyrazolyl exchange process attributed to intramolecular rotation of the $Tp^{t\text{-Bu}}$ ligands about the dicopper(I) core [$\Delta H^{\ddagger} = 11.7(5)$ kcal mol^{-1} and $\Delta S^{\ddagger} = -9(2)$ eu; Fig. 20] (60). Although the detailed mechanism of this fluxional process is not known, and the possible intermediacy of a trace amount of a mononuclear species cannot be ruled out, a route involving movement of a pyrazolyl ring from a terminal to a bridging and then to another terminal site on the other Cu^I ion is easily envisioned. It is interesting to view the crystal structures of the $[Tp^{RR'}Cu]_2$ complexes as representatives of intermediate structures that lie along the potential energy surface of this pyrazolyl ring exchange pathway (62). Thus, the differing steric demands of the R^3 groups in the dimers result in molecular "snapshots" in which terminal, semibridging, and bridging pyrazolyl binding modes mimic plausible intermediate geometries in a dynamic fluxional process.

A third illustration of ligand steric effects in a homologous set of complexes concerns $Tp^{RR'}Zn(NO_3)$ (R = R′ = Ph or H; R = *t*-Bu or R′ = H) (63–65).

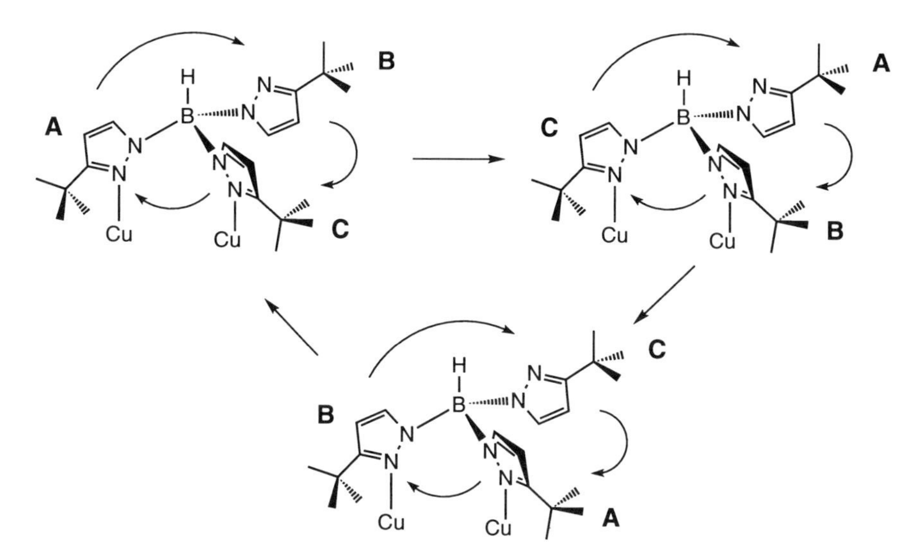

Figure 20. Intramolecular fluxional process postulated to occur in solution for $[Tp^{t\text{-Bu}}Cu]_2$ (only one $Tp^{t\text{-Bu}}$ ligand shown for clarity) (60).

Structural studies of these compounds were undertaken as part of an effort to unravel the tendencies of divalent metal ions to bind to nitrate and and/or bicarbonate ions in unidendate versus bidentate modes, ultimately in order to shed light on the nature of substrate adducts to the active sites of various metal-substituted forms of carbonic anhydrase (see Section IV.D.1) (43, 66–68). As shown by the X-ray crystal structural results for the set of nitrate complexes listed in Fig. 21, the nitrate binding mode shifts from essentially unidentate (Δ Zn—O = 0.60 Å) in the most hindered case ($Tp^{t\text{-Bu}}$) to anisobidentate (Δ Zn—O = 0.42 Å) for Tp. In agreement with the steric order of Eq. 1, the bonding in $Tp^{Ph2}Zn(NO_3)$ lies intermediate between these two extremes (Δ Zn—O = 0.53 Å). Since the $Tp^{RR'}$—Zn interactions (e.g., N_{pz}—Zn bond lengths and N_{pz}—Zn—N_{pz} angles) are similar among all three compounds, these differences in nitrate binding geometries can be viewed as a direct reflection of the divergent steric influences of the $Tp^{RR'}$ ligand substituents.

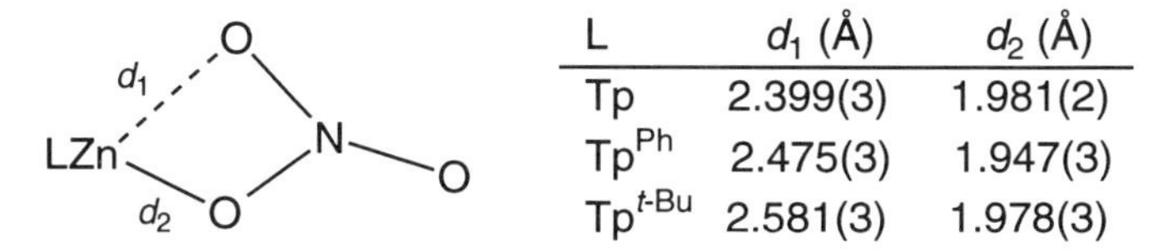

L	d_1 (Å)	d_2 (Å)
Tp	2.399(3)	1.981(2)
Tp^{Ph}	2.475(3)	1.947(3)
$Tp^{t\text{-Bu}}$	2.581(3)	1.978(3)

Figure 21. Comparison of X-ray structural data for $Tp^RZn(NO_3)$ (R = H, Ph, or *t*-Bu) (63–65).

2. *Electronic Effects*

Experimental data with which one might evaluate the relative electron-donating or -withdrawing capabilities of Tp ligands having pyrazolyl substituents other than H or Me are relatively scarce. However, insight into the probable stereoelectronic effects of larger substituents can be gained by considering the growing amount of data being acquired for Tp and Tp^{Me2}, much of which has been collected during efforts to compare these ligands with their formally isoelectronic analogues Cp and Cp*. A useful set of molecules to contemplate are the $LM(CO)_3$ radicals and anions (L = Tp, Tp^{Me2}, Cp, or Cp*; M = Cr, Mo, or W). The steric differences between the Tp and Cp ligands are evinced by the relative stability of the $TpM(CO)_3\cdot$ versus the $CpM(CO)_3\cdot$ radicals toward dimerization, the former being isolable as discrete species (69–72) while the latter either dimerize at close to diffusion-controlled rates (M = Mo or W) or are only identifiable in solution (M = Cr) (73–75). Electronic effect differences have been inferred from IR ν_{CO} values and oxidation potentials for the carbonyl anions (Table IV) (72, 76, 77). As recently discussed by Skagestad and Tilset (76), the lower oxidation potentials (by greater than 0.1 V) and the 15–20-cm^{-1} lower energy of the *E* symmetry ν_{CO} bands for the Tp versus the Cp complexes suggest that the Tp ligand is the stronger electron donor [but see (78) for conflicting ν_{CO} data for some Tp and Cp ruthenium complexes]. Importantly for the purposes of this discussion, substitution of Me for H in the pyrazolyl 3 and 5 positions results in further shifts of oxidation potentials and ν_{CO} bands of the anions to lower values (Table IV). These data, in combination with redox potential measurements acquired for other complexes, including $CpRuTp^{RR'}$ ($Tp^{RR'}$ = Tp and Tp^{Me2}) (79) and the set Cp_2Fe, Cp_2^*Fe, and Tp_2Fe (80), support the following trend in ligand electron-releasing capability: Tp^{Me2} > Tp ~ Cp* > Cp. This trend has been further corroborated by extended Hückel molecular orbital (EHMO) calculations for the LMo^- (L = Cp or Tp)

TABLE IV
Infrared ν_{CO} (cm^{-1}) and Oxidation Potentials (V vs. Ferrocene/Ferrocenium Couple) for $LM(CO)_3^-$ (L = Tp, Tp^{Me2}, or Cp; M = Cr, Mo, or W)[a]

M =	Cr	Mo	W
ν_{CO} (Tp)	1889, 1751	1888, 1752	1878, 1744
ν_{CO} (Tp^{Me2})	1881, 1742	1881, 1745	1872, 1733
ν_{CO}(Cp)	1888, 1768	1891, 1770	1886, 1765
E_{ox} (Tp)	−0.821	−0.521	−0.582
E_{ox} (Tp^{Me2})	−0.857	−0.583	−0.652
E_{ox} (Cp)	−0.688	−0.403	−0.397

[a]Table adapted from (76). All data measured in MeCN for Et_4N^+ salts. See (76) for other experimental details.

fragments (81), which have also been interpreted to indicate that Tp is a better π donor than Cp and that the σ donor orbitals of Tp are involved in π interactions with the metal center.

It is relevant to note that the thermodynamic acidities of $LM(CO)_3H$ species do *not* follow the trend predicted by the relative electron-donor capabilities of L for L = Tp^{Me2}, Tp, and Cp (76). Thus, the acidity order for a given M is $Tp^{Me2}M(CO)_3H > TpM(CO)_3H > CpM(CO)_3H$, as illustrated by the pK_a values in MeCN for M = Mo of 9.7, 10.7, and 13.9, respectively. There is a parallel trend in the metal-hydride bond dissociation energies, exemplified by the respective values for M = Mo of 248, 260, and 290 kJ mol^{-1}. Steric influences ordered as $Tp^{Me2} > Tp > Cp$ that disfavor the attainment of crowded seven-coordinate geometries in the hydride complexes have been suggested to be responsible for both trends (76). Another rationale for the predominance of six rather than seven coordination in $Tp^{RR'}M$ versus CpM complexes, which is based on EHMO calculations is that the more "localized" bonding of the Tp—N donors results in more efficient polarization of metal *d* orbitals into an octahedral geometry (81). Taken together, these data illustrate how the interplay of steric and electronic effects must be considered in evaluating variation of properties among related complexes and underscore the potential difficulties associated with identification of relative electronic effects between differently substituted ligands.

Insight into the relative electron-donating–electron-releasing properties of the more hindered Tp ligands can be gained by comparing ν_{CO} data listed in Table V for sets of like compounds that differ only in their pyrazolyl ring substituents. Higher values for the carbonyl stretching frequencies reflect less electron density at the metal center and decreased electron donation (or, conversely, greater electron withdrawal) by the $Tp^{RR'}$ ligand for compounds within each set. Although caution must be exercised in interpreting such limited data in the absence of corroborative experimental and/or theoretical information, one can distinguish trends that support the order of electron-donating ability of $Tp^{RR'}$ ligands shown in Eq. 2.

$$Tp^{R2}\ (R = \text{alkyl}) > Tp^{t\text{-Bu}} \approx Tp^{i\text{-Pr}} \approx Tp^{Me} \approx Tp^{Ms} > Tp$$
$$> Tp^{Ph2} \approx Tp^{Ph} > Tp^{CF_3,Tn} > Tp^{i\text{-Pr},4Br} > Tp^{CF_3,Me} \quad (2)$$

This order is generally that which is expected on the basis of the extensive knowledge of electronic effects of substituents in organic chemistry. Note that electron donation increases with the number of pyrazolyl alkyl substituents (Table V, Entries 1–6 and 16 and 17) and that the effect of a 4-Br group is substantial, causing $Tp^{i\text{-Pr},4Br}$ to have similar electron-withdrawing ability as Tp^{Ph} (Entries 12, 14, 22, and 23). As discussed in more detail below, the

TABLE V

IR Data for $Tp^{RR'}$ Metal Carbonyl Complexes

Entry	Compound	ν_{CO} (cm^{-1})	Reference
1	$Tp^{t\text{-Bu},i\text{-Pr}}CuCO$	2050	48
2	$Tp^{i\text{-Pr}2}CuCO$	2056	167
3	$Tp^{t\text{-Bu},Me}CuCO$	2059	48
4	$Tp^{Me2}CuCO$	2066	61
5	$Tp^{t\text{-Bu}}CuCO$	2069	188
6	TpCuCO	2083	*g*
7	$Tp^{Ph2}CuCO$	2086	168
8	η^2-$Tp^{t\text{-Bu}}Rh(CO)_2$	2083, 2017[a]	19
9	η^2-$Tp^{i\text{-Pr}}Rh(CO)_2$	2082, 2017[a]	19
10	η^2-$Tp^{b}Rh(CO)_2$	2084, 2020	40
11	η^2-$Tp^{Me}Rh(CO)_2$	2087, 2019[b]	24
12	η^2-$Tp^{Ph}Rh(CO)_2$	2088, 2026[a]	19
13	η^2-$Tp^{CF_3,Tn}Rh(CO)_2$	2080, 2005	19
14	η^2-$Tp^{i\text{-Pr},4Br}Rh(CO)_2$	2089, 2026[b]	24
15	η^2-$Tp^{CF_3,Me}Rh(CO)_2$	2103, 2040[a]	19
16	η^3-$Tp^{Me2}Rh(CO)_2$	2054, 1981[a]	124
17	η^3-$Tp^{i\text{-Pr}}Rh(CO)_2$	2058, 1987[a]	19
18	η^3-$Tp^{b}Rh(CO)_2$	2056, 1988	40
19	η^3-$Tp^{Me}Rh(CO)_2$	2060, 1988[b]	24
20	η^3-$Tp^{Ms}Rh(CO)_2$	2062, 1986[a]	25
21	η^3-$Tp^{Ms*}Rh(CO)_2$	2062, 1965[a]	25
22	η^3-$Tp^{i\text{-Pr},4Br}Rh(CO)_2$	2077, 2026[b]	24
23	η^3-$Tp^{Ph}Rh(CO)_2$	2079, 2015[a]	19
24	η^3-$Tp^{CF_3,Me}Rh(CO)_2$	2091, 2026[a]	19
25	$[NEt_4][TpW(CO)_3]$	1883, 1750[c]	47
		1878, 1744[c]	76
26	$[NEt_4][Tp^{Me2}W(CO)_3]$	1877, 1741[c]	47
		1872, 1733[c]	76
27	$[NEt_4][Tp^{Ph}W(CO)_3]$	1887, 1756[c]	47
28	$[NEt_4][Tp^{i\text{-Pr}}W(CO)_3]$	1882, 1746[c]	47
29	$[NEt_4][Tp^{t\text{-Bu}}W(CO)_3]$	1903, 1770[c]	47
30	$Tp^{t\text{-Bu}}Mo(CO)_2NO$	2000, 1908, 1885[d]	9
31	$Tp^{i\text{-Pr}}Mo(CO)_2NO$	2005, 1910[d]	9
32	$Tp^{i\text{-Pr},4Br}Mo(CO)_2NO$	2025, 1920, 1870[d]	17
33	$Tp^{Ph}Mo(CO)_2NO$	2000, 1910[d]	9
34	$Tp^{i\text{-Pr},Me}Mo(CO)_2NO$[e]	1990, 1900[d]	*h*
35	$Tp^{Me2}Mo(CO)_2NO$	2006, 1906[f]	*i*

[a]In hexanes or cyclohexane.

[b]In chloroform.

[c]In acetonitrile.

[d]Nujol mull; ν_{NO} not shown.

[e]Mixture of isomers.

[f]KBr pellet; ν_{NO} not shown.

[g]M. I. Bruce and A. P. P. Ostazewski, *J. Chem. Soc. Dalton Trans.*, 2433 (1973).

[h]M. Cano, J. V. Heras, S. Trofimenko, A. Monge, E. Gutierrez, C. J. Jones, and J. A. McCleverty, *J. Chem. Soc. Dalton Trans.*, 3577 (1990).

[i]S. J. Reynolds, C. F. Smith, C. J. Jones, and J. A. McCleverty, *Inorg. Synth.*, *23*, 4 (1985).

$Tp^{RR'}Rh(CO)_2$ complexes listed (Entries 8–24) can exist as isomers with either η^2- or η^3-$Tp^{RR'}$ ligation. Carbonyl stretching frequencies for the η^2 isomers are generally higher by 10–25 cm^{-1} than those for the η^3 forms, reflecting the lesser electron density at the metal center in the four-coordinate (16 e^-) versus the five-coordinate (18 e^-) complexes. Although it was described as an η^2-Tp^{Ms} complex in the original literature (25), we prefer assignment of $Tp^{Ms}Rh(CO)_2$ as an η^3-Tp^{Ms} species based on its observed ν_{CO} bands about 20 cm^{-1} lower than the most electron rich of the other η^2 structures and within the range exhibited by the η^3 isomers. The reported NMR data for the complex are consistent with this assignment. Parallel trends in carbonyl stretching frequencies as a function of pyrazolyl ring substitution are exhibited by the $Tp^{RR'}CuCO$, η^2-$Tp^{RR'}Rh(CO)_2$, and η^3-$Tp^{RR'}Rh(CO)_2$ sets, but close similarities among the ν_{CO} values (as well as some inconsistencies among the data acquired by different research groups) (47, 76) hinders interpretation of the electronic effects of substituents for the $Tp^{RR'}W(CO)_3^-$ and $Tp^{RR'}Mo(CO)_2NO$ compounds (Entries 25–35) and demonstrates the limitations of this analytical method. The apparent poor sensitivity of the ν_{CO} bands to the nature of the pyrazolyl substituent(s) in these cases is difficult to explain, although in the case of the tungsten anions steric influences on ion pairing in solution is one possibility (82).

III. ORGANOMETALLIC CHEMISTRY

A. Hydride and Alkyl Complexes of Main Group Elements

A principal objective in modern organometallic chemistry has been to understand the composition, structures, and reactivity of hydride and alkyl complexes of main group (*s* and *p* block) elements. Such species, those involving Mg, Al, Ga, In, Cd, Be, and the related *d*-block element Zn being of particular relevance here, are critically important reagents for organic synthesis (83–86), catalysis of polymerizations (87), chemical vapor deposition of thin films (88), and/or the controlled preparation of materials (89). Despite their obvious practical importance, however, the majority of hydride and alkyl complexes of the main group metals are complex, oligomeric, and sometimes quite unstable species that often participate in kinetically facile and solvent dependent ligand redistribution equilibria that hamper studies of their structures and reaction mechanisms (90, 91). There has therefore been a surge of interest in using suitably designed coligands to increase their stability, inhibit their oligomerization, and control their reactivity to gain more detailed mechanistic insights and to perhaps uncover novel reactivity patterns [for examples of relevant chemistry using less hindered poly(pyrazolyl)borates see (92–98)]. The high degree of steric hindrance inherent to $Tp^{t\text{-}Bu}$ and $Tp^{t\text{-}Bu,Me}$ has been used to particular advantage toward these ends.

1. Alkyl Complexes of Mg, Zn, Al, and Cd

The well-recognized significance of organomagnesium compounds in organic synthesis has provided the impetus for numerous investigations of their gas-phase, solution, and solid state composition, structure, and reactivity (90, 91). In particular, studies of Grignard reagents have shown that they consist of a complex mixture of species in solution, with structural and reactivity differences that markedly depend on the alkyl group(s), solvent(s), and/or additive(s) (91). The tendency of Grignard reagents to associate into oligomeric structures in solution and to participate in exchange processes, such as the Schlenk equilibrium (Eq. 3), in combination with the complexity conferred by solvation of all involved species, greatly confounds efforts to relate structure to reactivity and to unravel reaction mechanisms. As succinctly stated by Han and Parkin (99), "an investigation of magnesium alkyl derivatives that are both solvent-free and soluble in noncoordinating hydrocarbon solvents would be particularly helpful in simplifying mechanistic interpretations." Sterically hindered Tp^{Me2} and $Tp^{t\text{-}Bu}$ were found to be useful for the preparation of monomeric derivatives of this sort, the particular steric properties of these ligands being instrumental for the isolation of species distinct from other magnesium alkyls complexed to multidentate chelates (90).

$$2RMgX \rightleftharpoons MgR_2 + MgX_2 \qquad (3)$$

Treatment of a wide range of dialkyl magnesium reagents R_2Mg (R = Me, Et, $CH_2(CH_2)_2Me$, $CH(Me)_2$, $C(Me)_3$, $CH_2Si(Me)_3$, $C{=}CH_2$, Ph) with $TlTp^{t\text{-}Bu}$ or KTp^{Me2} resulted in the formation of the monomeric alkyls $Tp^{t\text{-}Bu}MgR$ and $Tp^{Me2}MgR$ in good yield (99). The use of R_2Mg instead of the Grignard reagents RMgX (X = I, Br or Cl) in the synthesis is critical, as MgR metathesis to afford $Tp^{RR'}MgX$ derivatives competes with the desired MgX exchange in the Grignard reactions (Eq. 4) (99, 100). This competition is apparently kinetic in origin, following the metathesis order Mg—I > Mg—Me > Mg—X (X = Cl or Br) for the reaction with $TlTp^{t\text{-}Bu}$. The $Tp^{t\text{-}Bu}MgR$ and $Tp^{Me2}MgR$ products were isolated as white crystalline solids that were amenable to characterization by X-ray crystallography and, in solution, by NMR spectroscopy. The structure of $Tp^{t\text{-}Bu}MgMe$ shown in Fig. 22 is representative of the class and demonstrates the η^3 coordination of the $Tp^{RR'}$ ligand as well as the monomeric nature of the compounds.

$$RMgX + TlTp^{RR'} \longrightarrow Tp^{RR'}MgX + [TlR] \qquad (4)$$

The increased steric bulk of $Tp^{t\text{-}Bu}$ compared to Tp^{Me2} is clearly responsible for distinct differences in the synthesis, stability, and reactivity of their respective Mg alkyl derivatives (99, 101). Both sets of alkyls are stable with respect

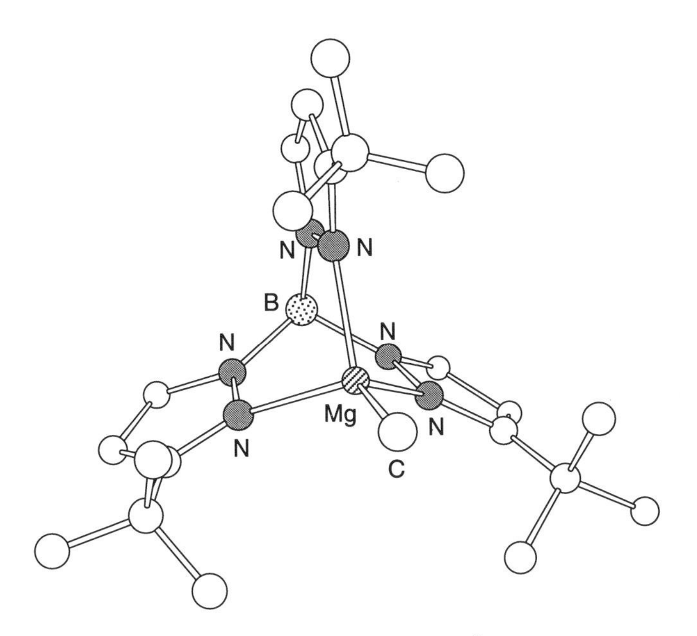

Figure 22. X-ray crystal structure of $Tp^{t\text{-}Bu}MgMe$ (99).

to inversion of configuration at the α carbon and to intermolecular alkyl exchange (cf. the lack of crossover products in the reaction depicted in Eq. 5), reactions known to occur with Grignard and dialkylmagnesium reagents [see literature listed in footnote 29 of (99)]. Differences are apparent, however, in the respective tendencies of the Tp^{Me2} and $Tp^{t\text{-}Bu}$ alkyls to undergo ligand redistribution reactions. Thus, whereas $Tp^{Me2}MgR$ complexes react in benzene (80–120°C) to form octahedral $Tp_2^{Me2}Mg$ and MgR_2 (Eq. 6), the $Tp^{t\text{-}Bu}MgR$ compounds are thermally stable, as typified by the lack of decomposition of a benzene solution of $Tp^{t\text{-}Bu}MgMe$ after 1 week at 120°C (101, 102). This increased stability is obviously a direct consequence of the greater steric demands of the $Tp^{t\text{-}Bu}$ ligand, which disfavors adoption of octahedral $Tp_2^{t\text{-}Bu}M$ structures.

$$Tp^{t\text{-}Bu}MgMe + Tp^{Me2}MgC(Me)_3 \not\longrightarrow Tp^{t\text{-}Bu}MgC(Me)_3 + Tp^{Me2}MgMe \quad (5)$$

$$Tp^{Me2}MgR \longrightarrow Tp_2^{Me2}Mg + MgR_2 \quad (6)$$

Additional aspects of the extensively studied reactivity of the $Tp^{t\text{-}Bu}MgR$ compounds that can be attributed to the steric features of the $Tp^{t\text{-}Bu}$ ligand have been reported. For example, $Tp^{t\text{-}Bu}MgR$ [R = Et, $CH(Me)_2$, $C(Me)_3$] react almost instantaneously with O_2 to quantitatively afford the alkylperoxo species $Tp^{t\text{-}Bu}MgOOR$ (101, 103). These novel products were identified by ^{17}O NMR and IR spectroscopic measurements on suitably isotopically labeled isomers, as well as by independent synthesis of $Tp^{t\text{-}Bu}MgOOC(Me)_3$ (with evolution of CH_4) from the metathesis reaction of $Tp^{t\text{-}Bu}MgMe$ with $(Me)_3COOH$. Evidence from trapping and crossover experiments implicated involvement of radicals in this O_2 insertion reaction mechanism. Radical species were also proposed as intermediates in the decomposition of $Tp^{t\text{-}Bu}MgCH_2Si(Me)_3$ by O_2, a reaction that took a different course due to the presence of oxophilic silicon (Eq. 7) (101, 104). The clean generation of alkylperoxo species in the reaction of O_2 with the simple magnesium alkyls is unusual, since the more commonly observed result in analogous reactions of metal alkyls is the isolation of alkoxo derivatives via bimolecular oxygen-atom abstraction from the initially formed alkylperoxo species by the starting metal alkyl (Eq. 8) (105). Such a bimolecular decomposition pathway is disfavored in this instance by the sterically encumbering $Tp^{t\text{-}Bu}$ ligand.

$$Tp^{t\text{-}Bu}MgCH_2Si(Me)_3 + O_2 \longrightarrow Tp^{t\text{-}Bu}MgOSi(Me)_3 + CH_2O \quad (7)$$

$$LMOOR + LMR \longrightarrow 2LMOR \quad (8)$$

Further unique reactivity features of the $Tp^{t\text{-}Bu}MgR$ complexes include (a) the formation of the magnesium enolate $Tp^{t\text{-}Bu}MgO(Me){=}CH_2$ instead of the alkoxide $Tp^{t\text{-}Bu}MgOC(Me)_3$ upon mixing of $Tp^{t\text{-}Bu}MgMe$ with acetone, and (b) the observation upon treatment of $Tp^{t\text{-}Bu}MgMe$ with $^{13}CH_3I$ of alkyl exchange (to afford $Tp^{t\text{-}Bu}Mg^{13}CH_3$ and CH_3I) in addition to the expected alkylation reaction (giving $^{13}CH_3CH_3$ and $Tp^{t\text{-}Bu}MgI$) (100). Both reactions, enolate formation with acetone (which is so susceptible to aldol condensations) and alkyl exchange [(7 and 8) in (101)], are uncommon for simple dialkylmagnesium and Grignard reagents (106–109) and also probably occur as a consequence of the specific sterically demanding coordination environment provided by the supporting $Tp^{t\text{-}Bu}$ chelate.

The synthesis and characterization of $Tp^{t\text{-}Bu}ZnR$ complexes was undertaken to directly compare the intrinsic reactivities of Mg—C and Zn—C bonds in a set of well-defined, monomeric, solvent free, and isostructural complexes (110). Major impetus for this investigation was provided by the importance of both types of alkyls in organic synthesis. As for the magnesium complexes discussed above, $Tp^{t\text{-}Bu}ZnR$ (R = Me and Et) were prepared by metathesis of the metal alkyl bonds of ZnR_2 with $TlTp^{t\text{-}Bu}$, the driving force presumably being the irreversible decomposition of TlR to Tl metal. The X-ray structure of $Tp^{t\text{-}Bu}ZnMe$ revealed it to be isomorphous with $Tp^{t\text{-}Bu}MgMe$, except the Zn—C bond length

[1.971(8) Å] is appreciably shorter than the Mg—C distance [2.182(8) Å] (110–112). These Zn alkyls were found to be remarkably resistant to oxidation by air, directly reflecting the stabilization provided by the sterically hindered $Tp^{t\text{-Bu}}$ ligand. Similar Zn alkyls were reported with Tp^{Ph} and Tp^{Tol} as supporting ligands (65). It is noteworthy in the current context that hydrolyses of some of these species led to $Tp^{R}_{2}Zn$ formation, again reflecting the steric trend of Eq. 1.

For the most part, the course of the reactions of the $Tp^{RR'}ZnR$ species with HX (X = halides, pseudohalides, —C≡CMe, $—O_2CMe$) were the same as those of their magnesium congeners, affording CH_4 and $Tp^{RR'}ZnX$. Important differences were uncovered, however, including (a) a faster rate for the reaction of $Tp^{t\text{-Bu}}MgEt$ with $PhCH_2I$ to give $Tp^{t\text{-Bu}}MgI$ and organic coupling products ($t_{1/2}$ = 0.23 h) than for the like reaction of $Tp^{t\text{-Bu}}ZnEt$ ($t_{1/2} = 2.3 \times 10^3$ h, both at 100°C), and (b) facile room temperature reaction of CO_2 with $Tp^{t\text{-Bu}}MgMe$ to afford $Tp^{t\text{-Bu}}Mg(\eta^1\text{-}O_2CMe)$ (100) versus no such insertion into the Zn—C bond of $Tp^{t\text{-Bu}}ZnMe$ at 140°C, despite the fact that the anticipated product $Tp^{t\text{-Bu}}Zn(\eta^1\text{-}O_2CMe)$ is a stable compound that could be synthesized independently (113). This data was cited as "good evidence for the intrinsic higher reactivity of the Mg—C versus the Zn—C bonds" (110).

Reactions of $Al(Me)_3$ with $Tp^{t\text{-Bu}}$ and Tp^{Me2} salts were also investigated in attempts to prepare monomeric species analogous to the Mg and Zn alkyls (114, 115). Current interest in gaining a detailed understanding of the organometallic chemistry of aluminum is derived by the need to control processes in which alkyl aluminums are used, such as catalysis of Ziegler–Natta alkene polymerization (87), chemical vapor deposition of thin films (88), and materials synthesis (89). Two types of $Tp^{RR'}Al(Me)_2$ complexes were synthesized, with their structures dependent on the steric demands of the $Tp^{RR'}$ used. Observation of a 2:1 ration of *tert*-butylpyrazolyl ring resonances and two inequivalent Al—Me peaks in its room temperature ^1H NMR spectrum indicated that with $Tp^{t\text{-Bu}}$ a four-coordinate complex **IX** with an η^2-$Tp^{t\text{-Bu}}$ was formed. A related complex

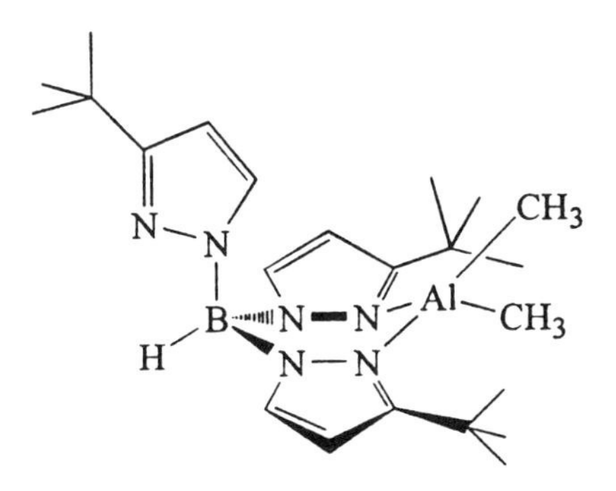

IX

η^2-[H_2B(3-*tert*-butylpyrazolyl)$_2$]Al(Me)$_2$ was also prepared from the reaction of Al(Me)$_3$ with K[H_2B(3-*tert*-butylpyrazolyl)$_2$]. In contrast, only a single set of resonances for the 3,5-dimethylpyrazolyl rings and one Al—Me signal were observed for Tp^{Me2}Al(Me)$_2$; the spectrum remained invariant to −90°C. These data are consistent with either a five-coordinate structure with an η^3-Tp^{Me2} ligand or a more highly fluxional Tp^{Me2} variant of the four-coordinate complex of $Tp^{t\text{-Bu}}$. That the former topology is most likely to be correct was implied by the fact that η^3-Tp^{Me2} ligands had been observed in another aluminum complex, [Tp_2^{Me2}Al][$AlCl_4$] (116), and that the barrier for intramolecular pyrazolyl ring exchange in the structurally defined and sterically *less* bulky analogue η^2-TpAu(Me)$_2$ was sufficiently large to enable its asymmetric static structure to be observed by ^{1}H NHR spectroscopy at −90°C (117).

Here again, the differences in steric demand between $Tp^{t\text{-Bu}}$ and Tp^{Me2} are directly indicated by the divergent structures of their complexes. In fact, the existence of steric strain intrinsic to the metal coordination of the 3-*tert*-butylpyrazolyl ring was revealed by the observation of a thermal isomerization of η^2-[H_2B(3-*tert*-butylpyrazolyl)$_2$]Al(Me)$_2$ to η^2-[H_2B(3-*tert*-butylpyrazolyl)(5-*tert*-butylpyrazolyl)]Al(Me)$_2$ (Fig. 23) (115). Such borotropic shifts have been seen for a number of complexes of $Tp^{RR'}$ ligands that have large R^3 substituents

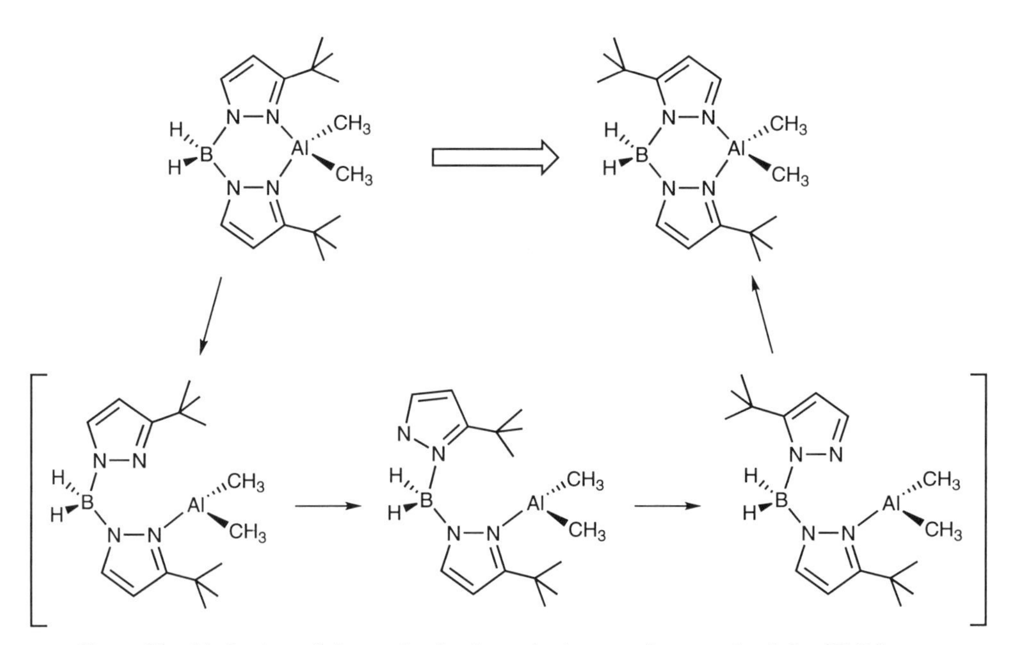

Figure 23. Mechanism of intramolecular isomerization reaction postulated for [H_2B(3-*tert*-Bupz)]Al(Me)$_2$ (115).

(see below) (18, 21, 25, 52). The commonly accepted mechanism is analogous to that shown in Fig. 23, which involves ring dissociation, a 1,2-borotropic shift, rotation, and ring reassociation to afford a less hindered and more thermodynamically stable structure. In this instance, activation parameters for the first-order process were obtained: $\Delta H^{\ddagger} = 34.5(8)$ kcal mol^{-1} and $\Delta S^{\ddagger} = 6(2)$ eu. Of note is the small positive $\Delta S^{\ddagger}$ value that is consistent with the dissociative nature of the rearrangement pathway (115).

Finally, we note that there is continuing interest in extending the aforementioned synthetic methodologies to the preparation of other main group alkyls. Recent work has focused on Cd, in part because of its utility as an NMR active ($I = \frac{1}{2}$) nucleus that can readily replace Zn in metalloproteins (118). As observed for Mg and Zn, reaction of $TlTp^{Me2}$ with CdR_2 yielded monomers $Tp^{Me2}CdR$, which were characterized by analytical, spectroscopic, and solution molecular weight methods (94). Also as seen for the $Tp^{Me2}MgR$ analogues, the cadmium complexes of the Tp^{Me2} ligand readily undergo ligand redistribution reactions to form $Tp_2^{Me2}Cd$ and CdR_2 (>80°C). The X-ray crystal structure of $Tp^{t\text{-Bu},Me}CdMe$ has been obtained (68); presumably, the more hindered $Tp^{t\text{-Bu},Me}$ enhances the stability of this complex towards such a ligand exchange process.

2. *Hydride Complexes of Be, Zn, and Cd*

The unique sterically hindered environment provided by $Tp^{t\text{-Bu}}$ has also allowed the isolation of novel monomeric hydride complexes of Be, Zn, and Cd. Reaction of ZnH_2 with $TlTp^{t\text{-Bu}}$ afforded a rare example of a structurally characterized zinc hydride, $Tp^{t\text{-Bu}}ZnH$ (112, 113), that is topologically analogous to its zinc alkyl congeners discussed above. Although the hydride ligand could not be located by X-ray crystallography, unequivocal evidence supporting its presence was evident from ^1H NMR ($\delta = 5.36$ ppm) and IR ($\nu_{ZnH} = 1770$ cm^{-1}) spectroscopic measurements of the complex and its deuterated analogue (cf. $\nu_{ZnD} = 1270$ cm^{-1}; $\nu_H/\nu_D = 1.39$). Loss of the hydride ligand is facile in its reactions with HX (X = SH, $OSi(Me)_3$, O_2CMe, C≡CMe) or RX (X = halide; R = CX_3, C(O)Me, etc.), the metal-containing product being $Tp^{t\text{-Bu}}ZnX$.

More detailed characterization of the Be—H moiety in $Tp^{t\text{-Bu}}BeH$ has been reported, including location of the hydride in its X-ray structure (119). This complex was prepared by the reaction of $Tp^{t\text{-Bu}}BeX$ (X = Cl or Br) with $LiAlH_4$ in Et_2O. It is a rare example of a monomeric beryllium hydride complex, all others beside CpBeH and Cp*BeH having oligomeric structures [(120); also see footnotes 9 and 17 in (119)]. Most importantly, it is only the second complex with a terminal Be—H bond to be examined by X-ray crystallography [Be—H distances = 1.23(7) Å] [see citations in (121)].

The Cd analogue of the above hydride complexes, $Tp^{t\text{-Bu}}CdH$, has also been

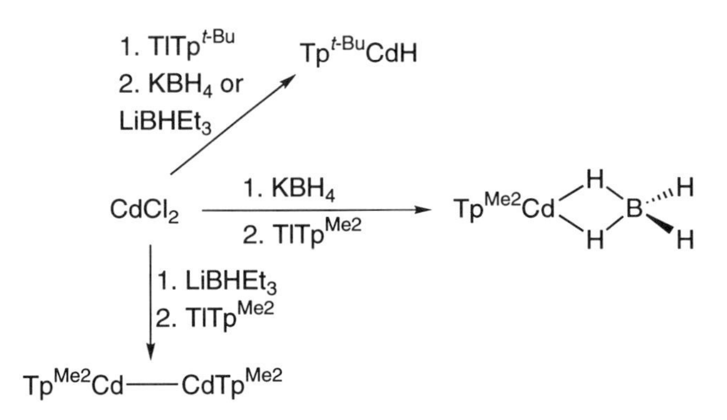

Figure 24. Reactions of $Tp^{RR'}Tl$, $CdCl_2$, and hydride reagents (122).

prepared from reactions of $CdCl_2$ with either KBH_4 or $LiBHEt_3$, followed by metathesis with $TlTp^{t\text{-Bu}}$ (Fig. 24) (122, 123). When $TlTp^{Me2}$ was used, however, different products were isolated. Thus, sequential treatment of $CdCl_2$ with KBH_4 and $TlTp^{Me2}$ yielded a crystallographically characterized η^2-BH_4 complex, $Tp^{Me2}Cd(\eta^2\text{-}BH_4)$, while use of $LiHBEt_3$ as the hydride source resulted in the formation of a novel dimer with a Cd—Cd bond, $[Tp^{Me2}Cd]_2$ (Fig. 24). The divergent results obtained for the $Tp^{t\text{-Bu}}$ and Tp^{Me2} ligands in these preparative reactions again illustrates the influence of the steric bulk of 3-pyrazolyl substituents on the chemistry of $Tp^{RR'}$ complexes.

B. Carbonyl and Hydrocarbon Complexes of Rhodium

Much of the interest in the synthesis and characterization of rhodium complexes of hindered Tp ligands has stemmed from a desire to augment and/or increase the selectivity of interesting reactions orginally uncovered for $Tp^{Me2}Rh^{I}$ and $Tp^{Me2}Rh^{III}$ fragments. An important example is the discovery that $Tp^{Me2}Rh(CO)_2$ will activate the C—H bonds of hydrocarbons upon photolysis (Fig. 25) (124). This reaction is distinguished by its high efficiency, as it generally proceeds to higher conversions upon irradiation with longer wavelength light than $Cp^*Ir(CO)_2$ or $Cp^*Ir[P(Me)_3]H_2$ (125–127) (see Note Added in Proof). Further functionalization of a hydrocarbon was achieved using $Tp^{Me2}Rh(CO)(CH_2{=}CH_2)$ as starting material; irradiation in benzene afforded two products (**X** and **XI**; Fig. 26), one of which (**XI**) yielded ethyl phenyl ketone upon successive carbonylation and $ZnBr_2$-promoted reductive elimination (128). Distinctive features that may be responsible for these unique organometallic transformations include the particular steric demands of Tp^{Me2}, its ability to readily interconvert between η^2 or η^3 binding modes [although $Tp^{Me2}Rh(CO_2)$ exists primarily in an η^3 form], and the presence of a low-

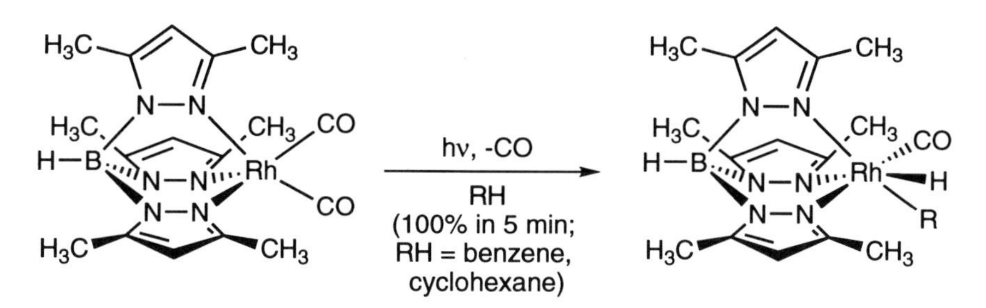

Figure 25. The C—H bond activation reactivity of $Tp^{Me2}Rh(CO)_2$ (124).

energy electronic absorption band at 353 nm in its Rh^I dicarbonyl complex. It was the profound effects of variations of R^3 substituents on the coordination chemistry of complexes of more hindered $Tp^{RR'}$ ligands that suggested that examination of rhodium complexes with these other chelates might lead to the discovery of more selective C—H bond activation reactions (e.g., size selection of hydrocarbon substrates) (19).

1. *Synthesis and Structures*

The compounds in the series $Tp^{RR'}Rh(CO)_2$ ($Tp^{RR'}$ = $Tp^{t\text{-Bu}}$, $Tp^{i\text{-Pr}}$, Tp^{Menth}, Tp^{Ph}, $Tp^{CF_3,Me}$, $Tp^{CF_3,Tn}$, Tp^{a}) were synthesized either from $[ClRh(CO)_2]_2$ and the appropriate ligand salt via a route analogous to that used to prepare the Tp^{Me2} derivative (19, 21) or via carbonylation of $Tp^{RR'}Rh(cod)$, where cod = cyclooctadiene (40, 129). Consideration of the structures and spectroscopic features of this set of complexes nicely demonstrates how the different steric and electronic effects of the variously substituted $Tp^{RR'}$ ligands can impact on the observed chemistry of derived metal complexes. A key finding was that there

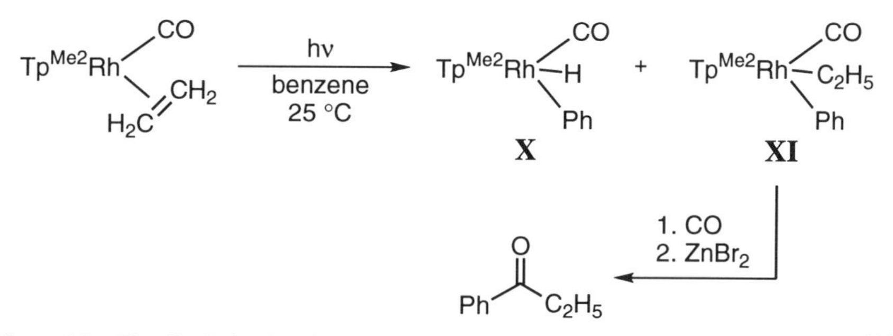

Figure 26. The C—H bond activation and subsequent functionalization reactions of Tp^{Me2}-$Rh(CO)(CH_2{=}CH_2)$ (128).

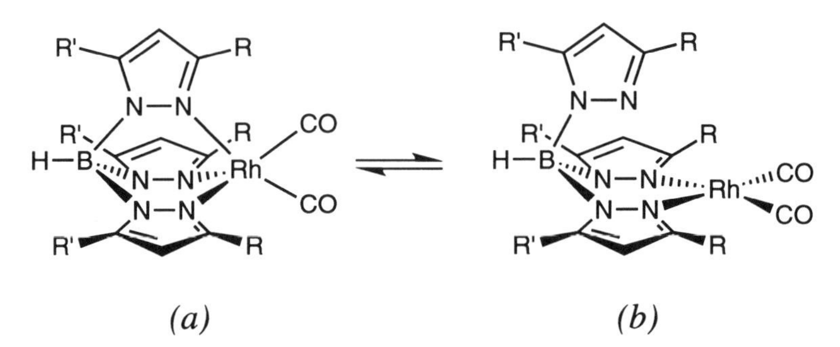

Figure 27. Equilibrium between 18 e$^-$, five-coordinate (*a*) and 16 e$^-$, four coordinate (*b*) forms of $Tp^{RR'}Rh(CO)_2$ (19).

are two accessible structural motifs for the $Tp^{RR'}Rh(CO)_2$ compounds, one with an η^2-$Tp^{RR'}$ coordinated to a four-coordinate, square planar, and formally 16 e$^-$ Rh^I ion, and the other having the third pyrazolyl ring bound so that the metal is five coordinate with a formal 18 e$^-$ count (Fig. 27). Identification of each isomeric form and observation of their interconversions were based on X-ray crystallographic studies of η^2-$Tp^{Ph}RH(CO)_2$ and η^2-$Tp^{CF_3,Me}Rh(CO)_2$ (Fig. 28) (130, 131), as well as on IR and variable temperature ^{1}H NMR spectroscopic measurements of the compounds in solution (19). The X-ray structures of η^2-$Tp^aRh(CO)_2$ and η^2-$Tp^{CF_3,Tn}Rh(CO)_2$ have also been obtained (40, 129). In the solution IR spectra of the compounds that exist as mixtures, two sets of carbonyl stretching bands were observed, of which the set of lower energy was assigned

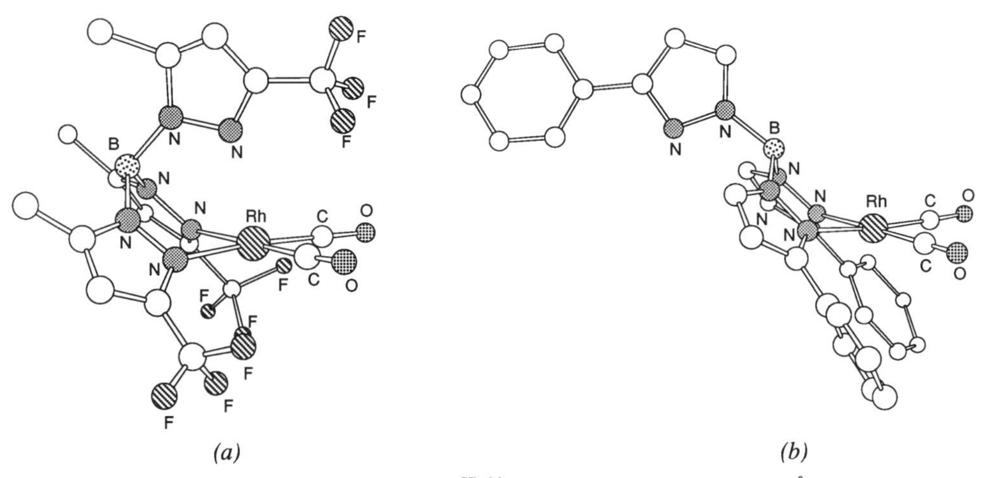

Figure 28. X-ray crystal structures of (*a*) $Tp^{CF_3,Me}Rh(CO)_2$ [$Rh-N_{ax}$ = 2.623(8) Å; (130)] and (*b*) $Tp^{Ph}Rh(CO)_2$ (131).

to the more electron rich four-coordinate structure (see Table V). In addition, analysis of the behavior of broad room temperature ^1H NMR resonances upon cooling allowed the relative rates of four to five-coordinate interconversions to be estimated in selected cases.

The relative proportions of the two isomeric structures were found to be dependent on solvent and, more importantly for this discussion, the nature of $Tp^{RR'}$ (Table VI). It has been suggested that the $\eta^2:\eta^3$ ratio is predominantly determined by steric, rather than electronic, influences of the $Tp^{RR'}$ ligands (19). Based on the steric order of Eq. 1 and the expectation that the greater the effective size of $Tp^{RR'}$ the more favored would be the four-coordinate η^2 structure, the trend in the $\eta^2:\eta^3$ ratios $Tp^{t\text{-}Bu} > Tp^{Ph} > Tp^{CF_3,Me} > Tp^{Me2}$ makes sense. Note that the rhodium dicarbonyl complex of the most hindered $Tp^{t\text{-}Bu}$ exists exclusively as the η^2 form in solution. Also, the differences in the orientations of the uncoordinated pyrazolyl ring in the X-ray structures of η^2-$Tp^{Ph}Rh(CO)_2$ and η^2-$Tp^{CF_3,Me}Rh(CO)_2$ (Fig. 28) (130, 131) reflect the greater steric forces exerted by the 3-phenyl versus the 3-CF_3 substituents. The $Tp^{Ms}Rh(CO)_2$ complex is an interesting case because it appears to exist solely in the η^3 form on the basis of its low carbonyl stretching frequencies (25). Its symmetric ^{1}NMR spectrum is consistent with this assignment if one postulates rapid Berry pseudorotation or turnstile rotation of the five-coordinate molecule. We speculate that exclusive observation of η^3 binding is due to a metal-centered cavity for Tp^{Ms} that is effectively larger than that provided by Tp^{Ph} because of the forced orthogonality of the mesityl and attached pyrazolyl groups.

A rationalization of the trend in $\eta^2:\eta^3$ ratios based on electronic structural arguments appears to be inappropriate, since both $Tp^{t\text{-}Bu}$ and Tp^{Me2} are more electron donating than Tp^{Ph} and $Tp^{CF_3,Me}$ (see IR data in Table V) but the $\eta^2:\eta^3$ ratios for the former bracket those of the latter. On the basis of the steric arguments alone, however, the $\eta^2:\eta^3$ ratio of 35:65 for $Tp^{i\text{-}Pr}Rh(CO)_2$ that is quite similar to that of $Tp^{CF_3,Me}Rh(CO)_2$ (32:68) appears anomalous, since most evidence (see above) suggests that $Tp^{i\text{-}Pr}$ is more sterically demanding than *both* Tp^{Ph} and $Tp^{CF_3,Me}$. Also, despite the relatively small size of its thienyl substituents, $Tp^{CF_3,Tn}Rh(CO)_2$ exhibits η^3 coordination in the solid state (129). Again,

TABLE VI
Isomer Ratios in Selected $Tp^{RR'}Rh(CO)_2$ Complexes[a]

Compound	$\eta^2:\eta^3$ ratio
$Tp^{t\text{-}Bu}Rh(CO)_2$	100:0
$Tp^{iPr}Rh(CO)_2$	35:65
$Tp^{Me2}Rh(CO)_2$	0:100
$Tp^{Ph}Rh(CO)_2$	85:15
$Tp^{CF_3,Me}Rh(CO)_2$	32:68

[a]Table adapted from (19). All values determined in cyclohexane.

a subtle interplay of steric and electronic influences may be at work for ligands of intermediate steric bulk.

The rate of interconversion between the η^2 and η^3 isomers of the rhodium dicarbonyl complexes in solution is too rapid for direct observation of both forms via NMR spectroscopy for most of the $Tp^{RR'}$ ligands, a single set of resonances for the pyrazolyl ring nuclei being typical even at temperatures as low as −90°C (19). For $Tp^{Ph}Rh(CO)_2$ and $Tp^{Menth}Rh(CO)_2$ (21), however, the process is sufficiently slowed to allow both isomers to be identified by ^1H NMR spectroscopy at low temperatures (−30–0°C). For example, for Tp^{Ph}-$Rh(CO)_2$, two sets of pyrazolyl 4-hydrogen signals were observed at −30°C, a pair of resonances with a 2:1 ratio of intensities that was assigned to the η^2 form and a single resonance that was assigned to the five coordinate η^3 isomer. Rapid averaging of the pyrazolyl ring environments via a Berry pseudorotation or turnstile process was postulated for the latter species. Thermodynamic parameters were calculated for the interconversion of the η^2- and η^3-$Tp^{Ph}Rh(CO)_2$ isomers from the temperature dependence of the equilibrium constants determined from the combined ^1H NMR and IR spectroscopic data: $\Delta H° = -2.95$ kcal mol^{-1} and $\Delta S° = -11.6$ eu (CH_2Cl_2/CD_2Cl_2). Negative values for both the enthalpic and entropic terms make intuitive sense since $\eta^2 \rightarrow \eta^3$ isomerization involves formation of a new Rh-pyrazolyl ring bond and concomitant ordering due to the loss of degrees of freedom upon coordination of the dangling ring.

The propensity for the rhodium dicarbonyl complexes of the more sterically hindered $Tp^{RR'}$ ligands to exist as 16 e^- four-coordinate species has direct consequences on their reactivity, which differs in certain respects from that exhibited by $Tp^{Me2}Rh(CO)_2$. For example, no incorporation of ^{13}CO (1 atm) into a solution of $Tp^{Me2}Rh(CO)_2$ in cyclohexane, where it exists solely as the η^3 isomer, is observed after 4 h, but incorporation into $Tp^{Ph}Rh(CO)_2$ ($\eta^2:\eta^3 = 85:15$) under the same conditions is complete after 5 min. Support for the idea that the rate difference is due to the divergent accessibilities of the η^2 structures comes from the known rapidity of ^{13}CO exchange reactions in square planar and 16 e^- complexes of Rh^I, which occur by an associative mechanism, compared to the correspondingly slower reactions in 18 e^- complexes (132).

The third, sometimes unbound pyrazolyl ring in $Tp^{Ph}Rh(CO)_2$ can bind a second metal center, as demonstrated by the synthesis of **XII**. This dimer is highly fluxional and, on the basis of NMR spectroscopic experiments on selectively ^{13}CO-substituted forms, two processes were identified (Fig. 29). In the fast process, both a pyrazolyl ring and the chloride exchange between Rh atoms, resulting in exchange of both ^{13}CO ligands from Rh^1 to Rh^2. The slower process involves exchange of only one ^{13}CO at a time. Either or both inter- and intramolecular pathways are possible for both exchanges, since scrambling was observed in crossover experiments in which fully ^{13}CO labeled and unlabeled

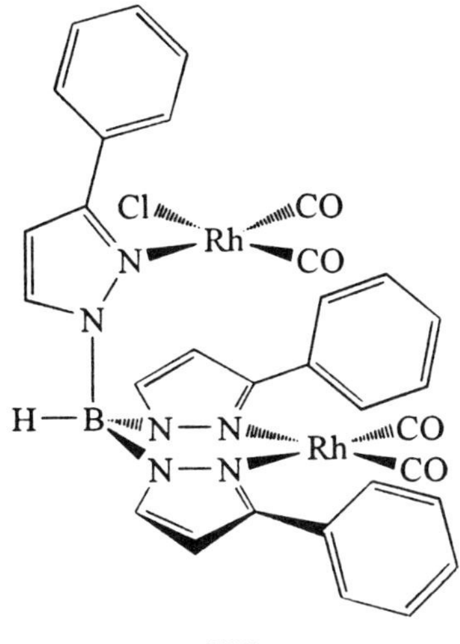

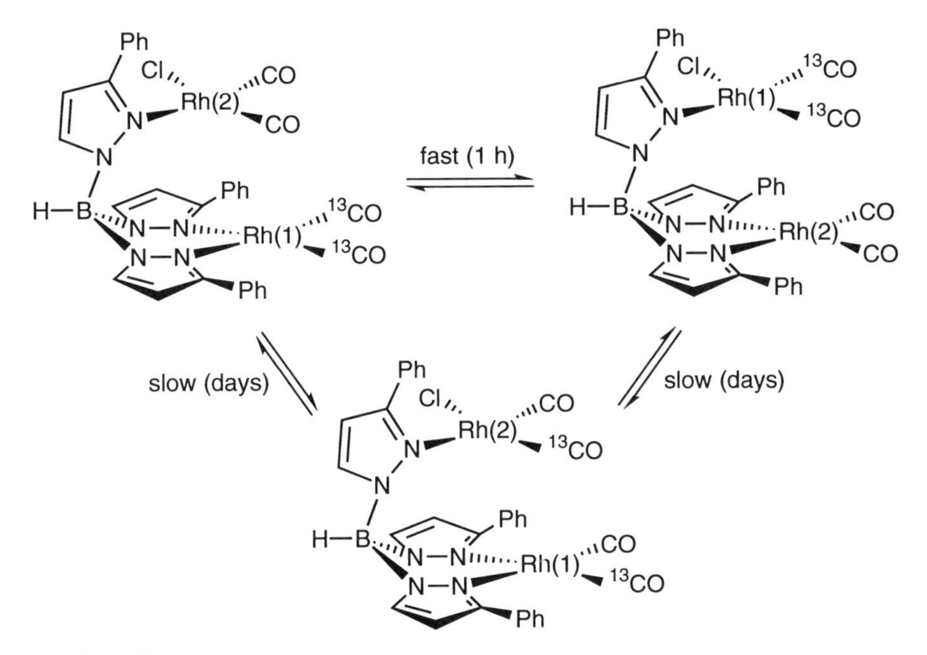

Figure 29. Intramolecular fluxional processes postulated to occur in solution for **XII** (19).

compounds were mixed. Parenthetically, it is interesting to note the close similarity of the above "fast" process with that proposed for the equilibration of the *tert*-butylpyrazolyl environments in $[Tp^{t\text{-}Bu}Cu]_2$ (Fig. 20).

2. *C—H Bond Activation Reactions*

Preliminary investigation of the C—H bond activation reactions of Rh complexes of $Tp^{RR'}$ (R, R′ ≠ Me) revealed significant differences in the pathways followed from those of the Tp^{Me2} systems (19). For example, whereas irradiation of $Tp^{Me2}Rh(CO)_2$ in benzene with a N_2 purge led to facile intermolecular C—H bond activation to afford a phenyl hydride complex (Fig. 25, R = Ph) (124), intramolecular attack occurred when the same reaction was performed using $Tp^{Ph}Rh(CO)_2$ as starting material to yield the cyclometalated species **XIII** (Fig. 30) (19). Intermolecular C—H bond activation was only observed with Tp^{Ph} when cyclopropane was the substrate; in this instance, the product isolated was the metallacyclobutane **XIV** that was presumably formed via rearrangement of the initial cyclopropyl hydride **XV** (133). Complex **XIV** was also produced in the thermal reaction of **XIII** with cyclopropane (Fig. 30). On the basis of this observation and the fact that the cyclometalation that afforded **XIII** was rapidly reversed upon exposure of this product to CO (1 atm), an equilibrium between **XIII** and a reactive, coordinatively unsaturated, 16 e^- species was proposed (Fig. 31). All C—H bond activation reactions in this system can be viewed to progress through this intermediate. In sum, in the Tp^{Ph} system intermolecular C—H bond activation appears viable, but because of the tendency

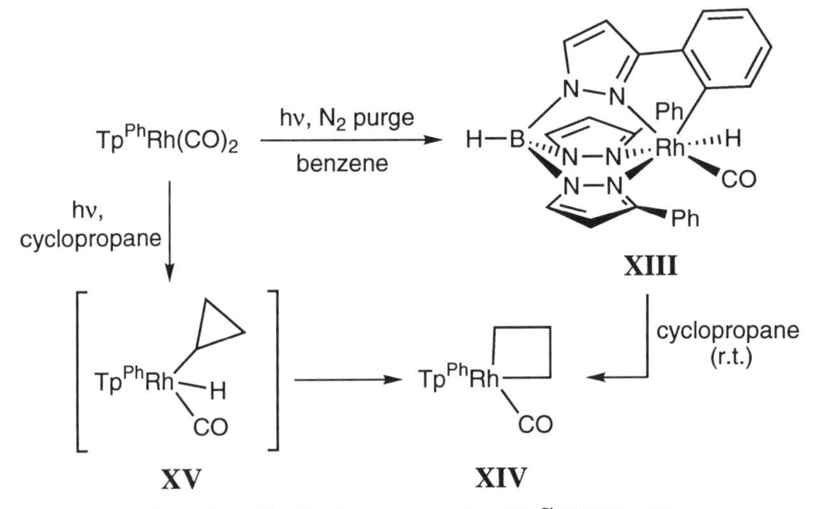

Figure 30. The C—H bond reactivity of $Tp^{Ph}Rh(CO)_2$ (19).

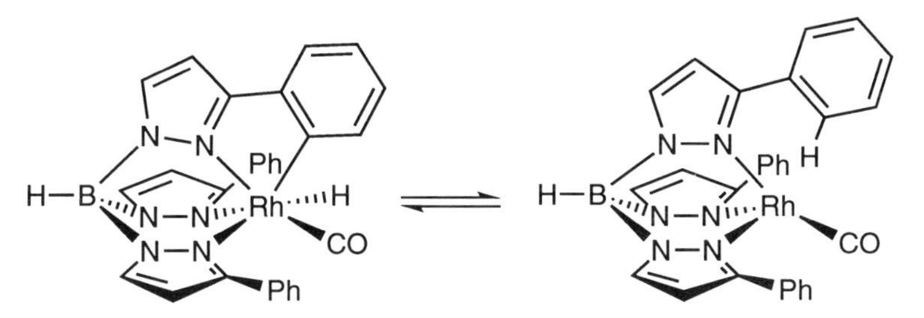

Figure 31. Postulated equilibrium for **XIII** in solution (19).

for the 16 e^- intermediate to cyclometalate, the product of intermolecular attack has been isolable only when its decomposition by reductive elimination is highly disfavored (i.e., as in metallacyclobutane **XIV**).

Intramolecular C—H bond activation also was identified for $Tp^{i\text{-}Pr}Rh(CO)_2$, but only when its photolysis was carried out in cyclohexane; in benzene, a phenyl hydride complex (**XVI**) formed (Fig. 32) (19). Cyclometalation for $Tp^{i\text{-}Pr}$ was shown by NMR spectroscopy to involve attack at an isopropyl methyl group to afford a six-membered ring in the product (**XVII**). This reaction is one of several reported cases of metal–ion promoted functionalization of isopropyl substituents on $Tp^{i\text{-}Pr_n}$ ($n = 1$ or 2; see below), but it is a rare instance involving attack at a methyl group rather than the methine hydrogen (see Section V). Like the cyclometalated Tp^{Ph} species **XIII**, Compound **XVII** was shown to react readily under thermolytic conditions with external substrates such as benzene

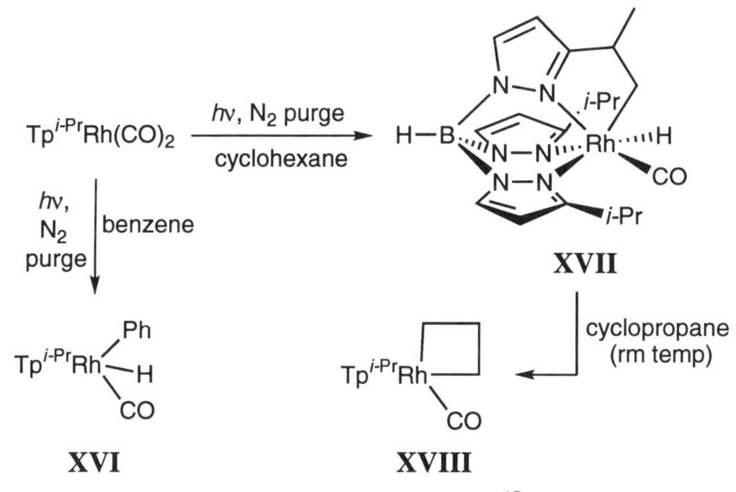

Figure 32. The C—H bond reactivity of $Tp^{i\text{-}Pr}Rh(CO)_2$ (19).

(to give **XVI**) or cyclopropane (to give the metallacyclobutane **XVIII**), thus implicating operation of an equilibria similar to that shown in Fig. 31, but involving isopropyl instead of phenyl substituents.

Further increasing the steric bulk of the supporting ligand by using $Tp^{t\text{-}Bu}$ effectively shut down the C—H bond activation chemistry; only decomposition of $Tp^{t\text{-}Bu}Rh(CO)_2$ was observed under the usual photolytic reaction conditions (19). Variation of the electron-donating properties of Tp^{Me2} by using $Tp^{CF_3,Me}$ gave interesting results, however. While the C—H bond activation reactivities of $Tp^{CF_3,Me}Rh(CO)_2$ and $Tp^{CF_3,Me}Rh(CO)(CH_2{=}CH_2)$ were similar in overall form to that of their Tp^{Me2} congeners, significant differences in reaction rates and product stabilities were observed. Thus, irradiation of $Tp^{CF_3,Me}Rh(CO)_2$ in cyclohexane with a H_2 purge or in benzene with an N_2 purge afforded dihydride **XIX** or phenyl hydride **XX,** respectively. In contrast to the formation of both phenyl hydride and phenyl ethyl products upon irradiation of Tp^{Me2}-$Rh(CO)(CH_2{=}CH_2)$ (Fig. 26) (124), the analogous $Tp^{CF_3,Me}$ complex yielded **XX** only. Like the Tp^{Me2} phenyl hydride, **XX** exchanged with deuterated solvent (toluene-d_8) and reverted to $Tp^{CF_3,Me}Rh(CO)_2$ under CO (1 atm), but the rates of both processes were significantly faster for the $Tp^{CF_3,Me}$ compounds. For the solvent exchange reaction, a $\Delta H^\ddagger \sim 10$ kcal mol^{-1} lower than that of the Tp^{Me2} phenyl hydride was measured for **XX**. This significantly lower barrier for reductive elimination is consistent with the electron-withdrawing properties of the CF_3-substituted ligand, which would be anticipated to preferentially stabilize the formally Rh^I intermediate formed upon loss of benzene from the formally Rh^{III} compound **XX**. Unfortunately, the greater tendency for $Tp^{CF_3,Me}$ species to undergo reductive elimination resulted in significantly decreased stability for other C—H bond activation products, such as those formed upon irradiation of $Tp^{CF_3,Me}Rh(CO)_2$ or dissolution of **XX** in alkanes.

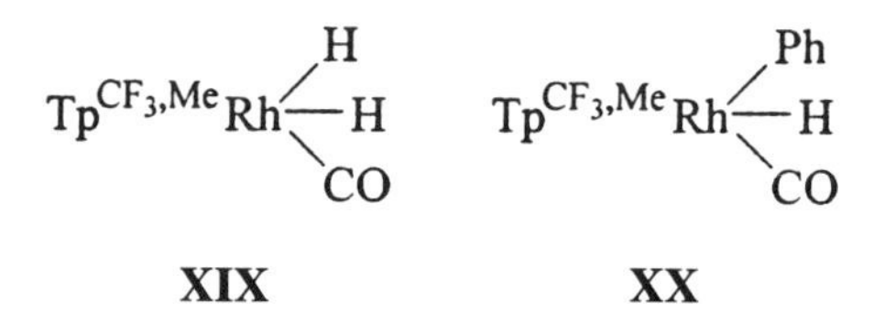

C. Complexes of Divalent Lanthanide Ions

The ever-increasing utility of Ln^{II} salts in organic synthesis (134, 135) combined with the unusual reactivity patterns reported for organometallic Ln^{II} complexes, such as $Cp^*_2Sm(thf)_{0\text{-}2}$ (136), has stimulated interest in developing the Tp chemistry of these ions (Ln = Sm and Yb). With $Tp^{RR'}$ ligands of low and intermediate steric bulk, such as Tp^{Ph}, Tp^{Me2}, and Tp^{Tn}, bis(ligand) complexes with the large Sm^{II} and Yb^{II} ions are formed predominantly [e.g., Tp^{Me2}_2Sm

(137, 138)]. Although expansion of the coordination number of such complexes to greater than 6 is possible (138), mono-Tp compounds of Ln^{II} ions would enable greater access to novel and potentially highly reactive derivatives. Such $(L)Ln^{II}$ species are rare (139, 140), and have been demonstrated to have a tendency to disproportionate to L_2Ln and a Ln-halide salt. Recently, the successful synthesis of a series of $(L)Ln^{II}$ compounds was achieved by using the highly hindered $Tp^{t\text{-Bu},Me}$ ligand to inhibit bis(ligand) complex formation (137, 138, 141–143) (also see Note Added in Proof).

The key starting materials $Tp^{t\text{-Bu},Me}LnI(thf)_n$ (Ln = Sm, $n = 2$; Yb, $n = 1$) were prepared by straightforward metathetical reactions between $LnI_2(thf)_2$ and $KTp^{t\text{-Bu},Me}$. The X-ray crystal structure of the trigonal bipyramidal Yb complex (Fig. 33) demonstrated the monomeric nature of the product, which showed no tendency to disproportionate in solution (142). Derivatization of these iodide compounds proceeded as indicated in Fig. 34 and each product (Ln = Yb) was crystallographically characterized. The structures of the $N(SiMe_3)_2$ and $CH(SiMe_3)_2$ complexes (Fig. 35), the latter of which apparently represents the first well-characterized hydrocarbyl complex of a divalent lanthanide, reveal metal ion geometries distorted from tetrahedral as shown by the presence of the amido N and hydrocarbyl C atoms displaced from the B· · ·Yb axis by 18.6°

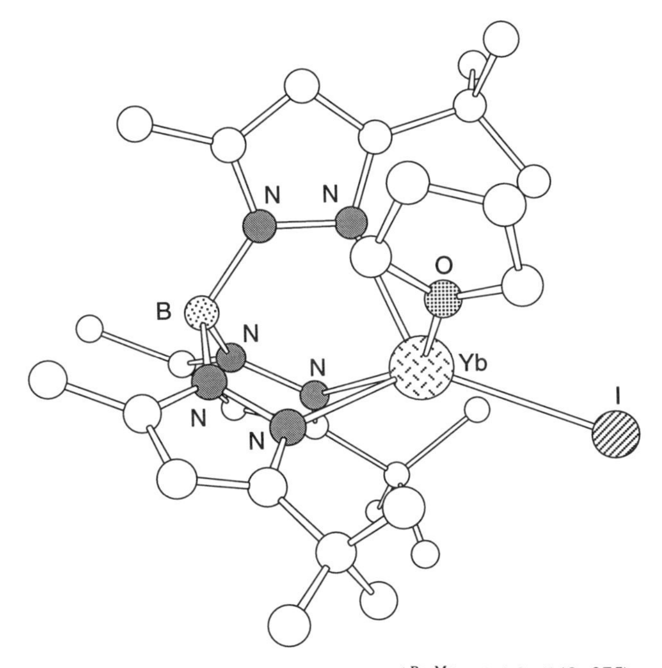

Figure 33. X-ray crystal structure of $Tp^{t\text{-Bu},Me}YBI(thf)$ (142, 275).

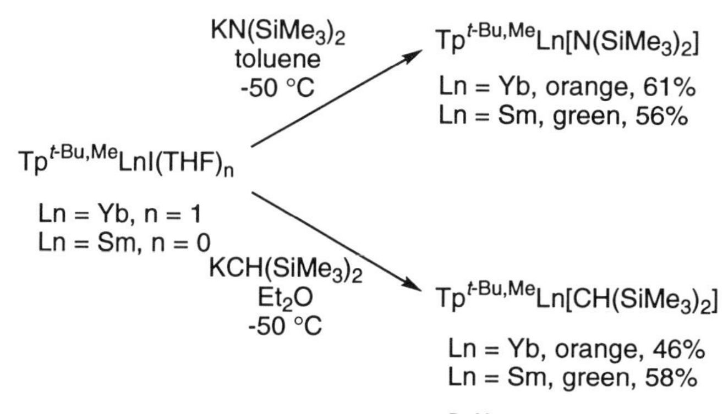

Figure 34. Some reactions of $Tp^{t\text{-Bu, Me}}LnI(thf)$ (141).

and 15.0°, respectively. In addition, severe distortions of the Yb-amido and Yb-hydrocarbyl moieties are evinced by the differences within the pairs of Yb—N—Si and Yb—C—Si angles [e.g., Yb—C—Si1 = 132(2)° and Yb—C—Si2 = 98(3)°]. These distortions, in combination with short $Yb\cdots C_{silyl}$ distances of about 3.1 Å, can be explained by the presence of

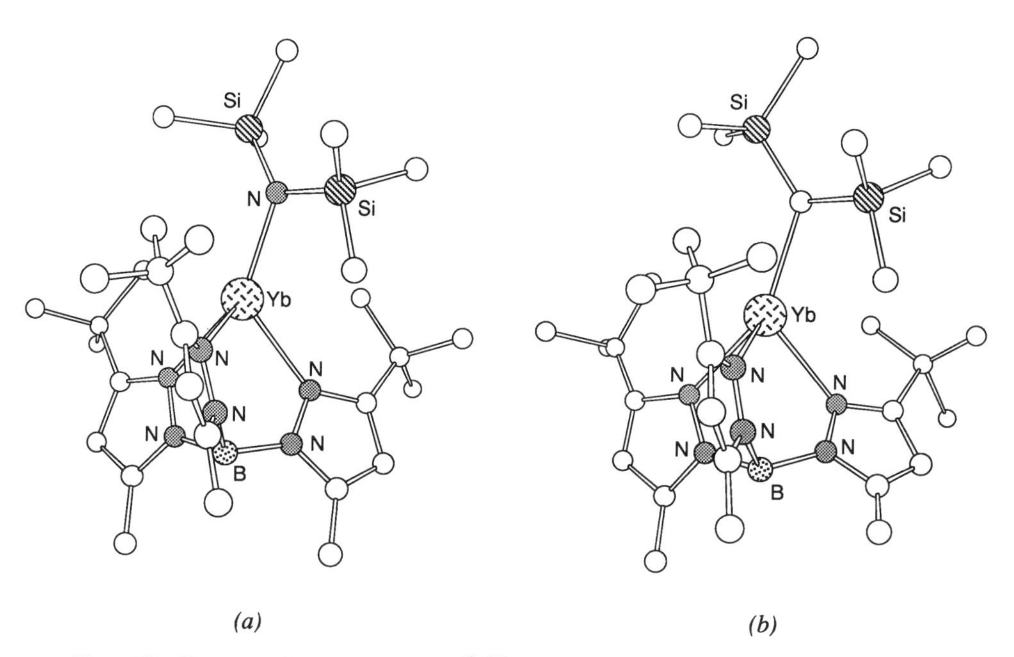

Figure 35. X-ray crystal structures of (*a*) $Tp^{t\text{-Bu, Me}}Yb[N(SiMe_3)_2]$ and (*b*) $Tp^{t\text{-Bu, Me}}Yb[CH(SiMe_3)_2]$ (142).

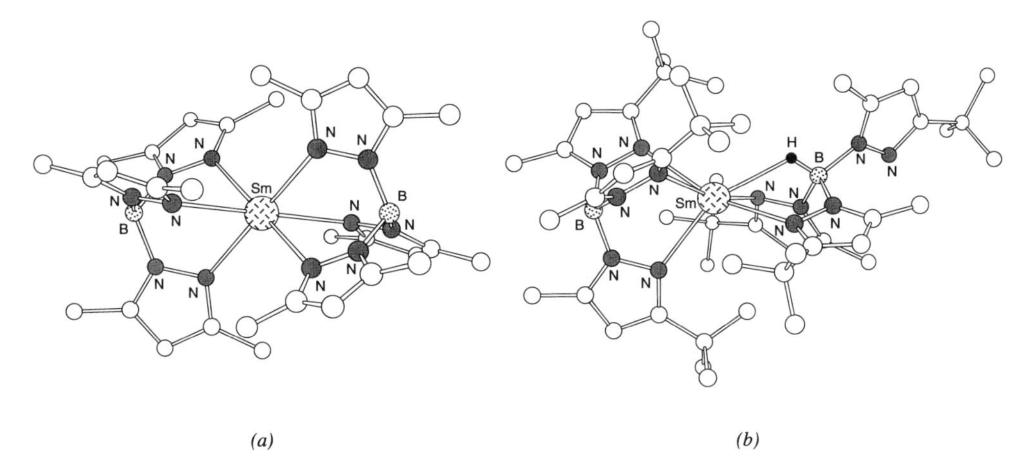

Figure 36. X-ray crystal structures (*a*) $Tp_2^{Me2}Sm$ (138) and (*b*) $Tp_2^{t\text{-Bu,Me}}Sm$ (142).

agostic interactions in the solid state analogous to ones previously observed in lanthanide chemistry (144, 145). Preliminary results of studies of the reactivity of these mono-$Tp^{t\text{-Bu,Me}}$ derivatives indicate that other interesting hydrido and hydrocarbyl derivatives are accessible and suggest great potential for future applications in synthetic and catalytic chemistry (142).

Finally, we note that disproportionation to $Tp_2^{t\text{-Bu,Me}}Sm$ was observed in the attempted preparation of $Tp^{t\text{-Bu,Me}}Sm(CH_2SiMe_3)$, but the structure for the product was shown to be different from that adopted by its analogue, $Tp_2^{Me2}Sm$ (Fig. 36) (137, 143). In contrast to the η^3-N_3 donation by both ligands in the latter compound, in the $Tp^{t\text{-Bu,Me}}$ complex the usual tridentate N-coordination occurs with one ligand, but the other binds via two pyrazolyl arms and a three-center B—H—Sm linkage, leaving the third pyrazolyl group uncoordinated. Note the similarity of this structure to that of $Tp^{i\text{-Pr,4Br}}Tp^{Ph}Co$ (Fig. 16). Steric differences between the 3-methyl and 3-*tert*-butyl-substituted ligands are, again, evidently responsible for the divergent structures adopted by the L_2Sm complexes.

IV. BIOINORGANIC CHEMISTRY

The indispensable physiological role of metal ions in biological systems has been the focus of growing interest in recent years (146, 147). Particular emphasis has been placed on understanding the structures and functions of metal-containing sites in enzymes and proteins, especially because of their often quite unusual properties. Synthetic compounds that model aspects of metalloprotein structure, spectroscopic signatures, and function have been demonstrated to be highly useful for deconvoluting the complex properties of many biological systems on a molecular level, as well as for extending our knowledge of fundamentally important chemical reactions such as oxygenations, electron transfers, and hydrolyses (148). Hydrotris(pyrazolyl)borates have been used with great success as supporting ligands for bioinorganic synthetic studies, their ability to mimic multiple histidine coordination to metal ions in proteins being particularly advantageous. Ligation of two or more histidine imidazoles to an active site metal ion has been recognized as a common biological motif, with examples identified by X-ray crystallography for the most common elements Cu [e.g., hemocyanin (149), ascorbate oxidase (150), superoxide dismutase (151)], Fe [e.g., hemerythrin (152)], and Mn [e.g., superoxide dismutase (153)], among others. Early successes in modeling the structural and spectroscopic properties of the active sites of metalloproteins such as hemerythrin with the unsubstituted, parent Tp as supporting ligand demonstrated that despite the obvious differences between imidazole and pyrazole, the tris(pyrazolyl) array in Tp was well suited for mimicking analogous polyimidazolyl units supplied by protein chains (cf. 154). More recently, significant inroads have been made in constructing complexes with novel features, including biomimetic reactivity, by appropriate use of substituted Tp ligands with differing steric and electronic properties. Although there exists an impressive body of bioinorganic modeling work centered on Tp and Tp^{Me2} complexes (cf. 154–156), the following discussion will remain within the stated scope of this chapter and thus will be limited to selected cases involving ligands with R^3 groups larger than Me, except when comparisons to more hindered ligands are deemed useful.

A. Copper Complexes

1. *Models of Hemocyanin and Tyrosinase*

Hemocyanins are copper-containing dioxygen-transport proteins found in arthropods and mollusks. The dioxygen binding site contains a pair of copper ions which, in their deoxygenated $Cu^{I}Cu^{I}$ form (deoxyhemocyanin, deoxyHc), are 3.54 Å apart and are each coordinated by three histidine imidazolyl N-donor ligands (157). Developing an understanding of how O_2 binds to this dicopper(I) site to form oxyhemocyanin (oxyHc) has been a long standing problem in the bioinorganic chemistry community, both because of the intrinsic importance of the dioxygen-binding phenomenon and because of the unusual spectroscopic and physical properties of the O_2 adduct. Key features of oxyHc that were identified in early work include its diamagnetism, the presence of intense charge-transfer transitions in its optical spectrum [λ_{max} (ϵ, M^{-1} cm^{-1})

340 (20,000), 480 (1000) nm], and a low ν_{O-O} in its resonance Raman spectrum ($\sim$750 cm^{-1}), analysis of which using isotope labeling indicated symmetrical coordination of O_2 as peroxide (158, 159). Similar properties have been identified for the dioxygen adduct of tyrosinase (oxyTyr), an enzyme that catalyzes the oxidation of phenols to catechols and benzoquinones (160), leading to the assumption that its structure is analogous to that of oxyHc. Substantial effort has been expended by synthetic chemists to construct dioxygen adducts that would replicate these features and shed light on the mode of O_2 binding in oxyHc and oxyTyr; this work has been reviewed extensively (161–164). Here we focus on the role of $Tp^{RR'}$ ligands in facilitating the construction and study of a set of model compounds that most accurately mimic the structural and spectroscopic properties of the oxyHc active site and that, in fact, presaged the recent definition of the site by X-ray crystallography (165, 166).

Initial synthetic efforts that focused on using Tp^{Me2} as supporting ligand in the attempted preparation of a dicopper(II)–peroxo model complex were plagued by difficulties associated with deleterious side reactions. Thus, the complex $[(Tp^{Me2}Cu)_2(O_2)]$ was identified by resonance Raman spectroscopy (ν_{O-O} = 731 cm^{-1}) and field desorption mass spectrometry as the product of the reaction of H_2O_2 with the μ-oxo complex $[(Tp^{Me2}Cu)_2(O)]$, but facile decomposition of this species to the Cu^{II} monomer $Tp_2^{Me2}Cu$ hindered more definitive characterization (167, 168). Evidence for the transient formation of the peroxo compound was obtained upon treatment of the dicopper(I) complex $(Tp^{Me2}Cu)_2$ with O_2 at low temperature, but decomposition to $Tp_2^{Me2}Cu$ was again problematic (168). It was clear that the 3,5-dimethyl substitution pattern on the Tp ligand was insufficient for inhibiting L_2M formation, so more hindered Tp^{Ph2} and $Tp^{i\text{-}Pr2}$ ligands were examined.

The model compounds $[(Tp^{R2}Cu)_2(O_2)]$ (R = Ph or *i*-Pr), which were prepared either from the reaction of O_2 with the appropriate $Tp^{R2}Cu^I$ precursor or via treatment of a dicopper(II)-bis(μ-hydroxo) species with H_2O_2 (Fig. 37), were significantly more stable than the Tp^{Me2} analogue (169, 170). The $Tp^{i\text{-}Pr2}$

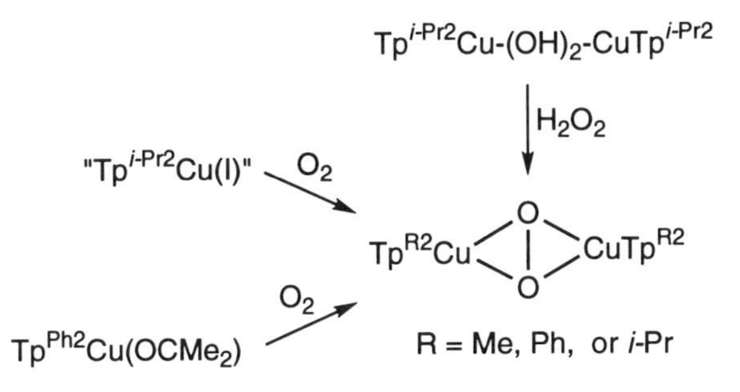

Figure 37. Syntheses of $[(Tp^{R2}Cu)_2(O_2)]$ (R = Me, Ph, or *i*-Pr) (168, 170).

complex persisted in solution below $-10°C$ and was sufficiently stable as a solid to allow characterization by X-ray crystallography at low temperature, while the Tp^{Ph2} compound was stable as a powder at room temperature. The X-ray structure of $[(Tp^{i\text{-}Pr2}Cu)_2(O_2)]$ exhibited a novel, planar, side-on-bridged peroxide (μ-$\eta^2:\eta^2$) bound between two square pyramidal Cu^{II} ions and encapsulated by a shroud of isopropyl groups from the $Tp^{i\text{-}Pr2}$ ligands (Fig. 38).

The μ-$\eta^2:\eta^2$-peroxo coordination mode is rare in transition metal chemistry, having only been observed previously in crystal structures of La (171), U (172), and V (173) complexes. Adoption of this bonding mode by these elements derives in part from their high oxophilicity, a rationalization that is inappropriate for Cu^{II}. Instead, the facial binding topology intrinsic to Tp, which prevents a square planar metal geometry, combined with the ligand field driven preference for tetragonal (square pyramidal) over tetrahedral coordination for the d^9 Cu^{II} ion favors the μ-$\eta^2:\eta^2$ mode over another reasonable possibility for a symmetric peroxo ligand, *trans*-μ-1,2 [a dicopper–peroxo complex with this mode has been structurally characterized (174)]. The recent identification of a similar μ-$\eta^2:\eta^2$ bonding arrangement for the disulfide bridge in $[(Tp^{i\text{-}Pr2}Cu)_2(S_2)]$ suggests that this rationalization may be general (175). It is noteworthy in the current context that the successful isolation of the μ-$\eta^2:\eta^2$ complexes was possible at least in part because of the intermediate steric properties of the $Tp^{i\text{-}Pr2}$ and Tp^{Ph2} li-

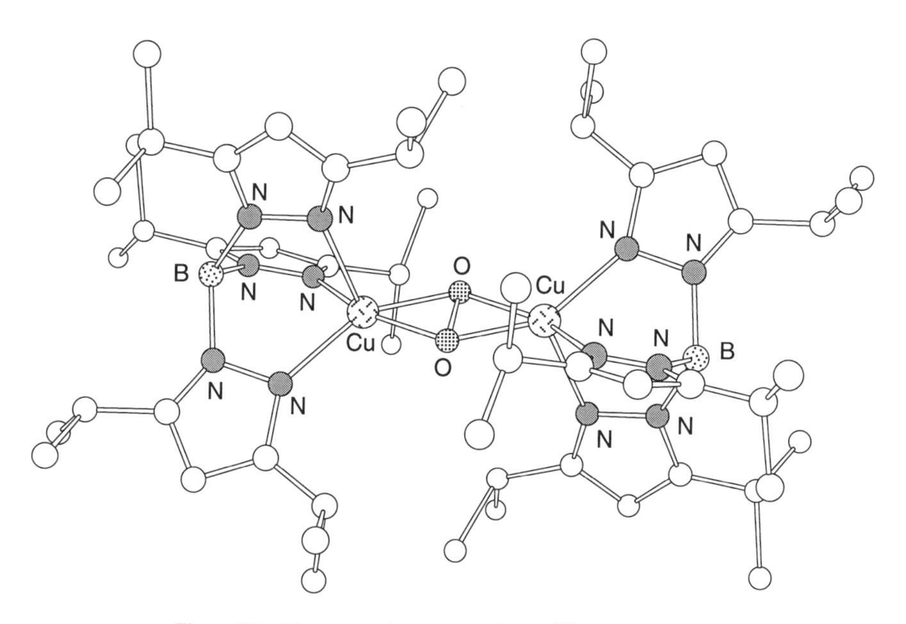

Figure 38. X-ray crystal structure of $[(Tp^{i\text{-}Pr2}Cu)_2(O_2)]$ (168).

gands. Both have the necessary high degree of steric hindrance with which to encapsulate the peroxo moiety and inhibit deleterious side reactions, but they also are able to support five-coordinate metal geometries. As discussed below, steric hindrance greater than that exhibited by $Tp^{i\text{-}Pr2}$ prevents dinuclear complex formation, allowing the isolation of a monomeric superoxo species.

The enhanced stability of the dicopper(II)-peroxo unit in these Tp^{R2} (R = Ph or *i*-Pr) complexes resulting from the steric properties of the polypyrazole ligands was critical in allowing the comprehensive definition of the physicochemical properties of these compounds, work that revealed their suitability as accurate models of the oxyHc and oxyTyr active sites. Thus, the diamagnetism, characteristic optical absorption bands, low ν_{O-O}, and Cu· · ·Cu separation (3.6 Å) of oxyHc and oxyTyr were closely replicated by the model complexes. Bonding models supported by calculations have been presented to explain the unique spectral features of the dicopper(II)-μ-$\eta^2:\eta^2$-peroxo unit (168, 176). The results indicate that the lowest unoccupied molecular orbital (LUMO) is comprised of an antibonding interaction between the Cu $d_{x^2-y^2}$ and the $O_2^{2-} - \pi^*$ orbitals, while the highest occupied molecular orbital (HOMO), which lies significantly lower in energy (5660 cm^{-1}), derives from bonding interactions that include the one between the same Cu $d_{x^2-y^2}$ orbitals and the $O_2^{2-} - \sigma^*$ level (Fig. 39) (158). Detailed correlation of the bonding description and the observed physicochemical features of oxyHc, oxyTyr, and the model complexes can be found elsewhere (170, 177); suffice it to say here that the combined spectroscopic, structural, and theoretical investigations have afforded a strikingly coherent and conceptually satisfying picture of the unique dicopper(II)-μ-$\eta^2:\eta^2$-peroxo moiety that was only accessible through parallel examination of both the protein and the synthetic systems.

The divergent steric properties of Tp^{Me2} and $Tp^{i\text{-}Pr2}$ have also influenced reactivity studies of their respective copper peroxo complexes, undertaken to provide detailed mechanistic information on important oxyHc and oxyTyr transformations. For example, addition of N_3^- to oxyHc results in the formation of metazidohemocyanin (metN_3Hc), a form of the protein that was shown by spectroscopic studies to contain a dicopper(II)-(μ-hydroxo)(μ-1,3-azido) core (178). The similarities between the typical binding modes adopted by azide and peroxide, combined with the EPR silence (diamagnetism) of metN_3Hc, had been cited as corroborative evidence for a similar dicopper(II)-(μ-hydroxo)(μ-1,2-peroxo) structure for oxyHc, with antiferromagnetic coupling between the Cu^{II} ions in both oxy- and metN_3Hc mediated by the hydroxo bridge (179). In accordance with this early view, the oxyHc → metN_3Hc conversion was perceived to involve simple ligand substitution of azide for peroxide. This pathway had to be reconsidered upon subsequent recognition of the μ-$\eta^2:\eta^2$-peroxo binding mode in oxyHc, so the reaction of N_3^- with $[(Tp^{i\text{-}Pr2}Cu)_2(O_2)]$ was investigated (Fig. 40) (180). The product, $[(Tp^{i\text{-}Pr2}Cu)_2(\mu\text{-}OH)(\mu\text{-}N_3)]$, exhibited sim-

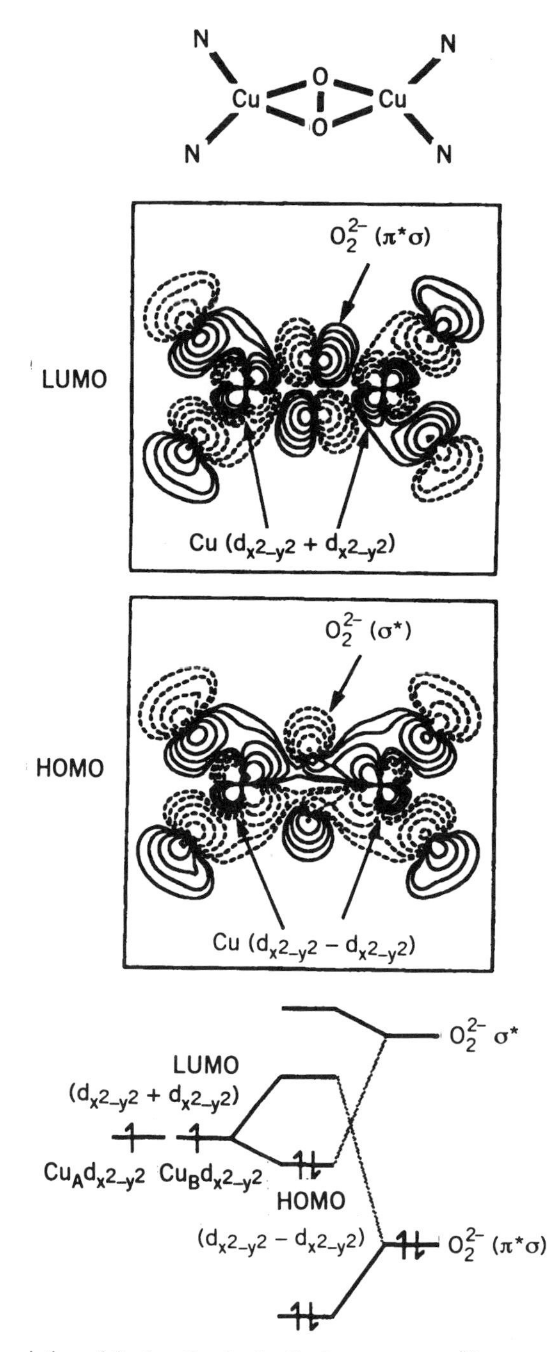

Figure 39. Description of the bonding in the Cu_2O_2 core of $[(Tp^{R2}Cu)_2(O_2)]$ (R = Me, Ph, or *i*-Pr). [Reprinted with permission from E. I. Solomon, M. J. Baldwin, M. D. Lowery, *Chem. Rev.*, *92*, 521 (1992). Copyright © 1992 American Chemical Society.]

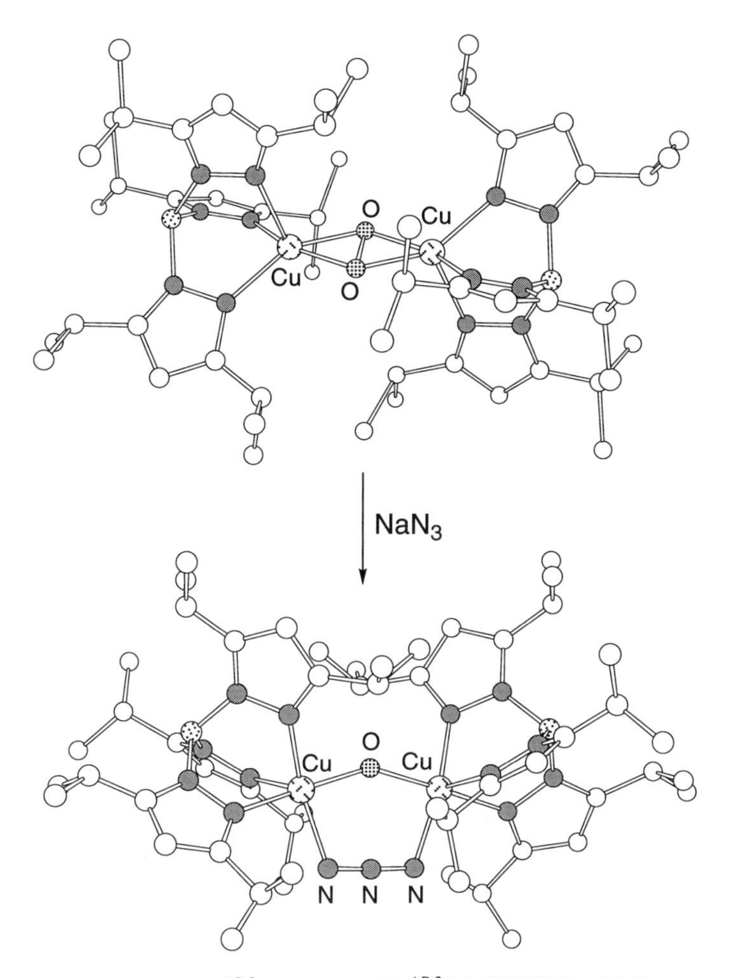

Figure 40. Conversion of $[(Tp^{i\text{-}Pr2}Cu)_2(O_2)]$ to $[(Tp^{i\text{-}Pr2}Cu)_2(OH)(N_3)]$, with X-ray crystal structures of both reactant and product shown (180).

ilar structural and spectroscopic features to metN$_3$Hc and its μ-hydroxo bridge was shown to derive from H_2O in the reaction medium. Thus, synthetic modeling of the oxyHc $\rightarrow$ metN$_3$Hc reaction was accomplished and the competence of the μ-$\eta^2:\eta^2$-peroxo $\rightarrow$ (μ-hydroxo)(μ-1,3-azido) bridge exchange in dicopper(II) chemistry was demonstrated.

Additional reactivity studies have focused on modeling the oxidation activity of Tyr by examining the interactions of $[(Tp^{R2}Cu)_2(O_2)]$ (R = Me or *i*-Pr) with phenols. Mechanistic studies have centered on the Tp^{Me2} complex because of its greater reactivity, but more hindered ligands have been used to help sup-

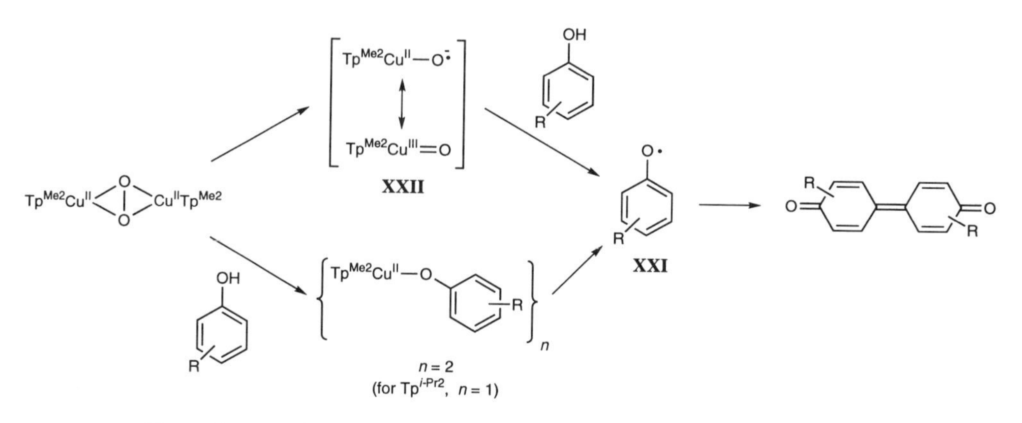

Figure 41. Postulated mechanisms of phenol oxidation by $[(Tp^{Me2}Cu)_2(O_2)]$ (181).

port the plausibility of key intermediates in the oxidations uncovered. Addition of hindered *tert*-butyl-substituted phenols to anaerobic solutions of $[(Tp^{Me2}Cu)_2(O_2)]$ yielded diphenoquinones (Fig. 41) (181). The same radical coupling products of these hindered phenols are generated by Tyr (182, 183). On the basis of kinetic evidence obtained for the model system, two separate reaction pathways were postulated, both of which result in the formation of phenyl radical **XXI**, which couples to give the final diphenoquinone. In one route, rate determining O—O bond cleavage yields monomeric "$Tp^{Me2}CuO$" (**XXII**), which then rapidly abstracts a hydrogen atom from the phenol to yield **XIX**. Corroboration of rate-determining O—O bond homolysis has come from separate kinetics investigations that showed a first-order dependence of the spontaneous decomposition rate on the concentration of the peroxo complex (181). In a second pathway, the peroxo moiety is envisioned to deprotonate the phenol to yield a Cu^{II}-phenoxide, which is susceptible to subsequent attack by O_2 (or a Cu^{II}-hydroperoxide) to afford the phenoxyl radical. The viability of this route was shown by the isolation and structural characterization of a mononuclear phenoxide complex of the more hindered $Tp^{i\text{-}Pr2}$ supporting ligand, $Tp^{i\text{-}Pr2}Cu(OC_6H_4\text{-}p\text{-}F)$, that affords diphenoquinone upon decomposition (Fig. 42) (183). Mechanisms for tyrosinase activity have been put forth based on these experimental results with the Tp^{R2} model systems (163), but alternative routes have been proposed based on data acquired for different synthetic compounds (158, 164), leaving many mechanistic issues unresolved and, therefore, open to further study.

2. *Models of Type 1 Copper Sites*

Type 1 copper proteins are ubiquitous electron carriers found in many bacteria and plants (184, 185). Unlike ordinary monomeric, tetragonal Cu^{II} centers

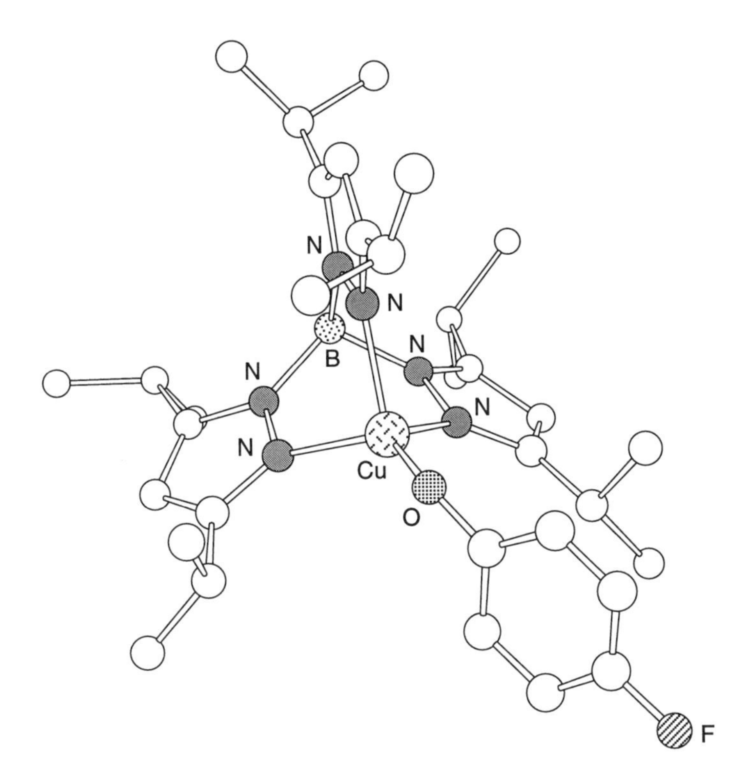

Figure 42. X-ray crystal structure of $Tp^{i\text{-}Pr2}Cu(O\text{-}C_6H_4\text{-}p\text{-}F)$ (48).

that are characterized by weak *dd* bands in the visible region of their absorption spectra (600–800 nm) and an axial signal ($g_{/\!/} > g_{\perp} > 2.0$) with $A_{/\!/}$ in the range 150–200 G in their EPR spectra indicative of a $d_{x^2-y^2}$ ground state, most type 1 copper sites derive their deep blue color from the presence of an intense (ϵ = 3000–7000 cm^{-1} M^{-1}) absorption at about 600 nm and exhibit an axial EPR signal with $A_{/\!/} < 70$ G (158). Some other "green" type 1 copper sites exhibit increased absorption intensity at about 480 nm and have a rhombic ($g_x \neq g_y \neq g_z$) signal in their EPR spectra (186). Other important characteristics of type 1 proteins include high and variable reduction potentials and complicated resonance Raman spectral features in the Cu—S vibration region (250–500 cm^{-1}) (186). X-ray crystal structures of a number of these proteins have been obtained and common to them all is tight coordination of two histidine imidazoles (~2.0 Å) and a thiolate sulfur (~2.1 Å) from a cysteine residue to a single copper ion in an approximately trigonal planar array (180). Additional weaker axial interactions with a methionine sulfur (2.6–3.1 Å) and/or a peptide carbonyl oxygen (>3 Å) also occur. The extent of these interactions varies among different proteins, with shorter $Cu-S_{met}$ distances resulting in copper ion geometries closer to tetrahedral.

A long standing research objective has been to correlate the unusual physicochemical properties of the type 1 copper centers with their structural features. Although significant insight into the electronic structures of these electron carriers has been obtained through combined theoretical and spectroscopic studies (158, 186), definitive integration of structural and spectral features has been hindered by a lack of suitable model complexes. The synthetic difficulties involved in the preparation of a Cu^{II} complex that simultaneously exhibits a trigonal planar or distorted tetrahedral geometry, thiolate ligation with a short $Cu-S$ bond length, and the appropriate electronic absorption and EPR spectroscopic features are substantial. Particularly problematic is the preference for Cu^{II} to adopt a tetragonal geometry and the facile nature of the reduction of Cu^{II} to Cu^{I} by thiolates to yield RSSR species via irreversible radical coupling of $RS\cdot$. The first problem can be overcome by using highly sterically hindered $Tp^{RR'}$ ligands that, as discussed in Section II.C.1, can act as "tetrahedral enforcers." A number of pseudotetrahedral Cu^{II} complexes of such ligands have been prepared, with X-ray crystal structures reported for $Tp^{t\text{-Bu}}CuX$ [X = Cl (43), Br (48), CF_3SO_3 (45), N_3 (48)], $Tp^{t\text{-Bu,Me}}CuX$ [X = Cl (48, 49), SCN (48)], and $Tp^{i\text{-Pr}2}CuX$ [X = Cl (51), Br (48), OC_6H_4-*p*-F (183), $OCH(CF_3)_2$ (48), $OOCMe_2Ph$ (187)]. As is generally found for $Tp^{RR'}MX$ species, most of the complexes have C_{3v}-distorted tetrahedral geometries, with acute $N_{pz}-Cu-N_{pz}$ and correspondingly obtuse $N_{pz}-Cu-X$ angles (91°–95° and 122–124°, respectively). A key spectroscopic signature of these compounds is a "reverse" EPR signal indicative of a d_z^2 ground state (cf. spectrum for $Tp^{t\text{-Bu}}CuCl$ in Fig. 43) (188). It is noteworthy that for $Tp^{i\text{-Pr}2}CuCl$ in *N,N*-dimethylformamide (solvent) (DMF), dimethyl sulfoxide (solvent) (DMSO), tetrahydrofuran (THF), or acetone at low temperature, an axial signal typical for tetragonal Cu^{II} is observed. This and other optical absorption data sugget that solvent coordination yields square pyramidal complexes, an hypothesis that was supported by the X-ray crystallographic characterization of $Tp^{i\text{-Pr}2}Cu(dmf)Cl$ (Fig. 7) (51).

In early work, Tp^{Me2} was used in attempted syntheses of type 1 copper protein models (189). A thermally unstable thiolate complex was generated which, on the basis of comparison to the X-ray structure of a related Co^{II} species $Tp^{Me2}Co(SC_6H_4\text{-}p\text{-}NO_2)$ (190) and observation of an intense electronic absorption band (λ_{max} = 588 nm, ϵ = 3900 M^{-1} cm^{-1}), was postulated to be monomeric, tetrahedral $Tp^{Me2}Cu(SC_6H_4\text{-}p\text{-}NO_2)$. Its EPR spectrum recorded in THF at 77 K is typical for a tetragonal complex, however. In light of the aforementioned results for $Tp^{i\text{-Pr}2}CuCl$, it is likely that the complex actually exists as a solvent-coordinated species, $Tp^{Me2}Cu(SC_6H_4\text{-}p\text{-}NO_2)(thf)$. Thus, it would appear that in order to isolate an accurate mimic of the type 1 site, relatively

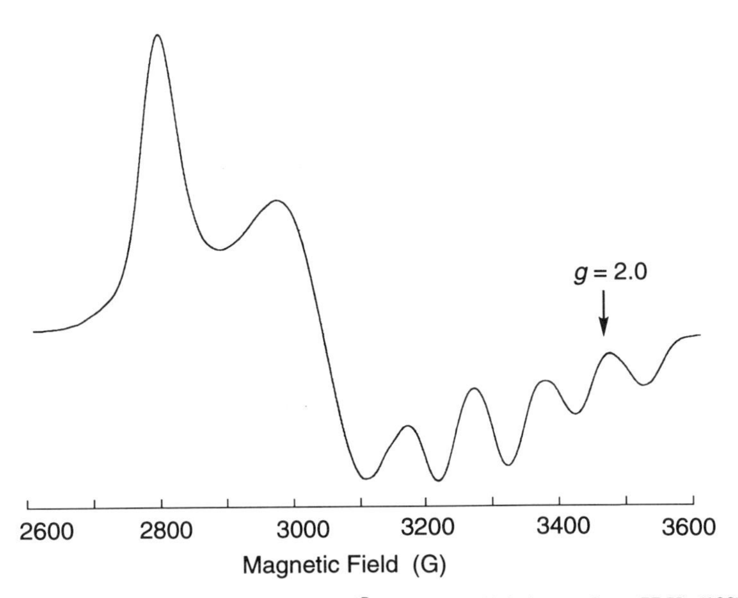

Figure 43. X-band EPR spectrum of $Tp^{t\text{-}Bu}CuCl$ (CH_2Cl_2/toluene glass, 77 K) (188).

noncoordinating solvents (e.g., CH_2Cl_2 or pentane) would have to be used in addition to using more hindered $Tp^{RR'}$ ligands to enhance product stability.

The ultimately successful strategy involved treating $[(Tp^{i\text{-}Pr2}Cu)_2(\mu\text{-}OH)_2]$ in CH_2Cl_2 or pentane with thiols having bulky or electron-withdrawing substituents (*t*-BuSH, *sec*-BuSH, Ph_3CSH, or C_6F_5SH), which would disfavor unwanted redox reactions (e.g., Cu^{II} reduction) (51, 191). An X-ray crystal structure was obtained for $Tp^{i\text{-}Pr2}Cu(SC_6F_5)$, which confirmed its monomeric formulation (Fig. 44). Interestingly, the most appropriate description of the geometry of the complex is trigonal pyramidal instead of tetrahedral, with the Cu^{II} ion residing 0.34 Å from the basal plane comprised of N_2S donor atoms (for a tetrahedron, this distance would be ~0.70 Å). This topology and a notably short Cu^{II}-S distance of 2.176(4) Å bear a striking resemblance to the analogous features of the copper sites of azurin and plastocyanin (185), although in the model compounds a pyrazolyl group occupies a position close to that of a weakly bound methionine sulfur in the proteins. The spectroscopic features of the model complexes are strikingly similar to those of the proteins, however, indicating that these properties derive primarily from the $CuN_2S_{thiolate}$ unit. Key features exhibited by the complexes that are analogous to those of the proteins include an axial EPR signal with a low $A_{\parallel}$ value (50–70 G), an intense $S_{thiolate} \rightarrow Cu^{II}$ charge-transfer band at 625–665 nm ($\epsilon \sim 6000\ M^{-1}\ cm^{-1}$), and complex resonance Raman spectral features in the Cu—S region due to coupling of Cu—S and C—C—S vibrational modes (192).

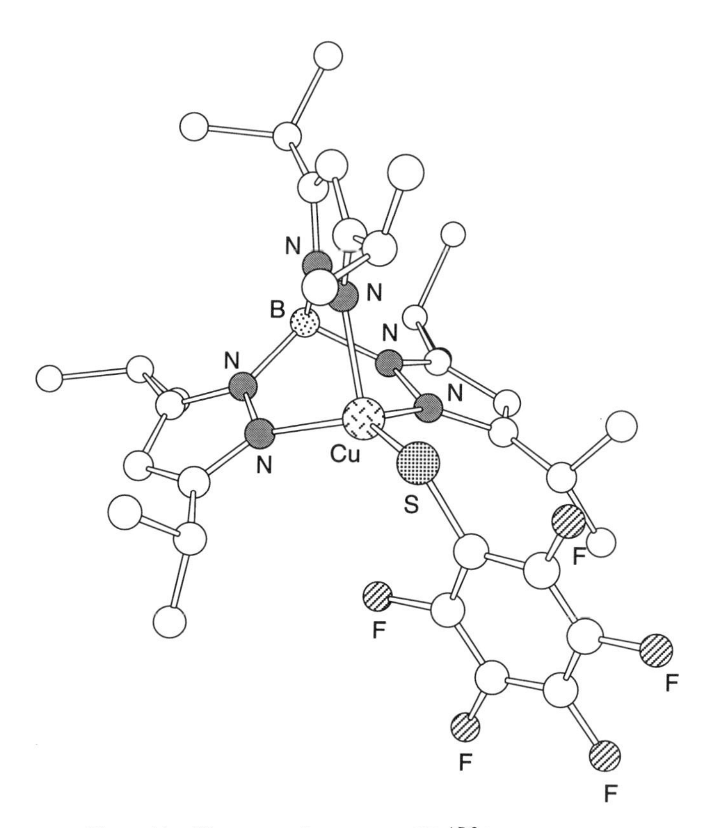

Figure 44. X-ray crystal structure of $Tp^{i\text{-}Pr2}Cu(SC_6F_5)$ (191).

A theoretical rationalization of these spectroscopic properties that emphasizes the importance of the strong, highly covalent $Cu^{II}—S_{thiolate}$ interaction has been provided and postulates relating the electronic structure of the unusual type 1 centers to pathways of electron transfer (i.e., function) have been put forth based on these conclusions (158). In addition, a relationship between the nature of the EPR signal (rhombic vs. axial), the relative intensity of the 460- and 600-nm absorption bands, and the Cu^{II} geometry (trigonal planar vs. distorted tetrahedral) has been noted (186). Interestingly, the model compounds fit into the category of sites that exhibit an axial EPR signal, a weak 460-nm absorption band, and a trigonal planar geometry, despite the more general tendency of $Tp^{RR'}CuX$ complexes to adopt more C_{3v}-symmetric structures. It appears that the combined effects of the unusually strong, covalent copper–thiolate interaction and the unique ability of the $Tp^{i\text{-}Pr2}$ ligand to allow distortions away from tetrahedral topology to occur (yet to prevent L_2M formation and enable isolation of sensitive species) are responsible for the observed chemistry.

3. *Models of Putative Monooxygenase Intermediates*

Monooxygenase activity has been characterized for a number of enzymes other than tyrosinase that use different numbers of copper ions to activate dioxygen. For example, although the active site of dopamine-β-monooxygenase (DβM), an important enzyme that catalyzes the benzylic hydroxylation of dopamine to yield the neurotransmitter norepinephrin, contains two copper ions, it is believed that the oxidation chemistry occurs at a single copper ion and that the other copper site located nearby functions to shuttle electrons (193–195). A single copper ion is also believed to perform the critical oxidation chemistry catalyzed by peptidylglycine-α-amidating enzyme (PAM) (196), while tricopper sites have been proposed to act in concert in particulate methane monooxygenase (pMMO) (197). Monocopper adducts of dioxygen, its redox partners superoxide and peroxide, as well as protonated variants, such as hydroperoxide, have been postulated to be important intermediates in the monooxygenase pathways of each of these enzymes, making the synthesis and characterization of complexes containing such moieties an important research goal.

Because an initial reaction of a protein-bound Cu^I ion with O_2 to yield a $Cu^{II}-(O_2^-)$ species is common to virtually all hypothesized reaction pathways, many attempts have been made to synthesize a Cu^{II} complex of superoxide (163). Such a complex was proposed to have been formed when the Cu^I species $Tp^{Me2}Cu(C_2H_4)$ was exposed to O_2 (198), but reanalysis of the published characterization data for the putative product, $Tp^{Me2}CuO_2$, in light of more recent work suggests that it is instead the dicopper(II)–peroxo complex $[(Tp^{Me2}Cu)_2(O_2)]$. The observation of such dinuclear species even with the $Tp^{i\text{-}Pr2}$ ligand implies that the secondary reaction of $Tp^{RR'}CuO_2$ with another $Tp^{RR'}Cu^I$ fragment is much faster than the initial dioxygen addition. A similar reaction pathway involving initial formation of a mononuclear superoxide complex followed by rapid trapping by a Cu^I fragment has been quantitatively delineated for oxygenation reactions that afford other peroxo dimers (199, 200). Recently, by using $Tp^{t\text{-}Bu,i\text{-}Pr}$ instead of $Tp^{i\text{-}Pr2}$, a subtle modification that nonetheless affords an increase in steric bulk sufficient to inhibit approach of the metal ions to closer than 4 Å (the Cu· · ·Cu distance in the μ-peroxo complexes is ~3.6 Å), the secondary reaction of a monomeric adduct with Cu^I starting material was prevented (201). Thus, reversible binding of 1 equivalent of O_2 by $Tp^{t\text{-}Bu,i\text{-}Pr}Cu(dmf)$ was observed and formulation of the product as superoxide complex $Tp^{t\text{-}Bu,i\text{-}Pr}CuO_2$ was confirmed by resonance Raman spectroscopy ($\nu_{^{16}O-^{16}O} = 1111\ cm^{-1}$; $\nu_{^{18}O-^{18}O} = 1062\ cm^{-1}$) and an X-ray crystal structure determination (Fig. 45). This first crystallographic characterization of a copper–superoxo complex (see Note Added in Proof) revealed a symmetric, side-on coordination of O_2^- [Cu—O(1) = Cu—O(1′) = 1.84(1) Å], and an O—O bond length [1.23(3) Å] typical for superoxide ion bound to

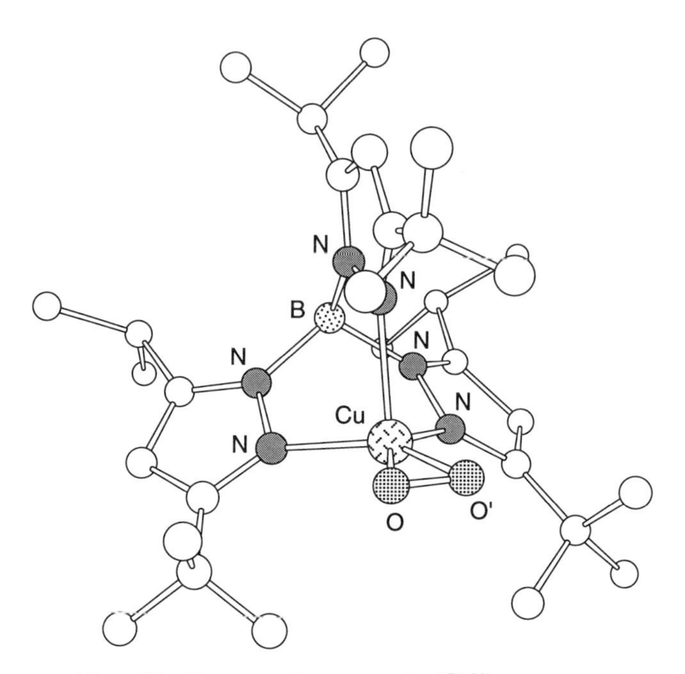

Figure 45. X-ray crystal structure of $Tp^{t\text{-Bu, Me}}Cu(O_2)$ (201).

a transition metal. Note that the symmetric binding of O_2^- in this complex is similar to the η^2 coordination of superoxide that occurs in the cobalt analogue $Tp^{t\text{-Bu, Me}}CoO_2$ [Co—O(1) = 1.816(5) Å, Co—O(2) = 1.799(6) Å] (202). A theoretical basis for these observations is eagerly awaited.

One-electron reduction and protonation of a Cu^{II}-(O_2^-) species to afford a Cu^{II}-(OOH) moiety has been postulated to occur in several monooxygenase mechanisms. In DβM, for example, the additional electron has been suggested to originate from a second nearby copper ion. According to an intriguing hypothesis, the resulting monocopper hydroperoxide oxidizes a nearby tyrosine residue to a tyrosyl radical, which then participates in the key benzylic C—H bond activation reaction step (Fig. 46) (195). Developing an understanding of the detailed chemistry of Cu^{II}—(OOH) compounds in thus an important goal. Since alkyl- or acylperoxo complexes would be expected to be more amenable to detailed study due to their enhanced stability compared to a hydroperoxo complex, yet should exhibit parallel reactivity patterns, they have also been targeted for study as models of putative monooxygenase intermediates.

Mononuclear alkyl- and acylperoxo complexes $Tp^{i\text{-Pr2}}CuOOR$ (R = alkyl or acyl) were isolated from the reaction of $[(Tp^{i\text{-Pr2}}Cu)_2(OH)_2]$ with the appropriate hydroperoxide (ROOH) (203, 187). A five-coordinate geometry was proposed

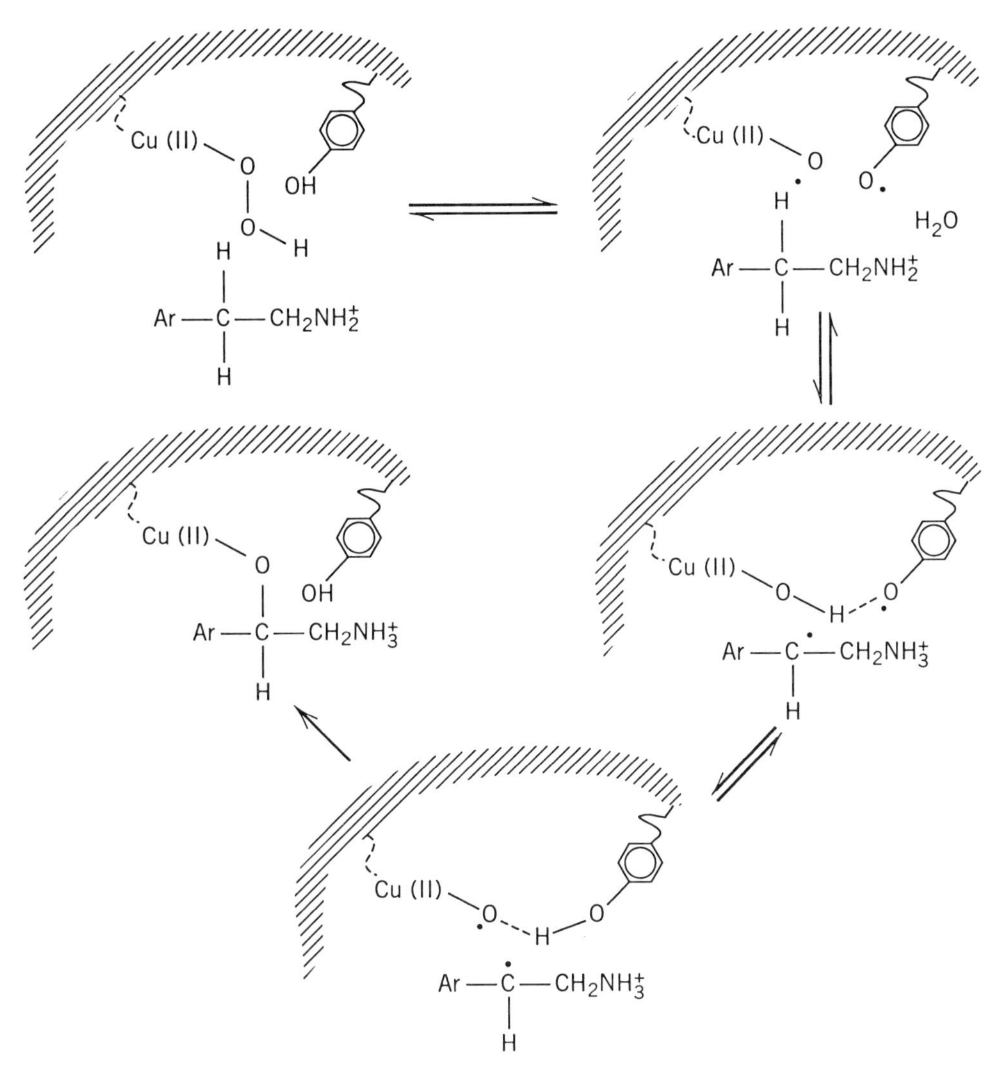

Figure 46. Proposed mechanisms for DβM-catalyzed benzylic hydroxylations. [Reprinted with permission from G. Tan, J. A. Bery, J. P. Keinman, *Biochemistry*, *33*, 226 (1994). Copyright © 1994 American Chemical Society.]

for the *m*-chloroperbenzoate complex on the basis of combined EPR (axial signal with $A_{\parallel} = 160$ G) and IR ($\nu_{CO} = 1640\ cm^{-1}$) data (203). An X-ray crystal structure obtained for the cumylperoxide complex [R = $C(Me)_2Ph$] indicated a four-coordinate, pseudotetrahedral geometry (Fig. 47) (187). Consistent with this structure, the complex exhibits a rhombic EPR signal similar to that observed for other $Tp^{RR'}CuX$ complexes (see above).

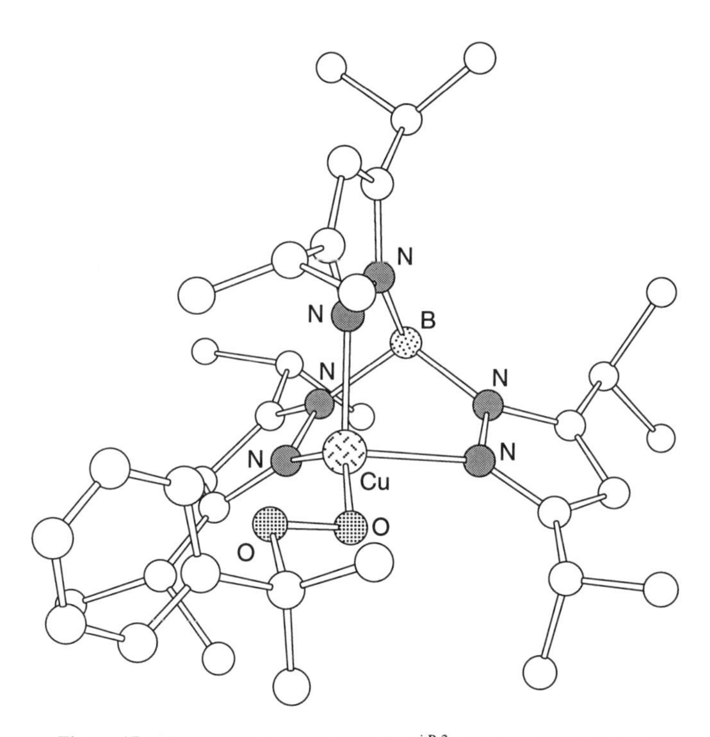

Figure 47. X-ray crystal structure of $Tp^{i\text{-}Pr2}Cu(OOCMe_2Ph)$ (187).

In more recent work using the extremely hindered $Tp^{t\text{-}Bu,i\text{-}Pr}$ ligand a novel monocopperhydroperoxo complex was identified (204). The synthesis utilized as starting material $Tp^{t\text{-}Bu,i\text{-}Pr}CuOH$, a rare example of a discrete, monomeric Cu^{II} complex with a terminal hydroxo group [with the smaller $Tp^{i\text{-}Pr2}$ ligand, only dinuclear μ-hydroxo-bridged dimers of divalent transition metals were obtained (205)]. Addition of H_2O_2 to this complex at $-78\,°C$ yielded a purple product which, although too unstable to be isolated, was identified as $Tp^{t\text{-}Bu,i\text{-}Pr}Cu(OOH)$ on the basis of spectroscopic measurements. Thus, a signal in its EPR spectrum closely analogous to that of tetrahedral $Tp^{i\text{-}Pr2}Cu(OO\textit{t}\text{-}Bu)$ and a ν_{OH} at 3611 cm^{-1} (ν_{OD} = 2688 cm^{-1}) different from that of the hydroxo precursor (ν_{OH} = 3655 cm^{-1}) in its Fourier transform infrared (FTIR) spectrum support the monomeric hydroperoxo formulation. Multiple lines in the resonance Raman spectrum in the region of an anticipated ν_{O-O} that shift upon oxygen isotope substitution were observed. While not yet interpreted definitively, the complicated resonance Raman spectral features are similar to those reported for a monomeric ferric alkyl peroxo complex (206).

The alkylperoxo, acylperoxo, and hydroperoxo complexes were found to be relatively unreactive oxidants; phosphines and sulfoxides are oxidized, but alkenes are not epoxidized. Analysis of the products of the spontaneous decomposition of the alkylperoxo complexes suggested that either Cu—O or O—O homolytic bond cleavage to yield HOO· or HO·, respectively, was the predominant pathway (203). This reactivity contrasts with that exhibited by a related (acylperoxo)(porphinato)iron(III) species, which undergoes O—O bond heterolysis to afford an electrophilic Fe^{IV}=O/porphyrin cation radical that rapidly epoxidizes cyclohexane (207). While the lack of reactivity of the $Tp^{i\text{-}Pr2}$CuOOR (R = acyl, alkyl, or H) complexes may be due to the protective steric bulk of the supporting $Tp^{i\text{-}Pr2}$ ligand, the observation of similar behavior for other copper hydroperoxo and acylperoxo complexes suggests that a more general rationale is warranted (208). A possible explanation for the intrinsically lower oxo-transfer reactivity of the copper species compared to the iron–porphyrin analogues may be that the copper(IV) oxo or copper(III) oxo radical cation products of O—O bond heterolysis are much less stable than the analogous iron–porphyrin species. More extensive mechanistic and reactivity studies of the novel $Tp^{RR'}$CuOOR complexes need to be completed before definitive conclusions can be reached, however.

4. *Models of Putative Nitrite Reductase Intermediates*

Copper-containing nitrite reductases (NiR) catalyze the dissimilatory reduction of NO_2^- to NO and, under certain conditions, N_2O, during biological denitrification, an important component of the global nitrogen cycle (209). An X-ray crystal structure of the enzyme from *Achromobacter cycloclastes* showed that each subunit in the trimeric protein contains two copper sites 12.5 Å apart (Fig. 48) (210). One (Cu—I) exhibits structural and spectroscopic parameters typical for type I copper sites, while the other (Cu—II) is coordinated to three histidine imidazolyl donors and a water molecule in an unusual, pseudotetrahedral geometry. It has been suggested that Cu—I acts to shuttle electrons and that substrate (NO_2^-) binding and subsequent reduction occurs at Cu—II (211, 212). Based on mechanistic work with this and other enzymes, a mechanism has been proposed that involves nitrite binding to the reduced copper ion, followed by dehydration to yield an electrophilic nitrosyl (Fig. 49) (213–215). Evolution of NO is then proposed to occur to afford this principal enzyme product and a Cu^{II} ion. The N—N bond formation route necessary for production of N_2O is a more complex issue. Pathways involving NO and/or NO_2^- attack on the nitrosyl intermediate have been proposed, although details of these processes are unknown. The lack of precedent in the form of synthetic model complexes for the various proposed monocopper intermediates has stimulated the recent interest in the study of Cu—NO_x compounds (45, 188, 216–222).

Sterically hindered $Tp^{RR'}$ ligands that mimic the tris(histidyl) ligand array in

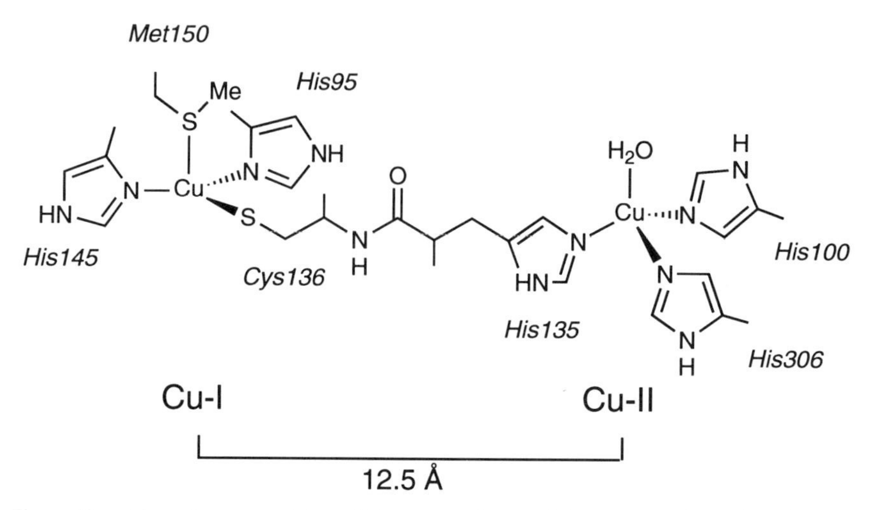

Figure 48. Schematic representation of the copper sites in nitrite reductase from *A. cycloclastes*. [Reprinted with permission from C. E. Ruggiero, S. M. Carrier, W. E. Antholine, J. W. Whittaker, C. J. Cramer, and W. B. Tolman, *J. Am. Chem. Soc.*, *115*, 11285 (1993). Copyright © 1993 American Chemical Society.]

NiR have been used to prepare $Cu^{I}{-}NO$ and $Cu^{II}{-}NO_2^-$ adducts, the study of which has provided insight into structural, spectroscopic, and reactivity features of possible NiR intermediates. A series of Cu^{II}–nitrite complexes $Tp^{RR'}CuNO_2$ (R = R′ = Me or Ph; R = *t*-Bu, R′ = H) have been synthesized and X-ray structures have been obtained for those supported by Tp^{Me2} and $Tp^{t\text{-}Bu}$ (Fig. 50) (45, 220). Both of these latter complexes exhibit η^2-*O*,*O*-nitrito coordination, but the differing steric influences of their respective ligands induces significantly

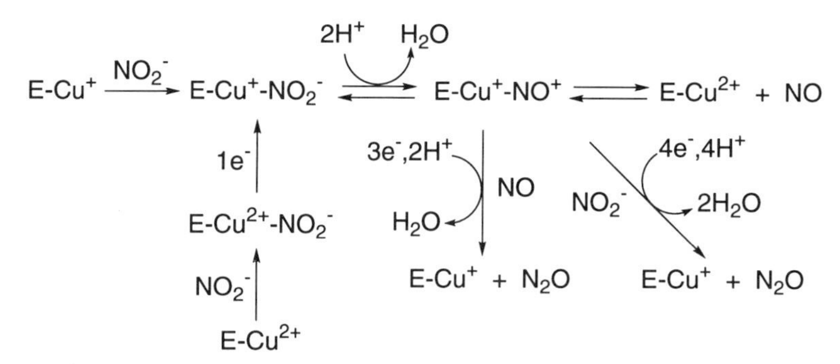

Figure 49. Proposed mechanisms of nitrite reduction by NiR. [Reprinted with permission from C. E. Ruggiero, S. M. Carrier, W. E. Antholine, J. W. Whittaker, C. J. Cramer, and W. B. Tolman, *J. Am. Chem. Soc.*, *115*, 11285 (1993). Copyright © 1993 American Chemical Society.]

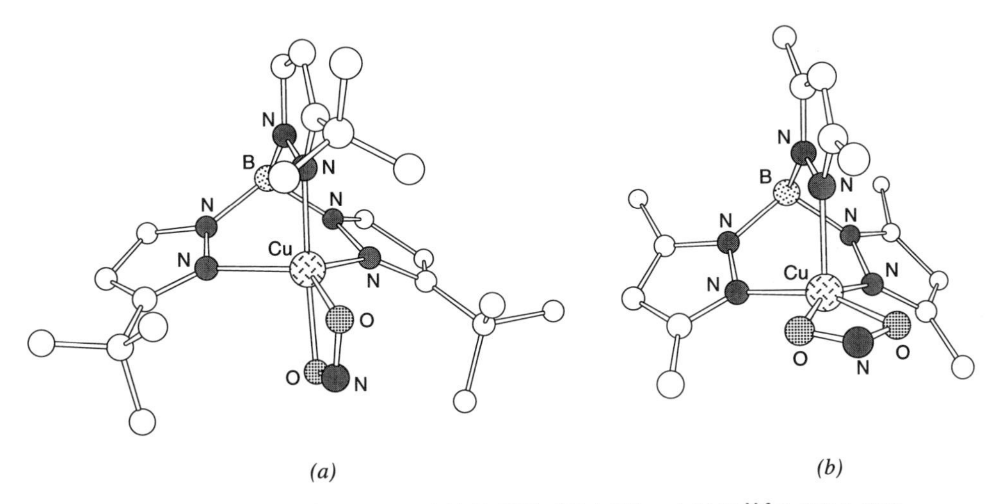

Figure 50. X-ray crystal structures of (*a*) $Tp^{t\text{-}Bu}Cu(NO_2)$ (45) and (*b*) $Tp^{Me2}Cu(NO_2)$ (220).

different copper ion geometries. While a square pyramidal topology with one pyrazolyl arm of the Tp^{Me2} ligand in the axial position is adopted by $Tp^{Me2}CuNO_2$ (and, by virtue of analogies between spectroscopic data, by $Tp^{Ph2}CuNO_2$ as well) (220), the $Tp^{t\text{-}Bu}$ complex is best described as trigonal bipyramidal (45). Important differences in spectroscopic properties, including EPR signal shapes (axial for Tp^{Me2} vs. rhombic for $Tp^{t\text{-}Bu}$), imply that these structural differences are retained in solution. We speculate that destabilizing intramolecular interactions between the nitrite oxygen atoms and the *tert*-butyl substituents that would be present if the square pyramidal structure were adopted by $Tp^{t\text{-}Bu}CuNO_2$ may cause the geometry switch in this complex (only one such interaction, instead of two, is present in the trigonal bipyramidal structure). Both $Cu^{II}{-}NO_2^-$ complexes, as well as others prepared using alternative supporting ligands, may be envisioned as models of the NiR substrate adduct, insofar as they have the appropriate molecular composition. Efforts to dehydrate the $Tp^{RR'}CuNO_2$ complexes by adding protonic acids have not led to NO evolution, however, suggesting that their functional relevance is limited (45). Reduction by an e^- may be a necessary prerequisite for dehydration of a nitrite adduct, a hypothesis supported by observation of clean NO evolution upon protonation of a $Cu^{I}{-}NO_2^-$ complex (N bound) supported by a hindered triazacyclononane ligand (222).

The postulate of a copper nitrosyl intermediate in denitrification is particularly intriguing because of the paucity of such species in the synthetic chemistry literature (216). In attempts to prepare a copper–NO adduct, the Cu^{I} dimers $[Tp^{RR'}Cu]_2$ (R = R′ = Me or Ph; R = *t*-Bu, R′ = H) were exposed to NO. The results of these reactions provide yet another illustration of the profound

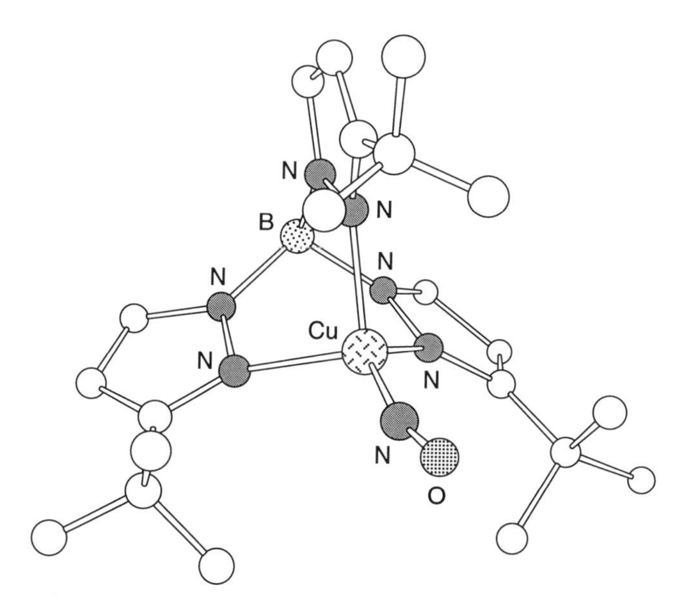

Figure 51. X-ray crystal structure of $Tp^{t\text{-}Bu}CuNO$ (188).

effects of differing Tp ligand steric bulk on the course of coordinated metal chemistry. When $Tp^{t\text{-}Bu}$ or Tp^{Ph2} were used, the respective $Tp^{RR'}CuNO$ complex was isolated and, for the $Tp^{t\text{-}Bu}$ case, characterized by X-ray crystallography (Fig. 51) (188). Consideration of the structural parameters of the CuNO unit in the complex and the ν_{NO} in the IR spectrum (1712 cm^{-1}) were insufficient for defining its electronic structure, which can be envisioned in valence bond terms as $Cu^0—NO^+$, $Cu^I—NO\cdot$, or $Cu^{II}—NO^-$. Detailed analysis of electronic absorption, magnetic circular dichroism (MCD), and multifrequency EPR spectroscopic data for the $Tp^{t\text{-}Bu}$ and Tp^{Ph2} complexes led to the conclusion that the most appropriate valence bond resonance description is $Cu^I—NO\cdot$. Alternatively, a formulation that more explicitly accounts for the covalency inherent in the bonding was provided based on ab initio calculations performed using the model fragment $(NH_3)_3CuNO$. According to this description, the molecular orbitals with mostly copper *d* character are filled and the unpaired electron resides in an orbital with significant contributions from a π^* orbital localized on NO, yet still having density on copper. Key supporting spectroscopic data include a lack of *dd* bands in the region where such bands are normally observed in pseudotetrahedral $Tp^{RR'}Cu^{II}X$ complexes (>800 nm) and unusual EPR features significantly different from those typical for Cu^{II} complexes ($g_e \sim g_\perp > g_{/\!/} = 1.83$, with large nitrogen and copper hyperfine couplings). Although the overall electron inventory of the nitrosyl complexes $\{[CuNO]^{11}$

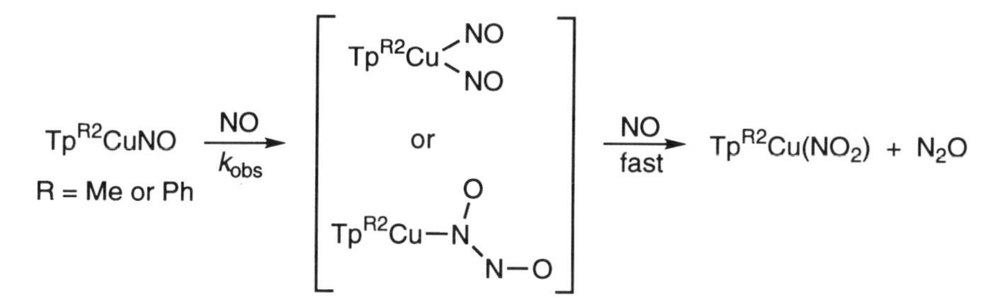

Figure 52. Proposed mechanisms for the disproportionation of NO by $Tp^{RR'}CuNO$ [adapted from (220)].

according to the Feltham and Enemark nomenclature (223)} is one electron less than that of the postulated product of dehydration of the enzyme nitrite adduct $\{[CuNO]^{10}\}$, a species in the biological system analogous to the one synthesized may result upon reduction of the dehydration product or by coordination of NO to the reduced Cu-II site.

When $Tp^{Ph2}CuNO$ was allowed to stand for extended periods (hours) under an NO atmosphere or when the NO reaction was carried out with $[Tp^{Me2}Cu]_2$, the nitrosyl adduct disappeared and one equivalent of N_2O was generated (220). Identification of the copper-containing products of these reactions as $Tp^{R2}CuNO_2$ (R = Me or Ph) and isotope labeling experiments indicated that an overall reductive disproportionation of NO had occurred (Eq. 9). Such a reaction, which may be involved in N_2O evolution by NiR, had not been definitively observed in copper chemistry previously. In accord with the relative size of the supporting Tp^{R2} ligands, the disproportionation was significantly faster for the complexes of Tp^{Me2} ($t_{1/2} \sim 8$ min) compared to Tp^{Ph2} ($t_{1/2} = 1.5$ h). Kinetic data acquired for the $Tp^{Ph2}CuNO$ decomposition indicated that the reaction was first order in the copper complex, ruling out rate-determining dimerization (which is not possible in the enzyme) and supporting attack of a second NO molecule, either at copper or at the bound NO, in the slow step (Fig. 52). The nature of the subsequent N—N bond formation, oxygen-atom transfer, and/or electron-transfer steps that afford the ultimate products (N_2O and $Cu^{II}—NO_2^-$ species) are not understood at present.

$$Tp^{R2}Cu^{I}(MeCN) + 3NO \longrightarrow Tp^{R2}Cu^{II}(NO_2) + N_2O \qquad (9)$$

B. Iron Complexes

The ubiquity of nonheme iron centers in biology is now well-recognized and extensive research efforts focused on understanding their wide-ranging structures and functions have been underway for some time (224–229). Noteworthy examples of proteins with such sites are the diiron-containing hemerythrin (O_2 carrier), ribonucleotide reductase (DNA biosynthesis), and methane monoxygenase ($CH_4 \rightarrow MeOH$), as well as those with monoiron active sites, such as extradiol catechol dioxygenases (aromatic ring cleavage), tyrosine hydroxylase (dopa formation), and isopenicillin N synthase (desaturative cyclization), among many others. Synthetic chemical approaches have been especially fruitful in providing structural, spectroscopic, and mechanistic insight into these proteins, with Tp ligands having been key building blocks useful for the construction of important active site model compounds. Indeed, one of the first accurate models of the diiron site of hemerythrin was constructed via self-assembly of KTp, Fe^{III}, and carboxylate ions (154). The product, $[(TpFe)_2(\mu\text{-O})(\mu\text{-O}_2CR)_2]$, was one of the first of which is now a large class of complexes with μ-oxo-bis(μ-carboxylato)diiron(III) cores that closely mimic the resting, diferric state of hemerythrin (Fig. 53) (228). More recently, emphasis has shifted towards the construction of Fe^{II} model complexes and investigation of their reactivity with dioxygen and its redox partners, since in most of the relevant proteins such interactions are central features of their mechanisms of action (229). The steric hindrance provided by large substituents in $Tp^{RR'}$ ligands has been critical in allowing the stabilization and isolation of several novel Fe^{II} complexes and their dioxygen adducts.

Attempts to construct analogues of $[(TpFe)_2(\mu\text{-O})(\mu\text{-O}_2CR)_2]$ using Tp^{Me2} or $Tp^{i\text{-Pr2}}$ instead of Tp were unsuccessful, presumably because of steric interactions that prohibited access to the compact μ-oxo-bis(μ-carboxylato)-

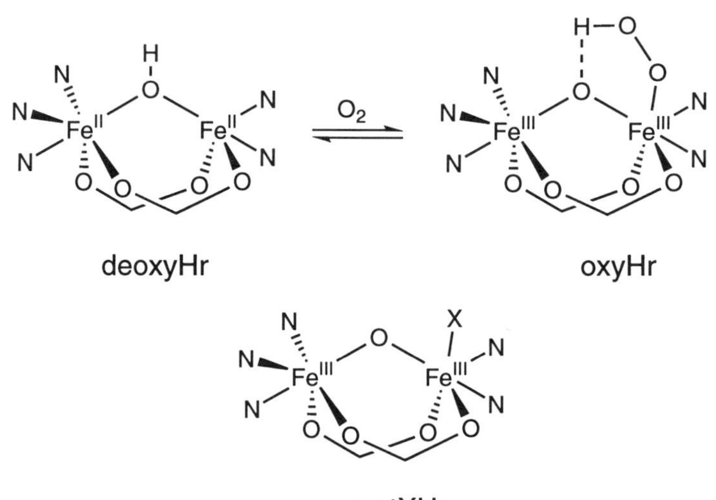

Figure 53. Schematic drawings of the active site structures of various forms of hemerythrin (Hr).

diiron(III) core (Fe· · ·Fe = 3.15 Å). However, the $Tp^{i\text{-}Pr2}$ ligand was used to construct a set of reactive mononuclear Fe^{II} complexes, $Tp^{i\text{-}Pr2}Fe(X)$ [X = OAc, OBz, O_2Ct-Bu, acetylacetonate (acac), OAr, SAr, or OMe] (230–233). The carboxylate and acetylacetonate complexes adopt similar structures (five-coordinate, bidentate, O,O-coordination of O_2CR or acac) and readily bind additional σ donors, such as pyridine or MeCN solvent (Fig. 54). The phenolate, thiolate, and alkoxide complexes are tetrahedral (231).

All of the $Tp^{i\text{-}Pr2}Fe(X)$ compounds react with dioxygen to give interesting Fe^{III} products. Reversible binding of O_2 at −20°C occurs with $Tp^{i\text{-}Pr2}Fe(OBz)$ to yield a product which, primarily on the basis of resonance Raman (ν_{O-O} = 876 cm^{-1}, ν_{Fe-O} = 418 cm^{-1}), extended X-ray absorption five structure (EXAFS) (Fe· · ·Fe = 4.3 Å), magnetic susceptibility (antiferromagnetic coupling with $-2J \sim 30\ cm^{-1}$), and manometric measurements (Fe/O_2 = 2 : 1), was assigned a dinuclear *trans*-1,2-peroxo differic structure (Fig. 55) (232). To explain the rather weak antiferromagnetic coupling and the short distance between the iron atoms, extended Hückel calculations were performed that supported a dihedral (Fe—O—O—Fe) angle of 90° [rather than the more usual 0°

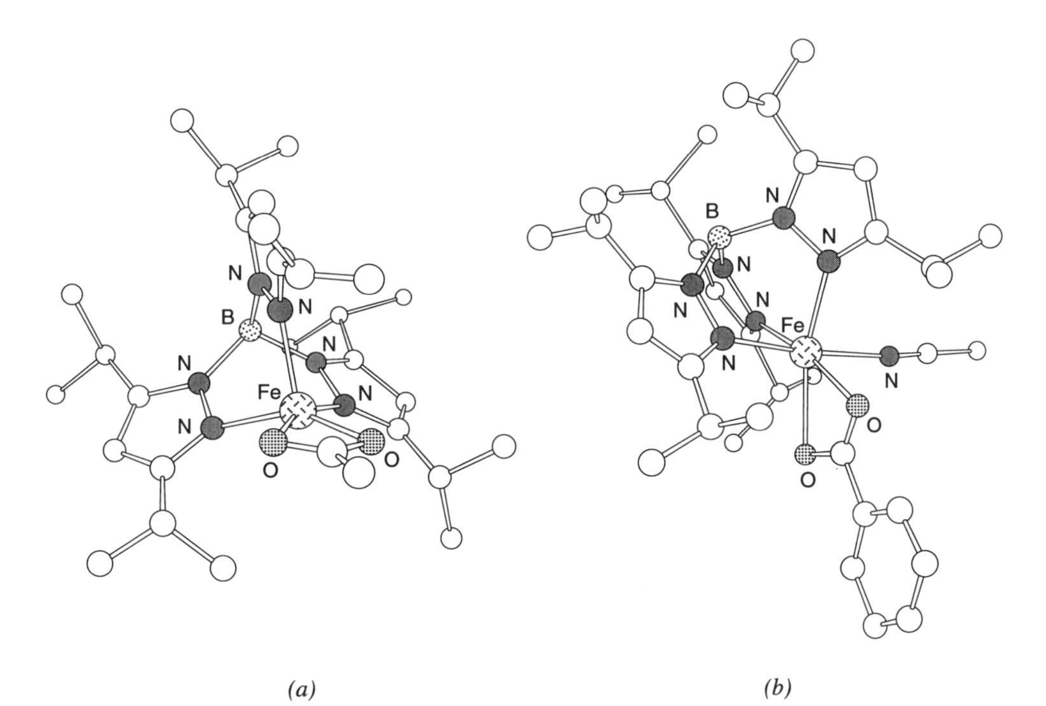

Figure 54. X-ray crystal structures of (*a*) $Tp^{i\text{-}Pr2}Fe(OAc)$ (232) and (*b*) $Tp^{i\text{-}Pr2}Fe(OBz)(MeCN)$, where Bz = benzyl ($CH_2Ph$) (230, 232).

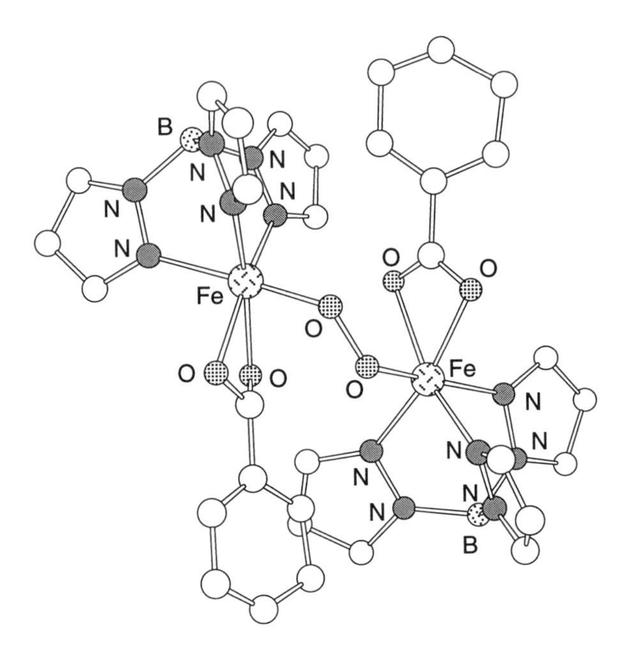

Figure 55. Proposed structure (not based on X-ray data) of the dioxygen adduct $[(Tp^{i\text{-}Pr2}Fe)_2(O_2)]$ [*i*-Pr groups omitted for clarity (232)].

(cis) or 180° (trans)]. This product is a rare example of a nonheme iron dioxygen complex (229) and thus represents an important model of such species in proteins, in particular those containing diiron active sites. The relatively high stability of this peroxo complex probably derives from the protective influence of the hindered $Tp^{i\text{-}Pr2}$ ligand, which inhibits deleterious autoxidation reactions. A similar peroxo species was noted for $Tp^{i\text{-}Pr2}$Fe(acac), but it was less robust and, thus, not characterized in detail.

Both peroxo species (acac and OBz) decomposed upon warming to a linear, trinuclear, Fe_3^{III} complex which, for the case of the product derived from $Tp^{i\text{-}Pr2}$Fe(acac), was characterized by X-ray crystallography (Fig. 56) (232, 233). Interestingly, oxidative cleavage of the acac ligands occurred to afford only acetato bridges between the Fe^{III} ions in the complex, with no acac ligands present. Although the structure contains a crystallographically imposed center of symmetry, analysis of combined Mössbauer, IR, and magnetic data indicated that the compound contains one μ-hydroxo and one μ-oxo bridge that are disordered in the crystal structure. A related triiron(III) complex with symmetrical μ-hydroxo bis(μ-carboxylato) bridges was also prepared using a different sterically hindered, tridentate N-donor capping ligand, **XXIII** (234). The isolation of trinuclear instead of dinuclear species with both $Tp^{i\text{-}Pr2}$ and **XXIII** reflects

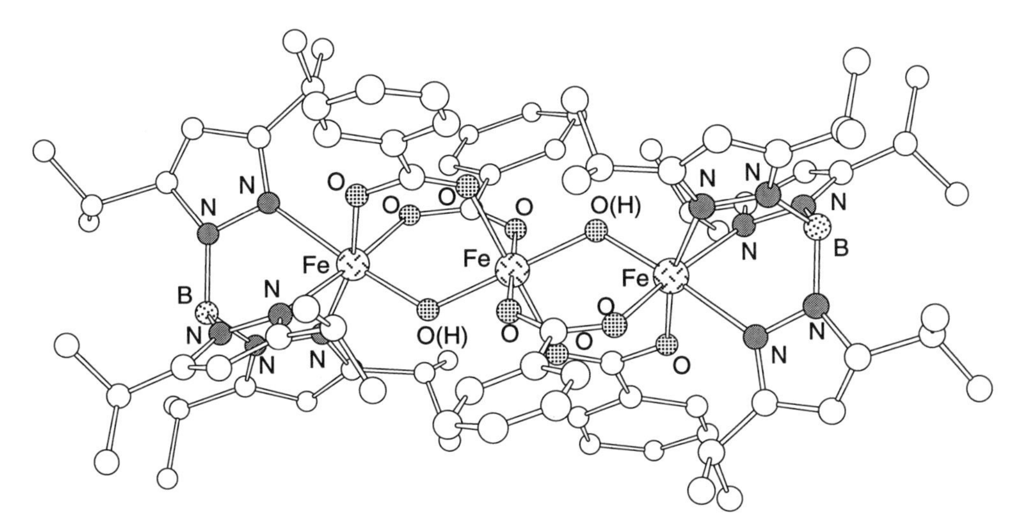

Figure 56. X-ray crystal structure of $[(Tp^{i\text{-}Pr}Fe(\mu\text{-}OH)(\mu\text{-}OBz)_2Fe(\mu\text{-}O)(\mu\text{-}OBz)_2FeTp^{i\text{-}Pr2}]$ (232, 233).

the steric bulk of these chelates, which effectively prevents the close approach necessary for formation of a tribridged diiron(III) complex.

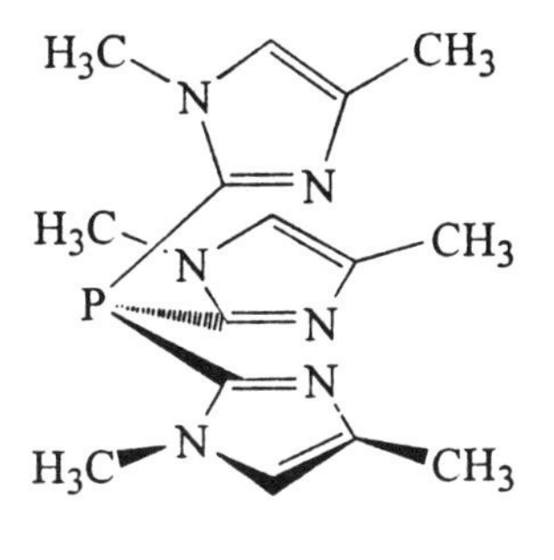

XXIII

Diiron(II) complexes were accessed using $Tp^{i\text{-}Pr2}$, however. Treatment of $Tp^{i\text{-}Pr2}Fe(OBz)$ with NaOH afforded a bis(μ-hydroxo)-bridged diiron(II) complex (Fig. 57) (235), which is but one member of a class of such complexes prepared with a range of divalent metal ions in an effort to model carbonic anhydrase activity (see Section IV.D.1) (205). Addition of one equivalent of benzoic acid to the bis(μ-hydroxo)diiron(II) complex yielded a novel μ-hydroxo μ-carboxylato complex with two five-coordinate Fe^{II} ions (Fig. 58) (235). Clearly, the particular steric properties of the $Tp^{i\text{-}Pr2}$ ligand were critical in en-

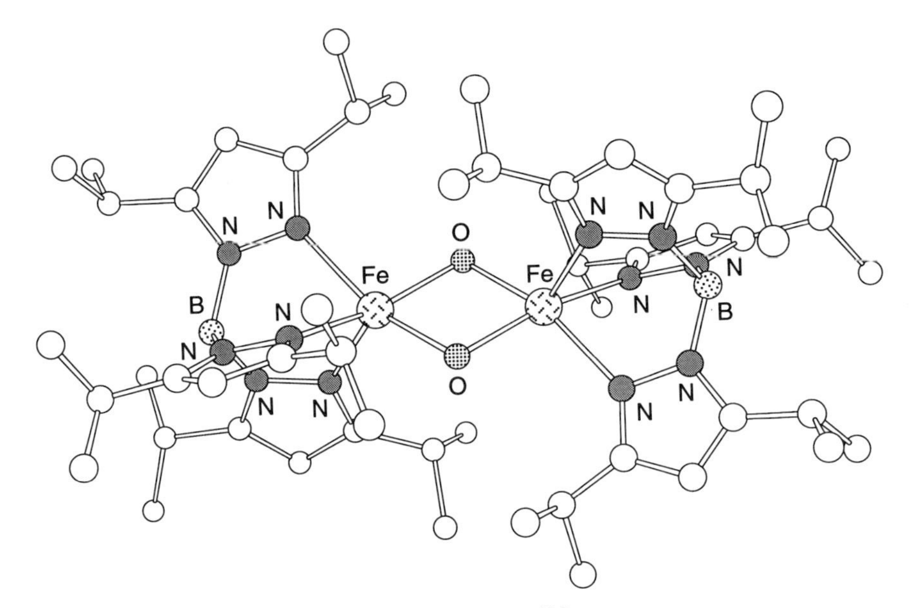

Figure 57. X-ray crystal structure of $[(Tp^{i\text{-}Pr2}Fe)_2(OH)_2]$ (205, 235).

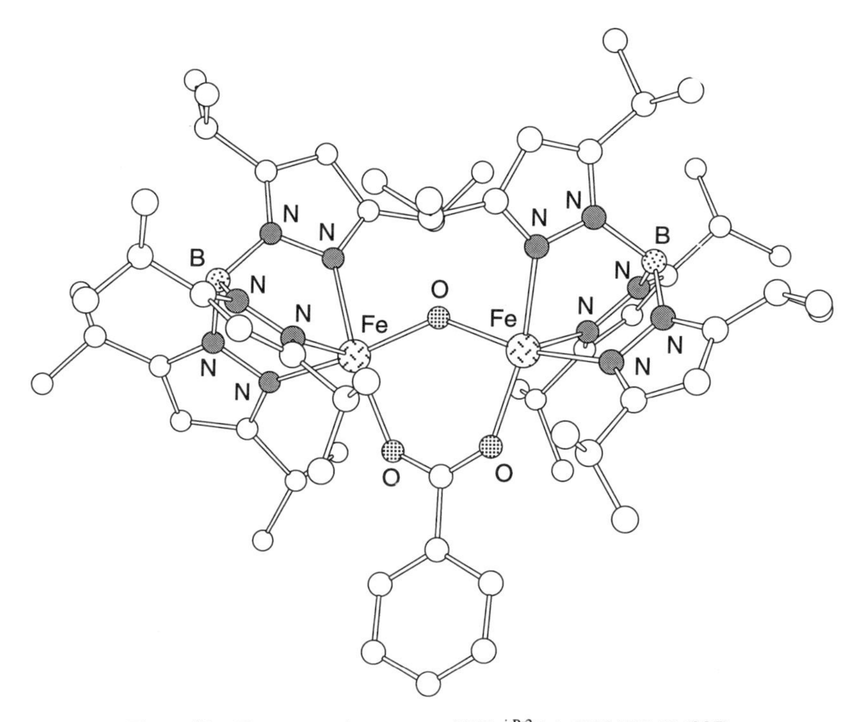

Figure 58. X-ray crystal structure of $[(Tp^{i\text{-}Pr2}Fe)_2(OH)(OBz)]$ (235).

abling the isolation of such a coordinatively unsaturated species. While preliminary experiments confirm that it is extremely oxygen sensitive and that the product of its reaction with dioxygen at low temperature is probably a peroxo species, full characterization of this molecule has yet to be completed. These studies of dinuclear species are particularly important in view of the current interest in understanding the dioxygen activation processes promoted by the diiron(II) sites in ribonucleotide reductase and methane monooxygenase (229).

Reactivity studies of the tetrahedral Fe^{II} complexes $Tp^{i\text{-}Pr2}Fe(X)$ (X = OAr, SAr, or OMe) with dioxygen have not been completed. However, functional mimicry of the monoiron enzyme tyrosine hydroxylase (TH) was accomplished starting from $Tp^{i\text{-}Pr2}Fe(OAr)$ (Ar = *p*-nitro-, chloro-, or methyl-phenyl) (236). Aerobic oxidation of these complexes afforded the Fe^{III} species $Tp^{i\text{-}Pr2}Fe(OAr)_2$, which were then treated with mCPBA at low temperature (Fig. 59). Hydroxylation of the coordinated phenoxide at the ortho position was confirmed by product isolation experiments and the identity of the proposed intermediate **XXIV** was corroborated by independent synthesis (236). The overall reaction closely resembles that promoted by TH [hydroxylation of tyrosine to 3,4-dihydroxyphenylalanine (dopa)] and suggests that a $Fe^{III}—OOR$ species is a competent phenol hydroxylation reagent. Again, the particular combination inherent to the

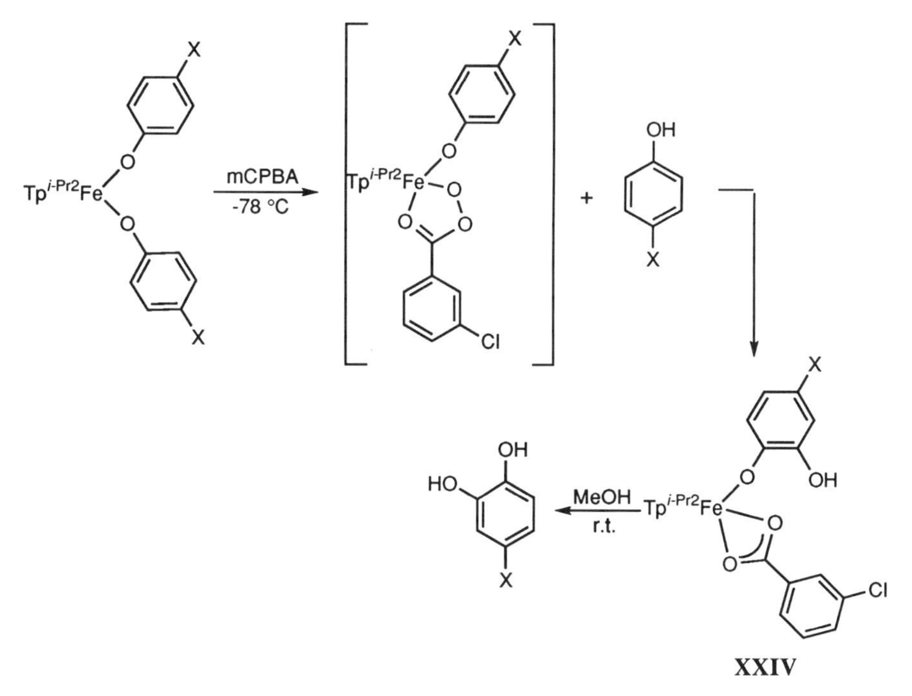

Figure 59. Synthetic modeling of tyrosine hydroxylase (TH) activity (236).

$Tp^{i\text{-}Pr2}$ ligand of steric bulk, which inhibits deleterious side reactions and formation of unwanted polynuclear species, and geometric flexibility, which allows varying coordination numbers to be adopted, was instrumental in enabling observation of novel biomimetic chemistry.

C. Manganese Complexes

The interactions of manganese centers in proteins with dioxygen and its reduced forms are also important, albeit less prevalent than those involving iron (237, 238). Examination of the structural and physical properties of redox active synthetic manganese complexes has been the focus of intense interest, with primary emphasis on modeling aspects of the chemistry of the mononuclear site of manganese superoxide dismutase (MnSOD), the dinuclear sites of catalase and manganese ribonucleotide reductase (MnRR), and the tetranuclear oxygen-evolving complex from photosystem II (237–239). Studies using $Tp^{RR'}$ ligands have paralleled those carried out with iron; for example, as in the case of Fe, self-assembly of KTp, Mn^{III}, and the appropriate carboyxlate afforded Tp-capped (μ-oxo)bis(μ-carboxylato)dimanganese(III) complexes (240). Also, as found for the case of iron the use of more sterically hindered Tp^{Me2} or $Tp^{i\text{-}Pr2}$ ligands has led to the isolation of complexes with different nuclearities and bridging motifs, particularly interesting results having been obtained when Mn^{II} species were treated with O_2^{n-} ($n = 0$, 1, or 2).

A structural and functional mimic of the mononuclear active site of MnSOD was prepared by treatment of $Tp^{i\text{-}Pr2}MnCl$ with equimolar NaOBz and 3,5-diisopropylpyrazole (3,5-*i*-Pr_2pzH) under anaerobic conditions (241). An X-ray crystal structure of the reaction product revealed a five-coordinate Mn^{II} ion supported by η^3-$Tp^{i\text{-}Pr2}$ and monodentate 3,5-*i*-Pr_2pzH and OBz in a trigonal bipyramidal geometry (Fig. 60; 3,5-*i*-Pr_2pzH and the pyrazolyl ring trans to it in the axial positions). The noncoordinated oxygen atom from the OBz ligand is hydrogen bonded to the pyrazole N—H. The congruence of this coordination geometry with that of the MnSOD active size is impressive, since the latter is comprised of a manganese ion with a N_3O_2 donor set in a trigonal bipyramidal array and a monodentate aspartate carboxylate in an equatorial position hydrogen bonded to the peptide chain (242, 243). Functional relevance of the synthetic model complex was revealed by assaying its SOD activity, which was shown to be greater than a number of other manganese compounds but still 100-fold less than that of the enzyme.

Other biologically relevant manganese-containing complexes of $Tp^{i\text{-}Pr2}$ were prepared from the bis(μ-hydroxo)-bridged dimanganese(II) compound $[(Tp^{i\text{-}Pr2}Mn)_2(OH)_2]$, a typical member of the class of such complexes prepared from a range of divalent metal ions (205). Permanganate oxidation of $[(Tp^{i\text{-}Pr2}Mn)_2(OH)_2]$ afforded the bis(μ-oxo)dimanganese(III) species

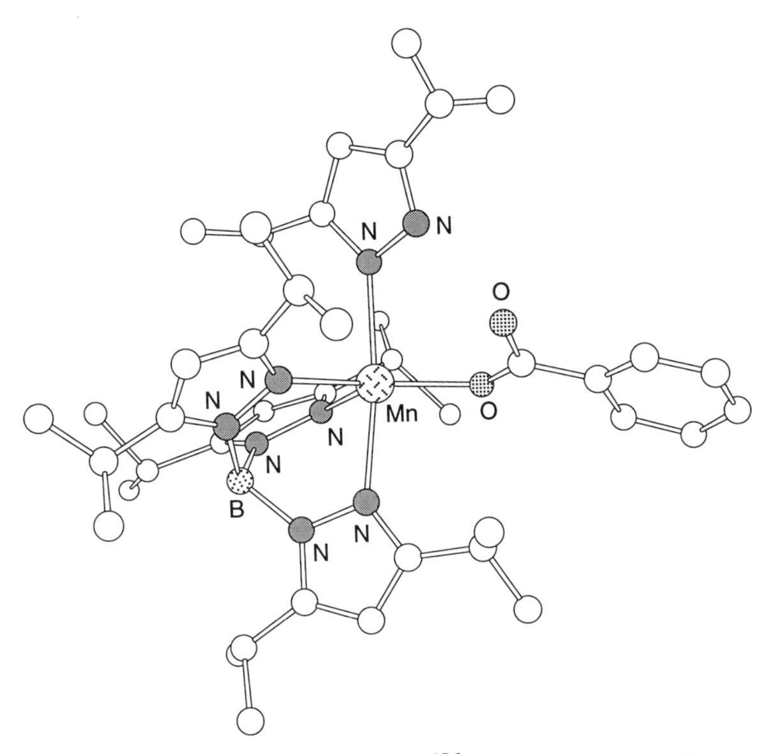

Figure 60. X-ray crystal structure of $Tp^{i\text{-}Pr2}Mn(3,5\text{-}i\text{-}Pr_2pz)(OBz)$ (241).

$[(Tp^{i\text{-}Pr2}Mn)_2(O)_2]$ in high yield (Fig. 61) (244). Such $Mn_2^{III}(\mu\text{-}O)_2$ cores are relatively rare, the other few examples having six-coordinate Mn^{III} ions (245). Presumably, the five-coordinate geometry exhibited by $[(Tp^{i\text{-}Pr2}Mn)_2(O)_2]$ arises as a result of the steric protection afforded by the substituent array of the $Tp^{i\text{-}Pr2}$ ligands. These substituents were attacked in the reaction of $[(Tp^{i\text{-}Pr2}Mn)_2(OH)_2]$ with O_2, which yielded a mixture of $[(Tp^{i\text{-}Pr2}Mn)_2(O)_2]$ and hydroxylated product **XXV** (Fig. 62) (246). Compound **XXV** was shown by X-ray crystallography to contain a single oxo bridge between two Mn^{III} ions, as well as two alkoxide ligands derived from hydroxylation of $Tp^{i\text{-}Pr2}$ isopropyl groups (Fig. 63). Isotope labeling studies showed that the two alkoxide oxygen atoms were derived from the same O_2 molecule, a result that can be explained by the mechanism proposed in Fig. 64. According to this proposal, dioxygen binding to yield a reactive Mn_2^{III}–peroxo complex occurs initially, followed by loss of water and O—O bond cleavage to afford the active oxidant **XXVI.** Related oxidations of $Tp^{RR'}$ substituents have been discovered recently, some of which have been argued to proceed via different pathway(s) (e.g., see discussion in Section V).

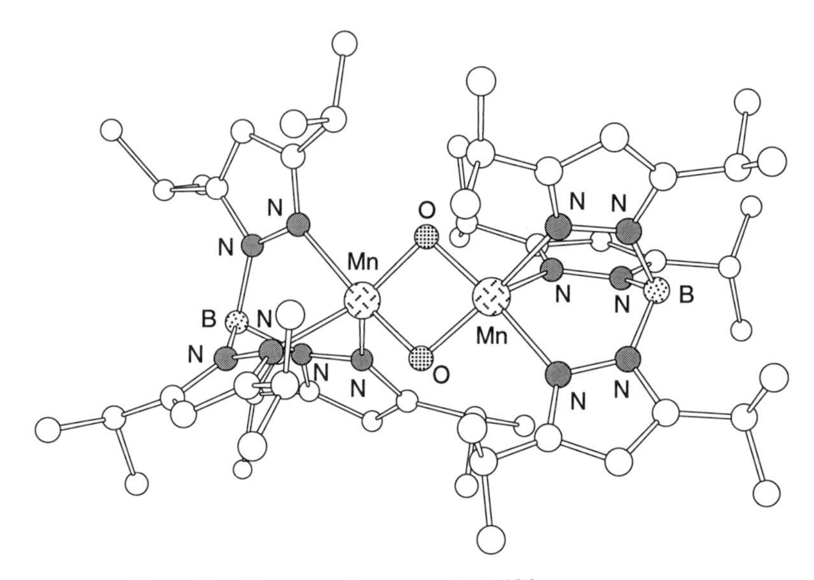

Figure 61. X-ray crystal structure of $[(Tp^{i\text{-}Pr2}Mn)_2(O)_2]$ (244).

An intriguing intermediate in the mechanism outlined in Fig. 64 is the primary dioxygen adduct (Mn_2^{III}–peroxo complex). As reviewed elsewhere (237), structurally characterized manganese dioxygen complexes are rare, yet they are of central importance in the mechanisms traversed by many manganese enzymes. Treatment of $[(Tp^{i\text{-}Pr2}Mn)_2(OH)_2]$ with an excess of H_2O_2 in the pres-

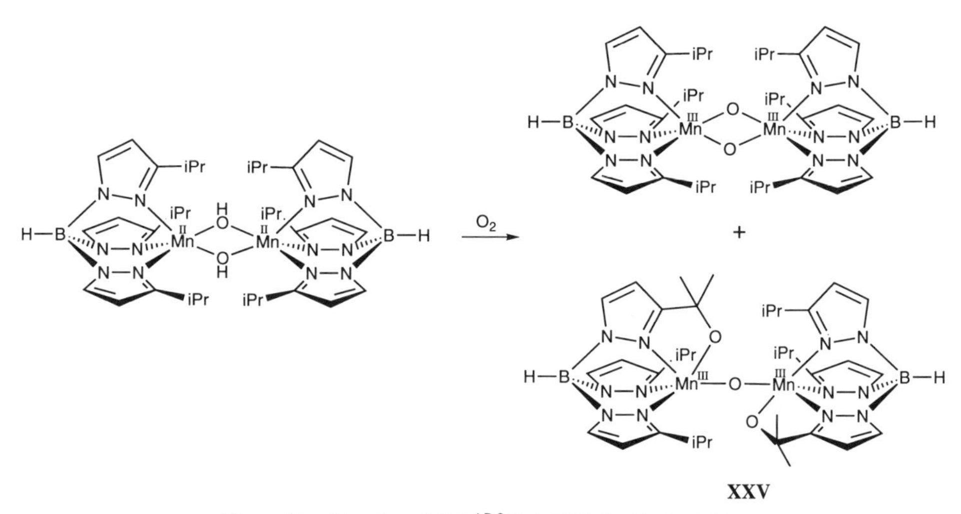

Figure 62. Reaction of $[(Tp^{i\text{-}Pr2}Mn)_2(OH)_2]$ with O_2 (246).

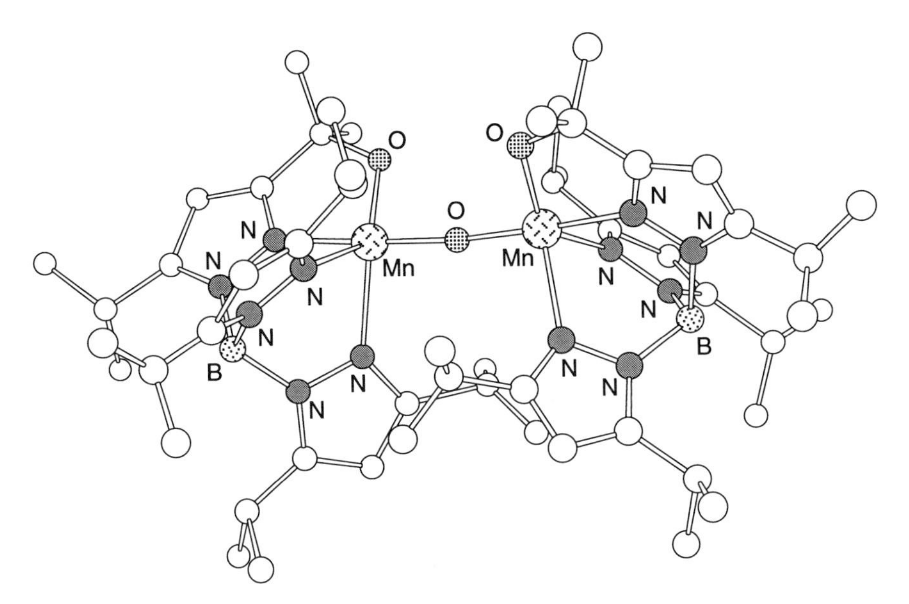

Figure 63. X-ray crystal structure of **XXV** (246).

ence of 3,5-*i*-Pr_2pzH yielded a novel dioxygen adduct, $Tp^{i\text{-}Pr2}$Mn(3,5-*i*-Pr_2pzH)(O_2), which was characterized by X-ray crystallography (Fig. 65) (247). Symmetric side-on coordination of a peroxo ligand to a Mn^{III} ion occurs in the complex, a structural motif similar to that observed in the porphyrin compound (TPP)Mn(O_2) (where TPP = dianion of *meso*-tetraphenylporphyrin) (248). A

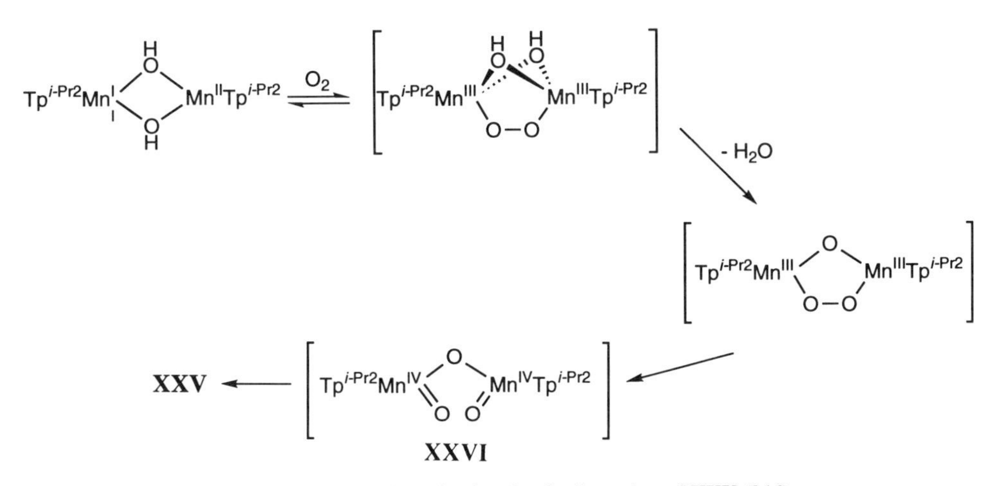

Figure 64. Proposed mechanism for the formation of **XXV** (246).

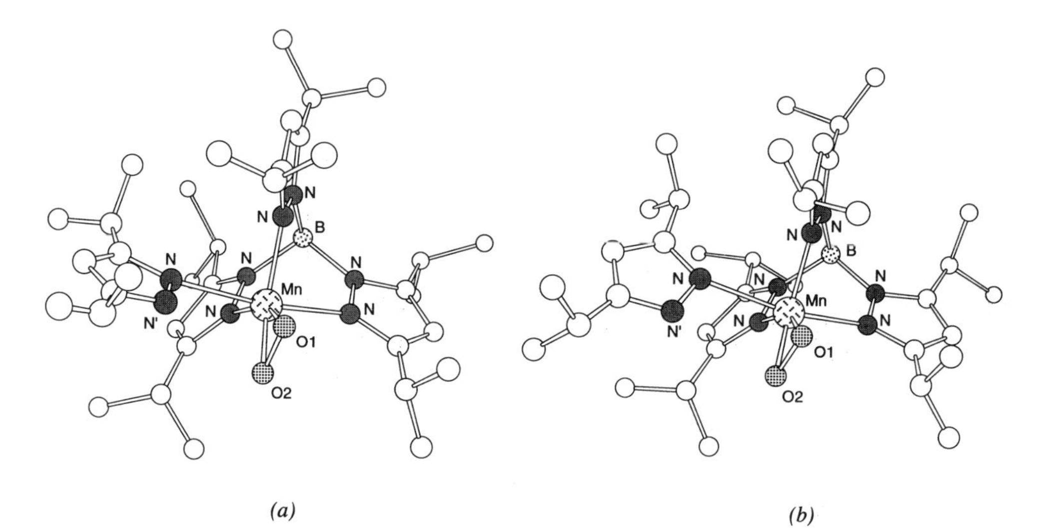

Figure 65. X-ray crystal structures of the two forms of $Tp^{i\text{-}Pr2}Mn(3,5\text{-}i\text{-}Pr_2pz)(O_2)$ (247): (*a*) blue, hydrogen-bonded form [N′—O(1) = 2.82(2) Å, N′—O(2) = 2.99(2) Å] and (*b*) brown, non-hydrogen-bonded form [N′—O(1) = 3.083(3) Å, N′—C(2) = 3.298(8) Å].

particularly interesting finding was that $Tp^{i\text{-}Pr2}Mn(3,5\text{-}i\text{-}Pr_2pzH)(O_2)$ exists in two forms that differ only with respect to the presence of a hydrogen bond between the 3,5-*i*-Pr_2pzH and O_2^{2-} moieties (Fig. 65). This observation of hydrogen bonding to a coordinated dioxygen molecule is of particular interest because of the known importance of such interactions in stabilizing dioxygen adducts in proteins (249, 250) and the possible involvement of such bonding in O—O bond cleavage mechanisms.

Finally, several other interesting polynuclear manganese complexes were obtained using $[(Tp^{i\text{-}Pr2}Mn)_2(OH)_2]$ as starting material. A possible model of the active site of reduced MnRR, $[(Tp^{i\text{-}Pr2}Mn)_2(\mu\text{-}OH)(\mu\text{-}OAc)]$, was prepared from $[(Tp^{i\text{-}Pr2}Mn)_2(OH)_2]$ plus 1 equivalent of acetic acid (251). Reaction of $[(Tp^{i\text{-}Pr2}Mn)_2(\mu\text{-}OH)(\mu\text{-}OAc)]$ with O_2 yielded $[(Tp^{i\text{-}Pr2}Mn)_2(\mu\text{-}O)_2(\mu\text{-}OAc)]$ (Fig. 66), a $Mn^{III}Mn^{IV}$ complex with a short Mn· · ·Mn distance (2.7 Å) (251). Although this distance is typical for such bridging motifs in $Mn^{III}Mn^{IV}$ chemistry (239), the fact that it can be supported by using highly hindered $Tp^{i\text{-}Pr2}$ caps is unusual, especially in view of the tendency to isolate trinuclear instead of dinuclear species with Fe^{III} or Mn^{III} and other combinations of oxo and carboxylato bridges with this supporting ligand. The course of the reaction of $[(Tp^{i\text{-}Pr2}Mn)_2(\mu\text{-}OH)(\mu\text{-}OAc)]$ with excess H_2O_2 is more typical, the isolated product being a trinuclear $Mn^{III}Mn^{II}Mn^{III}$ complex (252) topologically similar to other iron compounds discussed above (232–234) (Fig. 67).

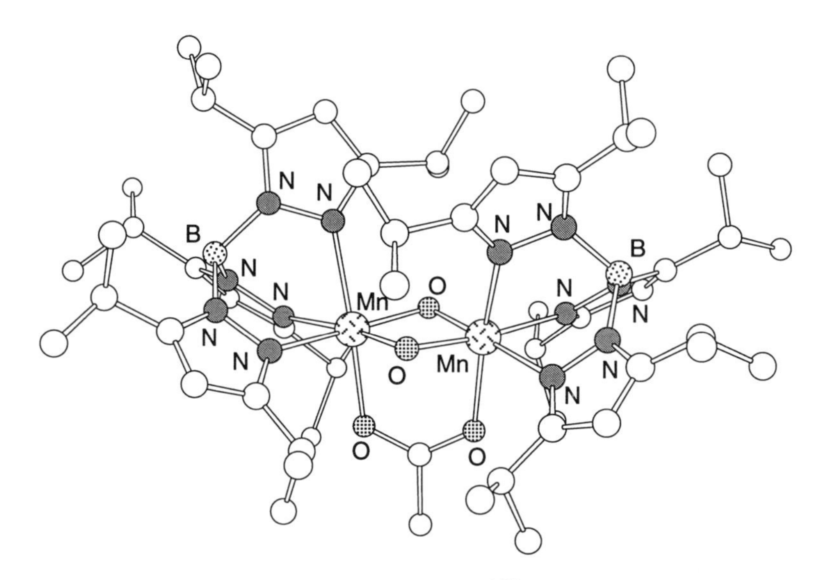

Figure 66. X-ray crystal structure of [(Tp$^{i\text{-}Pr2}$Mn)$_2$(O)$_2$(OAc)] (251).

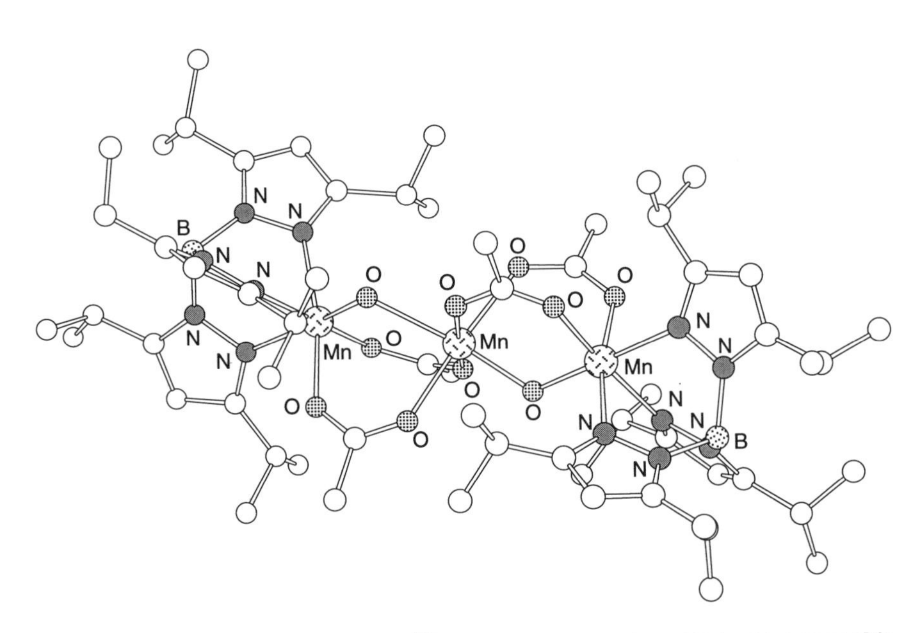

Figure 67. X-ray crystal structure of [Tp$^{i\text{-}Pr}$Mn(μ-OH)(μ-OAc)$_2$Mn(μ-OH)(μ-OAc)$_2$MnTp$^{i\text{-}Pr2}$] (252).

D. Zinc Complexes

1. Models of Carbonic Anhydrase

Carbonic anhydrase (CA), a ubiquitous enzyme found in animals, plants, and some bacteria, enhances the rate of hydration of CO_2 to bicarbonate by a factor of about 10^7 at physiological pH (253). The active site is composed of a single Zn^{II} ion coordinated to three histidine imidazolyl groups, with a water molecule or a hydroxide ion, depending on pH, occupying the fourth coordination position (254). It has been proposed that the catalytic cycle involves attack of the nucleophilic zinc hydroxide on CO_2 to yield a bicarbonate intermediate (Fig. 68). According to this simplified mechanism, subsequent displacement of bicarbonate by a water molecule followed by deprotonation of the zinc–water complex by an active site base then regenerates the key Zn—OH species. Synthetic modeling studies have focused on (a) preparing a mononuclear zinc–hydroxide complex with biologically relevant coligands that exhibits biomimetic pK_a behavior and CO_2 hydration activity, and (b) developing a rationale for the observed trend in activities of various metal(II)-substituted CA enzymes: Zn > Co >> Ni ~ Mn ~ Cd > Cu (253). Oligomerization has been the most vexing problem in attempted preparations of models for the zinc–hydroxide complex, as exemplified by the work of Kimura et al. (255) using the [12]aneN_3 ligand system (Fig. 69). Although exhibiting a pK_a = 7.30 at 25°C quite similar to that measured for CA (~7.5) and relevant hydrolysis activity, the LZnOH complex crystallized as a trimer with bridging hydroxide units [Zn—O = 1.944(5) Å].

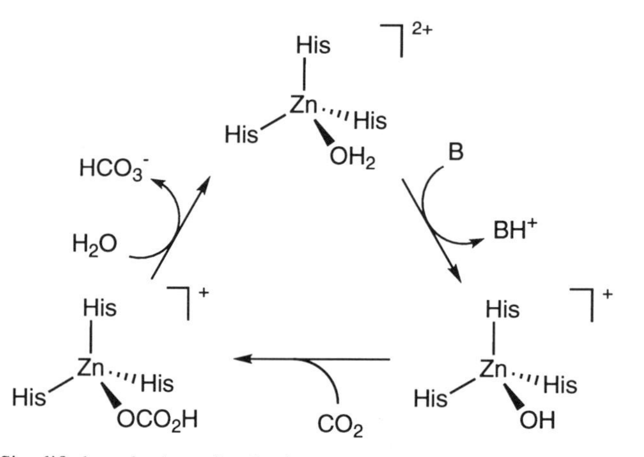

Figure 68. Simplified mechanism of carbonic anhydrase (CA) action (His = histidine imidazole groups, B = buffer or basic site in protein).

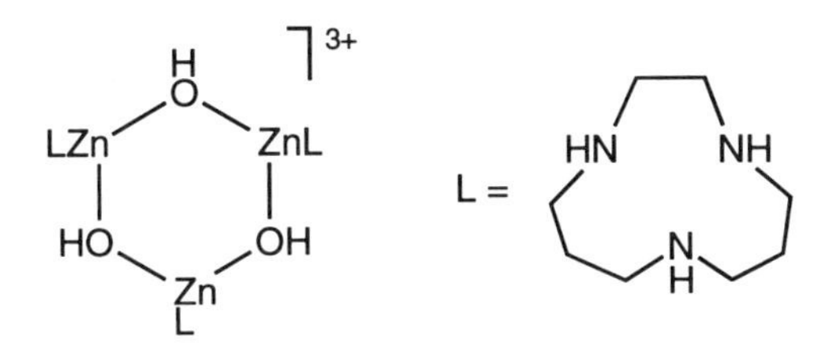

Figure 69. Trimeric zinc hydroxide complex supported by a [11]aneN_3 ligand (255).

By using the sterically hindered, encapsulating $Tp^{t\text{-Bu,Me}}$, $Tp^{i\text{-Pr2}}$, and $Tp^{\text{cumyl,Me}}$ ligands, the first monomeric zinc complexes with terminal hydroxide ligands, $Tp^{RR'}$ZnOH, were prepared (205, 256, 257). The presence of substituents in the 5 position of the pyrazolyl rings was critical in inhibiting hydrolytic cleavage of the ligands (B—N bond disruption), which was identified as a predominant decomposition pathway. The X-ray crystal structure of $Tp^{t\text{-Bu,Me}}$ZnOH confirmed its monomeric formulation (256). The Zn—O bond length [1.850(8) Å] is shorter than those of other zinc complexes with bridging hydroxides (63, 255, 258), as well as that observed in a recent 2.0-Å resolution X-ray analysis of CA (2.1 Å) (254). As previously noted (256), the longer distance in the enzyme probably results from the presence of hydrogen-bonding interactions not observed in the model system.

Studies of the reactivity of $Tp^{t\text{-Bu,Me}}$ZnOH and $Tp^{i\text{-Pr2}}$ZnOH with CO_2 demonstrated the suitability of these complexes as models for functional CA intermediates (66, 205, 259). For the $Tp^{t\text{-Bu,Me}}$ case, exposure of the hydroxide complex to 1 atm of CO_2 led to the formation of an equilibrium mixture of the starting materials and a bicarbonato complex, tentatively assigned as $Tp^{t\text{-Bu,Me}}Zn[\eta^1\text{-OC(O)OH}]$ on the basis of IR data (ν_{CO} = 1675 and 1302 cm^{-1}) (215, 259). Attempted isolation of this complex was unsuccessful; instead, the carbonato complex $[(Tp^{t\text{-Bu,Me}}Zn)_2(\mu\text{-}\eta^1:\eta^1\text{-}O,O\text{-}CO_3)]$ was obtained and subjected to an X-ray crystallographic analysis (Fig. 70) (43). Unidentate binding of the CO_3^{2-} ion to both symmetry-related Zn^{II} centers is evident (Zn—O = 1.85 and 3.20 Å). The structure of the analogous $Tp^{i\text{-Pr2}}$ complex was also determined, but a different carbonate binding mode, η^1 to one Zn^{II} ion and asymmetric η^2 to the other, was observed (Fig. 71) (205). These subtle geometric differences clearly result from differences in the steric properties of $Tp^{t\text{-Bu,Me}}$ and $Tp^{i\text{-Pr2}}$, where the larger *tert*-butyl substituents in the former ligand effectively prevent bidentate carbonate coordination to the Zn^{II} ion.

Interestingly, the structural divergence between the $Tp^{t\text{-Bu,Me}}$ and $Tp^{i\text{-Pr2}}$ carbonate complexes is paralleled by a key reactivity difference. While $[(Tp^{t\text{-Bu,Me}}Zn)_2(\mu\text{-}\eta^1:\eta^1\text{-}O,O\text{-}CO_3)]$ is readily hydrolyzed to its monomeric zinc hydroxide precursor, the $\mu\text{-}\eta^1:\eta^2\text{-}O,O\text{-}CO_3$ complex with $Tp^{i\text{-Pr2}}$ does not react with H_2O. Apparently, the bidentate coordination in the latter compound con-

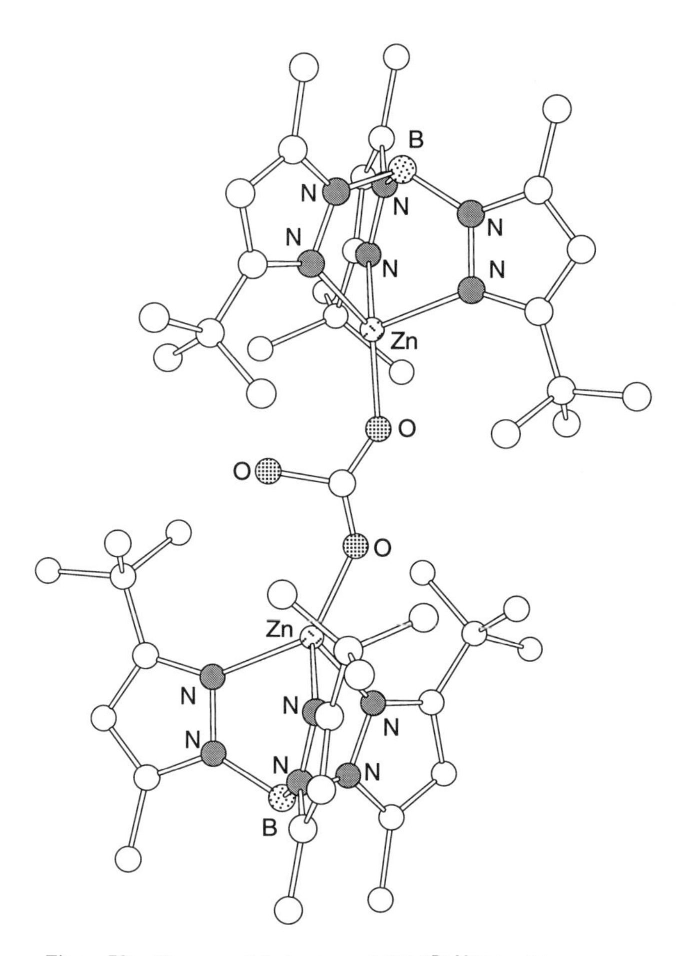

Figure 70. X-ray crystal structure of $[(Tp^{t\text{-Bu,Me}}Zn)_2(CO_3)]$ (43).

fers sufficient thermodynamic stability to prevent hydrolysis or simple water substitution, implying that such a binding mode in CA might be inhibitory. A related idea has been put forth by Parkin and co-workers (43, 63, 66–68) to explain the lower activities of the enzymes substituted by divalent Co, Ni, Cd, and Cu ions. They analyzed the crystal structures of a series of nitrate derivatives $Tp^{t\text{-Bu}}M(NO_3)$ (M = Co, Ni, Cu, Cd, or Zn) and argued that structural differences in this homologous set reflect the binding mode tendencies of the different metal ions with nitrate and, by analogy, bicarbonate. The nitrate-binding modes in the $Tp^{t\text{-Bu}}$ complexes vary from essentially unidentate for Zn (Fig. 21) to asymmetric bidentate for Ni, Cu, and Cd. An intermediate asymmetric bidentate geometry is adopted by the Co complex (Co—O = 2.001(3) and

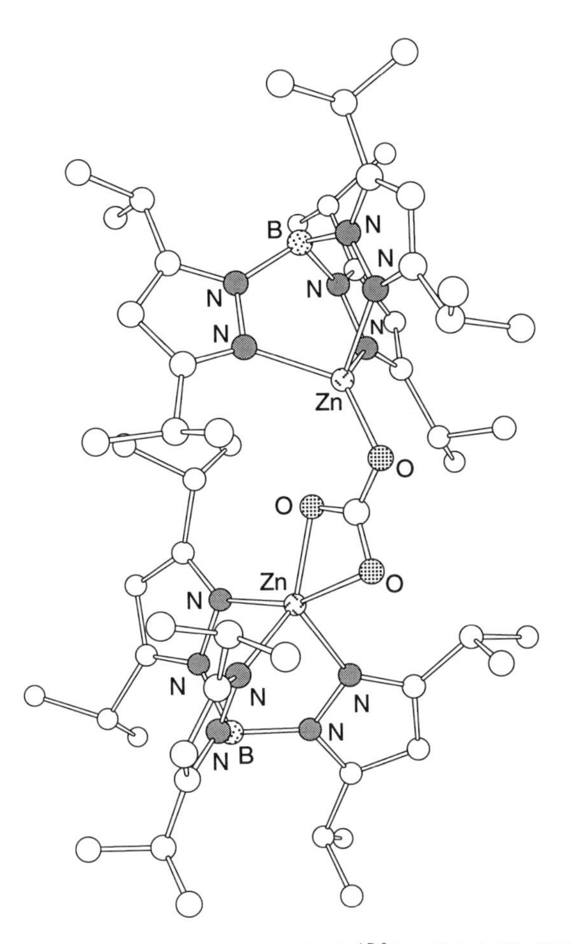

Figure 71. X-ray crystal structure of $[(Tp^{i\text{-}Pr2}Zn)_2(CO_3)]$ (43, 205).

2.339(3) Å]. The parallel between the unidentate → bidentate nitrate-binding geometry trend and the metal-substituted CA reactivity order led to the suggestion that the latter can be rationalized, at least in part, by the differences in the relative reactivity of the metal-bicarbonate intermediate. Thus, unidentate binding of bicarbonate, which is most favored for the most active Zn-substituted CA, was argued to be the most accessible to hydrolysis. Rates of water displacement and, thus, activity were argued to decrease in forms of CA substituted with metal ions that progressively favor bidentate coordination (Co > Cd, Ni, Cu). It is noteworthy that the trend in nitrate-binding modes in the $Tp^{t\text{-}Bu}M(NO_3)$ series is mirrored by the trend in carbonate-binding modes in the bridged, dinuclear set $[(Tp^{i\text{-}Pr2}M)_2(\mu\text{-}CO_3)]$ (M = Co, Ni, Cu, or Zn) (203).

Again, the order of increasing symmetric bidentate coordination to each metal ion by the bridging carbonate ligand is Zn < Co < Cu or Ni, further corroborating the arguments applied to the catalytically important (yet unobserved) bicarbonate species.

2. *Models of Biological Phosphate, Ester, and Amide Hydrolyses*

As described above, modeling studies have demonstrated how the Zn—OH unit may play a key role as a nucleophile in the hydration of CO_2 by CA. Similar Zn—OH units have also been proposed to be the central reactive species in other hydrolysis enzymes that catalyze phosphate, ester, and amide cleavage, such as alkaline phosphatase and carboxypeptidase, among others (260). To shed light on the molecular details of the reactions catalyzed by these enzymes, reactivity studies of the $Tp^{RR'}ZnOH$ complexes were extended to include reactions with phosphates, esters, and amides.

Stoichiometric O—P bond cleavage was observed when $Tp^{i\text{-}Pr2}ZnOH$ was treated with 0.5 equivalents of tris(*p*-nitrophenyl)phosphate to ultimately afford the dinuclear complex $\{(Tp^{i\text{-}Pr2}Zn)_2[\mu\text{-}O_2P(O)(OAr)]\}$ (Fig. 72) (261). The phosphate bridges the Zn^{II} ions in the product in a syn-anti mode, with a Zn· · ·Zn distance of 5.1 Å. The active site of alkaline phosphatase from *Escherichia coli* has been shown to have a similar overall structure [Zn(μ-phosphate)Zn], although in the enzyme the bridge is syn–syn with a metal–metal separation of 3.9 Å (263, 263). Based on the identification of various monomeric $Tp^{i\text{-}Pr2}Zn(OR)$ (R = Ar or phosphate derivative) species, a mechanism for the hydrolysis of tris(*p*-nitrophenyl)phosphate involving stepwise cleavages of P—O bonds was postulated (Fig. 73). Notwithstanding these results, the reactivity of $Tp^{i\text{-}Pr2}ZnOH$ is rather low, as other phosphate esters lacking activating *p*-nitrophenyl groups and amides are not cleaved. The related hydroxo complexes $[Tp^{i\text{-}Pr2}M]_2(OH)_2$ (M = Cu or Ni) react with *p*-nitroacetanilide to give four-membered chelated amido complexes, with no C—N bond scission (264).

Recently, $Tp^{Cum,Me}ZnOH$ has been shown to be a nucleophile of greater strength (257). In addition to attacking *p*-nitrophenyl phosphates, it hydrolyzes aliphatic phosphate esters and diphosphate esters. Moreover, activated esters and amides are cleaved as well in reactions that mimic postulated steps in, for example, peptidase activity. Since $Tp^{RR'}$ ligands with aryl substituents are electron withdrawing compared to alkyl groups, the greater nucleophilicity of the Zn—OH unit in the $Tp^{Cum,Me}$ complex compared to that of $Tp^{i\text{-}Pr2}$ was suggested to arise from a steric factor (“a higher degree of encapsulation” and “a more hydrophobic environment” (257).

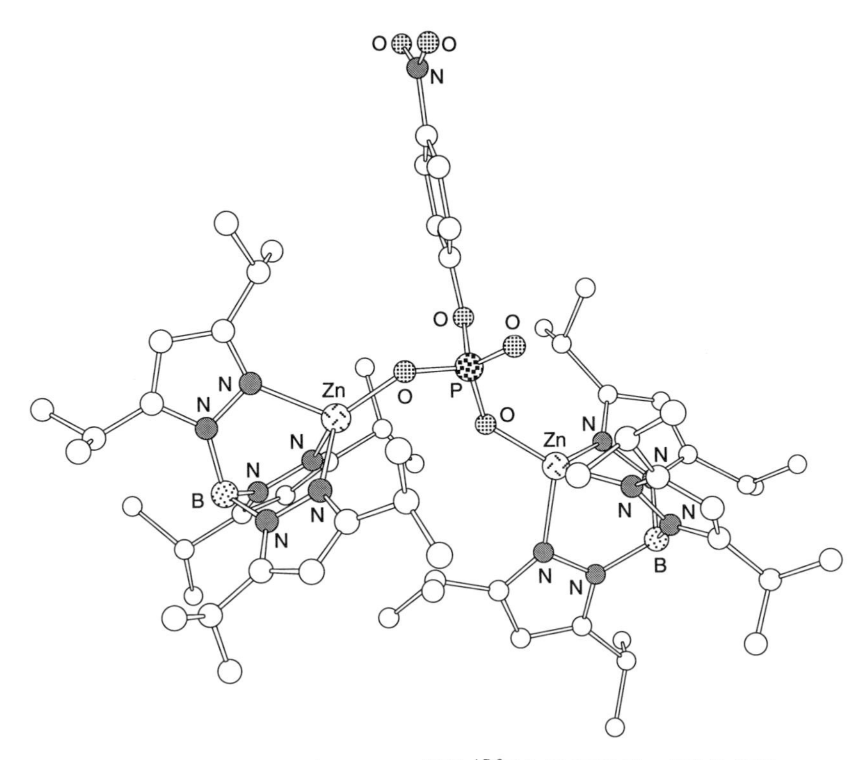

Figure 72. X-ray crystal structure of $[(Tp^{i\text{-}Pr2}Zn)_2(O_3POC_6H_4\text{-}p\text{-}NO_2)]$ (261).

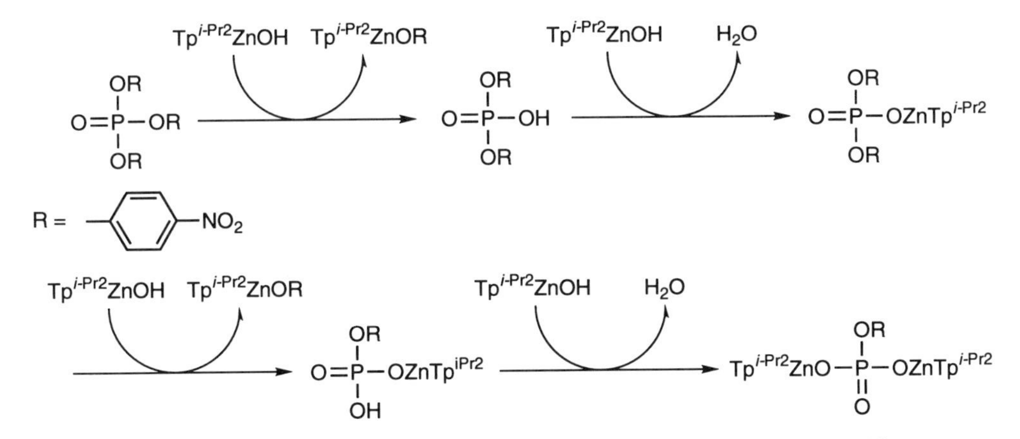

Figure 73. Proposed mechanism for hydrolysis of tris(p-nitrophenyl)phosphate by $Tp^{i\text{-}Pr2}ZnOH$ (261).

V. OTHER COMPLEXES

A particularly intriguing aspect of the chemistry of complexes of hindered Tp ligands that does not nearly fall into the organometallic or bioinorganic categories, yet which is relevant to processes important from many perspectives, concerns recently discovered oxidations of the 3-isopropyl substituents in $Tp^{i\text{-Pr},R'}$ (R′ = Me or *i*-Pr) (246, 265–268). Particularly significant insights into the reactivity of active metal–oxygen species with aliphatic C—H bonds have been obtained from studies of $Tp^{i\text{-Pr},R'}$ complexes, the success of this work being at least partially derived from the presence of relatively reactive methine hydrogen atoms in the 3-isopropyl groups held in close proximity to the encapsulated metal ion. Hydroxylation of the methine C—H bonds of the isopropyl substituents upon oxygenation of $[(Tp^{i\text{-Pr}2}Mn)_2(OH)_2]$ was discussed previously (Section IV.C).

A different reaction course was followed upon oxygenation of $[(Tp^{i\text{-Pr,Me}}Co)_2(\mu\text{-}N_2)]$ (Fig. 74) (265). The initially formed species was identified as the monomeric superoxo complex $Tp^{i\text{-Pr,Me}}CoO_2$ on the basis of spectroscopic comparisons to its crystallographically characterized $Tp^{t\text{-Bu,Me}}$ analogue (201). Whereas the steric protection afforded by the 3-*tert*-butyl groups allowed isolation of the latter complex, less hindered $Tp^{i\text{-Pr,Me}}CoO_2$ converted

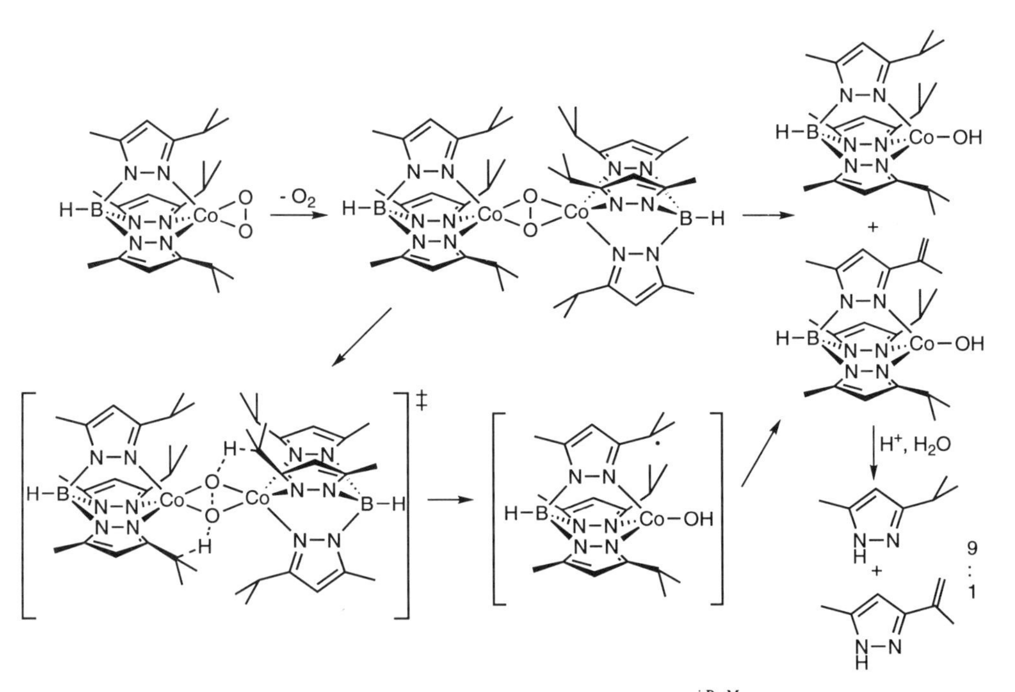

Figure 74. Dioxygen reactivity of cobalt complexes of $Tp^{i\text{-Pr,Me}}$ (265).

to [$(Tp^{i\text{-Pr,Me}}Co)_2(\mu\text{-}O_2)$] at $-78°C$, a reaction closely parallel to that proposed for related copper compounds (Sections IV.A1 and IV.A3). Most interesting was the finding that the dicobalt(II)–μ-peroxo complex decomposed upon warming to afford [$(Tp^{i\text{-Pr,Me}}Co)_2(\mu\text{-OH})_2$] in about 45% yield and that subsequent hydrolysis of the crude product mixture liberated 3-isopropenyl-5-methylpyrazole, a product derived from dehydrogenation of the $Tp^{i\text{-Pr,Me}}$ ligands. A mechanism for the dehydrogenation reaction was proposed (Fig. 74) that involves dual hydrogen-atom abstraction by the peroxo group followed by disproportionation of the isopropyl radicals to the indicated isopropyl- and isopropenyl-substituted cobalt hydroxo species. Observation of a large kinetic isotope effect [KIE = k_H/k_D = 22(1) at 281 K] provided key evidence in favor of isopropyl C—H bond breaking during the rate-determining step of the decomposition reaction and tunneling was proposed for the hydrogen-atom transfer on the basis of the temperature dependence of the KIE. The highly ordered transition state shown was proposed to be consistent with these results and to explain the measured negative $\Delta S^{\ddagger}$ values [$-12(1)$ eu for the protio-ligand]. This work, in conjunction with research on related chemistry involving hydrogen-atom abstraction from a proximate isopropyl group in a coordinated ligand by a dicopper-μ-peroxo complex (266), provides key insight into possible mechanisms of hydrocarbon functionalization by metal–dioxygen adducts.

The reaction of [$(Tp^{i\text{-Pr}2}Co)_2(\mu\text{-OH})_2$] with H_2O_2 also yielded a μ-peroxo complex with spectral features quite similar to those exhibited by its $Tp^{i\text{-Pr,Me}}$ analogue (267). When this complex was decomposed in the presence of excess H_2O_2 at room temperature, ligand hydroxylation instead of dehydrogenation occurred to afford two products, **XXVII** and **XXVIII**, both of which were structurally characterized by X-ray crystallography (Fig. 75). The dicobalt(II) com-

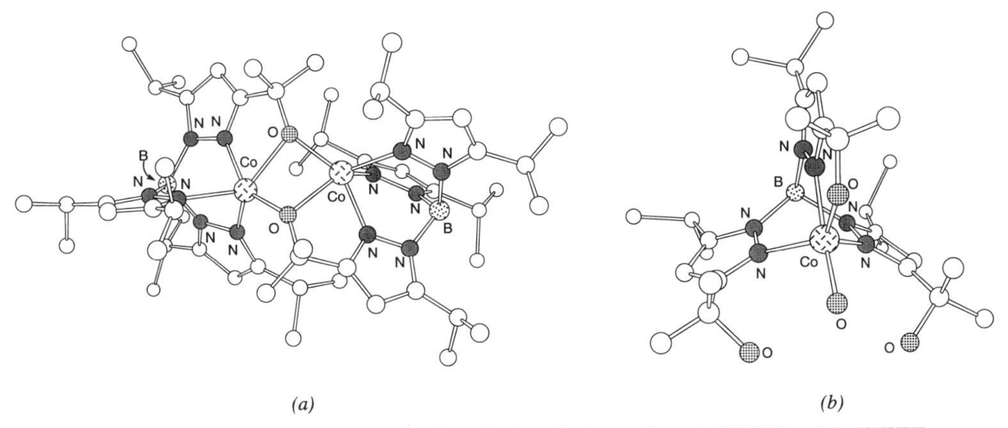

Figure 75. X-ray crystal structures of the ligand hydroxylation products (*a*) **XXVII** and (*b*) **XXVIII** (267).

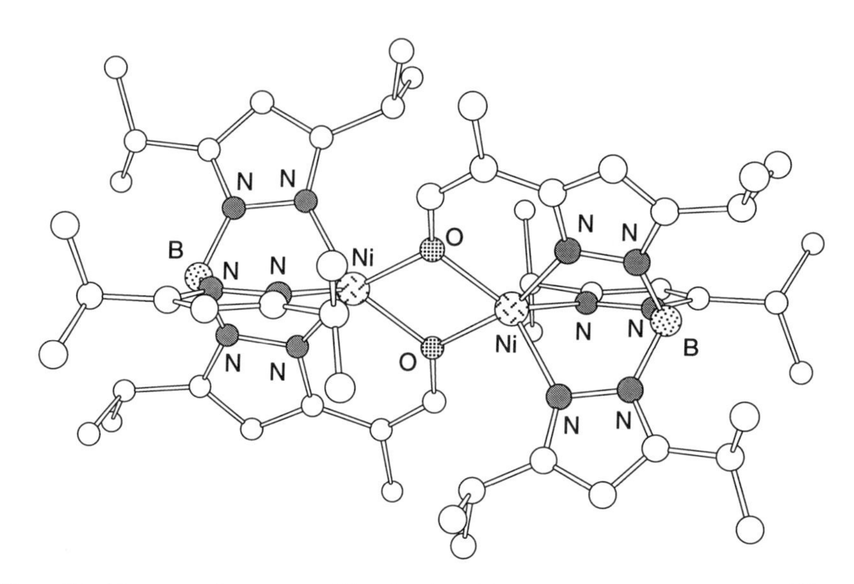

Figure 76. X-ray crystal structure and line drawing of the product of oxygenation of $[(Tp^{i\text{-Pr}2}Ni)_2(\mu\text{-OH})_2]$ (268).

plex **XXVII** is analogous to the Mn hydroxylation product **XXV** (Fig. 63), while **XXVIII** is monomeric with three hydroxylated 3-isopropyl groups on a single $Tp^{i\text{-Pr}2}$ ligand, one of which is coordinated to the Co^{III} center. It is notable that the hydroxylated isopropyl groups lie on the metal side (i.e., the 3 position), implying that the oxo-transfer reactions take place within the coordination sphere of the metal ion. Although details of the mechanism of these oxidations remain unclear, the fact that these hydroxylation products did not form from the μ-peroxo compound in the absence of H_2O_2 suggests that a different species, perhaps a hydroperoxo Co^{II} complex such as $Tp^{i\text{-Pr}2}CoOOH$, is the reactive intermediate responsible for the observed chemistry.

Finally, an additional reaction pathway resulting in an isopropyl group functionalized differently from the above examples was observed upon treatment of $[(Tp^{i\text{-Pr}2}Ni)_2(\mu\text{-OH})_2]$ with excess H_2O_2 (Fig. 76) (268). Not only were the methyl instead of the methine positions hydroxylated in the indicated product, but dehydrogenation also occurred. Mechanistic analysis of this and the above Co hydroxylation reactions have yet to be performed.

VI. SUMMARY AND PERSPECTIVES

Consideration of the chemistry discussed herein and in previous reviews (8–12) leads to the inescapable conclusion that Tp–metal complex properties are significantly affected by the nature of the substituents attached to the Tp ligand frame. The nuclearities, geometries, spectroscopic properties, and reactivities of main group, transition metal, and lanthanide complexes can be controlled by varying the size and shape of the metal-centered cavities generated by the $Tp^{RR'}$ scorpionates. In particular, the use of highly hindered ligands to enable the isolation of sensitive compounds with novel structures has been a common theme in both the organometallic and bioinorganic research arenas. More recently, subtle but important changes in metal complex structures and reactivity patterns have been uncovered through either the incremental decrease of the steric bulk of the $Tp^{RR'}$ substituents or the attenuation of their shape, both of which control the approach of substrates or additional TpM fragments.

It is also evident from the extant literature that synthesis and characterization of hindered $Tp^{RR'}$ complexes has been a primary emphasis of research, and that somewhat less attention has been directed toward detailed examination of their reactivity. Future exploitation of the unique properties of $Tp^{RR'}M$ complexes for discovering and understanding the mechanisms of organic synthetic and biomimetic reactions is anticipated. Thus, based on the precedent provided by the work discussed here, extension of the existing rich chemistry of organometallic compounds supported by Tp^{Me2} [cf. (269–273)] to complexes of more hindered ligands should lead to new reactivity patterns. With the construction and characterization of optically pure Tp ligands, enantioselective processes that take advantage of the versatile binding properties and C_3 symmetry of the Tp frame can be targeted. Moreover, the availability of encapsulating Tp ligands that afford metal-centered cavities of variable size and shape offers clear opportunities for further, more sophisticated modeling of reactions that take place within the coordination sphere of metal ions buried within proteins. In view of the breadth of the chemistry already uncovered by using $Tp^{RR'}$ scorpionates as supporting ligands, it would appear that the extent of future applications will be limited only by the imagination of the researcher.

NOTE ADDED IN PROOF

Trofimenko and co-workers have recently prepared a tris(indazolyl)-hydroborate ligand in which the indazolyl moiety bonds to the central B atom via its more hindered N atom, a situation opposite from the usual binding of B to the less hindered N atom of substituted pyrazolyl rings (274). Several new Yb^{II} complexes of $Tp^{t\text{-Bu},Me}$ have been structurally characterized (275). A report has appeared that describes X-ray crystallographic characterization of a Cu^{II}-superoxide complex supported by a tripodal ligand that exhibits an end-on superoxide binding mode different from the side-on geometry of $Tp^{t\text{-Bu},i\text{-Pr}}CuO_2$ (276). Some insights into the detailed photochemical processes involved in C—H bond activation by $Tp^{Me2}Rh(CO)_2$ have been published recently (277). Finally, by using Tp^{Ph2} as supporting ligand, novel mixed epoxide–carboxylate adducts of Cd^{II} that model possible intermediates in the catalytic copolymerization of CO_2 and epoxides were prepared and structurally characterized (278).

ACKNOWLEDGMENTS

Support for the research carried out by our groups are provided from the U.S. National Institutes of Health (GM47365 to WBT), the U.S. National Science Foundation (CHE9207152 and National Young Investigator Award to WBT), the Exxon Education Foundation (WBT), the Searle Scholars Award/Chicago Community Trust (WBT), and the Japanese Ministry of Education, Culture, and Science (Grants-in-Aid for Scientific Research 04225107 and 15235106 to NK). We thank Dr. S. Trofimenko and Professors Josef Takats, Daniel Reger, Gerard Parkin, William Graham, Luigi Venanzi, and Klaus Theopold for helpful discussions and for providing access to results prior to publication. The assistance of Michael Keyes in Chem3D picture generation is gratefully acknowledged.

REFERENCES

1. C. Janiak and H. Schumann, *Adv. Organomet. Chem.*, *33*, 291 (1991).
2. P. P. Power, K. Ruhlandt-Senge, and S. C. Shoner, *Inorg. Chem.*, *30*, 5013 (1991).
3. M. M. Olmstead, P. P. Power, and S. C. Shoner, *Inorg. Chem.*, *30*, 2547 (1991) and references cited therein.
4. For example, see K. J. Covert, P. T. Wolczanski, S. A. Hill, and P. J. Krusic, *Inorg. Chem.*, *31*, 66 (1992).
5. For example, see K. Ruhlandt-Senge and P. P. Power, *Inorg. Chem.*, *30*, 2633 (1991) and references cited therein.
6. B. Morgan and D. Dolphin, *Structure Bonding*, *64*, 115 (1987).
7. C. A. Tolman, *Chem. Rev.*, *77*, 313 (1977).
8. J. C. Calabrese, S. Trofimenko, and J. S. Thompson, *J. Chem. Soc. Chem. Commun.*, 1122 (1986).
9. S. Trofimenko, J. C. Calabrese, and J. S. Thompson, *Inorg. Chem.*, *26*, 1507 (1987).
10. S. Trofimenko, *Progress In Inorganic Chemistry*, Wiley-Interscience, New York, 1986, Vol. 34, p. 115.
11. K. Niedenzu and S. Trofimenko, *Top. Curr. Chem.*, *131*, 1 (1986).
12. A. Shaver, in *Comprehensive Coordination Chemistry*, G. Wilkinson, R. D. Gillard, and J. A. McCleverty, Eds., Pergamon, Oxford, UK, 1987, Vol. 2, pp. 245–259.
13. P. K. Byers, A. J. Canty, and R. T. Honeyman, *Adv. Organomet. Chem.*, *34*, 1 (1992).

14. S. Trofimenko, *Chem. Rev.*, *93*, 943 (1993).
15. S. Trofimenko, *Inorg. Synth.*, *12*, 99 (1970).
16. C. F. Beam, R. M. Sandifer, R. S. Foote, and C. R. Hauser, *Synth. Commun.*, *6*, 5 (1976).
17. S. Trofimenko, J. C. Calabrese, P. J. Domaille, and J. S. Thompson, *Inorg. Chem.*, *28*, 1091 (1989).
18. M. Cano, J. V. Heras, C. J. Jones, J. A. McCleverty, and S. Trofimenko, *Polyhedron*, *9*, 619 (1990).
19. R. Krentz, Ph.D. Thesis, ''Model Compounds in Carbon–Hydrogen Activation,'' University of Alberta, Canada, 1989.
20. O. M. Reinaud, A. L. Rheingold, and K. H. Theopold, *Inorg. Chem.*, *33*, 2306 (1994).
21. D. D. LeCloux, M. C. Keyes, M. Osawa, V. Reynolds, and W. B. Tolman, *Inorg. Chem.*, *33*, 6361 (1994).
22. J. Elguero, *The Structure, Reactions, Synthesis and Uses of Heteroaromatic Compounds*, Pergamon, Oxford, UK, 1984.
23. C. Lopez, R. M. Claramunt, S. Trofimenki, and J. Elguero, *Can. J. Chem.*, *71*, 678 (1993).
24. U. E. Bucher, Ph.D. Thesis, ''Poly(1-Pyrazolyl)borate Compounds with Rhodium,'' Swiss Federal Institute of Technology, Switzerland, 1993.
25. S. L. Rheingold, C. B. White, and S. Trofimenko, *Inorg. Chem.*, *32*, 3471 (1993).
26. E. Frauendorfer and G. Agrifoglio, *Inorg. Chem.*, *21*, 4122 (1982).
27. H. Brunner and T. Scheck, *Chem. Ber.*, *125*, 701 (1992).
28. T. Scheck, Ph.D. Thesis, ''New optisch aktive pyrazolhaltige Liganden für die enantioselektive Katalyse,'' Universität Regensburg, 1991.
29. P. J. Steel and E. C. Constable, *J. Chem. Res.*, 189 (1989).
30. A. A. Watson, D. A. House, and P. J. Steel, *J. Organomet. Chem.*, *311*, 387 (1986).
31. A. A. Watson, D. A. House, and P. J. Steel, *J. Org. Chem.*, *56*, 4072 (1991).
32. D. D. LeCloux and W. B. Tolman, *J. Am. Chem. Soc.*, *115*, 1153 (1993).
33. D. D. LeCloux, C. J. Tokar, M. Osawa, R. P. Houser, M. C. Keyes, and W. B. Tolman, *Organometallics*, *13*, 2855 (1994).
34. D. D. LeCloux and W. B. Tolman, unpublished results.
35. H. Brunner, U. P. Singh, T. Boeck, S. Altmann, T. Scheck, and B. Wrackmeyer, *J. Organomet. Chem.*, *443*, C16 (1993).
36. C. J. Tokar, P. B. Kettler, and W. B. Tolman, *Organometallics*, *11*, 2738 (1992).
37. H. C. Clark, *Isr. J. Chem.*, *15*, 210 (1977).
38. T. L. Brown and K. J. Lee, *Coord. Chem. Rev.*, *128*, 89 (1993).
39. J. C. Calabrese, P. J. Domaille, S. Trofimenko, and G. J. Long, *Inorg. Chem.*, *30*, 2795 (1991).
40. A. L. Rheingold, R. Ostrander, B. S. Haggerty, and S. Trofimenko, *Inorg. Chem.*, *33*, 3666 (1994).

41. A. L. Rheingold, B. S. Haggerty, and S. Trofimenko, *J. Chem. Soc. Chem. Commun.*, 1973 (1994).
42. S. Trofimenko, J. C. Calabrese, J. K. Kochi, S. Wolowiec, R. B. Hulsbergen, and J. Reedijk, *Inorg. Chem.*, *31*, 3943 (1992).
43. R. Han, A. Looney, K. McNeill, G. Parkin, A. L. Rheingold, and B. S. Haggerty, *J. Inorg. Biochem.*, *49*, 105 (1993).
44. J. J. W. Egan, B. S. Haggerty, A. L. Rheingold, S. C. Sendlinger, and K. H. Theopold, *J. Am. Chem. Soc.*, *112*, 2445 (1990).
45. W. B. Tolman, *Inorg. Chem.*, *30*, 4878 (1991).
46. R. Han and G. Parkin, *J. Am. Chem. Soc.*, *113*, 9707 (1991).
47. A. J. M. Caffyn, S. G. Feng, A. Dierdorf, A. S. Gamble, P. A. Eldredge, M. R. Vossen, P. S. White, and J. L. Templeton, *Organometallics*, *10*, 2842 (1991).
48. K. Fujisawa and N. Kitajama, unpublished results.
49. G. Parkin, unpublished results.
50. N. Kitajima, K. Fujisawa, and Y. Moro-oka, *J. Am. Chem. Soc.*, *112*, 3210 (1990).
51. J. C. Calabrese and S. Trofimenko, *Inorg. Chem.*, *31*, 4810 (1992).
52. N. Kitajima, *Adv. Inorg. Chem.*, *39*, 1 (1992).
53. D. M. Eichhorn and W. H. Armstrong, *Inorg. Chem.*, *29*, 3607 (1990).
54. R. Han, G. Parkin, and S. Trofimenko, *Polyhedron*, *14*, 387 (1995).
55. E. Libertini, K. Yoon, and G. Parkin, *Polyhedron*, *12*, 2539 (1993).
56. A. H. Cowley, R. L. Geerts, C. M. Nunn, and S. Trofimenko, *J. Organomet. Chem.*, *365*, 19 (1989).
57. J. C. Calabrese, P. J. Domaille, J. S. Thompson, and S. Trofimenko, *Inorg. Chem.*, *29*, 4429 (1990).
58. G. Ferguson, M. C. Jennings, F. J. Laor, and C. Shanahan, *Acta Crystallogr. Ser.*, *C47*, 2079 (1991).
59. A. Haaland, *Angew. Chem. Int. Ed. Engl.*, *28*, 992 (1989).
60. S. M. Carrier, C. E. Ruggiero, R. P. Houser, and W. B. Tolman, *Inorg. Chem.*, *32*, 4889 (1993).
61. C. Mealli, C. S. Arcus, J. L. Wilkinson, T. J. Marks, and J. A. Ibers, *J. Am. Chem. Soc.*, *98*, 711 (1976).
62. H. B. Bürgi and J. Dunitz, Eds. *Structure Correlation* VCH, Weinheim, 1994, Vols. 1 and 2.
63. A. Looney and G. Parkin, *Inorg. Chem.*, *33*, 1234 (1994).
64. R. Alsfasser, A. K. Powell, and H. Vahrenkamp, *Angew. Chem. Int. Ed. Engl.*, *29*, 898 (1990).
65. R. Alsfasser, A. K. Powell, S. Trofimenko, and H. Vahrenkamp, *Chem. Ber.*, *126*, 685 (1993).
66. R. Han and G. Parkin, *J. Am. Chem. Soc.*, *113*, 9707 (1991).
67. A. Looney, R. Han, K. McNeill, and G. Parkin, *J. Am. Chem. Soc.*, *115*, 4690 (1993).

68. A. Looney, A. Saleh, Y. Zhang, and G. Parkin, *Inorg. Chem.*, *33*, 1158 (1994).
69. J. H. MacNeil, W. C. Watkins, M. C. Baird, and K. F. Preston, *Organometallics*, *11*, 2761 (1992).
70. M. D. Curtis, K.-B. Shiu, W. M. Butler, and J. C. Huffman, *J. Am. Chem. Soc.*, *108*, 3335 (1986).
71. K.-B. Shiu and L.-Y. Lee, *J. Organomet. Chem.*, *348*, 357 (1988).
72. T. M. Bockman and J. K. Kochi, *New J. Chem.*, *16*, 39 (1992).
73. T. Madach and H. Vahrenkamp, *Z. Naturforsch.*, *33B*, 1301 (1978).
74. K. A. E. O'Callaghan, S. J. Brown, J. A. Page, M. C. Baird, T. C. Richards, and W. E. Geiger, *Organometallics*, *10*, 3119 (1991).
75. M. Tilset, *J. Am. Chem. Soc.*, *114*, 2740 (1992).
76. V. Skagestad and M. Tilset, *J. Am. Chem. Soc.*, *115*, 5077 (1993).
77. S. Trofimenko, *J. Am. Chem. Soc.*, *91*, 588 (1969).
78. N.-Y. Sun and S. J. Simpson, *J. Organomet. Chem.*, *434*, 341 (1992).
79. A. M. McNair, D. C. Boyd, and K. R. Mann, *Organometallics*, *5*, 303 (1986).
80. P. R. Sharp and A. J. Bard, *Inorg. Chem.*, *22*, 2689 (1983).
81. M. D. Curtis and K.-B. Shiu, *Inorg. Chem.*, *24*, 1213 (1985).
82. M. Y. Darensbourg, *Progress in Inorganic Chemistry*, Wiley-Interscience, New York, 1985, Vol. 33, pp. 221–274.
83. E.-I. Negishi, *Organometallics in Organic Synthesis*, Wiley, New York, 1980, Vol. 1.
84. G. Wilkinson, F. G. A. Stone, and E. W. Abel, Eds., *Comprehensive Organometallic Chemistry*, Pergamon, Oxford, UK, Vol. 7, 1982.
85. J. R. Hauske, in *Comprehensive Organic Synthesis*, B. M. Trost and I. Fleming, Eds., Pergamon, Oxford, UK, 1991, Vol. 1, pp. 77–106.
86. P. Knochel, in *Comprehensive Organic Synthesis*, B. M. Trost and I. Fleming, Eds., Pergamon, Oxford, UK, 1991, Vol. 1, pp. 211–232.
87. H. Sinn and W. Kaminsky, *Adv. Organomet. Chem.*, *18*, 99 (1980).
88. K. F. Jensen and W. Kern, in *Thin Film Processes II*, J. L. Vossen and W. Kern, Eds., Academic, Boston, MA, 1991, pp. 283–368.
89. For example, see L. V. Interrante, L. E. Carpenter, II, C. Whitmarsh, W. Lee, M. Garbuskas, and G. A. Slack, *Mater. Res. Soc. Symp. Proc.*, *73*, 359 (1986).
90. P. R. Markies, O. S. Akkerman, F. Bickelhaupt, W. J. J. Smeets, and A. L. Spek, *Adv. Organomet. Chem.*, *32*, 147 (1991).
91. See, for example, E. C. Ashby, *Pure Appl. Chem.*, *52*, 545 (1980) and references cited therein.
92. D. L. Reger, S. J. Knox, A. L. Rheingold, and B. S. Haggerty, *Organometallics*, *9*, 2581 (1990).
93. D. L. Reger, S. J. Knox, and L. Leboida, *Organometallics*, *9*, 2218 (1990).
94. D. L. Reger and S. S. Mason, *Organometallics*, *12*, 2600 (1993).
95. G. G. Lobbia, F. Bonati, P. Cecchi, A. Cingolani, and A. Lorenzotti, *J. Organomet. Chem.*, *378*, 139 (1989).

96. B. K. Nicholson, *J. Organomet. Chem.*, *265*, 153 (1984).
97. S. K. Lee and B. K. Nicholson, *J. Organomet. Chem.*, *309*, 257 (1986).
98. D. L. Reger, S. S. Mason, A. L. Rheingold, and R. L. Ostrander, *Inorg. Chem.*, *33*, 1803 (1994).
99. R. Han and G. Parkin, *Organometallics*, *10*, 1010 (1991).
100. R. Han, M. Bachrach, and G. Parkin, *Polyhedron*, *9*, 1775 (1990).
101. R. Han and G. Parkin, *J. Am. Chem. Soc.*, *114*, 748 (1992).
102. R. Han and G. Parkin, *J. Organomet. Chem.*, *393*, C43 (1990).
103. R. Han and G. Parkin, *J. Am. Chem. Soc.*, *112*, 3662 (1990).
104. R. Han and G. Parkin, *Polyhedron*, *9*, 2655 (1990).
105. See, for example, T. V. Lubben and P. T. Wolczanski, *J. Am. Chem. Soc.*, *109*, 424 (1987).
106. P. G. Williard and J. M. Salvino, *J. Chem. Soc. Chem. Commun.*, 153 (1986).
107. A. G. Pinkus, J. G. Lindberg, and A.-B. Wu, *J. Chem. Soc. Chem. Commun.*, 1350 (1969).
108. M. S. Kharasch and C. F. Fuchs, *J. Org. Chem.*, *10*, 292 (1945).
109. M. S. Kharasch, F. L. Lambert, and W. H. Urry, *J. Org. Chem.*, *10*, 298 (1945).
110. I. B. Gorrell, A. Looney, and G. Parkin, *J. Chem. Soc. Chem. Commun.*, 220 (1990).
111. K. Yoon and G. Parkin, *J. Am. Chem. Soc.*, *113*, 8414 (1991).
112. A. Looney, R. Han, I. B. Gorrell, M. Cornebise, K. Yoon, G. Parkin, and A. L. Rheingold, *Organometallics*, *14*, 274 (1995).
113. R. Han, I. B. Gorrell, A. G. Looney, and G. Parkin, *J. Chem. Soc. Chem. Commun.*, 717 (1991).
114. R. Han, A. Looney, and G. Parkin, *J. Am. Chem. Soc.*, *111*, 7276 (1989).
115. A. Looney and G. Parkin, *Polyhedron*, *9*, 265 (1990).
116. A. H. Cowley, C. J. Carrano, R. L. Geerts, R. A. Jones, and C. M. Nunn, *Angew. Chem. Int. Ed. Engl.*, *27*, 277 (1988).
117. A. J. Canty, N. J. Minchin, J. M. Patrick, and A. H. White, *Aust. J. Chem.*, *36*, 1107 (1983).
118. M. A. Kennedy and P. D. Ellis, *J. Am. Chem. Soc.*, *111*, 3195 (1989).
119. R. Han and G. Parkin, *Inorg. Chem.*, *31*, 983 (1992).
120. H. Schmidbaur, *Gmelin, Be-Org.*, 1, 155 (1987).
121. R. Han and G. Parkin, *Inorg. Chem.*, *33*, 3848 (1994).
122. D. Reger, S. Mason, and A. L. Rheingold, *J. Am. Chem. Soc.*, *115*, 10406 (1993).
123. D. Reger, S. Mason, and A. L. Rheingold, *J. Am. Chem. Soc.*, *116*, 2233 (1994).
124. C. K. Ghosh and W. A. G. Graham, *J. Am. Chem. Soc.*, *109*, 4726 (1987).
125. J. K. Hoyano and W. A. G. Graham, *J. Am. Chem. Soc.*, *104*, 3723 (1982).
126. A. H. Janowicz and R. G. Bergman, *J. Am. Chem. Soc.*, *104*, 352 (1982).
127. J. M. Buchanan, J. M. Stryker, and R. G. Bergman, *J. Am. Chem. Soc.*, *108*, 1537 (1986).

128. C. K. Ghosh and W. A. G. Graham, *J. Am. Chem. Soc.*, *111*, 375 (1989).
129. S. Trofimenko, unpublished results.
130. R. G. Ball, University of Alberta Structure Determination Laboratory Report No. SR: 071801-12-987. Quoted in (19).
131. M. Cowie and R. G. Ball, unpublished results. Quoted in (19).
132. F. Basolo and R. G. Pearson, *Mechanisms of Inorganic Reactions*, Wiley, New York, 1967, pp. 553–556.
133. R. A. Periana and R. G. Bergman, *J. Am. Chem. Soc.*, *108*, 7346 (1986).
134. T. Iwamoto, in *Comprehensive Organic Synthesis*, B. M. Trost, Eds., Pergamon, Oxford, UK, 1991, Vol. 1, pp. 231–250.
135. G. A. Molander, *Chem. Rev.*, *92*, 44 (1992).
136. See for example, W. J. Evans, S. L. Gonzales, and J. W. Ziller, *J. Am. Chem. Soc.*, *113*, 9880 (1991) and references cited therein.
137. J. Takats, *Twentieth Rare Earth Conference*, Monterey, California, September 1993.
138. J. Takats, X. Zhang, V. W. Day, and T. A. Eberspacher, *Organometallics*, *12*, 4286 (1993).
139. W. J. Evans, J. W. Grate, H. W. Choi, I. Bloom, W. E. Hunter, and J. L. Atwood, *J. Am. Chem. Soc.*, *107*, 941 (1985).
140. S. J. Swamy and H. Schumann, *J. Organomet. Chem.*, *334*, 1 (1987).
141. J. Takats, *77th CSC Conference*, Winnipeg, Manitoba, June 1994, Abstract 538.
142. L. Hasinoff, J. Takats, X. W. Zhang, A. H. Bond, and R. D. Rogers, *J. Am. Chem. Soc.*, *116*, 8833 (1994).
143. R. McDonald, X. Zhang, and J. Takats, *New J. Chem.*, submitted for publication.
144. H. J. Heeres, A. Meetsma, J. H. Teuben, and R. D. Rogers, *Organometallics*, *8*, 2637 (1989) and references cited therein.
145. J. M. Boncella and R. A. Andersen, *Organometallics*, *4*, 205 (1985).
146. I. Bertini, H. B. Gray, S. J. Lippard, and J. S. Valentine, Eds., *Bioinorganic Chemistry*, University Science Books, Mill Valley, CA, 1994.
147. S. J. Lippard and J. M. Berg, *Principles of Bioinorganic Chemistry*, University Science Books, Mill Valley, CA, 1994.
148. K. D. Karlin, *Science*, *261*, 701 (1993).
149. K. A. Magnus, H. Ton-That, and J. E. Carpenter, *Chem. Rev.*, *94*, 727 (1994).
150. A. Messerschmidt, in *Bioinorganic Chemistry of Copper*, K. D. Karlin and Z. Tyeklár., Eds., Chapman & Hall, New York, 1993, pp. 471–484.
151. See for example, K. Djinovic, G. Gatti, A. Coda, L. Antolini, G. Pelosi, A. Desider, M. Falconi, F. Marmocchi, G. Rotilio, and M. Bolognesi, *J. Mol. Biol.*, *225*, 791 (1992).
152. R. E. Stenkamp, *Chem. Rev.*, *94*, 715 (1994).
153. See for example, W. C. Stallings, K. A. Pattridge, R. K. Strong, and M. L. Ludwig, *J. Biol. Chem.*, *260*, 16424 (1985).

154. W. H. Armstrong, A. Spool, G. C. Papaefthymiou, R. B. Frankel, and S. J. Lippard, *J. Am. Chem. Soc.*, *106*, 3653 (1984).
155. M. Mohan, S. M. Holmes, R. J. Butcher, J. P. Jasinski, and C. J. Carrano, *Inorg. Chem.*, *31*, 2029 (1991) and references cited therein.
156. I. K. Dhawan, A. Pacheco, and J. H. Enemark, *J. Am. Chem. Soc.*, *116*, 7911 (1994).
157. A. Volbeda and W. G. J. Hol, *J. Mol. Biol.*, *209*, 249 (1989).
158. E. I. Solomon, M. J. Baldwin, and M. D. Lowery, *Chem. Rev.*, *92*, 521 (1992) and references cited therein.
159. J. Ling, L. P. Nestor, R. S. Czernuszewicz, T. G. Spiro, R. Frackzkiewicz, K. D. Sharma, T. M. Loehr, and J. Sanders-Loehr, *J. Am. Chem. Soc.*, *116*, 7682 (1994).
160. R. S. Himmelwright, N. C. Eickman, C. D. LuBien, K. Lerch, and E. I. Solomon, *J. Am. Chem. Soc.*, *102*, 7339 (1980).
161. K. D. Karlin and Y. Gultneh, *Prog. Inorg. Chem.*, *35*, 219 (1987).
162. T. N. Sorrell, *Tetrahedron*, *40*, 3 (1989).
163. N. Kitajima and Y. Moro-oka, *Chem. Rev.*, *94*, 737 (1994).
164. K. D. Karlin, Z. Tyeklár, and A. D. Zuberbühler, in *Bioinorganic Catalysis*, J. Reedijk, Eds., Marcel-Dekker, New York, 1993, pp. 261–315.
165. K. A. Magnus, H. Ton-That, and J. E. Carpenter, *Chem. Rev.*, *94*, 727 (1994).
166. N. Kitajima and Y. Moro-oka, *J. Chem. Soc. Dalton Trans.*, 2665 (1993).
167. N. Kitajima, T. Koda, S. Hashimoto, T. Kitagawa, and Y. Moro-oka, *J. Chem. Soc. Chem. Commun.*, 151 (1988).
168. N. Kitajima, T. Koda, S. Hashimoto, T. Kitagawa, and Y. Moro-oka, *J. Am. Chem. Soc.*, *113*, 5664 (1991).
169. N. Kitajima, K. Fujisawa, Y. Moro-oka, and K. Toriumi, *J. Am. Chem. Soc.*, *111*, 8975 (1989).
170. N. Kitajima, K. Fujisawa, C. Fujimoto, Y. Moro-oka, S. Hashimoto, T. Kitagawa, K. Toriumi, K. Tasumi, and A. Nakamura, *J. Am. Chem. Soc.*, *114*, 1277 (1992).
171. D. C. Bradley, J. S. Ghotra, F. A. Hart, M. B. Hursthouse, and P. R. Raithby, *J. Chem. Soc. Dalton Trans.*, 1166 (1977).
172. R. Haegele and J. C. A. Boeyens, *J. Chem. Soc. Dalton Trans.*, 548 (1977).
173. A. E. Lapsin, Y. J. Smolin, and Y. F. Shepelev, *Acta Crystallogr. Ser. C46*, 1755 (1990).
174. R. R. Jacobson, Z. Tyeklár, A. Farooq, K. D. Karlin, S. Liu, and J. Zubieta, *J. Am. Chem. Soc.*, *110*, 3690 (1988).
175. K. Fujisawa, Y. Moro-oka, and N. Kitajima, *J. Chem. Soc. Chem. Commun.*, 623 (1994).
176. F. Tuczek and E. I. Solomon, *J. Am. Chem. Soc.*, *116*, 6916 (1994).
177. M. J. Baldwin, D. E. Root, J. E. Pate, K. Fujisawa, N. Kitajima, and E. I. Solomon, *J. Am. Chem. Soc.*, *114*, 10421 (1992).

178. See J. E. Pate, P. K. Ross, T. J. Thamann, C. A. Reed, K. D. Karlin, T. N. Sorrell, and E. I. Solomon, *J. Am. Chem. Soc.*, *111*, 5198 (1989) and references cited therein.

179. E. I. Solomon, in *Metal Clusters in Proteins*, L. Que, Jr., Ed., American Chemical Society, Washington, DC, 1988, Vol. 372, pp. 116–150.

180. N. Kitajima, K. Fujisawa, S. Hikichi, and Y. Moro-oka, *J. Am. Chem. Soc.*, *115*, 7874 (1993).

181. N. Kitajima, T. Koda, Y. Iwata, and Y. Moro-oka, *J. Am. Chem. Soc.*, *112*, 8833 (1990).

182. G. Pandey, C. Muralikrishna, and U. T. Bhalerao, *Tetrahedron Lett.*, 3771 (1990).

183. N. Kitajima, T. Katayama, K. Fujisawa, and Y. Moro-oka, to be submitted.

184. A. G. Sykes, in *Advances in Inorganic Chemistry*, A. G. Sykes, Eds., Academic, New York, 1991, Vol. 36, pp. 377–408.

185. E. T. Adman, *Adv. Prot. Chem.*, *42*, 145 (1991).

186. J. Han, T. M. Loehr, Y. Lu, J. S. Valentine, B. A. Averill, and J. Sanders-Loehr, *J. Am. Chem. Soc.*, *115*, 4256 (1993) and references cited therein.

187. N. Kitajima, T. Katayama, K. Fujisawa, Y. Iwata, and Y. Moro-oka, *J. Am. Chem. Soc.*, *115*, 7872 (1993).

188. C. E. Ruggiero, S. M. Carrier, W. E. Antholine, J. W. Whittaker, C. J. Cramer, and W. B. Tolman, *J. Am. Chem. Soc.*, *115*, 11285 (1993).

189. J. S. Thompson, T. J. Marks, and J. A. Ibers, *J. Am. Chem. Soc.*, *101*, 4180 (1979).

190. J. S. Thompson, T. Sorrell, T. J. Marks, and J. A. Ibers, *J. Am. Chem. Soc.*, *101*, 4193 (1979).

191. N. Kitajima, K. Fujisawa, M. Tanaka, and Y. Moro-oka, *J. Am. Chem. Soc.*, *114*, 9232 (1992).

192. D. Qui, L. Kilpatrick, N. Kitajima, and T. G. Spiro, *J. Am. Chem. Soc.*, *116*, 2585 (1994).

193. L. C. Stewart and J. P. Klinman, *Annu. Rev. Biochem.*, *57*, 551 (1988).

194. B. J. Reedy and N. J. Blackburn, *J. Am. Chem. Soc.*, *116*, 1924 (1994) and references cited therein.

195. G. Tian, J. A. Berry, and J. P. Klinman, *Biochemistry*, *33*, 226 (1994).

196. D. J. Merkler, R. Kulathila, S. D. Young, J. Freeman, and J. J. Villafranca, in *Bioinorganic Chemistry of Copper*, K. D. Karlin, and Z. Tyeklár, Eds., Chapman & Hall, New York, 1993, pp. 196–209.

197. S. I. Chan, H.-H. T. Nguyen, A. K. Shiemke, and M. E. Lidstrom, in *Bioinorganic Chemistry of Copper*, K. D. Karlin and Z. Tyeklár, Eds., Chapman & Hall, New York, 1993, pp. 184–195.

198. J. S. Thompson, *J. Am. Chem. Soc.*, *106*, 4057 (1984).

199. K. D. Karlin, N. Wei, B. Jung, S. Kaderli, and A. D. Zuberbuhler, *J. Am. Chem. Soc.*, *113*, 5868 (1991).

200. K. D. Karlin, N. Wei, B. Jung, S. Kaderli, P. Niklaus, and A. D. Zuberbuhler, *J. Am. Chem. Soc.*, *115*, 9506 (1993).
201. K. Fujisawa, M. Tanaka, Y. Moro-oka, and N. Kitajima, *J. Am. Chem. Soc.*, *116*, 12079 (1994).
202. J. W. Egan, Jr., B. S. Haggerty, A. L. Rheingold, S. C. Sendlinger, and K. H. Theopold, *J. Am. Chem. Soc.*, *112*, 2445 (1990).
203. N. Kitajima, K. Fujisawa, and Y. Moro-oka, *Inorg. Chem.*, *29*, 357 (1990).
204. K. Fujisawa, N. Kitajima, and Y. Moro-oka, to be submitted.
205. N. Kitajima, S. Hikichi, M. Tanaka, and Y. Moro-oka, *J. Am. Chem. Soc.*, *115*, 5496 (1993).
206. Y. Zang, T. E. Elgren, Y. Dong, and L. Que, Jr. *J. Am. Chem. Soc.*, *115*, 811 (1993).
207. J. T. Groves and Y. Watanabe, *J. Am. Chem. Soc.*, *108*, 7834 (1986).
208. See for example, K. D. Karlin, P. Ghosh, R. W. Cruse, A. Farooq, Y. Gultneh, R. R. Jacobson, N. J. Blackburn, R. W. Strange, and J. Zubieta, *J. Am. Chem. Soc.*, *110*, 6769 (1988).
209. P. M. H. Kroneck, J. Beuerle, and W. Schumacher, in *Degradation of Environmental Pollutants by Microoganisms and their Metalloenzymes*, H. Sigel and A. Sigel, Eds., Marcel Dekker, New York, 1992, Vol. 28, pp. 455–505.
210. J. W. Godden, S. Turley, D. C. Teller, E. T. Adman, M. Y. Liu, W. J. Payne, and J. LeGall, *Science*, *253*, 438 (1991).
211. B. D. Howes, Z. H. L. Abraham, D. J. Lowe, T. Bruser, R. R. Eady, and B. E. Smith, *Biochemistry*, *33*, 3171 (1994).
212. Z. H. L. Abraham, D. J. Lowe, B. E. Smith, *Biochem. J.*, *295*, 587 (1993).
213. C. L. Hulse, B. A. Averill, and J. M. Tiedje, *J. Am. Chem. Soc.*, *111*, 2322 (1989).
214. M. A. Jackson, J. M. Tiedje, and B. A. Averill, *FEBS Lett.*, *291*, 41 (1991).
215. R. W. Ye, I. Toro-Suarez, J. M. Tiedje, and B. A. Averill, *J. Biol. Chem.*, *266*, 12848 (1991).
216. P. P. Paul, Z. Tyeklar, A. Farooq, K. D. Karlin, S. Liu, and J. Zubieta, *J. Am. Chem. Soc.*, *112*, 2430 (1990).
217. P. P. Paul and K. D. Karlin, *J. Am. Chem. Soc.*, *113*, 6331 (1991).
218. F. Jiang, R. R. Conry, L. Bubacco, Z. Tyeklár, R. R. Jacobson, K. D. Karlin, and J. Peisach, *J. Am. Chem. Soc.*, *115*, 2093 (1993).
219. W. B. Tolman, S. M. Carrier, C. E. Ruggiero, W. E. Antholine, and J. W. Whittaker, in *Bioinorganic Chemistry of Copper*, K. D. Karlin and Z. Tyeklár, Eds., Chapman & Hall, New York, 1993, pp. 406–418.
220. C. E. Ruggiero, S. M. Carrier, and W. B. Tolman, *Angew. Chem. Int. Ed. Engl.*, *33*, 895 (1993).
221. J. A. Halfen, S. Mahapatra, M. M. Olmstead, and W. B. Tolman, *J. Am. Chem. Soc.*, *116*, 2173 (1994).
222. J. A. Halfen and W. B. Tolman, *J. Am. Chem. Soc.*, *116*, 5475 (1994).

223. R. D. Feltham and J. H. Enemark, *Top. Stereochem.*, *12*, 155 (1981).
224. S. J. Lippard, *Angew. Chem. Int. Ed. Engl.*, *27*, 344 (1988).
225. J. Sanders-Loehr, in *Iron Carriers and Iron Proteins*, T. M. Loehr, Eds., VCH, New York, 1989, Vol. 5, pp. 373–466.
226. L. Que, Jr. and A. E. True, *Progress in Inorganic Chemistry*, Wiley-Interscience, New York, 1980, Vol. 38, pp. 97–200.
227. J. B. Vincent, G. L. Olivier-Lilley, and B. A. Averill, *Chem. Rev.*, *90*, 1447 (1990).
228. D. M. Kurtz, Jr., *Chem. Rev.*, *90*, 585 (1990).
229. A. L. Feig and S. J. Lippard, *Chem. Rev.*, *94*, 759 (1994).
230. N. Kitajima, H. Fukui, Y. Moro-oka, Y. Mizutani, and T. Kitagawa, *J. Am. Chem. Soc.*, *112*, 6402 (1990).
231. N. Kitajima, M. Ito, H. Amagai, H. Fukui, and Y. Moro-oka, unpublished results.
232. N. Kitajima, N. Tamura, H. Amagai, H. Fukui, Y. Moro-oka, Y. Mizutani, T. Kitagawa, R. Mathur, K. Heerwegh, C. A. Reed, C. R. Randall, J. L. Que, and K. Tatsumi, *J. Am. Chem. Soc.*, *116*, 9071 (1994).
233. N. Kitajima, H. Amagai, N. Tamura, M. Ito, Y. Moro-oka, K. Heerwegh, A. Pénicaud, R. Mathur, C. A. Reed, and P. D. W. Boyd, *Inorg. Chem.*, *32*, 3583 (1993).
234. V. A. Vankai, M. G. Newton, and D. M. Kurtz, Jr., *Inorg. Chem.*, *31*, 341 (1992).
235. N. Kitajima, N. Tamura, M. Tanaka, and Y. Moro-oka, *Inorg. Chem.*, *31*, 3342 (1992).
236. N. Kitajima, M. Ito, H. Fukui, and Y. Moro-oka, *J. Am. Chem. Soc.*, *115*, 9335 (1993).
237. V. L. Pecoraro, M. J. Baldwin, and A. Gelasco, *Chem. Rev.*, *94*, 807 (1994).
238. V. L. Pecoraro, Ed., *Manganese Redox Enzymes* VCH, New York, 1992.
239. K. Wieghardt, *Angew. Chem. Int. Ed. Engel.*, *28*, 1153 (1989).
240. J. E. Sheats, R. S. Czernuszewicz, G. C. Dismukes, A. L. Rheingold, V. Petrouleas, J. Stubbe, W. H. Armstrong, R. H. Beer, and S. J. Lippard, *J. Am. Chem. Soc.*, *109*, 1435 (1987).
241. N. Kitajima, M. Osawa, N. Tamura, Y. Moro-oka, T. Hirano, M. Hirobe, and T. Nagano, *Inorg. Chem.*, *32*, 1879 (1993).
242. W. C. Stallings, K. A. Pattridge, R. K. Strong, and M. L. Ludwig, *J. Biol. Chem.*, *260*, 16424 (1985).
243. M. W. Parker and C. C. F. Blake, *J. Mol. Biol.*, *199*, 649 (1988).
244. N. Kitajima, U. P. Singh, H. Amagai, M. Osawa, and Y. Moro-oka, *J. Am. Chem. Soc.*, *113*, 7757 (1991).
245. See for example, P. A. Goodson, A. R. Oki, J. Glerup, and D. J. Hodgson, *J. Am. Chem. Soc.*, *112*, 6248 (1990).
246. N. Kitajima, M. Osawa, M. Tanaka, and Y. Moro-oka, *J. Am. Chem. Soc.*, *113*, 8952 (1991).

247. N. Kitajima, H. Komatsuzaki, S. Hikichi, M. Osawa, and Y. Moro-oka, *J. Am. Chem. Soc.*, *116*, 11596 (1994).

248. R. B. VanAtta, C. E. Strouse, L. K. Hanson, and J. S. Valentine, *J. Am. Chem. Soc.*, *109*, 1425 (1987).

249. B. A. Springer, S. G. Sligar, J. S. Olson, and J. G. N. Phillips, *Chem. Rev.*, *94*, 699 (1994).

250. R. E. Stenkamp, *Chem. Rev.*, *94*, 715 (1994).

251. N. Kitajima, M. Osawa, S. Imai, and Y. Moro-oka, unpublished results.

252. N. Kitajima, M. Osawa, S. Imai, K. Fujisawa, Y. Moro-oka, K. Heeregh, C. A. Reed, and P. Boyd, *Inorg. Chem.*, *33*, 4613 (1994).

253. S. J. Dodgson, R. E. Tashian, G. Gros, and N. D. Carter, Eds., *The Carbonic Anhydrases*, Plenum, New York, 1991.

254. A. E. Eriksson, T. A. Jones, and A. Liljas, *Proteins*, *4*, 274 (1988).

255. E. Kimura, T. Shiota, T. Koike, M. Shiro, and M. Kodama, *J. Am. Chem. Soc.*, *112*, 5805 (1990).

256. R. Alsfasser, S. Trofimenko, A. Looney, G. Parkin, and H. Vahrenkamp, *Inorg. Chem.*, *30*, 4098 (1991).

257. M. Ruf, K. Weis, and H. Vahrenkamp, *J. Chem. Soc. Chem. Commun.*, 135 (1994).

258. I. B. Gorrell, A. Looney, G. Parkin, and A. L. Rheingold, *J. Am. Chem. Soc.*, *112*, 4068 (1990).

259. A. Looney, G. Parkin, R. Alsfasser, M. Ruf, and H. Vahrenkamp, *Angew. Chem. Int. Ed. Engl.*, *31*, 92 (1992).

260. T. G. Spiro, Ed., *Zinc Enzymes*, Wiley, New York, 1983.

261. S. Hikichi, M. Tanaka, Y. Moro-oka, and N. Kitajima, *J. Chem. Soc. Chem. Commun.*, 814 (1992).

262. J. M. Sowadski, M. D. Handschumacher, H. M. K. Murthy, B. A. Foster, and H. W. Wyckoff, *J. Mol. Biol.*, *186*, 417 (1985).

263. E. E. Kim and H. W. Wyckoff, *Clin. Chem. Acta*, *186*, 175 (1989).

264. S. Kikichi, M. Tanaka, Y. Moro-oka, and N. Kitajima, *J. Chem. Soc. Chem. Commun.*, 1737 (1994).

265. O. M. Reinaud and K. H. Theopold, *J. Am. Chem. Soc.*, *116*, 6979 (1994).

266. S. Mahapatra, J. A. Halfen, E. C. Wilkinson, L. Que, Jr., and W. B. Tolman, *J. Am. Chem. Soc.*, *116*, 9785 (1994).

267. N. Kitajima and S. Hikichi, Y. Moro-oka, to be submitted.

268. N. Kitajima, S. Hikichi, H. Komatsuzaki, and Y. Moro-oka, to be submitted.

269. P. J. Pérez, M. Brookhart, and J. L. Templeton, *Organometallics*, *12*, 261 (1993).

270. P. J. Pérez, L. Luan, P. S. White, M. Brookhart, and J. L. Templeton, *J. Am. Chem. Soc.*, *114*, 7928 (1992).

271. S. G. Feng and J. L. Templeton, *Organometallics*, *11*, 1295 (1992).

272. M. A. Collins, S. G,. Feng, P. A. White, and J. L. Templeton, *J. Am. Chem. Soc.*, *114*, 3771 (1992).

273. L. L. Blosch, K. Abboud, and J. M. Boncella, *J. Am. Chem. Soc.*, *113*, 7066 (1991).

274. A. L. Rheingold, G. Yap, and S. Trofimenko, *Inorg. Chem.*, *34*, in press.

275. G. H. Mamder, A. Sella, and D. A. Tocher, *J. Chem. Soc. Commun.*, 2689 (1994).

276. M. Harata, K. Jitsukawa, H. Masuda, and H. Einaga, *J. Am. Chem. Soc.*, *116*, 10817 (1994).

277. A. A. Purwoko and A. J. Lees, *Inorg. Chem.*, *34*, 424 (1995).

278. D. J. Darenslfourg, M. W. Holtcamp, B. Khandeewal, K. K. Klausmeyer, and J. H. Reibenspies, *J. Am. Chem. Soc.*, *117*, 538 (1995).

Metal Complexes of Calixarenes

D. MAX ROUNDHILL

Department of Chemistry
Tulane University
New Orleans, LA

CONTENTS

Progress in Inorganic Chemistry, Vol. 43, Edited by Kenneth D. Karlin.
ISBN 0-471-12336-6

I. INTRODUCTION

This chapter is written to introduce the reader to a class of ligands that are becoming increasingly used as complexants for metal ions. This class of ligands is the calixarenes. At present only a rather limited number of metal complexes of this class of ligands have been prepared, but it is likely that an increasing number of such complexes will be prepared in the near future. This prediction is made because calixarenes possess many of the features that are desirable in a series of compounds for them to become broadly useful as a class of new ligands.

Calixarenes are a family of macrocycles that are prepared by condensation reactions between para-substituted phenols and formaldehyde (1, 2). Synthetic procedures have been developed for selectively obtaining these macrocycles with either 4-, 6-, or 8-phenolic residues in the ring (3, 4). Examples with 5- and 7-phenolic residues are also known, but at present they are much less studied. The number of phenolic residues in the macrocycle is designated by the value of n (4, 6, or 8) in the general term calix[n]arene. The structure of such a calix[4]arene with R groups as substituents in the para positions is shown in Fig. 1. Figure 2 shows the numbering systems, the abbreviations that are used for designating the different calixarenes, and the four principal conformers of the calix[4]arenes. The parent calixarenes are synthesized from phenols that have an alkyl (usually a tertiary butyl) group substituent at the para position of the phenolic ring.

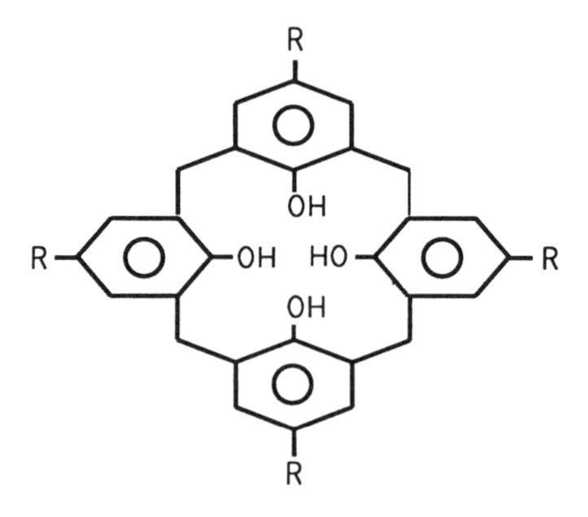

Figure 1. Calix[4]arene structure.

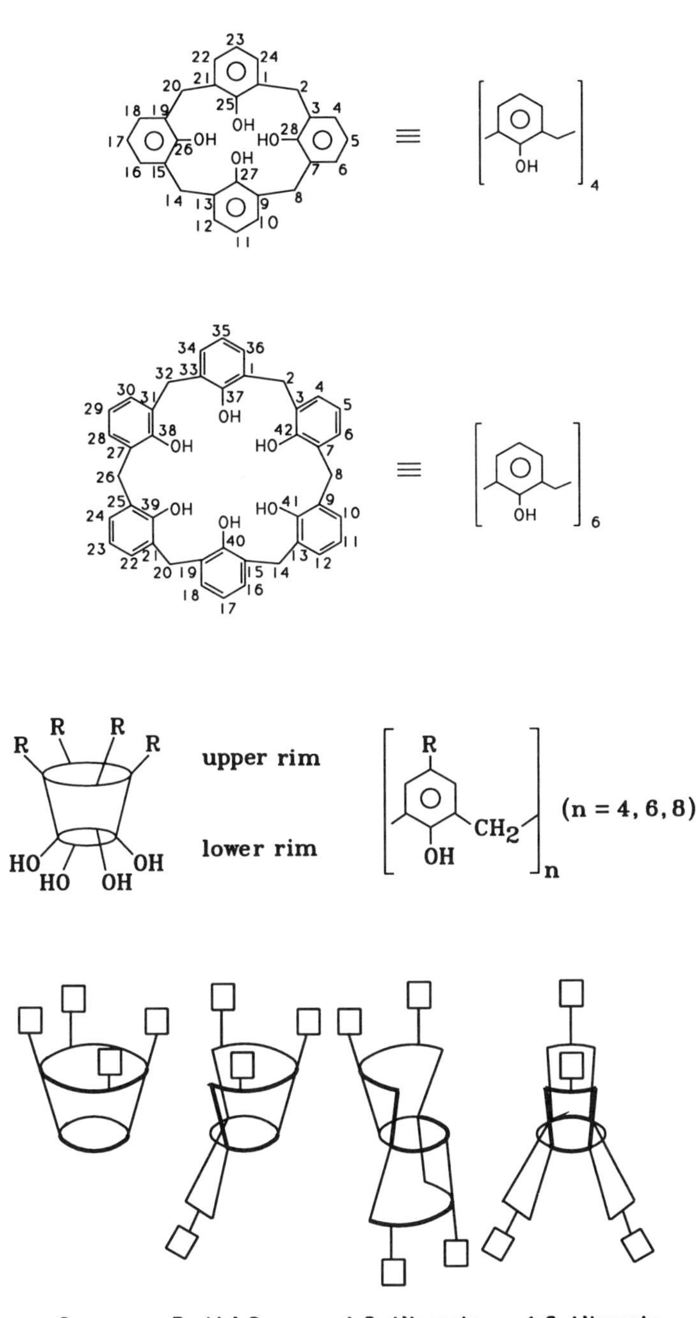

Figure 2. Numbering system, abbreviations, and conformers of calix[4]arenes.

Calixarenes are particularly attractive as ligands for metal ions because of their chemical architecture. The following list summarizes some of the more important general features of calixarenes that can be used to design ligands for individual applications.

- The calixarene framework can be regarded as a platform with two rims that can be separately modified with different functional groups.
- The cavity size for intercalation of the metal ion can be modified by changing the value of *n* in the calix[*n*]arene.
- Both the number and the geometrical arrangement of the ligating groups can be varied to accommodate the particular coordination properties of the metal ion.
- The calixarene framework has sufficient flexibility that it can conformationally adapt to a preferred cavity size and shape.
- Separate chemical modifications to each rim can be used to incorporate ligating groups on one rim and groups that confer desired solubility characteristics on the other.
- The selective incorporation of different functional groups on each rim can be used to synthesize calixarenes that can selectively complex dissimilar metals on these rims.
- The noncentrosymmetric structure of calixarenes and the ease with which enantiomeric congeners can be synthesized, makes calixarenes attractive as synthons for a wide range of ligand applications.

II. UPPER RIM MODIFICATION

Calix[*n*]arenes are generally prepared with tertiary butyl groups on the para positions in order to induce condensation reactions at the ortho positions of the phenol, and prevent the formation of insoluble polymers. If desired, subsequent removal of these tertiary butyl groups can be achieved by reacting the para substituted calixarenes with aluminum trichloride (Eq. 1) (5, 6). The "naked" calix[*n*]arene can then be used to prepare a broad range of derivatives. For

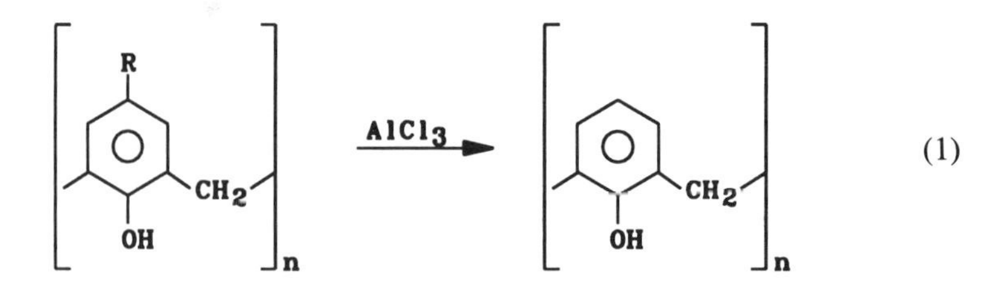

(1)

example, chemical modification of the upper rim can be used to prepare sulfonate and phosphonate ligands and complexes that are soluble in aqueous solution, or to prepare chloromethylated derivatives that can be used as precursors for a wide range of functionalized calixarenes.

A. Water Soluble Sulfonates and Phosphonates

Sulfonation of the unsubstituted calixarenes can be achieved by reaction sith concentrated sulfuric acid at 80°C for approximately 4 h (Eq. 2) (7–10). An

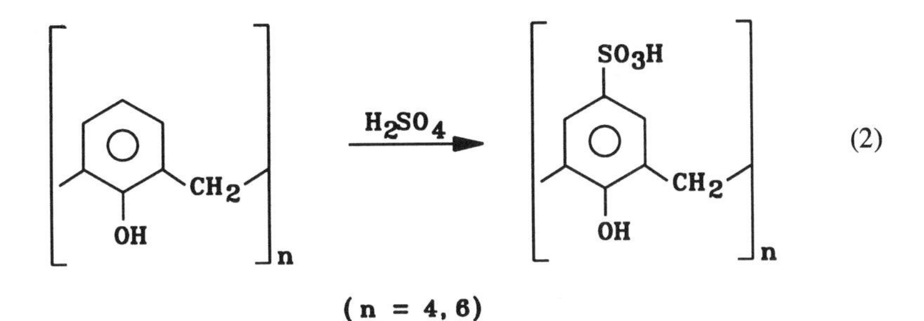

(2)

alternative route to synthesizing water soluble calixarenes involves first treating the unsubstituted calixarene with octyl chloromethyl ether and tin(IV) chloride, followed by reaction with triethyl phosphite. Subsequent hydrolysis of this phosphonated calixarene gives the arylphosphonic acid derivative (Scheme 1) (11).

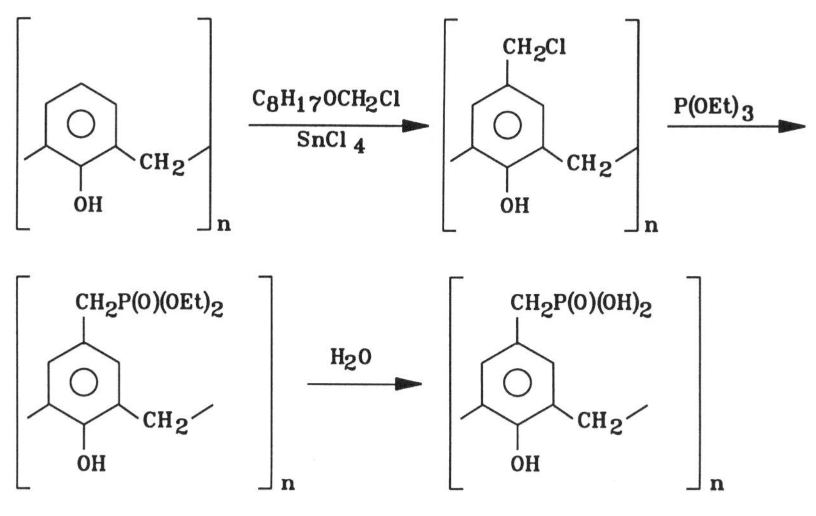

Scheme 1

B. Chloromethylation

These calixarenes with chloromethyl groups on the upper rim can also be used as synthetic precursors for a broad range of other compounds. Since this group is a benzylic chloride, it is reactive to nucleophilic reagents. Examples of such potentially useful reactions are shown in Eqs. 3 and 4.

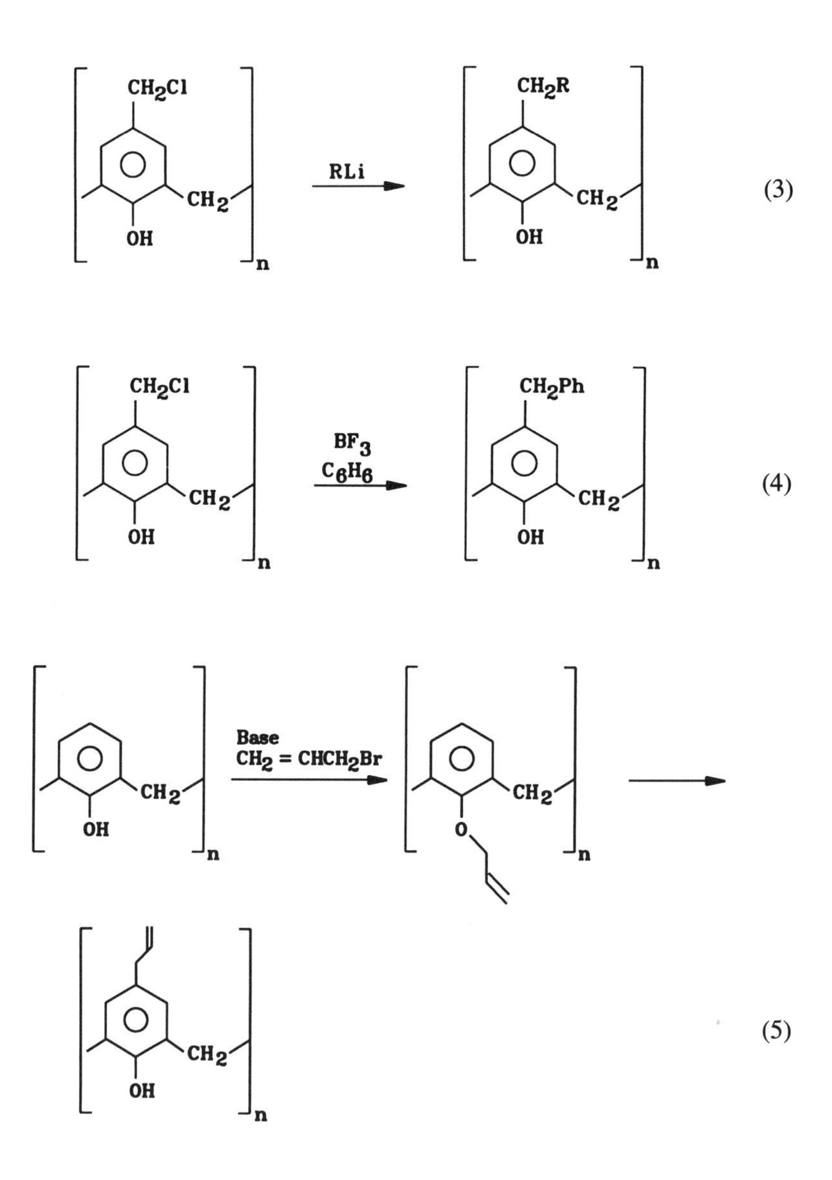

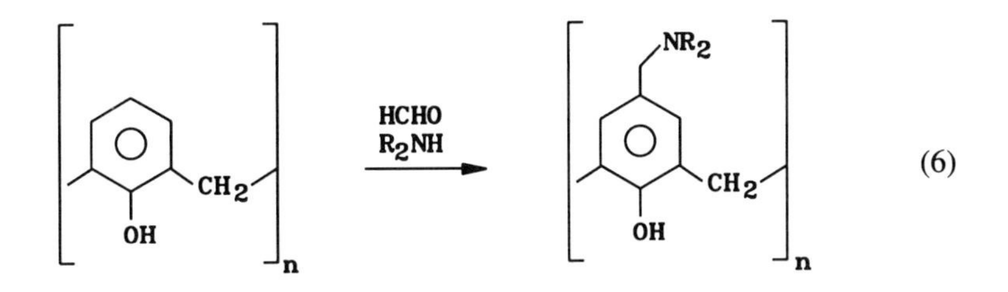

In Eqs. 5 and 6 are shown examples where the upper rim can be directly modified by either a Cope rearrangement or by a Mannich reaction. Since these examples are either one- or two-step reactions, they offer relatively simple approaches to introducing a broad range of different substituents onto the upper rim (12).

III. LOWER RIM MODIFICATION

In addition to carrying out chemical changes to the upper rim, the lower rim with phenolic oxygen atoms can be independently modified. Each rim can therefore be considered to be a separate region of the molecule that can be tailored for a specific purpose or application. Examples of such reactions that have been carried out at the lower rim phenolic positions are shown in Eqs. 7–10.

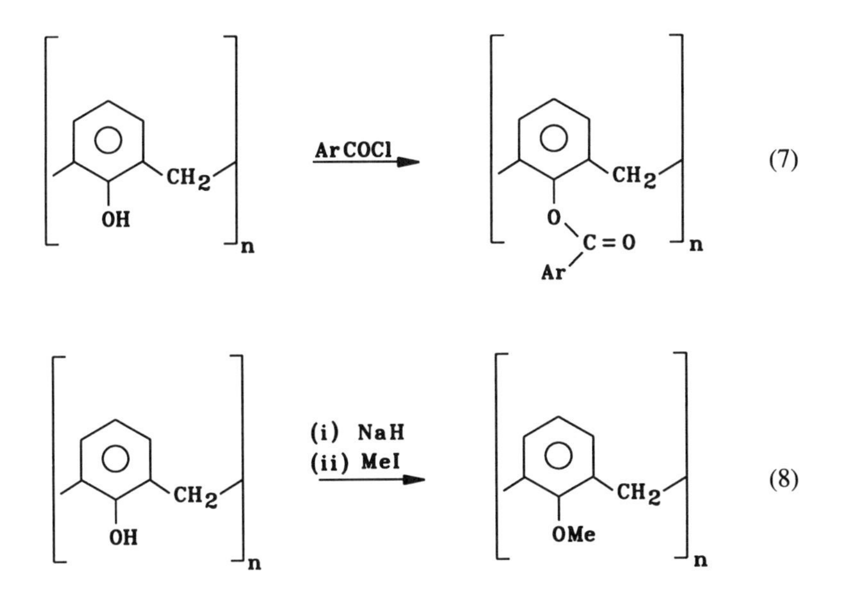

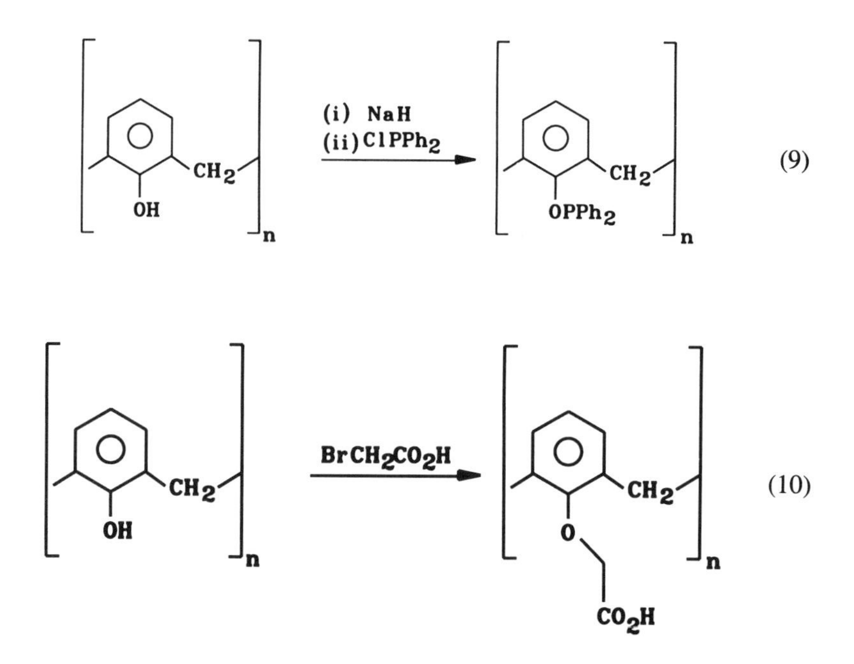

These reactions show that the lower rim can be used to incorporate ester, ether, phosphinite, or carboxylate moieties.

IV. ATTACHMENT TO POLYMERS AND RECEPTORS

In addition to chemically modifying the calixarene rims to modify the solubility characteristics of the calixarene, such chemical modifications can be used to effect their immobilization on polymeric supports. For certain applications such as the separation of heavy metal ions from aqueous media or for the fabrication of devices, it is advantageous for these calixarene-impregnated polymers to be available in an insoluble form. For other applications such as the synthesis of imaging agents it is preferable for the modified polymers to be water soluble. The preparation of insoluble products with calixarenes appended is relatively easy to achieve since such materials can be obtained by cross-linking the calixarene to the polymer via several attachment points. Since reactions leading to the chemical modification of calixarenes usually lead to multiple site functionalization, precursor calixarenes for such uses are readily available. Several examples of such cross-linked calixarene modified polymers are reported in the literature. An example by Harris and co-workers (13, 14) involves the synthesis of silicone bound calixarenes where the attachment is via multiple ester functionalities between the polymer and the calixarene.

Harris has also reported the synthesis of a new calixarene bearing a single methacrylate functionality that is potentially suitable for use as a reagent to give living polymers and copolymers. Attempted homopolymerization of his compound, however, yielded an oligomer that only had about six calixarene units in the chain (15). An alternate procedure for the synthesis of soluble calixarene modified polymers involves the alkylation of the nitrogen atoms of polyethyleneimine by a single haloalkyl substituent on the upper rim (16).

Calixarenes can be used for molecular recognition by incorporating complementary binding groups within a synthetic cavity. If such binding sites have both hydrogen-bonding donors and acceptors the molecule tends to associate intramolecularly, therefore resulting in a closed receptor. Such molecules have been transformed into open receptors for guests by synthesizing a calixarene that has a metal-binding site in close proximity to the hydrogen-bonding guest-binding site. Such a compound is shown in Fig. 3, which binds the guest molecule only when the metal ion bound to the metal-binding site disrupts the intramolecular hydrogen bonds in the guest-binding site to generate the open guest-binding site (17). In the side chain the ketonic and ether oxygen atoms of the calix[4]arene act as the metal-binding site, and the amino and pyridinium nitrogen atoms act as the guest-binding site (18).

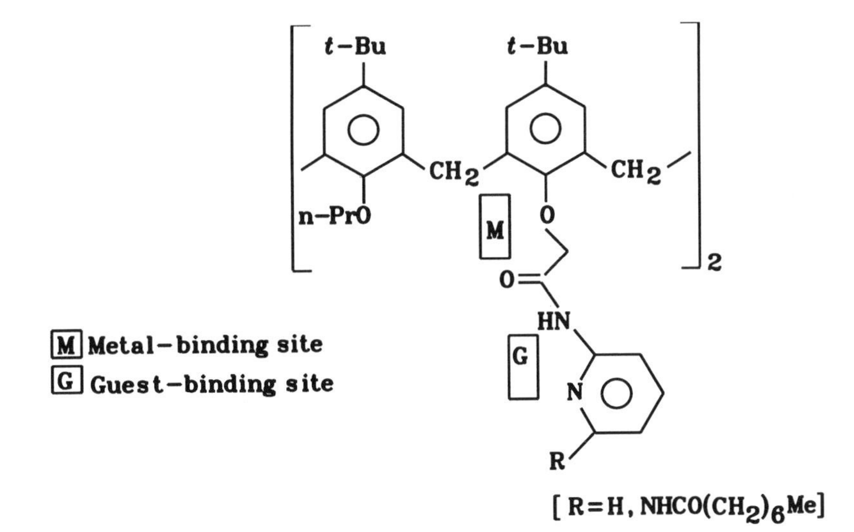

Figure 3. Calixarene having both a metal-binding site and a hydrogen-bonding site.

V. ALKALI AND ALKALINE EARTH METAL COMPLEXES

A. Unsubstituted Calixarenes

Calixarenes, like crown ethers and cyclodextrins, are a class of compounds that can be potentially used to transport cations across a liquid membrane. Calixarenes have the advantage of other molecules that both the size and shape of the lower rim cavity can be adjusted to accommodate metal ions of differing sizes. The size of the cavity can be changed by using the four-, six-, or eight-membered calixarene ring, and the shape of the cavity can be changed by varying the steric effects of the substituents on the para positions on the lower rim. Early predictions that calixarenes may be effective extractants for alkali metal ions come from studies on polymerized para octylphenol, where it was found that the pH dependent extractability of alkali metal ions followed the sequence $Cs^+ > K^+ > Rb^+ > Na^+ > Li^+$ (19).

The unsubstituted calixarenes themselves are ineffective as carriers of alkali metal cations in neutral solution, but in basic solution they can be used for the transport of these ions (20, 21). The data in Table I show that the transport effectiveness for alkali metal ions is dependent on the ring size of the *p-tert*-butylcalix[*n*]arene. The cation diameters range from 1.52 Å for Li^+ to 3.40 Å for Cs^+, and the calix[*n*]arene monoanion diameters are approximately 1.0, 2.4, and 4.8 Å for the tetramer ($n = 4$), hexamer ($n = 6$), and octamer ($n = 8$), respectively. Linked calixarenes (Fig. 4) show an unexpectedly high transport ability for Cs^+, which is likely due to the possibility that an endo complex is formed with Cs^+ being occluded within the cavity formed by aromatic groups (22). Support for such a proposal comes from the crystal structure of the Cs^+ complex of the *p-tert*-butylcalix[4]arene monoanion where the Cs^+ is bound in the larger cavity offered between the *tert*-butyl substituents on the upper rim. The metal–ligand interaction is likely with the π electrons of the aromatic groups of the deprotonated calix[4]arene ligand (23). The calixarenes also exhibit synergism for the transport of these ions. At a low Cs^+/Rb^+ ratio, *p-tert*-butylcalix[6]arene transports Rb^+ more rapidly than Cs^+, but as this ratio increases it

TABLE I
Cation Transport (in mol s^{-1} m^{-2} $\times$ 10^8) by *p-tert*-Butylcalix[*n*]arenes from Basic Solution

Cation	Diameter (Å)	$n = 4$	$n = 6$	$n = 8$
Li^+	1.52		10	2
Na^+	2.04	2	22	9
K^+	2.76	<0.7	13	10
Rb^+	3.04	6	71	340
Cs^+	3.40	260	810	996

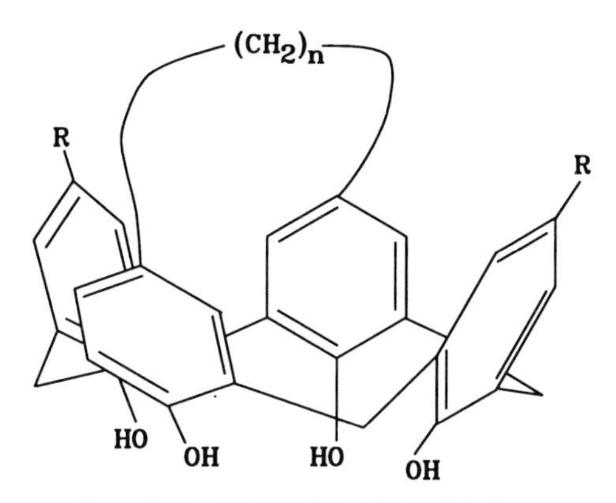

Figure 4. Structure of a linked calix[4]arene.

is Cs^+ that is transported more rapidly. Furthermore the transport of Na^+ is more rapid in the presence of Cs^+ than it is with Na^+ alone.

Water soluble calix[4]arenes having sulfonate substituents in the para positions of the upper rim undergo proton-transfer reactions in the presence of Na^+ ions (24). X-ray crystallography of the salt shows the calix[4]arene to be associated with five Na^+ ions (25).

B. Lower Rim Modified Calixarenes

Calixarenes that are chemically modified on the lower phenolic rim can be used as complexants for alkali [Group 1 (IA)] and alkaline earth [Group 2 (IIA)] metal ions. Among the most studied calixarenes are those that have uncharged ester, ketone, or ether groups appended to this rim. These functionalities are particularly attractive as complexants because they have oxygen atoms as donor ligands. Such atoms are preferable for coordination to alkali and alkaline earth metal ions because both these metal ions and this donor atom are classified as "hard." Many of the complexation studies with these metals have been carried out with the picrate salts because this anion has a chromophore that can be used to analyze the salts as they are extracted between different solution phases.

The following generalizations can be made from a series of extraction experiments involving the complexation of Group 1 (IA) metal ions with calixarene ketones and esters. When using these "rules," however, it should be recognized that the differential selectivities can be quite different between the various systems, and that in many cases the solvent can play an important role.

1. Tetramers show a preference for the extraction of Na^+, regardless of whether the substituent on the calixarene is an ester or a ketone.

2. Generally, tetraketones are more effective extractants for Li^+, Rb^+, and Cs^+ than are the tetraesters.
3. Hexamer ketones and esters show a preference for the larger cations. These compounds have a higher affinity for K^+ than Na^+, and their maximum preference is for Rb^+ and Cs^+, although little selectivity between these two latter ions is observed.
4. Octamers show low levels of phase-transfer effectiveness for Group 1 (IA) cations. The larger cations are favored but the selectivity between them is poor.

Similar conclusions can be drawn from stability constant data. These stability constant data also provide the additional information that Na^+ binds more strongly to a tetramethyl ketone substituted calix[4]arene than to a tetraethyl ester substituted calix[4]arene. Calix[4]arene tetraamides, like tetraketones and tetraesters, also show a very high selectivity for Na^+ as compared to the larger alkali metal ions. The binding constants for amides are significantly greater than those found for tetraketones or tetraester substituted calix[4]arenes. For the *p-tert*-butylcalix[4]arene tetramide the K^+ ion is encapsulated within the lower rim enclosure. The coordination sphere of the metal ion is that of an antiprism with eight oxygen atoms coordinated (26).

Calixarenes with mixed substituents on a rim have been studied (27). Replacement of an ethyl ester group on a tetraethyl ester substituted calix[4]arene by a methyl ester group results in extractant characteristics intermediate between that of the tetraethyl and tetramethyl esters. The peak selectivity for Na^+ is retained but complexation with Li^+ is reduced. Replacement of a tetraethyl ester group with an amide retains the Na^+ selectivity. The compound does not show an increased binding constant for Na^+, despite the tetramide showing such a higher binding for Na^+. Since the mixed-substituted calixarene shows a higher binding for K^+, this calixarene has a lower Na^+/K^+ selectivity ratio than does the tetraester (2).

Much of the more recent work directed toward the synthesis of calixarenes as selective complexants for alkali and alkaline earth metal ions has focused on the smaller calix[4]arenes. These compounds frequently have uncharged methyl ether or ester substituents on the lower rim (27–34). The structures of the complexes do not necessarily have a monomeric 1:1 stoichiometry, since 2:2 complexes can also be formed (35). Other lower rim modified calixarenes can also be used as selective complexants for alkali metal ions. One such group of calixarenes are those that have acetate functionalities as the peripheral ligating moieties for the ions. The data for the extraction of alkali metal ions by these compounds are collected in Table II (36). These data show that the calix[4]arenes show the highest selectivity for Na^+, the calix[6]arenes have less affinity for Na^+ than K^+, with the highest selectivity being for Rb^+ and Cs^+.

TABLE II
Extraction of Alkali Metal Ions by Acetato Ester Substituted Calixarenes

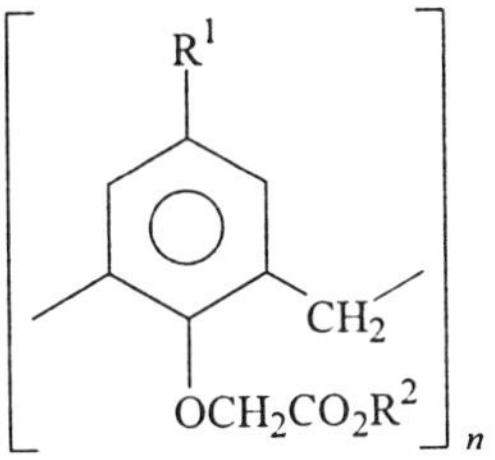

(A) R^1 = *tert*-butyl, R^2 = Et
(B) R^1 = *tert*-butyl, R^2 = Me
(C) R^1 = H, R^2 = Et
(D) R^1 = H, R^2 = Me

	$n = 4$				$n = 6$				$n = 8$			
Ion	A	B	C	D	A	B	C	D	A	B	C	D
Li^+	15.0	6.7	1.8	1.1	11.4	1.7	4.7	2.6	1.1	0.9	0.8	0.4
Na^+	94.6	85.7	60.4	34.2	50.1	10.3	10.4	6.7	6.0	8.3	7.5	4.1
K^+	49.1	22.3	12.9	4.8	85.9	29.1	51.3	25.2	26.0	25.5	20.2	12.1
Rb^+	23.6	9.8	4.1	1.9	88.7	41.2	94.1	77.7	20.3	29.8	28.9	17.5
Cs^+	48.9	25.5	10.6	10.8	100.0	54.8	94.6	94.6	24.5	20.1	30.1	27.0

The calix[8]arenes are the least effective extractants for alkali metal ions, and no calixarenes are effective extractants for Li^+. The presence of a *p-tert*-butyl group on the upper rim increases the selectivity for the transport of Na^+ ions, which may result from the smaller lower rim cavity size in the *p-tert*-butyl derivatives.

A comparison has been made between amide, ester, and ketone derivatives of *tert*-butyl calix[4]arenes (n = 4, 6, or 8) for the extraction of alkali and alkaline earth metal ions as 1 : 1 complexes. The extraction data in dichloromethane solvent for a series of ligands are collected in Table III (37–39). In general, the ketone C.1 shows greater complexation than does the ester B.1. The calix[4]arene A.1 selectively complexes Na^+ in preference to Li^+ and K^+, but the calix[n]arenes (n = 6 or 8) A.2 and A.3 show lower selectivities. The amides (A.1–A.3) show greater complexation for the alkaline earths than do the esters and ketones, with the calix[6]arene A.2 being a strong complexant for Ca^{2+} and Sr^{2+}. The stability constants in methanol and acetonitrile solvents for a similar series of ester and ketone substituted calixarenes shown in Fig. 5 are collected in Table IV (40). In methanol solvent the calix[4]arenes show a greater selectivity for Na^+, with the *tert*-butyl showing a slight preference for K^+. In acetonitrile solvent the greater preference with the calix[4]arene ethyl ester and the ketones with R = Me and Ph is found for Li^+ and Na^+, with the calix[6]arene ethyl ester showing an overall lower complexing ability. These

TABLE III
Extractions of Picrate Salts with Calixarene Derivatives[a]

t-Bu, CH_2, OR, *n*

(A) R = $CH_2CONHBn$-Bu (1) $n = 4$
(B) R = CH_2CO_2Et (2) $n = 6$
(C) R = CH_2COMe (3) $n = 8$

	Fraction of Picrate Salt Extracted (%)								
Ligand	Li^+	Na^+	K^+	Rb^+	Cs^+	Mg^{2+}	Ca^{2+}	Sr^{2+}	Ba^{2+}
A.1	<1	2.7	<1	5.2	3.1	<1	5.8	4.6	4.8
A.2	5.8	3.1	<1	4.1	2.9	11.8	33.4	56.8	30.2
A.3	5.9	1.4	5.4	4.7	3.6	14.2	16.4	28.6	37.0
B.1[b]	15.0	94.6	49.1	23.6	48.9				
B.2	6.7	15.6	66.5	60.5	88.9		5.3		8.2
B.3	<1	4.5	21.5	16.4	17.0	<1	6.4		17.9
C.1[b]	31.4	99.2	84.1	53.7	83.8				

[a] H_2O/CH_2Cl_2.
[b] CH_2Cl_2.

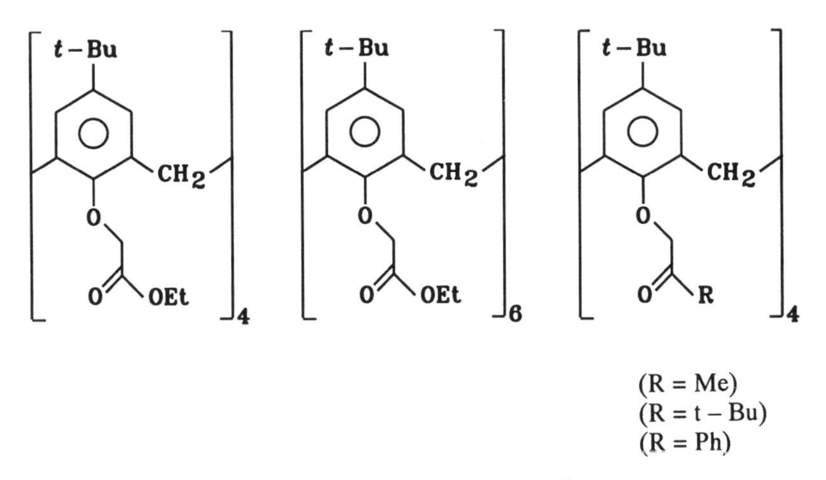

Figure 5. Ester and ketone substituted calixarenes.

TABLE IV
Logarithm of the Stability Constants (β) of the Alkali Metal Cation Complexes of Calixarene Derivatives **A**, **B**, **C**, **D**, and **E** at 25°C

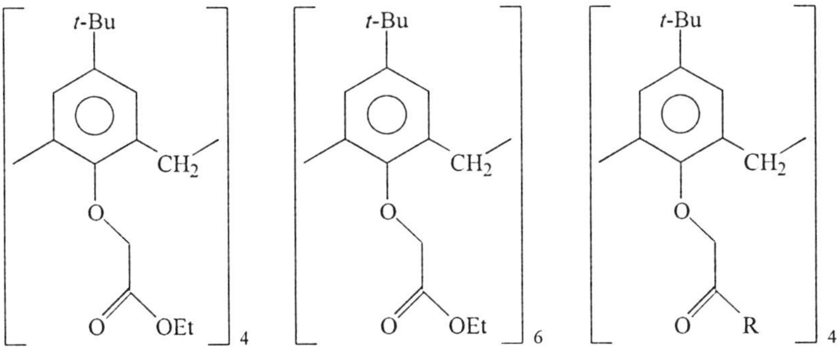

A ($n = 4$, R = OEt); B ($n = 6$, R = OEt)

C (R = Me)
D (R = *t*-Bu)
E (R = Ph)

Solvent	Cation	A	C	D	
MeOH	Li^+	2.6	2.7	1.8	
	Na^+	5.0	5.1	4.3	
	K^+	2.4	3.1	5.0	
	Rb^+	3.1	3.6	1.6	
	Cs^+	2.7	3.1	<1	
	Ag^+	4.0			
	Tl^+	1.6			
	Cation	**A**	**C**	**E**	**B**
MeCN	Li^+	6.4	5.8	6.3	3.7
	Na^+	5.8	5.6	6.1	3.5
	K^+	4.5	4.4	5.1	5.1
	Rb^+	1.9	1.7	4.5	4.8
	Cs^+	2.8	3.7	5.6	4.3
	Ag^+	2.4			

studies have been extended to include Ag^+ and Tl^+ ions. The logarithms of these stability constants are also shown in Table IV. From these data it is apparent that the binding of Ag^+ is comparable to that of Na^+, with the binding of Tl^+ being more closely similar to that of K^+ (41). One factor that leads to solvent effects in complex formation with calixarenes is that whereas the smaller

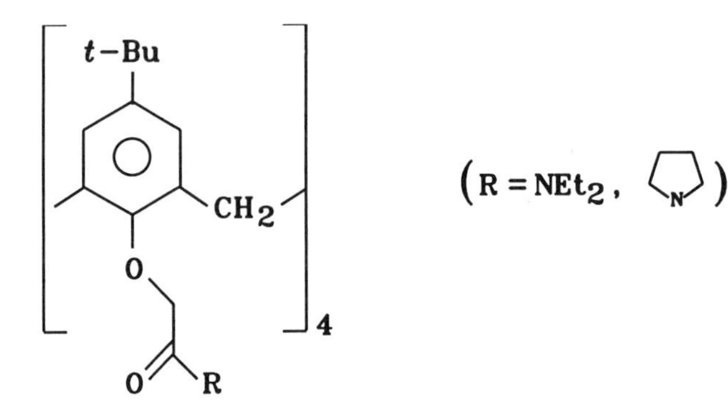

Figure 6. Amide substituted calixarenes.

Na^+ ion is encapsulated to give solvent-separated ion pairs, the larger Cs^+ ion is exposed to the solvent, resulting in the formation of contact ion pairs (42).

The amides shown in Fig. 6 have also been used as selective complexants for alkali and alkaline earth ions. Again, by contrast with the tetraester and tetraketone analogues, these amides show substantial complexation for alkaline earth ions. The amides show a strong preference for Ca^{2+} and Sr^{2+}, resulting in a high Ca^{2+}/Mg^{2+} selectivity ratio (43). Since the Na^+ and Ca^{2+} ions have very similar ionic radii, it is to be expected that the calix[4]arene cavity is the preferred size for selective extraction of both of these ions. Charge effects must also be considered. For extraction of metal ions into an organic medium, uncharged complexes are preferable. For a divalent cation such as Ca^{2+}, therefore, selective phase-transfer reagents used to extract this ion from aqueous solution should be dianionic. Such a calix[4]arene is shown in Fig. 7. This particular calix[4]arene is a highly selective extractant for Ca^{2+} ion in the presence of a mixture of Mg^{2+}, Ca^{2+}, Sr^{2+}, and Ba^{2+} ions (44).

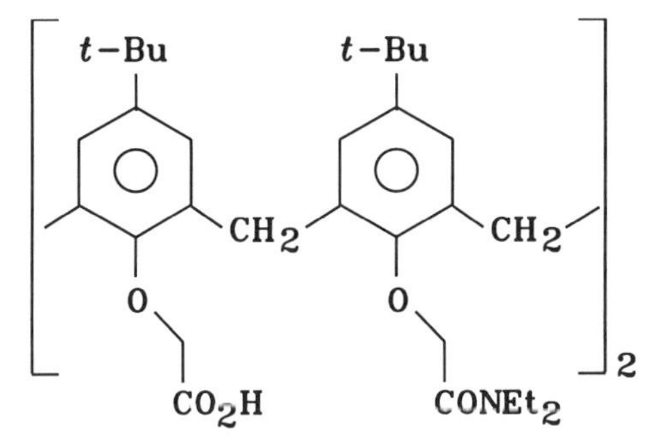

Figure 7. Mixed amide–carboxylic acid substituted calixarene.

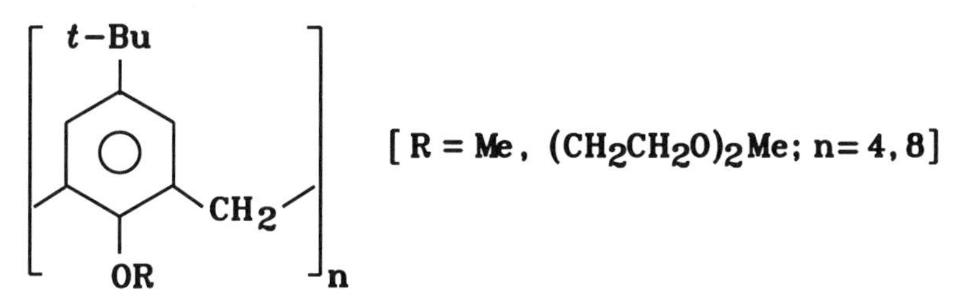

Figure 8. Oligoethylene glycol substituted calixarenes.

Calix[4]arenes and calix[8]arenes with short oligoethylene glycol units attached to the lower rim (Fig. 8) have also been used for cation binding. The calix[4]arene shows little or no complexation with alkali metal ions, but the calix[8]arene strongly binds Cs^+ (45).

The conformational properties of the calixarene must also be considered when the stability of the alkali and alkaline earth metal complexes are being evaluated. An example is the finding that for a calix[4]arene, which can be separately isolated in both its cone and partial cone conformers, different selectivities are observed for each. Such a compound is 5, 11, 17, 23-tetra-*tert*-butyl-25, 27-bis[(ethoxycarbonyl)methoxy]-26, 28-bis(2-pyridyl methoxy) calix[4]arene (Fig. 9). Solvent extraction data show that the cone conformer binds both Na^+ and Li^+, whereas the partial cone conformer has little affinity for any metal ions.[46,47] Molecular dynamics simulations have been carried out with *tert*-butylcalix[4]arenes having amide functional groups on the lower rim. The uncomplexed calixarene is not preorganized to coordinate metal ions in vacuo or in an aqueous environment. A detailed analysis of hydration patterns and dynamics features of the apolar cone and of the hydrophilic moiety shows a guest-dependent dynamic coupling between the motions of the cone and of the lower rim oxygen atoms (48).

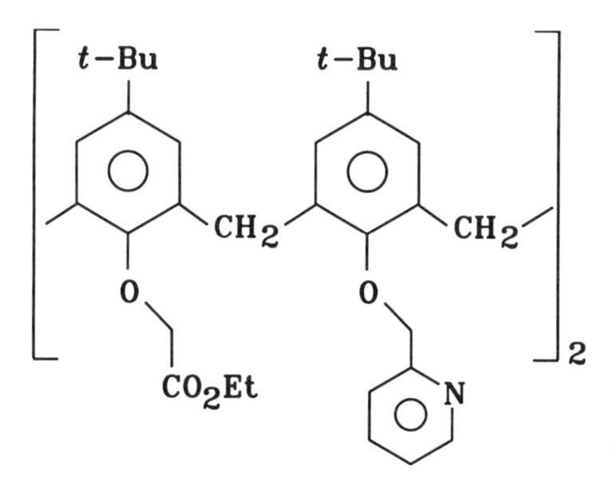

Figure 9. Calixarene with pyridyl functionality attached.

TABLE V
Thermodynamics Parameters for the Complexation of 5, 11, 17, 23-Tetra-*tert*-butyl-25, 26, 27, 28-ethyl acetato calix[6]arene with Group 1 (IA) Metal Ions

Solvent	Cation	$\Delta G°$ (kJ mol^{-1})	$\Delta H°$ (kJ mol^{-1})	$\Delta S°$ (JK^{-1} mol^{-1})
MeOH	Li^+	−14.84	5.05	66.7
	Na^+	−28.54	−45.60	−57.2
	K^+	−13.70	−14.22	−1.7
MeCN	Li^+	−36.53	−48.78	−41.1
	Na^+	−33.11	−61.55	−95.4
	K^+	−25.69	−43.85	−60.9
	Rb^+	−10.85	−18.67	−26.2
	Cs^+	−15.98	−11.48	15.1

The thermodynamics of binding of alkali metal ions into an ester derivatized calixarene have been investigated in both methanol and acetonitrile solvent. The data collected in Table V for 5, 11, 17, 23-tetra-*tert*-butyl-25, 26, 27, 28-tetrakis ethyl acetato calix[4]arene show that the complexation of alkali metal ions is enthalpy controlled except for the case of Li^+ in methanol, where it is entropy controlled (49). The results suggest that acetonitrile enters the hydrophobic cavity of the ligand to produce a synergistic effect that makes the hydrophilic cavity better preorganized to interact with cations in this solvent.

The selective extraction of alkali metal ions by different calixarene conformers has been used to probe the site preferance for ion binding. Comparative data for the extraction of alkali metal picrates by both the cone and 1,3-alternate conformations of calix[4]arene ethers are shown in Table VI. From the higher binding of these ions to the 1,3-alternate calix[4]arenes it has been concluded that π-donor effects of the phenyl ring may be an important factor in causing an increase in the stability constant for metal ion binding to calixarenes (50). The calixarene conformation can also affect the mode of coordination of a metal ion. An example of such behavior is suggested in the complex formed between

TABLE VI
Percent Extraction of Alkali Metal Picrates

	Extraction (%)			
Calix[4]arene	Li^+	Na^+	K^+	Cs^+
Cone-1	0.7	0.8	1.9	5.2
1,3-Alternate-1	4.3	1.6	16.2	11.8
Cone-2	1.6	3.9	3.9	7.7
1,3-Alternate-2	5.3	8.8	56.8	44.7
Cone-3	0.0	9.0	3.5	0.0
1,3-Alternate-3	4.8	0.0	38.6	41.4
1,3-Alternate-4	1.1	0.0	88.8	95.7

25, 27-di(allyloxy)-5, 17-di-*tert*-butyl-26, 28-dimethoxycalix[4]arene and Ag^+. It is proposed that Ag^+ coordinates to the alkene double bonds of the allyloxy functionalities rather than into the calixarene cavity because the flattened partial cone conformation of the calixarene causes one of the methoxy groups to restrict access of a metal ion into the lower rim calixarene host (51). Further work on this system with a metal ion that does not have such a high preference for binding to an alkene as does the Ag^+ ion would be interesting.

Charge effects can also affect ion binding. Calix[4]arenes having either one, two, three, or four redox active quinone residues have been prepared (Fig. 10). The neutral calix[4]arenemonoquinone binds Na^+ and shows a binding enhancement on the order of 10^6 when it is reduced to the monoanionic state. The other calix[4]arenequinones show less enhancement (52).

C. Protonation and Stability Constants

For certain biological applications it is useful to have available alkali or alkaline earth metal ion carriers that are soluble in aqueous solution. A series of water soluble alkali and alkaline earth metal calix[4]arene complexes have recently been obtained with carboxylate substituted calixarenes. The introduction of a carboxylic acid substituent onto the lower rim results in increased stability for the complexes, but in a decrease in the Na^+/K^+ selectivity. These ligands with either carboxylic acid or mixed carboxylic acid–ester functionalities

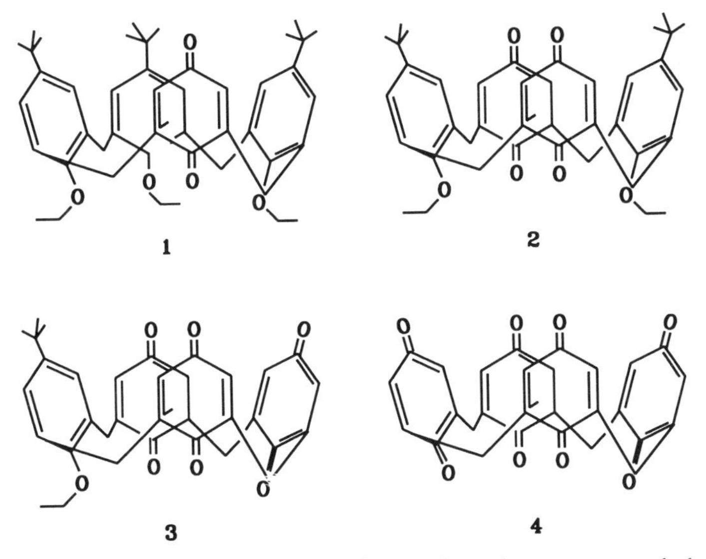

Figure 10. Calixarenes with one, two, three, or four quinone groups attached.

TABLE VII
Protonation K and Stability β (MLH) Constants of the Ligand L and the K^+, Rb^+, and Cs^+ Complexes[a]

System	log K_{011}	log K_{012}	log K_{013}
L	6.72 (0.02)	5.95 (0.03)	5.58 (0.02)
System	log β_{110}	log β_{111}	log β_{112}
K^+ + L	2.1 (0.1)	9.2 (0.2)	14.90 (0.06)
Rb^+ + L	3.2 (0.3)	9.0 (0.3)	14.9 (0.3)
Cs^+ + L	~1	8.0 (0.2)	14.9 (0.2)

[a] $\log \beta_{110} = \log K_{110}$; $\log \beta_{111} = \log K_{110} + \log K_{111}$; $\log \beta_{112} = \log K_{110} + \log K_{111} + \log K_{112}$.

all show very large Ca^{2+}/Mg^{2+} selectivities (53). The carboxylic acid substituted calix[6]arene, 5, 11, 17, 23, 29, 35-hexakis-*tert*-butyl-37, 39, 41-trimethoxy-38, 40, 42-tris-oxoacetic acid calix[6]arene (L) with alternating carboxylic acid and methoxy groups on the lower rim has been used as a complexant for alkali metal ions (54–56). The protonation and stability constants for the reactions shown in Scheme 2 are shown in Table VII (57). The p*K* values for

$$ML^{2-} \underset{K_{110}}{\overset{H^+}{\rightleftharpoons}} MLH^- \underset{K_{111}}{\overset{H^+}{\rightleftharpoons}} MLH_2 \underset{K_{112}}{\overset{H^+}{\rightleftharpoons}} MLH_3^+$$

Scheme 2

L vary over the 6.72–5.58 range, and show little change with successive protonations. The data fitting routines to obtain the stability (β_{110}) and protonation (β_{111} and β_{112}) constants show that K^+, Rb^+, and Cs^+ bind to L in a 1 : 1 ratio for M/L even with water as a competing ligand. In the presence of water, β_{110} for the Rb^+ complex ML is slightly higher than those for K^+ or Cs^+, but the stability constants β_{112} for MLH_2 are essentially identical for all three metal ions. At high pH the predominant solution species is ML, with MLH_2 becoming the major species in solution at lower pH. At a pH of below 4, dissociation into the free ligand and metal ion occurs. The abundance of the species MHL in the pH range of 6–7 decreases along the series K^+, Rb^+, Cs^+, until for Cs^+ the only complex being present in significant concentration is MLH_2.

D. Oxacalix[3]arenes and Dioxacalix[4]arenes

Oxacalix[3]arenes have also been used as extractants for alkali metal ions (58, 59). The compounds that have been used for this purpose are shown in Fig. 11. The compound with R = Et cannot be used to investigate the effect of

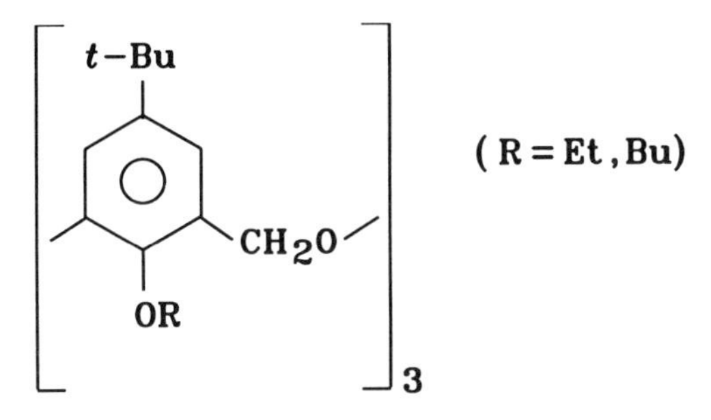

Figure 11. Oxacalix[3]arenes.

conformational structure of ion binding because it undergoes interconversion between the cone and partial cone forms. In solution the partial cone is the predominant conformer. For the compound with R = Bu, however, the ring inversion is inhibited, and complexation with each conformer can be separately studied. From the extraction data shown in Table VIII for this series of compounds it is apparent that the highest selectivity is found for K^+, with the cone conformers generally having the higher extractability. Oxacalix[3]arenes have also been used as complexants for both alkali and alkaline earth metals. The complexation properties of this series of compounds is particularly dependent on the substituent in the para position of the upper rim. When this substituent is a tertiary butyl group, no binding of Li^+, Na^+, K^+, Ca^{2+}, Y^{3+}, or La^{3+} to the oxacalix[3]arene occurs. By contrast, when the oxacalix[3]arene has a chloro

TABLE VIII
Percent Extraction of Alkali Metal Ions

R = Et or Bu

R	Li^+	Na^+	K^+	Cs^+
Et	1.0	14.0	49.6	36.7
Bu (cone)	0.0	5.7	58.8	35.0
Bu (partial-cone)	0.0	11.9	34.9	23.6

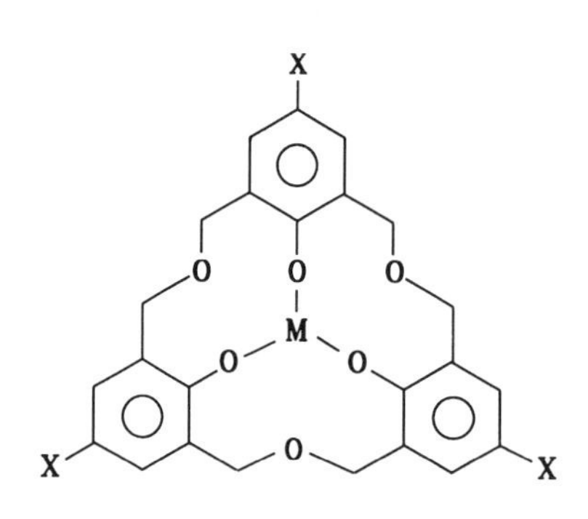

Figure 12. Oxacalix[3]arene complexes with trivalent metals.

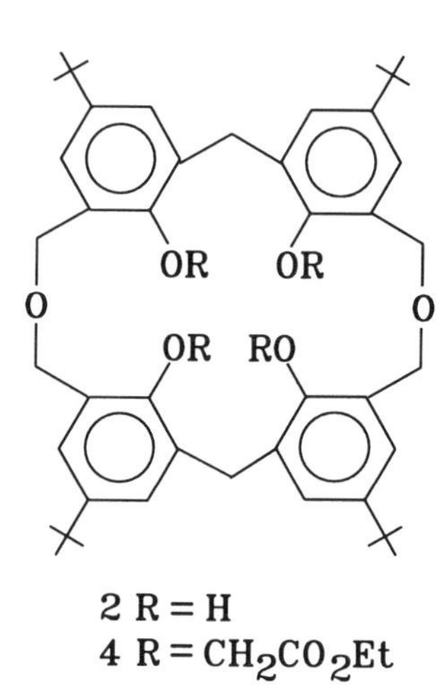

Figure 13. Tetraester derivatives of dioxocalix[4]arenes.

substituent in this position, the order of binding of the alkali metal ions follows the sequence $Na^+ > K^+ > Li^+$. The addition of Et_3N leads to the formation of the deprotonated oxacalix[3]arene anion, and now in contrast with the protonated form the *tert*-butyl substituted derivative coordinates to Sc^{3+}, Y^{3+}, and Ti^{4+} (Fig. 12) (60).

In addition to oxacalix[3]arenes, dioxacalix[4]arenes have also been synthesized. When these compounds are modified with ester groups ($R = CH_2CO_2Et$) on their lower rim positions, the resulting tetraester (Fig. 13) can be used as a selective complexant for metal ions. Since these dioxacalix[4]arenes have a larger cavity size than the more conventional calix[4]arenes, their complexation properties more closely resemble those of the calix[6]arenes. Thus they show a preference for complexing the larger alkali metal cations. As a result, Ca^{2+} and Ba^{2+} are more strongly complexed than the alkali metal ions, and Eu^{3+} is complexed more strongly than Na^+ (61).

E. Influence of Complexation on Calixarene Properties

In the temperature-dependent ^1H NMR spectrum of sulfonated calix[4]arenes and their methoxylated derivatives, the coalescence temperature is affected by the presence of cations in the solution. The effect for alkylammonium ions is greater than that for Group I (IA) metal ions, which is likely due to electrostatic bridge formation occurring on the cavity edge (62, 63). The conformer distribution in calix[4]arenes is also affected by the presence of Group I (IA) metal ions. For the ethyl acetato derivatives in the absence of any metal ion, either the partial cone or the 1,3-alternate conformation usually predominates. In the presence of a metal ion, the predominant conformation is now the cone (64, 65). This finding has been confirmed with the methoxy calix[4]arene derivative, where the interaction of metal ions such as Li^+ and Na^+ result in a shift of the equilibrium from a partial cone to a cone conformation (66). A double calixarene having two "lower rim" metal ion binding sites has been synthesized. By variable temperature NMR spectroscopy it has been shown that when the metal ion is Na^+ or K^+ two exchange processes occur. One of these processes is a fast intramolecular exchange of the metal ion between the two binding sites, and the other is a slow intermolecular metal ion exchange between adjacent calixarene channels (Fig. 14) (67, 68). Extension of these studies using ^1H NMR spectroscopy with complexed Na^+, K^+, or Ag^+ has shown that binding occurs asymmetrically to one of two binding sites in conformationally immobilized calix[4]arenes. For the 1,3-alternate conformation of the calix[4]arene, the Ag^+ ion is bound to one of two binding sites. Thus in addition to an intermolecular complexation–decomplexation process, the bound Ag^+ tunnels through the π-basic tube to another binding site. These competing processes are shown diagrammatically in Fig. 15 (69).

Alkali metal ion complexation can also affect chemical reactions that occur at the substituent groups on the calixarene rim. An example is the conversion of calix[4]arene tetraethyl ester into the triester with trifluoroacetic acid (70). This reaction is inhibited by the presence of Na^+. This result can be explained

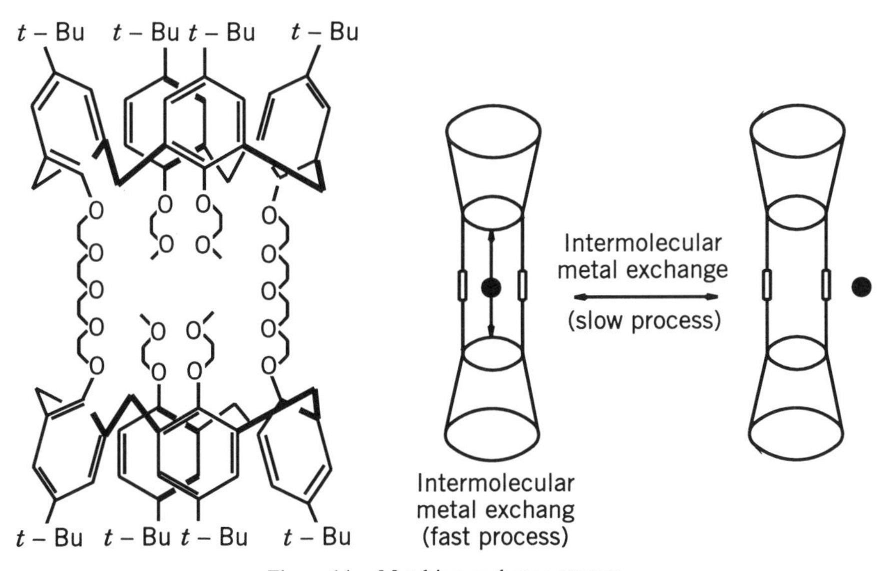

Figure 14. Metal ion exchange process.

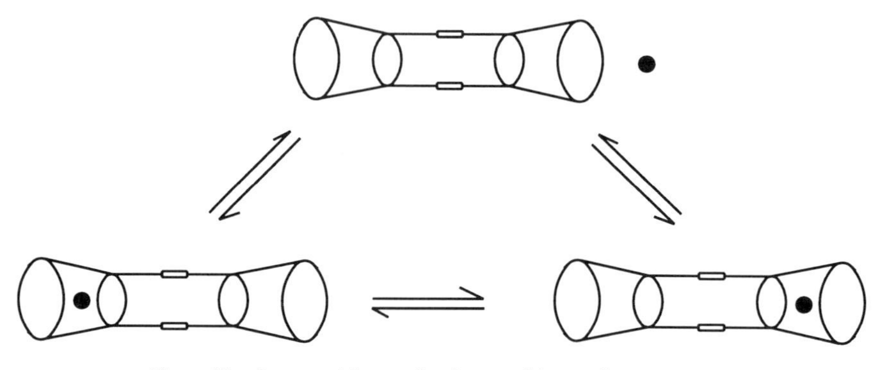

Figure 15. Inter- and intramolecular metal ion exchange processes.

if the reaction occurs via a proton bound into the lower rim cavity of the tetraethyl ester, a process that is impeded by the occlusion of Na^+ into that site. A further example involves a calix[4]arene that has been prepared with 2-[2-(2-methoxyethoxy)ethoxy]ethyl groups appended to the lower rim (Fig. 16). This compound behaves as an ionophore for alkali or alkali metal ions. The compound acts as a catalyst for the Finkelstein reaction involving halide replacement reactions with octyl halides. Acceleration is observed for the K^+, Rb^+, and Cs^+ salts (71).

Calixarene esters show monolayer behavior that is characteristic of their ring size. Unsubstituted calixarenes do not form monolayers, but their ester derivatives do. These monolayers show responsive behavior to monovalent cations by undergoing an expansion upon binding of alkali metal ions (72). The selectivity of alkali metal ion complexation by calixarenes at the air–water interface follows the following sequences:

Calix[4]arene	$Li^+ < Na^+ > K^+ > Rb^+$
Calix[6]arene	$K^+ > Rb^+ > Na^+ > Li^+$
Calix[8]arene	$Rb^+ > K^+ > Na^+ > Li^+$

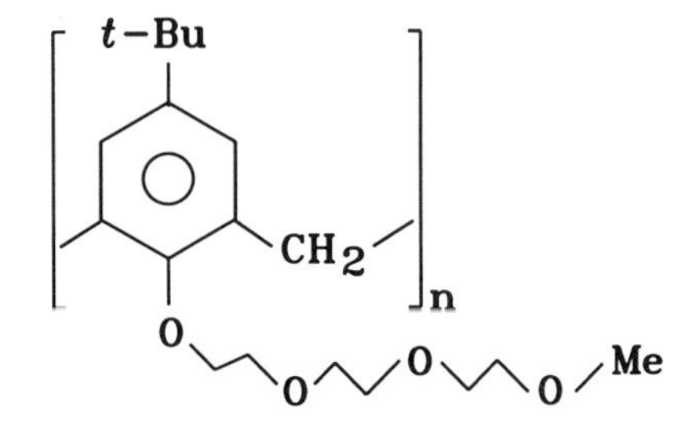

Figure 16. Polyether substituted calixarenes.

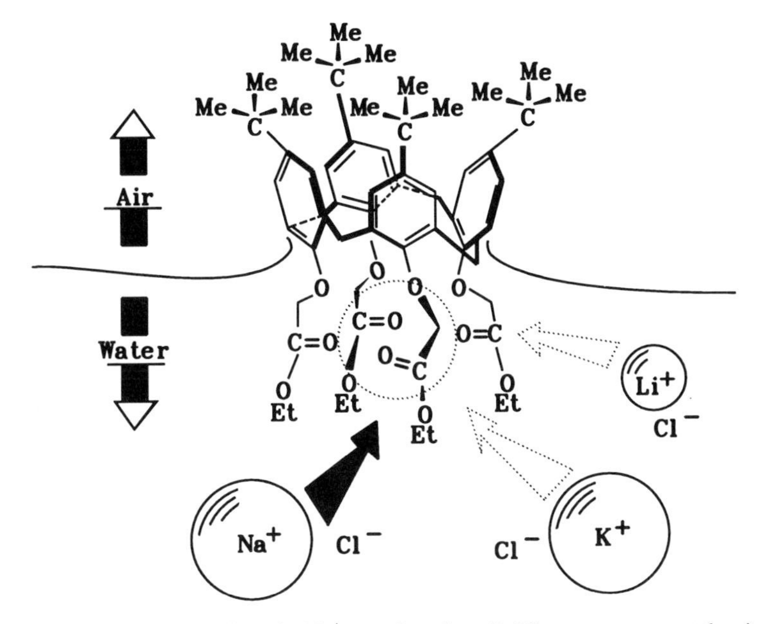

Figure 17. A selective formation of a Na^+ complex of a calix[4]arene tetraester at the air–water interface, resulting in a modification to the usual monolayer behavior.

This relative ordering of binding closely corresponds with those obtained for calixarenes that are contained entirely within the solution phase. Figure 17 shows a representation of the selective formation of a Na^+ complex of a calix[4]arene tetraester at the air–water interface, resulting in a modification to the usual monolayer behavior.

F. Chromogenic Calixarenes as Ion Sensors

The combination of selective ion-binding sites and conformational flexibility of the calixarenes makes them potentially useful as chromogenic ion sensors. If the ion-bound complex has a different conformation than the uncomplexed form, then the emissive properties of a chromophore within the molecule can be potentially modified. Such effects have been used in the design of ion sensors with calixarenes.

Chromogenic ionophoric calixarenes have been synthesized by incorporating a chromophoric functional group onto the upper calixarene rim, along with functional groups on the lower rim that selectively bind metal ions. Such a compound is shown in Fig. 18 where an azophenol moiety is used as the chromophoric group on an upper rim of the calix[4]arene, and an ethyl acetate functionality is used for the selective binding of the Group 1 (IA) metal ion (73).

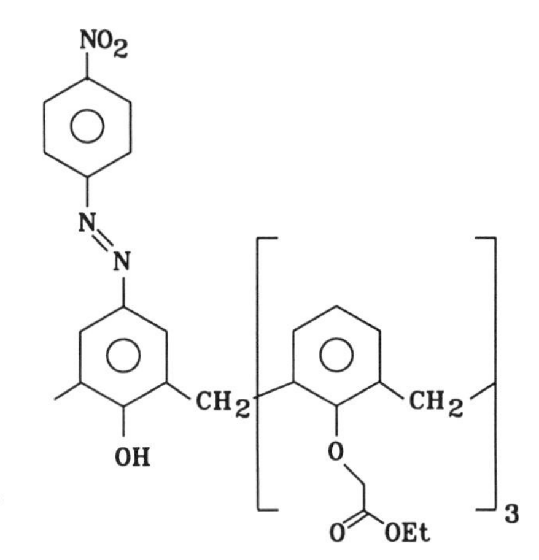

Figure 18. Diaza chromophore bound to a calixarene.

This compound shows a very high selectivity for Li^+, leading to a new absorbance band in the vicinity of 550 nm in the presence of this ion. A similar approach has bccn followcd with anothcr calix[4]arcnc that has bccn synthcsized by attaching two methoxy and two pyrene functionalities onto the lower calixarene rim. The ratio of the monomer versus excimer emission from this host molecule is sensitive to both the solvent polarity and the presence of added Group 1 (IA) metal ions (74). The attachment of a 5-benzothiazyl moiety can also be used to synthesize a fluorogenic calix[4]arene that has a very high selectivity for the Li^+ ion (75). When a pyrene is introduced as a fluorophore along with nitrobenzene as a quencher, the fluorescence increases some six to seven times over that observed in the absence of a metal ion (76). A chromogenic receptor has been prepared having a 1,3-bis(indoaniline) functional group attached to the upper calixarene rim (Fig. 19) (77). The lower rim binding site has ethyl acetate groups that show high association constants for binding Ca^{2+} ions. The binding of Ca^{2+} correlates with a chromophoric shift from 609 nm ($\varepsilon = 3.5 \times 10^4\ M^{-1}\ cm^{-1}$) to 719 nm.

Fluorescent response to alkali metal ions is observed with a calix[4]arene that has four anthracene moieties connected to the lower calixarene rim (Fig. 20) (78). Addition of Li^+ or Na^+ results in a decrease of the fluorescence intensity as the ion concentration increases. By contrast, the addition of K^+ leads to a decrease in the emission band at about 418 nm, and an increase in the band at 443 nm. Analysis of the ^1H NMR spectra suggests the formation of a 1:2 (L/M) complex for the case of added K^+. The fluorescence quenching by the presence of Na^+ may result from interactions that result from the four anthracene groups being bought into closer proximity. A similar type fluorescent calix[4]arene has been prepared with a pyrene substituent (Fig. 21) (31). This compound acts as a Na^+ ion sensor showing dual emission from the pyrene.

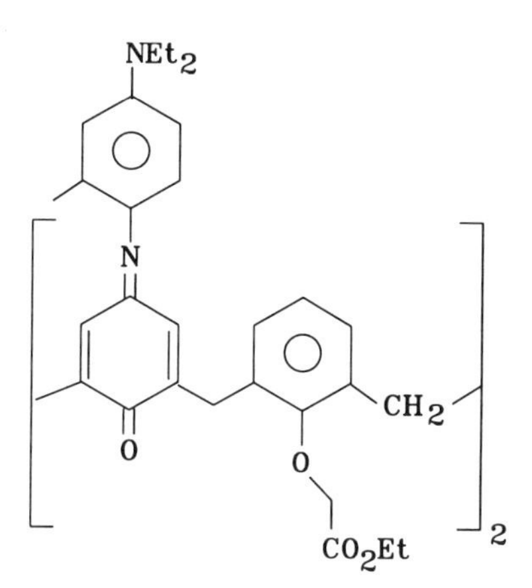

Figure 19. Indoaniline chromophore bound to a calixarene.

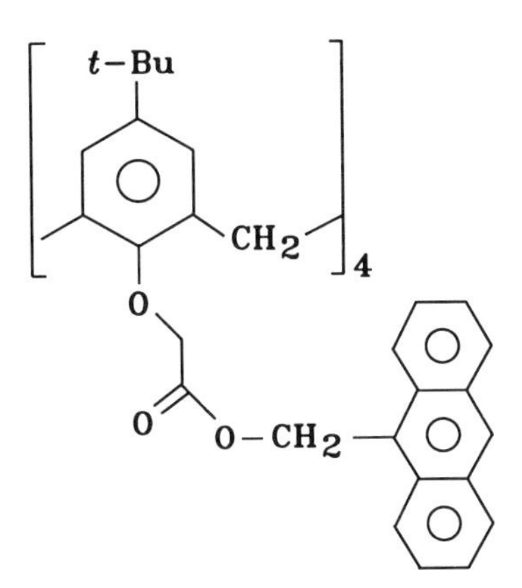

Figure 20. Anthracene chromophore bound to a calixarene.

The two pyrene fluorophores form an intramolecular pyrene excimer. When this fluorescent calix[4]arene complexes with Na^+ the excimer/monomer intensity ratio is altered by changes in the relative configurations of the two pyrene moieties. This change arises from the reorientation of the four carbonyl groups upon binding of these oxygen atoms to Na^+. This compound acts as a fluorescent sensor for the selective detection of Na^+ in nonaqueous solution.

Two selective chromoionophores have also been developed that employ an aromatic nitro group as substituent on a calix[4]arene. These two new chromoionophores show a high selectivity for K^+ in the presence of other ions such as Na^+, Mg^{2+}, and Ca^{2+}. These calix[4]arenes are potentially useful for use in optical fiber sensors for K^+ in biological fluids. One of these chemically modified calix[4]arenes having all of the lower rim positions substituted by a nitrophenyl group, is shown in Fig. 22. The second chromoionophore has a single aromatic nitro group appended to one of the lower calix[4]arene rims.

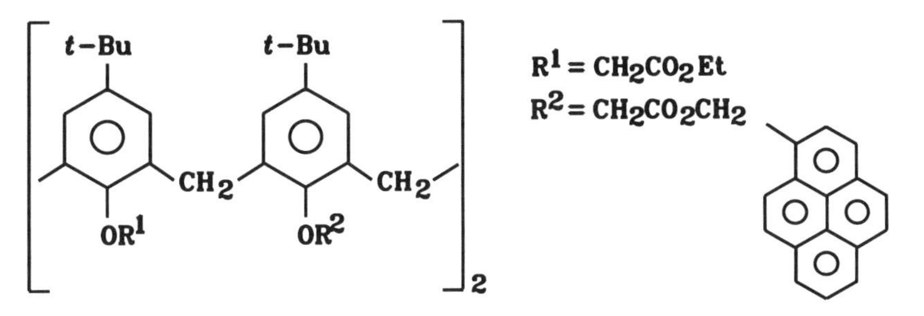

Figure 21. Pyrene chromophore bound to a calixarene.

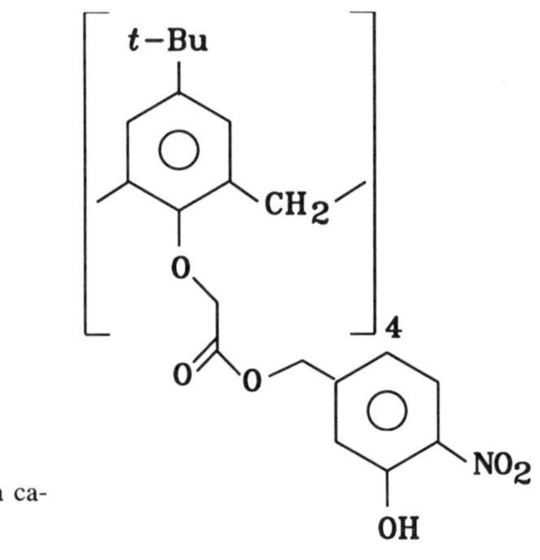

Figure 22. Nitrophenyl chromophore bound to a calixarene.

This compound is shown in Fig. 23. This latter compound exhibits a color change from colorless to yellow upon the addition of Li^+, with the color intensity being dependent on the ion concentration (33). A lesser effect is observed with the addition of Na^+.

By using a reverse strategy, conformational changes in calixarene ligands can be monitored by following their fluorescence changes. Such a compound with a pteridine substituent on the lower rim has been synthesized (Fig. 24) (79). This compound exists in a "closed" form with intramolecular hydrogen bonds, which then converts to an "open" form with intermolecular hydrogen bonds upon binding of Na^+. This interconversion can be monitored by following fluorescence changes in the flavin guest.

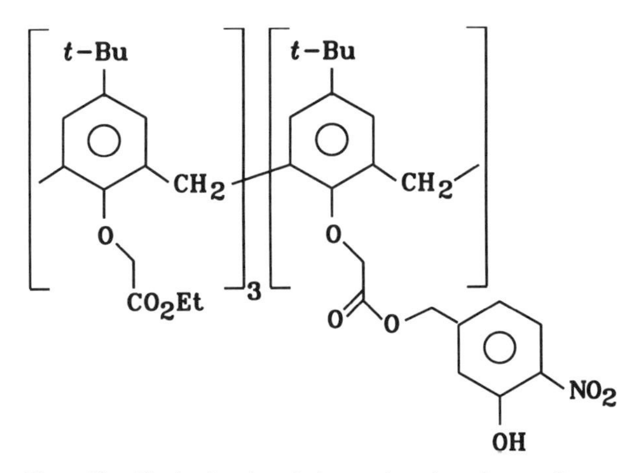

Figure 23. Single nitrophenyl chromophore bound to a calixarene.

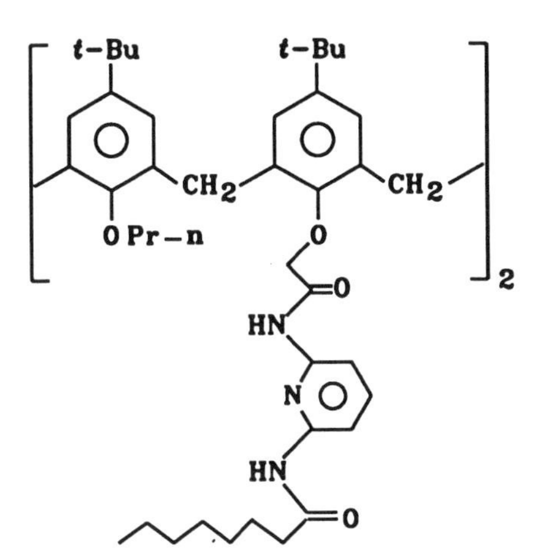

Figure 24. Pteridine chromophore bound to a calixarene.

G. Metal Ion Encapsulation in Calixspherands and Calixcrowns

The three-dimensional shape of calixarenes is particularly conducive for the introduction of metal ions as guests. However, although the calixarenes can be chemically modified to have a relatively large coordination number as a ligand, the metal ion can migrate out of the host in a manner analogous to the way in which it was initially incorporated. One manner in which this migration of a metal ion out of its host lattice can be retarded or eliminated is to chemically attach a cap above the calixarene rim after the metal ion has been complexed as a guest. Such complexes have been termed calixspherands. An alternative approach involves incorporating a crown ether metallocycle above the calixarene rim. Such complexes have been termed calixcrowns. Many of the complexes that have been prepared using these encapsulation techniques have either a Ag^+ or an alkali metal ion bound as guest.

A light switched ionophoric calixspherand has been synthesized whereby two anthracene moieties attached to the lower calixarene rim can reversibly form a "photochemical lid" (80, 81). Photolysis at wavelengths greater than 350 nm leads to closure, whereas thermal conditions or photolysis at wavelengths less than 280 nm leads to opening of the calixspherand (Fig. 25). The formation of a closed dimer is verified by the significantly reduced extractability of Na^+ from this dimer form.

Kinetically stable complexes with alkali metal ions have been prepared using calixspherands and calixcrowns. One series of calix[4]crowns that have been used are shown in Fig. 26 (82). These ligands have the 1,3 positions on the lower rim connected by bridging groups X that have different types and numbers of O- and N-functional groups within the spanning chain. The association con-

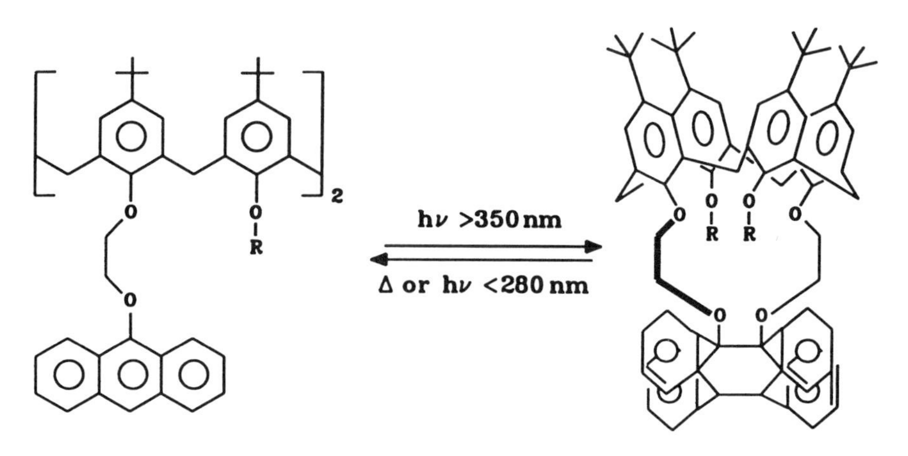

Figure 25. Calixspherand with "photochemical lid".

stants and free energy changes for binding alkali metal ions to these calix[4]crowns are given in Table IX. All ligands are selective toward K^+ with K^+/Na^+ selectivities up to 1.18×10^4 being obtained. The free energies of complexation range from -6 to -13.5 kcal mol^{-1}. Further studies with this family of calix[4]crowns have shown that in these highly preorganized ligands

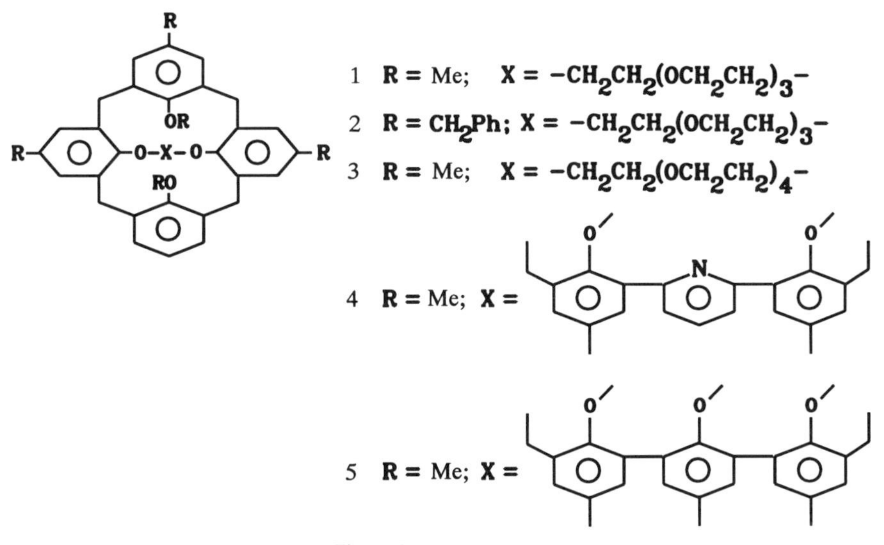

Figure 26. Calixcrowns.

TABLE IX
Association Constants and Free Energies of Complexation of Alkali Metal Ions with Calixspherands

	K_a (M^{-1})					$-\Delta G°$ (k cal mol^{-1})				
Compound[a]	Li^+	Na^+	K^+	Rb^+	Cs^+	Li^+	Na^+	K^+	Rb^+	Cs^+
1	3.8×10^4	1.1×10^5	3.0×10^8	1.1×10^8	4.7×10^5	6.3	6.7	11.4	10.8	7.6
2	$<10^4$	4.3×10^4	1.2×10^6	5.9×10^4	$<10^4$	<6	6.3	8.2	6.4	<6
3	5.5×10^4	4.2×10^4	1.5×10^5	2.1×10^5	3.2×10^6	6.5	6.3	7.1	7.3	8.9
4	2.1×10^5	3.1×10^6	8.2×10^8	6.2×10^8	1.8×10^6	7.1	8.7	12.0	11.9	8.4
5		2.1×10^{12}	2.2×10^{13}	3.6×10^9			16.8	18.1	13.0	

[a]The compounds **1–5** are the ones shown in Fig. 26.

the stability of the complexes can be strongly affected by subtle changes in the geometry in the vicinity of the binding region. The partial cone arrangement of the calix[4]spherand subunit is the preferred conformation for the selective binding of K^+. These calixcrown derivatives transport K^+ ions with a good K^+/Na^+ selectivity. In particular, the lipophilic 1,3-dimethoxy calix[4]arene crown-5 shown in Fig. 27 has a lower selectivity than does valinomycin, but a higher selectivity than does dibenzo-18-crown-6 (83, 84). In experiments with supported liquid membranes, however, it has been found that the transport selectivity does not correlate with the transport rates in single cation experiments or to association constants. This calixcrown has a high extraction selectivity but a low transport selectivity. This combination of properties leads to complex saturation at the source-phase side of the membrane. Calixspherands can be used to form Rb^+ complexes of sufficient stability that they have the potential

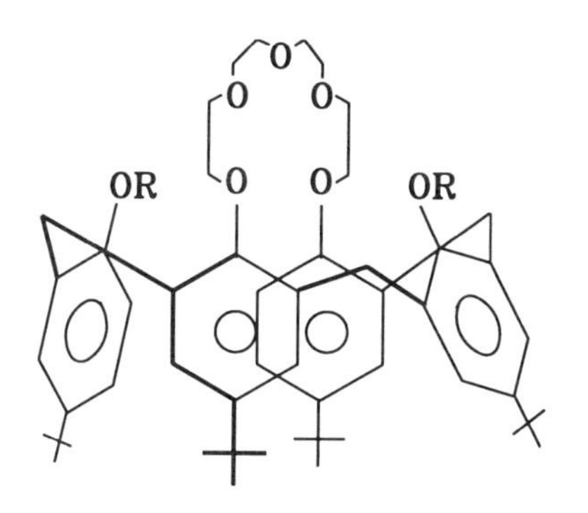

Figure 27. 1,3-Dimethoxy calix[4]arene crown-5.

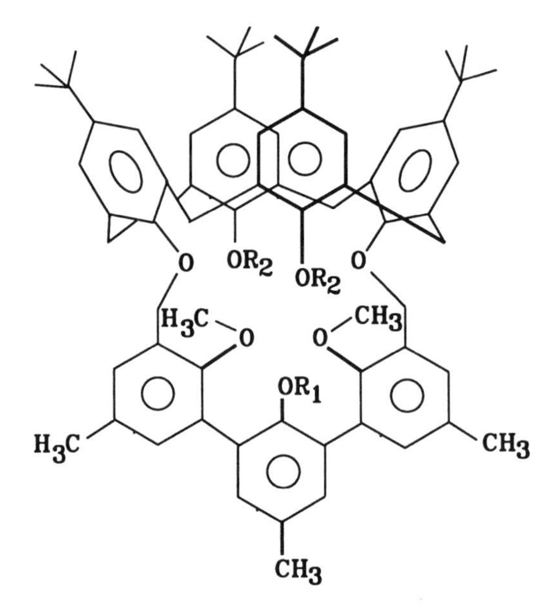

Figure 28. Calixspherand for Rb^+.

for use in blood flow monitoring using the isotope ^{81}Rb (85). In the series of calixspherands shown in Fig. 28 where R_1 and R_2 are a combination of the H, Me, Et, and *i*-Pr groups, it has been found that the half-lifetimes for decomplexation of Rb^+ ion can be varied by changing the substituents on the calixspherand cap. As R is varied from Me to Et to *i*-Pr, this half-life changes from 2.8 to 139 h and then to 180 days (86). These calixspherands also form kinetically stable Ag^+ complexes (87).

Molecular dynamics with the thermodynamic perturbation method has been used to probe the binding of alkali metal cations to calixspherands. This method gives the calculated relative free energies of the complexes. The most successful model included solvation parameters. The model reproduces the observed selectivity of the binding of the K^+ ion, and for the case of Rb^+, the value of the absolute binding free energy (-11 to -13 kcal mol^{-1}) is found to be in good agreement with the experimental value of -12 to -13 kcal mol^{-1} (82, 88).

The barium complex of the calixarene-crown-5 shown in Fig. 29 acts as a nucleophilic catalyst with transacylase activity (89). It is proposed that the function of the Ba^{2+} ion is to act as an internal electrophilic catalyst that favors nucleophilic addition to the carbonyl group in both the acylation and deacylation steps. This concept is supported from observations using the *O*, *O*′-oxybis(ethyleneoxyethylene)calix[4]arene monoacetate derivative shown in Fig. 30,

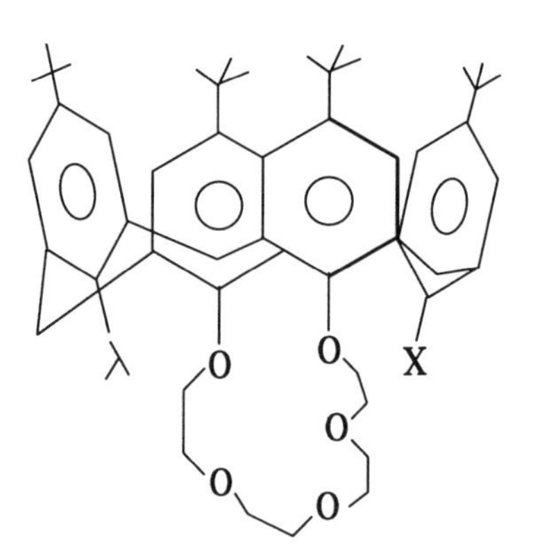

Figure 29. Calixarene-crown-5 for transacylase activity.

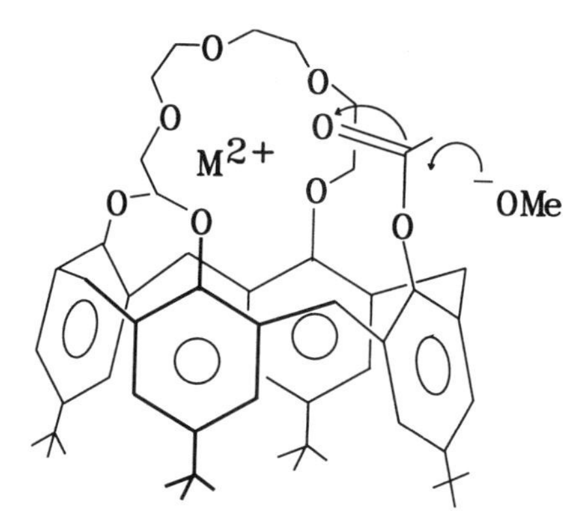

Figure 30. O,O′-Oxybis(ethyleneoxyethylene)calix[4]arene.

which shows an acceleration in the base methanolysis of the cone isomer of more than 10^6-fold in the presence of Ba^{2+} and Sr^{2+} (90). The metal ion occluded within the calix[4]arene cavity is proposed to activate the carbonyl oxygen of the ester group toward addition of the methoxide nucleophile.

Ion sensitive field-effect transistors (ISFETs) or chemically modified field-effect transistors (CHEMFETs) have been synthesized using derivatized calix[4]arenes. These devices have been fabricated by incorporating the calix-[4]arenes into a polyvinyl chloride (PVC) membrane directly attached to the SiO_2 gate surface. Such a membrane-modified FET is an integrated device of an ion-sensing membrane and an ISFET transducer that functions as a chemical sensor. Both Na^+ and K^+ sensitive FETs have been prepared that show near-Nernstian behavior. Either Na^+ or K^+ selectivity can be achieved by the use of different calixspherands and calixcrowns (91, 92).

VI. LANTHANIDE COMPLEXES

The simple unsubstituted calixarenes will themselves form complexes with lanthanides (2). The majority of such complexes are formed with europium, terbium, and cerium. Europium complexes with both *p-tert*-butyl calix[4]arene and *p-tert*-butyl calix[8]arene have been prepared. These complexes have structures containing two europium ions coordinated to the calixarene either as a 2:2 metal/ligand (LH) calix[4]arene complex (93) or as a 2:1 metal/ligand (LH_2) calix[8]arene complex (94). The 2:2 europium calix[4]arene complex (Eq. 11) has a stoichiometry corresponding to the formula $Eu_2(LH)_2(dmf)_4$, with the europiums being bridged by a phenoxide group from each calixarene ligand.

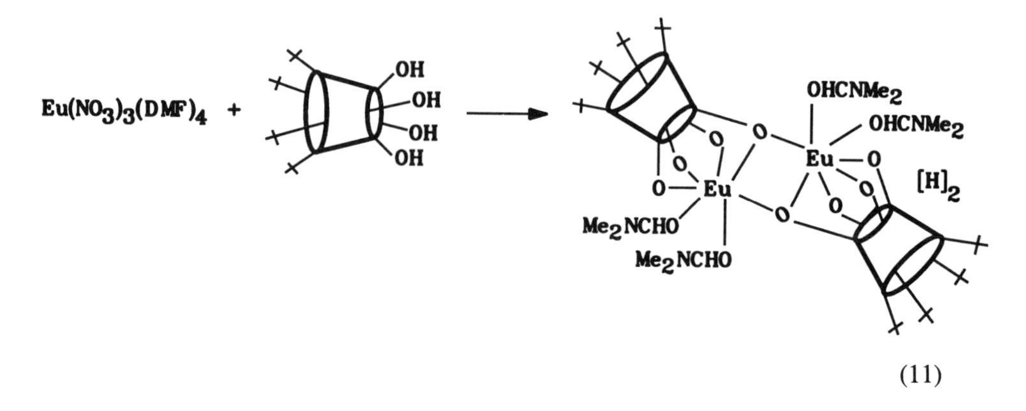

(11)

The overall neutral change corresponds to the complex having three of the four calixarene ligands deprotonated, but the specific oxygen atoms that are protonated have not been identified from the crystal structure. The 2 : 1 europium calix[8]arene complex $Eu_2(LH_2)(dmf)_5$ has been prepared in a similar manner to its calix[4]arene congener (Eq. 12). For charge balance, two of the phenolic

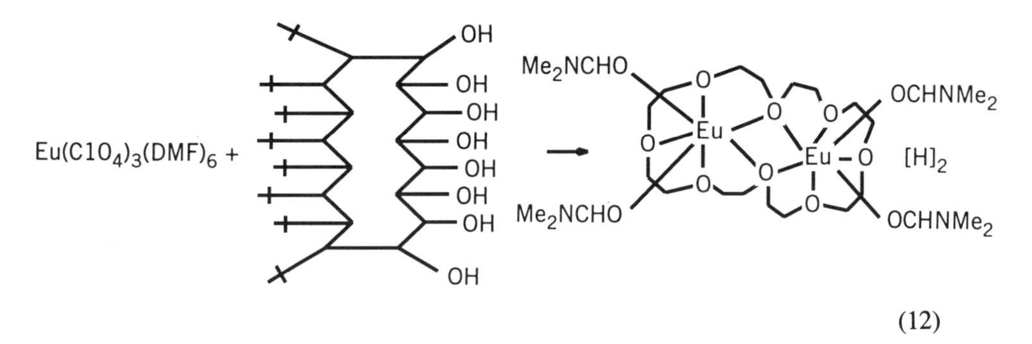

(12)

oxygen atoms of the calix[8]arene ligand are protonated, although again these have not been crystallographically identified. The 1 : 1 europium calix[8]arene complex $Eu(H_6L)(NO_3)(dmf)_4$ has been prepared by treating $Eu(NO_3)_3(dmso)_4$ with *p-tert*-butylcalix[8]arene (LH_8) in DMF as solvent (Eq. 13). The ligating

$$Eu(NO_3)_3(dmso)_4 + LH_8 \xrightarrow{DMF} Eu(LH_6)NO_3(dmf)_4 - 3dmf \qquad (13)$$

calixarene now has six of its eight phenolic groups in the complex protonated (95). These *p-tert*-butyl substituted calixarene lanthanide complexes have rather low solubilities, and in the presence of water they have a tendency to release the calixarene ligand and precipitate the lanthanide trihydroxide (2).

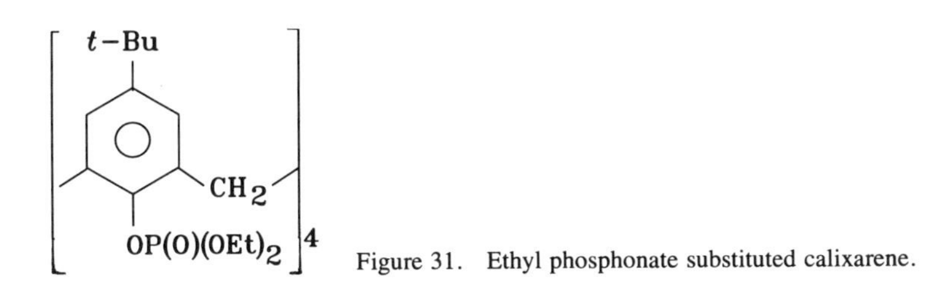

Figure 31. Ethyl phosphonate substituted calixarene.

The complex stability can be increased by introducing functional groups onto the lower rim of the calixarenes that can encapsulate the metal ion into a multidentate cavity. Ethyl phosphonate substituted calixarenes have been synthesized (96) and the example shown in Fig. 31 has been used as an extractant for lanthanides in both neutral and acidic media (97, 98). The lanthanides coordinate to the phosphonate oxygen atoms. A further example of the complexation of lanthanides into a preorganized receptor is found in the binding of europium, gadolinium, and terbium into a *p-tert*-butyl calix[4]arene tetramide ligand (99, 100). The complexes are soluble in water, and no hydrolysis to the hydroxide is observed. The europium and terbium complexes of this ligand (Fig. 32) are particularly useful because their highly luminescent properties in aqueous solution make them potentially valuable in fluoroimmunoassay applications. The emission from the free lanthanide ions is rapidly quenched by water. The luminescence properties of these calix[4]arene tetramide complexes have been investigated, and by using the Horrocks and Sudnick equation it has been concluded that a single water molecule is coordinated to the central metal ion in the calixarene complex (101).

Both carbamoyloxy and 2-aminoethoxy substituted calixarene hosts have been used to encapsulate the Tb^{3+} ion (Fig. 33). Long-lived emission is observed from the Tb^{3+} center after absorption of the light into a ligand band followed

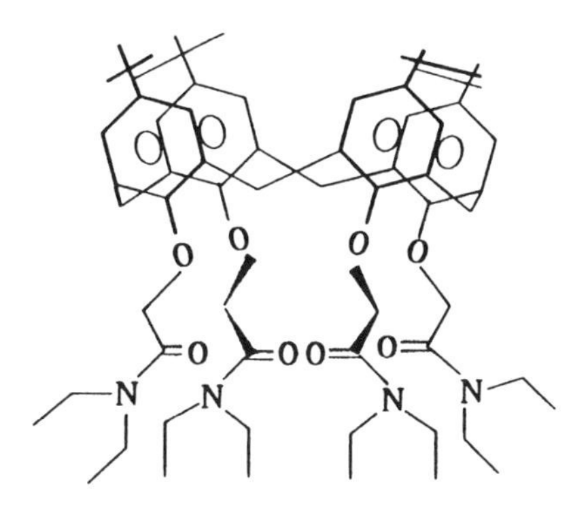

Figure 32. Calixarene amide.

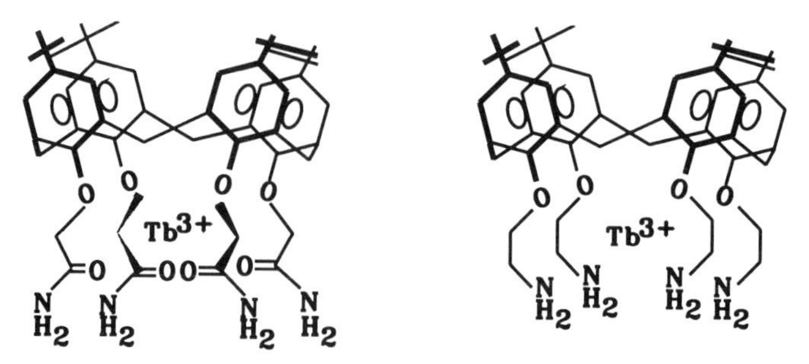

Figure 33. Fluorescent Tb complexes.

by intersystem crossing. This emission is partially quenched by the addition of water to a solution of the complex in methanol solvent (Table X) (102). To further induce the light-emitting properties of Tb^{3+} and Eu^{3+} centers, calix[4]arene ligands have been synthesized that have three amide groups on the lower rim that are involved in metal binding, and one group on that rim that

TABLE X
Excited State Lifetime Measurements for Tb^{3+} Ions Encapsulated in Calixarene Hosts[a]

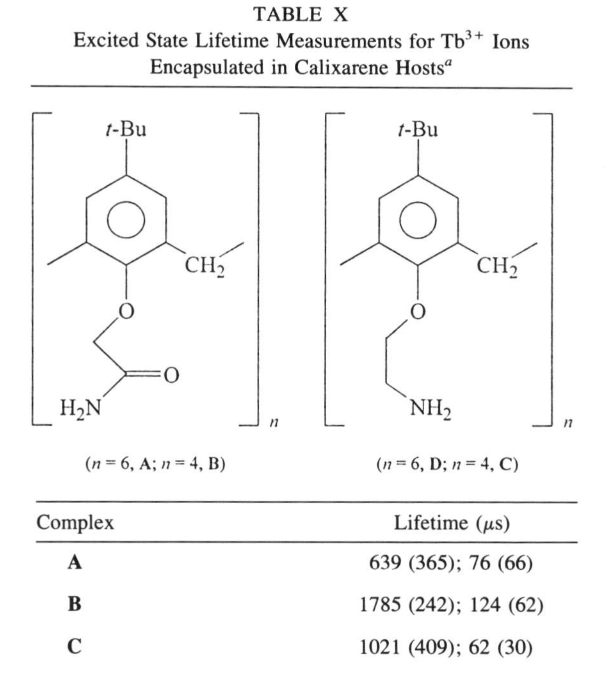

Complex	Lifetime (μs)
A	639 (365); 76 (66)
B	1785 (242); 124 (62)
C	1021 (409); 62 (30)
D	554 (456); 66 (42)

[a]Lifetimes measured as 5×10^{-3} *M* solutions in methanol [values in parentheses are in methanol/water (95:5)]. The excitation is into the ligand band at λ = 266 nm.

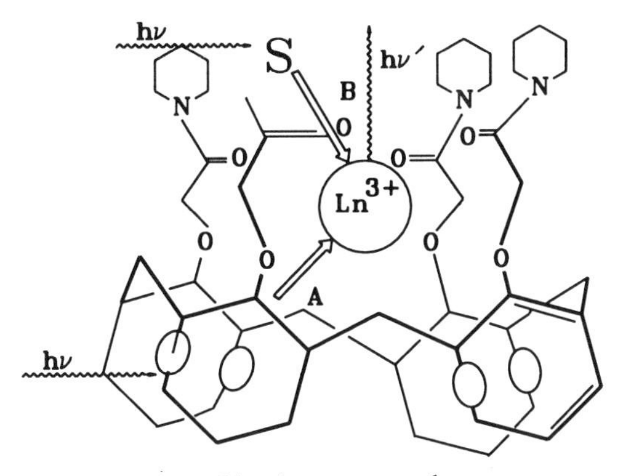

Figure 34. Antennae complexes.

can act as a sensitizer (Fig. 34). This sensitizer S is either a phenyl or a biphenyl group (103).

The coordination of a water molecule to a lanthanide coordinated into a calix[4]arene host makes such complexes potentially useful as magnetic resonance imaging (MRI) agents. The carbamoyloxy ligand in Fig. 33 has been used as a host for such an application. Although the relaxivity of the Gd^{3+} complex (0.8 s) is suitable for MRI applications, its stability constant ($\beta = 1.0 \times 10^3\ M^{-1}$) is too low for the complex to be useful (104). A cerium calix[8]arene complex $[Ce(L)(dmso)_5]\cdot(dmso)_2$ that contains two cerium(IV) centers encapsulated within the host (Fig. 35) has been used in conjunction with hydrogen peroxide for the regioselective hydroxylation of phenols (Eq. 14) (105).

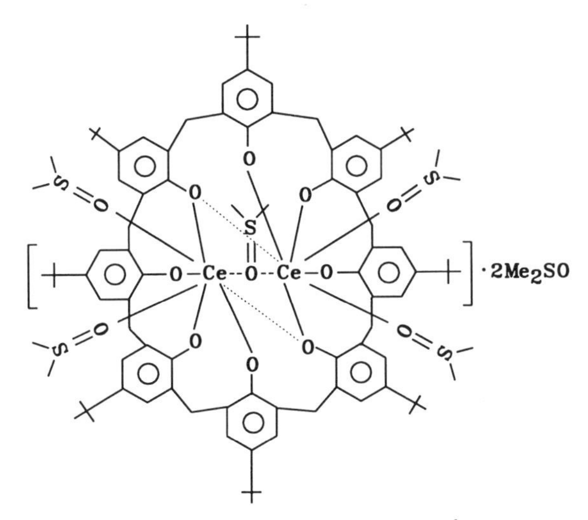

Figure 35. A Ce_2 calix[8]arene complex.

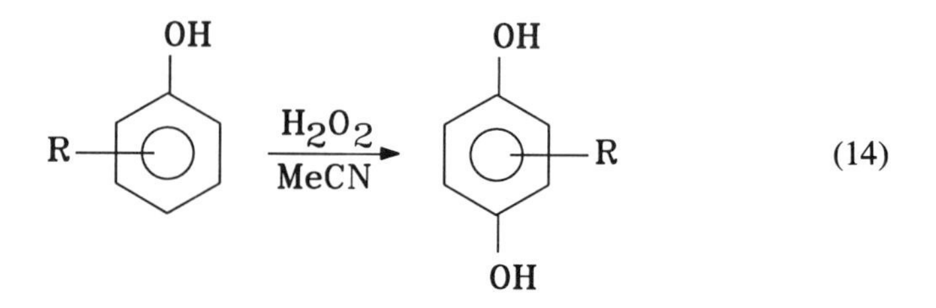

(14)

VII. ACTINIDE COMPLEXES

A. Uranyl Complexes

The complexation of the uranyl ion UO_2^{2+} is of interest because of its importance as an essential element in the nuclear industry, and also because of the need to extract the ion from waste fission products. In each case it is necessary to have available an extractant that both strongly binds this ion, and is also highly selective for it. Several calix[6]arenes have been synthesized that strongly bind the uranyl ion. Among these are 37, 38, 39, 40, 41, 42-hexahydroxycalix[6]arene-5, 11, 17, 23, 29, 35-hexasulfonate (106), 37, 38, 39, 40, 41, 42-hexamethoxy-5, 11, 17, 23, 29, 35-hexakis(phosphonomethyl)calix[6]arene, and calix[6]arenes that have either carboxylate or hydroxamate groups substituted at each of the positions on the lower rim (29). The phosphonomethyl derivative has been prepared by the Arbuzov reaction between a phosphite and a chloromethyl substituted calixarene (18). These calixarenes have high-binding constants for uranyl (log $K \approx$ 18–20), and high selectives for complexation of this ion over other metal ions (107). The binding selectivity is sensitive to the size of the calixarene cavity. An example of this selectivity is shown in the data in Table XI for a series of unsubstituted and carboxylate substituted calixarenes from which it is apparent that whereas UO_2^{2+} strongly binds to the calix[5]arenes and calix[6]arenes, there is weaker binding to the smaller calix[4]arenes. This selectivity may not simply be a reflection of the size difference between the cavities, but may also arise because the greater conformational flexibility of the larger calixarenes. This greater flexibility may allow the calixarene to more readily adopt the preferred pseudoplanar penta- or hexacoordination complexation of UO_2^{2+} (108, 109). The selectivity for UO_2^{2+} extraction is also sensitive to the number of carboxylate substituents on the lower calixarene rim. The tricarboxylate substituted calix[6]arene shown in Fig. 36 has a higher selectivity for binding UO_2^{2+} than does the corresponding hexacarboxylate in the presence of Mg^{2+}, Ni^{2+}, or Zn^{2+} ions dissolved in an acetate buffer solution (110). The extraction of uranyl ion is also sensitive to the nature of the ligating groups. Two calix[6]arenes that have been used for the extraction of

TABLE XI
Binding Constants (log K_{assoc}) for the Complexation of *p*-sulfonatocalix[*n*]arenes with the UO_2^{2+} Ion

Compound	R	*n*	log K_{assoc} [a]
$[SO_3H, CH_2, OR]_n$	H	4	3.2
	CH_2CO_2H	4	3.1
	H	5	18.9(6)
	CH_2CO_2H	5	18.4(1)
	H	6	19.2(1)
	CH_2CO_2H	6	18.7(1)

[a]Error values are in parentheses.

UO_2^{2+} are shown in Fig. 37. The lower rim substituted calix[6]arene hydroxamate has a sufficiently high-binding constant for UO_2^{2+} that it can effectively complete with carbonate ion as a ligand (111). The upper rim substituted calix[6]arene methylphosphonate has a stability constant (log K) of 17.5 for UO_2^{2+}, which makes it competitive with the calix[6]arene hexacarboxylate as a ligand for forming a 1:1 complex with UO_2^{2+} (112).

A potentially quinquedentate ligand is bis(homo-oxa)-*p*-*tert*-butylcalix-[4]arene (Fig. 38). This compound also forms a complex with UO_2^{2+}. Both the free ligand and the UO_2^{2+} complex adopt the cone conformation (113). In the UO_2^{2+} complex the primary ligation is via the four phenoxides, with only a long "bond" being formed between UO_2^{2+} and the ether oxygen of the oxacalix[4]arene. This high selectivity for UO_2^{2+} is considered to be an example of shape selectivity whereby the uranyl ion complexes with the oxygen atoms of the sulfonates, phosphonates, and carboxylates coordinating to the uranyl ion in the equatorial coordination positions.

If a hydrophobic substituent such as a *tert*-butyl or a hexyl group is introduced into the para positions instead of a sulfonate group, these chemically modified calixarenes can be used as extractants to transfer UO_2^{2+} from the

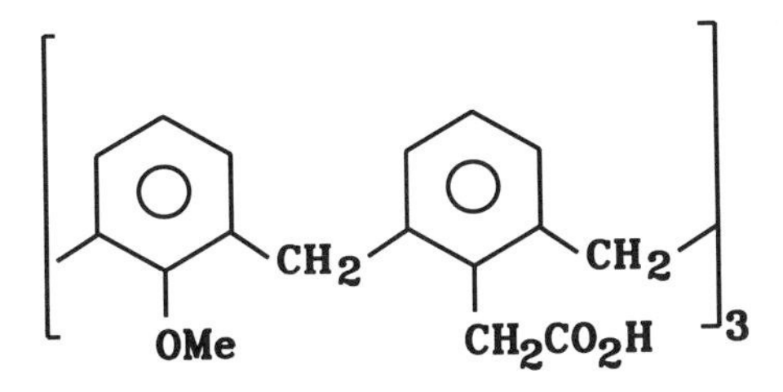

Figure 36. Uranyl selective calix[6]arene tricarboxylate.

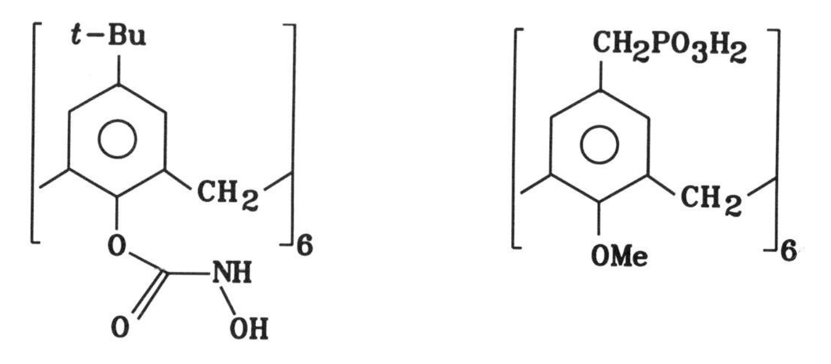

Figure 37. Hydroxamate and phosphonate substituted calix[6]arenes.

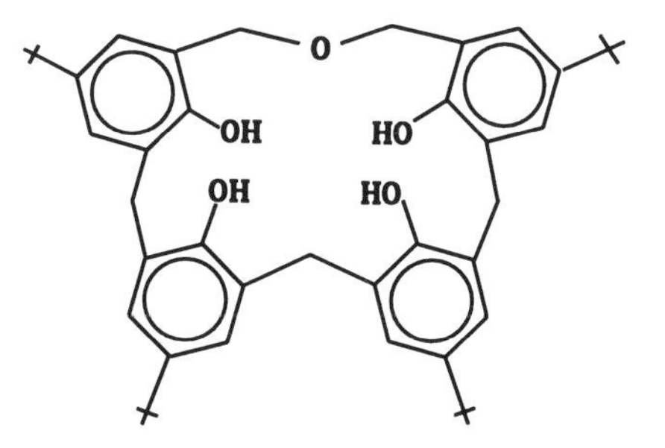

Figure 38. Homo-oxa-calixarene.

aqueous phase into an organic phase (114). Alternately these calixarenes can be anchored onto a polymer and used as a selective resin for UO_2^{2+}. In one approach the calix[6]arene hexacarboxylate has been first partially nitrated, and the nitro groups reduced to amines. These functionalities are then fixed onto a cross-linked polystyrene and the resulting calixarene impregnated resin used to selectively adsorb UO_2^{2+} from seawater (115). A further cross-linked calixarene modified polymer has been prepared by treating polyethyleneimine with a chlorosulfonated calixarene (116). These calixarene impreganted resins have also been used for the extraction of UO_2^{2+} from seawater.

B. Thorium Complexes

Tetravalent thorium complexes of calixarenes have been synthesized because such organic-soluble complexes are of interest as materials for use in the ''sol–gel'' production of metal–oxide coatings (117). When $Th(ClO_4)_4 \cdot 12dmso$ and *p-tert*-butylcalix[8]arene (H_8L) are reacted together, a complex of stoichiometry $Th_4(HL)(H_2L)(OH)_3H_2O(dmso)_4$ is obtained (Eq. 15). Each ligating calixarene

$$Th(ClO_4)_4.12dmso + 2H_8L \longrightarrow Th_4(HL)(H_2L)(OH)_3H_2(dmso)_4 \quad (15)$$

is coordinated to two Th atoms, resulting in the coordination sphere of each Th having five bonded phenolic oxygen atoms, three of these five have monoden-

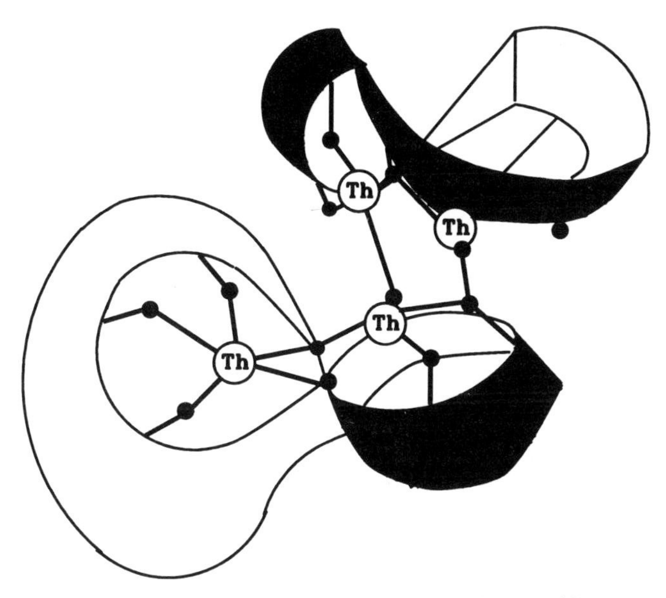

Figure 39. Tetrathorium supramolecular assembly.

tate coordination and two are bridging thoriums. The monodentate oxygen-bonded dmso ligands enter through the cup of each ligand. In addition hydroxide ligands bridge to the other Th within the same ligand, and also bind the two Th_2L systems together (118). A representation of the folding of the complexed calix[8]arenes in this complex is shown in Fig. 39.

C. Chromogenic UO_2^{2+} Sensors

A UO_2^{2+} selective chromophore has been synthesized by attaching a single indoaniline functionality onto the upper rim of a calix[6]arene (Fig. 40). This compound in the presence of base has an absorption band at 628 nm. Addition of UO_2^{2+} to this basic solution results in a large bathochromic shift of this band, along with a corresponding increase in absorption intensity due to the formation of a 1 : 1 complex. The addition of other metal ions causes little or no changes in the absorptioin spectrum of this indoaniline functionalized calixarene (119).

VIII. TRANSITION AND POSTTRANSITION METAL COMPLEXES

Both simple and chemically modified calixarenes can act as ligands to transition metal ions. The simple calixarenes with phenolic groups on their lower rim can bind to metal ions via the deprotonated phenolate ligand. This is a hard

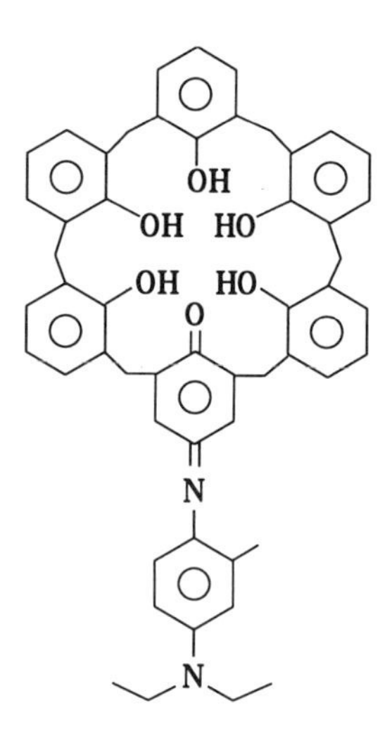

Figure 40. Indoaniline uranyl selective chromophore.

donor ligand that is expected to show a preference for hard metal ions. Such metal ions are first-row or highly charged metal ions of the early transition elements. For chemically modified calixarenes, however, it is possible to incorporate functional groups that are designed for complexation with the characteristics of specific metal centers.

A. Complexes with Simple Calixarenes

Calixarene complexes of Ti^{IV}, Fe^{III}, and Co^{II} have been prepared by reaction between the calixarene and the amide derivatives of the individual metal ions. This synthetic approach is analogous to that used to prepare aluminum calix[4]arene complexes where the adduct $[AlH_3{\cdot}NMe_3]_2$ is the synthetic reagent of choice (120). For the reaction between *p-tert*-butylcalix[4]arene and titanium amide, the product is the 1 : 1 dimer [Ti(*p-tert*-butylcalix[4]arene)]$_2$ (Eq. 16),

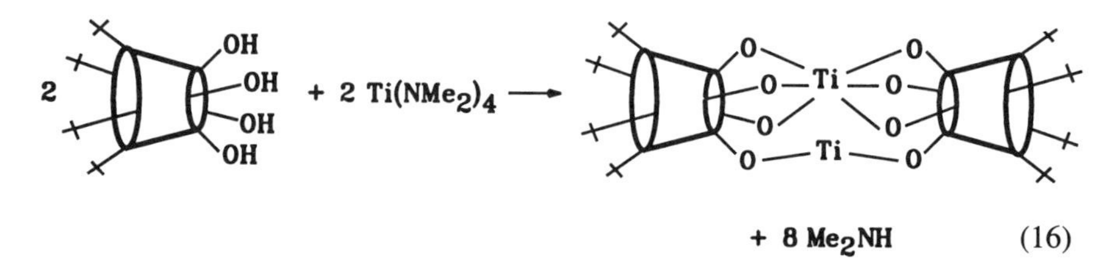

(16)

the structure of which has been characterized by X-ray crystallography (121). The reactions with iron and cobalt are more complicated. For the case of iron, a dimer $[Fe(NH_3)($*p-tert*-butylcalix[4]arene-$OSiMe_3]_2$ is again formed, but for cobalt the tricobalt(II) complex Co_3(*p-tert*-butylcalix[4]arene-$OSiMe_3$)$_2$.thf is obtained as the major product (Eqs. 17 and 18). The ammonia molecule that is coordinated to the Fe^{III} center is formed by the reaction between the free calixarene and the bis(trimethylsilyl) amide ligand. Again, both of these iron and

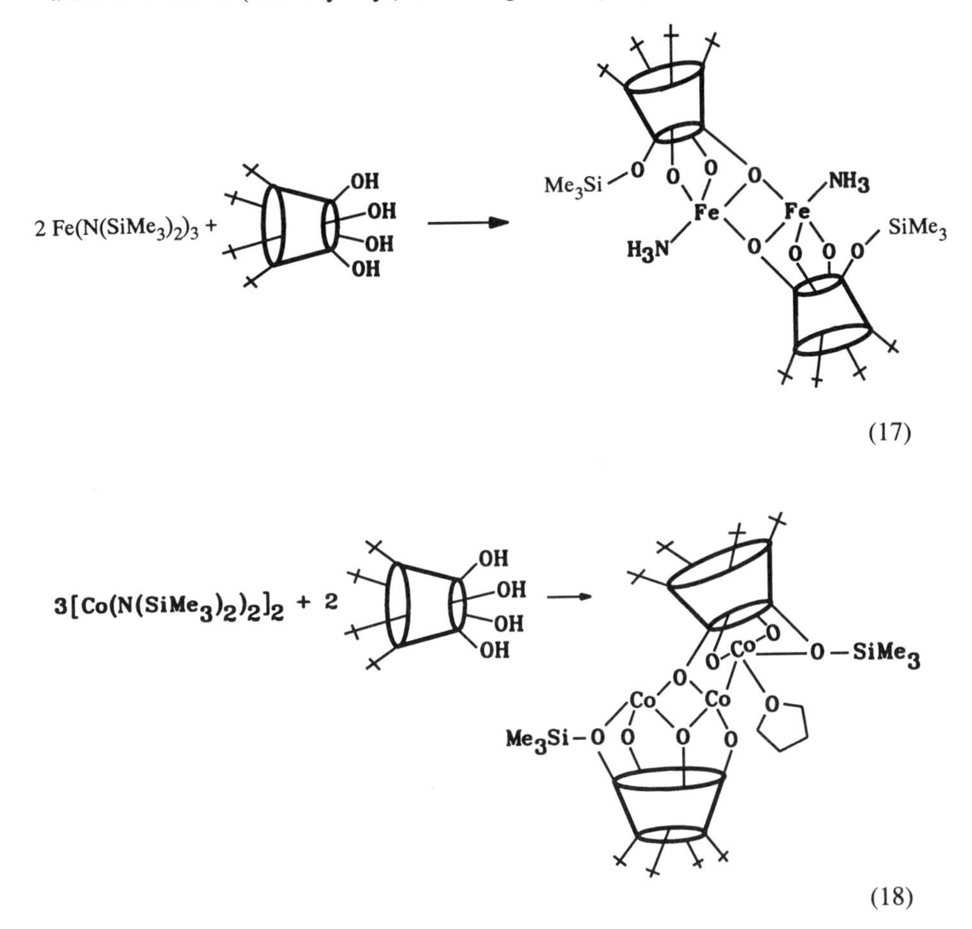

cobalt complexes have been structurally characterized by X-ray crystallography. Lower aggregate complexes such as these can be extended into the construction of larger solid state complex networks. An example of such a network is the synthesis of a noncentrosymmetric koiland consisting of two *p-tert*-butylcalix[4]arene units fused by both Si and Ti atoms. Such a double-hollow koiland can then be potentially assembled into a linear network by the interpenetration of a hydrogen-bonded connector (Fig. 41) (122). The range of transition metal complexes that can be obtained between calixarenes and high-va-

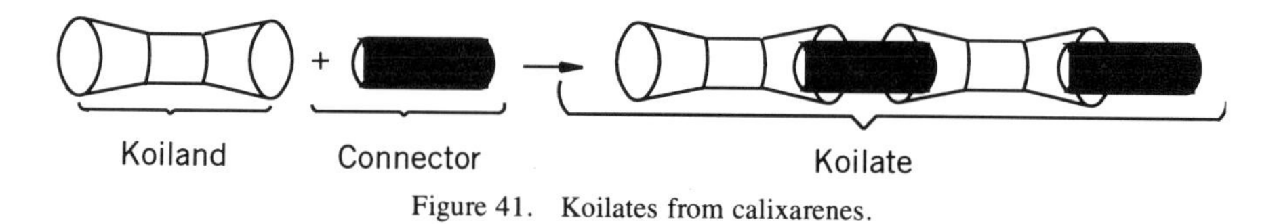

Figure 41. Koilates from calixarenes.

lent early transition metal ions is exemplified by the reactions of niobium(V) and tantalum(V). Treating $NbCl_5$ or $TaCl_5$ with *p-tert*-butylcalix[4]arene (LH_4) leads to the formation of the monomeric complex $MCl_2(LH)$ (Eq. 19). For the

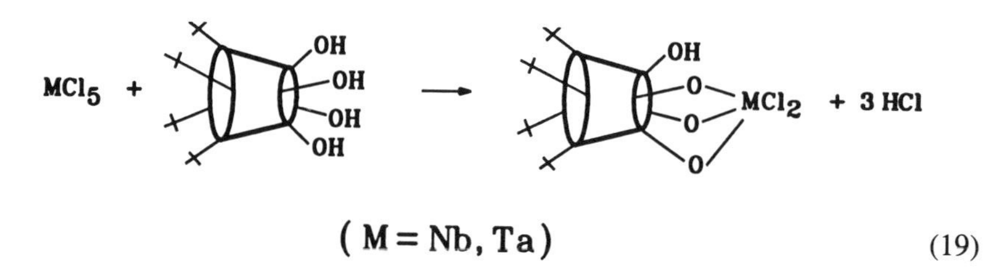

(M = Nb, Ta) (19)

case of niobium, however, when the reaction mixture contains acetic acid or water, the tetrametallic complex $Nb_4Cl_8O_2(L_2)$ is formed (Eq. 20) (123).

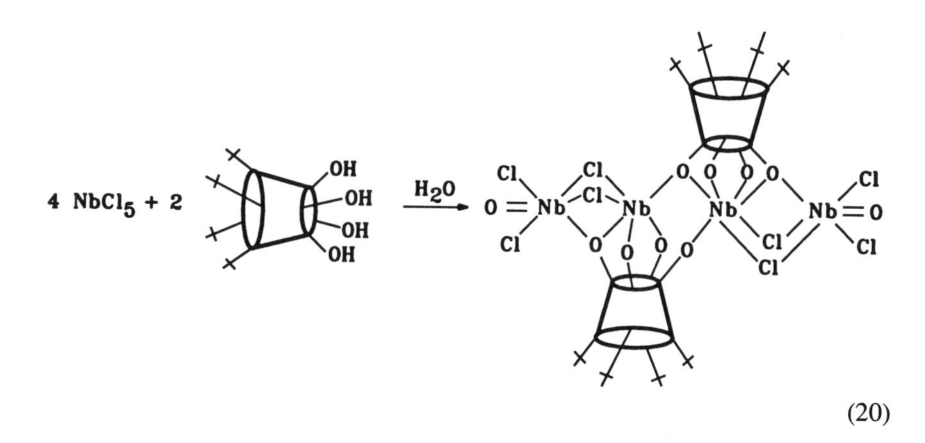

(20)

Calix[8]arenes have also been used as oxygen-donor ligands for early transition metal ions (124, 125). For the cases of Ti and Zr the transition metal center has been introduced into the reaction mixture as its alkoxide. The synthesis of the complexes $[(MOR)_2(p\text{-}tert\text{-butylcalix[8]arene})]^-$ (M = Ti, R = *i*-Pr, or *t*-Bu; M = Zr, R = *i*-Pr) (Eq. 21) and $[(VO)_2(p\text{-}tert\text{-butylcalix[8]arene})]^-$ (Eq. 22) have been achieved by this procedure.

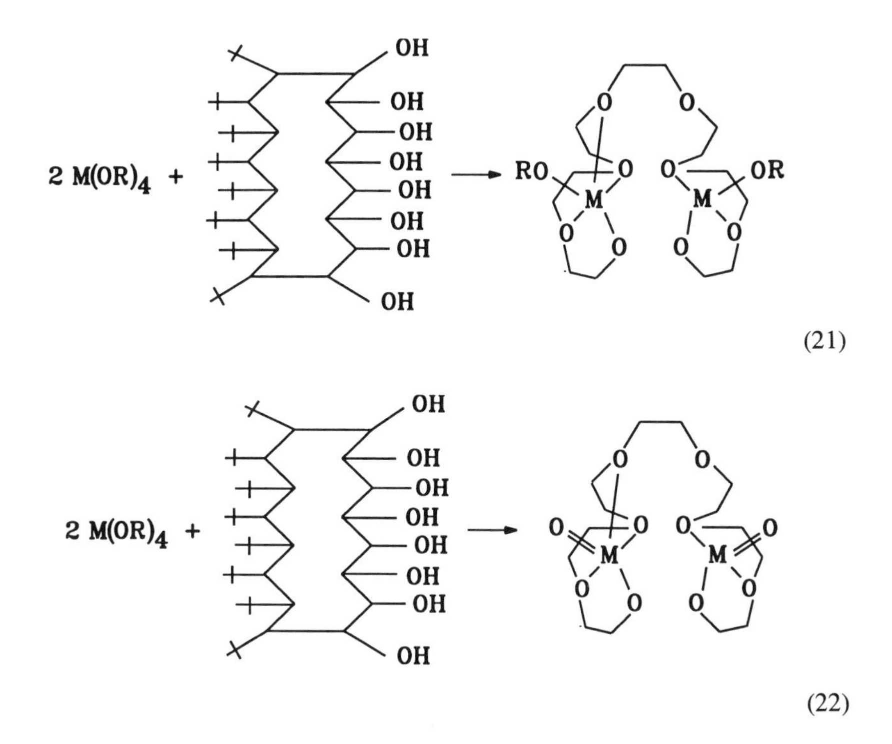

Calixarenes will also form water soluble complexes with the mid-transition metals. Thus the water soluble sulfonated calix[6]arene ($H_{12}A$) forms complexes with aqueous solutions of Fe^{III} over a pH range of 2.5–5.5 to give characteristic violet-colored solutions. By using the continuous variation method, the stoichiometry of the complex is believed to be $[Fe(OH)]_2H_6A$. The sequence of reactions leading to the formation of this complex are shown in Scheme 3 (126). Water soluble calix[*n*]arenes have also been used for different applications that involve metal complexes. Once such application is to use the inclusion properties of the sulfonated calix[*n*]arenes (n = 4 or 8) to modify the kinetics of ligand substitution and electron-transfer reactions of transition metal complexes (127). Modifications include the acceleration and inhibition of reactions by both ion-pairing effects and the shielding of the reactants in the calixarene cavity.

$$2\ Fe(H_2O)_6^{3+} \rightleftharpoons 2\ Fe(OH)(H_2O)_5^{2+} + H^+$$

$$2\ Fe(OH)(H_2O)_5^{2+} + H_6A^{6-} \rightleftharpoons [Fe(OH)]_2(H_6A)^{2-}$$

Scheme 3

In addition to forming transition metal complexes where the calixarene is bound into the first coordination sphere of the metal ion, examples exist where a hydrated metal ion is intercalated within the hydrophilic layer of a calixarene. Such a case is found with the sulfonated calix[4]arenes where the metal complexes $Cr(H_2O)_6^{3+}$, $Yb(H_2O)_7^{3+}$, and $Cu(H_2O)_4^{2+}$ with water molecules in the first coordination sphere interact with sulfonate oxygen atoms in different molecular layers of the calixarene structure (128).

B. Complexes with Chemically Modified Calixarenes

The examples in Section VIII.A show that the coordination of transition metal ions to simple unmodified calixarenes can lead to the formation of complexes having a wide range of different structural types. This structural diversity can be potentially reduced or controlled by using chemically modified calixarenes that are designed to favor certain coordination numbers and geometries. Calixarenes can be chemically modified at either the upper or lower rim to incorporate functional groups that can coordinate with selected metal ions. For such designed complexes the calixarene framework can be considered to be a platform whereby one rim is used to attach the ligating groups, and the other rim is used to introduce functional groups that modify the solubility characteristics of the complex, or immobilize the calixarene by covalently binding it to a polymeric support.

Upper rim substituted aminocalixarenes have been used as complexants for transition metal ions such as nickel^{II}, copper^{II}, cobalt^{II}, iron^{II}, and palladium^{II} (129). An example of such an aminocalixarene is 5, 11, 17, 23-tetrakis(2-aminoethyl)-25, 26, 27, 28-tetrakis[[*p*-bromophenyl)sulfonyl]oxy]-calix[4]-arene. This compound forms octahedral metal complexes with the four amino groups of the calix[4]arene coordinated, along with two supporting ligands. These supporting ligands may be in an exo position, or one of them may be in an endo site within the calixarene cavity (Structure A in Fig. 42). For the case of nickel^{II}, the complex with this aminocalix[4]arene and imidazole likely has an imidazole ligand in the octahedral coordination position within the cavity. An octahedral complex is also obtained by treating $CuCl_2{\cdot}H_2O$ with this aminocalix[4]arene, followed by treatment with imidazole. For the case of cobalt^{II} a mixture of tetrahedral and octahedral complexes are formed. This aminocalix[4]arene forms an iron^{II} complex, but since it does not coordinate molecular oxygen it cannot be used as an oxygen carrier. When this aminocalix[4]arene is reacted with $PdCl_2(MeCN)_2$ a planar four-coordinate complex is obtained where the amine nitrogen atoms act as donor ligands to Pd^{II}.

Complexes of 1 : 1 stoichiometry between Fe^{III} and the calix[4]arene have been obtained from reacting $FeCl_3{\cdot}6H_2O$ with upper rim functionalized calix[4]arenes having acetyl or carboxylate groups (130). These ligands can be

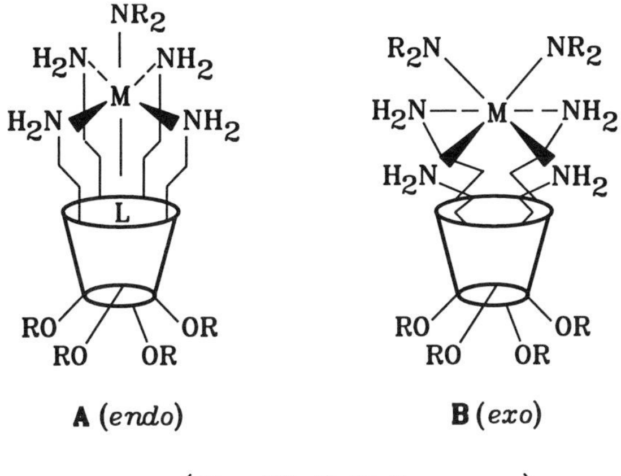

Figure 42. Metal complexes of aminocalixarenes.

used as precipitating agents for Fe^{III}. A calix[6]arene with a *vic*-dioxime substituent attached to the upper rim has been prepared and used as a complexant for Cu^{II}, Ni^{II}, and Co^{II}. The ligand has been synthesized from *anti*-chloroglyoxime and 5, 11, 17, 23, 29, 35-hexamino-37, 38, 39, 40, 41, 42-hexamethoxycalix[6]arene (Eq. 23). This hexaglyoxime calix[6]arene gives square planar complexes with the Cu^{II} and Ni^{II}, and an octahedral complex with Co^{II} (131).

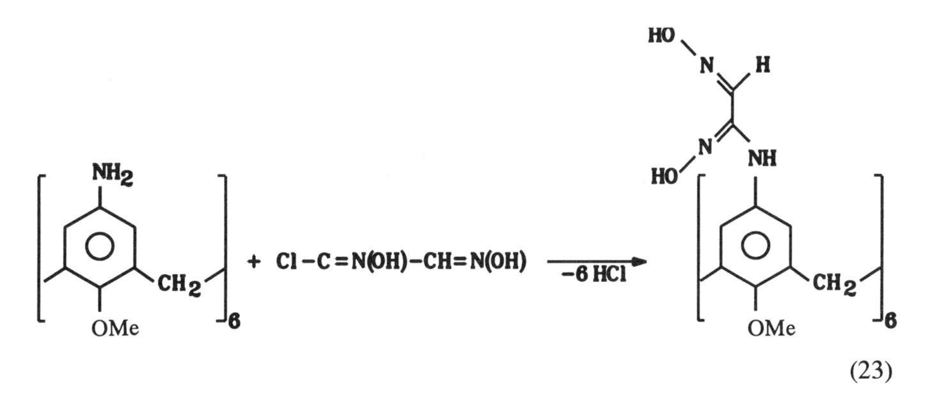

(23)

Transition metal calixarene complexes can also be used in molecular recognition. Substituted calix[4]arenes have been modified with $Ru(bpy)_2^{2+}$ (where bpy = 2,2′-bipyridine) derivatived functionalities bound via one or two linkages to one or two $Ru(bpy)_2^{2+}$ moieties. This group of complexes (Figs. 43 and 44) recognize halide, $H_2PO_4^-$ and HSO_4^- anions (132). The monotopic anion

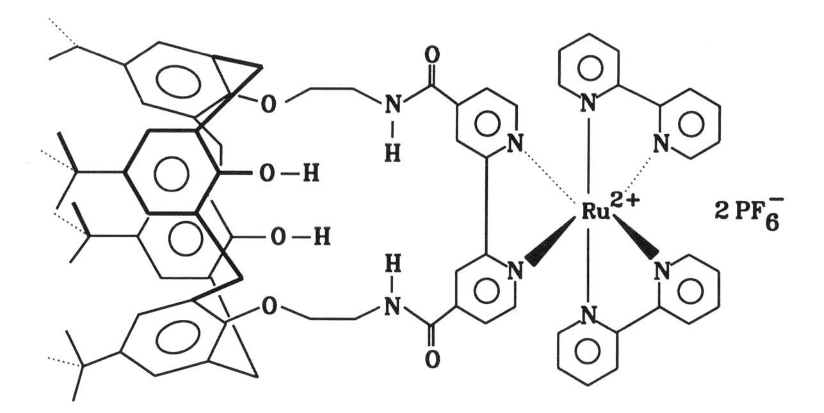

Figure 43. A Ru(bpy)-substituted calixarene.

receptor in Fig. 43 electrochemically recognizes $H_2PO_4^-$ in the presence of a 10-fold excess of HSO_4^- and Cl^-. The bimetallic complex in Fig. 44, by contrast, has the potential to exhibit ditopic behavior with two anionic molecules occluded into the host. Tris(bpy)ruthenium(II) centers covalently attached to a calix[4]arene can also be used as luminescent pH sensors. In the complex shown

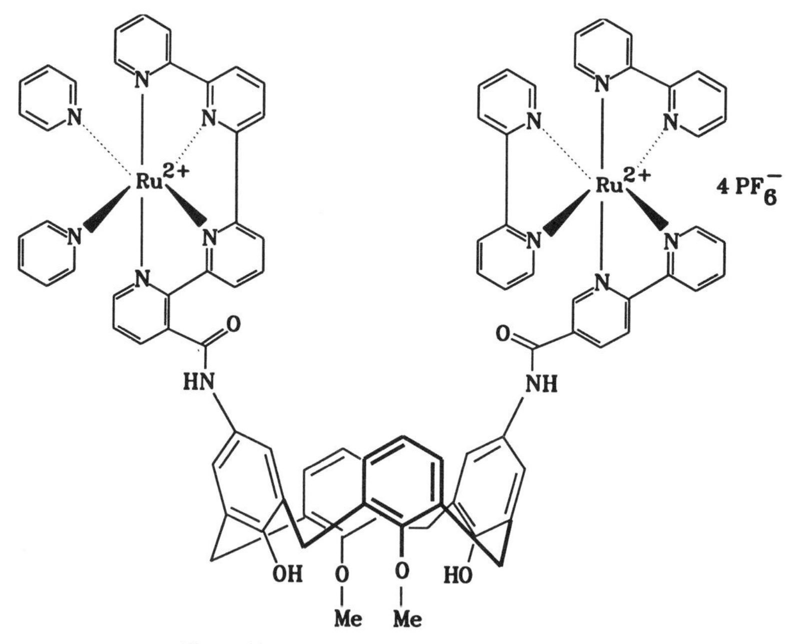

Figure 44. A $[Ru(bpy)]_2$-substituted calixarene.

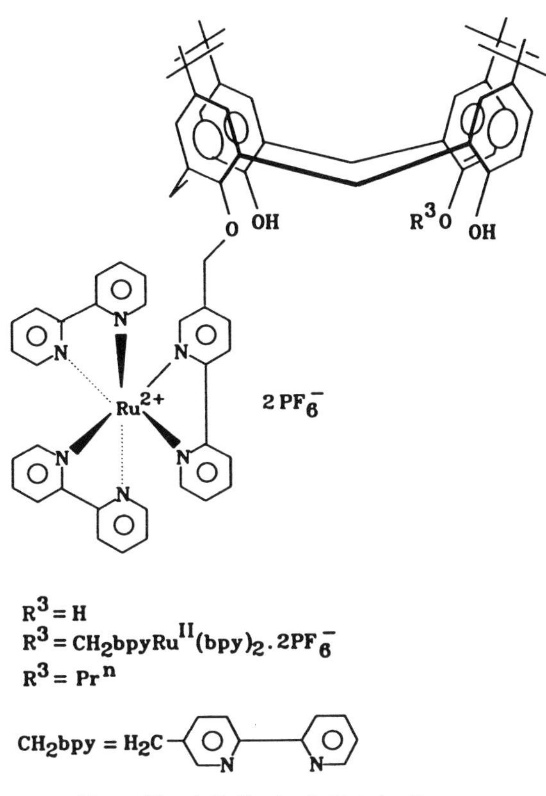

Figure 45. A Ru(bpy)-substituted calixarene.

in Fig. 45 the $Ru(bpy)_3^{2+}$ moiety acts as the luminophore, and the free phenolic units of the calixarene act as the acid–base sites (133). This complex has a methylene group as a spacer between the bpy edge and the lower calixarene rim. This group is introduced both to maximize the electron-transfer rate and to buffer the two units from each other. As the solution pH is raised the formation of phenolate anion(s) at the calixarene center causes photoinduced intramolecular electron transfer to occur from this ion to the $Ru(bpy)_3^{2+}$ center, thus causing quenching of the luminescence.

Transition metal complexes of calix[4]arenes can be used for the directed synthesis of liquid crystals. This opportunity arises because calix[4]arenes have a greater upper rim than lower rim diameter, and are therefore a class of bowlic compounds that can undergo a head-to-tail aggregation. Such aggregated bowlic liquid crystals are natural noncentrosymmetric building blocks since a head-to-tail organization maximizes the interactions between the bowlic cores. A modified calixarene with long-chain alkyloxy groups attached to an upper rim ary-

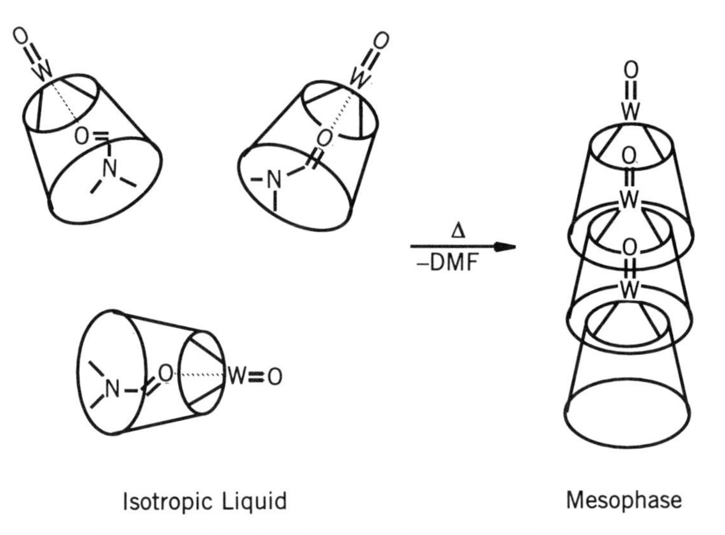

Figure 46. Mesophase formation from a tungstyl calixarene.

lazo sustituent has been prepared that forms a pyramidal complex with a tungstyl group (Fig. 46). At ambient temperature in DMF solvent a solvent molecule is occluded into the calix[4]arene cavity. Upon heating the isotropic liquid eliminates the solvent DMF molecule, and an ordered mesophase is obtained (134). Such organized intermolecular associations of calixarene complexes are increasingly attracting the interest of chemists who are working to correlate the properties of solid state materials with supramolecular organization (135).

Phosphorus donor atoms attached to the lower rim of a calix[4]arene have been used as donor ligands for transition metals. For example, an octacopper(I) complex has been prepared by treating $[Cu(CO)Cl]_n$ with 5, 11, 17, 23-tetra-*tert*-butyl-25, 26, 27, 28-tetra-diphenylphosphinito calix[4]arene (Eq. 24) (136). This diphenylphosphinito substituted calix[4]arene also reacts with $Fe(CO)_3(\eta^2\text{-}C_8H_{14})_2$ in the presence of excess cyclooctene at low temperature to give a complex having two $Fe(CO)_3$ moieties spanning pairs of diphenylphosphinito moieties (Eq. 25) (137). Calix[4]arenes with diphenylphosphino groups attached to the upper rim have been used for the liquid–liquid extraction of both transition metal and alkali metal cations. The diphenylphosphino groups have been introduced by the reaction shown in Eq. 26. The extractability of metal cations by this diphenylphosphino calix[4]arene decreases in the sequence: $Hg^{2+} > Cu^{2+} > Cd^{2+} > Zn^{2+} > Ni^{2+} > Al^{3+} > Na^{+} > K^{+}$. This selectivity is different from that of triphenylphosphine, which implies that the cyclic structure of the calix[4]arene may be playing an important role in influencing the selectivity (138).

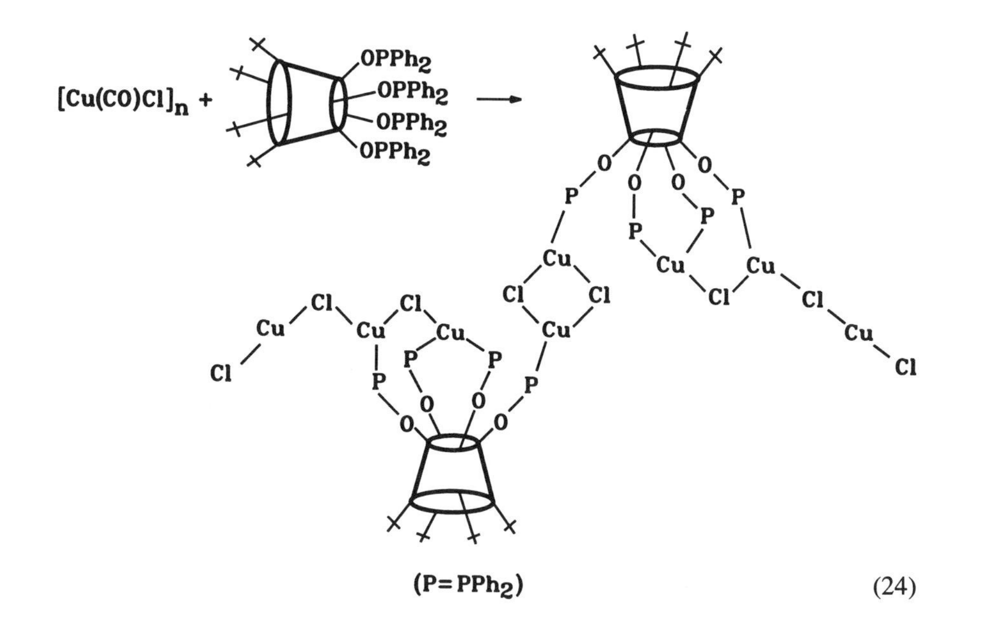
[Cu(CO)Cl]n +
OPPh2
OPPh2
OPPh2
OPPh2
Cu
Cl
Cl
Cu
Cu
Cu
P
O
(P= PPh2)

(24)

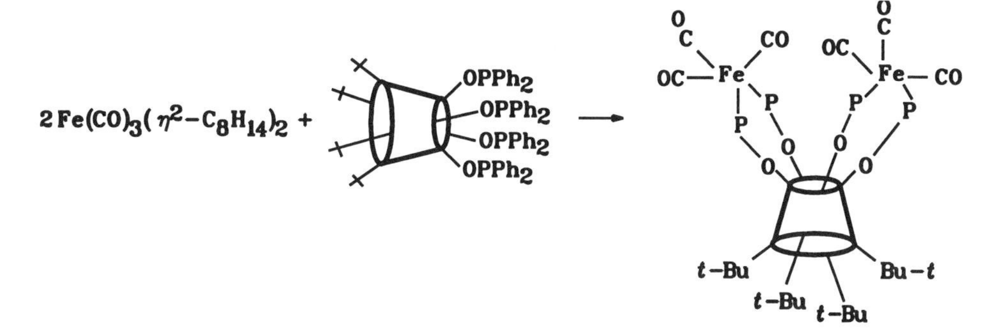
2 Fe(CO)3(η2–C8H14)2 +
OPPh2
OPPh2
OPPh2
OPPh2
OC
CO
Fe
P
O
t-Bu
Bu-t
(P = PPh2)

(25)

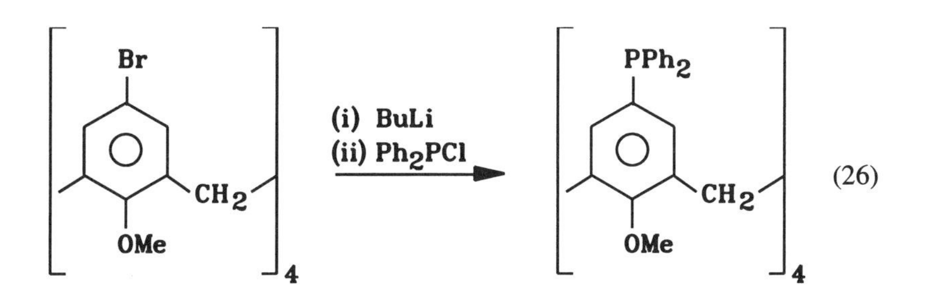
Br
CH2
OMe
4
(i) BuLi
(ii) Ph2PCl
PPh2
CH2
OMe
4

(26)

Sulfur-donor atoms attached to the lower calixarene rim have also been used as ligands for soft metal ions. A series of calix[4]arenes derivatized with thioether, thioamide, and dithiocarbamoyl functionalities show selective complexation of heavy metal ions. Chemically modified field effect transistors based on these derivatized calixarenes have been synthesized that respond selectively to the presence of Ag^+, Hg^{2+}, Cd^{2+}, and Pb^{2+} (139). Sulfur-donor ligands have also been used in the design of a combination hard–soft ditopic metal-binding calix[4]arene (Fig. 47). The preferred ligands have a chain length with $n = 6$, which is sufficiently long that the lower rim can accept two metal ions in different coordination sites. For the case of $M = Ag^+$, both it and Na^+ can be coexistent in the ionophoric cavity. For the case of $M = Cd^{2+}$, however, its complexation into the soft sulfur-donor binding site excludes Na^+ from the hard oxygen-donor binding site (140). A series of mercapto, thiocarbamates, and dithiocarbamates have been synthesized for complexation with heavy metals (Fig. 48). It has been found that conformational structure effects can lead to the calixarene acting as a ligand for two metal ions rather than for just a single one. An example of such behavior is found with the 1,3-alternate conformer of *p-tert*-butyltetramercaptocalix[4]arene that coordinates to two Hg^{2+} ions, one across each pair of 1,3-alternate sulfurs as shown in the cartoon in Fig. 49 (141).

A series of calixarene-based ligands have been synthesized with N, S- and

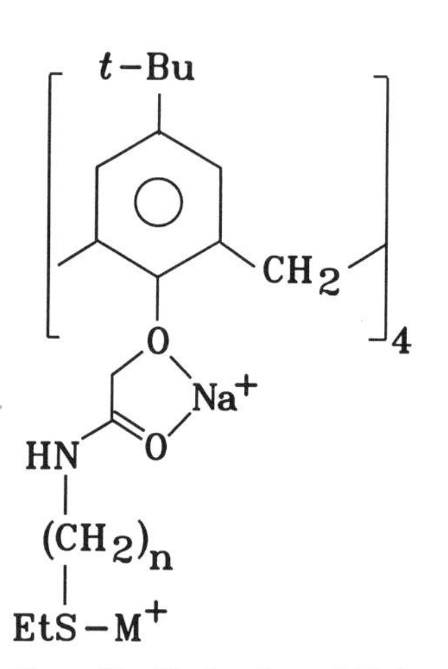

Figure 47. Hard soft metal binding calix[4]-arene.

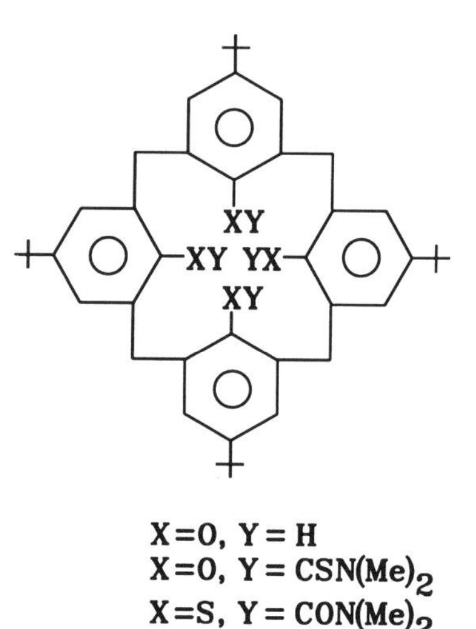

Figure 48. Sulfur-donor derivatized calix[4]arenes.

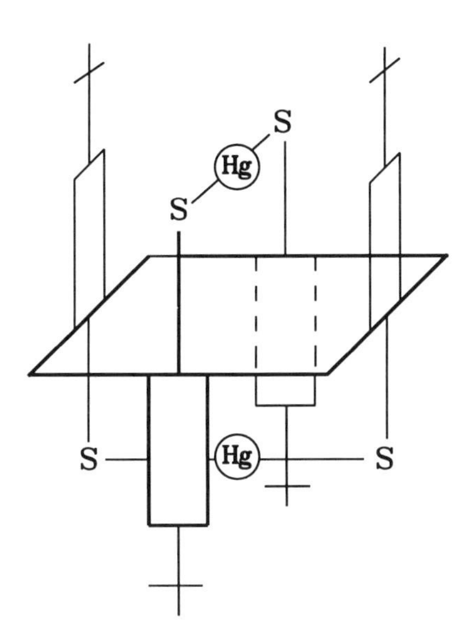

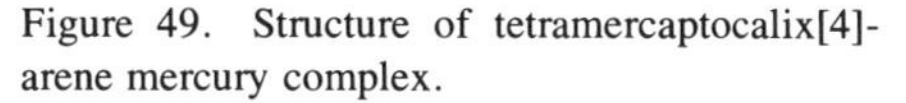
Figure 49. Structure of tetramercaptocalix[4]-arene mercury complex.

N, O-donor atom combinations on the upper rim. Examples include the calix[4]arene shown in Fig. 50 (142). These modified calixarenes show a high selectivity for complexation with heavy metals or transition metals.

Calix[5]arene cations functionalized with organometallic substituents have been synthesized. Complexes of Ru^{II}, Rh^{III}, and Ir^{III} have been synthesized, and the complex $[\{Ir(Cp^*)\}_3(\eta^6:\eta^6:\eta^6$-*p-tert*-butylcalix[5]arene-H]$[BF_4]_5$, where Cp* = η^5-pentamethylcyclopentadienyl, has been characterized crystallographically. These complexes have been used as complexant hosts both for anions and for polyphosphate residues of adenosine triphosphate (ATP) (143).

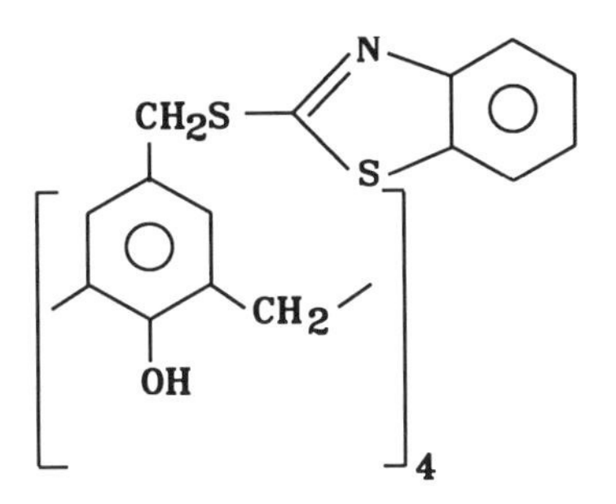

Figure 50. Sulfur-donor derivatized calix[4]arene.

ABBREVIATIONS

acac	Acetylacetonate
Ar	Aryl
bpy	2,2′-Bipyridine
Bu	Butyl
dmf	*N*,*N*-Dimethylformamide (ligand)
DMF	*N*,*N*-Dimethylformamide (solvent)
dmso	Dimethyl sulfoxide (ligand)
DMSO	Dimethyl sulfoxide (solvent)
Et	Ethyl
Me	Methyl
Ph	Phenyl
Pr	Propyl
thf	Tetrahydrofuran (ligand)

REFERENCES

1. C. D. Gutsche, *Calixarenes*, Royal Society of Chemistry, Cambridge, UK, 1989.
2. J. Vicens and V. Böhmer, *Calixarenes: A Versatile Class of Macrocyclic Compounds*, Kluwer, Dordrecht, The Netherlands, 1991.
3. C. D. Gutsche and M. Iqubal, *Org. Synth.*, *68*, 234 (1989).
4. J. H. Munch and C. D. Gutsche, *Org. Synth.*, *68*, 243 (1989).
5. V. Bocchi, D. Foina, A. Pochini, R. Ungaro, and G. D. Andreetti, *Tetrahedron*, *38*, 373 (1982).
6. C. D. Gutsche and L.-G. Lin, *Tetrahedron*, *42*, 1639 (1986).
7. J.-P. Scharff and M. Mahjoubi, *New J. Chem.*, *15*, 883 (1991).
8. S. Shinkai, S. Mori, T. Tsubaki, T. Sone, and O. Manabe, *Tetrahedron Lett.*, *25*, 5315 (1984).
9. S. Shinkai, K. Araki, T. Tsubaki, T. Arimura, and O. Manabe, *J. Chem. Soc., Perkin Trans 1*, 2297 (1987).
10. J. L. Atwood, D. L. Clark, R. K. Juneja, G. W. Orr, K. D. Robinson, and R. L. Vincent, *J. Am. Chem. Soc.*, *114*, 7558 (1992).
11. M. Almi, A. Arduini, A. Casnati, A. Pochini, and U. Ungaro, *Tetrahedron*, *45*, 2177 (1989).
12. C. D. Gutsche and I. Alam, *Tetrahedron*, *44*, 489 (1988).
13. S. J. Harris, M. A. McKervey, D. P. Melody, J. G. Woods, and J. M. Rooney, *Eur. Patent Appl.*, 151, 527 (1985); *Chem. Abstr.*, *103*, 216,392x (1985).
14. B. Kneafsey, J. M. Rooney, and S. J. Harris, *U. S. Patent*, 4, 912, 183 (1990); *Chem. Abstr.*, *114*, 123,273 (1991).
15. S. J. Harris, G. Barrett, and M. A. McKervey, *J. Chem. Soc., Chem. Commun.*, 1224 (1991).

16. E. M. Georgiev, K. Troev, and D. M. Roundhill, *Supramol. Chem.*, *2*, 61 (1993).
17. H. Murakami and S. Shinkai, *Workshop on Calixarenes and Related Compounds*, Fukuoka, Japan, 1993, P5/A-39.
18. H. Murakami and S. Shinkai, *Tetrahedron Lett.*, *34*, 4237 (1993).
19. V. I. Bukin, V. E. Plyushchev, and A. M. Reznik, *Russ. J. Inorg. Chem.*, *18*, 1011 (1973).
20. R. M. Izatt, J. D. Lamb, R. T. Hawkins, P. R. Brown, S. R. Izatt, and J. Christensen, *J. Am. Chem. Soc.*, *105*, 1782 (1983).
21. R. M. Izatt, R. T. Hawkins, J. J. Christensen, and S. R. Izatt, *J. Am. Chem. Soc.*, *107*, 63 (1985).
22. H. Goldman, W. Vogt, E. Paulus, and V. Böhmer, *J. Am. Chem. Soc.*, *110*, 6811 (1988).
23. J. M. Harrowfield, M. I. Ogden, W. R. Richmond, and A. H. White, *J. Chem. Soc. Chem. Commun.*, 1159 (1991).
24. G. Arena, R. Cali, G. G. Lombardo, E. Rizzarelli, D. Sciotto, R. Ungaro, and A. Casnati, *Supramol. Chem.*, *1*, 19 (1992).
25. A. W. Coleman, S. G. Bott, S. D. Morley, C. M. Means, K. D. Robinson, H. Zhang, and J. L. Atwood, *Agnew. Chem. Int. Ed. Engl.*, *27*, 1361 (1988).
26. G. Calestani, F. Ugozzoli, A. Arduini, E. Ghidini, and R. Ungaro, *J. Chem. Soc. Chem. Commun.*, 344 (1987).
27. E. M. Collins, M. A. McKervey, and S. J. Harris, *J. Chem. Soc. Perkin Trans.*, *1*, 372 (1989).
28. A. Arduini, A. Pochini, S. Reverberi, and R. Ungaro, *Tetrahedron*, *42*, 2089 (1986).
29. A. M. King, C. P. Moore, K. R. A. S. Sandanayaka, and I. O. Sutherland, *J. Chem. Soc. Chem. Commun.*, 582 (1992).
30. F. Arnaud-Neu, G. Barrett, S. Cremin, M. Deasy, G. Ferguson, S. J. Harris, A. J. Lough, L. Guerra, M. A. McKervey, M. J. Schwing-Weill, and P. Schwinte, *J. Chem. Soc. Perkin Trans.*, *2*, 1119 (1992).
31. T. Jin, K. Ichikawa, and T. Koyama, *J. Chem. Soc. Chem. Commun.*, 499 (1992).
32. S. J. Harris, G. Barrett, and M. A. McKervey, *J. Chem. Soc. Chem. Commun.*, 1224 (1992).
33. M. McCarrick, B. Wu, S. J. Harris, D. Diamond, G. Barrett, and M. A. McKervey, *J. Chem. Soc. Chem. Commun.*, 1287 (1992).
34. R. Assmus, V. Böhmer, J. M. Harrowfield, M. I. Ogden, W. R. Richmond, B. W. Shelton, and A. H. White, *J. Chem. Soc. Dalton Trans.*, 2427 (1993).
35. M. A. McKervey, *Workshop on Calixarenes and Related Compounds*, Fukuoka, Japan, 1992, 1L-2.
36. M. A. McKervey, E. M. Seward, G. Ferguson, B. Ruhl, and S. J. Harris, *J. Chem. Soc. Chem. Commun.*, 388 (1985).
37. G. Ferguson, B. Kaitner, M. A. McKervey, and E. M. Seward, *J. Chem. Soc. Chem. Commun.*, 584 (1987).
38. S.-K. Chang, S.-K. Kwon, and I. Cho, *Chem. Lett.*, 947 (1987).

39. S.-K. Chang and I. Cho, *J. Chem. Soc. Perkin Trans.*, *1*, 211 (1986).
40. F. Arnaud-Neu, E. M. Collins, M. Deasy, G. Ferguson, S. J. Harris, B. Kaitner, A. J. Lough, M. A. McKervey, E. Marques, B. L. Ruhl, M. J. Schwing-Weill, and E. M. Seward, *J. Chem. Soc.*, *111*, 8681 (1989).
41. M.-J. Schwing, F. Arnaud, and E. Marques, *Pure Appl. Chem.*, *61*, 1597 (1989).
42. T. Arimura, M. Kubota, T. Matsuda, O. Manabe, and S. Shinkai, *Bull. Chem. Soc. Jpn.*, *62*, 1674 (1989).
43. F. Arnaud-Neu, M. J. Schwing-Weill, K. Ziat, S. Cremin, S. J. Harris, and M. A. McKervey, *New J. Chem.*, *15*, 33 (1991).
44. M. Ogata, K. Fugimoto, and S. Shinkai, *J. Am. Chem. Soc.*, *116*, 4505 (1994).
45. V. Bocchi, D. Forina, A. Pochini, R. Ungaro, and G. D. Andretti, *Tetrahedron*, *38*, 373 (1982).
46. S. Shinkai, T. Otsuka, K. Fujimoto, and T. Matsuda, *Chem. Lett.*, 835 (1990).
47. S. Shinkai, K. Fujimoto, T. Otsuka, and H. L. Ammon, *J. Org. Chem.*, *57*, 1516 (1992).
48. P. Guilbaud, A. Varnek, and G. Wipff, *J. Am. Chem. Soc.*, *115*, 8298 (1993).
49. A. F. D. de Namor, N. A. de Sueros, M. A. McKervey, G. Barrett, F. Arnaud-Neu, and M. J. Schwing-Weill, *J. Chem. Soc. Chem. Commun.*, 1546 (1991).
50. A. Ikeda and S. Skinkai, *Tetrahedron Lett.*, *33*, 7385 (1992).
51. J. M. Harrowfield, M. Mocerino, B. W. Skelton, C. R. Whitaker, and A. H. White, *Aust. J. Chem.*, *47*, 1185 (1994).
52. M. Gomez-Kaifer, P. A. Reddy, C. D. Gutsche, and L. Echegoyen, *J. Am. Chem. Soc.*, *116*, 3580 (1994).
53. F. Arnaud-Neu, G. Barrett, S. J. Harris, M. Owens, M. A. McKervey, M.-J. Schwing-Weill, and P. Schwinte, *Inorg. Chem.*, *32*, 2644 (1993).
54. J. K. Moran and D. M. Roundhill, *Inorg. Chem.*, *31*, 4213 (1993).
55. A. Casnati, P. Minari, A. Pochini, and R. Ungaro, *J. Chem. Soc. Chem. Commun.*, 1413 (1991).
56. R. G. Janssen, W. Verboom, D. N. Reinhoudt, A. Casnati, M. Freriks, A. Pochini, F. Ugozzoli, R. Ungaro, P. M. Nieto, M. Carramolino, F. Cuevas, P. Prados, and J. de Mendoza, *Synthesis*, 380 (1993).
57. J. E. Bollinger, J. K. Moran, E. M. Georgiev, and D. M. Roundhill, *Supramolecular Chem.*, in press.
58. K. Araki, K. Inada, H. Otsuka, and S. Shinkai, *Tetrahedron*, *49*, 9465 (1993).
59. K. Araki, N. Hashimoto, H. Otsuka, and S. Shinkai, *J. Org. Chem.*, *58*, 5958 (1993).
60. P. D. Hampton, C. E. Daitch, Z. Bencze, T. M. Alam, W. Tong, and S. Wu, *XIXth International Symposium on Macrocylic Chemistry*, Lawrence, KS, 1994, Abstr. ST 24.
61. F. Arnaud-Neu, S. Cremin, D. Cunningham, S. J. Harris, P. McArdle, M. A. McKervey, M. McManus, M.-J. Schwing-Weill, and K. Ziat, *J. Incl. Phenom. Mol. Recogn. Chem.*, *10*, 329 (1991).

62. T. Arimura, M. Kubota, K. Araki, S. Shinkai, and T. Matsuda, *Tetrahedron Lett.*, 2563 (1989).
63. S. Shinkai, K. Araki, M. Kubota, T. Arimura, and T. Masuda, *J. Org. Chem.*, *56*, 295 (1991).
64. K. Iwamoto and S. Shinkai, *J. Org. Chem.*, *57*, 7066 (1992).
65. T. Arimura, M. Kubota, K. Araki, S. Shinkai, and T. Matsuda, *Tetrahedron Lett.*, *30*, 2563 (1993).
66. K. Iwamoto, A. Ikeda, K. Araki, T. Harada, and S. Shinkai, *Tetrahedron*, *49*, 9937 (1993).
67. F. Ohseto and S. Shinkai, *Workshop of Calixarenes and Related Compounds*, Fukuoka, Japan, 1993, Abst. PS/B-4.
68. F. Ohseto, T. Sakaki, K. Araki, and S. Shinkai, *Tetrahedron Lett.*, *34*, 2149 (1993).
69. A. Ikeda and S. Shinkai, *J. Am. Chem. Soc.*, *116*, 3102 (1994).
70. V. Böhmer, W. Vogt, S. J. Harris, R. G. Leonard, E. M. Collins, M. Deasy, M. A. McKervey, and M. Owens, *J. Chem. Soc. Perkin Trans. 1*, 431 (1990).
71. Y. Okada, Y. Sugitani, Y. Kasai, and J. Nishimura, *Bull. Chem. Soc. Jpn.*, *67*, 586 (1994).
72. Y. Ishikawa, T. Kunitake, T. Matsuda, T. Otsuka, and S. Shinkai, *J. Chem. Soc. Chem. Commun.*, 736 (1989).
73. H. Shimizu, K. Iwamoto, K. Fujimoto, and S. Shinkai, *Chem. Lett.*, 2147 (1991).
74. I. Aoki, H. Kawabata, K. Nakashima, and S. Shinkai, *J. Chem. Soc. Chem. Commun.*, 1771 (1991).
75. K. Iwamoto, K. Araki, H. Fujishima, and S. Shinkai, *J. Chem. Soc. Perkin Trans. 1*, 1885 (1992).
76. I. Aoki, T. Sakaki, and S. Shinkai, *J. Chem. Soc. Chem. Commun.*, 730 (1992).
77. Y. Kubo, S.-I. Hamaguchi, A. Niimi, K. Yoshida, and S. Tokita, *J. Chem. Soc. Chem. Commun.*, 305 (1993).
78. C. Perez-Jimenez, S. J. Harris, and D. Diamond, *J. Chem. Soc. Chem. Commun.*, 480 (1993).
79. H. Murakami and S. Shinkai, *J. Chem. Soc. Chem. Commun.*, 1533 (1993).
80. G. Deng, T. Sakaki, Y. Kawahara, and S. Shinkai, *Tetrahedron Lett.*, *33*, 2163 (1992).
81. G. Deng, K. Sakaki, K. Nakashima, and S. Shinkai, *Chem. Lett.*, 1287 (1992).
82. P. J. Dijkstra, J. A. J. Brunink, K.-E. Bugge, D. N. Reinhoudt, S. Harkema, R. Ungaro, F. Ugozzoli, and E. Ghidini, *J. Am. Chem. Soc.*, *111*, 7567 (1989).
83. E. Ghidini, F. Ugozzoli, R. Ungaro, S. Harkema, A. A. El-Fadl, and D. N. Reinhoudt, *J. Am. Chem. Soc.*, *112*, 6979 (1990).
84. W. F. Nijenhuis, E. G. Buitenhuis, F. de Jong, E. J. R. Sudholter, and D. N. Reinhoudt, *J. Am. Chem. Soc.*, *113*, 7963 (1991).
85. W. I. Iwema Bakker, M. Haas, H. J. den Hertog, Jr., W. Verboom, D. de Zeeuw, A. P. Bruins, and D. N. Reinhoudt, *J. Org. Chem.*, *59*, 972 (1994).

86. W. I. Iwema Bakker, M. Haas, C. Khoo-Beattie, R. Ostaszewski, S. M. Franken, H. J. den Hertog, Jr., W. Verboom, D. de Zeeuw, S. Harkema, and D. N. Reinhoudt, *J. Am. Chem. Soc.*, *116*, 123 (1994).
87. W. I. Iwema Bakker, W. Verboom, and D. N. Reinhoudt, *J. Chem. Soc. Chem. Commun.*, *71* (1994).
88. S. Miyamoto and P. A. Kollman, *J. Am. Chem. Soc.*, *114*, 3668 (1992).
89. R. Cacciapaglia, A. Casnati, L. Mandolini, and R. Ungaro, *J. Am. Chem. Soc.*, *114*, 10956 (1992).
90. R. Cacciapaglia, A. Casnati, L. Mandolini, and R. Ungaro, *J. Chem. Soc. Chem. Commun.*, 1291 (1992).
91. E. J. R. Sudhölter, P. D. van der Waal, M. Skowronska-Ptasinska, A. van den Berg, P. Bergveld, and D. N. Reinhoudt, *Recl. Trav. Chim. Pays-Bas*, *109*, 222 (1990).
92. J. A. J. Brunink, J. R. Haak, J. G. Bomer, D. N. Reinhoudt, M. A. McKervey, and S. J. Harris, *Anal. Chim. Acta*, *254*, 75 (1991).
93. B. M. Furphy, J. M. Harrowfield, M. I. Ogden, B. W. Skelton, A. H. White, and F. R. Wilner, *J. Chem. Soc. Dalton Trans.*, 2217 (1989).
94. J. M. Harrowfield, M. I. Ogden, A. H. White, and F. R. Wilner, *Aust. J. Chem.*, *42*, 949 (1989).
95. J. M. Harrowfield, M. I. Ogden, W. R. Richmond, and A. H. White, *J. Chem. Soc. Dalton Trans.*, 2153 (1991).
96. J. K. Moran and D. M. Roundhill, *Phosphorus Sulfur*, *71*, 7 (1992).
97. L. T. Byrne, J. M. Harrowfield, D. C. R. Hockless, B. J. Peachey, B. W. Skelton, and A. H. White, *Aust. J. Chem.*, *46*, 1673 (1993).
98. J. M. Harrowfield, M. Mocerino, and B. J. Peachey, *XIXth. International Symposium on Macrocyclic Chemistry*, Lawrence, KS, 1994, Abstr. ST 25.
99. N. Sabbatini, M. Guardigli, A. Mecati, V. Balzani, R. Ungaro, E. Ghidini, A. Casnati, and A. Pochini, *J. Chem. Soc. Chem. Commun.*, 878 (1990).
100. M. F. Hazenkamp, G. Blasse, N. Sabbatini, and R. Ungaro, *Inorg. Chim. Acta*, *172*, 93 (1990).
101. W. D. Horrocks and M. Albin, *Progress in Inorganic Chemistry*, Wiley-Interscience, New York, 1984, Vol. 31, p. 1.
102. E. M. Georgiev, J. Clymire, G. L. McPherson, and D. M. Roundhill, *Inorg. Chim. Acta*, *227*, 293 (1994).
103. N. Sato and S. Shinkai, *Workshop on Calixarenes and Related Compounds*, Fukuoka, Japan, 1993, Abstr. PS/B-13.
104. E. M. Georgiev and D. M. Roundhill, unpublished results.
105. H. M. Chawla, U. Hooda, and V. Singh, *J. Chem. Soc. Chem. Commun.*, 617 (1994).
106. Y. Kondo, T. Yamamoto, O. Manabe, and S. Shinkai, *Jpn. Kokai Tokkyo Koho* JP 63/7837 A2 [88/7837] (1988); *Chem. Abstr.*, *109*, 137,280b (1988).
107. S. Shinkai, *J. Incl. Phenom. Mol. Recogn. Chem.*, *7*, 193 (1989).

108. S. Shinkai, H. Koreishi, K. Ueda, and O. Manabe, *J. Chem. Soc. Chem. Commun.*, 233 (1986).
109. S. Shinkai, H. Koreishi, K. Ueda, T. Arimura, and O. Manabe, *J. Am. Chem. Soc.*, *109*, 6371 (1987).
110. K. Araki, N. Hashimoto, H. Otsuka, T. Nagasaki, and S. Shinkai, *Chem. Lett.*, 829 (1993).
111. T. Nagasaki, S. Shinkai, and T. Matsuda, *J. Chem. Soc. Perkin Trans. 1*, 2617 (1990).
112. T. Nagasaki, T. Arimura, and S. Shinkai, *Bull. Chem. Soc. Jpn.*, *64*, 2575 (1991).
113. J. M. Harrowfield, M. I. Ogden, and A. H. White, *J. Chem. Soc. Dalton Trans.*, 979 (1991).
114. S. Shinkai, Y. Shirahama, H. Satoh, and O. Manabe, *J. Chem. Soc. Perkin Trans. 2*, *2*, 1167 (1989).
115. Y. Kondo, T. Yamamoto, O. Manabe, and S. Shinkai, *Jpn. Kokai Tokkyo Koho* JP 62/210055 A2 [87/210055] (1987); *Chem. Abstr.*, *108*, 116,380b (1988).
116. S. Shinkai, H. Kawaguchi, and O. Manabe, *J. Polym. Sci. Part C: Polym. Lett.*, *26*, 391 (1988); *Chem. Abstr. 109*, 171425r (1988).
117. H. Reuter, *Adv. Mater.*, 351 (1989).
118. J. M. Harrowfield, M. I. Ogden, and A. H. White, *J. Chem. Soc. Dalton Trans.*, 2625 (1991).
119. Y. Kubo, S. Maeda, M. Nakamura, and S. Tokita, *J. Chem. Soc. Chem. Commun.*, 1725 (1994).
120. J. L. Atwood, S. G. Bott, C. Jones, and C. L. Raston, *J. Chem. Soc. Chem. Commun.*, 1349 (1992).
121. M. M. Olmstead, G. Sigel, H. Hope, X. Xu, and P. P. Power, *J. Am. Chem. Soc.*, *107*, 8087 (1985).
122. X. Delaigue, M. W. Hosseini, E. Leize, S. Kieffer, and A. Van Dorseelaer, *Tetrahedron Lett.*, *34*, 7561 (1993).
123. F. Corazza, C. Floriani, A. Chiesi-Villa, and C. Guastini, *J. Chem. Soc. Chem. Commun.*, 1083 (1990).
124. G. E. Hofmeister, F. E. Hahn, and S. F. Pedersen, *J. Am. Chem. Soc.*, *111*, 2318 (1989).
125. G. E. Hofmeister, E. Alvarado, J. A. Leary, D. I. Yoon, and S. F. Pedersen, *J. Am. Chem. Soc.*, *112*, 8843 (1990).
126. J.-P. Scharff, M. Mahjoubi, and R. Perrin, *New J. Chem.*, *17*, 793 (1993).
127. S. Bartlett, J. A. Imonigie, and D. H. Macartney, *XIXth. International Symposium on Macrocyclic Chemistry*, Lawrence, KS, 1994, Abstr. ST 27.
128. J. L. Atwood, J. W. Orr, N. C. Means, F. Hamada, H. Zhang, S. G. Bott, and K. D. Robinson, *Inorg. Chem.*, *31*, 603 (1992).
129. C. D. Gutsche and K. C. Nam, *J. Am. Chem. Soc.*, *110*, 6153 (1988).
130. M. Yilmaz and U. S. Vural, *Synth. React. Inorg. Met.-Org. Chem.*, *21*, 1231 (1991).

131. M. Yilmaz and H. Deligoz, *Synth. React. Inorg. Met.-Org. Chem.*, *23*, 67 (1993).
132. P. D. Beer, Z. Chen, A. J. Goulden, A. Grieve, D. Hesek, F. Szemes, and T. Wear, *J. Chem. Soc. Chem. Commun.*, 1269 (1994).
133. R. Grigg, J. M. Holmes, S. K. Jones, and W. D. J. Amilaprasach Norbert, *J. Chem. Soc. Chem. Commun.*, 185 (1994).
134. B. Xu and T. M. Swager, *J. Am. Chem. Soc.*, *115*, 1159 (1993).
135. T. Bein, *Nature (London)*, *31*, 207 (1993).
136. C. Floriani, D. Jacoby, A. Chiesi-Villa, and C. Guastini, *Agnew. Chem. Int. Ed. Engl.*, *28*, 137 (1989).
137. D. Jacoby, C. Floriani, A. Chiesi-Villa, and C. Rizzoli, *J. Chem. Soc. Dalton Trans.*, 813 (1993).
138. F. Hamada, T. Fukugaki, K. Murai, G. W. Orr, and J. L. Atwood, *J. Incl. Phenom. Mol. Recogn. Chem.*, *10*, 57 (1991).
139. P. L. H. M. Cobben, R. J. M. Egberink, J. G. Bomer, P. Bergveld, W. Verboom, and D. N. Reinhoudt, *J. Am. Chem. Soc.*, *114*, 10573 (1992).
140. K. N. Koh, T. Imada, T. Nagasaki, and S. Shinkai, *Tetrahedron Lett.*, *35*, 4157 (1994).
141. X. Delaigue, J. M. Harrowfield, M. W. Hosseini, A. De Cian, J. Fischer, and N. Kyritsakis, *J. Chem. Soc. Chem. Commun.*, 1579 (1994).
142. F. Hamada, Y. Kondo, S. Ohnuki, and J. L. Atwood, *Workshop on Calixarenes and Related Compounds*, Fukuoka, Japan, 1993, Abstr. PS/B-9.
143. J. W. Steed and J. K. Atwood, *XIXth. International Symposium on Macrocyclic Chemistry*, Lawrence, KS, 1994, Abstr. B 20.

Subject Index

Cumulative Index, Volumes 1–43